The Eponym Dictionary of Mammals

The Eponym Dictionary of Mammals

Bo Beolens
Michael Watkins
Michael Grayson

THE JOHNS HOPKINS UNIVERSITY PRESS
Baltimore

Printed in the United States of America on acid-free paper
9 8 7 6 5 4 3 2 1

The Johns Hopkins University Press
2715 North Charles Street
Baltimore, Maryland 21218-4363
www.press.jhu.edu

Library of Congress Cataloging-in-Publication Data

Beolens, Bo.
The eponym dictionary of mammals / by Bo Beolens, Michael Watkins, and Mike Grayson.
p. cm.
Includes bibliographical references.
ISBN-13: 978-0-8018-9304-9 (hardcover : alk. paper)
ISBN-10: 0-8018-9304-6 (hardcover : alk. paper)
1. Mammals—Dictionaries. 2. Eponyms—Dictionaries. I. Watkins, Michael, 1940– II. Grayson, Mike. III. Title.
QL701.2.B46 2009
599.03—dc22 2008046475

A catalog record for this book is available from the British Library.

The title page illustration is a tarsier from the nineteenth-century engraving; he also appears on the opener pages of the alphabetic groups and on page 164.

Special discounts are available for bulk purchases of this book. For more information, please contact Special Sales at 410-516-6936 or specialsales@press.jhu.edu.

The Johns Hopkins University Press uses environmentally friendly book materials, including recycled text paper that is composed of at least 30 percent post-consumer waste, whenever possible. All of our book papers are acid-free, and our jackets and covers are printed on paper with recycled content.

Contents

Preface

Two of us, Bo Beolens and Mike Watkins, wrote *Whose Bird?* which was published in November 2003. A review of *Whose Bird?* was written by Nicholas Gould for the journal *International Zoo News*. Gould suggested that there could be a need for similar volumes on other animal classes, and among them he suggested mammals. We wish to give credit and thanks here to the person whose suggestion began the conversations that led to our writing of this book.

As there are only about half as many mammal species as bird species, we assumed our new book would not be as long as *Whose Bird?* How wrong we were! It turns out that there are a very large number of people who have only one mammal species named after them.

Given that one man, Oldfield Thomas of the British Museum of Natural History, seemed to have described half of the mammal species ever discovered, we also assumed our research task would prove much easier than it had been for *Whose Bird?* Wrong again. But in the end it paid off, and the book you hold is the result not only of our perseverance, but of the labors of many people who assisted us without hesitation.

We are deeply indebted to the following people and organizations for their generous help with research and, where needed, translations: Mark A. Adams, Researcher, Evolutionary Biology Unit, South Australian Museum, Australia; Cleber J. R. Alho, Conservação e Uso Sustentável da Biodiversidade, Brasília, Brazil; Mike Archer, Australian Museum, Sydney, Australia; Dickon and Ito Corrado, Tokyo, Japan; Sylvie Coten-Watkins, Montmorency, France; Gabor Csorba, Deputy Director, Curator of Mammals, Hungarian Natural History Museum, Budapest, Hungary; Ross Cunningham, Canberra ACT, Australia; Fritz Dieterlen, Staatlisches Museum für Naturkunde, Stuttgart, Germany; Peter D. Dwyer, Research Fellow, University of Melbourne, Australia; Louise H. Emmons, Research Associate at the Smithsonian Division of Mammals, Washington DC, USA; Tim Flannery, Macquarie University, New South Wales, Australia; Pavel German, Wildlife Images, New South Wales, Australia; Nicholas Gould, *International Zoo News*, Orkney, Scotland; David L. Harrison, Harrison Zoological Museum Trust, England; Lawrence R. Heaney, Curator of Mammals, Department of Zoology, University of Chicago, USA; Colin Higgins, Bat Conservation Trust, London, UK; Geoffrey Hope, Department of Archaeology and Natural History, Australian National University, Canberra, Australia; Kim M. Howell, Professor of Zoology and Marine Biology, University of Dar es Salaam, Tanzania; Tony M. Hutson, East Sussex, England, UK; Rainer Hutterer, Museum Alexander Koenig, Bonn, Germany; Paula D. Jenkins, Collections Manager, Mammal

Curation Group, Natural History Museum, London; Viner Khabibullin, Bashkir State University, Ufa City, Bashkortostan, Russia; Rael and Helena Loon, South Africa; Tim May, London, England; David Minter, Buenos Aires, Argentina; Philip Myers, Associate Professor and Associate Curator of Mammals, Museum of Zoology, University of Michigan, USA; Bruce Patterson, MacArthur Curator, Department of Zoology (Mammals), Field Museum, Chicago, USA; Heather Prestridge, Assistant Curator, Texas Cooperative Wildlife Collection, Department of Wildlife and Fisheries, Texas A&M University, Texas, USA; Gavin J. Prideaux, Research Fellow, Western Australian Museum, Perth, Australia; Eric Rickart, Curator of Vertebrates, Utah Museum of Natural History, University of Utah, USA; Jevgeni Shergalin, Tallinn, Estonia; Steve Van Dyck, Senior Curator of Vertebrates, Queensland Museum, Brisbane, Australia; Manfred Warth, Staatlisches Museum für Naturkunde, Stuttgart, Germany; Charles Watkins, Montmorency, France; Nicholas Watkins, Oxford, England; Suzanne Watkins, Bushey Heath, Hertfordshire, England; Chris Watts, South Australian Museum, Adelaide, Australia.

Introduction

Who Is It For?

Much as birders often come across bird names that include the name of a person (such names are properly called "eponyms"), and their curiosity is aroused just as ours was, so will people come across similar eponyms for mammals. We have all heard of Przewalski's Horse or Thomson's Gazelle, but how familiar is Nolthenius' Long-tailed Climbing Mouse or Bannister's Melomys? This book is for the curious mammalogist as much as it is for the student of zoology.

How to Use This Book

This book is arranged alphabetically by the names of the people after whom mammals have been named. Generally, the easiest way to find your animal is to look it up under the personal name that is apparently embedded in the animal's common or scientific name. We say "apparently," as things are not always as simple as they seem. In some names, for example, the apostrophe implying ownership is a transcription error; in other instances the animal may have been named after a place, not a person. We have included any such names where we think confusion might arise, but we do not promise to have been comprehensive in that respect. You should also beware of spelling. Surf the Internet, and you may well find animals' names spelled in a number of different ways; the greatest resource there has ever been is also full of inaccuracies and misinformation, so beware. We have tried to include entries on those alternatives, if we have ourselves come across them.

Each entry follows a standard format. First, you will find the name of the person honored. Next, there follows a list of animals named after that person, arranged in order of the year in which they were described. This list gives common names, scientific names, names of the people who first described each species, and the date of the original descriptions—in that sequence. Alternative English names follow in brackets and are each preceded by the abbreviation Alt. Alternative scientific names (in cases where taxonomists are not in agreement) are preceded by the abbreviation Syn. (synonym). Finally, there is a brief biography of that individual.

To assist you in your search, we have cross-referenced the entries by highlighting (in bold) the names of those describers who also appear in the book. Some mammals are named in different ways after the same person, and we have also tried to marry these up using cross-references. So, for example, a species named after Queen Victoria might be called Queen Victoria's Shrew or Victoria's Shrew, or Queen's Shrew or even Empress's Shrew. Interestingly, this is most often the case where aristocratic titles are concerned. For example, the Earl of Derby, whose family name is Stanley, has mammals named after him in at least three different ways.

The greatness of a person's fame does not correspond to the length of the entry—in fact, often the opposite. Very famous people such as Queen Victoria (1819–1901) and President Theodore Roosevelt (1858–1919) have fairly brief write-ups. They are, after all, so well known, and so much has been written about them, that it is unnecessary for us to reiterate it.

We have provided lists of vernacular and scientific names at the back; these include two animals for which there is a vernacular name but no scientific name. These lists often provided the only way in which we could cross-reference the various personal names that had at different times been given to the same animal. Mammals may be named in the vernacular after the finder, after the person who wrote the description, or after some other person of the latter's choice. When more than one person has thought a species new, the mammal may get more than one set of names. An animal could thus warrant an entry in as many as six different places!

There are a great number of recent namings of fossil mammals. As the rate at which fossil remains are discovered and described seems to be increasing so rapidly, and the disagreement among the paleontologists appears epidemic, we decided that we would ignore anything that became extinct in prehistoric times—that is, more than about 500 years ago (in simplistic terms, before Columbus discovered America).

What's in a Name?

Tracking down the provenance of eponymous mammal names, and finding out about the individuals responsible for them, proved to be fraught with difficulties. Our final list contains 2,351 entries. However, this may be misleading, since these entries actually cover only 2,310 animals. The names honor 1,388 individual people, but there are also 47 that sound like people's names but in fact are not. Additionally we can add 6 tribes of indigenous people, 3 fictional characters, 8 biblical references, and 45 references to classical mythology or literature. Most annoying of all, there are entries for 10 names of people whom we have been unable to identify.

Describers and Namers

New species are first brought to the notice of the scientific community in a formal, published description of a type specimen—essentially a dead example of the species—which will eventually be lodged in a scientific collection. The person who describes the species will give it its scientific name, usually in Latin but sometimes in Latinized ancient Greek. Sometimes the "new" animal is later reclassified, and then the scientific name may be changed. This frequently applies to generic names (the first part of a binomial name), but specific scientific names (the second part of a binomial), once proposed, usually cannot be amended or replaced; there are precise and complicated rules governing any such name changes. Conventionally, a changed name is indicated by putting parentheses around the describer's name. Let us use a classic ornithological example from *Whose Bird?* to illustrate this. The Grey Heron was named *Ardea cinerea* by Linnaeus in 1758, and since that name remains recognized to this day, the bird is officially named *Ardea cinerea* Linnaeus, 1758. Linnaeus also described the Great Bittern as *Ardea stellaris* in the same year. However, bitterns have since been awarded their own genus, and we now officially call the bird *Botaurus stellaris* (Linnaeus, 1758), the parentheses indicating that the name was not the namer's original choice. The scientific names used in this book are largely those used by Duff and Lawson in their *Mammals of the World—A Checklist,* published in 2004. Occasionally we have been persuaded to swap a name from this checklist for one used in the Smithsonian database or in other recent authorities. We may have missed a few recently published taxonomic changes, but we have put the name of the original describer after every entry; hence the normal convention regarding such parentheses does *not* apply here. Because alterations to taxonomy have been so radical, and so often swiftly changing, we decided we would never get the parentheses around the right entries and so have omitted them entirely.

Although we have used current scientific names as far as possible, these are not always as universal as the casual observer might suppose. In addition to some animals having been reclassified since they were first discovered and described, there are other cases where various authorities and textbooks do not agree on which name to use; this invariably applies to generic names, where there may be disagreement on an animal's taxonomic affinities. There is no "world authority" on such matters.

There are no agreed-upon conventions for English names, and indeed the choice of vernacular names is often controversial. Very often the person who coined the scientific name will also have given it a vernacular name, which may not be an English name if the describer was not an English-speaker. On the other hand, vernacular names—including English ones—have often been added afterward, frequently by people other than the describers. In this book, therefore, when we refer to an animal as having been *named* by someone, we mean that that person gave it the *English* name in question. We refer to someone as a *describer* when that person was responsible for the original description of the species and hence for its scientific or Latin name. As we said above, it is the describer's name that is given after the scientific name in the entries.

Animals Named after More Than One Person

Throughout the text you may come across several different names for the same species. In some cases these names are honorifics; for example, Andrews' Beaked Whale is the same species as Bowdoin's Beaked Whale. This peculiarity has sometimes come about through simple mistakes or misunderstandings, such as believing juveniles or females to be a different species from the adult male. In some cases the same animal was found at about the same time in two different places, and only later has it emerged that this is the same animal named twice. Some of these duplications persist even today, with the same mammal being called something different in different places or by different people.

Recently there has been an example of an animal changing not only its vernacular and scientific names but its describer and date too. Huet's Dormouse *Graphiurus hueti* Rochebrune, 1883, has now been changed to Nagtglas' Dormouse *Graphiurus nagtglasii* Jentink, 1888. It seems that no type specimen exists for *hueti*, and as it was said to have come from Senegal and has never been seen in that country since, a number of authorities have decided to use the next oldest synonym for which a type specimen does exist; thus *nagtglasii* gets promoted. The reader will find the animal entered under both Huet and Nagtglas but should remember that whatever it is called, it is still just one species of dormouse.

Unidentified Persons

Unfortunately we have not been able to identify everyone whose name appears in that of an animal. There are, as we observed above, 10 names on our list that are just not traceable. For example, who was the Fellows of *Protochromys fellowsi?* Some Victorians seem to have had the sentimental notion of naming animals after female relatives or mistresses. In some cases we suspect the author has named the animal after a woman but has deliberately (gallantly?) withheld her full name—for example, Dorothy's Slender Mouse-Opossum *Marmosops dorothea* as described by Thomas in 1911. Such people clearly had no regard for those of us doing this kind of research!

Male or Female?

In some cases we know that an animal is named after a man, even though its scientific name is in the feminine. This seems to occur only when a name ends in the letter *a*. Presumably the reason for this is that many singular Latin nouns ending in *a* are feminine; for example, *mensa* means "table" (nothing very feminine about that), and the possessive/genitive case is *mensae,* not *mensai.* There are a number of masculine Latin nouns ending in *a* (e.g. *agricola,* meaning "farmer"), but they are declined as though they were feminine. Thus the convention is that the feminine form is adopted in such cases. For example, Olalla's Titi is named after Alfonso M. Olalla, but the scientific name is *Callicebus olallae.* This convention seems to have been falling into disuse in recent years. It is quite striking how many modern namings ignore it.

Red Herrings

Further confusion arises from the number of animals that appear to be named after people but upon closer examination turn out to be named after a place, such as an island, that was itself named after a person. We have included these with an appropriate note, as other sources of reference will not necessarily help the enquirer.

Interestingly, this volume contains a couple of examples where, because the common local name has been incorrectly assumed to be a person, a misleading possessive apostrophe has been added.

Weighing the Evidence

Ultimately, our decisions on what to include in this book depended upon the weight of available evidence. Wherever there is any doubt, we have made this clear. In some cases we have had to reject a possible attribution when the evidence is just too nebulous. For example, we might have attributed Therese's Shrew to Theresa Clay, the niece and companion of Richard Meinertzhagen, as it became clear that Heim de Balsac, who described and named the animal in 1968, knew them both. However, the animal in question proved to be named after someone entirely different.

The Eponym Dictionary of Mammals

Abbott

Abbott's Duiker *Cephalophus spadix* **True,** 1890

Abbott's Grey Gibbon *Hylobates muelleri abbotti* **Kloss,** 1929

Dr. William Louis Abbott (1860–1936) was a student, naturalist, and collector. He initially qualified as a medical doctor at the University of Pennsylvania and worked as a surgeon at Guy's Hospital in London. However, he decided not to pursue medicine but instead to use his private wealth for scientific exploration. In 1880, as a student, he had collected in Iowa and Dakota, and in 1883 in Cuba and San Domingo. In 1887 he went to East Africa, spending two years there. From 1891 he studied the wildlife of the Indo-Malayan region, using his Singapore-based ship *Terrapin,* and made large collections of mammals from Southeast Asia for the National Museum of Natural History (Smithsonian) in Washington, DC, USA. In 1897 he switched to Siam (Thailand) and spent 10 years exploring and collecting in and around the China Sea. He provided much of the Kenya material in the Smithsonian and was the author of "Ethnological Collections in the United States National Museum from Kilima-Njaro, East Africa," published in the museum's Annual Report for 1891. In 1917 he returned to San Domingo (Hispaniola), exploring the interior. He retired to Maryland but continued his lifelong study of natural history until his death. He is also commemorated in the names of several birds, such as Abbott's Babbler *Malacocincla abbotti* and Abbott's Starling *Cinnyricinclus femoralis,* and of a lizard, Abbott's Day Gecko *Phelsuma abbotti.* The duiker is found in the highlands of Tanzania and the gibbon in western Borneo.

Abe

Abe's Whiskered Bat *Myotis abei* Yoshikura, 1944 [Alt. Sakhalin Myotis]

Yoshio Abe (1883–1945) was Professor of Zoology at Karahuto Normal University. Makoto Yoshikura, who described the bat, studied there under him. In his description Yoshikura says that the bat is named to commemorate Professor Abe and to express sincere gratitude for the guidance and instruction he provided. In 1930 Professor Abe was the first Japanese scientist to study and publish on kinorhynchs (microscopic marine invertebrates), and one, *Dracoderes abei,* was named after him as late as 1990 in recognition of his studies. The bat is known only from the type specimen found on the island of Sakhalin (now part of Russia but occupied by Japan in 1944). It was only in 1956 that Yoshikura decided on the English name Abe's Whiskered Bat in his paper "Insectivores and Bats of South Sakhalin." A study published in 2004 concluded that this "species" is not valid and should be regarded as a junior synonym of Daubenton's Bat *Myotis daubentoni.*

Abel

Sumatran Orangutan *Pongo abelii* **Lesson,** 1827

Dr. Clarke Abel (1780–1826) was a British physician and naturalist. He was Chief Medical Officer and Naturalist to the Embassy of Lord Amherst to the Court of Peking from 1816 to 1817; this was Britain's second attempt at establishing relations with the Emperor of China. After detailed observation and collection of assorted cultivated and wild plants on the way to and back from the capital, he wrote *Narrative of a Journey in the Interior of China and of a Voyage to and from That Country, in 1816 and 1817.* Returning from China in 1818, Abel was subsequently appointed Physician to Lord Amherst in India, where he later died. At some point he was shipwrecked and all of his original collections went down with the ship, but he continued to collect while in Batavia (now Jakarta), where he acquired the orangutan skin. The familiar garden plant abelia is named after him. The orangutans of Borneo and Sumatra were formerly regarded as a single species, but the Sumatran form is now usually accorded the status of a full species.

Abert

Abert's Squirrel *Sciurus aberti* **Woodhouse,** 1853

Colonel John James Abert (1788–1863) was an American military man and engineer. He studied at West Point from 1808 until 1811 but resigned from the army on the very day that he graduated. He was then employed by the War Office and studied law, being admitted to the Bar in the District of Columbia in 1813. During the War of 1812 he served as a private for the defense of the capital, which was sacked and burned by the British army. In 1814 he was reappointed to the army with the rank of Major of Topographical Engineers. From 1816 to 1824 he was based mainly on the Atlantic coast. In 1824 he was promoted to Lieutenant Colonel in recognition of 10 years' service. He worked on the Ohio Canal from 1824 to 1825 and then worked in Maine until 1827. In 1828 he became Chief of the Topographical Bureau in Washington, being promoted Colonel in charge of the newly formed Corps of Topographical Engineers in 1838, a position he held until his retirement on the grounds of ill-health in 1861. He belonged to a number of scientific societies and was among those who organized the National Institute of Science that was later merged in to the Smithsonian Institution. The squirrel inhabits the pine forests of the southwest USA and northwest Mexico.

Abid

Northern Glider *Petaurus abidi* **Ziegler,** 1981

Abid Beg Mirza (dates not found) was a collector employed by the University of Maryland Pakistan Medical Centre in Lahore in 1964. He was involved in a survey of Pakistani mammals and their parasites. In an article, "Additions to the Avifauna of the Adelbert Range, Papua New Guinea," published in *The Emu*, he is mentioned as having "organized the trip and supervised the survey of mammals and preparation of all specimens." He collected the type specimen of the glider, perhaps on that trip to New Guinea. This marsupial is found in the Torricelli Mountains of northern New Guinea.

Achates

Brown-bearded Sheath-tailed Bat *Taphozous achates* **Thomas,** 1915 [Alt. Indonesian Tomb Bat]

Achates is a character in Virgil's *Aeneid*. He is always referred to as Aeneas' faithful friend. Thomas was fond of using names from classical literature and mythology when describing mammal species. Often there seems to be no obvious reason for his choice of name. The bat is found on some small islands in eastern Indonesia, including Savu, Roti, and the Kei Islands.

Adam

Adam's Horseshoe Bat *Rhinolophus adami* **Aellen** and **Brosset,** 1968

François Adam is (or was) a French agronomist and zoologist who collected bats in West Africa and co-wrote a number of articles on them in the early 1970s including, in 1975, jointly with Aellen (one of the describers of the bat) "Présence de *Glauconycteris beatrix* (Chiroptera, Vespertilionidae) en Côte d'Ivoire." He became Director of the National Agronomic

Research Institute in Lomé, Togo. The horseshoe bat is found in the Republic of Congo.

Adams, C. D.

Ear-spot Squirrel *Callosciurus adamsi* **Kloss,** 1921

C. D. Adams (dates not found) was the District Officer at Baram in North Sarawak on the island of Borneo during the late 19th and early 20th centuries. Major J. C. Moulton (1886–1926) collected the type specimen and specifically asked Kloss to name the animal after Adams as a mark of his gratitude for the exceptional help Adams had given him during his expedition. Moulton had been the Curator of the Sarawak Museum from 1909 to 1915 and was Director of the Raffles Museum in Singapore from 1919 to 1923, during which period Kloss was a member of his staff. Kloss succeeded him as Director. The squirrel is found only on Borneo.

Adams, M. A.

Cape York Pipistrelle *Pipistrellus adamsi* Kitchener, Caputi, and Jones, 1986

Dr. Mark Andrew Adams (b. 1954) is Senior Researcher at the Evolutionary Biology Unit, South Australian Museum. He is the Evolutionary Biology Unit's longest-serving researcher, with over 30 years' experience in the molecular systematics of Australian fauna. He has published over 130 scientific papers and is a leading authority on allozyme electrophoresis, having co-authored a major reference work on the subject in 1986, *Allozyme Electrophoresis: A Handbook for Animal Systematics and Population Studies*. He is also a very experienced fieldworker, having undertaken trips to every part of Australia and to New Guinea. The pipistrelle is found in northern Australia (Queensland and the Northern Territory).

Aders

Black-and-Rufous Elephant-Shrew *Rhynchocyon petersi adersi* **Dollman,** 1912

Aders' Duiker *Cephalophus adersi* **Thomas,** 1918

Zanzibar Leopard *Panthera pardus adersi* **Pocock,** 1932

Dr. William Mansfield-Aders D.Sc. (dates not found) was an entomologist and physician. He was living and working in Zanzibar as Government Zoologist between the 1900s and late 1920s. There is a collection of beetles, found by him in Zanzibar, now in the British Museum of Natural History. He also wrote articles on diseases and the invertebrates that carry them, as well as on other aspects of natural history, for example "Economic Zoology Report for the Year 1913. Section III Zanzibar Protectorate," "Entomology in Relation to Agriculture and Insects Injurious to Man and Stock in Zanzibar" (1914), and "Insects Injurious to Man and Stock in Zanzibar" (1917). He conducted a survey in 1928 regarding the prevalence of sleeping sickness there and how it could be controlled by the control of insect carriers. He is noted as having photographed someone climbing into palm trees to collect malarial mosquitoes in 1921. He is also commemorated in the scientific names of other taxa including a blackfly, *Simulium adersi*, and a nematode, *Cylicocyclus adersi*. All three mammals named after Aders are found on Zanzibar; the duiker is also found in one locality in coastal Kenya. The leopard has not been scientifically recorded since the early 1980s, but a few may still be extant based on unconfirmed local sightings.

Adolf Friedrich

Adolf Friedrich's Angolan Colobus *Colobus angolensis ruwenzorii* **Thomas,** 1901

Adolf Friedrich Albrecht Heinrich, Duke of Mecklenburg-Schwerin (1873–1969), led an expedition to German East Africa from 1907 to

1908 and also led the German Central Africa expedition of 1910–1911. He served as Governor of Togoland (then a German colony, now called Togo) from 1912 to 1914. Between 1926 and 1956 he was a member of the German Olympic Committee, being President of it from 1949 to 1951. He wrote a number of books about his travels in Africa including, in 1912, *Von Kongo zum Niger und Nil* (From the Congo to the Niger and the Nile). In 1914, Matschie named a species of colobus as *Colobus adolfi-frederici*. This is now regarded as a junior synonym of *Colobus angolensis ruwenzorii* (Thomas, 1901), but the synonym has survived as a rarely used common name. The colobus is found in eastern DRC (Zaire), Rwanda, Burundi, and southwest Uganda.

Aeecl

Aeecl's Sportive Lemur *Lepilemur aeeclis* Andriaholinirina et al., 2006

This lemur is named not after a person but, possibly uniquely, after the acronym of an organization, Association Européenne pour l'Étude et la Conservation des Lémuriens (AEECL). The lemur is endemic to a small area of western Madagascar.

Aellen

Aellen's Roundleaf Bat *Hipposideros marisae* Aellen, 1954
Aellen's Pipistrelle *Pipistrellus inexspectatus* Aellen, 1959
Southern Myotis *Myotis aelleni* Baud, 1979

Professor Villy Aellen (1926–2000) was a Swiss zoologist. His first degree was from the University of Neuchâtel, and his doctorate, which he earned from a study of bats from Cameroon, was awarded by that city's Zoological Museum in 1952. His first job was as a conservator in the Vertebrates Section of the Geneva Museum of Natural History, where he went on to become Director in 1969; he remained an Honorary Director until his death. He taught at the University of Geneva from 1966 until 1989. He was recognized internationally as an expert on cave-dwelling fauna of all kinds. He wrote a great many papers about bats in particular, very often those in his native Switzerland, but he also made field trips elsewhere, for example to Morocco. He wrote a paper in 1965, "Les rongeurs de basse Côte-d'Ivoire." The pipistrelle has been recorded in Benin, Cameroon, DRC (Zaire), Uganda, and Kenya; the roundleaf bat in Ivory Coast, Liberia, and Guinea; and the myotis in southwest Argentina. See also **Marisa.**

Aello

Broad-striped Tube-nosed Bat *Nyctimene aello* **Thomas,** 1900

Aello, which means "whirlwind," was one of the mythical Harpies, creatures usually depicted as birds with women's upper bodies. They are found in Greek mythology (in Homer's *Odyssey*, they are regarded as storm winds), and three of them—Aello, Ocypete, and Celaeno, who were the daughters of the Nereid Electra and Thaumas—were very ugly. Their sister Iris was, in contrast, very beautiful. Thomas was fond of choosing names from mythology and using them as binomials, as in this example, without putting them into the genitive. The bat is found in New Guinea.

Agag

Agag Gerbil *Gerbillus agag* **Thomas,** 1903

Agag is a biblical character, King of the Amalekites. He was defeated in battle and captured by King Saul, who spared his life, but unfortunately for Agag, the Prophet Samuel disagreed and hacked him to death. A fuller account can be read in the Bible, in the Old Testament (1 Sam. 15). There is another reference in the Old Testament to an Agag, and it may be that it was actually the Amalekite word for "king." Thomas seems to have been inspired to use this choice of scientific name by the type locality where the gerbil was taken: Agageh Wells in Kordofan (Sudan). Its taxonomy is not well defined, and its distribution—sometimes said to extend from southern Mauritania and northern Nigeria to Sudan—may involve more than one taxon.

Agricola

Agricola's Gracile Opossum *Cryptonanus agricolai* **Moojen,** 1943 [Syn. *Gracilinanus agricolai*]

Dr. Ernani Agricola (1883–1978) was a prominent Brazilian physician who held positions in several government organizations and international bodies such as the International Leprosy Society. In addition to the opossum he is commemorated in the names of hospital departments and a street in his home province of Minas Gerais. In his description of the opossum, Moojen explains that in 1936 Dr. Agricola encouraged a man by the name of Antenor Leitão de Carvalho to collect mammal specimens in northeast Brazil. This work resulted in the incorporation into the Brazilian National Museum of a very significant collection, accompanied by excellent field notes. The opossum is found in eastern Brazil. It was long regarded as being synonymous with *Gracilinanus emiliae* but is now again treated as a valid species.

Ahmanson

Ahmanson's Sportive Lemur *Lepilemur ahmansoni* Louis et al., 2006

Robert H. Ahmanson (1927–2007) and the Ahmanson Foundation, of which Robert was President and Trustee, give a lot of support to Malagasy students both in Madagascar and at Henry Dorley Zoo's Center for Conservation and Research at Omaha, Nebraska. He was a right-wing businessman from the finance world before his retirement in 1995. The lemur is endemic to western Madagascar.

Aitken

Kangaroo Island Dunnart *Sminthopsis aitkeni* Kitchener, Stoddart, and Henry, 1984

Peter F. Aitken was an expert on Australian mammals who was Senior Curator of Mammals at the South Australian Museum from 1972 to 1986. He was a prolific publisher of scientific papers from the 1960s to the 1980s, mostly on marsupials, often in the *South Australian Naturalist* or the *Records of the South Australian Museum.* He co-wrote *Marine Mammals in South Australia* with J. K. Ling. There is a Peter F. Aitken Medal that has been awarded annually since 1989 for "Contributions to the Conservation of Australian Mammals." The dunnart lives on Kangaroo Island in South Australia and was first found in the early 1980s by the museum's researchers when Aitken was Curator of Mammals.

Ajax

Kashmir Grey Langur *Semnopithecus (entellus) ajax* **Pocock,** 1928

Ajax was a hero in Greek mythology who is mentioned by Homer in his account of the Trojan War and who is also the subject of a tragedy by Sophocles. There was something of a fad for naming Indian langurs after characters from Homer and Virgil (see also **Entellus, Hector,** and **Priam**). This monkey is found in northern India.

Alberico

Alberico's Broad-nosed Bat *Platyrrhinus albericoi* Velazco, 2005

Professor Dr. Michael Alberico (1937–2005) was an American biologist and zoologist who was a Professor at Del Valle University, Colombia. He moved to Colombia in 1980, after graduating in biology from the University of Illinois and taking his master's degree and doctorate in zoology at the University of New

Mexico in 1979. He was murdered, shot dead after withdrawing money from an ATM in the city of Cali. In the citation he is described as one "who devoted his scientific career to the study of Colombian mammals." The bat is found on the eastern slope of the Andes from Ecuador to Bolivia.

Alcathoe

Alcathoe's Myotis *Myotis alcathoe* Helversen and Heller, 2001

In Greek myth, the Minyads were the daughters of Minyas, at the time when the worship of Dionysus was introduced into Boeotia. They became insane and conceived a craving for human flesh. One of the Minyads was Alcathoe. According to one myth, Alcathoe and her sister Arsinoe, while the other women and maidens were indulging in Bacchic frolics, remained at home devoting themselves to their usual occupations—thus profaning the days sacred to the god. Dionysus punished them by changing them into bats. This species of bat was first identified in Greece and Hungary but has since been found in several other European countries including Slovakia, Bulgaria, France, and Spain.

Alcorn

Alcorn's Pocket Gopher *Pappogeomys alcorni* Russell, 1957

J. Ray Alcorn (dates not found) spent most of his life and career devoted to wildlife biology, with much of it spent working for the U.S. Fish and Wildlife Service. In 1943 he published an article entitled "The Introduced Fishes of Nevada with a History of Their Introduction" and in 1946, as an Assistant District Agent of the Fish and Wildlife Service, he wrote an article for the *Journal of Mammalogy* entitled "On the Decoying of Coyotes." He also spent over 50 years gathering data for *The Birds of Nevada*, which was eventually published in 1988. In 1947 and 1948 he collected small mammals along the Alaska Highway in British Columbia, southern Yukon, and southern Alaska. In 1949 and 1954–1956 he was collecting in Mexico. In 1956 he and his family sent specimens from Nicaragua to the Mammalogy Division, Museum of Natural History, University of Kansas. There is a J. R. Alcorn Collection held at the University of Nevada, made up of specimens he collected and donated. The gopher is found in west-central Mexico.

Alecto

Small Asian Sheath-tailed Bat *Emballonura alecto* Eydoux and **Gervais,** 1836
Black Flying Fox *Pteropus alecto* **Temminck,** 1837
Short-eared Bat *Cyttarops alecto* **Thomas,** 1913
Pygmy Fruit Bat *Aethalops alecto* Thomas, 1923

Alecto, in Greek mythology, was one of the Furies; the others were Tisiphone and Megaera. They had snakes for hair, eyes that dripped blood, heads of dogs, and wings of bats. They persecuted men and women who committed crimes against the moral and natural order, such as patricide or fratricide. They worked by driving their victims mad. The ancient Greeks believed in the power of euphemism, and so to propitiate the Furies they often referred to them as the Eumenides—the kind-hearted ones. The sheath-tailed bat is found in the Philippines and parts of Indonesia; the flying fox is found in Sulawesi, southern New Guinea, and northern Australia; the short-eared bat's range extends from Nicaragua to northeast Brazil; and the fruit bat is found in Malaysia and Indonesia (Sumatra, Borneo, Java, Bali, and Lombok).

Alexander

Alexander's Bush Squirrel *Paraxerus alexandri* **Thomas** and **Wroughton,** 1907

Alexander's Cusimanse *Crossarchus alexandri* Thomas and Wroughton, 1907 [Alt. Alexander's Mongoose]

Captain Boyd Alexander (1873–1910) was an African traveler and ornithologist. He was educated at Radley College from 1887 until 1891, joined the army in 1893, and was at Kumasi in 1900. He explored Lake Chad from 1904 until 1905 and made a geographical survey of West Africa between 1905 and 1906. He spent some time on the island of Fernando Pó (now Bioko), and many of the taxa that he described have *poensis* in their binomial, referring to that island. He was a Royal Geographical Society medalist in 1908. He continued his African explorations from 1908 until 1910, ending in Chad where local people killed him. He published *From the Niger to the Nile* in two volumes in 1907. Alexander studied birdlife across a large part of West Africa and is also commemorated in the common names of two birds: Alexander's Akalat *Sheppardia poensis* and Alexander's Swift *Apus alexandri*. The squirrel and mongoose are found in DRC (Zaire) and Uganda.

Alexandra

Gebe Cuscus *Phalanger alexandrae* **Flannery** and **Boeadi,** 1995

Dr. Alexandra Szalay is an Australian anthropologist. In private life she is Mrs. Flannery and so was commemorated by her husband in the name of this cuscus, which is found on the island of Gebe (North Moluccas, Indonesia).

Alfaro

Alfaro's Rice Rat *Oryzomys alfaroi* **J. A. Allen,** 1891

Alfaro's Pygmy Squirrel *Microsciurus alfari* J. A. Allen, 1895 [Alt. Central American Dwarf Squirrel]

Alfaro's Rice Water Rat *Sigmodontomys alfari* J. A. Allen, 1897 [Alt. Cana Rice Rat]

Dr. Don Anastasio Alfaro (1865–1951) was an archeologist, geologist, ethnologist, zoologist, and famous Costa Rican writer. From a young age he collected birds, insects, minerals, and plants. He took his first degree at the University of Santo Tomás in 1883. In 1885 he asked the President of Costa Rica to create a National Museum, and then he dedicated much of his life to it, becoming Director not long after it was established in 1887. He spent his life teaching and exploring as well as continuing to collect, thereby discovering a number of new taxa that carry his name. He wrote a number of books including one on Costa Rican mammals, but he also wrote poetry. He was much admired throughout Europe and the Americas and corresponded with all the leading naturalists of his day. He is also commemorated in the names of other taxa, from ants, such as *Cephalotes alfaroi* and *Pheidole alfaroi,* to amphibians such as *Oedipina alfaroi*. The rice rat is found from eastern Mexico to northwest Ecuador; the squirrel from southern Nicaragua to northern Colombia; and the water rat from eastern Honduras to Colombia and northwest Venezuela.

Allen, G. M.

Allen's Big-eared Bat *Idionycteris phyllotis* G. M. Allen, 1916

Professor Dr. Glover Morrill Allen (1879–1942) was an American collector, curator, editor, librarian, mammalogist, ornithologist, scientist, taxonomist, teacher, and writer. Between 1901 and 1927 he was Librarian at the Boston Society of Natural History. In 1907 he was hired to oversee the mammal collection at the Museum of Comparative Zoology, Harvard University, having taken his Ph.D. there in 1904. He became Curator of Mammals in 1925, staying in that post until 1938 and then becoming Professor of Zoology, in which capacity he served until his death in 1942. One of his collaborators on *Narrative of a Trip to Bahamas* (1904) was Thomas Barbour, whom he had met at Harvard and who went on to become Director of the Museum in 1927. Allen was keen on all vertebrates, particularly birds (he edited *The Auk* from 1939 to 1942) and mammals (he was President of the American Society of Mammalogists from 1927 until 1929). He took part in the Harvard African expedition to Liberia in 1926, as well as collecting trips to the Bahamas in 1903, Labrador in 1906, East Africa in 1909, the West Indies in 1910, Africa again in 1912 and 1926, Brazil in 1929, and Australia in 1931. He wrote a great many scientific papers and articles and a number of books. Early works include *Mammals of the West Indies* (1911). Later works include the two-volume *Mammals of China and Mongolia*, published between 1938 and 1940. His *Bats* of 1939 is considered a classic. His last book, *Extinct and Vanishing Mammals of the Western Hemisphere*, was published posthumously. The bat is found from southern Utah and southern Nevada, USA, to central Mexico. See also **Glover** and **Gloverallen.**

Allen, H.

Allen's Yellow Bat *Rhogeessa alleni* **Thomas,** 1892

Dr. Harrison Allen (1841–1897) was regarded by Thomas as "the chief authority on North-American bats," and so he named the bat in Allen's honor. Allen was a physician who graduated in Philadelphia in 1861 and spent the years 1862 to 1865 as a surgeon in the U.S. Army during the American Civil War. After the war he was made Professor of Comparative Anatomy and Medical Zoology at Pennsylvania University Medical School in Philadelphia. In 1875 he transferred to the Chair of Physiology and stayed in that post until 1895. His interest in bats arose as a result of a chance meeting in 1861 with Spencer Fullerton Baird, who noticed the young Allen reading, in translation, the chapter on bats in Cuvier's *Regne animal*—instead of what he should have been doing, revising for his final medical exams. Baird advised him to forget the book and look at the specimens instead and went on to use his influence to have Allen's military postings to be as convenient as possible for pursuing his natural history studies at the Smithsonian. Allen's best-known work on a zoological subject is *A Monograph of the Bats of North America*, which appeared in 1893. The bat is endemic to Mexico.

Allen, J. A.

Allen's Olingo *Bassaricyon alleni* **Thomas,** 1880
Round-tailed Muskrat *Neofiber alleni* **True,** 1884
Antelope Jackrabbit *Lepus alleni* **Mearns,** 1890
Allen's Woodrat *Hodomys alleni* **Merriam,** 1892
Allen's Squirrel *Sciurus alleni* **Nelson,** 1898
Allen's Cotton Rat *Sigmodon alleni* **Bailey,** 1902
Allen's Mastiff Bat *Molossus sinaloae* J. A. Allen, 1906
Allen's Swamp Monkey *Allenopithecus nigroviridis* **Pocock,** 1907
Allen's Round-eared Bat *Lophostoma carrikeri* J. A. Allen, 1910 [Alt. Carriker's Round-eared Bat]
Allen's Hutia *Isolobodon portoricensis* J. A. Allen, 1916 extinct
Allen's Striped Bat *Glauconycteris alboguttatus* J. A. Allen, 1917

Joel Asaph Allen (1838–1921) was an America zoologist with interests in both mammals and birds. He studied under Agassiz and accompanied him to Brazil in 1865. He made a number of field trips in North America and in 1873 became chief of an expedition sent out by the Northern Pacific Railroad. He became an assistant in ornithology at the Museum of Comparative Zoology at Cambridge, Massachusetts, in 1870. In 1885 he became Curator of the Department of Mammals and Birds in the American Museum of Natural History in New York, a post he held until his death in 1921. He wrote many scientific papers and edited the ornithological journal *The Auk*. Allen also wrote a number of monographs, including one with Dr. Elliott Coues. He organized the American Ornithologists' Union with Coues and Brewster, serving as its first President, and was also a founding member of the National Audubon Society. In addition to naming many species, he made important studies on geographic variation relative to climate. Allen's recognition of "variation within populations and intergradation across geographic gradients" helped to overturn the typological species concept current in the mid-1800s, setting out the principle that intergrading populations should be treated as subspecies instead of separate species. This idea led to the widespread adoption of trinomials by American zoologists, a practice that Allen helped to spread through his editorship of *The Auk* and through the American Ornithologists' Union (AOU) code of nomenclature. He established what was later called "Allen's rule," the observation that animals in cold climates had small extremities: so, for example, from the northern Arctic Hare *(Lepus arcticus)* through the more southerly Antelope Jackrabbit *(L. alleni)*, members of the genus show progressively longer extremities (legs and ears) and leaner bodies. The olingo is found in Ecuador, Peru, and Bolivia. The muskrat is found in Florida and extreme southeast Georgia, USA. The jackrabbit occurs in southern Arizona and northwest Mexico. The woodrat and the cotton rat are both found in western Mexico, while the squirrel is endemic to northeast Mexico. The mastiff bat occurs from Mexico south to Suriname and Trinidad, and the round-eared bat in northern South America. The striped bat is known from DRC (Zaire) and Cameroon. The swamp monkey dwells in riparian forests along the Congo River and its tributaries. The hutia, from Hispaniola and Puerto Rico, is generally regarded as extinct, although some authorities believe that populations may yet persist and await rediscovery.

Allen, W. A.

Allen's Squirrel Galago *Galago alleni* **Waterhouse,** 1838 [Alt. Allen's Bushbaby; Syn. *Galagoides alleni*]
Allen's Wood Mouse *Hylomyscus alleni* Waterhouse, 1838

Rear Admiral William Allen (1793–1864) was an English naval officer who was involved in fighting the African slave trade. He led three expeditions up the Niger River in West Africa, two in 1832 and one in 1841, where he collected the type specimen of a bird that commemorates his name, Allen's Gallinule *Porphyrio alleni*. The type specimens of both mammals were collected on the island of Fernando Pó (now Bioko). The galago is endemic to the island, with similar populations found from southeast Nigeria to Gabon and the Congo Republic now being regarded as separate species. The wood mouse also occurs on the West African mainland, from Guinea east to Gabon and the Central African Republic.

Allenby

Allenby's Gerbil *Gerbillus allenbyi* **Thomas,** 1918

Field Marshall Sir Edmund Henry Hynman, Viscount Allenby of Megiddo (1861–1936), was a career soldier. He served in Africa before and during the Boer War. He went on to command the Cavalry Division, the Cavalry Corps, V Corps, and the Third Army on the Western Front. From June 1917 he was Commander in Chief, Egyptian Expeditionary Force. He is most famous for capturing Jerusalem from Turkish occupation in December 1917. He is also known to be the only Christian General to have succeeded in capturing both Jerusalem and the strategic fortress of Acre—a feat that was beyond the early Crusaders such as Richard Coeur-de-Lion and St. Louis. Thomas was commemorating what the western world of the time saw as a great victory; his original description of the gerbil reflects the chauvinist sentiments of his era. He wrote, "I have named it in honour of the general to whose forces the country where it occurs (i.e. Palestine) owes release from the barbarian domination under which it has suffered for so many centuries." Allenby was also a keen ornithologist, extremely interested in bird migration, and among his staff in Palestine and Egypt toward the end of and immediately after WW1 was Richard Meinertzhagen. The two men are known to have been on very good terms and remained friends, even after Meinertzhagen was sacked by Allenby. The gerbil is endemic to Israel, though it may be conspecific with the more widespread *Gerbillus andersoni*.

Aloysius

Duke of Abruzzi's Free-tailed Bat *Chaerephon aloysiisabaudiae* Festa, 1907

The binomial *aloysiisabaudiae* is just a Latinized way of saying Luigi of Sabaudia: Aloysius is the Latin for Luigi, and the House of Savoy (sometime Kings of Italy) traced its descent from the medieval Counts of Sabaudia. See **Duke of Abruzzi** for biographical details.

Alston

Alston's Brown Mouse *Scotinomys teguina* Alston, 1877

Alston's Cotton Rat *Sigmodon alstoni* **Thomas,** 1881

Mexican Volcano Mouse *Neotomodon alstoni* **Merriam,** 1898

Alston's Woolly Mouse-Opossum *Micoureus alstoni* **J. A. Allen,** 1900 [Alt. Alston's Opossum; formerly *Marmosa alstoni*]

Edward Richard Alston F.Z.S. (1845–1881) was Secretary of the Linnean Society. He wrote a number of scientific papers, including, with C. Danford, "On the Mammals of Asia Minor," which appeared in the *Proceedings of the Zoological Society of London* in 1877, and at least one longer work, *Biologia Centrali-American: Mammalia*, published between 1878 and 1882. In 1871 he visited Norway and, in 1872, Archangel in Russia. He was collecting in Central America, particularly Guatemala and Belize, from 1879 until 1881. The brown mouse occurs from southern Mexico to western Panama. The cotton rat is found from northeast Colombia to Suriname and northern Brazil. The volcano mouse is confined to central Mexico. The mouse-opossum is found from Belize south to Colombia.

Ammon

Argali *Ovis ammon* **Linnaeus,** 1758

Ammon was an Egyptian god usually depicted as a human with a ram's head—an appropriate deity to use for this sheep, which is found in central Asia, with scattered populations from Kazakhstan and Kyrgyzstan to Tibet, Mongolia, and southern Siberia.

Anak

Giant Naked-tailed Rat *Uromys anak* **Thomas,** 1907

Thomas sometimes used biblical or mythological names but did not explain why. Such is the case with *Uromys anak,* but we can be fairly sure that the large size of this rat led him to name it after Anak, the biblical progenitor of the giants, as stated in Numbers 13.33: "And there we saw the giants, the sons of Anak, which come of the giants: and we were in our own sight as grasshoppers, and so we were in their sight." It is possible that Thomas may have had an actual man in mind, as the Victorians loved freak shows, and very tall men would be exhibited as giants. For instance, Joseph Brice, a Frenchman who claimed to be 244 cm (8 feet) tall (but was probably a bit shorter) exhibited himself at St. James Hall in Piccadilly in 1865 as "Anak, King of the Anakims, or the Giant of Giants." The rat comes from the highlands of New Guinea.

Anchieta

Anchieta's Antelope *Cephalophus (melanorheus) anchietae* **Bocage,** 1878 [Alt. Angolan Blue Duiker; now *Cephalophus monticola anchietae*]

José (Alberto) de (Oliveira) Anchieta (1832–1897) was an independent Portuguese naturalist and collector. The common name Anchieta's Antelope is seldom used nowadays, as this taxon is regarded as the Angolan subspecies of the Blue Duiker. See **D'Anchieta** for biographical details and other eponymous species.

Andersen

Andersen's Bare-backed Fruit Bat *Dobsonia anderseni* **Thomas,** 1904
Andersen's Horseshoe Bat *Rhinolophus anderseni* **Cabrera,** 1909
Andersen's Fruit-eating Bat *Artibeus anderseni* **Osgood,** 1916

Dr. Knud Andersen (d. 1918) was a Danish zoologist specializing in Chiroptera (bats). He was employed by the British Museum of Natural History from 1904 but mysteriously disappeared in 1918 and is presumed to have died then. He collected in the Pacific in the 1900s and in Queensland in 1912. He wrote *Catalogue of the Chiroptera in the Collection of the British Museum,* published in 1912, which remains the most comprehensive treatment of the Megachiroptera, and he also published a number of descriptions of new species. The bare-backed bat is found in the Bismarck Archipelago (Papua New Guinea). The horseshoe bat occurs in the Philippines and the fruit-eating bat in Ecuador, Peru, Bolivia, and western Brazil. See also **Canut.**

Anderson, J.

Anderson's Squirrel *Callosciurus quinquestriatus* Anderson, 1871
Anderson's Shrew *Suncus stoliczkanus* Anderson, 1877
Anderson's Gerbil *Gerbillus andersoni* **de Winton,** 1902

Professor Dr. John Anderson F.R.S. (1833–1900) was a Scottish naturalist who was Professor of Natural History at Free Church College in Edinburgh. He became Curator of the Indian Museum in Calcutta in 1865 and collected for the trustees. He went on scientific expeditions to Yunnan in 1867, to Burma between 1875 and 1876, and to the Mergui Archipelago between 1881 and 1882. In 1885 he became Professor of Comparative Anatomy at the Calcutta Medical School, then returned to London in 1886. He wrote *Guide to the Calcutta*

Zoological Gardens and began another work, *Zoology of Egypt—Mammalia,* which was completed by de Winton and published in 1902. The squirrel is found in northeast Myanmar and western Yunnan (China). The shrew occurs in deserts and arid country in Pakistan, Nepal, India, and Bangladesh. The gerbil is found in coastal areas of North Africa from Tunisia to Sinai.

Anderson, M. P.

Anderson's Red-backed Vole *Myodes andersoni* **Thomas,** 1905 [Alt. Japanese Red-backed Vole; Syn. *Phaulomys andersoni*]
Anderson's Shrew-Mole *Uropsilus andersoni* Thomas, 1911
Anderson's White-bellied Rat *Niviventer andersoni* Thomas, 1911
Anderson's Four-eyed Opossum *Philander andersoni* **Osgood,** 1913 [Alt. Black Four-eyed Opossum]

Malcolm Playfair Anderson (1879–1919) was born in Indianapolis but was educated at secondary level in Germany, returning to the USA to study zoology at Stanford University and graduating in 1904. From the age of 15 he took part in collecting expeditions to Arizona, Alaska, and California. In 1901 he joined the Cooper Ornithological Club and wrote a number of articles on ornithology, yet did not confine himself to that subject. In 1904 he was chosen to conduct the Duke of Bedford's Exploration of Eastern Asia for the Zoological Society of London; he took photographs and extensive notes on the collections and wrote several short stories about the people with whom he lived and worked in the Orient. He was again in western China in 1909 and 1910. In 1912 he found the type specimen of the four-eyed opossum during a trip to Peru with Osgood. He died in 1919 after falling from scaffolding at the shipyards in Oakland, California. The vole is endemic to Honshu, Japan, and the shrew-mole to Sichuan, China. The rat dwells in the highlands of central and southern China. The opossum's range extends from Peru to western Brazil and southern Venezuela.

Anderson, S.

Anderson's Mouse-Opossum *Marmosa andersoni* **Pine,** 1972 [Alt. Heavy-browed Mouse-Opossum]
Anderson's Rice Rat *Cerradomys andersoni* Brooks et al., 2004
Anderson's Oldfield Mouse *Thomasomys andersoni* Salazar-Bravo and Yates, 2007

Sydney Anderson is an American mammalogist who is Emeritus Curator of Mammals at the American Museum of Natural History. He has devoted his life to the study of the mammals of Bolivia and has written widely on the subject, in books such as *Mammals of Bolivia, Taxonomy and Distribution* (1997) and articles published in the *Bulletin of the American Museum of Natural History.* He was President of the American Society of Mammalogists between 1974 and 1976. The mouse-opossum is critically endangered and occupies a small range in southern Peru. The rice rat comes from the Cerrado region of Brazil. This species may be a junior synonym of *Cerradomys scotti,* named two years earlier. The oldfield mouse is found on the eastern slopes of the central Bolivian Andes.

Andrews

Andrews' Beaked Whale *Mesoplodon bowdoini* Andrews, 1908 [Alt. Bowdoin's Beaked Whale]
Andrews' Hill Rat *Bunomys andrewsi* **J. A. Allen,** 1911
Andrews' Three-toed Jerboa *Stylodipus andrewsi* J. A. Allen, 1925

Roy Chapman Andrews (1884–1960) was a larger-than-life American who became an explorer, collector, and curator. Many believe him to have been the real-life model for the hero of the Indiana Jones movies. He always maintained that from his earliest childhood he had a desire for travel and adventure. "I was born to be an explorer," he wrote in 1935 in *The Business of Exploring*. "There was never any decision to make. I couldn't do anything else and be happy." He said too that his only ambition was to work at the American Museum of Natural History. He first worked as a taxidermist. After graduating in 1906 he went to New York City and applied for a job at the museum. The Director told him there were no jobs, but Andrews persisted, saying "You have to have somebody to scrub floors, don't you?" The Director took him on, and from this humble beginning he went on to become the museum's most famous explorer. As a taxidermist he developed an interest in whales and traveled to Alaska, Japan, Korea, and China to collect various marine mammals. (We do not know if he ever tried to combine his interests in whales with his skills as a taxidermist.) From 1909 to 1910 he was naturalist on the USS *Albatross* voyage to the Dutch East Indies. From 1921 to 1923 he led an expedition to China and Outer Mongolia, where he collected both live specimens and fossils, including the first eggs to be positively identified as those of a dinosaur. He continued to make further expeditions over a number of years until 1930. He returned to the USA and four years later became Director of the museum. He retired in 1942, moved to California, and spent the rest of his life writing about his exploits in works including an autobiography, *Under a Lucky Star*. The jerboa is found in Mongolia and northern China, and the rat on the Indonesian islands of Sulawesi and Butung. The whale is recorded mainly in the cold-temperate waters off southern Australia and New Zealand.

Angas

Nyala *Tragelaphus angasii* **Gray,** 1849

George Francis (French) Angas (1822–1886) was an English explorer, zoologist, and artist. He painted African animals as well as other subjects. He certainly traveled widely, having painted in New Zealand in 1844, Australia in 1846, and South Africa in 1847. He published *The Kafirs Illustrated* in 1849. He served as Secretary of the Australian Museum, Sydney, in the 1850s. His *Savage Life and Scenes in Australia and New Zealand* was published in 1847 in two volumes with numerous illustrations. He was also a noted conchologist, and several molluscs were named after him, including the cone shell *Conus angasi*. The nyala is found in southern Malawi, Mozambique, eastern Zimbabwe, Swaziland, and KwaZulu-Natal and eastern Transvaal (South Africa).

Anita

Anita's Leaf-eared Mouse *Phyllotis anitae* Jayat, D'Elía, Pardiñas, and Namen, 2007

Anita Kelley Pearson (dates not found) was the wife of Oliver Payne Pearson (q.v.), whom she married in 1944. She was active in the same fields of zoology as her husband and traveled widely with him; often they were also accompanied by their young children. The mouse comes from the province of Tucuman, northwest Argentina.

Annandale

Annandale's Rat *Rattus annandalei* **Bonhote,** 1903

Dr. (Thomas) Nelson Annandale (1876–1924) was a zoologist who was Director of the Indian Museum during the British Raj. He published a number of scientific papers during the 1900s through the 1920s—for example, in 1915, "Fauna of the Chilka Lake: Mammals, Reptiles and Batrachians." He was instrumental in separating out a zoological survey (as opposed to one combined with anthropology) and undertook several expeditions. Most notable of these was the Annandale-Robinson expedition, which collected the type specimen of the rat, in Malaya from 1901 to 1902. The rat is found on the Malay Peninsula, Singapore, eastern Sumatra, and the islands of Padang and Rupat off the Sumatran coast.

Ansell, P. D. H.

Ansell's Shrew *Crocidura ansellorum* **Hutterer** and Dippenaar, 1987

P. D. H. Ansell is the son of W. F. H. Ansell (see below), and it was he who collected the type specimen of the shrew in Zambia, although the species is named after both father and son (*ansellorum* means "of the Ansells"). He has written a number of articles, including "More Light on the Problem of Hartebeest Calves" (1970) and, with his father, "Mammals of the Northeastern Montane Areas of Zambia" (1973).

Ansell, W. F. H.

Ansell's Wood Mouse *Hylomyscus anselli* Bishop, 1979
Ansell's Shrew *Crocidura ansellorum* **Hutterer** and Dippenaar, 1987
Ansell's Mole-Rat *Fukomys anselli* Burda et al., 1999 [Alt. Zambian Mole-Rat; formerly *Cryptomys anselli*]
Ansell's Epauletted Fruit Bat *Epomophorus anselli* Bergmans and Van Strien, 2004
Upemba Lechwe *Kobus anselli* Cotterill, 2005

Dr. William Frank Harding Ansell (1923–1996) was a zoologist who studied the mammals of southeast Africa. He obtained his Ph.D. at Liverpool University in 1960 and in the same year was appointed as Provincial Game Officer in Northern Rhodesia (now Zambia). He wrote *Mammals of Northern Rhodesia* (1960), *The Mammals of Zambia* (1978), and *African Mammals* (1989), and co-wrote *Mammals of Malawi* (1988). He studied south-central African lechwe antelopes intensively throughout the 1960s. He is also commemorated in the scientific name of a bird, a subspecies of the Cloud Cisticola *Cisticola textrix anselli*. The wood mouse is found in the highlands of northern Zambia and southwest Tanzania. It was formerly treated as conspecific with *Hylomyscus denniae*. The shrew and mole-rat are both endemic to Zambia. The fruit bat was identified from two museum specimens from Malawi. The lechwe was also identified as a new species from museum specimens and comes from the Upemba wetlands of the southeast Congo basin.

Ansorge

Southern Giant Pouched Rat *Cricetomys ansorgei* **Thomas,** 1904
Ansorge's Cusimanse *Crossarchus ansorgei* Thomas, 1910
Ansorge's Grass Rat *Arvicanthis ansorgei* Thomas, 1910
Ansorge's Free-tailed Bat *Chaerephon ansorgei* Thomas, 1913

Dr. William John Ansorge (1850–1913) was an English explorer and collector who was active in Africa in the second half of the 19th century. He wrote *Under the African Sun* in 1899. He is also commemorated in the common names of three birds, Ansorge's Crombec *Sylvietta rufescens ansorgei*, Ansorge's Greenbul *Andropadus ansorgei*, and Ansorge's Robin–Chat *Xenocopsychus ansorgei*. Many African fish are named after him, including one with the splendid common name of Slender Stonebasher *Hippopotamyrus ansorgii*. His son, Sir Eric Cecil Ansorge (1887–1977), continued the family tradition of interest in natural history: he was a lepidopterist and was also interested in beetles. The pouched rat is found from Uganda and Kenya south to northeast South Africa, extending westward through parts of Zambia and Zimbabwe to Angola. The cusimanse (mongoose) occurs in northern Angola and southern DRC (Zaire). The grass rat is a West African species, found from Gambia and Senegal east to southern Chad. The bat's distribution extends from Cameroon east to Ethiopia and south to Angola and Natal (South Africa).

Anthony, A. W.

Anthony's Woodrat *Neotoma anthonyi* **J. A. Allen,** 1898
Anthony's Pocket Mouse *Chaetodipus (fallax) anthonyi* **Osgood,** 1900

Alfred Webster Anthony (1865–1939) was an American collector and ornithologist (and father of H. E. Anthony, q.v.). His primary interest was birds, and in 1886 he published *Field Notes on the Birds of Washington County, Oregon.* He was President of the Audubon Society in Portland in 1904. He collected birds for years in the Tualatin Valley, his specimens now being in the Carnegie Museum in Pittsburgh, Pennsylvania. However, he also collected mammals, and this collection can also be found in the American Museum of Natural History and in the San Diego Museum of Natural History. His collections also included reptiles, invertebrates, plants, and minerals. His interest in natural history was firmly rooted in its conservation. His first trip out of the USA was to North Coronado Island, Mexico. He later traveled to Alaska, during the gold rush, and went on a collecting trip to Guatemala. He is commemorated in the common names of five birds, including Anthony's Towhee *Pipilo crissalis senicula* and Anthony's Vireo *Vireo huttoni obscurus,* at least one plant, *Dudleya anthonyi,* and a crab, *Cancer anthonyi.* Both of the rodents named after him are confined to small islands off Baja California, Mexico: the woodrat to the Todos Santos Islands, and the pocket mouse to Cedros Island.

Anthony, H. E.

Anthony's Bat *Sturnira ludovici* Anthony, 1924 [Alt. Highland Yellow-shouldered Bat]
Anthony's Puma *Puma concolor anthonyi* **Nelson** and **Goldman,** 1931
Anthony's Pipistrelle *Hypsugo anthonyi* **Tate,** 1942 [formerly *Pipistrellus anthonyi*]

Dr. Harold Elmer Anthony (1890–1970) was Curator of Mammals at the American Museum of Natural History and a noted collector of animals, especially in the Neotropics. He was the son of A. W. Anthony (q.v.), but his prime interest was mammals rather than the birds favored by his father. Early papers include "A New Rabbit and a New Bat from Neotropical Regions" and "Two New Fossil Bats from Porto Rico," both published in the *Bulletin of the American Museum of Natural History* in 1917. He was President of the American Society of Mammalogists from 1935 to 1937. He took part in the Vernay-Cutting expedition to Burma in 1938, where "he was delighted with the native Lisu who brought him many species of voles and mice. Delighted, that is, until he realised they were selling him animals that they had stolen from his own traps." The yellow-shouldered bat is found from Mexico south to Ecuador and Guyana. The pipistrelle is known only from the type locality in northern Myanmar. The puma subspecies occurs in southern Venezuela.

Anubis

Olive Baboon *Papio anubis* **Lesson,** 1827

Anubis was a god of the ancient Egyptians usually shown as a man with a jackal's head. So this name could possibly be inspired by the doglike physiognomy of baboons, or perhaps Lesson might have had his gods mixed up. Was he thinking of Thoth (sometimes depicted as a baboon)? The species is found from Mali eastward to Kenya and northwest Tanzania.

Aplin

Arfak Pygmy Bandicoot *Microperoryctes aplini* Helgen and **Flannery,** 2004

Dr. Ken P. Aplin works as a herpetologist and general zoologist at the Museum of Western Australia. He co-authored a "Checklist of the Frogs and Reptiles of Western Australia." The bandicoot comes from the Arfak Mountains of western New Guinea.

Arata

Arata and Thomas' Yellow-shouldered Bat *Sturnira aratathomasi* **Peterson** and Tamsitt, 1968 [Alt. Aratathomas's Yellow-shouldered Bat, Giant Andean Fruit Bat]

Dr. Andrew A. Arata is a tropical disease expert who retired in March 1999. From 1968 to 1985 he was a scientist/ecologist for the World Health Organization and Pan American Health Organization in Mexico, Venezuela, and Geneva, and from 1985 to 1992 he was associated with the Vector Biology and Control (VBC) project of the U.S. Agency for International Development. He was the overall Director and Manager of this activity. He wrote *The Anatomy and Taxonomic Significance of the Male Accessory Reproductive Glands of Muroid Rodents* in 1964. The bat is known from Colombia, northwest Venezuela, Ecuador, and northern Peru. It was named after Andrew Arata and Dr. Maurice Thomas (see **Thomas, M.**), a Professor of Biology, who collected the holotype.

Archbold

The shrew-mouse genus *Archboldomys* **Musser,** 1982 [3 species: *kalinga*, *luzonensis*, and *musseri*]

Richard Archbold (1907–1976) was an American patron of science who became a zoologist at the American Museum of Natural History. He financed and led expeditions, particularly to Australasia. He also set up a permanent research station at Lake Placid in Florida. Seven bird species from the early expeditions to New Guinea and Madagascar are named in his honor in their scientific names, as well as 18 insects, 3 spiders, and no fewer than 44 plants. The shrew-mice named after him are found in the Philippines.

Archer

Green Ringtail Possum *Pseudochirops archeri* **Collett,** 1884

The original Green Ringtail Possum specimen was collected by Carl Lumholtz when he was the guest of the Archer family, who were pastoralists at a property called Gracemere in Queensland. The Archer family originally came to Australia from Scotland via Larvik in Norway, to which they had moved in the early 19th century. There were many children in the Archer family, and nine of them either visited or settled in Australia. Unfortunately the original description of the possum does not mention which member of the family was intended to be the recipient of the dedication. Eliminating those who had already died or who had left Australia before 1881, we are left with four brothers, so you can take your pick. The four were William (1818–1896), Archibald (1820–1902), Thomas (1823–1905), and Colin (1832–1921), and any or all of those could have been, and almost certainly were, at Gracemere when Lumholtz stayed there for about 10 months in 1881. The possum is found only in northeast Queensland.

Archer, M. A.

Chestnut Dunnart *Sminthopsis archeri* Van Dyck, 1986

Professor Michael Archer (b. 1945) was born in Sydney, Australia, with dual Australian and American citizenship. His career in vertebrate paleontology began when he was 11. He graduated from Princeton University in 1967 and was Fulbright Scholar at the Western Australian Museum from 1967 to 1968. He obtained his Ph.D. in zoology at the University of Western Australia in 1976. From 1972 to 1978 he was Curator of Mammals, Queensland Museum. In 1978 he joined the University of New South Wales. After a notable academic career with many Ph.D. students, he was appointed Director of the Australian Museum, in which position he served from 1999 to 2004. In 2004 he was appointed Dean of the Faculty of Science at UNSW. He has spent equal amounts of time on vertebrate paleontology and modern mammalogy, producing hundreds of books, research papers, CDs, and documentaries in both areas, including *Carnivorous Marsupials* (1982), *Vertebrate Zoogeography and Evolution* (1984), *Possums and Opossums: Studies in Evolution* (1987) and *Going Native* (2005). He has been awarded many academic honors, such as the Eureka Prize for Promotion of Science and a range of professional fellowships including the Australian Academy of Science. He has researched and published in all fields of vertebrate paleontology (most taxonomic groups), stratigraphy, paleoecology, and biocorrelation. The dunnart lives in northeast Australia and southern New Guinea.

Arends

Arends' Golden-Mole *Carpitalpa arendsi* Lundholm, 1955 [formerly *Chlorotalpa arendsi*]

Nicholas P. Arends (dates not found) assisted Guy Shortridge, who was a collector for the British Museum of Natural History. He also helped Shortridge in the collecting of specimens of African mammals for the Kaffrarian Museum in King William's Town, South Africa, during the years 1922 to 1947, with some specimens being sent to the New York Museum of Natural History as well as the British Museum of Natural History. The golden-mole is found in eastern Zimbabwe and a small adjacent area of western Mozambique.

Arianus

Persian Field Mouse *Apodemus arianus* **Blanford,** 1881

Ariana (as mentioned by Strabo) was an area close to the River Indus, which was part of the Persian Empire. We believe that this, rather than a person, is what the scientific name refers to. However, in the unlikely event that it was named after an individual, then a possible candidate would be Arianus (d. ca. A.D. 311), a Roman official in Egypt who became governor of Thebes. He is known only from literary and documentary sources. With Theoticus and three others he was converted to Christianity on witnessing at Alexandria the martyrdom of Apollonius and Philemon. The field mouse is found from Iran westward through Iraq to Israel and Lebanon. (The taxonomy of this genus is complex; it has been argued that the holotype of *Apodemus arianus* is actually an example of the Yellow-necked Field Mouse *A. flavicollis*.)

Arianus, M.

Arianus' Rat *Stenomys omichlodes* **Misonne,** 1979 [Syn. *Rattus omichlodes*]

Arianus Murip (1980–1993), who came from the village of Ilaga, was a victim of the extreme inhumanity of the Indonesian military against the people of Irian Jaya. With his sister, he suffered from frostbite and exposure on the flanks of Mount Carstensz—at 4,884 m (16,024 feet), the highest mountain in New Guinea—and when they descended to find help and treatment they were detained by security guards at a mine. Arianus was beaten and left to die but was found and taken to hospital. Sadly he died there: he was almost certainly never given treatment, as his injuries should not have been fatal had he been properly cared for. Tim Flannery coined the common name of this rodent. In his book *Mammals of New Guinea* he writes, "The common name, Arianus's Rat, is in memory of 13 year old Arianus Murip of Ilaga, and the many others who have lost their lives on Mount Carstensz on the long walk home." This species of rat is common in Meren Valley, Mount Carstensz.

Armandville

Flores Giant Tree Rat *Papagomys armandvillei* **Jentink**, 1892

Father Cornelis J. F. le Cocq d'Armandville (1846–1896) was a Dutch Jesuit missionary in the Dutch East Indies (now Indonesia). He was ordained a priest in 1876 and sent to the Indies, where he was stationed in East Ceram. In 1894 he landed at Kapaur, becoming the first Christian missionary to be in that part of New Guinea.In 1896 he was due to return to Java for medical treatment but went instead on a trip to the southern coast of New Guinea. The ship returned without him, with the report that he was dead, but there were no clear facts or coherent story as to how he died. The rat is endemic to the island of Flores.

Arnoux

Arnoux's Beaked Whale *Berardius arnuxii* Duvernoy, 1851

Dr. Maurice Arnoux (dates not found) was a French ship's surgeon who presented the skull of this whale to the Paris Museum of Natural History in 1846. He had been aboard the corvette *Le Rhin* on its journey to Australia and New Zealand, and he had found the skull on a beach close to Akaroa, New Zealand. (Unfortunately Duvernoy made a spelling error when coining the scientific name, omitting the *o* from Arnoux.) The whale is found in the Southern Hemisphere in circumpolar and cold-temperate waters.

Ascanius

Red-tailed Guenon *Cercopithecus ascanius* Audebert, 1799 [Alt. Black-cheeked White-nosed Monkey]

Ascanius was the son of Aeneas, the hero of Virgil's *Aeneid*, and considering that it was popular to name primates after figures in classical mythology and literature, we believe that this is who Audubert had in mind. The other possibility is that he was thinking of Professor Peder Ascanius (1723–1803), a Norwegian zoologist and mineralogist who gave his very extensive collections and his fortune to Det Kongelige Norske Videnskabers Selskab (Royal Norwegian Society of Sciences). During the period 1768 to 1770 he traveled along the west coast of Norway and collected marine fauna for his work *Icones rerum naturalium, ou Figures enluminées d'histoire naturelles du Nord* (Illustrations of the Natural History of the North). On balance we think that Audubert would have used *ascanii* in the genitive to show that he had the man in mind, rather than the nominative *ascanius*, which is not unexpected for a mythological personage. The guenon is found from the Central African Republic eastward through Uganda and western Kenya and south into the northern parts of Angola and Zambia.

Attenborough

Attenborough's Echidna *Zaglossus attenboroughi* **Flannery** and Groves, 1998 [Alt. Sir David's Long-beaked Echidna, Cyclops Long-beaked Echidna]

Sir David Frederick Attenborough (b. 1926) is famous as a maker of wildlife television programs. He studied natural sciences at Cambridge and in 1950 joined a firm of publishers in London, staying only briefly before joining the BBC in the early days of its postwar television service. He has been associated with the BBC, first as an employee and later as a freelance journalist, virtually ever since. He rose high in the organization's ranks, becoming Controller of BBC2 and introducing color television to Britain, yet his first love was not administration but photojournalism. He has made some of the most stunning series of nature programs ever produced and, through the BBC as publishers, excellent books to accompany them: *The Living Planet* (1984), *The Life of Birds* (1998), and *The Life of Mammals* (2002), to name a few. His elder brother, Richard (Lord) Attenborough is a well-known actor and film director. A prehistoric plesiosaur was named *Attenborosaurus conybeari* by Bakker in 1993 with the words "in honour of the naturalist and filmmaker David Attenborough, whose childhood fascination with Liassic plesiosaurs sparked a brilliant career in scientific journalism." David Attenborough has said that Flannery was "in the league of the all-time great explorers like Dr. David Livingstone." Attenborough also wrote the introduction in Flannery's 1995 book *Mammals of New Guinea*. This species of echidna was known from only one (damaged) specimen, collected around 1961 in the Cyclops Mountains of New Guinea. It was over 35 years before the specimen was recognized as a distinct species. It had been feared that the echidna might have become extinct, but in July 2007 it was reported that researchers had discovered evidence of its continued existence in the Cyclops Mountains.

Attila

Hun Shrew *Crocidura attila* **Dollman,** 1915

Attila (ca. A.D. 406–453) the Hun, also known as "the Scourge of God," was a conqueror who, having become King or General of his people in 433, came out of the plains of central Asia to conquer half the known world. "Attila" was first used zoologically by Lesson in 1830, for a genus of New World flycatchers named after the same historical figure, an example of which is the Rufous-tailed Attila *Attila phoenicurus*. The shrew is found from southeast Nigeria and Cameroon to eastern DRC (Zaire).

Attwater

Attwater's Pocket Gopher *Geomys attwateri* **Merriam,** 1895

Attwater's White-footed Mouse *Peromyscus attwateri* **J. A. Allen,** 1895 [Alt. Texas Mouse]

Henry Philemon Attwater (1854–1931) was a naturalist and conservationist. He was born in Brighton, England, emigrating in 1873 to Ontario, Canada, where he farmed and kept bees. He became interested in natural history and, together with John A. Morden, prepared and exhibited natural history specimens in 1883. During 1884 the two men collected specimens in Bexar County, Texas. During the latter part of 1884 and early 1885 Attwater and Toudouze were employed to prepare and exhibit natural history specimens in the Texas Pavilion at the New Orleans World's Fair. Attwater's major contributions to natural history were in the areas of ornithology, mammalogy, and conservation. He also contributed specimens to the Smithsonian. However, it is in the field of ornithology that he remains best known, and his three ornithological papers deal with the nesting habits of 50 species of birds in Bexar County, the occurrence of 242 species of birds near San Antonio, and the deaths of thousands of warblers in 1892. He was elected a Director of the National Audubon Society in about 1900 and again in 1905. Through his influence with farmers, by 1910 the Texas Audubon Society had gained affiliation with the Texas Farmers' Congress, the Texas Cotton Growers' Association, and the Texas Corn Growers' Association. He was also active in the promotion of legislation to protect the Mourning Dove *Zenaida macroura*, which was rapidly declining during the early 1900s. His most important conservation works include *Boll Weevils and Birds* (1903), *Use and Value of Wild Birds to Texas Farmers and Stockmen and Fruit and Truck Growers* (1914), and *The Disappearance of Wild Life* (1917). He is also commemorated in the name of a bird, Attwater's Prairie Chicken *Tympanuchus cupido attwateri.* The gopher is restricted to south-central Texas. The mouse is found from southeast Kansas and southwest Missouri to central Texas.

Aubinn

Slender-tailed Giant Squirrel *Protoxerus aubinnii* **Gray,** 1873

Mr. Aubinn (dates not found) was a collector. The type specimen was received by the British Museum in "a series of skins of Mammalia that were collected by Mr. Aubinn at Fantee" (Ghana). Gray says in his description, "This species I have named after Mr. Aubinn, who has sent many good specimens of Mammalia and birds from Fantee, and is a very intelligent native collector." No other name(s) or details of Mr. Aubinn are recorded. The squirrel comes from the forests of Liberia, Ivory Coast, and Ghana.

Audebert

Lowland Red Forest Rat *Nesomys audeberti* **Jentink,** 1879

Josef-Peter Audebert (1848–1933) was a German naturalist who was collecting in Madagascar during the period 1876 to 1882. In 1878 he was employed as a collector by Hermann Schlegel, who at the time was in charge of the museum at Leiden (where Jentink worked as a zoologist), which is why his collections were all sent to Holland. The rat is endemic to Madagascar.

Audubon

Desert Cottontail *Sylvilagus audubonii* **Baird,** 1858

Audubon's Bighorn Sheep *Ovis canadensis auduboni* **Merriam,** 1901 extinct [Alt. Badlands Bighorn Sheep]

John James Audubon (1785–1851) is remembered as the father of U.S. ornithology. He gave several different accounts of his birth, but he was the son of a French naval captain and a French girl who worked at his sugar plantation in San Domingo (Haiti). Audubon's mother died within a short time of his birth, so Audubon's father took him back to France, where he was adopted by the captain and his legal wife. In 1803 Captain Audubon sent him to manage his plantation near Philadelphia (and so, also, to avoid conscription into Napoleon's army). He became an American citizen in 1812. In Philadelphia, Audubon met and married his wife, Lucy, whose support was critical in his success. He succeeded only because he went in 1826 to England, where his work was appreciated and subscribers made possible the long publication of his 435 bird prints (1826–1838). In the 1830s Audubon also wrote his *Ornithological Biography,* which describes the habits of the birds he drew. Audubon made a trip to the western regions of North America in 1843, his last great adventure prior to his death in 1851. While remembered for his love of, and paintings of, birds, he was interested in all aspects of natural history. He even painted all the plates for a book he conceived on American mammals, which he asked Bachman to write the text for, and together they created *Viviparous Quadrupeds of North America,* which Bachman and his sons eventually published in 1851 after Audubon's death. He spent weeks in the woods studying birds and other animals, and his spectacular drawings, which were criticized as overimaginative by some, were scenes he actually witnessed. Several birds are named after him, including Audubon's Shearwater *Puffinus lherminieri*, Audubon's Warbler *Dendroica coronata auduboni*, and Audubon's Woodpecker *Picoides villosus audubonii*. The cottontail is found in the southwestern and central USA and northern Mexico. The sheep formerly inhabited North Dakota, South Dakota, Montana, Nebraska, and Wyoming but was pushed to extinction by 1925. However, modern research suggests that the subspecies *auduboni* was not distinct from the Rocky Mountain Bighorn *Ovis canadensis canadensis*.

Avila Pires

Ávila Pires' Saddle-back Tamarin *Saguinus fuscicollis avilapiresi* **Hershkovitz,** 1969

Professor Dr. Fernando Dias de Avila Pires (b. 1933) is a Brazilian zoologist. He joined the National Museum of Natural History, Rio de Janeiro, as a voluntary assistant in 1954, and in 1957 he published his first paper on mammalogy. In 1958 he became Assistant Professor at the Federal University of Rio de Janeiro. In 1959 he became a Fellow of the John Simon Guggenheim Memorial Foundation, and between 1959 and 1961 he was a Research Associate at the American Museum of Natural History. After his return to Brazil, although permanently on the staff of the National Museum, he undertook other assignments such as being Director of the newly established Museum of Natural History at the University of Minas Gerais and also working at the universities of Campinas, São Paulo, and Rio Grande do Sul. In 1986 he became a Professor at the National School of Public Health, Oswaldo Cruz Foundation, Ministry of Health, and from 1991 to 1993 was Vice President of the Foundation; he later moved to the Department of Tropical Medicine, Oswaldo Cruz Institute. He has published on mammal taxonomy, biogeography, ecology, conservation, ecology of nonhuman reservoirs of diseases, and human ecology. His publications include "Mamíferos colecionados na região do Rio Negro, Amazonas, Brazil" (1964) and "Taxonomia e zoogeografia do gênero *Callithrix* Erxleben, 1777 (Primates, Callithricidae)" (1969). The tamarin is endemic to the Brazilian state of Amazonas.

Ayres

Ayres' Uacari *Cacajao ayresi* Boubli et al., 2008 [Alt. Aracá Uakari]

Dr. José Márcio Ayres (1954–2003) was a Brazilian biologist, conservationist, and primatologist whose bachelor's degree was awarded by the University of São Paulo in 1976, as was his master's in 1981, and his doctorate by Cambridge University in England in 1986. He worked at both the São Paulo and the National Institute for Amazonian Research. He founded the Mamirauá Sustainable Development Reserve in 1996 and the Amaña Sustainable Development Reserve a year later, both in the Brazilian state of Amazonas. His thesis at Cambridge was *The White Uakaris and the Amazonian Flooded Forest.* He died of lung cancer. The monkey lives in a small area that encompasses the Rio Curuduri basin and adjacent areas, Amazonas, Brazil.

Azara

Azara's Night Monkey *Aotus azarai* **Humboldt,** 1811

Azara's Dog *Pseudalopex gymnocercus* **Fischer,** 1814 [Alt. Azara's Fox, Pampas Fox; Syn. *Lycalopex gymnocercus*]

Azara's Agouti *Dasyprocta azarai* **Lichtenstein,** 1823

Azara's Grass Mouse *Akodon azarae* Fischer, 1829

Azara's Tuco-tuco *Ctenomys azarai* **Thomas,** 1903

Féliz de Azara (1746–1811) was a Spanish naturalist. He started his working life as an engineer, distinguished himself on various expeditions, and so rose to the eventual rank of

Brigadier General in the Spanish army. He was appointed to the Spanish commission and sent to South America, in 1781, to try to settle the boundaries between Portuguese and Spanish colonies. He stayed in South America until 1801. It was there that he started to study mammals, particularly as an observer of the behavior of quadrupeds. His notes are generally acknowledged to be meticulous, but they also contained reports from others that were not as accurate as his own. He sent his observations to his brother, who was the Spanish Ambassador in Paris, and Moreau de Saint-Méry published them in Paris in 1801, under the title *Essai sur l'histoire naturelle des quadrupèdes du Paraguay* (also published in Madrid in 1802 as *Apuntamientos para la historia natural de los cuadrúpedos del Paraguay y Río de la Plata*). In 1809 he published *Voyage dans l'Amérique Méridionale depuis 1781 jusqu'en 1801*. The monkey comes from Bolivia south of Rio Madre de Dios, Paraguay, and northern Argentina. The dog is found in eastern Bolivia, Paraguay, southeast Brazil, Uruguay, and north and central Argentina (south to Rio Negro Province). The grass mouse has a similar range to the dog but in Argentina is confined to the northeast of the country. The agouti is found in central and south Brazil, Paraguay, and northeast Argentina. The tuco-tuco is endemic to La Pampa Province, Argentina.

Babault

Babault's Mouse-Shrew *Myosorex babaulti* **Heim de Balsac** and **Lamotte,** 1956

Guy Babault (1888–ca. 1932) was a French traveler, naturalist, conservationist, and collector. He was collecting in British East Africa from 1912 until 1920, at which time he is known to have been in India and Ceylon (now Sri Lanka). He wrote about his extensive collecting missions and published, among others, *Chasses et recherches zoologiques en Afrique Orientale Anglaise* in 1917 and *Recherches zoologiques dans les provinces centrales de l'Inde et dans les régions occidentales de l'Himalaya* in 1922. Many of the animals he collected can be seen in the Bourges Museum, gifted in 1927 on his return from another trip to East Africa. At least one book was written about his journeys, *Voyage de M. Guy Babault dans l'Afrique Orientale Anglaise 1912–1913.* There were further specimens gifted to the museum by his wife in 1932, and we can find no record of him after that date. He is also commemorated in the scientific name of a fish found in Lake Tanganyika, *Simochromis babaulti.* The shrew is confined to the mountains west and east of Lake Kivu, in DRC (Zaire), Rwanda, and Burundi.

Baber

Afghan Flying Squirrel *Eoglaucomys baberi* **Blyth,** 1847 [Syn. *E. fimbriatus baberi*]

Emperor Zahiruddin Muhammad Baber (1483–1530) was the founder of the Mughal Empire that ruled over Afghanistan. He was born in what is today Uzbekistan and conquered India in 1526, founding the dynasty that ruled there until 1857. One of his descendants built the Taj Mahal. It was said of him that the oath of temperance weighed heavily on him (he made a late and ecstatic discovery of wine): "His drinking parties took on an epic quality, on boats, in subterranean chambers of delight, in the violet garden or the garden of fidelity, because of a victory or a good harvest or a haircut." Nevertheless, he died sober. Baber wrote his memoirs, and the "Flying Fox" mentioned in them was equated with this flying squirrel; hence the scientific name *baberi.* Zoologists disagree on its taxonomic status, as some consider it to be only a subspecies of the Kashmir Flying Squirrel *Eoglaucomys fimbriatus.* It is found in the mountains of northern Afghanistan and Pakistan.

Bachman

Bachman's Shrew *Sorex longirostris* Bachman, 1837 [Alt. Southeastern Shrew]

Bachman's Hare *Sylvilagus bachmani* **Waterhouse,** 1839 [Alt. Brush Rabbit]

Dr. John Bachman (1790–1874) grew up on a farm and was always a great outdoorsman. On a trip to Philadelphia he met Alexander Wilson and Humboldt, and it was these illustrious friends who convinced him that the study of nature was a pursuit he should follow. When he was a teacher and preacher (he was a Lutheran minister of Charleston, South Carolina) he took up the study of birds and small mammals, especially rabbits, a project he continued for the rest of his life. He remained an active churchman and helped establish the Lutheran Synod of South Carolina, serving twice as its President (1824–1833 and 1839–1840). Although he owned slaves, he bucked the trend by educating them and was said to have baptized more than 90 African Americans in one year. Bachman wrote *Characteristics of Genera and Species, as Applicable to the Doctrine of Unity in the Human Race* (1864). This was a radical yet scientifically accurate tract that took the position that master and slave were the same species, providing a scientific rationale against slavery. Bachman had eight children by his first wife Harriet Martin and, we think, none by his second wife Maria, Harriet's sister, who became a very fine illustrative artist. One of his daughters, also called Maria, married John Woodhouse Audubon (1812–1862), the younger son of John James Audubon (Bachman's close friend—two of his daughters were the first wives of Audubon's two sons). When Audubon conceived the idea of a book on American mammals he asked Bachman to write the text, and together they created *Viviparous Quadrupeds of North America*, which Bachman and his sons eventually published in 1851 after Audubon's death. Audubon named a number of bird species after Bachman, such as Bachman's Sparrow *Aimophila aestivalis* and Bachman's Warbler *Vermivora bachmanii*, and also immortalized his friend in the scientific name of the American Black Oystercatcher *Haematopus bachmani*. The shrew is found in the southeastern USA. The rabbit occurs from western Oregon south to Baja California in Mexico.

Baer

Baer's Wood Mouse *Hylomyscus baeri* **Heim de Balsac** and **Aellen,** 1965

Professor Jean George Baer (1902–1975) was a Swiss parasitologist. He has been described as one of the great pioneer conservationists and environmentalists. He worked for many years at the Zoology Institute, University of Neuchâtel , Swiss Center for Scientific Research in Ivory Coast, serving as its head from 1941 until 1972. He published more than 260 scientific papers and books. Much of his work was on the parasites of rodents, particularly those in Ivory Coast. (Aellen, who named the mouse, undertook his doctorate at Neuchâtel when Baer was Professor there.) The wood mouse is confined to Ivory Coast and Ghana.

Bahamonde

Bahamonde's Beaked Whale *Mesoplodon bahamondi* Reyes, Van Waerebeek, Cardenas, and Yañez, 1995 [Alt. Spade-toothed Beaked Whale; Syn. *M. traversii* **Gray,** 1874]

Professor Nibaldo Bahamonde Navarro (b. 1924) is a Chilean marine biologist. He graduated in 1943 and then took a teaching qualification. By 1946 he was Professor of Biological and Chemical Sciences at the University of Chile. During the 1950s he carried out research on marine fauna in Europe, including at the University of Bergen, Norway. During the 1960s he continued to teach, with his interests turning to ecology and conservation. It was in 1965 that he went to the Juan Fernandez Islands to investigate a rumor, which fortunately turned out to be true, that the Juan Fernandez Fur Seal *Arctocephalus philippii*, then thought extirpated in the middle of the 19th century, was in fact still extant. He has published over 200 books, articles, and papers, including, with G. R. Pequeño in 1975, *Peces de Chile: Lista sistematica*. In 2004 he was made Professor Emeritus of the University of Chile, Santiago, where he had long worked. The whale was described in 1995 from remains found in the Juan Fernández Archipelago, Chile, but it was later shown through DNA analysis that a species named much earlier, *Mesoplodon traversii* (Gray, 1874), is the same taxon. Thus the first description and name must take precedence (see **Travers**).

Bailey, A. M.

Bailey's Shrew *Crocidura baileyi* **Osgood,** 1936

Dr. Alfred Marshall Bailey (1894–1978) was an American zoologist. He was Curator of Birds and Mammals at the Louisiana State Museum from 1916 to 1919, and from 1919 until 1921 he was part of the survey of Alaska undertaken by the U.S. Fish and Wildlife Service. He worked for the Field Museum of Natural History in Chicago from 1926 to 1927, and during those years he was a member of the expedition to Abyssinia organized by the Field Museum in conjunction with the *Chicago Daily News*. From 1927 until 1936 he was Director of the Chicago Academy of Science, and from 1936 until 1969 he was at the Denver Museum of Natural History. He was also a leading light in the American Ornithologists' Union, as well as being a notable early photographer and cinematographer. He wrote many articles, particularly on ornithological topics, such as "Birds of Arctic Alaska" in (1948), and books such as his major work, written with R. J. Niedrach, *The Birds of Colorado* (1965). However, he also published more generalist and popular works such as *Fieldwork of a Naturalist* about his time in Alaska. He collected the type specimen of the shrew in 1927 in the Simien Mountains of Ethiopia, where it is endemic. He is also commemorated in the name of the Sierra Madre (or Bailey's) Sparrow *Xenospiza baileyi*.

Bailey, F. M.

Red Goral *Naemorhedus baileyi* **Pocock,** 1914

Lieutenant Colonel Frederick Markham Bailey (1882–1967) was a British army officer, spy, and explorer. He was known as "Eric," to differentiate him from his father, also an army officer whose first name was Frederick. "Eric" was a spy and butterfly collector. He was born at Lahore (in what is now Pakistan but was then part of India) and educated at Sandhurst. He was commissioned in 1901 and posted to the Nilgiri Hills, where he met Richard Meinertzhagen, who was convalescing after a bout of enteric fever; they were to become lifelong friends. Bailey was proficient in Tibetan and so accompanied Younghusband's invasion of Tibet in 1904, staying on until 1909. He later traveled alone in then-unknown parts of Tibet and China and was awarded the Gold Explorer's Medal by the Royal Geographical Society in London. Among his more famous discoveries was the Blue Poppy *Meconopsis betonicfolia baileyi*. During WW1 he fought in France with the Indian Expeditionary Forces and later at Gallipoli. In 1918 he was sent to Tashkent in order to discover the new Bolshevik government's intentions, particularly in relation to India. His presence became known, and he had to flee for his life; he made his escape by disguising himself as an Austrian prisoner of war and joining the Cheka (Russian secret police), where his task was to track down a dangerous British agent, namely himself. He later recorded his experiences in *Mission to Tashkent*, which was published in 1946. He was Political Officer in Sikkim from 1921 to 1928. In 1957 he published *No Passport to Tibet* about his experiences before 1914. The red goral is found in southeast Tibet and Yunnan (China), northern Myanmar, and Assam (India).

Bailey, V. O.

Bailey's Bobcat *Lynx rufus baileyi* **Merriam,** 1890

Bailey's Pocket Mouse *Chaetodipus baileyi baileyi* Merriam, 1894

Vernon Orlando Bailey (1864–1942) was a U.S. naturalist and ethnographer. As a young Minnesota farmer he sent many natural history specimens to C. Hart Merriam (q.v.), the then head of the U.S. Biological Survey. Bailey joined the survey in 1887, serving as its chief from 1889 until 1902. He retired in 1933. During his time with the survey he undertook many field trips, some with his wife, including six to Texas. In 1899 he married Florence Augusta Bailey (née Merriam), who also worked for the survey. He wrote *The Mammals and Life Zones of Oregon* (1936). A "Transcription of Vernon Bailey's Field Notes for His 1909 Trip to Lincoln

Co., Coos Co., and Curry Co., Oregon . . ." was published in the *Journal of Oregon Ornithology* in 1996. The subspecies of bobcat is found in the southwest USA and adjacent northern Mexico. The pocket mouse is found from southern California, Arizona, and southwest New Mexico in the USA, south to northern Sinaloa and Baja California in Mexico.

Bailward

Iranian Mouse-like Hamster *Calomyscus bailwardi* **Thomas,** 1905

Brigadier General Arthur Charles Bailward (1855–1923) was a British soldier and amateur collector. In 1905 Oldfield Thomas presented a paper entitled "On a Collection of Mammals from Persia and Armenia Presented to the British Museum by Col. A. C. Bailward" in the *Proceedings of the Zoological Society of London.* Gertude Bell, who was in Syria at the time, in her diary entry of 13 February 1909 wrote, "An English Colonel has arrived here. His name is Bailward and he had stayed with the Spring Rices at Tehran [Teheran]. He is deaf and not very enlightened." The hamster is found in central and southwest Iran.

Baird

Baird's Pocket Gopher *Geomys breviceps* Baird, 1855
Baird's Tapir *Tapirus bairdii* **Gill,** 1865
Baird's Dolphin *Delphinus bairdi* **Dall,** 1873 [Alt. Pacific Common Dolphin; Syn. *D. capensis*]
Baird's Beaked Whale *Berardius bairdii* **Stejneger,** 1883
Baird's Shrew *Sorex bairdi* **Merriam,** 1895

Spencer Fullerton Baird (1823–1887) was an American zoologist who became a giant of American ornithology, and sufficient has been written elsewhere to make a long entry here redundant. The young Baird became a friend of John James Audubon in 1838 and sent him specimens. After studying medicine for a time, Baird became Professor of Natural History at Dickinson College, Pennsylvania, in 1845. From 1850 until 1878 he was Assistant Secretary of the U.S. National Museum, the Smithsonian Institution, and then became Secretary. He organized a number of expeditions from 1850 until 1860, some with the SS *Albatross.* He wrote copiously: among his major works are *Catalogue of North American Reptiles* (1853), *Catalogue of North American Birds* (1858), and *Mammals of North America: Descriptions Based on Collections in the Smithsonian Institution* (1859). He was also commemorated in the names of no fewer than eight birds, including Baird's Sandpiper *Calidris bairdii,* Baird's Sparrow *Ammodramus bairdii, and* Baird's Trogon *Trogon bairdii.* The gopher is found in eastern Texas, western Louisiana, southwest Arkansas, and eastern Oklahoma. The tapir is found from southern Mexico to Colombia west of the Rio Cauca and Ecuador west of the Andes. The dolphin, from the eastern Pacific, is now usually regarded as a junior synonym of *Delphinus capensis.* The whale is found in the North Pacific, from the Bering Sea to Japan and California. The shrew has a limited distribution in the moist coniferous forests of Oregon.

Baker

Baker's Harvest Mouse *Reithrodontomys bakeri* Bradley et al., 2004

Dr. Robert J. Baker (b. 1942) is an American biologist. He is Horn Professor of Biological Sciences, Director Natural Science Research Laboratory, and Curator of Mammals at Texas Tech University. Dr. Baker has played a major role in investigating chromosomal evolution, systematics, and molecular evolution in *Reithrodontomys.* He was a research associate at the Carnegie Museum in Pittsburgh from 1975 to 1990. He has visited many countries in the neotropics and spent 20 months studying fauna at the Chernobyl site following the nuclear accident there. He has written many papers and described a number of species. He is also honored in the trinomials of three subspecies: two bats, *Glossophaga commissarisi bakeri* and *Lophostoma saurophila bakeri*, and one pocket gopher, *Geomys texensis bakeri*. The mouse is found in the mountains of Guerrero State, Mexico.

Balston

Inland Broad-nosed Bat *Scotorepens balstoni* **Thomas,** 1906 [Alt. Western Broad-nosed Bat; Syn. *Nycticeius balstoni*]
Tube-nosed Bat sp. *Murina balstoni* Thomas, 1908 [Syn. *M. suilla* **Temminck,** 1840]

William Edward Balston (1848–1918) was a successful British businessman who was also interested in ornithology. He sponsored collecting expeditions to Indonesia and some of the first collecting expeditions to Western Australia. His brother, Richard James Balston (1839–1916), was also a patron of the sciences, using wealth acquired in the business of papermaking. One or the other is also commemorated in the scientific names of three bird subspecies (the African Swift *Apus barbatus balstoni*, White-browed Scrubwren *Sericornis frontalis balstoni*, and Australian Yellow White-eye *Zosterops luteus balstoni*) and a fish (Balston's Pygmy Perch *Nannatherina balstoni*). As Oldfied Thomas wrote a paper in 1906 entitled "On Mammals Collected in Southwest Australia for Mr. W. E. Balson," we surmise that the bats are named after William. The broad-nosed bat comes from Australia. The tube-nosed bat comes from Java, but it is now often regarded as a junior synonym of the Brown Tube-nosed Bat *M. suilla*.

Bangs

Bangs' Mountain Squirrel *Syntheosciurus brochus* Bangs, 1902

Outram Bangs (1862–1932) was a zoologist and prolific collector born in Watertown, Massachusetts. He attended Harvard from 1880 to 1884. In 1890 he began a systematic study of the mammals of eastern North America and also collected bird specimens. Bangs was appointed Assistant in Mammalogy at Harvard and became Curator of Mammals at the Harvard Museum of Comparative Zoology in 1900. He visited Jamaica in 1906 and collected over 100 birds there, but his trip was cut short by dengue fever. His collection of over 10,000 mammalian skins and skulls, including over 100 type specimens, was presented to Harvard College in 1899. In 1908 his collection of over 24,000 bird skins was presented to the Museum of Comparative Zoology, and he went on to increase it. In 1925 he went to Europe, visiting museums and ornithologists and arranging scientific exchanges. He wrote over 70 books and articles, 55 of them on mammals, including an article on "The Florida Puma" that was published in the *Proceedings of the Biological Society of Washington* in 1899. Bangs is also commemorated in the names of Bangs' Black Parrot *Coracopsis nigra libs* and Bangs' Sparrow *Amphispiza bilineata bangsi*, and in the tanager genus *Bangsia*. The squirrel is found in Costa Rica and western Panama.

Bannister

Bannister's Melomys *Melomys bannisteri* Kitchener and Maryanto, 1993

Dr. John Leonard Bannister is a former Director and now Honorary Researcher of the Western Australian Museum, Perth. He has written a great many scientific papers on mammals in general and cetaceans in particular. The rodent is known only from the small island of Kai Besar in the Kai Islands, eastern Indonesia.

Baoule

Baoule's Mouse *Mus baoulei* Vermeiren and **Verheyen,** 1980

The mouse is named not after a person but after a people who live where it occurs: the Baoule, originally part of the Ashanti people. The mouse is known only from Ivory Coast and eastern Guinea.

Barbara

Tayra *Eira barbara* **Linnaeus,** 1758

Barbara is Latin for "wild," so this mammal is not named after a person called Barbara. This member of the weasel family (Mustelidae) is found in Central and South America.

Barbara Brown

Barbara Brown's Titi *Callicebus (personatus) barbarabrownae* **Hershkovitz,** 1990 [Alt. Northern Bahian Blond Titi, Blond Titi]
Barbara Brown's Brush-tailed Rat *Isothrix barbarabrownae* **Patterson** and Velazco, 2006

Dr. Barbara E. Brown (b. 1929) is an Associate in the zoology department and a benefactor of the Field Museum in Chicago, where there is a specific Barbara Brown Fund for Mammal Research. She was trained as an economist but was working as a helper at a children's nature camp when she was recruited in about 1970 to the Field Museum by the then Curator of Mammals, Philip Hershkovitz. The titi monkey was first described as a race of the Masked Titi *Callicebus personatus,* and the citation reads that "the subspecies is named in honor of Associate Barbara E. Brown for her many years of active support of the Field Museum of Natural History and valuable contributions to research by the staff of the Division of Mammals." She has written some papers on mammals, including "Atlas of New World Marsupials" (2004). The titi, now sometimes elevated to a full species, is found in the state of Bahia, eastern Brazil. The brush-tailed rat is from the Peruvian Andes.

Barbe

Barbe's Leaf Monkey *Trachypithecus barbei* **Blyth,** 1847 [Alt. Tenasserim Leaf Monkey, Tenasserim Langur]

The Rev. J. Barbe (fl. 1860) was a traveler and collector. In addition to collecting the type specimen of this monkey, Barbe traveled in pursuit of natural history specimens. He is known to have been on board the Danish steamer *Ganges* and briefly visited the Nicobar Islands in 1846. Blyth in 1852 listed five specimens of orioles from the Nicobars "presented by Capt. J. Lewis and the Rev. J. Barbe in 1846." Barbe also published a paper in 1847 on the ambergris that was collected in the Nicobars and used by the Burmese and Chinese for medicinal purposes. The monkey is found in a narrow strip of land in the peninsular area of Myanmar and Thailand.

Barbour

Barbour's Rock Mouse *Petromyscus barbouri* **Shortridge** and Carter, 1938
Barbour's Vlei Rat *Otomys barbouri* **Lawrence** and Loveridge, 1953

Dr. Thomas Barbour (1884–1946) was an American zoologist. He graduated from Harvard in 1906 and obtained his Ph.D. there in 1910. The following year he became an Associate Curator of Reptiles and Amphibians at the Harvard Museum of Comparative Zoology, and in 1927 he was appointed as its Director, staying until his death in 1946. In 1927 he became Custodian of the Harvard Biological Station and Botanical Garden Soledad, Cuba. He was Executive Officer in charge of Barro Colorado Island Laboratory in Panama from 1923 until 1945. During his time at the museum he explored in the East Indies, the West Indies, India, Burma, China, Japan, and South and Central America. He was famously jovial good company and would invite all and sundry to eat and converse next door to his office in the "Eateria" in which his secretary, Helen Robinson, prepared the food for his thousands of guests. Something of an all-rounder, he wrote many articles and books, including *The Birds of Cuba* (1923) and *Naturalist at Large* (1943). He also co-wrote *Checklist of North American Amphibians and Reptiles.* His special area of interest was the herpetology of Central America, and many amphibian and reptile species are named after him, including Barbour's Map Turtle *Graptemys barbouri* and the Streamside Salamander *Ambystoma barbouri.* The mouse is found in a small area of northwest Cape Province, South Africa. The vlei rat is confined to Mount Elgon on the Kenya-Uganda border.

Barnard

Barnard's Hairy-nosed Wombat *Lasiorhinus latifrons barnardi* **Longman,** 1939 [Alt. Northern Hairy-nosed Wombat, Yaminon; Syn. *L. krefftii*]

Henry "Harry" Greensill Barnard (1869–1966) was an Australian zoologist, naturalist, and grazier in Queensland. His father, George, was an entomologist and an oologist whose collection got so large that he had to build himself a private museum at Coomooboolaroo in 1891. On George's death in 1894 the collection was sold to Lord Rothschild's private museum at Tring, England (now part of the British Museum of Natural History). Many naturalists and zoologists, such as Carl Lumholtz in 1883, stayed at Coomooboolaroo. In 1888 Henry joined a government expedition to explore the Bellenden Ker Range. Albert Stewart Meek, who was a family friend of the Barnard family, collected in Northern Queensland for Rothschild in 1894, and Barnard accompanied him. In 1896 and 1899 Barnard was in Cape York collecting for a number of different people. He spent his working life as a grazier, alternating with collecting trips, until he retired to Brisbane. He is honored in the scientific names of other taxa, such as the Yellow-naped Snake *Furina barnardi.* The Northern Hairy-nosed Wombat is today restricted to one population in the Epping Forest National Park, central Queensland. Due to wombat taxonomy having been revised since Longman described this form, the correct scientific name for the species is now *Lasiorhinus krefftii.*

Barnes

Barnes' Mastiff Bat *Molossus barnesi* **Thomas,** 1905 [Alt. Barnes' Free-tailed Bat]

Mr. Barnes (dates and forenames not found), from Cayenne, French Guiana, sent the type specimen of this bat to Oldfield Thomas at the British Museum of Natural History. Unfortunately Thomas recorded no other details. The bat is known only from French Guiana.

Barnes, C. T.

Barnes' Pika *Ochotona princeps barnesi* Durrant and Lee, 1955 [Alt. American Pika]

Claude Teancum Barnes (1884–1968) was a naturalist in Utah, USA. He published widely on a variety of subjects in many magazines, such as *Utah Mammals,* in which he published an essay on badgers in 1927. Stephan Durrant, one of the describers of the pika, refers to this essay in his publication of 1952, *Mammals of Utah: Taxonomy and Distribution.* Barnes, whose last publication in 1966 seems to have been a *Dictionary of Utah Slang,* co-written with Dorothy B. Jensen, is the subject of a biography by Davis Bitton. About 35 different subspecies of the American Pika have been described; this one is found in Utah.

Barrera

Red Viscacha Rat *Tympanoctomys barrerae* **Lawrence,** 1941

Dr. J. M. de la Barrera (dates not found) was a parasitologist. Given his profession, it is not surprising that he is also noted for having collected some new types of flea in Argentina in the 1930s–1950s. The viscacha rat is found in west-central Argentina and was named in his honor because he studied the parasites associated with this species and its nests, finding a species of flea *Parapsyllus barrerai* exclusively associated with it.

Barrett

Red Rock Hare sp. *Pronolagus barretti* **Roberts,** 1948 [Alt. Hewitt's Red Rock Hare]

Dudley A. Barrett (dates not found) was the person who supplied the type specimen of the hare in response to a request. The describer, Roberts, suspected that a population of rock hares in Natal would prove to be taxonomically distinct and asked Barrett to get a specimen for him from that area. Barrett succeeded, and Roberts named the hare in his honor. The "species" is currently regarded as being a junior synonym of Hewitt's Red Rock Hare *Pronolagus saundersiae*, although a 1995 study by Whiteford indicated that it might still deserve to be recognized as a separate species.

Bartels

Bartels' Spiny Rat *Maxomys bartelsii* **Jentink,** 1910

Bartels' Rat *Sundamys maxi* **Sody,** 1932 [Alt. Bartels' Giant Sunda Rat]

Bartels' Flying Squirrel *Hylopetes bartelsi* Chasen, 1939

Max Eduard Gottlieb Bartels (1871–1936) was a Dutch plantation owner and naturalist who lived in Java from 1896 until his death. Bartels named a rat after Sody in 1937, so returning the favor of Sody's naming the giant rat after him in 1932. He is also commemorated in the common name of a bird, Bartels' Wood Owl *Strix (leptogrammica) bartelsi*, and in the scientific name of the Javan Hawk-Eagle *Spizaetus bartelsi*. All three mammals are endemic to Java.

Barton

Barton's [Long-beaked] Echidna *Zaglossus bartoni* **Thomas,** 1907

Captain Francis Rickman Barton (1865–1947) was a colonial administrator in New Guinea whose main interests were botany, anthropology, and early pioneering photography. He went to Papua in 1899 as Commandant of the Armed Native Constabulary. He was Resident Magistrate of the Central Division from 1902 to 1904 and was Acting Administrator for the colony between 1904 and 1907. He went on leave in 1907 and resigned from his post in 1908, becoming Vizier or First Minister in Zanzibar and holding that post until 1913. He also "collected" a human skull in New Guinea that became part of the British Museum of Natural History collections in 1919. He accompanied Captain Charles A. W. Monckton on some of his expeditions. He is noted for taking a remarkable series of photographs of the native tribes between 1897 and 1907 and may have shot some cinema film as early as 1904. He corresponded widely with anthropologists and photographic enthusiasts. In 1920 he was made President of the Royal Geographical Society. The echidna is found in eastern New Guinea.

Basilio

Father Basilio's Striped Mouse *Hybomys basilii* **Eisentraut,** 1965

Father Aurelio Basilio (dates not found) was a priest who lived on the island of Fernando Pó (now Bioko) from about 1944 to about 1972. He was the parish priest of Sácriba Pámue, his religious order being the Congregation of the Missionary Brothers of the Immaculate Heart of the Blessed Virgin Mary. During that period Eisentraut also spent some time on the island, and there is no doubt that the two men knew each other. Basilio published "La vida animal en la Guinea Española" (*Instituto de Estudios Africanos,* 1962) and "Aves de la isla de Fernando Poo" (*Editorial Coculsa,* 1963). He also jointly published, with D. Amadon, "Notes on the Birds of Fernando Po Island, Spanish Equatorial Guinea" (*American Museum Novitates*, 1957). The mouse is endemic to Bioko and is an endangered species.

Bastard

Bastard Big-footed Mouse *Macrotarsomys bastardi* **Milne-Edwards** and **Grandidier,** 1898

Eugène Joseph Bastard (1865–1910) was a French colonial administrator, traveler, zoologist, paleontologist, and collector. He worked at the National Museum of Natural History in Paris alongside Grandidier and was particularly connected with Madagascan fauna. He undertook a collecting expedition to Madagascar, returning to France in 1898, and published "Exploration au sud de l'Onilahy" (*Notes, Reconnaissances et Explorations,* 1899) and "Mission chez les Mahafalys" (*Revue de Madagascar,* 1900), having visited Madagascar again in 1899. He is also remembered in the binomial of a reptile, Mocquard's Madagascar Ground Gecko *Paroedura bastardi.* The mouse is from west and south Madagascar.

Bates

Bates' Slit-faced Bat *Nycteris arge* **Thomas,** 1903
Bates' Dwarf Antelope *Neotragus batesi* **de Winton,** 1903
Dollman's Tree Mouse *Prionomys batesi* **Dollman,** 1910
Bates' Shrew *Crocidura batesi* Dollman, 1915

George Latimer Bates (1863–1940) was an American ornithologist and botanist who traveled in West Africa between 1895 and 1931, residing in Cameroon for some years. He wrote *Handbook of the Birds of West Africa* (1930) and a number of articles that were published in *The Ibis,* notably "Birds of the Southern Sahara and Adjoining Countries" (1933). He also left unpublished manuscripts on the birds of Arabia, which were subsequently utilized by Meinertzhagen for his 1950s work on the subject. The results of his collecting forays were sent to the British Museum of Natural History, the Philadelphia Academy, and other institutions. An African member of the nettle family, *Pouzolzia batesii,* is also named after him, as are at least five birds including Bates' Weaver *Ploceus batesi* and Bates' Nightjar *Caprimulgus batesi.* The bat can be found from Sierra Leone east to western Kenya, and on Bioko. The dwarf antelope has a range extending from southeast Nigeria and Cameroon to Equatorial Guinea, DRC (Zaire), and Uganda. The tree mouse inhabits forest-savannah mosaic in Cameroon, the Central African Republic, and the Congo Republic. The shrew is restricted to lowland forest in southern Cameroon and Gabon.

Bauer

Van Gelder's Bat *Bauerus dubiaquercus* Van Gelder, 1959

Harry J. Bauer (1886–1960) of Los Angeles was a prominent entrepreneur who in 1933 became President of Edison International. He was cosponsor of the Puritan-American Museum of

Natural History expedition to Baja California in 1958. Bauer also owned the *Puritan*, which ship was used for the expedition. There is a "Harry J. Bauer Collection in the History of Science" at the California Institute of Technology, in Pasadena. He is also remembered in the name of Bauer's Nightsnake *Hypsiglena torquata baueri*. The bat was first described from the Tres Marias Islands, off western Mexico, but has also been found on the mainland of Central America from southern Mexico to Costa Rica.

Baverstock

Baverstock's Forest Bat *Vespadelus baverstocki* Kitchener, Jones, and Caputi, 1987 [Alt. Inland Forest Bat]

Professor Peter R. Baverstock (b. 1948) is an Australian zoologist. He graduated in 1969, obtained his Ph.D. in 1972, and received a D.Sc. from the University of Adelaide in 1989. He has a long and distinguished academic career and is currently Pro Vice Chancellor (Research) at Southern Cross University. He has published numerous academic papers and holds a number of patents and directorships in industry. The bat, an Australian endemic, was named in recognition of Baverstock's contributions to systematics and study of the evolution of Australian fauna.

Bayon

Bocage's Tree Squirrel *Funisciurus bayonii* **Bocage,** 1890 [Alt. Lunda Rope Squirrel]

Dr. Enrico Pietro Bayon (b. 1856) was a physician and naturalist working in West Africa in the latter part of the 19th century and in Uganda in the early part of the 20th. His father was in the Swiss Consular Service in Genoa, where Dr. Bayon was born. He collected for the Natural History Museum in Genoa and the Regional Museum of Natural Science in Turin. He practiced medicine and collected ants, fish, amphibians, reptiles, and basically anything else that was collectable. He is also commemorated in the names of a number of reptiles, including Bayon's Skink *Sepsina bayoni* and an amphibian, Bayon's Common Reed Frog *Hyperolius viridiflavus bayoni*. The squirrel is found in northeast Angola and southern DRC (Zaire).

Beatrix

Beatrix Oryx *Oryx beatrix* **Gray,** 1857 [Alt. Arabian Oryx; Syn. *O. leucoryx* **Pallas,** 1777]

Beatrix's Bat *Glauconycteris beatrix* **Thomas,** 1901

HRH the Princess Beatrice (1857–1944), Princess Henry of Battenberg, was the fifth daughter and youngest child of Queen Victoria and Prince Albert of Saxe-Coburg-Gotha. In 1885 Princess Beatrice married His Serene Highness Prince Henry of Battenberg (1858–1896). The antelope, from the Arabian Peninsula, was almost certainly named after the Princess, as it was described in the year of her birth. Some years later it was realized that the species had been described by Pallas 80 years earlier, and so the name *Oryx beatrix* was discarded as a junior synonym. The bat, which ranges from Ivory Coast to Angola, was collected in 1898 by George Latimer Bates (1863–1940), who traveled and collected in West Africa. Perhaps Thomas had the Princess in mind when he named the species, but it may just have been a whimsical choice of name; he provided no explanation in his description.

Beaufort

Beaufort's Bare-backed Fruit Bat *Dobsonia beauforti* Bergmans, 1975

Professor Lieven Ferdinand de Beaufort (1879–1968) was a Dutch zoologist whose main interests were birds and fish. He was one of the founders of the Dutch Ornithological Association and was its President from 1924 to 1956. He undertook his first expedition to the Indo-Australian archipelago in 1900 and collected extensively with Weber in New Guinea from 1907 until 1922, obtaining specimens of 474 birds. In 1922 he succeeded Weber as Director of the Zoological Museum in Amsterdam and stayed in that post until 1949. Bergmans, who named the fruit bat, became an Assistant in the Mammals Department there in the 1970s. From 1911 until 1964 Beaufort published, with Weber, *The Fishes of the Indo-Australian Archipelago* in six volumes. He is also commemorated in the common name of a bird, Beaufort's Black-capped Lory *Lorius lory viridicrissalis.* The bat is found on small islands off western New Guinea.

Beccari

Beccari's Margareta Rat *Margaretamys beccarii* **Jentink,** 1880
Beccari's Mastiff Bat *Mormopterus beccarii* **Peters,** 1881
Beccari's Sheath-tailed Bat *Emballonura beccarii* Peters and **Doria,** 1881
Beccari's Shrew *Crocidura beccarii* **Dobson,** 1886

Dr. Odoardo Beccari (1843–1920) was an Italian botanist. He explored the Arfak Mountains of New Guinea during extensive zoological exploration with d'Albertis in 1872 and 1873. He also explored and collected on various islands of the East Indies, including Sumatra, where he found the Titan Arum or Corpse Flower *Amorphophallus titanium,* the world's largest flower. Seeds of it were sent to the Royal Botanic Gardens at Kew and were successfully grown, flowering for the first time in cultivation in 1889. At some stage of his career Beccari also collected in Ethiopia. He is also commemorated in the names of four birds: Beccari's Ground Dove *Gallicolumba beccarii,* Beccari's Pygmy Parrot *Micropsitta pusio beccarii,* Beccari's Scops Owl *Otus beccarii,* and Beccari's Scrubwren *Sericornis beccarii.* The rat is found in lowland tropical rainforest of Sulawesi. The mastiff bat is found in the Moluccas, New Guinea, and northern Australia. The sheath-tailed bat is found in New Guinea and adjacent islands. The shrew is confined to Sumatra.

Bechstein

Bechstein's Bat *Myotis bechsteini* **Kuhl,** 1817

Johann Matthäus Bechstein (1757–1822) was a German naturalist. His interests ranged from ornithology to forestry, and he was one of the first to advocate conservation, even of species then considered to be pests, such as bats. From 1810 until 1818 he was Director of the Herzoglichen Academy of Forestry. Among his writings was *Gemeinnutzige Naturgeschichte Deutschlands* (1789). In 1810, after the death of his son, he adopted Ludwig Bechstein, who became famous for writing fairy tales. Bechstein's Violet-necked Lory *Eos squamata riciniata* is also named after him. The bat ranges across Europe, as far north as England and southern Sweden, and east to the Caucasus and Iran.

Beddard

Beddard's Olingo *Bassaricyon beddardi* **Pocock,** 1921

Frank Evers Beddard M.A., F.R.S. (1858–1925) was a British anatomist and zoologist. He was a lecturer in biology at Guys Hospital, London, and also held the post of "Prosector" at the Zoological Society of London (meaning that he dissected and studied the animals that died at the zoo). He published a number of articles, such as, in 1889, in the *Proceedings of the Zoological Society of London,* "Notes upon the Anatomy of the American Tapir." He also wrote larger works such as *A Text-book of Zoogeography* (1895), *A Book of Whales* (1900), and *Mammalia* (1902), a definitive early work on mammalian anatomy. He also *Animal Coloration,* largely attacking some of Darwin's ideas on sexual selection, as well as books on everything from bird anatomy to the life history of worms. The olingo is known from Guyana and possibly extends its range into neighboring countries.

Beddome

Lesser Woolly Horseshoe Bat *Rhinolophus beddomei* **Andersen,** 1905

Colonel Richard Henry Beddome (1830–1911) was a British military officer in India who was also a keen naturalist with a particular interest in botany. He joined the army in India in 1848 and in 1857 became Assistant Conservator of Forests in the Madras presidency. He rose to Chief Conservator in 1860, a post he held until 1882. He became a member of the University of Madras in 1880. He conducted many studies of the flora of India and Sri Lanka (then Ceylon). He wrote *Trees of the Madras Presidency* (1863) and *Handbook of the Ferns of British India, Ceylon, and Malaya Peninsula* (1892). His other interests included herpetology (many reptiles and amphibians are named after him) and the study of molluscs. The bat is found in India and Sri Lanka.

Bedford (Duchess of)

Lesser Striped Shrew *Sorex bedfordiae* **Thomas,** 1911

Mary, Duchess of Bedford (1865–1937), was a naturalist, an ornithologist, an aviatrix, and a radiographer. She was the wife of Herbrand Arthur Russell, 11th Duke of Bedford (see below), and was known as "the Flying Duchess." She had a particular interest in bird migration. Between 1909 and 1914 she spent much of her time on Fair Isle, and her journals were published posthumously as *A Bird-watcher's Diary* in 1938. She became very interested in aviation and made record flights from the United Kingdom to India and back in 1929 and in the following year from the United Kingdom to Cape Town and back. She crashed into the North Sea in 1937, and her body was never recovered. The shrew is found from central China south to Nepal, Assam, and northern Myanmar (Burma).

Bedford (Duke of)

Duke of Bedford's Vole *Proedromys bedfordi* **Thomas,** 1911

Bedford Takin *Budorcas taxicolor bedfordi* Thomas, 1911 [Alt. Golden Takin, Shensin Takin]

Herbrand Arthur Russell, 11th Duke of Bedford (1858–1940), was a career soldier and amateur naturalist. His wife was Mary, Duchess of Bedford (see above). After graduating in 1877 he became an officer in the Grenadier Guards and fought in the Egyptian campaign of 1882. He held the office of Aide-de-Camp to the Viceroy of India between 1885 and 1886 and Colonel of the 3rd Battalion, Bedfordshire Regiment, between 1897 and 1908. Between 1908 and 1920 he was Militia Aide-de-Camp to King Edward VII and King George V, and he also saw distinguished service in WW1. He held a number of aristocratic titles and was Deputy Lieutenant of Bedfordshire. In addition he was a Fellow of the Royal Society, and in 1899 he became President of the Zoological Society, a post he held until 1936. He was also a trustee of the British Museum. He is commemorated in the name of a bird, Bedford's Paradise-flycatcher *Terpsiphone bedfordi,* and in the scientific name of an eastern race of the Eurasian Nuthatch *Sitta europaea bedfordi.* The vole and the takin are both found in central China.

Beechey

California Ground Squirrel *Spermophilus beecheyi* **Richardson,** 1829

Captain Frederick William Beechey (1796–1856) was a noted British geographer and mapmaker. From 1825 to 1828 he led an expedition to the Pacific and the Bering Strait in HMS *Blossom,* which made significant discoveries in the Arctic, California, and the Pacific islands. During that voyage Beechey took a formal pardon to John Adams, the last survivor of the mutiny on the *Bounty,* on Pitcairn Island. Also during the voyage, in 1825, he named Point

Barrow after Sir John Barrow of the British Admiralty. *The Zoology of Captain Beechey's Voyage* was published in 1839 with colored plates by George B. Sowerby. A further book, *The Botany of Captain Beechey's Voyage*, was published in 1841. Since many specimens were collected on that voyage, it is no surprise that some would be named after the captain; he is also commemorated in the name of a bird, Beechey's Jay *Cyanocorax beecheii*. He made further voyages such as that on HMS *Sulphur* in 1836. The ground squirrel is found from southwest Washington State, USA, south to northern Baja California, Mexico.

Beecroft

Beecroft's Scaly-tailed Squirrel *Anomalurus beecrofti* **Fraser,** 1853
Beecroft's Tree Hyrax *Dendrohyrax dorsalis* Fraser, 1855 [Alt Western Tree Hyrax]

Captain John Beecroft (1790–1854) was an English adventurer, trader, explorer, and naturalist. The French captured him in 1814 while he was serving on a coastal vessel. By 1832 he had become Acting Governor, then in 1844 Governor, on the Spanish island of Fernando Pó (now Bioko). He served as British Consul from 1849 until 1854 for the Bights of Benin and Biafra. His job was to regulate commercial relations with the coastal city-states, this "regulation" being backed up with gunboats enabling him to interfere in local internal affairs and engineer the eventual imposition of colonial rule. Previously, in 1840, he had explored the Benin River with a view to its strategic importance, rather than for any purely scientific pursuit. The "squirrel"—more correctly an anomalure, a uniquely African family of rodents—is found from Senegal to Uganda and northern Zambia. The tree hyrax's range extends from Gambia to northwest Angola, DRC (Zaire), and Uganda. Both species also occur on the island of Bioko.

Behn

Behn's Big-eared Bat *Glyphonycteris behnii* **Peters,** 1865 [formerly *Micronycteris behnii*]

William Friedrich Georg Behn (1808–1878) was a German explorer famed for his crossing of South America in 1847. He was the Director of the Zoological Museum of the Christian Albrechts University of Kiel from 1836 until 1868. He is also commemorated in the common names of two birds, Behn's Parakeet *Brotogeris chiriri behni* and Behn's Thrush *Turdus subalaris*. The bat is known from central Brazil and southern Peru.

Bélanger

Northern Tree-Shrew *Tupaia belangeri* **Wagner,** 1841

Charles Paulus Bélanger (1805–1881) was a French traveler. His voyage is commemorated in his published journal, *Voyage aux Indes-Orientales, par le nord de l'Europe, les provinces du Caucase, la Géorgie, l'Armenie et la Perse, suivi de détails topographiques, statistiques et autres surle Pégou, les îles de Java, de Maurice et de Bourbon, sur le Cap-de-Bonne-Espérance et Sainte Hélène, pendant les années 1825, 1826, 1827, 1828 et 1829*, which has illustrations, by Geoffroy St. Hilaire, of the mammals he collected. His trip was part of the French expedition across Europe to India undertaken in order to make a botanical garden at Pondicherry. The expedition collected vast numbers of specimens of dried and living plants and seeds as well as fish, birds, crustaceans and molluscs, and a few mammals. Bélanger became the Director of the Botanical Gardens in Martinique in 1853. He is also commemorated in the names of other taxa, such as the fish *Osteobrama belangeri*. The tree-shrew is found from eastern Nepal and northeast India to Thailand, Laos, Vietnam, and southeast China.

Belding

Belding's Ground Squirrel *Spermophilus beldingi* **Merriam,** 1888

Lyman Belding (1829–1917) was a professional animal collector who specialized in birds. He wrote a series of articles: "Collecting in the Cape Region of Lower California, West" (1877); "Catalogue of a Collection of Birds Made Near the Southern Extremity of the Peninsula of Lower California" (1882); "Second Catalogue of a Collection of Birds Made Near the Southern Extremity of Lower California" (1883); "List of Birds Found at Guaymas, Sonora" (December 1882 and April 1883); "Land Birds of the Pacific District" (1890); and "A Part of My Experience in Collecting" (1900). Not surprisingly, he is also remembered in the names of several birds, such as Belding's Plover *Charadrius wilsonia beldingi,* Belding's Rail *Rallus longirostris beldingi,* and Belding's Yellowthroat *Geothlypis beldingi.* The squirrel is found in the USA, in eastern Oregon, northeast California, northern Nevada, southwest Idaho, and northwest Utah.

Bellier

Bellier's Striped Grass Mouse *Lemniscomys bellieri* Van der Straeten, 1975

Louis Bellier is a French researcher and agronomist. He collected the type specimen of the mouse while Head of Research (1963–1967) at the Laboratoire d'Ecologie des Mammifères et des Oiseaux, in Abidjan (part of the Institut Français de Recherche Scientifique pour le Développement en Coopération in Ivory Coast). The mouse is found in "Doka" woodland (where the dominant tree is *Isoberlinia doka*) of Ivory Coast and Guinea.

Belzebul [Belzebuth]

Red-handed Howler Monkey *Alouatta belzebul* **Linnaeus,** 1766
White-fronted Spider Monkey *Ateles belzebuth* **Geoffroy,** 1806 [Alt. White-bellied Spider Monkey, Long-haired Spider Monkey]

Both scientific names are variants of Beelzebub, or Baalzebûb, the Philistine god of Accaron (Ekron) west of Jerusalem. According to 2 Kings 1.2, King Ochozias (Ahaziah) attempted to consult Beelzebub as an oracle in his last illness. We know of the god only as an oracle; no other mention of him occurs in the Old Testament, although many people think his name is synonymous with that of the devil. The name is commonly translated as "the Lord of the Flies." The howler monkey is endemic to Brazil. The spider monkey is found in northwest Brazil, northern Peru, Ecuador and Venezuela.

Bemmelen

Gland-tailed Free-tailed Bat *Chaerephon bemmeleni* **Jentink,** 1879

Adriaan Anthoni van Bemmelen (1830–1897) was a Dutch naturalist. He was appointed to be the Director of the Zoological Gardens at Rotterdam in 1866. Jentink said in his description that van Bemmelen "has been so kind as to present this curious Bat to the Leyden Museum: I therefore propose to name it in honour of its donor." The type specimen of the bat was supposedly found in Liberia, but its true distribution seems to be central African, from Cameroon east to Sudan, Uganda, Kenya, and Tanzania.

Bendire

Bendire's Shrew *Sorex bendirii* **Merriam,** 1884 [Alt. Marsh Shrew]

Major Charles Emil Bendire (1836–1897) was an oologist, zoologist, and army surgeon who was born Karl Emil Bendire in Germany, from where he emigrated to the USA. He collected birds' eggs in the 1860s and 1870s while stationed at frontier posts throughout the Department of the Columbia (now the states of Oregon, Idaho, and Washington) and was famous for the copious notes he made of everything he observed. Fellow officers sent Bendire feathers and eggs from other posts in the West. He became Honorary Curator of Oology at the Smithsonian Institution in 1883 and compiled a two-volume work entitled *Life Histories of North American Birds*. He personally oversaw the watercolor illustrations to ensure accuracy. This work was curtailed when he died of Bright's disease at the age of 60. A lake and a mountain in Oregon are named in Bendire's honor. His remarkable collection of 8,000 eggs remains an exhibit at the American Natural History Museum in Washington, DC. Fans of Westerns might like to know that he also once argued Chief Cochise into a truce. He is also commemorated in the common names of three birds: Bendire's Crossbill *Loxia curvirostra bendirei*, Bendire's Summer Sparrow *Aimophila carpalis*, and Bendire's Thrasher *Toxostoma bendirei*. The shrew is found along a narrow coastal area from northwest California to Washington State, USA, with a few records from southeast British Columbia, Canada.

Ben Keith

Keith's Short-tailed Bat *Carollia benkeithi* Solari and **Baker,** 2006

Benjamin Ellington Keith (1882–1959) was an industrialist who founded a company that bears his name and is the ninth largest food distributor in the USA, distributing over 35 million cases of beer annually. It is based in Forth Worth, Texas, where the founder was born. He was also considered a pioneer in the field of employee relations, establishing both pension funds and profit sharing. A longtime advocate of higher education, he was instrumental in establishing no fewer than three Texas universities. Over the years his company has given many millions of dollars to the Museum of Texas Tech University. The original etymology reads, "The specific epithet *benkeithi* is a modified Latin genitive after Mr. Ben E. Keith, a long-time benefactor of the Natural Science Research Laboratory (NSRL) of the Museum of Texas Tech University." The bat is found in Peru and Bolivia.

Bennett, E. T.

Indian Gazelle *Gazella bennetti* **Sykes,** 1831

Bennett's Chinchilla-Rat *Abrocoma bennetti* **Waterhouse,** 1837

Otter-Civet *Cynogale bennetti* **Gray,** 1837

Bennett's Spear-nosed Bat *Mimon bennettii* Gray, 1838 [Alt. Golden Bat]

Bennett's Wallaby *Macropus rufogriseus fruticus* **Ogilby,** 1838 [Alt. Red-necked Wallaby]

Edward Turner Bennett (1797–1836) was a British zoologist. His brother John Joseph Bennett (1801–1876) was a botanist. In 1822 Edward Bennett proposed the setting up of an entomology society under the auspices of the Linnean Society, which soon became a zoological club and, in 1826, the Zoology Society of London. He was its first Vice Secretary, then served as Secretary from 1831 until his untimely death in 1836. A number of his papers are recorded in the society's *Proceedings* of the 1820s and 1830s, such as "Characters of a New Species of Otter (Lutra, Erxl), and of a New Species of Mouse (Mus L.), Collected in Chile by Mr. Cuming" (1832) and "On the Family of Chinchillidae, and on a New Genus Referrible to It" (1833). He also wrote two longer works, *The Tower Menagerie* (1829) and *The Gardens and Menagerie of the Zoological Society* (1831). The gazelle is found from northwest India to Iran. The chinchilla-rat is endemic to Chile.

The otter-civet comes from Thailand, Malaysia, Sumatra, and Borneo. The bat is found from southeast Mexico south to the Guianas and eastern Brazil. The wallaby comes from Tasmania.

Bennett, G.

Bennett's Tree Kangaroo *Dendrolagus bennettianus* **De Vis,** 1887

Dr. George Bennett (1804–1893) was a British surgeon, botanist, and zoologist. He took passage to the South Seas and Australia as surgeon-naturalist on board the *Sophia.* He visited Australia around 1834, then returned to Britain, but settled permanently in Australia in 1836. He wrote a scientific paper, "A Recent Visit to Several of the Polynesian Islands" (1830), as well as *Wanderings in New South Wales, Batavia* (1834) and *Gatherings of a Naturalist in Australia—Being Observations Principally on the Animal and Vegetable Productions of New South Wales, New Zealand, and Some of the Austral Islands* (1860). He was the first Curator and Secretary of the Australian Museum (from 1835) and an early conservationist. In 1860 he wrote, "Many of the Australian quadrupeds and birds are not only peculiar to that country, but are, even there, of comparatively rare occurrence: and such has been the war of extermination recklessly waged against them, that they are in a fair way of becoming extinct. Even in our own time, several have been exterminated; and unless the hand of man be stayed from their destruction, the Ornithorhynchus and the Echidna, the Emeu and the Megapodius, like the Dodo, Moa and Notornis, will shortly exist only in the pages of the naturalist. The Author hopes that what he has been induced to say with reference to this important subject will not be without weight to every thoughtful colonist." The fact that all four survive today in good numbers may well owe a great deal to Bennett's timely campaign. He spent 50 years unsuccessfully trying to fully understand monotreme and marsupial biology. He was a frequent correspondent of both Gould and Gilbert and offered both advice and support to their various Australian sojourns. He is also commemorated in the common names of two birds, Bennett's Bird-of-Paradise *Drepanornis albertisi cervinicauda* and Bennett's Cassowary *Casuarius bennetti.* The tree kangaroo is found in northeast Queensland, Australia.

Berard

The beaked whale genus *Berardius* Duvernoy, 1851 [2 species: *arnuxii* and *bairdii*]

Admiral Auguste Bérard (1796–1852) was the captain of the French corvette *Le Rhin,* from 1842 to 1846, on board of which Arnoux was the ship's surgeon. Arnoux discovered the skull of what is now known as Arnoux's Beaked Whale on a beach close to Akaroa, New Zealand, and Bérard graciously agreed to keep it on board and transport it to France. Duvernoy then named the genus after the then-Commodore Bérard. He, his crew, and his ship made an excellent impression on everyone they encountered in Australia and New Zealand. Commodore Bérard seems to have been most hospitable and to have thrown more than one good party. In his early career he served with such well-known explorers and naturalists as Captain Louis Claude Desaules de Freycinet, Jean Paul Gaimard, and Jean René Constant Quoy on board the corvette *L'Uranie,* which sailed from France in 1820 and was wrecked on the Falkland Islands in 1820, and with Prosper Garnot and René Primevère Lesson on board *La Coquille,* which circumnavigated the globe between 1822 and 1825.

Berdmore

Berdmore's Palm Squirrel *Menetes berdmorei* **Blyth,** 1849 [Alt. Indochinese Ground Squirrel]
Small White-toothed Rat *Berylmys berdmorei* Blyth, 1851

Captain Thomas Matthew Berdmore (1811–1859) amassed, with W. Theobald, a significant natural history collection that he presented to the Asiatic Society of Bengal in 1856, which Blyth later reported on. He is also commemorated in the names of other taxa, including fish such as Berdmore's Loach *Syncrossus (Botia) berdmorei* and amphibians such as Berdmore's Narrow-mouthed Frog *Microhyla berdmorei.* The two rodents named after him have a broadly similar distribution, from Myanmar east to Vietnam.

Berezovski

Chinese Forest Musk Deer *Moschus berezovskii* Flerov, 1929

Michael Michaelovitch Berezovski (d. 1911) was a Russian zoologist and traveler who was in Mongolia between 1876 and 1878, and in the Kansu area of China between 1892 and 1894. Between 1905 and 1907 he led an expedition to Chinese Turkestan. On this trip he was accompanied by his artist brother N. Michaelovitch Berezovski. He is also commemorated in the name of a bird, Berezowski's Blood Pheasant *Ithaginis cruentus berezowskii.* The deer is found in central and southern China and in northern Vietnam.

Berg

Berg's Tuco-tuco *Ctenomys bergi* **Thomas,** 1902

Dr. Frederico Guillermo Carlos Berg (1843–1902) was an entomologist and naturalist. He was a Latvian of Baltic German extraction and until he went to South America he was known as Friedrich Wilhelm Carl Berg. After a number of years working in a commercial company, he got a job as a conservator of entomological specimens at the museum in Riga. In 1873 he was invited by the Director of the Museum of Buenos Aires, Hermann Burmeister, to join his staff. In 1874 he undertook a collecting expedition to Patagonia, and in 1879 he visited Chile and collected insects with Rodolfo Amando Philippi. He worked at the Montevideo National Museum in Uruguay from 1890 until 1892, and in that year he succeeded Burmeister as Director of the Buenos Aires Museum, retiring in 1901. The tuco-tuco comes from the Argentine province of Córdoba.

Bergman

Bergman's Bear *Ursus arctos piscator* **Pucheran,** 1855 [Alt. Kamchatka Bear]

Dr. Sten Bergman (1895–1975) was a Swedish zoologist, biologist, writer, traveler, and explorer. He was a member of the Swedish Kamchatka expedition of 1920–1922, during which he first came across the skin of an enormous bear. The skin seemed different from that of normal brown bears from that region. Unfortunately there are now various websites and articles claiming that Bergman named this "new" bear as *Ursus arctos piscator.* In fact that name had been coined for Kamchatkan bears by Pucheran some 70 years earlier, and it has nothing to do with the skin seen by Bergman. Whether the latter represents a valid taxon is still uncertain, as no further specimens seem to have come to the attention of scientists. In the 1930s Bergman traveled in the Kurile Islands and Korea. He was in Dutch New Guinea (now Irian Jaya or West Papua) on three separate expeditions after the end of WW2, and in the 1960s he visited Japan.

Beringe

Eastern Gorilla *Gorilla beringei* **Matschie,** 1903

Major Friedrich Robert von Beringe (1865–1940) was a German military officer. From 1894 to 1906 he belonged to a Hussar regiment and was posted to German East Africa as a lieutenant. In 1898 he conducted a successful punitive expedition against the Watumbi tribe. Between 1902 and 1904 he commanded the Usumbura Military Post, and in 1902 he led an expedition to the Virungas as a way of keeping in touch with German outposts and establishing the boundaries of German East Africa. On day two of the expedition one of the team spotted "a herd of big, black monkeys [trying] to climb the crest of the volcano" near their camp. Beringe shot and killed two of the "monkeys" for closer examination. One was recovered and dispatched to the Berlin Museum, although en route from the camp one hand and part of the skin were eaten by a hyena. In Berlin it was named in his honor *Gorilla beringei.* (Later taxonomists tended to treat all gorilla populations as subspecies of a single species, *Gorilla gorilla.* It is only quite recently that the eastern form of gorilla has been reinstated as a full species.) During his 1902 expedition Beringe had the offer (according to his grandson's biography of him) of a young woman from a tribal chief (history does not record his reaction) and a tankard of beer from another chief; he suspected that the tankard was poisoned (which it was) and declined the gift. In 1903, by which date he had been promoted to Captain, Beringe led an expedition against a rebellious tribal chief called Muezi Kisabo, who finally surrendered and submitted to German rule. In 1906 Beringe returned to Germany, transferring to a Dragoon regiment, and in 1908 was promoted to Major, which rank he kept until his retirement in 1913. He was forced to retire by a fall from horseback caused by diabetes, which disease led to his death in 1940 while staying in Stettin where his daughter's parents-in-law lived. Two races of Eastern Gorilla are recognized. The Mountain Gorilla *(Gorilla beringei beringei)* lives in isolated populations in the Virunga volcano range on the DRC (Zaire)–Rwanda–Uganda border. For more on the second subspecies, see under **Grauer.**

Bernard

Black Wallaroo *Macropus bernardus* **Rothschild,** 1904

Bernard Woodward (1846–1912). The wallaroo inhabits the Northern Territory, Australia. See **Woodward** for full biographical details.

Bernard, P. and J.

Bernard's Wolf *Canis lupus bernardi* **Anderson,** 1934 extinct [Alt. Banks Island Tundra Wolf]

This canid is named after not one person but two (so the trinomial really ought to be *bernardorum*): Peter Bernard (d. 1917) and his nephew, Joseph F. Bernard (1878–1972). Peter Barnard was the sailing master of the gas schooner *Mary Sachs,* which he had sold to the Canadian government in 1913 for the Canadian Arctic expedition of 1913–1918. He had the option to buy her back at the end of the expedition but died while on a sledging trip to Banks Island, and the ship was frozen into the ice pack. Joseph Bernard was master and owner of the gas schooner *Teddy Bear* and made many trips to the Arctic. He started out as a hunter and fur trader and moved his operations from Prince Edward Island to Alaska in 1901. Between 1916 and 1920 he gathered an important collection of ethnographical and archeological objects, mainly from the Copper Eskimos of Coronation Gulf. In 1959 he was appointed to be Harbor Master at Cordova in Alaska. It is worth recording just how small these schooners were: *Mary Sachs* was all of 36 tons, and the *Teddy Bear* a mere 13. Anderson was in the high Arctic in 1916, so he had the chance to see the wolf—which was described as white with black-tipped hair along the ridge of the back—before it became extinct in about 1920. Its range was very limited, being only Banks and Victoria Islands in the high Arctic. In his description Anderson specifically mentions the two men with these words: "With this well-marked subspecies is joined the subspecific name *bernardi* in recognition of two natives of Tignish, Prince Edward Island, Canada, well-known for many years in Alaska and Western Arctic—Peter Bernard, sailing master of the gas schooner *Mary Sachs* of the Canadian Arctic expedition, 1913–18, who collected the type and four other specimens of *Canis lupus bernardi* and lost his life in the line of duty on a sledge trip in northern Banks Island in spring of 1917, and his nephew, Joseph F. Bernard, who made many voyages into the Arctic as master and owner of the gas schooner *Teddy Bear*, and collected many scientific specimens of mammals and birds, as well as ethnological material in the Northwest Territories, Alaska, and eastern Siberia."

Berthe

Berthe's Mouse-Lemur *Microcebus berthae* Rasoloarison, **Goodman** and Ganzhorn 2000 [Alt. Madame Berthe's Mouse Lemur]

Professor Berthe Rakotosamimanana (d. 2005) was the head of the Department of Paleontology and Biological Anthropology, University of Antananarivo, Madagascar. She made significant contributions to the study of lemurs and wrote a number of scientific papers and contributed to collective works, as well as editing *New Directions in Lemur Studies* (1999). The lemur is found in the dry deciduous forests of western Madagascar.

Beryl

The white-toothed rat genus *Berylmys* **Ellerman,** 1947 [4 species]

Ellerman makes no mention as to what led him to use the name Beryl in the genus; it may or may not be eponymous. We speculate one possible source may be someone whom Ellerman certainly would have heard of, and who was much in the public eye in 1947: Professor Bernard Beryl Brodie (1907–1989), who was born in England but lived and worked most of his life in North America. Brodie's undergraduate degree was awarded by McGill University in Canada and his doctorate by New York University in 1935. A leading figure in the study of how drugs interact in the body, he brought the era of modern pharmacology into prominence in the 1940s and 1950s. He was an Associate Professor at New York University from 1935 until 1950, when he joined the National Institute of Health, heading up the pharmacology laboratory there until he retired in 1979, after which he worked as a consultant and as a Professor of Pharmacology at Pennsylvania State University. As sole or co-author he produced over 350 scientific papers.

Betsileo

Betsileo Sportive Lemur *Lepilemur betsileo* Louis et al., 2006
Betsileo Woolly Lemur *Avahi betsileo* Andriantompohavana et al., 2007

These primates are named after a people rather than a person. Betsileo is the name of a tribe who occupy the area of eastern Madagascar where the animals are endemic.

Bibi

The genus of crimson-nosed rats *Bibimys* Massoia, 1979 [3 species: *chacoensis, labiosus*, and *torresi*]

Bibiana Massoia is the only daughter of the late Argentinean zoologist Elio Massoia (1936–2001), who named this genus after her. The rodents occur in Argentina and southeast Brazil.

Biet

Chinese Desert Cat *Felis bieti* **Milne-Edwards,** 1892 [Alt. Chinese Mountain Cat]
Black Snub-nosed Monkey *Rhinopithecus bieti* Milne-Edwards, 1897

Monsignor Felix Biet (1838–1904) was a French missionary on the Burmese-Chinese border. He was posted to Bhamo in 1873 and became Bishop of Diana. In 1891 he was based in Tatsienlu (China), and among his helpers was Père Jean André Soulie, who introduced the butterfly bush or *Buddleia davidii* into Europe. In 1891 he was visited by the naturalist Antwerp Edgar Pratt (q.v.). Biet's name is also commemorated in the name of a bird, Biet's (or White-speckled) Laughing-thrush *Garrulax bieti.* The cat is endemic to China, occurring mainly on the eastern edge of the Tibetan Plateau. The monkey is found only in Yunnan Province, China.

Bilarni [Bill Harney]

Sandstone Pseudantechinus *Pseudantechinus bilarni* **Johnson,** 1954 [Alt. Northern Dibbler; Syn. *Parantechinus bilarni*]

Bill Harney (1895–1962) was an Australian who had many roles: drover, soldier, laborer, fencer, sailor, fisherman, and ranger. The Aboriginals called him Bilarni, which was the nearest they could get to the correct pronunciation. He left home at the age of 12 to go droving. He served in the Australian army as an infantryman in Egypt and France from 1914 to 1918. He returned to Australia and was accused of cattle rustling in the Northern Territory and was held on remand in prison, where he read the classics and acquired the education he had missed as a child. The case was dismissed, and after he was released he spent seven years deep-sea fishing. The late 1920s and the 1930s he spent wandering the Northern Territory, where he spent much time with the Aboriginal tribes learning their languages, rites and customs, and tribal law. His expertise was recognized when he was made Protector of Aborigines in the Northern Territory in 1940. He retired in 1948 and wrote a number of books and became a well-known personality as a speaker on the radio and at conferences. In the 1950s, tourism as an industry expanded and people started to visit Ayres Rock (now called Uluru). The local Aborigines were concerned as they regard the site as sacred and requested that "Bilarni" help them. He was made first Ranger of Ayres Rock and Mount Olga in 1956. The pseudantechinus was first discovered in 1948 in the Northern Territory by an expedition of which Bill Harney was a member.

Billardiere

Tasmanian Pademelon *Thylogale billardierii* **Desmarest,** 1822 [Alt. Red-bellied Pademelon]

Jacques Julien Houton de La Billardière (1755–1834) was the botanist and explorer on Bruni d'Entrecasteaux's voyage of 1791–1794 that visited Western Australia and Tasmania on *La Recherche* and *L'Espérance,* and that collected many species, especially plants. Billardière acted as naturalist and collected not only plants but also mammals, fish, birds, and insects. He was author of *Novae Hollandiae plantarum specimen,* published between 1804 and 1806. He is commemorated in all sorts of taxa but is perhaps best known for the shrub genus *Billardiera,* otherwise known as appleberries. The pademelon is a kind of wallaby found in southeast Australia and Tasmania.

Bindi

Kakadu Dunnart *Sminthopsis bindi* Van Dyck, Woinarski, and Press, 1994

The binomial is not, as one might suppose, eponymous. Dr. Steve Van Dyck confirmed that "Bindi is the name for small dasyurids (carnivorous marsupials) in the language of the Jawoyn people, traditional owners of the land from which most specimens have been recorded." The dunnart lives in Arnhem Land, Northern Territory, Australia.

Bishop

Bishop's Fossorial Spiny Rat *Clyomys bishopi* **Ávila-Pires** and Wutke, 1981
Bishop's Slender Mouse-Opossum *Marmosops bishopi* **Pine,** 1981

Dr. Ian R. Bishop was Assistant Keeper of Zoology at the British Museum, Natural History. He led the Guy's Hospital Amazon expedition in 1964, and in 1966 he was a lecturer in zoology at Leicester University. He was a member of the group that made a preliminary visit to Brazil for what became the 1967–1969 joint Royal Society–Royal Geographical Society Brazil expedition, of which Bishop was the leader. He wrote "An Annotated List of Caviomorph Rodents Collected in Northeastern Mato Grosso Brazil" (*Mammalia,* 1974) and, with Pine and Jackson, "Preliminary List of Mammals of the Xavantina/Cachimbo Expedition (Central Brazil)" (*Transactions of the Royal Society of Tropical Medicine and Hygiene,* 1970). In 1995 he gained public attention when he identified a skull found on Bodmin Moor, Cornwall, as being that of a male leopard. Locals thought it might belong to the mysterious "Beast of Bodmin Moor," but Bishop determined that the leopard had not died in Britain and concluded that the skull had probably been imported as part of a leopard-skin rug. The spiny rat is endemic to São Paulo State, Brazil. The opossum comes from Bolivia and the Mato Grosso region of Brazil.

Bismarck

Bismarck Flying Fox *Pteropus capistratus* **Peters,** 1876
Bismarck Blossom Bat *Melonycteris melanops* **Dobson,** 1877 [Alt. Black-bellied Fruit Bat]
Bismarck Bare-backed Fruit Bat *Dobsonia praedatrix* **K. Andersen,** 1909
Bismarck Trumpet-eared Bat *Kerivoula myrella* **Thomas,** 1914

We believe that none of these mammals are named directly after the famous German Chancellor, but that they take their name from the Bismarck Archipelago, Papua New Guinea, as they are all found in various islands of that archipelago.

Biswas

Namdapha Flying Squirrel *Biswamoyopterus biswasi* Saha, 1981

Dr. Biswamoy Biswas (1923–1994) was an Indian ornithologist. In 1947 he was awarded a three-year fellowship by Sunderlal Hora, then Director of the Zoological Survey of India (ZSI), that enabled him to study at the British Museum, the Berlin Zoological Museum under Stresemann, and the American Museum of Natural History under Ernst Mayr. He was awarded his Ph.D. in 1952 at the University of Calcutta. In 1954 he was part of the famous *Daily Mail* expedition sent to look for the Yeti around Mount Everest. He was elected Corresponding Fellow of the American Ornithologists' Union in 1963 and later headed the Bird and Mammal Section of the Zoological Survey of India, serving as Joint Director until his retirement in 1981. He remained Emeritus Scientist until 1986. The common name comes from the Namdapha National Park, northeast India, where this endangered species is found.

Blainville

Blainville's Beaked Whale *Mesoplodon densirostris* **Blainville,** 1817
Antillean Ghost-faced Bat *Mormoops blainvillii* **Leach,** 1821
Golden Atlantic Tree Rat *Phyllomys blainvilii* Jourdan, 1837
Franciscana *Pontoporia blainvillei* **Gervais** and **d'Orbigny,** 1844 [Alt. La Plata Dolphin]
Blainville's Spotted Dolphin *Stenella pernettensis* **Desmarest,** 1817 [Alt. Pernetty's Dolphin, Atlantic Spotted Dolphin; Syn. *S. frontalis* Cuvier, 1829]

Professor Henry Marie Ducrotay de Blainville (1777–1850) was a French zoologist and anatomist. He was one of Cuvier's bitterest rivals and his successor to the Chair of Comparative Anatomy at the French Museum of Natural History and in the Collège de France. He applied his principles of anatomy to classification and had much influence in establishing the skeletal evolution as one determinant of classification. He wrote *Cours de physiologie générale et comparée* (1829). He is also remembered in the scientific name of a bird, the Lowland Peltops *Peltops blainvilli.* The whale occurs worldwide in temperate to tropical waters. The bat is from the Greater Antilles. The tree rat comes from southeast Brazil. The Franciscana is found from Brazil to Argentina in the coastal waters from Doce River, Espírito Santo, to the Valdez Peninsula. The spotted dolphin is found in the tropical and warm-temperate zones of the Atlantic Ocean. (The name *Stenella pernettensis,* though older than *S. frontalis,* has been suppressed by the so-called Taxonomy Police, the International Commission of Zoological Nomenclature, and *frontalis* is now the correct name to use.)

Blandford

See below. This is a common misspelling of Blanford.

Blanford

Blanford's Rat *Cremnomys blanfordi* **Thomas,** 1881
Blanford's Jerboa *Jaculus blanfordi* Murray, 1884
Blanford's False Serotine *Hesperoptenus blanfordi* **Dobson,** 1877
Blanford's Fox *Vulpes cana* Blanford, 1877
Blanford's Urial *Ovis vignei blanfordi* **Hume,** 1877
Blanford's Fruit Bat *Sphaerias blanfordi* Thomas, 1891 [Alt. Mountain Fruit Bat]
The vole genus *Blanfordimys* Argyropulo, 1933 [2 species: *afghanus* and *bucharicus*]

William Thomas Blanford (1832–1905) was a British geologist and zoologist. He studied at the Royal School of Mines between 1852 and 1854 and at Freiberg in Saxony before obtaining a post in the Indian Geological Survey in 1854, in which position he investigated coal mines at Talchir from 1854 until 1857. In 1860 he undertook a geological survey of Burma and was appointed Deputy Superintendent. He surveyed in Bombay from 1862 until 1866 and was then attached to the Abyssinian expedition in 1867. Blanford published works on the geology of Abyssinia (now Ethiopia) in 1870 and of India in 1879, before settling in London in 1881. He edited works for the government on Indian fauna, contributing two volumes on mammals in 1888 and 1891. He was elected to Fellowships of the Royal Society and the Royal Geographical Society in 1874 and was President of the latter from 1888 until 1890. Blanford wrote *The Scientific Results of the Second Yarkand Mission: Mammalia,* published by the Indian government in 1879.

He is also commemorated in the common names of six birds, including Blanford's Lark *Calandrella blanfordi,* Blandford's Warbler *Sylvia leucomelaena,* and Blanford's Olive Bulbul *Pycnonotus blanfordi.* The rat is found in India and Sri Lanka. The jerboa comes from Iran, Afghanistan, and western Pakistan. The false serotine bat is found in Thailand, southeast Myanmar (Burma), the Malay Peninsula, and Borneo. The fox ranges from Egypt and Israel east to Afghanistan, Turkmenistan, and Pakistan. The urial is found in Baluchistan (this race of wild sheep may not be separable from the Afghan Urial *Ovis vignei cycloceros*). The fruit bat lives in the montane forests of northern India, Bhutan, Nepal, southern China, Myanmar, and northwest Thailand.

Blasius

Blasius' Horseshoe Bat *Rhinolophus blasii* **Peters,** 1866

Johann Heinrich Blasius (1809–1870) was a German zoologist. Both his sons, Wilhelm August Heinrich Blasius (1845–1912) and Rudolf Heinrich Blasius (1842–1907), were ornithologists. For much of his adult life he was a professor of natural sciences at the Collegium Carolinum, Brunswick, Lower Saxony. Along with his friend Alexander Graf Keyserling, Blasius co-authored *Die Wirbeltiere Europas* (The Vertebrates of Europe, 1840). The bat is found from Italy and the Balkans through the Middle East to Afghanistan and Iran; also in North Africa and from Ethiopia to Swaziland and the Transvaal (South Africa).

Bleyenbergh

Bleyenbergh's Lion *Panthera leo bleyenberghi* Lönnberg, 1916 [Alt. Angolan Lion, Katanga Lion]

Lieutenant Van Bleyenbergh (dates not found) was a Belgian soldier. The citation is disappointingly brief: "The type specimen was a male shot in Katanga by Lieutenant Van Bleyenbergh." No further details of the man are given. Katanga is in the south of present-day DRC (Zaire). This subspecies of lion is also found in Angola.

Blick

Blick's Grass Rat *Arvicanthis blicki* Frick, 1914

Dr. John Charles Blick (fl. 1896–1971) was an American physician. He accompanied his friend Henry Childs Frick on an expedition to Abyssinia (now Ethiopia) and Lake Rudolf that Frick had organized with Mearns in 1911, the express intension of which was to collect birds and small mammals. The rat has a limited range in Ethiopia.

Blosseville

Western Red Bat *Lasiurus blossevillii* **Lesson** and Garnot 1826

Baron Jules Alphonse Rene Poret de Blosseville (1802–1834) was a French navigator and astronomical observer. He was born in Rouen and died somewhere in the Arctic Ocean around February 1834. He entered the navy as a volunteer in 1818, visiting the Antilles and Cayenne that year and South America in 1819. He accompanied Dumont d'Urville and René Primevère Lesson aboard *La Coquille* on her voyage of discovery from 1822 to 1825 under the command of Louis Isidor Duperrey, the three being there to make scientific observations. While in Australia in 1824 he worked with Governor Brisbane in his observatory. In 1826 he made a report for the French government on the suitability of western New Holland as a French penal colony. In 1827 he went on to explore Asian waters and was on an expedition to Algiers in 1830. In 1833 he was appointed to command the brig *La Liloise* and sent to the Arctic Ocean. He sailed from France in May 1833, visited Iceland and Greenland (where he made astronomical observations), and drew up a chart of the western coast of Greenland. The expedition had reached latitude 83°N when they became stuck in the ice, but they managed to send news of their predicament to France via a passing whaler. This was the last that was heard of Blosseville or the ship, and several French and English expeditions failed to find traces of him. The expedition of *La Recherche* and *L'Aventare* ascertained through Inuits that he had advanced farther than latitude 84°N, but nothing more was ever known. He published *Histoire des découvertes faites e diverses epoques par les navigateurs* in 1826 and *Histoire des explorations de l'Amerique du Sud* in 1832. The bat has a range from southern British Columbia, Canada, south through the western USA, Mexico, and Central America, all the way to Argentina and Chile.

Blyth, A. C.

Brush-tailed Mulgara *Dasycercus blythi* Waite, 1904

A. C. Blyth (dates not found) collected live specimens of this small carnivorous marsupial in the Pilbara district of Western Australia. He provided some to Edgar Waite of the Australian Museum, Sydney, who named the animal after him. Waite returned the animals to Blyth, from whose care they apparently escaped. The exact range of this species in the arid regions of Australia is not fully known, due to earlier confusion with the Crest-tailed Mulgara *Dasycercus cristicauda* (until 2005 the two forms were regarded as conspecific).

Blyth, E.

Blyth's Horseshoe Bat *Rhinolophus lepidus* Blyth, 1844
Blyth's Pouched Bat *Saccolaimus saccolaimus crassus* Blyth, 1844
Lesser Mouse-eared Bat *Myotis blythii* **Tomes,** 1857
Blyth's Clubfooted Bat *Tylonycteris pachypus fulvida* Blyth, 1859 [Alt. Lesser Bamboo Bat]
Blyth's Vole *Phaiomys leucurus* Blyth, 1863

Edward Blyth (1810–1873) was an English zoologist and author. He was Curator of the Museum of the Asiatic Society of Bengal from 1842 to 1864 and author of its catalogue. He wrote *The Natural History of Cranes*, published in 1881. Hume said of him, "Neither neglect nor harshness could drive, nor wealth nor worldly advantages tempt him, from what he deemed the nobler path. He was] ill paid and subjected . . . to ceaseless humiliations." In a similar tribute Arthur Grote wrote, "Had he been a less imaginative and more practical man, he must have been a prosperous one. . . . All that he knew was at the service of everybody. No one asking him for information asked in vain." He is also commemorated in no fewer than 19 common bird names, including Blyth's Pipit *Anthus godlewskii* and Blyth's Tragopan *Tragopan blythii*. The horseshoe bat ranges from Afghanistan across northern India, Burma, and Yunnan (China) to the Malay Peninsula and Sumatra. The pouched bat occurs in Sri Lanka and from India east to Thailand. The mouse-eared bat has a discontinuous distribution, being found in the Mediterranean zone of Europe (these populations are sometimes separated as *Myotis oxygnathus*) and from Turkey east to Iran, in northwest India and the Himalayas, and also in Inner Mongolia and Shensi (China). The club-footed bat is found from India east to Indochina. The vole inhabits high altitudes of the Himalayas and the Tibetan plateau.

Bobrinski

Bobrinski's Serotine *Eptesicus bobrinskoi* Kuzyakin, 1935
Bobrinski's Jerboa *Allactodipus bobrinskii* Kolesnikov, 1937

Professor Count Nikolay Alekseyevich Bobrinski (1890–1964) was a Russian zoologist. He worked at the Zoological Museum in Moscow in 1930 and despite being an aristocrat was regarded as one of the most prominent Soviet zoologists. (The present count, Nicholas, is a geographer and lives in Moscow.) He wrote, with others, *Mammals of the USSR* (published in 1944) and *The Animal World and Nature of the USSR* and *Animal Geography* (published in 1951). The serotine bat is found in the North Caucasus, Kazakhstan, Uzbekistan, and Turkmenistan. The jerboa is found in the Kyzylkum and Karakumy deserts of Turkmenistan and western Uzbekistan.

Bocage

Rufous Mouse-eared Bat *Myotis bocagei* **Peters,** 1870
Bocage's Tree Squirrel *Funisciurus bayonii* Bocage, 1890 [Alt. Lunda Rope Squirrel]
Bocage's Gerbil *Gerbilliscus validus* Bocage, 1890 [Alt. Southern Savannah Gerbil]
Bocage's Mole-Rat *Fukomys bocagei* **de Winton,** 1897 [Alt. Angolan Mole-Rat; formerly *Cryptomys bocagei*]
Bocage's Rock Rat *Aethomys bocagei* **Thomas,** 1904

José Vicente Barboza du Bocage (1823–1907) was Director of the National Zoological Museum of Lisbon, Portugal, which is now named in his honor. He became known as the father of Angolan ornithology and wrote *Ornithologie d'Angola.* He also collected sponges and other specimens. He is commemorated in the common names of six birds, including Bocage's Sunbird *Nectarinia bocagii* and Bocage's [Golden] Weaver *Ploceus temporalis.* The bat is found from Senegal to Yemen and south to Angola, Zambia, and Transvaal (South Africa). The squirrel comes from northeast Angola and southern DRC (Zaire). The gerbil ranges from Angola east through Zambia and southern DRC to western Tanzania. The mole-rat is found in Angola, northwest Zambia, and southern DRC, while the rock rat has a restricted range in western and central Angola.

Bodenheimer

Bodenheimer's Pipistrelle *Hypsugo bodenheimeri* **Harrison,** 1960 [formerly *Pipistrellus bodenheimeri*]

Professor Frederick Simon (Fritz Shimon) Bodenheimer (1897–1959) was an Israeli zoologist, entomologist, and pioneer ecologist. He was a professor at the Hebrew University of Jerusalem. He wrote many scientific papers, such as "Studies on the Ecology and Control of the Moroccan Locust *(Dociostaurus maroccanus)* in Iraq" (1944) and "The Present Taxonomic Status of the Terrestrial Mammals of Palestine" (1958), as well as works of more popular science such as *Insects as Human Food* (1951) and *Animal and Man in Bible Lands,* published posthumously in 1960. Another work was *History of Biology* (1958). He identified many of the species mentioned in the Bible. The bat has been found in Israel, Egypt (Sinai), and the Arabian Peninsula.

Boeadi

Biak Giant Rat *Uromys boeadii* Groves and **Flannery,** 1994
Boeadi's Roundleaf Bat *Hipposideros boeadii* Bates et al., 2007

Dr. D. R. S. "Pak" Boeadi is an Indonesian zoologist. He works in the Institute of Environmental Sciences Department of the Museum Zoologicum Bogoriense in Indonesia, as Curator of Mammals. In 2001 he was the project leader of taxonomy of mammal species on an expedition to Sulawesi. He has written many articles since the 1970s, including at least one with Tim Flannery, and has collected zoological specimens in New Guinea. The rat is restricted to the island of Biak, off northern New Guinea. The bat is found on Sulawesi.

Bogdanov

Balkan Snow Vole *Dinaromys bogdanovi* **Martino,** 1922 [Alt. Martino's Snow Vole]

Unfortunately Martino's original brief citation does not say who Bogdanov was, and so we have to speculate and suggest two prominent scientists. One possibility is Professor Modest Nikolaevich Bogdanov (1841–1888), a Russian zoologist and ornithologist who between 1879 and 1885 was supervisor of the ornithological collection at the St. Petersburg museum. He carried out a meticulous inventory, identifying all the species, and as a result produced a monograph, *Shrikes of Russian Fauna and Allies* (1881). He also prepared the first issue of the *Checklist of the Birds of the Russian Empire,* which appeared in 1884. Alternatively, the vole may have been named after Professor Anatoly Petrovich Bogdanov (1834–1896), a Russian anthropologist and zoologist who was the Director of the Moscow Zoological Museum and a professor at the Moscow State University. The Russian imperial government wanted to increase museum collections and to encourage popular interest in science through public lectures, and Bogdanov was appointed Chairman of a committee to achieve this. He traveled to England, Germany, and France to study physics and mechanics. He was especially impressed by the Crystal Palace, which had been re-erected at Sydenham, and from its exhibits he developed the idea of an exhibition in Russia to show off the diversity of the peoples of the Russian Empire. In 1867 it was decided to sponsor a National Ethnographic Exposition combined with an industrial exhibition. In 1883 Bogdanov organized an anthropological exhibition, and from that sprang the Anthropological Museum of the Russian State University. The vole is found in the Balkans. The type specimen was found in the Montenegrin town of Cetinje, in what was then called Yugoslavia (and, coincidentally, there is a street called "Bogdanov kraj" in that town).

Böhm

Böhm's Bush Squirrel *Paraxerus boehmi* Reichenow, 1886

Böhm's Gerbil *Gerbilliscus boehmi* **Noack,** 1887

Böhm's Zebra *Equus quagga boehmi* **Matschie,** 1892 [Alt. Grant's Zebra]

Dr. Richard Böhm (1854–1884) was a German traveler and zoologist who worked in Tanganyika (now Tanzania) during the late 1800s until his premature death. He was reportedly murdered by "the tribes westward of Lake Tanganyika." He wrote *Von Sansibar zum Tanganjika,* which was published in 1888. He is commemorated in the common names of five birds, including Böhm's Bee-eater *Merops boehmi,* Böhm's Flufftail *Sarothrura boehmi,* and Böhm's Flycatcher *Myiornis boehmi.* The squirrel is found from southern Sudan and western Kenya to southern DRC (Zaire) and northern Zambia. The gerbil occurs in Angola, southern DRC (Zaire), Zambia, Malawi, Tanzania, Kenya, and Uganda. The zebra's range covers southeast Sudan, Uganda, Kenya, and Tanzania.

Bokermann

Bokermann's Nectar Bat *Lonchophylla bokermanni* Sazima et al., 1978

Dr. Werner Carlos Augusto Bokermann (1929–1995) was a leading Brazilian herpetologist. He was born in São Paulo and received his doctorate in zoology at the Bioscience Institute of São Paulo University. He became head of the Bird Section of the São Paulo Zoo and remained there the whole of his working life. He described a great number of new species, including at least 61 frogs and toads from Brazil, many of them with Sazima. At least 13 such species have been given his name in their binomials. He published a great many scientific papers, too. The bat is found in southeast Brazil.

Bolam

Bolam's Mouse *Pseudomys bolami* **Troughton,** 1932

Anthony Gladstone Bolam (1893–1966) was an Australian railwayman and amateur naturalist. He collected the type specimen of the mouse along the transcontinental railway. He worked for Commonwealth Railways from 1918 until 1926 and was a stationmaster in Ooldea from 1920. He also wrote *The Trans-Australian Wonderland,* published in 1927, which brought him to public notice. In the book he reported his observations of contact between whites and Aboriginal people. Bolam thought that contact with whites brought diseases like VD and measles, as well as alcoholism. He described Native Australians who had walked from hundreds of kilometers away, surviving by getting water from wells and from roots of trees, and they arrived, in his words, "wonderfully bright and clean." They then donned "filthy cast-off" clothes, became "dirty," and succumbed to disease. Forced by "missionary benevolence" to wear clothes, they got wet when it rained, were afraid to remove the wet clothes, then got chilled and died of pneumonia. He also reported that some "Whites" shot "Blacks" for target practice. The mouse is found in arid and semi-arid regions of southern Australia.

Bole

Bole's Douroucouli *Aotus bipunctatus* Bole, 1937 [Syn. *A. (lemurinus) griseimembra*]

Benjamin Patterson Bole Jr. (1908–1980) was an ornithologist and zoologist at the Cleveland Museum of Natural History. He was related by marriage to the Holden family, who founded the Holden Trust, which in turn provided funds and land for the Holden Arboretum in Cleveland. Bole published quite widely, often in collaboration with others—for example, with John W. Aldrich in 1937, *The Birds and Mammals of the Western Slope of the Azuero Peninsula (Republic of Panama).* The monkey was described as a full species, coming from the Azuero Peninsula of Panama, but many authorities regard it as a junior synonym of *Aotus griseimembra.*

Bonaparte

Bonaparte's Weasel *Mustela cicognanii* Bonaparte, 1838 [Alt. North American Ermine; Syn. *M. erminea cicognanii*]

Prince Charles Lucien Bonaparte, originally Jules Laurent Lucien (1803–1857), was a nephew of the famous statesman, the Emperor Napoleon Bonaparte. He was a renowned ornithologist, both in Europe and the USA. Bonaparte was much traveled, but he spent many years in the USA cataloguing birds, and he has been described as the "father of systematic ornithology." While in America, at the age of 21, he met Audubon and much admired his work. He put Audubon forward for membership of the Philadelphia Academy of Natural Sciences. Bonaparte eventually settled in Paris and commenced his *Conspectus generum avium,* a catalogue of every bird species in the world. He died before finishing it, but its publication was heralded as a major step forward in accomplishing one of the most important goals of ornithology: a complete list of the world's birds. He also wrote *American Ornithology* (1825) and *Iconografia della fauna Italica—Uccelli* (1832). Swainson described Bonaparte as "destined by nature to confer unperishable benefits on this noble science." He is commemorated in the common names of at least a dozen birds, including Bonaparte's Gull *Larus philadelphia* and Bonaparte's Blackbird *Sturnella supercilia-*

ris. While clearly known as an ornithologist, he was interested in all of natural life, as evidenced by the fact that he described this species of weasel. The weasel is found in North America.

Bonetto

Bonetto's Tuco-tuco *Ctenomys bonettoi* **Contreras** and Berry, 1982

Dr. Argentino Aurelio Bonetto (1920–1998) was an Argentinean naturalist, ichthyologist, and ecologist who attended the University of Córdoba and took a doctorate in natural sciences. He was a specialist in the study of river systems and flood plains, their flora, fauna, and ecology. He published quite widely on his subject, including, with Wais, *Southern South American Streams and Rivers* (1995) and, with others, parts of *River and Stream Ecosystems* (1990). He also wrote "Wetland Management in the Gran Pantanal, Paraná Basin, South America," published in *The People's Role in Wetland Management: Proceedings of the International Conference on Wetlands* (1989). The tuco-tuco is confined to Chaco Province, northern Argentina.

Bonhote

Bonhote's Mouse *Mus famulus* Bonhote, 1898 [Alt. Servant Mouse]
Bonhote's Gerbil *Gerbillus bonhotei* **Thomas,** 1919

John Lewis James Bonhote M.A., F.Z.S. (1875–1922) was a British zoologist, ornithologist, and writer. After graduating from Trinity College, Cambridge, in 1897, he took a post as Private Secretary to the Governor of the Bahamas. He collected birds and spiders there, both then and also on a second visit from 1901 to 1902. He described a number of small mammals during the 1890s and 1900s, and was involved in the 1901 collecting expedition to the Malay States. He wrote *Birds of Britain and Their Eggs* (1907) and a number of reports, scientific papers, and articles, such as "The Study of Bird Migration" published in the *Strand Magazine* in 1912. He worked at the Zoological Gardens at Giza, Egypt, from 1913 until 1919. While there he collected in, and studied, the Nile Delta and played a large part in the reestablishment of the egret colony there. He also edited the *Journal of the Avicultural Society* and was Secretary, Treasurer, and Council Member at different times. He bred captive birds, notably ducks. He was one of the secretaries of the Fourth International Ornithological Congress, served in various posts in the British Ornithologists Club, and for many years was on the council of the Royal Society for the Protection of Birds. While he was primarily interested in birds, the fact that he described the mouse is evidence that he was interested in other taxa too, particularly mammals. The mouse occurs in the Nilgiri Hills of southwest India. The gerbil is only found in the northeast Sinai Peninsula, Egypt. It is sometimes regarded as conspecific with *Gerbillus andersoni*.

Bosman

Bosman's Potto *Perodicticus potto* Müller, 1766 [Alt. Potto]

Willem Bosman (1672–1702) was a Dutch sea captain who is credited as the supposed discoverer of the potto. He mapped the coast of Guinea and settled there for 14 years as an agent for the Dutch East India Company based at the fort at Elmina. His account *A New and Accurate Description of the Coast of Guinea, Divided into the Gold, the Slave, and the Ivory Coasts . . . with a Particular Account of the Rise, Progress and Present Condition of All the European Settlements upon That Coast; and the Just Measures for Improving the Several Branches of the Guinea Trade,* was published posthumously in 1704. Less than a year later it had been translated into both French and English and published in London, in an edition "To which is prefix'd, an exact map of the whole coast of

Guinea that was not in the original." The book took the form of letters to his friends and is a snapshot of the times, and particularly of the slave trade. The potto is found from Sierra Leone to Equatorial Guinea, eastern DRC (Zaire), and Uganda.

Botta

Botta's Pocket Gopher *Thomomys bottae* Eydoux and **Gervais,** 1836
Botta's Serotine *Eptesicus bottae* **Peters,** 1869
Botta's Gerbil *Gerbillus bottai* **Lataste,** 1882 [Alt. Botta's Dipodil; Syn. *Dipodillus bottai*]

Paolo Emilio (Paul-Emile) Botta (1802–1870) was an Italian explorer and a doctor. Early in his career he spent a year on board the French ship *Héros* as Ship's Surgeon and Naturalist. Around 1827 the vessel traded on the Californian coast under the command of Captain Auguste Duhaut-Cilly, who recorded expeditions ashore with Botta. Botta wrote *Observations on the Inhabitants of California* in the 1820s and *Notes on a Journey in Arabia* and *Account of a Journey in Yemen* in 1841. He was in Arabia from 1832 until 1846, excavating near Nineveh, the ancient capital of Assyria (in modern-day Iraq), from 1842 until 1845. He is also commemorated in the common name of a bird, Botta's Wheatear *Oenanthe bottae,* and in the scientific name of a snake, the Rubber Boa *Charina bottae.* The gopher is found in the southwest USA and northern Mexico. The serotine bat is found from Yemen, Turkey, and Egypt east to Mongolia and Pakistan. The gerbil is found only in Sudan and Kenya.

Bottego

Bottego's Shrew *Crocidura bottegi* **Thomas,** 1898

Vittorio Bottego (1860–1897) was an Italian explorer. He was an artilleryman and a skilled horseman who wanted adventure and to become a hero, so he arranged a transfer to Eritrea in 1887. He set out on a journey of exploration with Captain Matteo Grixoni, leaving Berbera in 1892. They reached the upper flow of the Giuba River and had penetrated as far as its source by March 1893. After parting ways with Grixonit, Bottego reached Daua Parma and discovered the Barattieri waterfalls, finally reaching Brava in September 1893. The expedition lost 35 men en route. In 1895 Bottego set off again under the auspices of the Italian Geographical Society with a contingent of 250 local troops. He later tried crossing Ethiopia and was offered a truce but turned it down and was killed in fighting. The King of Ethiopia kept his men imprisoned for two years, and only when they were released did word of Bottego's fate reach the Italian colonial regime. All the expeditions collected natural history specimens along the way. The shrew is found from Guinea to Ethiopia and northern Kenya.

Bougainville

Western Barred Bandicoot *Perameles bougainville* Quoy and **Gaimard,** 1834
Bougainville's Melomys *Melomys bougainville* **Troughton,** 1936 [Alt. Bougainville Mosaic-tailed Rat]

Admiral Baron Hyacinthe Yves Philippe Potentien de Bouganville (1781–1846) was a French scientist and mariner who was in command of the French corvette *L'Espérance.* René Primevère Lesson was on board and described the Guanay Cormorant *Phalacrocorax bougainvilli,* which he named after the Admiral. The bandicoot survives only on Bernier and Dorre Islands (off Western Australia), having become extinct on the Australian mainland. The melomys is found on the islands of Bougainville and Choiseul in the western Solomon Island chain. It presumably takes its name from the island, rather than directly from Admiral de Bougainville.

Bourret

Bourret's Horseshoe Bat *Rhinolophus paradoxolophus* Bourret, 1951

René Leon Bourret (1884–1957) was a French zoologist. He undertook a comprehensive herpetological survey of Vietnam before WW2 and spent two decades studying the fauna of Indochina from 1922 until 1942. He described a number of species in different taxa and had other taxa named after him, such as a subspecies of turtle, Bourret's Box Turtle *Cuora galbinifrons bourreti*. In 1941 he published *Les tortues de l'Indochine,* which was the first detailed monograph to deal with all the turtles and tortoises of Southeast Asia. The bat is found in Vietnam and Thailand.

Bourlon

Bourlon's Genet *Genetta bourloni* Gaubert, 2003

Phillipe Bourlon (1978–2001) was a zookeeper at the Vincennes Zoo in Paris. On 24 September 2001 an accident occurred with a trapdoor in a lion enclosure when he was feeding the lions. He was attacked by a lion and died of his injuries in hospital. The genet was described in Gaubert's Ph.D. thesis and relates to a skull collected in 1958. He dedicated the new taxa thus: "I dedicate this new species as a token of grateful friendship to Philippe Bourlon, keeper at the Vincennes Zoo, Paris, who lost his life in the course of his duties." The genet is known from Guinea and probably also occurs in Sierra Leone and Liberia.

Boutourline

Boutourlini's Blue Monkey *Cercopithecus mitis boutourlinii* Giglioli, 1887

Count Augusto Boutourline (dates not found) was a descendant of a Russian noble family. The family had lost most of its possessions in the great Moscow fire of 1812, when the city was occupied by the troops of Napoleon's army. They thus decided to rebuild in the more friendly surroundings of Italy, emigrating there in 1818 and using the Palazzo Buturlin in Florence as their family home. Boutourline traveled widely in Asia and Africa. In 1883 he was in Ceylon (now Sri Lanka) and is recorded as having paid a visit to eminent Egyptian exiles there. Giglioli wrote an account of his journey to Africa, published as *Note intorno agli animali vertebrati raccolti dal Conte Augusto Boutourline e dal Dr. Leopoldo Traversi ad Assab e nello Scioa anni 1884–87* (Notes on the Vertebrate Animals Collected by Count Augusto Boutourline and Dr. Leopoldo Traveling to Assab and in Scioa in the Years 1884–1887). Scioa is an area of Abyssinia (Ethiopia) close to the border with Sudan. The monkey is found in southern Ethiopia.

Bouvier

Bouvier's Red Colobus *Piliocolobus pennantii bouvieri* Rochebrune, 1887

Aimé Bouvier (d. 1919) was a French collector and zoologist who appears to have been active around 1876, in which year he became Secretary of the French Zoological Society. He and other committee members were forced to resign in 1880 after it was discovered that about 5,000 francs of the society's funds were missing and not accounted for. Bouvier is remembered in the scientific names of other animals including insects, fishes, and birds, among them Bouvier's Fishing Owl *Scotopelia bouvieri* and Bouvier's Sunbird *Nectarinia bouvieri.* He made two trips to Australia to collect birds, the first to Cairns between 1884 and 1885 and the second, with the taxidermist Walter Burton, to northwestern Australia in 1886. Thereafter he collected on Thursday Island, off Cape York, and at Palmerston in the Northern Territory in 1886. Many of the bird skins were presented to the British Museum in 1887, with the remainder given to G. M. Mathews. The colobus is found in a small area of the Congo Republic, if indeed it is not extinct. There have been no confirmed sightings for 25 years at the time of writing.

Bowdoin

Bowdoin's Beaked Whale *Mesoplodon bowdoini* **Andrews,** 1908 [Alt. Andrews' Beaked Whale]

George Sullivan Bowdoin (1833–1913) was a direct descendant of Alexander Hamilton, one of George Washington's confidants and a leading light in the American Revolution, so his social credentials were perfect for the age in which he lived. He was a trustee of the American Museum of Natural History (AMNH) from 1883 until 1913. He was also an important member of the Board of the Metropolitan Opera in New York. He was a lawyer by training and worked with various law firms associated with J. P. Morgan; he was also a director of the Guaranty Trust from 1896 to 1902. The whale was first recorded on New Brighton Beach, New Zealand, in 1904. It was described in 1908 by R. C. Andrews, who worked at the AMNH. The species is known from strandings of only about 35 animals, mainly in southern Australia and New Zealand. They appear to avoid vessels and are rarely seen at sea.

Bowers

Bowers' White-toothed Rat *Berylmys bowersi* **Anderson,** 1879

Captain Alexander Bowers (d. 1887) was a British merchant navy master mariner who commanded fast-sailing clippers and penetrated further up the Yangtze-Kiang than any other westerner. He was known to be very religious and gave free passage on his ships to missionaries, helping to set up mission stations at places like Swatow, which was opened up to foreign trade in 1869, when he was master of *Geelong*, a passenger/cargo ship built in 1866 and owned by P&O. Bowers must have been involved with Edward Sladen's expedition in 1868 that went through Burma into China, as in 1869 he wrote a *Report on the Practicability of Re-opening the Trade Route, between Burma and Western China*. Anderson, who described this rodent, was a member of the Sladen expedition. Bowers died in Rangoon, leaving a widow and young children, one of whom was Henry Robertson Bowers (1883–1912), who died on Scott's last expedition on their way back from the South Pole. Surprisingly, Anderson's original description of the rat says nothing about his choice of name, and we could find no mention of Bowers in the text. The evidence points to Captain Alexander Bowers, but Anderson didn't make things clear for posterity. The type specimen of the rat was taken in Yunnan Province (China), and the animal is distributed widely from northeast India to southern China, Thailand, and northern Indochina, and on the Malay Peninsula and northern Sumatra.

Boyle

Brush Deer Mouse *Peromyscus boylii* **Baird,** 1855

Dr. Charles Elisha Boyle (1821–1870) was an American physician and amateur naturalist. After working at a printing press and teaching, he entered medical college, graduating in 1847. Between 1848 and 1849 he caught "gold rush fever," joining the Columbus and California Industrial Association party as its official physician and traveling the Oregon Trail to California in the summer—one of the 49-ers. By 1850 he had settled in California, opening a practice as a physician. In his spare time he collected natural history specimens, particularly reptiles and amphibians, which he sent to the Smithsonian. He collected the type specimen of the Foothill Yellow-legged Frog *Rana boylii*, which Baird named after him. Between 1850 and 1852 he and a friend built a boat and then sailed it around Cape Horn. He settled down again but served as a military surgeon during the Civil War from 1860 until 1865, attaining the rank of Captain. After the war he returned to private practice. "Dr. Boyle was an accomplished linguist (fluent in 32 languages) and speaker. Although largely self educated, he was much in demand for local meetings and clubs due to his vast knowledge, phenomenal memory, and often accurate predictions of future events. He also gave much of his time and practice to the poor of the city and as a result never amassed much money and died poor himself." The mouse is found from the southern USA to central Mexico.

Brahms

Brahms' Wallaby

This is not a real mammal species but the title of a short story written by E. W. Smith and published in 1986.

Brandt

Brandt's Yellow-toothed Cavy *Galea flavidens* **Brandt,** 1835
Brandt's Hedgehog *Hemiechinus hypomelas* Brandt, 1836
Brandt's Bat *Myotis brandti* **Eversmann,** 1845
Brandt's Vole *Lasiopodomys brandtii* **Radde,** 1861
Brandt's Hamster *Mesocricetus brandti* **Nehring,** 1898

Johann Friedrich (Fedor Fedorovich) von Brandt (1802–1879) was a German who emigrated to Russia; he was a zoologist, surgeon, pharmacologist, and botanist. He explored Siberia, having settled in Russia in 1831 when he founded and become the first Director of the Zoological Museum of the Academy of Science in St. Petersburg. He produced works on systematics, zoogeography, comparative anatomy, and the paleontology of mammals. Between 1829 and 1833 he co-wrote the two-volume *Medical Zoology*. His book *Descriptiones et icones animalium rossicorum novorum vel minus rite cognitorum*, dealing in particular with Russian fauna, was published in St. Petersburg in 1836. He is also commemorated in the common names of three birds (Brandt's Cormorant *Phalacrocorax penicillatus*, Brandt's Jay *Garrulus glandarius brandtii*, and Brandt's Rosy Finch *Leucosticte brandti*) and other taxa as diverse as a woodlouse and a piranha. The cavy is a Brazilian endemic. The hedgehog is found in the arid steppes and desert zones of Iran and Turkmenistan east to Pakistan, with isolated populations in the Arabian Peninsula. The bat is found across Europe and in eastern Russia, Mongolia, Korea, and Japan. The vole is found in Mongolia and adjacent parts of China and Transbaikalia (Russia). The hamster is found from Turkey and Lebanon east through Syria and northern Iraq to northwest Iran.

Brants

Brants' Climbing Mouse *Dendromus mesomelas* Brants, 1827
Brants' Whistling Rat *Parotomys brantsii* **A. Smith,** 1834
Highveld Gerbil *Gerbilliscus brantsii* A. Smith, 1836

Anton Brants (dates not found) was a Dutch zoologist and author at Leiden. He published in a number of languages, as was common with savants of his time: in Dutch, *Het Geslacht der Muizen* (1827), a commentary on the mice described by Linnaeus; in Latin, *Dissertatio zoological inauguralis de tardigradis* (1828); and in French, *Observations sur les yeux des animaux* (1838). The climbing mouse has a discontinuous distribution from South Africa north to Tanzania and DRC (Zaire). The rat is found in Cape Province, South Africa, to southern Botswana and southeast Namibia. The gerbil ranges from South Africa north to western Zimbabwe, Botswana, southern Angola, and southwest Zambia.

Brauer

Brauer's Gerbil *Desmodilliscus braueri* Wettstein, 1916 [Alt. Pouched Gerbil]

Professor Dr. August Bernhard Brauer (1863–1917) was a German zoologist whose primary areas of interest were herpetology and ichthyology. He graduated from the Humboldt University in Berlin in natural sciences in 1885 and took his doctorate in 1892. He collected in the Seychelles in 1897. In 1898 he took part in the *Valdivia* expedition and described the fish they collected in a paper that appeared in 1908. He was appointed a Professor at Berlin University in 1905 and also, in the following year, Director of the university's Zoological Museum. He was appointed a Professor of the Zoological University in Berlin in 1914. The describer, Otto von Wettstein, Ritter von Westersheimb (1892–1967), was Curator of the Herpetological Collection of the Natural History Museum in Vienna from 1920 to 1945, so we assume they would have had correspondence at least. The gerbil is found in the Sahel savannah, south of the Sahara, from Senegal east to Sudan.

Bregulla

Fijian Mastiff Bat *Chaerephon bregullae* Felten, 1964

Heinrich L. Bregulla (b. 1930) is a German naturalist and ornithologist who is an expert on the biology of parrots. He first went to the Pacific in 1959 and settled in the New Hebrides (now Vanuatu), where he collected the type specimen of the mastiff bat. In 1992 he published *Birds of Vanuatu*. The bat is also found in Fiji.

Brelich

Brelich's Snub-nosed Langur *Rhinopithecus brelichi* **Thomas,** 1903 [Alt. Grey Snub-nosed Monkey]

Henry Brelich (dates not found) was working in the Chinese province of Guizhou, where this monkey has a very limited range, at the end of the 19th century and the beginning of the 20th. We know very little about him except that he may have been a metallurgist or mining engineer, as he published "Chinese Methods of Mining Quicksilver" in the *Proceedings of the Institute of Mining and Metal* in 1904–1905. He is quoted as writing in 1904, "Cinnabar was mined at the turn of this century in the province of Kweichow, and produced quicksilver from it in iron and clay retorts. It was exported to different parts of China where it is used for the manufacture of vermilion, for which there is great demand throughout the Chinese empire." In the *Proceedings of the Zoological Society of London,* 1903, it is written that "Mr. Oldfield Thomas exhibited the skin of a Chinese Monkey, which had been obtained from a hunter by Mr. Henry Brelich, and presented by him to the National Museum. . . . This magnificent Monkey, one of the largest in the world apart from the anthropoids, is a very remarkable discovery, and one on which we may congratulate Mr. Brelich, who obtained and sent it to the Museum on the suggestion of Mr. Herbert Ingram, himself a frequent contributor to the National collections."

Brewer

Beach Vole *Microtus breweri* **Baird,** 1858

Brewer's [Shrew] Mole *Parascalops breweri* **Bachman,** 1842 [Alt. Hairy-tailed Mole]

Dr. Thomas Mayo Brewer (1814–1880) was an American naturalist and ornithologist who co-wrote the *History of North American Birds* with Baird and Ridgway. He was also the author of *North American Oology* (1857). He was very involved in politics—a family trait, as his father had actually taken part in the Boston Tea Party. He had contempt for those studying animals in the field, preferring to study their skins in museum collections. He also defended the House Sparrow *Passer domesticus* against moves to eliminate this introduced species from the USA. He is commemorated in the common names of two birds, Brewer's Blackbird *Euphagus cyanocephalus* and Brewer's Sparrow *Spizella breweri*. The vole has a tiny distribution, being found in beach grassland on Muskeget Island, Massachusetts. The mole occurs in the northeast USA and southeast Canada.

Briceño

Merida Brocket *Mazama bricenii* **Thomas,** 1908 [Alt. Grey Dwarf Brocket]

Salomón Briceño Gabaldón (1826–1912) was a Venezuelan collector. Among his clients was Lord Lionel Walter Rothschild, for whom he supplied many specimens, particularly of birds. Thomas wrote that he named this species of deer after S. Briceño "in recognition of the immense number of mammals which he has been instrumental in discovering." A subspecies of the Flammulated Treehunter *Thripadectes flammulatus* is named *bricenoi* after him in the trinomial. The brocket is found in western Venezuela.

Bridges

Bridges' Degu *Octodon bridgesi* **Waterhouse,** 1845

Dr. Thomas Charles Bridges (1807–1865) was a traveler and collector in tropical America from 1822 until 1865. His particular targets were birds, and there are more than three dozen specimens of bird species in the British Museum collected by Bridges in Chile, Bolivia, and Guatemala and bought by, or sent to, a number of different ornithologists. Not surprisingly, he is commemorated in the names of three birds: Bridges' Antshrike *Thamnophilus bridgesi*, Bridges' Parrot *Pionus maximiliani siy*, and Bridges' Woodhewer *Drymornis bridgesii*. The degu is found in the Chilean Andes.

Bright

Bright's Gazelle *Gazella granti brighti* **Thomas,** 1900

Lieutenant Colonel R. G. T. Bright C.M.G. (dates not found) of the Rifle Brigade was involved in, and wrote the summary report of, the Uganda-Congo Boundary Commission in 1908. In 1900 through 1901 he was a member of a party that explored the northern part of the Sudan, close to the border with Ethiopia. In 1903, at that time a Captain, he reported on a journey he had made through eastern Northern Rhodesia (now Zambia). In 1910 he published *An Exploration in Central Equatorial Africa*, the text of a talk he had given to the Royal African Society when it held its dinner at the Trocadero. When he was still a Lieutenant he supplied the first known example of this subspecies of Grant's Gazelle to the British Museum. The gazelle is found in northwest Kenya, southeast Sudan, and adjacent parts of Uganda.

Broadbent

Giant Bandicoot *Peroryctes broadbenti* Ramsay, 1879

Kendall Broadbent (1837–1911) was born in Yorkshire, England, but went to Australia in 1852 and became a collector of natural history specimens for museums and wealthy private individuals. While collecting in the 1880s and 1890s he kept detailed diaries, which are now held at the Queensland Museum, for which he worked as a collector and later as a museum attendant. He is also commemorated in the scientific name of the Rufous Bristlebird *Dasyornis broadbenti*. The bandicoot is found in southeast New Guinea.

Brock

Brock's Yellow-eared Bat *Vampyressa brocki* **Peterson,** 1968

Stanley E. Brock (b. 1937) was born in England and went to what is now Guyana as a young teenager, when his father was on a posting there for the British government. He did not go home to England when his parents did and grew up with the Wapishana Indians in the central Amazon basin. He spent 15 years as General Manager of the Rupununi Development Company Dadanawa Ranch in Guyana. This was the world's largest tropical cattle ranch, with an area of 10,360 sq km (4,000 sq miles) and 40,000 cattle, 2,000 horses, and 1,200 sheep. He acted as a collector for the Royal Ontario Museum and made "a rather extensive collection of mammal specimens" for that museum in southern Guyana and surrounding regions. The type specimen of this bat was found in that collection, and Peterson wrote, "I take pleasure in naming it after Mr. Brock in recognition of his outstanding efforts." The museum's Department of Mammalogy made him a Research Associate. He became famous as a star in a number of wildlife series on American television, the best known being NBC's *Wild Kingdom*. He was an early bush pilot in the Amazon basin and worked as flight instructor. He now lives in Knoxville, Tennessee, and has written a number of books of memoirs about his time in the Amazon basin, including *All the Cowboys Were Indians*, *Leemo*, *More about Leemo*, and *Hunting in the Wilderness*. Brock is founder and current volunteer Director of Operations of the global medical relief charity, Remote Area Medical Volunteer Corps. The bat is found in Amazonian Brazil, Colombia, and the Guianas.

Brockman

Brockman's Mouse *Myomyscus brockmani* **Thomas,** 1908
Brockman's Gerbil *Gerbillus brockmani* Thomas, 1910

Lieutenant Colonel Ralph Evelyn Drake-Brockman (1875–1952) was a British zoologist and fellow of the Zoological Society of London. He served in Somalia as part of the British Colonial Service. He wrote *The Mammals of Somaliland* (1910) and *British Somaliland* (1912), as well as a number of papers, mostly about antelopes, published in the *Proceedings of the Zoological Society* from the 1900s through to 1930. The mouse is found in southern Sudan, Uganda, Somalia, Kenya, and Tanzania, and the gerbil is only found in Somalia.

Brooke, C. J.

Brooke's Squirrel *Sundasciurus brookei* **Thomas,** 1892

Sir Charles Johnson Brooke (1829–1917) was originally plain Charles Johnson but took the name Brooke when he became the second "White Rajah" of Sarawak in 1868, on the death of his uncle, James Brooke, the original "White Rajah." Rajah James invited Wallace to collect in Sarawak, but it isn't clear if Wallace went. Rajah Charles was a notable eccentric who lost an eye in a riding accident and replaced it with a false one that had originally been destined for a stuffed albatross. It was he who placed Sarawak under British protection, giving the British government full control over itsforeign relations; this led to the British government proclaiming a formal protectorate overthe states of Sarawak, Brunei, and North Borneo. He is also commemorated in the name of a bird, Brooke's Scops Owl *Otus brookii*. The squirrel is endemic to Borneo.

Brooke, V. A.

Brooke's Duiker *Cephalophus brookei* **Thomas,** 1903 [Syn. *C. ogilbyi brookei*]

Sir Victor Alexander Brooke, 3rd Baronet (1843–1891), was a naturalist and hunter. He traveled widely and, at the time of his death from pneumonia at Pau in France, where he had settled, was writing a work on antelopes. The plates that had been intended to illustrate this book were later used by Thomas (who described the duiker) and by PhilipSclaterin their publication *The Book of Antelopes*, which appeared in four volumes between 1894 and

1900. Brooke's youngest child was the distinguished soldier Field Marshall Sir Alan Francis Brooke, 1st Viscount Alanbrooke (1883–1963), and his grandson was Basil Brooke, Lord Brookeborough (1888–1973), who was Prime Minister of Northern Ireland for 20 years. This West African duiker is often regarded as a subspecies of Ogilby's Duiker, but some authorities elevate it to the rank of a full species.

Brooks, A. C.

Delany's Swamp Mouse *Delanymys brooksi* **Hayman,** 1962

Allan Cecil Brooks (1926–2000) was a Canadian biologist, teacher, and naturalist who worked for the Department of Game and Fisheries in Uganda. Michael Delany (q.v.) collected the type specimen of the mouse in 1961, in southwest Uganda, and sent it to Hayman at the British Museum (Natural History). Delany suspected, correctly, that it represented a new species. He must have mentioned Brooks' name to Hayman, saying how indebted he was to him for help and encouragement, as Hayman honored Brooks in the binomial. Brooks left Uganda and returned to Canada, settling in British Columbia in 1981. The mouse is also found in Rwanda and adjacent parts of DRC (Zaire).

Brooks, C. J.

Brooks' Large-headed Fruit Bat *Dyacopterus brooksi* **Thomas,** 1920 [Alt. Brooks' Dayak Fruit Bat]

Cecil Joslin Brooks (1875–1953) was a collector, mainly of botanical specimens, in Borneo and Sumatra in the early part of the 20th century.His wife, Alida C. De Jongh, assisted him. He was a metallurgical chemist in the employ of the Borneo Co. Ltd. in Sarawak from 1900 to 1910. From 1912 until 1923 he was employed at the Simau goldmine in Benkoelen, Sumatra, and traveled subsequently in the Dutch East Indies, then sailed via Australia and New Zealand to Europe in 1924. Once back in England he devoted himself to the study of his collections and of the butterfly collection in the British Museum. He wrote a number of scientific papers, mainly on ferns (several species of which are also named after him, such as *Pteris brooksianus* and *Protolindsaya brooksii*). He is also remembered in the name of a lizard, Brooks' Wolf Gecko *Luperosaurus brooksi.* Brooks collected the type specimen of the fruit bat, which comes from Borneo and Sumatra.

Brosset

Brosset's Big-eared Bat *Micronycteris brosseti* Simmons and **Voss,** 1998

Dr. André Brosset (1926–2004) was a French zoologist. He was Assistant Director of the Museum of Natural History, Paris. In 1962 he became the Director of the Charles Darwin Research Station in the Galápagos Islands. Over a number of years he collected and identified bats from French Guiana. He published numerous articles both in scientific journals and in popular science journals such as *Nature* and was also a keen ornithologist. The bat is known from Brazil, French Guiana, Guyana, and eastern Peru.

Brower

Alaska Marmot *Marmota broweri* Hall and Gilmore, 1934 [Alt. Brooks Range Marmot]

Charles Dewitt Brower (1886–1945) was a whaler, trapper, trader, and postmaster in Barrow, Alaska, and for many years held the title of "northernmost white man" in Alaska. He was brought up in New York but went to Alaska in search of coalfields. Beginning with his first voyage in 1884, Brower became an important citizen of the north, studying the ways of the indigenous peoples and their natural environment. He wrote, but never published, *My Arctic*

Outpost and an autobiography, *The Northernmost American*. But *Fifty Years below Freezing* was published 1942, as well as several articles. He married an Inuit girl, Asiannataq from Shishmaref, and they had 20 children. He counted among his friends the Arctic explorers and experts Vilhjalmur Steffansson, Knud Rassmussen, George Wilkins, and Roald Amundsen. He established a whaling company with a friend, and the Cape Smythe Whaling and Trading Company opened the first store in Barrow. The marmot is, as its name implies, confined to the north of Alaska. There is also a village in Alaska named after (or possibly by) Brower.

Brown

Brown's Pademelon *Thylogale browni* Ramsay, 1877 [Alt. New Guinea Pademelon]

The Rev. George Brown D.D. (1835–1917) was a missionary to Melanesia. He was a Methodist, and in 1875 he seems to have become the first Christian missionary to have landed in New Ireland, an island to the east of the mainland of Papua New Guinea and part of that present-day country. He spent much time exploring in the company of Cockerell. In 1905 he published *George Brown, D.D., Pioneer Missionary and Explorer—An Autobiography*. A paper by E. R. Alston entitled "On the Rodents and Marsupials Collected by the Rev. G. Brown in Duke-of-York Island, New Britain, and New Ireland" appeared in the *Proceedings of the Zoological Society* in 1877. Brown is also commemorated in the names of two birds, Brown's Long-tailed Pigeon *Reinwardtoena browni* and Brown's Monarch *Monarcha browni*. The pademelon (a type of wallaby) is found in northern New Guinea and on some of the islands of the Bismarck Archipelago. It is presumed to have been introduced to these islands by humans, since no land bridge ever existed between them and the New Guinea mainland.

Browne

Brown's Hutia *Geocapromys brownii* J. Fischer, 1829 [Alt. Jamaican Hutia]

Dr. Patrick Browne (1720–1790) was an Irish physician, botanist, and historian. He studied medicine in Paris and Leiden and graduated from the University of Rheims in 1742. After a short period of further study he took up a post at St. Thomas's Hospital, London. After this he lived for many years in the Caribbean, in Antigua (to which he had made an earlier visit in 1737), Jamaica, Saint Croix, and Montserrat, but when he retired in 1771 it was to his native County Mayo. He published *Civil and Natural History of Jamaica* in 1756, which included new names for 104, mostly botanical, genera. The hutia (the spelling of whose common name as "Brown's" without an *e* represents a "mistranslation" of the scientific name) is a Jamaican endemic.

Bruce

Yellow-spotted Hyrax *Heterohyrax brucei* **Gray,** 1868 [Alt. Bush Hyrax, Small-toothed Rock Hyrax]

James Bruce (1730–1794) was a British traveler and explorer. He went to Edinburgh University in 1746 to study for the bar, but ill health forced him to go back to the family farm. He married in 1753 and was starting work in his in-laws' wine business when his wife died. He wanted nothing more to do with the wine business and decided to learn Spanish and Portuguese in order to travel to Iberia for his health. In 1757 he sailed to Portugal to study its society, art, and science. In 1758, following visits to France, Germany, Brussels, and Holland, he heard of his father's death and returned to Scotland. Later living in Spain as war with England approached, he left with intelligence that led to him being favored by the English government, and was appointed to the post of British Consul in Algiers, a post he held from 1763 until 1765. He then took to traveling, visiting Tunis and Syria before heading for Upper Egypt, Abyssinia (now Ethiopia), and Sennaar. From 1769 he undertook an ambitious journey looking for the source and the course of the Blue Nile, which he achieved by the following year. He eventually made it home in 1774. The epitaph on his tomb reads, "In this tomb are deposited the remains of James Bruce Esq. of Kinnaird who died on the 27th of April 1794 in the 64th year of his age. His life was spent in performing useful and splendid actions. He explored many distant regions; he discovered the Fountains of the Nile; he traversed the deserts of Nubia." The hyrax is found from Egypt to northern South Africa and Angola.

Bruce Patterson

The rodent genus ("brucies") *Brucepattersonius* **Hershkovitz,** 1998 [8 species]

See **Patterson, B. D.**

Bruijn, A. A.

Bruijn's Echidna *Zaglossus bruijni* **Peters** and **Doria,** 1876 [Alt. Western Long-beaked Echidna]

Lowland Brush Mouse *Pogonomelomys bruijni* Peters and Doria, 1876 [Alt. Large Pogonomelomys]

Anton August Bruijn (d. 1885) was a Dutch plumassier, or feather merchant, who was the son-in-law of Duivenbode, the so-called King of Ternate. He exhibited many natural history specimens during the large international Colonial Trade Exhibition in Amsterdam in the summer of 1883. Others have described him as a botanist, explorer, and zoologist and use a different initial, which might mean there were two men of the same name, perhaps brothers, who were operating in the same area. Bruijn's Brush Turkey *Aepypodius bruijnii* and Bruijn's Pygmy Parrot *Micropsitta bruijnii* were also named after him. Both mammals are found in New Guinea.

Bruijn, C.

Bruijn's Pademelon *Thylogale brunii* **Schreber,** 1778 [Alt. Dusky Pademelon]

Cornelis de Bruijn (1652–1727) was a Dutch painter who in 1714 provided the first description to Europeans of a kangaroo. He had a very adventurous life, traveling widely and being suspected of trying to murder Johan de Witt, who was later lynched by a mob. Bruijn traveled to Italy in 1674 and stayed in Rome for more than two years. After traveling around Italy, he went to Greece and Turkey in 1678 and from there to Egypt and Palestine, where he visited Jerusalem. At the end of 1684 he arrived in Venice, where he remained for eight years, returning to the Netherlands in 1693 after an absence of 19 years. In 1700 he visited London, and the next year he set out for Russia. Since the Great Northern War (between Sweden and Russia) had broken out, he was forced to travel by sea via the North Cape and Barents Sea and landed at Archangel. He then traveled to Moscow, arriving there in 1702. He was introduced to the Tsar, Peter I "the Great." In 1703 he left Moscow, traveling toward the Volga, which he sailed down to Astrakhan and on to Persia (now Iran). In 1705 he sailed from Gamron (now Bandar Abbas) to Cochin in India, then to Ceylon (now Sri Lanka) and on to Batavia (now Jakarta) in Java in 1706. In 1707 he returned to Persia and from there overland to Moscow, where he arrived in 1708. He tried to return to the Netherlands via Smolensk and Vilnius but was caught up in the fighting between the Russians and the Swedes and was arrested as a spy. He was forced to return to Moscow and finally returned home to The Hague by sea from Archangel. He stayed at home after that and published a number of outstanding works on his travels, including, in 1711, *Reizen over Moskovie, door Persie en Indie* (translated in 1720 into English as *Voyage to the Levant and Travels into Moscovy, Persia, and the East Indies*). He saw the pademelon in 1706 in the Dutch Governor's garden in Batavia where it had been brought as a curiosity, but it comes from southern New Guinea and the Aru Islands.

Brumback

Brumback's Night Monkey *Aotus brumbacki* **Hershkovitz,** 1983 [Alt. Brumback's Owl Monkey]

Professor Dr. Roger Alan Brumback (b. 1948) is David Ross Boyd Professor and Chairman of the Department of Pathology in the University of Oklahoma College of Medicine, specializing in the study of Alzheimer's disease. He did not know a monkey had been named after him for nearly 20 years and discovered this fact only after a visit to the San Diego Zoo had got him thinking about the group of monkeys on which he had undertaken chromosome analysis to identify evolutionary trends in primates when he was a freshman at Pennsylvania State Medical College. To his surprise, he found one named after him by the person he had handed his research findings over to upon completion of his medical training in Pennsylvania. It is not known if Hershkovitz tried to contact him to let him know he had honored him thus. Brumback went on to a distinguished career. He was at Washington University School of Medicine in St. Louis in 1975, the National Institutes of Health to 1977, the University of Pittsburgh to 1978, the University of North Dakota to 1982, the University of Rochester to 1986, and finally University of Oklahoma, where he has been since 1986. As a self-described "born-again conservationist," Brumback has given the American Society of Primatologists' Conservation Fund a substantial donation, targeted for the study and protection of Brumback's Owl Monkey, which is under threat in its Colombian range.

Bryant

Bryant's Woodrat *Neotoma bryanti* **Merriam,** 1887

Dr. Henry Bryant (1820–1867) of Boston, Massachusetts, was a physician and naturalist who trained in Paris and served a year in the French army in North Africa before returning to Boston. As he was in poor health, he did not practice as a doctor but spent his time bird collecting instead, traveling widely in North America and the Caribbean. During the American Civil War he was the surgeon of the 20th Massachusetts Regiment. In 1866 LaFresnaye's collection was put up for sale by his widow, and Bryant bought the lot and donated it to the Boston Natural History Society. He is also commemorated in the common names of two birds, Bryant's Grassquit *Tiaris olivacea bryanti* and Bryant's Sparrow *Passerculus sandwichensis.* The endangered rat is confined to Cedros Island, off Baja California, Mexico.

Bryde

Pygmy Bryde's Whale *Balaenoptera edeni* **Anderson,** 1879 [Alt. Eden's Whale]
Bryde's Whale *Balaenoptera brydei* Olsen, 1913

Johan Bryde (1858–1925) was a Norwegian master mariner and shipowner who started up the whaling industry in South Africa and became Norwegian Consul in South Africa. In 1883 he was part owner and master of a sailing vessel called *Kommandør Sven Føyn,* which gives a clue to his future, as the vessel appears to have been named after her major shareholder Sven Føyn Bruun, who invented the explosive harpoon and revolutionized whaling. The vessel was sold in 1889 to a Sandefjord Whaling Company and sold again in 1892—

but this time to Bryde, who by now had an office in Sandefjord. In 1907 the Norwegian Consul in Durban, Jacob Egeland, returned to Norway and persuaded Bryde to join him in a whaling venture in South Africa. Bryde went to Durban in 1908, and the partners established the first whaling station in 1909. Later, Egeland and Bryde dissolved their partnership and were then in competition with each other. Bryde financed the first scientific whale investigation off South Africa. Eden's Whale and Bryde's Whale were long considered to be one and the same species but are now regarded as separate. The "pygmy" whale seems to be confined to the coastal waters of the eastern Indian Ocean and western Pacific Ocean, whereas the larger Bryde's Whale occurs worldwide in tropical and warm-temperate waters.

Budgett

Budgett's Tantalus [Monkey] *Chlorocebus tantalus budgetti* **Pocock,** 1907

John Samuel Budgett (1872–1904) was an embryologist and ichthyologist, a naturalist, an artist, and an explorer employed by Trinity College, Cambridge. He made at least four expeditions to West Africa and made a very important and extensive collection of African fishes. He suffered so badly from malaria that, although a Lieutenant in the mounted infantry regiment raised at Cambridge by the University, he was not passed as fit for service in the Boer War, so he went exploring again instead. He is associated with the attempts to document the embryonic development of *Polypterus*, a genus of freshwater fish found in tropical Africa. He succeeded in the task but died of malaria during the attempt. The monkey is found in Uganda and northeast DRC (Zaire).

Budin

Budin's Tuco-tuco *Ctenomys budini* **Thomas,** 1913
Budin's Grass Mouse *Akodon budini* Thomas, 1918
Budin's Chinchilla-Rat *Abrocoma budini* Thomas, 1920
Pampas Cat ssp. *Oncifelis colocolo budini* **Pocock,** 1941 [Syn. *Leopardus pajeros budini*]

Señor Emilio Budin (1877–1935) was a collector and explorer in Argentina and Bolivia. He was born in Geneva, but his parents emigrated from Switzerland to Argentina when he was only a few months old. He collected in Patagonia and other areas of Argentina at the direction of Oldfield Thomas of the British Natural History Museum, to whom he sent many specimens in the 1890s and 1900s. Oldfield Thomas was a man of many parts, and he and Budin carried on their correspondence in Spanish. Budin also collected for Jose Yepes at the Argentine Museum of Natural Sciences as well as other museums in Argentina and Europe. In addition to writing notes on his finds, he provided route maps and described the local people and topography. All four of the mammals named after Budin are found in northwest Argentina. The cat has, in turn, had an Argentinean wine named after it.

Buettikofer

Buettikofer's Monkey *Cercopithecus petaurista buettikoferi* **Jentink,** 1886 [Alt. Western Lesser Spot-nosed Monkey]
Buettikofer's Shrew *Crocidura buettikoferi* Jentink, 1888
Buettikofer's Epauletted Bat *Epomops buettikoferi* **Matschie,** 1899

Johan Büttikofer (1850–1929) was a Swiss zoologist. He made two collecting trips to Liberia, where the mammals were collected, one from 1879 to 1882, and the second from 1886 to 1887, when he returned home due to ill health. He did, however, make one more trip there in 1888. He wrote "Zoological Researches in Liberia: A List of Birds, Collected by the Author and Mr. F. X. Stampfli during Their Last Sojourn in Liberia" as a note for the Leyden Museum. He is also commemorated in the names of two birds, Büttikofer's Babbler *Trichastoma buettikoferi* and Büttikofer's Warbler *Buettikoferella bivittata,* and in the scientific names of fish such as *Clarias buettikoferi.* The monkey is found in Sierra Leone, Liberia and western Ivory Coast. The shrew and the bat are found from Guinea to Nigeria.

Buettner

Groove-toothed Forest Mouse *Leimacomys buettneri* **Matschie,** 1893

Oscar Alexander Richard Buettner (1858–1927) was a German collector in Africa. He is commemorated in the scientific names of all sorts of taxa from insects to lichens. He collected in Ghana for the Humboldt University Natural History Museum, Berlin. The mouse may be extinct, as it is known only from two specimens found in central Togo in 1890.

Buffon

Buffon's Tarsier *Tarsius syrichta* **Linnaeus,** 1758 [Alt. Philippine Tarsier]
Buffon's Kob *Kobus kob* **Erxleben,** 1777

Count Georges Louis Leclerc de Buffon (1707–1788) is one of the giants of zoology, and sufficient is written about him elsewhere to allow this entry to be brief. He is perhaps most famous for having developed the species concept, the basis of all taxonomy. Louis XV appointed him to the Academy of Sciences in 1734 and, five years later, to the directorship of the Jardin du Roi, which Buffon transformed into what we now know as the Jardin des Plantes in Paris. His *Histoire naturelle générale et particulière* was published between 1749 and 1804. He had aimed to publish 50 volumes, but "only" 36 had been produced by the time of his death, and a further 8 thereafter. A number of birds are also named after him, for instance Buffon's Macaw *Ara ambigua.* The tarsier is found in the Philippines, and the kob (an antelope) occurs in the savannah zones of Africa, from Senegal to western Kenya.

Buller

Buller's Chipmunk *Tamias bulleri* **J. A. Allen,** 1889
Buller's Pocket Gopher *Pappogeomys bulleri* **Thomas,** 1892

Dr. Audley Cecil Buller (1853–1894) was a collector of mammals and reptiles. He co-wrote a number of articles with J. A. Allen, mostly about the collection he made in Mexico. He collected the type specimens of a bat, a spiny lizard, and the chipmunk and gopher—all of which are found in Mexico. His epic collecting trip for the American Museum of Natural History took him 1,610 km (1,000 miles) across the Sierra de Nayarit and ranges of the Sierra Madre to Zacatecas, then the least known area of Mexico. He is also remembered in the name of a reptile, Buller's Spiny Lizard *Sceloporus bulleri.*

Bulmer

Bulmer's Fruit Bat *Aproteles bulmerae* **Menzies,** 1977

Dr. Susan Bulmer (b. 1945) is an archeologist from New Zealand who excavated the bones of this bat in 1976, in a cave in Papua New Guinea. It was assumed to be an extinct taxon, but living specimens were then discovered in an enormous and remote cave called Luplupwintem. The species is now regarded as critically endangered. Dr. Bulmer is currently the Senior Regional Archeologist for the New Zealand Historic Places Trust. She has published numerous papers, mostly on Maori settlement history.

Bumback

Bumback's Owl Monkey *Aotus bumbacki* **Hershkovitz,** 1983

This is a transcription error. See **Brumback.**

Bunker

Bunker's Woodrat *Neotoma bunkeri* **Burt,** 1932

Charles Dean Bunker (1870–1948) was an American zoologist. On the advice of the family physician, his childhood was mostly spent outdoors, which led to his interest in natural history. He collected specimens and gave them to a local taxidermist in exchange for being taught how to preserve skins. In 1895 he went to work at the University of Kansas as a taxidermist. In 1904, after a brief time at the University of Oklahoma, he returned to Kansas. In 1907 he became Assistant Curator of Birds and Mammals, then in 1909 Assistant Curator in Charge at the Museum of Natural History, University of Kansas. He subsequently became Curator, a post he held from 1912 until he retired in 1942. One of his students was William H. Burt. Bunker was keen on fieldwork and collected vertebrates, specialized in birds, and concentrated on the preparation of whole skeletons. He developed innovative techniques for cleaning bones, using *dermestid* beetles, the larvae of which were already known for their ability to clean bones precisely and without damage. His fame, however, lies chiefly in his teaching ability, the achievements of many of his students, and his delight in their success. The endangered woodrat is confined to Coronados Island, Mexico.

Bunni

Bunn's Short-tailed Bandicoot Rat *Nesokia bunni* **Khajuria,** 1981

Munir K. Bunni is an Iraqi zoologist. He works at the Iraq Natural History Research Center. He co-authored many articles from the 1950s through to the 1980s, such as "A New Blind Cyprinid Fish from Iraq" (1980) and "The Breeding Habits of the Iraqi Babbler, *Turdoides altirostris*" (1981). The rat is found in the marshes of southern Iraq.

Bunting

Bunting's Thicket Rat *Grammomys buntingi* **Thomas,** 1911

Robert Hugh Bunting (fl. 1911–1966) was a British biologist and natural history collector who wrote on plant diseases in the 1920s and 1930s. He seems to have been a leading light in the Dorset Natural History and Archaeological Society for about 10 years from 1939. Thomas's original citation reads, "I have named it after Mr. R. H. Bunting, its captor, in whose collection there are several interesting Liberian species not previously possessed by the Museum." While we can find no direct evidence, we believe this to refer to Robert Hugh Bunting. The rat is found in Sierra Leone, Guinea, Ivory Coast, and Liberia.

Burbidge

Monjon *Petrogale burbidgei* Kitchener and Sanson, 1978 [Alt. Warabi]

Dr. Andrew A. Burbidge is an Australian zoologist. He works as a Research Fellow for the Department of Conservation and Land Management, Western Australia, and is Chairman of the Australasian Marsupial and Monotreme Specialist Group. He has been involved in conservation since the 1960s. In 2004 he published *Threatened Animals of Western Australia.* He is also remembered in the scientific name of a lizard, the Plain-backed Kimberley Ctenotus *Ctenotus burbidgei.* The monjon (a small species of rock wallaby) is found in the Kimberley area of northern Western Australia.

Burchell

Burchell's Zebra *Equus quagga burchelli* **Gray,** 1823

William John Burchell (1781–1863) was an English explorer-naturalist. He went to the Cape of Good Hope in 1810 and undertook a major exploration of the interior of South Africa between 1811 and 1815, during which time he traveled more than 7,000 km (4,000 miles) through largely unexplored country. He published his two-volume work, *Travels in the Interior of Southern Africa,* in 1822 and 1824. He was renowned as a meticulous collector, botanist, and artist. He returned to London in 1815 to work on his collections. In 1825 he spent two months in Lisbon and then proceeded to Brazil where he collected extensively, not returning again to England until 1830. He became increasingly reclusive and in the last two years of his life became seriously ill, eventually taking his own life. Among the taxa named in his honor are the Wild Pomegranate *Burchellia bubalina* and at least six birds, including Burchell's Courser and Burchell's Sandgrouse. Burchell was also the first person to describe the White Rhinoceros *Ceratotherium simum.* The "typical" form of Burchell's Zebra, from South Africa, is often listed as extinct. However, in 2004 Groves and Bell concluded after a study of zebras in Natal, Swaziland, and Zululand that a small proportion of the existing population resembles typical *burchelli.*

Burmeister

Burmeister's Porpoise *Phocoena spinipinnis* Burmeister, 1865

Professor Karl Hermann Konrad Burmeister (1807–1892) was a German ornithologist who was Director of the Institute of Zoology of Martin Luther University at Halle Wittenberg, Germany, from 1837 to 1861. He was in the Prussian civil service but won his release from it by using the very inventive excuse that a persistent stomach complaint was caused by arsenic emissions in the museum and by the drinking water in Halle, which had high sulphate content. He sent many specimens to the zoological collections at the Institute. These were largely collected during his two expeditions to South America: to Brazil from 1850 until 1852 and to the La Plata region of Argentina from 1857 to 1860. Subsequently he was resident in Argentina from 1861 until his death. He founded the Institute at the Museo Nacional in Buenos Aires and became its first Director, remaining in that post until his retirement in 1880. He wrote *Reise nach Brasilien* in 1853. He is also remembered in the name of a bird, Burmeister's Seriema *Chunga burmeisteri.* Burmeister first described the porpoise based on the skin and skull of a specimen that he came across in the museum collection. He named it in the binomial from the fin that uniquely had a thornlike protrusion. The porpoise is found in the coastal temperate waters of South America.

Burt

Burt's Deer Mouse *Peromyscus caniceps* Burt, 1932

Sonoran Harvest Mouse *Reithrodontomys burti* Benson, 1939

Professor William Henry Burt (1903–1987) was an American zoologist. He studied at the University of Kansas under Charles Dean Bunker. He acted as assistant to, and collector for, Donald Ryder Dickey in his field surveys of mammal species in Nevada, Arizona, and New Mexico from 1928 to 1931. He was awarded his doctorate at the University of California in 1930 and then worked until 1935 as a research fellow at the California Institute of Technology. In 1935 he became Professor of Zoology at the University of Michigan and doubled up as Curator of Mammals at that university's Museum of Zoology. He retired from the university in 1969. He wrote *Mammals of the Great*

Lakes Region (1957) and, with others, *Field Guide to the Mammals—North America North of Mexico.* He was President of the American Mammalogists Society from 1953 to 1955. He also described a number of new mammals, mostly from Mexico. There is a William H. Burt Memorial Fund at the University of Colorado Museum, set up "to encourage significant contributions to natural history, especially mammalogy." The deer mouse is confined to Monserrate Island in the Gulf of California, Mexico. The harvest mouse is found in northwest Mexico (Sonora and Sinaloa).

Burton, E.

Burton's Gerbil *Gerbillus burtoni* **Cuvier,** 1838

Cuvier's original description of this gerbil tells us only that the type specimen was brought alive from Darfur (Sudan) by a M(onsieur?) Burton. We must therefore guess at the gentleman's identity. It may refer to Major Edward Burton (1790–1867), a British army surgeon. He was stationed at Chatham, UK, from 1829 to 1837, and in 1838 he published *A Catalogue of the Collection of Mammalia and Birds in the Museum at Fort Pitt, Chatham.* He wrote a paper for the Zoological Society of London in 1835 entitled "A Specimen of the Species *Ratelus*" and in a separate article commented on fishes described by Cuvier. The gerbil is known only from the type locality in Sudan.

Burton, R.

Burton's Vlei Rat *Otomys burtoni* **Thomas,** 1918

Sir Richard Francis Burton (1821–1890) was a noted British explorer, linguist, author, and devotee of erotica. He was originally an army officer. Having previously been Consul in Fernando Pó in 1872, he was appointed to be British Consul in Trieste (at that time part of the Austro-Hungarian Empire) and lived there for the rest of his life. He was fluent in over 20 languages and devoted much of his time to literature, translating into English the *Kama Sutra* in 1883, *The Arabian Nights* in 1885, and *The Perfumed Garden* in 1886. Immediately after his death his widow (see entry for Lady Burton) burned all his papers—an action that has been condemned as one of the greatest acts of literary vandalism of all time. The rat comes from Mount Cameroon and was presumably collected by Burton before 1872. It was thus getting on for half a century before someone got around to describing it. Thomas used the following words in his description: "I have named it in honour of its famous collector, Sir Richard Burton, to whose ability and energies as a naturalist too little credit has been generally given." Burton is also remembered in the name of a bird, Burton's Grosbeak *Serinus burtoni.*

Burton, W.

Burton's Melomys *Melomys burtoni* **Ramsay,** 1887 [Alt. Grassland Mosaic-tailed Rat]

Walter Burton (dates not found) was a taxidermist. The etymology by Ramsay says that he received the type specimen from the late Thomas Boyer-Bower. Ramsey then writes, "I have named this species after Mr. Burton who accompanied the late Thos. Boyer-Bower, Esq., as taxidermist to North West Australia." The melomys is found in northern and eastern Australia, New Guinea, and islands in the Torres Strait.

Butler

Butler's Dunnart *Sminthopsis butleri* **Archer,** 1979 [Alt. Carpentarian Dunnart]

Dr. William Henry (Harry) Butler C.B.E. (b. 1930) was born in Perth and trained as a teacher but in 1963 began working for corporate and government bodies as an environmental consultant and collector and undertook a major study of Western Australian animals. He collected more than 2,000 examples of mammals, 14 of which were new to science. A passionate conservationist, he presented the popular ABC television series *In the Wild* that began in 1976. He received the Australian of the Year award in 1979, and in 2003 he was awarded an honorary Doctorate of Science by the Edith Cowan University in Perth. The dunnart is found on Bathurst and Melville islands, Northern Territory, Australia. It was formerly also found in the Kimberley area of northern Western Australia but has not recently been recorded there. Populations in New Guinea formerly assigned to this species are now believed to be a different form.

Buxton, E. N.

Buxton's Jird *Meriones sacramenti* **Thomas,** 1922 [Alt. Negev Jird]

Sir Edward North Buxton (1840–1923) was a British hunter and a conservationist. In 1903 he set up the Society for the Preservation of the Wild Fauna of the Empire, of which Oldfield Thomas became an ardent supporter. This gerbil has a small range in Israel.

Buxton, I.

Mountain Nyala *Tragelaphus buxtoni* **Lydekker,** 1910

Major Ivor Buxton (1884–1969) discovered this antelope in 1909 and collected the type specimen, which was first recognized as a new species in 1910 by Rowland Ward, when he was asked to mount it (i.e. carry out taxidermy on it). Ward sent it to Lydekker for a proper scientific description. The species is found only in the Ethiopian highlands.

Byatt

Byatt's Bush Squirrel *Paraxerus (vexillarius) byatti* Kershaw, 1923

Sir Horace Archer Byatt (1875–1933) was a colonial administrator. He was a Commissioner in British Somaliland from 1911 to 1914, and in 1916 he was appointed Administrator of the northern half of Tanganyika (now called Tanzania), which had been taken from Germany. The country was made a British Protectorate in 1919, and Byatt served as its first British Governor from 1920 until 1924. In 1924 he became Governor of Trinidad and Tobago, holding that post until 1930. While in Tanganyika he did much to promote the study of the country's natural history, and this is stressed by Kershaw in his description of the squirrel. A Tanzanian endemic, this taxon has variously been treated as a subspecies of *Paraxerus vexillarius*, as a subspecies of *P. lucifer*, or as a distinct species.

Bynoe

Agile Wallaby *Halmaturus binoe* **Gould,** 1842 [Syn. *Macropus agilis*]

Benjamin Bynoe (1804–1865) was a British naval surgeon. In 1831 he was appointed as Assistant Surgeon on HMS *Beagle*. His superior, Robert McCormick, was so annoyed that Darwin, instead of himself, was treated as the ship's naturalist that he resigned from the expedition in 1832 and went back to England. Bynoe was promoted to Surgeon and served in the position for the rest of that voyage, which ended in 1836. He was given the same position on the *Beagle*'s third voyage from 1837 to 1843. He was a great success also as a naturalist and collector of both mammals and birds. He wrote the first description of the birth of marsupials. He was greatly used by John Gould in the preparation of his *Birds of Australia*, and it was Bynoe who first collected the finch *Chloebia gouldiae* that Gould named in honor of his wife. Bynoe Harbour in Australia was named after him in 1839. The name *Halmaturus binoe* is a junior synonym of *Macropus agilis*, which is found in northern Australia and southern New Guinea.

Byrne

Byrne's Marsupial Mouse *Dasyroides byrnei* **Spencer,** 1896 [Alt. Kowari; Brush-tailed Marsupial Rat, Crested-tailed Marsupial Rat, Kawiri]

P. M. (Paddy) Byrne (dates not found) was an Australian telegraph official living in frontier settlements hundreds of kilometers from the nearest Europeans. He met the pioneering anthropologist and biologist Baldwin Spencer during the 1894 Horn scientific expedition to central Australia. From 1894 to 1925 he corresponded with Spencer about the Aboriginal people, the landscape in which they lived, and the unusual flora and fauna of their habitat. We assume Spencer named the marsupial mouse to honor his correspondent. The Kowari is found in central Australia.

Byron

Southern Sea-Lion *Otaria byronia* **Blainville,** 1820 [Alt. Patagonian Sea-Lion; Syn. *O. flavescens* **Shaw,** 1800]

The Hon. John Byron (1723–1786) was captain of HMS *Dolphin* from 1764 to 1766 and was known as "Foul-Weather Jack," as he always seemed to attract storms and disaster. He took the Falkland Islands as a British possession and in 1769 was appointed Governor of Newfoundland, then becoming an admiral in 1775 before retiring into private life in 1779. His grandson was the famous poet. He wrote *The Narrative of the Honourable John Byron (Commodore in a Late Expedition Round the World) Containing an Account of the Great Distresses Suffered by Himself and His Companions on the Coast of Patagonia, from the Year 1740, till Their Arrival in England, 1746. With a Description of St. Jago de Chili, and the Manners and Customs of the Inhabitants: Also a Relation of the Loss of the Wager, Man of War, One of Admiral Anson's Squadron.* He was a midshipman aboard the *Wager*, which was wrecked off Chile and the survivors made prisoners of the local "Indians" and then handed over to the Spanish authorities. In "Epistle to Augusta" Lord Byron wrote, "A strange doom is thy father's son's, and past Recalling as it lies beyond redress, Reversed for him our grandsire's fate of yore, He had no rest at sea, nor I on shore." There is disagreement among zoologists as to whether *byronia* or *flavescens* is the valid scientific name for this sea-lion. It is found on the coasts of Peru, Chile, Argentina, and the Falkland Islands.

Cabrera

Cabrera's Vole *Microtus cabrerae* **Thomas,** 1906
Cabrera's Hutia *Mesocapromys angelcabrerai* **Varona,** 1979

Dr. Angel Cabrera Latorre (1879–1960)—his full name is hardly ever used, and the "Latorre" is mostly omitted—was one of the foremost Spanish-speaking zoologists of his era. He was born in Madrid, Spain, and graduated from the university there with a doctorate in 1900. He published his first article, "Observations on a Chimpanzee of White Haunches," at the age of just 18—the first of over 200 articles produced during his lifetime. He worked at the Madrid Natural History Museum for 25 years, ending as Chief Curator of Mammals, and was a frequent correspondent with Oldfield Thomas. He participated in a number of expeditions to Morocco in 1913, 1919, 1921, and 1923, during the last of which he became firm friends with the English ornithologist Hubert Lynes. He published a series of books, starting with one on the mammals of the Iberian Peninsula and including another on the mammals of Morocco; by the time he left Spain he had already published 17 books. In 1925 he was offered the post of Director of the Department of Paleontology of the National Museum of Argentina, a post that he occupied until 1947. He spent the rest of his life in Argentina. Up to 1954 he published at least 13 more books including, in 1940, with Jose Yepes, *Mammals of South America,* which remains a seminal work. With the same co-author he also wrote *Catálogo de los mamíferos de América del Sur,* published between 1957 and 1961. He undertook a number of expeditions collecting both live specimens and fossils in Patagonia and Catamarca, as well as around Buenos Aires. The vole is found in Spain and Portugal. The hutia is endemic to the small islands of the Cayos de Ana Maria, off south-central Cuba.

Cadena

Cadena's Nectar Bat *Lonchophylla cadenai* Woodman and Timm, 2006
Cadena's Tailless Bat *Anoura cadenai* Mantilla-Meluk and **Baker,** 2006

Dr. Augusto Alberto Cadena-Garcia is a distinguished Colombian zoologist who is the Curator of Mammals at the Institute of Natural Sciences, National University of Colombia in Bogota. In 2006 a prize named after him was established for the best zoology student in Colombia. The original description of the tailless bat includes the following words about him: "[He] has dedicated his life to the study of the Colombian mammalian fauna. Dr. Cadena has not only contributed to the knowledge of Colombian mammals but also has mentored several generations of Colombian mammalogists." The tailless bat is known only from the Calima River basin in western Colombia. The nectar bat comes from western Colombia and northwest Ecuador.

Cadorna

Cadorna's Pipistrelle *Hypsugo cadornae* **Thomas,** 1916 [Alt. Thomas' Pipistrelle; formerly *Pipistrellus cadornae*]

General Luigi Cadorna (1850–1928) was an Italian WW1 military leader. In 1908 he was offered the post of Chief of Staff but rejected it because of his antipathy to the political control exercised during wartime. However, it was offered again in July 1914, and he accepted. Although Italy declared its intention to be neutral at the outbreak of WW1, Luigi Cadorna expected that Italy *would* enter the war and so began building up the army. He decided to concentrate his forces on the borders with Austria-Hungary. After a series of losing engagements he had a famous victory in August 1916 at the Battle of Gorizia. We speculate that this might have prompted Thomas' use of his name in the binomial that year. The original citation reads, "It is named in honour of General Count Luigi Cadorna, the Commander-in-Chief of our Italian Allies." The bat is found in northeast India, Myanmar, Thailand, and Laos.

Calaby

Calaby's Pademelon *Thylogale calabyi* **Flannery,** 1980 [Alt. Alpine Wallaby]
Calaby's Pebble-mound Mouse *Pseudomys calabyi* Kitchener and Humphreys, 1987

Dr. John Henry Calaby (1922–1998) was recognized as one of Australia's greatest biologists, known for his encyclopedic knowledge of Australian fauna and his willingness and enthusiasm for sharing it. He was a world-renowned mammalogist and natural historian. After graduating he worked in the WW2 munitions industry but in 1945 began his career in zoology. In 1950 he became the first Curator of Mammals for the Australian National Wildlife Collection, which is now located at the CSIRO (Commonwealth Scientific and Industrial Research Organisation) Discovery Centre in Canberra, where one of the buildings that houses it has been named after him. He retired in 1987. He was the author of many articles appearing in such publications as *Australian Natural History* from the 1960s through the 1990s. Dr. Brian Walker, Chief, CSIRO Wildlife and Ecology once said, "John is a great collector of wildlife literature and has compiled a treasure trove of wildlife books and journals. These formed the basis of Wildlife and Ecology's library—one of the leading ecological libraries in the world, and definitely the front-runner in the southern hemisphere." Thirty species across the animal kingdom are named after him. The (endangered) pademelon occurs only in southeast Papua New Guinea. The mouse is found in Arnhem Land, Northern Territory, Australia.

Callewaert

Callewaert's Mouse *Mus callewaerti* **Thomas,** 1925

Monsignor Richard Callewaert (1866–1943) was a Belgian missionary and amateur naturalist who collected many different taxa in the then Belgian Congo (now Zaire). There is a record of an R. P. Callewaert collecting the type specimen of an ant in Zaire in 1912, another in 1913, and yet another ant type in 1935 as well as a beetle type specimen in 1923 and small fish in 1932. The zoologist who described the ants reported Callewaert as "noting he found the [ant] colony in his room, going from here to there, through a hole in the wall whence they disappeared one after the other. . . . The ants did not climb the wall itself but ascended between wall and the wood of the door." We think he was the Callewaert who wrote the tract "Rapport sur l'éducation des Mulâtres," which appeared in *Compte-rendu du Congrès International pour l'étude des problèmes résultant du mélange des races* in 1935. The mouse has been recorded from Angola and southern DRC (Zaire).

Campbell, C. W.

Campbell's Hamster *Phodopus campbelli* **Thomas,** 1905

Charles William Campbell (1861–1927) was a member of the British Consular Service and served in China. In 1889 he traveled from China and spent two months in what is now North Korea, publishing a report of this journey in 1891. He is also commemorated in the common name of a bird, Campbell's Hill Partridge *Arborophila (orientalis) campbelli*. Campbell collected a specimen of the hamster on the Sino-Russian border in 1904. The hamster is found in Transbaikalia (Russia), Mongolia, and adjacent parts of China.

Campbell, H. D.

Campbell's Monkey *Cercopithecus campbelli* **Waterhouse,** 1838 [Alt. Campbell's Guenon]

Major Henry Dundas Campbell (d. 1837) was Governor of Sierra Leone from 1833 until his death four years later. Campbell presented the type specimen of the monkey to the Zoological Society of London. It is found from Gambia east to the Sassandra River (western Ivory Coast).

Cansdale

Cansdale's Long-eared Flying Mouse *Idiurus macrotis cansdalei* **Hayman,** 1946 [Alt. Long-eared Scaly-tailed Flying Squirrel]
Cansdale's Swamp Rat *Malacomys cansdalei* **Ansell,** 1958

George Soper Cansdale (1909–1993) was a noted British zoologist and TV broadcaster. He was an officer in the Colonial Forest Service of the Gold Coast (now Ghana) from 1934 to 1948. Then he was the Superintendent of the London Zoo from 1948 until 1952, in which year he seems to have been dismissed. (He was extremely unpopular with his staff. There had been keepers parading up and down in front of the office with banners saying "Cansdale must go.") He made his television debut with a squirrel in 1948 on *Picture Page*, transmitted from the old studios at London's Alexandra Palace. This was followed by further children's programs in which he anthropomorphized animals, such as in *Looking at Animals* and *All about Animals*, both in 1955. He enabled the Brooke Bond Tea Company to make advertisements with chimpanzees when he introduced them to a private owner. In 1952 the Television Society awarded him its medal for the best programs of that year. He also acted as a host in outdoor broadcasts from the London Zoo. He wrote a number of books such as *Animals and Man* (1952), *Reptiles of West Africa* (1955), *West African Snakes* (1961), and *All the Animals of the Bible Lands* (1970). He also wrote scientific papers and popular articles such as "Some Gold Coast Lizards," published in *Nigerian Field* in 1951. The "flying mouse" (more correctly an anomalure) is found from Sierra Leone to eastern DRC (Zaire); *cansdalei* is the west African subspecies (though the race is not considered valid by some mammalogists). The rat occurs in the forests of southern Ghana, Ivory Coast, and Liberia.

Cantor

Cantor's Dusky Leaf Monkey *Trachypithecus obscurus halonifer* Cantor, 1845
Cantor's Roundleaf Bat *Hipposideros galeritus* Cantor, 1846

Dr. Theodore Edward Cantor (1809–1860) was the Danish Superintendent of European Asylum at Bhowanipur, Calcutta. This was part of the Bengal Medical Service, which in turn was part of the Honourable East India Company. Cantor was also an amateur zoologist. He published "Catalogue of Reptiles Inhabiting the Malayan Peninsula and Islands, Collected or Observed by Theodore Cantor, Esq., M.D. Bengal Medical Service" (*Journal of the Asiatic Society,* 1847) and *Catalogue of Malayan Fishes* (1850). He was interested in tropical fish, and in about 1840 the King of Siam (now Thailand) gave some Bettas (commonly called fighting fish) to Doctor Cantor. He published an article about them, and this led to the so-called Betta fever, a fashion for keeping these fish. He also published *List of Malayan Birds Collected by Theodore Cantor, Esq. MD, with Descriptions of the Imperfectly Known Species*. The leaf monkey is found in the Malay Peninsula and offshore islands, with the subspecies *halonifer* coming from Penang Island. The bat has a wide distribution from Sri Lanka and India through Southeast Asia to Java and Borneo.

Canut

Canut's Horseshoe Bat *Rhinolophus canuti* **Thomas** and **Wroughton,** 1909

Dr. Knud Andersen (d. 1918) has been described as the greatest early exponent of panbiogeography. He disappeared in 1918 and probably died during WW1. His work on bats was published in articles in the *Proceedings of the Zoological Society of London*. The name *canuti* is a Latinized form of Knud, and a better common name would be Knud's Horseshoe Bat. Thomas and Wroughton say in their original description, "We have named this striking species of Rhinolophus . . . in honour of Dr. Knud Andersen, in recognition of the exhaustive work he has done on this complicated and difficult group." The bat is known only from Java and Timor, Indonesia. See **Andersen** for more biographical detail.

Caovansung

Van Sung's Shrew *Chodsigoa caovansunga* Lunde, **Musser,** and Nguyen, 2003

Professor Dr. Cao Van Sung (d. 2002) of the University of Hanoi was Emeritus Director of the Institute for Ecology and Biological Resources in Hanoi. He was a zoologist and biologist who received his training in Russian universities. A species of frog, *Leptolalax sungi*, and a species of gecko, *Cyrtodactylus caovansungi*, are also named after him. The shrew comes from Mount Tay Con Linh II in northern Vietnam.

Capaccini

Long-fingered Myotis *Myotis capaccinii* **Bonaparte,** 1837

Monsignor Francesco Capaccini (1784–1845) was a Roman Catholic priest who was also an astronomer. He became Director of the Observatory at Naples. For much of his life he was involved in Vatican administration and diplomacy, being, for example, the Papal Nuncio (Ambassador) to Holland from 1828 to 1831. He was made a cardinal in 1845 but died before he could receive the red hat and title. The bat named after him was first found in Sicily. It ranges through the Mediterranean zone from Spain and northwest Africa east to Turkey, Israel, Iraq, Iran, and Uzbekistan.

Carol, Pratt

Weyland Ringtail Possum *Pseudochirulus caroli* **Thomas,** 1921

Charles B. Pratt (dates not found) collected the type specimen of the possum in the Weyland Mountains in west-central New Guinea, where this marsupial is endemic. See **Pratt** for more biographical details.

Carol, Rothschild

Ryukyu Mouse *Mus caroli* **Bonhote,** 1902

Nathaniel Charles Rothschild (1877–1923) was a banker and entomologist. His elder brother was Lord Lionel Walter Rothschild, and his daughter was Miriam Rothschild, the famous parasitologist. He worked with Karl Jordan, and together they discovered the rat flea *Xenopsylla cheopis* that was the main vector of bubonic plague in Egypt. He suffered from encephalitis and committed suicide in 1923. The name *caroli* is a derivative of Carolus, which is the Latin for Charles. The mouse was first discovered in the Ryukyu Islands, south of Japan, but also occurs in Taiwan, southern China, Thailand, and Indochina. Records from Indonesian islands probably reflect accidental human introduction.

Caroline

Halmahera Blossom Bat *Syconycteris carolinae* **Rozendaal,** 1984

Caroline Rozendaal is the wife of Dutch ornithologist Frank G. Rozendaal. She is also commemorated in the scientific name of a bird, the Tanimbar Bush Warbler *Cettia carolinae,* which her husband named and described in 1987. The bat is found on the islands of Halmahera and Batjan in the Moluccas (Indonesia).

Carriker

Carriker's Round-eared Bat *Lophostoma carrikeri* **J. A. Allen,** 1910 [Alt. Allen's Round-eared Bat]

Melbourne Armstrong Carriker Jr. (1879–1965), born in Illinois, was one of the great early naturalists, primarily an ornithologist and entomologist, of Central America and northern South America. In 1910 he published the first modern systematic catalogue of the birds of Costa Rica, listing 713 species for the country. He was sent on several expeditions by the Philadelphia Academy of Natural Science to Peru and Bolivia in 1929–1930, 1935, and 1938. He greatly enhanced the bird collections of the Carnegie Museum and the Smithsonian Institution. He is the subject of a biography written by his son, Professor Melbourne Romaine Carriker. He is also commemorated in the common names of three birds: Carriker's Antpitta *Grallaria carrikeri,* Carriker's Conure *Aratinga wagleri minor,* and Carriker's Mountain Tanager *Dubusia taeniata carrikeri.* The bat ranges across Colombia, Venezuela, Suriname, northern Brazil, Bolivia, and Peru.

Carruthers

Carruthers' Mountain Squirrel *Funisciurus carruthersi* **Thomas,** 1906

Carruthers' Juniper Vole *Neodon (juldaschi) carruthersi* Thomas, 1909

Alexander Douglas Mitchell Carruthers (1882–1962) was an explorer and naturalist. He was educated at Haileybury and Trinity College, Cambridge, and then trained in land survey and taxidermy. He went with the British Museum expedition to Ruwenzori from 1905 to 1906, and traveled in Russian Turkestan and along the borders of Afghanistan from 1907 until 1908. There he did research on wild sheep. In 1910 he explored the deserts of Outer Mongolia and the upper Yenisey River with John H. Miller and Morgan Philips Price. Carruthers was Honorary Secretary of the Royal Geographical Society between 1916 and 1921. His publications include *Unknown Mongolia* (1913), *Arabian Adventure* (1935), and *Beyond the Caspian: A Naturalist in Central Asia* (1949). He is also commemorated in the name of a bird, Carruthers' Cisticola *Cisticola carruthersi.* The squirrel is found in montane forests of western Uganda, Rwanda, Burundi, and adjacent parts of DRC (Zaire). The vole is found in the Gissar Mountains, east of Samarkand, and was originally described as a full species under the name *Pitymys carruthersi.*

Carter

Carter's Myotis *Myotis carteri* **LaVal,** 1973

Dr. Dilford Campbell Carter (b. 1930) is an American biologist at the Museum and Department of Biology, Texas Tech. In 1964 he was working in the Amazon rainforest when four German girls suddenly appeared. One was called Sigrid, and later she married him; this may sound like a soap opera but is perfectly true. The bat is found in western Mexico. It is often regarded as a subspecies of the Black Myotis *Myotis nigricans,* of Central America and tropical South America.

Castroviejo

Broom Hare *Lepus castroviejoi* Palacios, 1976

Dr. Javier Castroviejo Bolibar is a Spanish zoologist and ecologist who has created a number of biological stations, including one in the Doñana, Spain. He is President of the Spanish Committee of the UNESCO Man and Biosphere Program. He has also written a number of scientific articles such as "Premières données sur l'ecologie hivernale des vertébrés de la Cordillière Cantabrique" (1970). The hare is confined to the Cantabrian Mountains of northern Spain. Recent studies confirm that its status as a full species, rather than a form of the European Hare, is deserved.

Celaeno

Dark Tube-nosed Bat *Nyctimene celaeno* **Thomas,** 1922

Celaeno was one of the Harpies, mythical creatures that looked like birds with the upper bodies of women. In Greek mythology there were three of them: Aello, Ocypete, and Celaeno. They were the daughters of the Nereid Electra and Thaumas, and were extremely ugly. In Homer's *Odyssey* they are regarded as storm winds. Their sister was Iris, who in contrast was beautiful. Thomas quite often chose names from mythology and used them as binomials without putting them into the genitive, and this is one example. The bat is found in western New Guinea. It is sometimes regarded as conspecific with *Nyctimene aello.*

Chaerephon

Bat genus *Chaerophon* **Dobson,** 1874 [ca. 18 species]

Chaerephon (ca. 450–400 B.C.) was a Greek philosopher and one of Socrates' closest friends. His nickname was "the Bat," as he never seemed to come out by day. He appears in the works of both Plato and Aristophanes.

Champion

Champion's Tree Mouse *Pogonomys championi* **Flannery,** 1988

Ivan Francis Champion O.B.E. (1904–1988) was a British explorer. He was born in Port Moresby, Papua New Guinea (where his father was Government Secretary), was educated in Queensland, and became a Cadet Patrol Officer in 1923. In 1927 he traversed New Guinea and Papua via the Fly River, hiking the highlands and ascending the Sepik River. He wrote an account of this trip, *Across New Guinea from the Fly to the Sepik.* He served in the navy during WW2, in New Guinea waters, charting parts of the coast for Allied shipping, and commanded HMAS *Laurabada,* which was originally the Papua New Guinea government motor yacht. In the 1950s he was Chief Commissioner of the Native Land Commission, and he retired in 1964. He was awarded the Gill Medal of the Royal Geographical Society as well as the John Lewis Gold Medal, and a DC-3 aircraft was named after him. He died in Canberra in 1988, the year James Sinclair's book *Last Frontiers: The Explorations of Ivan Champion of Papua* was published. Flannery's citation reads, "I take great pleasure in naming this species in honour of Mr. Ivan Champion who has contributed so much to the development of Papua New Guinea. Champion was the first European to cross New Guinea from the Fly to the Sepik, entering the Telefomin Valley in 1926. Thus it is fitting that one of the region's most attractive murids should bear his name." The tree mouse is known from just two high valleys in Papua New Guinea, including the Telefomin valley where the type specimen was taken.

Chanler

Chanler's Mountain Reedbuck *Redunca fulvorufula chanleri* **Rothschild,** 1895

William Astor Chanler (1867–1934) was an American explorer and fellow of the Royal Geographical Society. He explored the area around Mount

Kilimanjaro in 1889. He was the first westerner to explore the northern Uaso Nyiro River from where it passes through Samburu Reserve to where it mysteriously disappears into the Lorian Swamp. One of his explorations in 1893 was in the company of Lieutenant von Hohnel. Hohnel was seriously injured by a rhinoceros and consequently sent down to the coast. Shortly afterward Chanler's men deserted him and returned to the coast, too. Fortunately for Chanler, as he was returning to the coast he met A. H. Neumann (author of *Elephant Hunting in East Equatorial Africa*) coming up the river. Chanler wrote *Through Jungle and Desert: Travels in Eastern Africa*, published in 1896. During the Spanish-American War he was appointed Captain and Assistant Adjutant General of Volunteers, and served as the Acting Ordnance Officer, Cavalry Division, Fifth Army Corps, taking part in the Battle of Santiago. He went on to become a U.S. Congressman when elected to represent New York's 14th District in the House of Representatives, serving from 1899 to 1901. Most interestingly, one British historian named him as a suspect in the bombing of the U.S. battleship *Maine*. Chanler, with his brothers, was involved in smuggling arms to the Cuban insurrectionists. He reportedly claimed responsibility for the explosion on the *Maine* in a conversation with U.S. Ambassador William C. Bullitt in the early 1930s. Chanler died shortly afterward in Paris. The reedbuck is found in southeast Sudan, Ethiopia, Uganda, Kenya, and northern Tanzania.

Chapin

Chapin's Free-tailed Bat *Chaerephon chapini* **J. A. Allen,** 1917
Matadi Hyrax *Heterohyrax chapini* **Hatt,** 1933

Dr. James Paul Chapin (1889–1964) was an American ornithologist. He was joint leader of the Lang-Chapin expedition, which made the first comprehensive biological survey of what was then the Belgian Congo (now Zaire) from 1909 until 1915. He was Ornithology Curator for the American Museum of Natural History and President of the Explorers' Club from 1949 to 1950. In 1932 he wrote *Birds of the Belgian Congo*, which largely earned him the Daniel Giraud Elliot Gold Medal that year. He is also commemorated in the common names of at least eight birds, including Chapin's Flycatcher *Muscicapa lendu* and Chapin's Spinetail *Telacanthura melanopygia*. The bat has been recorded in Ghana, northern DRC (Zaire), Uganda, Kenya, and Ethiopia (more southerly populations are now regarded as a separate species, *Chaerephon shortridgei*). The hyrax has a very limited distribution near the mouth of the Congo River. It may be an isolated subspecies of the widespread *Heterohyrax brucei*.

Chapman, A.

Western Pebble-mound Mouse *Pseudomys chapmani* Kitchener, 1980

Andrew Chapman worked for the Mammal Department, Western Australian Museum, when this Australian mouse was first described. He publishes often on a wide variety of subjects in the *Western Australian Naturalist* (e.g. "Frost Damage to Vegetation" in 1998). The mouse is found in the Pilbara district of northern Western Australia.

Chapman, F. M.

Chapman's Rice Rat *Oryzomys chapmani* **Thomas,** 1898
Chapman's Prehensile-tailed Hutia *Mysateles gundlachi* Chapman, 1901 [Alt. Gundlach's Hutia; Syn. *M. prehensilis gundlachi*]

Frank Michler Chapman (1864–1945) was Curator of Ornithology for the American Museum of Natural History, New York, from 1908 until 1942. He photographed and collected data on North American birds for over 50 years and did much to popularize birdwatching in the USA in the 20th century. In 1899 he began publishing *Bird Lore* magazine, which became a unifying national forum for the Audubon movement. He had been an enthusiastic collector but became a leading light in the conservation movement. His interest in protection can be traced to 1886 when, during a walk in New York City, he observed that three-quarters of the ladies wore hats with feathers in them. Chapman sponsored the first national Christmas Bird Count in 1900. He wrote *Handbook of Birds of Eastern North America, with Keys to the Species and Descriptions of Their Plumages, Nests, and Eggs* (1903), *The Distribution of Bird Life in Colombia* (1917), and *The Distribution of Bird Life in Ecuador* (1926). He is also commemorated in the common names of no fewer than 13 birds, including Chapman's Warbler *Xenoligea montana*, Chapman's Trogon *Trogon massena australis*, and Chapman's Swift *Chaetura chapmani*. The rice rat is found in the cloud forests of eastern and southern Mexico. The hutia is endemic to the Isle of Pines (now Isla de la Juventud), Cuba.

Chapman, J.

Chapman's Zebra *Equus quagga chapmanni* **Layard,** 1865

James Chapman (1831–1872) was an English naturalist, hunter, and trader in ivory and cattle. Between 1861 and 1862 he explored, with his brother Henry Samuel Chapman (1834–1922), Thomas Baines (1822–1875), and others, from German Southwest Africa (now Namibia) to Victoria Falls. (Chapman had been meant to accompany David Livingstone in 1858 as his photographer, but they had a disagreement and he did not go.) He said of the Victoria Falls that they make "one's hair stand on end." His aim was to "establish a line of commercial stations across Southern Africa from sea to sea." It was said of him that he "had a habit of 'knocking up' [taming] quagga and 'jolling' [partying heartily] with them in town, along with the rest of his menagerie of four young lion, two leopards, two meerkats, two springbucks and one weasel." In 1868 he wrote *Travels in the Interior of South Africa*. An enormous tree, nearly 25 m (90 feet) in girth and the largest tree in Africa, is named Chapman's Baobob after him. Other famous explorers, Livingstone and Selous among them, all camped under its branches. One of Chapman's sons, Charles Henry, born in 1860, drowned when the *Titanic* sank in 1912. The zebra is found in Zimbabwe and in Mozambique south of the Zambezi, though taxonomists disagree on how many subspecies of *Equus quagga* should be recognized, and where exactly to draw the lines between them.

Chapman, J. W.

Chapman's Bare-backed Fruit Bat *Dobsonia chapmani* **Rabor,** 1952 [Alt. Negros Naked-backed Fruit Bat]

Dr. James W. Chapman (dates not found) was a Presbyterian missionary and Professor of Biology at Silliman University in the Philippines, where there is now a Research Foundation named in his honor. Rabor gives the etymology thus: "Named after Dr. James W. Chapman, retired Presbyterian Missionary and Professor of Biology, who unselfishly spent 35 years in Silliman University, teaching biology to Filipino students and doing his best toward developing their interest in natural history." At the end of the 20th century this bat was feared to be extinct, but populations have now been found on the Philippine islands of Cebu and Negros (although it remains critically endangered).

Cheesman

Cheesman's Gerbil *Gerbillus cheesmani* **Thomas,** 1919

Colonel Robert Ernest Cheesman C.B.E. (1878–1962) was a British army officer, field naturalist, explorer, and ornithologist. He was Private

Secretary to Sir Percy Sykes, High Commissioner in Iraq. In 1921 Cheesman mapped the Arabian coast from Uqair to the head of the Gulf of Salwa. From 1923 to 1924 he traveled to Hufuf and mapped 240 km (150 miles) of desert, identified the site of ancient Gerra, and corrected maps of the wadi system. He spent several months at al-Ahsa collecting specimens of local plants, insects, and other animals, and drawing up maps. He wrote an account of that journey, *In Unknown Arabia*, published in 1926, as well as an article, "Zoological Investigations in the Persian Gulf and Iraq" (1924). There is also a paper by D. R. E. Cheesman and M. A. C. Hinton entitled "On the Mammals Collected in the Desert of Central Arabia by Major R. E. Cheesman," published in 1924. He went on to serve as British Consul for Northwest Ethiopia from 1925 to 1934 and in that period explored and mapped Lake Tana and the Blue Nile, a journey he described in *Lake Tana and the Blue Nile: An Abyssinian Quest* (1936). He said of his explorations of the Blue Nile, "It proved to be a venture not to be lightly undertaken, and I understood why the secrets of the Nile Valley remained so long unrevealed." He was given the Gill Memorial Award by the Royal Geographical Society. The original citation states that the gerbil was named after "Major R. E. Cheesman of the Mesopotamian Expeditionary Force." The gerbil lives in Iraq, where the type specimen was collected, southwest Iran, and throughout the Arabian Peninsula.

Cheng

Cheng's Jird *Meriones chengi* Wang, 1964

Professor Tso-hsin Cheng (1906–1998) was a distinguished Chinese zoologist and ornithologist. He was educated at the Fujian Christian University in Fuzhou City, from where he received the degree of B.Sc. in 1926. He went to the USA and received his doctorate from the University of Michigan in 1930, then returned to his old university in Fuzhou as Professor and Director of the Department of Biology. In 1934 he was one of the founders of the China Zoological Society. In 1950 he transferred to Peking (Beijing) to take charge as Curator of Bird Specimens in the Institute of Zoology, Academia Sinica, and in 1951 he founded the Peking Natural History Museum. For over 60 years he undertook fieldwork and research in ornithology and conservation, publishing more than 10 million words in 30 books, 20 monographs, 150 scientific papers, and 260 popular articles. The original citation of the jird, which is known only from the type locality in Sinkiang, China, reads, "The new species is named in honor of Prof. Tso-hsin Cheng of the Institute of Zoology, Academia Sinica for his contributions to systematic ornithology."

Cherrie

Cherrie's Pocket Gopher *Orthogeomys cherriei* J. A. Allen, 1893

George Kruck Cherrie (1865–1948) was an American naturalist and ornithologist who accompanied Theodore Roosevelt on a trip to Brazil in 1913 to find the source of one of the tributaries of the Amazon. He was Assistant Curator to the Department of Ornithology at the Field Museum, Chicago, in the 1890s. He also collected extensively in Costa Rica between 1894 and 1897 and, with his wife Stella M. Cherrie, in Colombia in 1898. In *Through the Brazilian Wilderness*, published in 1914, Roosevelt described him as an "efficient and fearless man; and willy-nilly he had been forced at times to vary his career by taking part in insurrections. Twice he had been behind the bars in consequence, on one occasion spending three months in a prison of a certain South American state, expecting each day to be taken out and shot. In another state he had, as an interlude to his ornithological pursuits, followed the career of a gun-runner, acting as such off and on for two and a half years. The particular revolutionary chief whose fortunes he was following finally came into power, and Cherrie immortalized his name by naming a new species of ant-thrush after him—a delightful touch, in its practical combination of those not normally kindred pursuits, ornithology and gun-running." Cherrie published *Ornithology Orinoco* in 1916. He published *Dark Trails* in 1930, as well as *Adventures of a Naturalist.* He is commemorated in the scientific names of such birds as the Chestnut-throated Spinetail *Synallaxis cherriei* and the Spot-fronted Swift *Cypseloides cherriei.* The gopher has a limited range in north-central Costa Rica.

Child

Child's Rice Rat *Nephelomys childi*
Thomas, 1895

George D. Child (dates not found). Thomas writes, "The name of this species is given in honour of Mr. George D. Child, through whose kind instrumentality the specimens have been obtained." We have been unable to find any more about him. The rat is found in cloud forests of Colombia (the type specimen was a "Bogota" skin). It has sometimes been treated as a subspecies of *Nephelomys albigularis* but is now considered a full species.

Christie

Christie's Long-eared Bat *Plecotus christiei*
Gray, 1838

Dr. Alexander Turnbull Christie (d. 1832) was a Scottish physician. After working as an assistant surgeon with the Honourable East India Company, he returned home to Scotland in 1828 to study geology and meteorology. He was appointed to be Geological Surveyor at the Company's Madras establishment in 1930 and returned to India in the following year. In 1828 he wrote *A Treatise on the Epidemic Cholera: Containing Its History, Symptoms, Autopsy, Etiology, Causes, and Treatment.* He donated the holotype of the bat, which was found in Egypt. This bat has only recently been rerecognized as a valid species. It is found in Egypt (including Sinai) and northern Sudan.

Christy

Christy's Dormouse *Graphiurus christyi*
Dollman, 1914

Dr. Cuthbert Christy (1863–1932) qualified as a physician at Edinburgh University. In the early 1890s he traveled in the West Indies and South America. He then joined the army as a doctor and served in northern Nigeria between 1898 and 1900 and in India for a short time thereafter. His travels took him to Uganda and the Congo in 1902 and 1903, and between then and the outbreak of WW1 he visited Ceylon (now Sri Lanka), Kenya, Uganda, southern Nigeria, the Gold Coast (now Ghana), Cameroon, and Sudan. During WW1 he served in Africa and Mesopotamia (now Iraq). After the war he explored in the Sudan, Nyasaland (now Malawi), and Tanganyika (now Tanzania) and was a member of a League of Nations Commission enquiring into slavery and forced labor in Liberia. He was at some time a Director of the Congo Museum in Tervuren, Belgium. He was on a zoological expedition to the Congo in 1932 when he was killed by being gored by a buffalo. The dormouse is known from southern Cameroon and northern DRC (Zaire).

Chudeau

Chudeau's Spiny Mouse *Acomys chudeaui*
Kollman, 1911

René Chudeau (1864–1921) was a French explorer and geologist who worked in French West Africa, particularly in the Sahara and Sahel. A number of other taxa, including a species of toad, *Bufo chudeaui,* are also named after him. The spiny mouse is found in the Western Sahara (Mauritania and Morocco).

Churchill

Winston Churchill's Flying Squirrel *Hylopetes winstoni* **Sody,** 1949 [Alt. Sumatran Flying Squirrel]

Sir Winston Churchill (1874–1965) was the famous British Prime Minister during WW2. In a poll carried out by the BBC in 2004 he was voted the greatest Briton of all time. We feel it unnecessary to give a longer biography here. The squirrel is found only in northern Sumatra. See also **Winston.**

Cinderella

Cinderella Fat-tailed Opossum *Thylamys cinderella* **Thomas,** 1902
Cinderella's Shrew *Crocidura cinderella* Thomas, 1911 [Alt. Cinderella Shrew]

Cinderella is a fictional character, best known in the United Kingdom as a pantomime, though the original story may have originated in China (pre–A.D. 900). There have been many versions of it including the best known, "Cendrillon" by Charles Perrault (it was one of his *Fairy Tales* published in 1697) and an opera by Rossini called La Cenerentola that had its premiere in 1817. In the shrew's case, at least, we think the inspiration for use of this name must lie in the animal's coloration, described as "drab-grey" (like cinders). The opossum is found in northern Argentina. The shrew is known from Gambia, Senegal, Mali, and Niger.

Cirne

Checkered Elephant-Shrew *Rhynchocyon cirnei* **Peters,** 1847

Manuel Joaquim Mendes de Vasconcelos e Cirne (1784–1832) was Governor of Mozambique when it was a Portuguese colony. Using documents from the archives in Lisbon, a study of his administration was published in 1890 under the title *Memória sobre a província de Moçambique.* His provincial capital was at Quelimane in the Bororo district of Zambezia Province, where the type specimen was collected. Another source states that "Herr Cirne stayed for two months in the Bororo district of Mozambique." We do not regard these two possible explanations of the name as being mutually exclusive, and we think that Peters may have used "Herr Cirne" as a convenient piece of shorthand for a long and complicated Portuguese name. The elephant-shrew is found from northeast DRC (Zaire) and Uganda to Malawi and Mozambique.

Claire

Claire's Mouse-Lemur *Microcebus mamiratra* Andriantompohavana et al., 2006

Mrs. Claire M. Hubbard is the widow of Dr. Theodore F. Hubbard M.D., who was a member of the first cardiovascular surgical team in Omaha and also perfected dry-field heart surgery (a process that uses a machine to perform the functions of a patient's heart and lungs so that doctors may complete heart operations free of blood). He dedicated his entire life to the research and practice of cardiology. The etymology reads, "The name *mamiratra* is derived from the Malagasy language and means 'clear and bright' and is proposed for Claire Hubbard and the Theodore F. and Claire M. Hubbard Family Foundation. The Hubbard Foundation has provided generous support over the past five years, providing the Malagasy graduate students the opportunity to conduct conservation genetics projects in the field and in the laboratory." This lemur is found on the island of Nosy Be, off the northwest coast of Madagascar.

Clara

Clara's Echymipera *Echymipera clara* **Stein,** 1932 [Alt. White-lipped Bandicoot]

Mrs. Clara Stein (dates not found) was the wife of George H. W. Stein and accompanied him on his expedition to the Dutch East Indies between 1931 and 1932. This bandicoot is endemic to New Guinea.

Clarke, S. R.

Clarke's Vole *Volemys clarkei* Hinton, 1923 [Syn. *Microtus clarkei*]

Colonel Stephenson Robert Clarke (1862–1948) was a British naturalist, a great traveler, and a keen hunter of big game. He was also interested in botany and wrote descriptions of a great number of birds, too. In 1893 he bought Borde Hill House, a Tudor mansion in West Sussex, and created the famous garden there that remains in the care of his descendants. The garden is based on the wonderful plant and seed collections from China, Burma, Tasmania, the Himalayas, and the Andes brought there by collectors including Francis Kingdon-Ward. A misprint has reached the literature, and Clarke is often referred to erroneously as Colonel "Robert Stephenson" Clarke. One of the Colonel's great-grandsons provided us full details. He is also remembered in the name of a subspecies of bird, Clarke's Blood Pheasant *Ithaginis cruentus clarkei*. The vole is found in the mountains of northern Myanmar and Yunnan (China).

Clarke, T. W. H.

Clarke's Gazelle *Ammodorcas clarkei* **Thomas,** 1891 [Alt. Dibatag]

Thomas William Henric Clarke (1860–1945) was the person who discovered and collected the type specimen of the gazelle in 1890, doing so in the Marehan area of southern Somalia. He was an Australian pastoralist, with an estate in Tasmania, as well as an avid big-game hunter. He hunted in North America and in East Africa, and amassed the largest private collection of game trophies in the Southern Hemisphere. This collection is still housed at Quorn Hall, Campbell Town, Tasmania, where his family has lived since 1844. His expeditions were notable in that he never took any European companions with him. He was a formidable shot, twice winning at Bisley, competitions for which he had entered as an "Unknown." His gravestone bears an epitaph that harks back to an age long past: "One of England's Empire sons. Be Faithful to your Country and your King." The gazelle is found only in Somalia and the Ogaden region of eastern Ethiopia.

Cleber

Cleber's Arboreal Rice Rat *Oecomys cleberi* Locks, 1981

Professor Dr. Cleber J. R. Alho (b. 1937) is a Brazilian conservation biologist. He was born in the Amazon region, where he was imprinted in childhood by the surrounding images of forest, rivers, and wildlife, especially the Amazon turtle, object of his scientific investigations. He moved to Rio as a teenager, and his first degree was in natural history, earned in 1963 from the Federal University of Rio de Janeiro (formerly the University of Brazil). In 1964 he received his master's degree from the Oswaldo Cruz Institute, working in parasitology *(helminths)*. His first professional position was back in the Amazon, as an Assistant Researcher at the Emilio Goeldi Museum, Belém, in 1964. He moved to Brasília in 1965 to join the new Institute of Biology of the recently created University of Brasília, where he remained until retiring, in 1990, as a Full Professor of Ecology. In 1977 he obtained his Ph.D. in ecology from the University of North Carolina at Chapel Hill. Dr. Alho concluded his postdoctoral studies in museology (museum studies) at the Museum of Natural History of the Smithsonian Institution in Washington, DC. Over his career he published more than a hundred scientific papers on ecology, particularly on small mammals of the Cerrado biome of central Brazil. In addition he published papers on the reproductive biology, conservation, and management of the Amazon turtle. He served in a number of organizations including the World Wildlife Fund, the Brazilian National Research Council, EMBRAPA (Brazilian Agency for Agricultural Research), FUNATURA (Pro-Nature Foundation), and IBAMA (Brazilian Institute for the Environment and Renewable Resources) and as a consultant for various national and international institutions. Dr. Alho's current research interest is the conservation of nature. The rice rat was named in his honor by a researcher at Brazil's National Museum and is found only in a small area of Brazil, in the vicinity of Brasília.

Cleese

Cleese's Woolly Lemur *Avahi cleesei* Thalmann and Geissmann, 2005 [Alt. Bemaraha woolly lemur]

John Marwood Cleese (b. 1939) is a British actor, writer, and comedian most famed for his appearances in the television programs *Monty Python's Flying Circus* and *Fawlty Towers*, and in the film *A Fish Called Wanda* (another eponymous animal?). The lemur was discovered in a western Madagascar nature reserve in 1990 by a team of scientists from Zurich University, but was only named 15 years later as a tribute to Cleese's promotion of the plight of lemurs in the film *Fierce Creatures* and in the documentary *Operation Lemur with John Cleese*. The alternative name comes from the Bemaraha Strict Nature Reserve where the species was discovered.

Cloet

Cloet's Vervet *Chlorocebus aethiops cloetei* **Roberts,** 1931 [now *C. pygerythrus*]

W. M. G. Cloete (dates not found) was a South African farmer near Fort Beaufort in the Eastern Cape. He had a large collection of skulls of medium-sized mammals, which his widow donated to the Transvaal Museum. He had skulls of 45 vervet monkeys, most of which he had procured on his own land. This race of monkey is not now generally regarded as being distinct from the typical form of Vervet Monkey *Chlorocebus pygerythrus* from southern Africa.

Clymene

Clymene Dolphin *Stenella clymene* **Gray,** 1850 [Alt. Atlantic Spinner Dolphin]

Clymene was a Greek sea-goddess, the daughter of Oceanus and Tethys. This dolphin occurs in tropical and subtropical areas of the Atlantic Ocean.

Cockrum

Cockrum's Desert Shrew *Notiosorex cockrumi* **Baker,** 2004

Professor E. Lendell Cockrum (b. 1920) is Emeritus Professor of Ecology and Evolutionary Biology at the University of Arizona. He took his first degree at the Southern Illinois University in 1942 and was awarded his doctorate by the University of Kansas in 1951. The shrew was discovered in 1966 by the describer Baker, who was at the time doing his doctorate under the tutelage of Cockrum at the University of Arizona, but it was identified as a new species only in 2003. Baker named it after his professor "for his lifetime of research on mammals and for his commitment to students in mammalogy and general biology." Cockrum has written numerous papers and books such as the textbook *Introduction to Mammalogy* (1962) and *Zoology* (1965). The shrew lives in southern Arizona (USA) and northern Sonora (Mexico).

Coimbra-Filho

Coimbra-Filho's Titi *Callicebus coimbrai* Kobayashi and **Langguth,** 1999

Professor Dr. Aldemar F. Coimbra-Filho is a Brazilian zoologist and biologist. In 1962, together with Alceo Magnanini, he began a program to breed endangered Lion Tamarins in captivity and then reintroduce them into the wild. The titi is found in the coastal forests of the state of Sergipe, eastern Brazil.

Coke

Coke's Hartebeest *Alcelaphus buselaphus cokii* **Günther,** 1884 [Alt. Kongoni]

Colonel the Hon. Wenman Clarence Walpole Coke (1828–1907) was the younger brother of the 2nd Earl of Leicester and played in one first-class cricket match for the Marylebone Cricket Club in 1851 against Oxford University. He was Member of Parliament at Westminster for Norfolk East from 1858 to 1865. He collected the type specimen of the hartebeest, which is found in southern Kenya and northern Tanzania.

Colburn

Colburn's Tuco-tuco *Ctenomys colburni* **J. A. Allen,** 1903

E. A. Colburn (1842–1935) was a zoologist and an associate of J. A. Allen, who described the tuco-tuco. They were joint authors, with Joseph H. Batty and Melbourne Armstrong Carriker, of *Mammals from Southern Mexico and Central and South America,* published in 1904 under the auspices of the American Museum of Natural History. The tuco-tuco is endemic to a small area of Santa Cruz Province, southern Argentina.

Collett

Dusky Rat *Rattus colletti* **Thomas,** 1904

Professor Dr. Robert Collett (1842–1913) was a Norwegian zoologist. He started his working life as an Assistant Curator in 1871, rose to Curator, and finally became Director of the Christiania Museum, a post he held from 1882 until his death. He was also Professor of Zoology at Christiania University from 1884. His main interest was ichthyology. When Thomas wrote the original citation, which reads, "I have named this species in honour of my friend Dr. Robert Collett, of Christiania, the author of the chief paper on the mammals of the region in which it occurs," he was referring to Norway's chief town, now known as Oslo; but at that time Norway had not yet become independent from Sweden. Collett described a number of Australian mammals including Lumholtz's Tree Kangaroo *Dendrolagus lumholtzi.* He also has an Australian snake, *Pseudechis colletti,* named after him. The rat is found in the Northern Territory of Australia.

Collie

Collie's Squirrel *Sciurus colliaei* **Richardson,** 1839

Lieutenant Dr. Alexander Collie (1793–1835) was a British physician, naturalist, and explorer. He was the naval surgeon and naturalist on an expedition from 1825 until 1828 led by Captain Frederick Beechey on HMS *Blossom,* which made some significant zoological findings during the voyage from Chile to Alaska. Collie collected many specimens that did not survive the return journey to England in good condition, but he made some colored drawings of birds he thought were new and also took extensive notes. From these the British ornithologist Nicholas Vigors wrote his *Zoology of Captain Beechey's Voyage, 1839.* Dr. Collie later went to Perth as a colonial administrator and died there before Vigors' work was published. When aboard HMS *Sulphur* he discovered what is now called the Collie River in Western Australia. A town in Australia was also named after him. The squirrel is found in western Mexico.

Commerson

Commerson's Dolphin *Cephalorhynchus commersonii* **Lacépède,** 1804
Commerson's Roundleaf Bat *Hipposideros commersoni* **Geoffroy,** 1813

Dr. Philibert Commerson (1727–1773) was known as "doctor, botanist, and naturalist of the King." He accompanied the French explorer Louis Antoine de Bougainville on his round-the-world expedition of 1766–1769 on *La Boudeuse* and *L'Etoile*. Commerson was primarily a botanist, but he also had a wide diversity of animal species named after him, including Commerson's Frogfish *Antennarius commersoni* and Commerson's Cornetfish *Fistularia commersonii*. He also discovered the vine Bougainvillea in the 1760s, naming it for the expedition leader. He is commemorated in the name of a now-extinct bird, Commerson's Owl *Otus commersoni (Mascarenotus sauzieri)* from Mauritius. The dolphin is found in waters off central and southern Argentina, the Straits of Magellan, and the South Shetland and Falkland Islands. Unusually, no type specimen was taken, but the description was based on specimens seen and described by Commerson in a manuscript addressed to Compte de Buffon. The bat is found from Gambia east to Ethiopia and south to Namibia, Botswana, and Transvaal (South Africa), and also on Madagascar.

Commissaris

Commissaris' Long-tongued Bat *Glossophaga commissarisi* **Gardner,** 1962

Larry R. Commissaris (d. 1962) was a graduate student working with Gardner at the University of Arizona when he suffered an untimely death. The bat is found from Mexico to eastern Peru and northwest Brazil.

Conover

Conover's Tuco-tuco *Ctenomys conoveri* **Osgood,** 1946

Henry Boardman Conover (1892–1950) was a soldier and amateur ornithologist. He served in the U.S. Army during WW1. Throughout his life he was an enthusiast for field sports. Around 1920 he became interested in scientific ornithology, particularly in relation to game birds. He was a Trustee of the Field Museum of Chicago, to which he donated his collection of ornithological texts. He made significant contributions to Hellmayr's *The Catalogue of Birds of the Americas* and published articles in *The Auk*, such as "Game Birds of the Hooper Bay Region, Alaska" in 1926 and "A New Species of Rail from Paraguay" in 1934. He is also commemorated in the name of a bird, Conover's Dove *Leptotila conoveri*. The tuco-tuco is found in the Chaco region of Paraguay and adjacent northern Argentina.

Constance

Pale-bellied Woolly Mouse-Opossum *Micoureus constantiae* **Thomas,** 1904

Constance Sladen, nee Anderson (1848–1906), was the wife of Walter Percy Sladen (1849–1900), the self-taught biologist and leading light of the Linnean Society. She was an artist of some repute whose works were exhibited in galleries in London and the provinces. She was also an authority on the archeology of Yorkshire and contributed articles on York Minster, Selby Abbey, and Castle Howard to various collaborative works. She married "Percy" in 1890, 20 years after first meeting him. It was said of them, "Community of tastes in literature and art, together with that rare sympathy which instinctively avoids all sources of difference, rendered their union one of exceptional harmony; the more so, perhaps, because, owing to Sladen's stern sense of duty, the consummation of their devotion had been postponed for nearly twenty years." Alphonse Robert collected the type specimen of the opossum on an expedition financed by Constance Sladen, who was by then widowed. Thomas says, "I have named this pretty species in honour of the donor of the present most valuable accession to the National Collection, in recognition of her enlightened method of commemorating her late husband's memory" (i.e. financing Robert to collect more specimens). She also made an initial £20,000 endowment to a trust fund, the Percy Sladen Trust, for the furtherance of scientific research. When the Linnean Society changed its admissions policy in 1904. Constance was one of the first women to be admitted. The opossum is found in the Mato Grosso area of Brazil and in eastern Bolivia.

Contreras

Contreras' Juliomys *Juliomys pictipes* **Osgood,** 1933 [Alt. Lesser Wilfred's Mouse; Syn. *Wilfredomys pictipes*]

Professor Dr. Julio Rafael Contreras Roque (fl. 1950–2007) is an Argentine biologist who was at the Bernardino Rivadavia National Museum of Natural Sciences in Buenos Aires and is now President of the Felix de Azara Natural History Foundation. In 1975 he was Director of the Biological Station at Isla Victoria in Bariloche. Since 2003 he has lived in Paraguay, where he is a member of the Scientific Society and the Academy of Paraguayan History. He also teaches and conducts research at the National University of Pilar and directs the Felix de Azara Bioecological and Subtropical Research Institute of that University. The mouse was first described by Osgood under the name *Thomasomys pictipes*. It was later moved into the genus *Wilfredomys* and is now placed in *Juliomys*. It is found in southeast Brazil and northeast Argentina.

Cook, J. A.

Cook's Hocicudo *Oxymycterus josei* Hoffmann, Lessa, and Smith, 2002 [Alt. José's Hocicudo]

Professor Dr. Joseph Anthony Cook. See **José** for biographical details.

Cook, J. P.

Cook's Mouse *Mus cookii* Ryley, 1914

J. Pemberton Cook (1865–1924) worked for the Burma Teak Corporation. The type specimen was collected by G. C. Shortridge, but Ryley wrote, "I have named this species after Mr. J. P. Cook who assisted Mr. Shortridge by sending in a small collection of specimens from Moulmein." The mouse has a wide distribution across India, Nepal, Myanmar, southwest Yunnan (China), Thailand, Laos, and northern Vietnam.

Cookson

Cookson's Wildebeest *Connochaetes taurinus cooksoni* Blaine, 1914 [Alt. Cookson's Blue Wildebeest]

Harold Cookson (1876–1969) was an Englishman who between 1898 and 1907 traveled widely in Asia and Africa. He also traveled in the Pacific, bringing a number of artifacts back to England from Samoa and presenting them to the Hancock Museum in Newcastle-upon-Tyne in 1913. He prospected for minerals in Northern Rhodesia (now Zambia) and hunted big game. His major interest in natural history was lepidoptery. He and his two sons bought a fruit farm in Natal in 1946 but sold it in 1957 and moved to the Vumba Mountains in what is now Zimbabwe. This race of wildebeest is endemic to the Luangwa valley of Zambia.

Cooper, J. G.

Southern Bog Lemming *Synaptomys cooperi* **Baird,** 1858

James Graham Cooper (1830–1902) was an American surgeon and naturalist whose main interests were ornithology and botany. He took part in the Pacific Railroad Surveys, and in 1860 he wrote *Botanical Report: Explorations and Surveys for a Railroad Route from the Mississippi River to the Pacific Ocean.* The surveys were government sponsored, but all the naturalists appointed to accompany them were selected by Baird (who described the Lemming) at the Smithsonian. In 1861 Cooper made a significant botanical collection from the area between San Diego, California, and Fort Mohave in Arizona. He was an active member of the California Academy of Science and became Director of the museum attached to it. His father, William Cooper (1798–1864), was also a famous American naturalist. He is also remembered in the scientific name of the Pacific Screech Owl *Otus cooperi*, and the Cooper Ornithological Society is named after him. The lemming is found in southeast Canada and the eastern USA, as far south as northeast Arkansas and North Carolina.

Cooper, L. G.

Cooper's Mountain Squirrel *Paraxerus cooperi* **Hayman,** 1950 [Alt. Cooper's Green Squirrel]

L. G. Cooper (dates not found) sent to the British Museum (Natural History) a collection of small mammals from the Kumba Division of British Cameroon. Nothing more is recorded about him. The squirrel is known only from Cameroon.

Cooper, N. K.

Cooper's Melomys *Melomys cooperae* Kitchener and Maryanto, 1995 [Alt. Yamdena Island Melomys]

Norah K. Cooper is an Australian zoologist who, at the time of writing, is Curator of Mammals at the Museum of Western Australia. She has written quite extensively on rodents. The melomys is found only on Yamdena Island, Tanimbar Group, eastern Indonesia.

Cooper, S.

Cooper's Margay *Leopardus wiedii cooperi* **Goldman,** 1943

General Samuel Cooper (1798–1876) was an American professional soldier. He was commissioned into the artillery in 1815 and served as an artillery officer until 1837, when he was made Chief Clerk at the War Department. The next year he was made Assistant Adjutant General of the U.S. Army. He served in the field during the Seminole War from 1841 to 1842. He was Assistant Adjutant General of the U.S. Army again in 1844, and in 1852 he was promoted to colonel. Upon the outbreak of the American Civil War in 1861, his sympathies being with the Confederacy, he resigned his commission and traveled to the South and joined the Army of the Confederacy as a brigadier general. He served as Adjutant General and Inspector General of the Confederate army until the end of the war. He was promoted to full general in May 1861 and so became the most senior general, outranking more famous officers such as Robert E. Lee. His last official act was to preserve the official records of the Confederate army and turn them over to the U.S. government. He then returned to his home in Virginia, farmed his land, and lived quietly until his death. The margay is known to have been taken only once in the USA, and that was by Colonel Cooper in southern Texas in 1852. The validity of the subspecies is uncertain.

Coquerel

Coquerel's Mouse-Lemur *Mirza coquereli* **Grandidier,** 1867
Coquerel's Sifaka *Propithecus coquereli* **Milne-Edwards,** 1867

Dr. Charles Coquerel (1822–1867) was a surgeon in the French Imperial Navy and an entomologist who was involved in collecting expeditions to Madagascar and its neighboring islands. He was the first to identify the screwworm fly in 1858 and gave it the name *hominivorax*, which literally means "man-eater." Many of the specimens he collected, mostly of insects, were studied only after his death. He was author of many scientific papers such as "Orthoptères de Bourbon et de Madagascar" and "Sur les Monandroptères et Raphiderus." He is also commemorated in the name of a bird, Coquerel's Coua *Coua coquereli*. Both lemurs are endemic to western Madagascar.

Corbett

Corbett's Tiger *Panthera tigris corbetti* Mazak, 1968 [Alt. Indochinese Tiger]

Edward James "Jim" Corbett (1875–1955) was an Englishman born in India, where his father was a postmaster. Originally he was a hunter and was widely called upon in his home territory of Kumaon to hunt down and kill man-eating tigers and leopards. In the 1920s he became worried by the increasing numbers of hunters and by the exploitation of the forests for timber rather than as sanctuary for wild animals. In 1934 he was responsible for the creation of India's first national park, in the Kumaon hills. About this time he entirely abandoned hunting with a rifle in favor of recording tigers on film. When WW2 broke out, he started to train Allied troops in jungle survival techniques, but the strain on a man in his mid-60s was too much and he became extremely ill. While convalescing from his illness he wrote *Man-eaters of Kumaon*, which became an international best-seller. After India became independent in 1947 he retired to Kenya. He suffered a heart attack, from which he died in 1955, and is buried in Africa. In 1957 the national park he established in India was renamed in his honor as Corbett National Park. It is now nearly doubled in size and still holds a small population of tigers. This subspecies of tiger is found in eastern Myanmar, Thailand, and Indochina.

Cordeaux

Cordeaux's Dik-dik *Madoqua (saltiana) cordeauxi* Drake-Brockman, 1909 [Alt. Salt's Dik-dik]

Captain Sir Harry Edward Spiller Cordeaux C.B., C.M.G. (1870–1943) was a career diplomat. He was Consul General, and then Commissioner, for the Somaliland Protectorate. The original etymology calls him one "whose interest in the fauna of Somaliland is well known." He served in Somaliland from 1889 until 1909, supervising the construction of a railway from Jinja to Kakindu. In 1910 he was made Governor of Uganda. He became Governor of St. Helena in 1912 and was then Governor of the Bahamas from 1920 until 1926. The dik-dik named after him is usually now regarded as a subspecies of Salt's Dik-dik. It is found in the Awash River basin of Ethiopia.

Cordier

Cordier's Angolan Colobus *Colobus angolensis cordieri* Rahm, 1959

Charles Cordier (dates not found) was a professional animal collector who, with his wife Emy, collected for zoos and museums during the 1940s, 1950s, and 1960s. He is most famous for trying to track down the "Kakundakari" in the Congo; this is Africa's equivalent of the Yeti. Once, said Cordier, a Kakundakari had become entangled in one of his bird snares: "It fell on its face, turned over, sat up, took the noose off its feet, and walked away before the nearby African could do anything." Zoologists have yet to see a specimen of Kakundakari. He wrote a number of articles about his adventures such as "Further Adventures of Charles Cordier," published in 1949 in *Animal Kingdom*. The colobus subspecies is found in the Manyema region of eastern DRC (Zaire).

Corinna

Plush-coated Ringtail Possum *Pseudochirops corinnae* **Thomas,** 1897

Corinna Loria (dates not found) was the sister of Lamberto Loria (1855–1913), an Italian naturalist and ethnologist. It was he who collected the type specimen and additional co-types from the central highlands of New

Guinea. Thomas named the possum after Loria's deceased sister.

Corrie

Fynbos Golden Mole *Amblysomus corriae* **Thomas,** 1905 [Alt. Western Cape Golden-Mole]

Mrs. Corrie (sometimes Cornie) Maria Rudd (dates not found). The original etymology says that the type specimen was "obtained by Mr. Grant in connection with Mr. C. D. Rudd's exploration of South Africa." Thomas named it in honor of Mrs. Rudd, "who had taken much interest in the results of the exploration." The golden mole is endemic to South Africa.

Cosens

Cosens' Gerbil *Gerbillus cosensi* **Dollman,** 1914

Lieutenant Colonel G. P. L. Cosens (1884–1928) was in Kenya from 1912 to 1913, in which year he entered the Egyptian army. The gerbil is known from only one locality in Kenya. It is sometimes regarded as being conspecific with *Gerbillus agag.*

Cotton

Lado Giraffe *Giraffa camelopardalis cottoni* **Lydekker,** 1904

Cotton's Colobus *Colobus angolensis cottoni* Lydekker, 1905

Cotton's Oribi *Ourebia ourebi cottoni* **Thomas** and **Wroughton,** 1908

Cotton's Wide-lipped Rhinoceros *Ceratotherium simum cottoni* Lydekker, 1908 [Alt. Northern White Rhinoceros]

Major Percy Horace Gordon Powell-Cotton F. G.S., F.Z.S. (1866–1940) was the archetypal British gentleman hunter and explorer. Unlike his contemporaries, he did not take the "Grand Tour of Europe" but ventured in the early 1880s first to India, then to Tibet and the Himalayas, climbing to 6,100 m (20,000 feet). He also visited Singapore and other parts of Southeast Asia. From there he traveled to Somalia for his first expedition to Africa—and a lifelong fascination with the African continent and all its wonders ensued. He spent more than 26 years on African soil during 25 expeditions. He collected more than 6,000 mammal specimens and, with his family, collected another 18,000 artifacts from the many different cultures he encountered. His further acquisitions of furniture, Chinese imperial porcelain, glassware, and hundreds of other decorative *objets d'art* are today housed in his former home, Quex House in Kent, UK, where they may be viewed. The "Lado" referred to in the vernacular name of the giraffe is the Lado enclave in northeastern Congo, of which Powell-Cotton made an extensive survey in 1905 through 1906, also collecting many specimens and artifacts there. "After the success of his sporting trip to Abyssinia, Powell-Cotton was determined to attempt another hunting expedition, this time to British East Africa, not only for sport, but also to prove the existence of the five-horned giraffe, which Harry Johnston had discovered. . . . Crossing the Tana, they spent time with the Kikuyu. . . . On the Baringo Plain, trophy kudu was collected as was lion. Powell-Cotton eventually bagged a five-horned giraffe before entering Masai territory." Three of the mammals named after Cotton are found in northeast DRC (Zaire). The Northern White Rhinoceros is, at the time of writing, on the edge of extinction. Poachers have decimated the last known wild population in the Garamba National Park. The oribi is found in Tanzania.

Coues

Coues' Deer *Odocoileus virginianus couesi* Coues and Yarrow, 1875 [Alt. Arizona White-tailed Deer]
Coues' Rice Rat *Oryzomys couesi* **Alston,** 1877
Coues' Climbing Mouse *Rhipidomys couesi* **J. A. Allen** and **Chapman,** 1893

Dr. Elliott Coues (pronounced "cows") (1842–1899) was a U.S. Army surgeon and one of the founders of the American Ornithologists' Union. Among other works, he wrote *Fur-Bearing Animals* (1877). He also wrote on birds, publishing such works as *Handbook of Field and General Ornithology* (1890), in which he set out in meticulous detail how to "collect" and preserve birds. He is commemorated in the common names of three birds: Coues' Flycatcher *Contopus pertinax,* Coues' Gadwall *Anas strepera couesi* (extinct), and Coues' Redpoll *Carduelis hornemanni exilipes.* The deer occurs in southeast California, central and southern Arizona, southwest New Mexico (USA), and northern Mexico. The rice rat is found from southern Texas to northwest Colombia. The climbing mouse occurs on Trinidad and in Colombia, Venezuela, Ecuador, and Peru.

Count Branicki

Count Branicki's Terrible Mouse *Dinomys branickii* **Peters,** 1873 [Alt. Pacarana]

Hieronim Florian Radziwill Konstanty, Count Branicki (1823–1884), was a wealthy Polish nobleman whose son and nephew co-founded the Branicki Museum of Warsaw. The Pacarana is not a mouse, but the scientific name *Dinomys* means "terrible mouse." It is the only member of the family Dinomyidae and is found in northwest Venezuela, Colombia, Ecuador, Peru, western Brazil, and Bolivia.

Cowan

Cowan's Shrew-Tenrec *Microgale cowani* **Thomas,** 1882

The Rev. William Deans Cowan (1844–1923) was a missionary in Madagascar for 10 years in the late 1800s. He made a geological expedition in south-central Madagascar and was the author of *The Bara Land: A Description of the Country and People.* The shrew-tenrec is found in north and east Madagascar.

Cox

Cox's Roundleaf Bat *Hipposideros coxi* Shelford, 1901

E. A. W. Cox (dates not found) was a British colonial civil servant. Shelford wrote in his formal description, "I have much pleasure in naming this species . . . after Mr. E. A. W. Cox, of the Sarawak Service, to whom I am indebted for the unique specimen." The bat is known only from Sarawak, East Malaysia.

Coxing

Coxing's White-bellied Rat *Niviventer coxingi* **Swinhoe,** 1864 [Syn. *N. coninga*]

Zheng Chenggong (ca. 1624–1662) was a Chinese general at the end of the Ming dynasty who succeeded in expelling the Dutch from Taiwan. He maintained Taiwan as an independent country of which he was ruler until it was taken from his son by the Qing dynasty in 1683. He is regarded as a hero by Taiwanese nationalists of all centuries since his time. "Coxinga" was what the Dutch called him—a corruption of his nickname in the Fujian dialect, Koku sen-ya. Swinhoe, as a diplomat who spent many years in China, would have known all about "Coxinga." Swinhoe originally spelled the scientific name *coninga,* but he says, "I propose to name the animal after the powerful pirate chief who seized the island [Formosa] from the Dutch," so it is clear that he meant to write *coxinga* rather than *coninga*. It seems that Thomas at some point "corrected' the spelling of the rat's scientific name to *coxingi*, though *coxinga* would have been the more correct amendment, and in fact the species can be found on some lists as *Niviventer coxinga*. Thus the animal's common name should really be Coxinga's White-bellied Rat.The rat is found, naturally, in Taiwan.

Cranbrook

Burmese Red Goral *Naemorhedus baileyi cranbrooki* **Hayman,** 1961

John David Gathorne-Hardy, 4th Earl of Cranbrook (1900–1978), was an amateur zoologist who became a Trustee of the Natural History Museum in London in 1963. He was in Burma (currently called Myanmar) in 1930 through 1931, and it was then that he collected the type specimen of this goat-antelope; but the taxon had to wait another 30 years before it was officially described. When Hayman named it as a full species, *Naemorhedus cranbrooki*, he raised the possibility that it was the same species as *baileyi* (brown goral), given the proximity in their ranges. He at first suggested, then dismissed, the idea that the deeper red coloration of *cranbrooki*, from specimens collected from February to April, might represent the winter pelage, and the browner coat of *baileyi* (the type specimen of which was collected in July) might be the summer coat. Today *cranbrooki* is sometimes regarded as a race of *baileyi* and sometimes as a mere synonym of the latter species. The goral is found in the state of Arunachal Pradesh, northeast India, and in northern Myanmar.

Crandall

Crandall's Saddle-back Tamarin *Saguinus fuscicollis crandalli* **Hershkovitz,** 1966 [Syn. *S. melanoleucus*]

Lee Saunders Crandall (1887–1969) was an American naturalist who spent most of his working life at the Bronx Zoo, New York. His most influential publication was *The Management of Wild Mammals in Captivity* (1964). The type specimen of *crandalli* was exhibited at the Bronx Zoo from 1951 to 1954. The tamarin came from an unknown location in South America, and the status of this subspecies is still unclear. It has been suggested that it could represent a color-morph of *Saguinus (fuscicollis) melanoleucus*.

Crawford

Desert Shrew *Notiosorex crawfordi* **Coues,** 1877

S. W. Crawford (dates not found) collected the holotype of the shrew, which is found in desert area of northern Mexico and the adjacent southern USA. The type specimen was obtained not far from El Paso, Texas.

Crawford-Cabral

Crawford-Cabral's Marsh Rat *Dasymys cabrali* **Verheyen** et al., 2003

Dr. João Crawford-Cabral is a Portuguese zoologist. He is Director of the Tropical Scientific Research Institute, Center for Zoology, in Lisbon. He has published widely on African mammals, in such works as *The Angolan Rodents of the Superfamily Muroidea: An Account of Their Distribution*. The marsh rat is found in the Okavango region of southern Africa.

Crawshay

Crawshay's Zebra *Equus guagga crawshaii* **de Winton,** 1896

Crawshay's Hare *Lepus crawshayi* de Winton, 1899 [Syn. *L. microtis* **Heuglin,** 1865]

Captain Richard Crawshay (1862–1958) was a hunter and collector who became an agent for the African Lakes Company. The company sent him to set up a permanent post on the shores of Lake Mweru in 1890. Crawshay found a spot at Puta, where the Chienge stream enters Lake Mweru, and set up residence there. He stayed in Rhodesia (now Zambia) from 1890 to 1891, when he abandoned the station and returned to Blantyre in Nyasaland (now Malawi), from where he sent collections of ants to the museums of South Africa. He also collected butterflies in the Kikuyu Country of British East Africa in 1899 and 1900. He presented the type specimen of the zebra, which is a subspecies of Plains Zebra, to the British Museum. The zebra is found in Mozambique north of the Zambezi, Malawi, and eastern Zambia. The hare is usually now regarded as a synonym of the widespread African Savanna Hare, *Lepus microtis*.

Creagh

Creagh's Horseshoe Bat *Rhinolophus creaghi* **Thomas,** 1896

Charles Vandelleur Creagh C.M.G. (1842–1917) was a civil servant. Having earlier been an administrator in Perak and the Malay States, he was Governor of North Borneo (now part of Malaysia) from 1888 to 1895, which is where the type specimen of the bat was collected. Thomas wrote that the specimen was "presented and collected by Governor C. V. Creagh, C.M.G., in whose honour I have ventured to name the species." The bat is found on Borneo, Madura Island (northeast of Java), and Timor.

Creighton

Voss' Slender Opossum *Marmosops creightoni* **Voss** et al., 2004

Dr. G. Ken Creighton works as Global Coordinator of the Biodiversity Planning Support Programme, part of the United Nations Development Programme. Creighton collected, but did not identify, this species of opossum in Bolivia in 1979. According to the etymology given by Voss it was named for "G. Ken Creighton, collector of the first known specimens of this species and author of the first modern analysis of didelphid phylogeny based on morphological characters." This was actually the subject of his Ph.D. thesis at the University of Michigan, published in 1984 as *Systematic Studies on Opossums (Didelphidae) and Rodents (Cricetidae).* He also co-wrote, with Sydney Anderson and Karl F. Koopman, *Bats of Bolivia—An Annotated Checklist*, published in 1982. The opossum is known from montane forest in the valley of the Rio Zongo, Bolivia.

Crespo

Crespo's Pampas Cat *Oncifelis colocolo crespoi* **Cabrera,** 1957 [Syn. *Leopardus pajeros crespoi* or *L. pajeros budini*]

Dr. Jorge A. Crespo (b. 1915) is a Brazilian biologist, zoologist, and paleontologist. He was born in Rio de Janeiro and became Chief of the Department of Mammals at the Bernardino Rivadavia National Museum of Natural Sciences in Buenos Aires, where he worked for 41 years including 20 years as Professor of Animal Ecology. He followed Cabrera, the describer of the cat, as Director. In 1954 he was elected to be a Fellow of the John Simon Guggenheim Memorial Foundation, and in 1956 he won a scholarship for postgraduate study at the University of Michigan. He published quite widely, including, in 1950, *Nota sobre mamíferos de misiones nuevos para Argentina* and, in 1982, *Introducción a la ecología de los mamíferos del Parque Nacional El Palmar, Entre Ríos.* This subspecies of Pampas Cat is found in northern Argentina. It is now usually not considered distinct from the subspecies *budini.*

Cross

Cross's Guenon *Cercopithecus preussi* **Matschie,** 1898 [Alt. **Preuss's** Guenon]

William Cross (1840–1900) and his sons William Simpson Cross and James Conrad Cross were famous importers of, and dealers in, wild animals. They had a shop called Cross'

Menagerie and Museum in Earle Street, Liverpool, England. The common name Cross's Guenon has fallen out of use; it was given to a live monkey, supplied by Cross and described by Forbes as *Cercopithecus crossi* in 1905, but this was later shown to be the same species as *C. preussi*, described by Matschie seven years earlier. It could be that the animal is named after the firm, but in all probability it is after William Simpson Cross, who became the leading light in the business after his father's death. This primate is, coincidentally, found in the Cross River area of western Cameroon and southeast Nigeria.

Crosse

Crosse's Shrew *Crocidura crossei* **Thomas,** 1895

Dr. William Henry Crosse (1859–1903) was a traveler and collector who took the type specimen of the shrew. He found it at Asaba "150 miles [240 km] up the River Niger." Crosse is also remembered as the author of a work entitled *Notes on the Malarial Fevers Met with on the River Niger, West Africa* (1892). He once said, "The ideal traveller is a temperate person, with a sound constitution, a digestion like an ostrich, a good temper, and no race prejudice." The shrew occurs in lowland forest from Sierra Leone to western Cameroon.

Crossley

Crossley's Dwarf Lemur *Cheirogaleus crossleyi* **Grandidier,** 1870 [Alt. Furry-eared Dwarf Lemur]

Alfred Crossley (dates not found) collected in Madagascar and Cameroon in the 1860s and 1870s. Many specimens of birds collected by him are now in the British Museum of Natural History, and three bird species are named after him: Crossley's Babbler *Mystacornis crossleyi*, Crossley's Ground Roller *Atelornis crossleyi*, and Crossley's Ground Thrush *Zoothera crossleyi*. By virtue of the other fauna commemorating him, it is clear he did not confine his efforts to collecting birds. For example, there are butterflies and moths that bear his name, such as Crossley's Forest Queen *Euxanthe crossleyi* and *Godartia crossleyi*. The lemur is endemic to eastern Madagascar.

Crowther

Crowther's Bear *Ursus crowtheri* **Blyth,** 1841 extinct [Alt. Atlas Bear; Syn. *U. arctos crowtheri*]

Crowther (dates not found) was an officer in the 63rd Regiment of Foot. He killed a specimen of this bear in the Tétouan, in Morocco, in 1834. It may be that his regiment was posted to India and that he encountered Blyth there, as Blyth wrote in 1841 to the Zoological Society of London that he had received a description of the bear from Crowther and that it was definitely a different species from the one he had expected *(Ursus syriacus)*. The Atlas Bear was Africa's only native bear, and the last individual is thought to have been killed by hunters in northern Morocco sometime in the 1870s.

Crump

Squirrel sp. *Callosciurus crumpi* **Wroughton,** 1916 [now *Callosciurus erythraeus crumpi*]

Crump's Mouse *Diomys crumpi* **Thomas,** 1917

C. A. Crump (dates not found) was a prolific collector of small mammals, working in the Indian subcontinent, where he arrived in 1910 and was taken on the payroll of the Bombay Natural History Society as a paid collector. The squirrel, from Sikkim, was originally described as a full species but has now been "demoted" to a race of *Callosciurus erythraeus*. The mouse comes from northeast India and western Nepal.

Cruz Lima

Cruz Lima's Saddle-back Tamarin *Saguinus fuscicollis cruzlimai* **Hershkovitz,** 1966

Eladio Da Cruz Lima (1900–1943) was a Brazilian zoologist associated with the Goeldi Museum in Pará. He wrote *Mammals of Amazonia—Volume I: General Introduction and Primates,* published in 1943. The citation reads, "Dr. Eladio Da Cruz Lima, in whose memory this tamarin is named, was a Justice of the Supreme Court of the State of Pará, an accomplished artist, literary critic, archeologist, and zoologist. His untimely death in 1943 at the age of 43 during the printing of his first volume of the *Mammals of Amazonia* deprived Brazil of a ranking scholar and one of its finest naturalist-artists." The distribution of this tamarin subspecies is uncertain; the type specimen was said to come from the upper Rio Purus, Amazonian Brazil.

Csorba

Csorba's Myotis *Myotis csorbai* Topal, 1997

Dr. Gabor Csorba (b. 1961) is a Hungarian zoologist. He graduated from Eötvös Loránd University in 1983 and went on to József Attila University, Szeged, where he took his master's in 1986 and was awarded his doctorate in 1995. He has written and published widely about a wide range of zoological subjects. Among his first publications was, with J. Török as co-author, "Táplálékszegregáció négy, fatörzsön tálálkozó madárfajnál" (Food Segregation among Four Bark-foraging Bird Species), published in 1986 in a larger work. One of his most recent publications was a lecture with V. Matveev, "Bat Fauna of Cambodia" (2007). He has been employed by the Hungarian Natural History Museum in Budapest, Department of Zoology. since 1983 and is currently Deputy Director and Curator of Mammals. He has undertaken many field trips since 1988, mostly to Asian countries. His main interests are the taxonomy and systematics of bats (he has described 10 new bat taxa) and conservation biology of Hungarian mammals. The bat named after him is known from just two localities in Nepal.

Cuming

Southern Luzon Cloud Rat *Phloeomys cumingi* **Waterhouse,** 1839

Hugh Cuming (1791–1865) was an English naturalist and conchologist who has been described as "the Prince of Collectors." He was a sailmaker who was living in Valparaíso, Chile, in 1819. He changed profession and became a collector, working inthe Neotropics from 1822 until 1826 and again from 1828 to 1830. He also collected in Polynesia between 1827 and 1828 and in the East Indies from 1836 until 1840. He preceded Darwin in having collected in the Galápagos, in 1829. His shell collection is housed in the Linnean Library in London. He is also remembered in the names of various other taxa, from birds such as Cuming's Scrubfowl *Megapodius cumingii* to marine molluscs such as *Cribrarula cumingii.* The cloud rat is found on Luzon, Marinduque, and the Catanduanes Islands (Philippines).

Cunningham

Southern Mountain Brushtail Possum *Trichosurus cunninghamii* Lindenmayer, Dubach, and Viggers, 2002

Ross Benjamin Cunningham (b. 1946) is an Associate Professor of Statistics at the Australian National University in Canberra and has over 35 years' experience in statistical science, having started off as a teacher of mathematics. He has interests not only in ecology but also in the application of statistics to sports and psephology (the study of elections). He has written 230 scientific papers published in diverse fields and has co-authored two books.

The possum was discovered by a team of scientists from the Australian National University who had been measuring mountain possums, all populations of which had been considered to be *Trichosurus caninus*. The scientists named this newly recognized species in honor of Cunningham, their biostatistician, who was the first to recognize the differences in size and facial shape between the northern and southern mountain possums. It is found in eastern Australia.

Curio

Curio's Giant Rat *Megaoryzomys curioi* **Niethammer,** 1964 extinct

Professor Dr. Eberhard Curio (b. 1932) is a zoologist and ornithologist who graduated from university in Berlin in 1957 and afterward became a research associate at the Max Planck Institute for Behavioral Physiology at Seewiesen. From 1964 to 1967 he was an Assistant Professor at Tübingen University, and in 1968 he was a Lecturer in the Biology Department at Ruhr University Bochum, where he became a Professor in 1971. He appears to have held that position until his retirement in 1997. He made research trips to southern Macedonia in 1958, Spain in 1960, the Galápagos Islands from 1962 to 1963, Jamaica in 1969, Panama in 1979, Tonga in 1990, Fiji from 1990 to 1991, and the Philippines from 1993 onward; he founded and became the Director of the Philippine Endemic Species Conservation Project in 1995. He is a member of many organizations and has received honors and much acclaim for his work. Since his official retirement from employment in Germany he has become a Professor of Biodiversity in the Philippines. The rat was from Santa Cruz Island in the Galápagos. It may have survived until around 1900 but could not survive the onslaught of introduced feral cats, dogs, and pigs.

Curry, A. W.

Curry's Red Rock Rabbit *Pronolagus rupestris curryi* **Thomas,** 1902

A. W. Curry (dates not found). In his description Thomas says that the type specimen was taken "some years ago" at Boshof, Orange River Colony, by Mr. A. W. Curry. No other details of Curry are given. The rock rabbit is found in the Orange Free State (South Africa).

Curry, N.

Curry's Bat *Glauconycteris curryae* Eger and Schlitter, 2001

Noreen Curry (1915–2004) was a well-known Canadian collector most famed for her extensive library. She was also passionately interested in nature and wildlife conservation. She helped fund, and participated in, two bat-collecting expeditions to Africa in the 1970s led by Dr. Randolph Peterson, who was Curator of Mammals at the Royal Ontario Museum at that time. She was also responsible for obtaining partial funding for construction of the "Bat Cave" at the Royal Ontario Museum, a gallery she considered an important means of educating children about bats. She was living in California at the time of her death. The type specimen of the bat was collected in Cameroon.

Curzon

Black-lipped Pika *Ochotona curzoniae* **Hodgson,** 1858

Mrs. Curzon (dates not found). Hodgson's etymology reads, "This beautiful little animal is appropriately dedicated to the Honourable Mrs. Curzon." Unfortunately this is as full a description of the lady as we can find. The pika inhabits eastern Nepal, Sikkim, the Tibetan Plateau, and adjacent areas of Gansu, Qinghai, and Sichuan (China).

Cuvier, F.

Cuvier's Vervet *Chlorocebus pygerythrus pygerythrus*,F. Cuvier, 1821 [Alt. South African Vervet Monkey]
Cuvier's Fire-footed Squirrel *Funisciurus pyrropus* F. Cuvier, 1833 [Alt. Fire-footed Rope Squirrel]
Cuvier's Hutia *Plagiodontia aedium* F. Cuvier, 1836 [Alt. Hispaniolan Hutia]
Cuvier's Gazelle *Gazella cuvieri* **Ogilby,** 1841

Frédéric Cuvier (1773–1838) was a zoologist like his brother Georges (see next entry), though not as well known. He was the head keeper of the menagerie at the Museum of Natural History in Paris from 1804 to 1838. He is regarded as the first of the modern zookeepers in that he believed that animals should receive a good diet and be treated with kindness. The Chair of Comparative Physiology at the museum was especially established for him in 1837. He was co-author, with Étienne Geoffroy Saint-Hilaire, of *Histoire naturelles des mammifères,* which appeared in four volumes between 1819 and 1842. He must have felt overshadowed by his famous older brother, as on his deathbed he asked that his tombstone contain nothing beyond the words "Brother of Georges Cuvier." The vervet monkey comes from southern Africa. The squirrel is widespread in western and central Africa, from Gambia east to Uganda and south to northwest Angola. The hutia is found on the Caribbean island of Hispaniola (Haiti and Dominican Republic), and the gazelle occurs in Morocco and northern Algeria.

Cuvier, G.

Cuvier's Spotted Dolphin *Stenella dubia* G. Cuvier, 1812
Cuvier's Beaked Whale *Ziphius cavirostris* G. Cuvier, 1823 [Alt. Goose-beaked Whale]
Cuvier's Spiny Rat *Proechimys cuvieri* **Petter,** 1978

Georges Léopold Chrétien Frédéric Dagobert, Baron Cuvier (1769–1832), is better known by his pen name Georges Cuvier. He was a French naturalist and one of the scientific giants of his age. He believed that paleontological discontinuities were evidence of sudden and widespread catastrophes—that is, that mass extinctions can happen suddenly. He is also famed for managing to stay in a top government post, as Permanent Secretary of the Academy of Sciences, through three regimes, including Napoleon's. He was born in eastern France in what was then part of the Duchy of Württemberg. In 1788 he took a job as a private tutor to an aristocratic family and moved to Caen in Normandy. He stayed in Normandy until 1795, when he went to Paris at the invitation of Étienne Geoffroy Saint-Hilaire. He taught natural history in a number of establishments before becoming Assistant to the Professor of Animal Anatomy at the Museum of Natural History in Paris. Among his achievements, he coined the term *Pterodactyle* for a genus of prehistoric flying reptiles. Audubon said of a kinglet he collected in June 1812, "I named this pretty and rare species after Baron Cuvier, not merely by way of acknowledgment for the kind attentions which I received at the hands of that deservedly celebrated naturalist, but as a homage due by every student of nature to one unrivalled in the knowledge of General Zoology." Cuvier wrote *Tableau élémentaire de l'histoire naturelle des animaux* (1798); *Mémoires sur les espèces d'éléphants vivants et fossils* (1800); *Leçons d'anatomie comparée* (1801–1805);

Récherches sur les ossements fossiles des quadrupeds (1812); and *Le règne animal destribué d'après son organisation* (1817). The whale has a worldwide marine distribution from tropical to cold-temperate waters. The spiny rat comes from eastern Venezuela, the Guianas, and northern Brazil. The taxonomy of spotted dolphins is confused (and confusing), and various scientific names can be found in the literature. *Stenella dubia* is what is known as a *nomen nudum*, which means that it is not considered a valid scientific name because it was published without an adequate description of the species it was meant to refer to. Current taxonomy recognizes two species of spotted dolphin under the names *Stenella attenuata* and *S. frontalis*.

Cyclops

Cyclops Roundleaf Bat *Hipposideros cyclops* **Temminck,** 1853

Cyclops Long-beaked Echidna *Zaglossus attenboroughi* **Flannery** and Groves, 1998 [Alt. Attenborough's Echidna, Sir David's Long-beaked Echidna]

In mythology, the Cyclops was a giant with one eye in the middle of his forehead. His most famous appearance is in Homer's *Odyssey*. The bat has a glandular sac on its forehead which, with a little bit of fancy, can be compared to an eye. It ranges from Guinea-Bissau east to southern Sudan and Kenya, and on the island of Bioko. The echidna is named for its geographical distribution (the Cyclops Mountains in New Guinea) and not for its physical appearance.

Dabbene

Dabbene's Mastiff Bat *Eumops dabbenei* **Thomas,** 1914 [Alt. Big Bonneted Bat]

Roberto Dabbene (1864–1938) was an Italian-Argentine ornithologist. Born in Turin, he moved to Argentina in 1887. For a time he worked in the Buenos Aires Zoo. In 1916 he became the first president of the Del Plata Ornithological Association. He wrote many articles on birds in journals such as *El Hornero* and *Revista Chilena de Historia Natural.* He is also commemorated in the name of a bird, Dabbene's (or Red-faced) Guan *Penelope dabbenei.* The bat has a curious distribution, being found in the Chaco region of northern Argentina and Paraguay and also in Colombia and northern Venezuela.

Dahl

Rock Ringtail Possum *Petropseudes dahli* **Collett,** 1895

Dahl's Jird *Meriones dahli* Shidlovsky, 1962

Professor Dr. Knut Dahl (1871–1953) was a Norwegian naturalist, explorer, and collector. He was in Australia from 1894 to 1896, collecting specimens for the Zoological Museum of the Norwegian University. He wrote an account of this expedition, *In Savage Australia: An Account of a Hunting and Collecting Expedition to Arnhem Land and Dampier Land,* first published in English in 1926. The ringtail possum was first collected by Dahl and is found in northern Australia. The jird is endemic to sandy habitats in Armenia.

D'Albertis

D'Albertis' Ringtail Possum *Pseudochirops albertisii* **Peters,** 1874

Cavaglieri Luig Maria D'Albertis (1841–1901) was an Italian botanist, ethnologist, and zoologist. He was in New Guinea from 1871 to 1877, where he ventured further than any European had before, using a steamboat furnished by the New South Wales government that enabled him to explore and chart the Fly River. He had an adventurous time of it, being attacked by indigenous people who fired arrows at him. He is reported to have taken a number of human skulls and even the recently severed head of an elderly woman. His behavior toward the local people probably contributed considerably to their hostility toward later European explorers. Presumably he also collected natural history specimens, as he is commemorated in the names of a number of different taxa, including D'Albertis' Python *Leiopython albertisii* and birds such as D'Albertis' Mountain Pigeon *Gymnophaps albertisii.* The ringtail possum is found in northern and western New Guinea.

Dall

Dall's Sheep *Ovis dalli* **Nelson,** 1884

Dall's Porpoise *Phocoenoides dalli* **True,** 1885

Alaskan Brown Bear *Ursus arctos dalli* **Merriam,** 1896

William Healey Dall (1845–1927) was an American naturalist. His interests were wide-ranging, including anthropology, meteorology, and oceanography as well as paleontology, zoology, ornithology, and in particular malacology. He was a pupil of Louis Agassiz of Harvard's Museum of Comparative Zoology. He undertook fieldwork in Alaska and along the coasts of the USA, initially as an assistant to Kennicott. His findings were published as *Alaska and Its Resources.* In 1870 he was appointed Acting Assistant to the U.S. Coast Survey. He continued to collect and send his specimens to Agassiz at Harvard. In 1877 he published *Tribes of the Extreme Northwest.* In 1879 or 1880 he first spotted the sheep that was later named after him, on the slopes of Mount McKinley. He is also commemorated in the scientific names of two fish found off the Californian coast: the Blue-banded Goby *Lythrypnus dalli* and the Calico Rockfish *Sebastes dallii.* He published over 1,600 papers, reviews, and commentaries describing 5,302 species, many of them molluscs. The sheep is found from Alaska to northern British Columbia (Canada). The porpoise occurs in cold-temperate waters of the northern Pacific Ocean. The bear is from coastal Alaska. (Merriam was notorious for describing new "species" and "subspecies" of bear at every opportunity, and most of them are no longer considered valid, but this one is still accepted by some authorities.)

Dallon

Dallon's Gerbil *Gerbillus dalloni* **Heim de Balsac,** 1936

Marius-Gustave Dalloni (1880–1959) was Professor of Geology at the University of Algiers between 1925 and 1950. He made a collecting expedition to the Spanish Pyrenees in 1910 to survey the geology of the area. He published a great number of papers and other works, notably, in 1935, *Mission au Tibesti zoologie: Étude préliminaire de la faune du Tibesti* (Tibesti, northern Chad, is where the type specimen was collected). This is the record of an expedition he led to the area in 1930 through 1931, principally to collect mineral specimens. The gerbil is known only from the Tibesti area.

Dalquest

Dalquest's Pocket Mouse *Chaetodipus dalquesti* Roth, 1976

Dr. Walter Woelber Dalquest (1917–2000) was a zoologist at Midwestern State University, Wichita Falls, Texas. He also had connections with the American Museum of Natural History and the Museum of Comparative Zoology at Harvard. He published both on mammals (e.g. in 1950, with E. R. Hall, *Geographic Range of the Hooded Skunk, Mephitis macroura*) and on birds (e.g. in 1967, with R. W. Storer, *Birds from the Save River Area of Mozambique*). The mouse is found in southern Baja California, Mexico.

Dalton

Dalton's Mouse *Praomys daltoni* **Thomas,** 1892 [Syn. *Myomys daltoni*]

J. T. Dalton (dates not found) was a collector who took the first known specimen of this mouse. We could find nothing more recorded about him, not even the locality where the type specimen was collected. The mouse ranges from Gambia and Senegal east to southern Chad, the Central African Republic, and southwest Sudan.

Dammerman

Western White-eared Giant Rat *Hyomys dammermani* **Stein,** 1933

Dr. Karel Willem Dammerman (1888–1951) was a Dutch field zoologist, botanist, and collector who worked in the East Indies from the beginning of the 20th century. He wrote *Preservation of Wildlife and Nature Reserves in the Netherlands Indies* and an article entitled "The Orang Pendek or Ape-man of Sumatra" (both published in 1929). The existence of the Orang-Pendek, a sort of mini-Yeti, is still questioned by zoologists. Dammerman is also commemorated in the name of a subspecies of bird, Dammerman's Moustached Parakeet *Psittacula alexandri dammermani.* The rat is found in the mountains of west and central New Guinea.

D'Anchieta

Angolan Vlei Rat *Otomys anchietae* **Bocage,** 1882
D'Anchieta's Fruit Bat *Plerotes anchietai* **Seabra,** 1900
D'Anchieta's Pipistrelle *Hypsugo anchietai* Seabra, 1900 [formerly *Pipistrellus anchietai*]

José (Alberto) de (Oliveira) Anchieta (1832–1897) was an independent Portuguese naturalist and collector who traveled in Africa. He left Lisbon in 1857 to join a close friend who had settled in Cabo Verde off the West African coast. Here he practiced some medicine, self-taught. A cholera outbreak killed many locals and nearly killed him, and he returned to Portugal in 1859. He studied medicine in Lisbon but before completing his studies left again for Africa, this time Angola. Here he collected many types of animals and other specimens that he donated to a museum upon his return to Portugal, which he left for the last time in 1866. Unfortunately little is known about the next period of Anchieta's life because most of the museum specimens, as well as his letters to Bocage, disappeared in a fire in 1978. We do know that he died, probably from chronic malaria, while returning from a zoological expedition to Caconda, and he was recorded at various locations between 1866 and 1897 in Angola and Mozambique. He was responsible for identifying at least 25 new mammals, 46 birds, and as many amphibians and reptiles. He is commemorated in the names of many taxa, such as the Angolan Python *Python anchietae,* the dune lizard *Meroles anchietae,* and Anchieta's Sunbird *Anthreptes anchietae.* All three mammals listed above are found in Angola. The fruit bat and the pipistrelle also extend their range into Zambia and southern DRC (Zaire). See also **Anchieta.**

Danfoss

Danfoss Mouse-Lemur *Microcebus danfossi* Olivieri et al., 2007

Danfoss is not a person but a Danish engineering company that has subsidiaries in many countries. The German-based Heating Division used lemurs in a marketing campaign and has sponsored field research on these primates. The mouse-lemur is found in northwest Madagascar.

Dang

Dang's Giant Squirrel *Ratufa indica dealbata* **Blanford,** 1897 extinct [Alt. Indian Giant Squirrel]

The squirrel is named after Surat Dangs, a locality in Gujerat (India) to which it was restricted, and not after a person. This albinistic subspecies has not been seen since 1945 and is believed to be extinct.

Daniel

Daniel's Tufted-tailed Rat *Eliurus danieli* Carleton and **Goodman,** 2007

Professor Daniel Rakotondravony is Professor of Animal Biology at the University of Antananarivo, Madagascar. He has published a number of studies and articles, often with Steve Goodman as co-author. The citation states that the animal is named after him "especially for his contributions to our knowledge on Malagasy rodents and for helping to foster zoological research on Madagascar for numerous national and foreign scientists." The rat is endemic to south-central Madagascar.

Dao Van Tien

Tonkin Limestone Rat *Tonkinomys daovantieni* **Musser,** Lunde and Truong Son, 2006

Professor Dao Van Tien (1917–1995) was a Professor of Biology at the National University of Hanoi. He was a primatologist, although he is probably best known for asserting his belief in the existence of "Forest Man"—a supposed primitive hominid reported from remote parts of Asia. Professor Tien was educated in Hanoi under the French colonial administration. He taught several generations of Vietnamese scientists, heading the Biology Faculty through the difficult years after the war. His principal interest, and the subject of many of his scientific papers, was mammalian zoology. The importance of this gentle and thoughtful man in the history of Vietnam's science is widely acknowledged, and he was considered to be the father of his field in Vietnam. He described a number of new taxa, including two langurs, *Trachypithecus francoisi hatinhensis* and *Trachypithecus cristatus caudalis* (the latter described from two individuals that were in the Hanoi Zoo). He was also the first person to record the presence of all-black gibbons, which he identified as *Hylobates concolor hainanus,* between the Red River delta and the Chinese border to the northeast. The original etymology reads, "We are honored to name this distinctive gray-furred and short-tailed limestone endemic after the late Dao Van Tien, who when he died on May 3, 1995, at the age of 78, was emeritus professor of biology at the National University of Hanoi." The rat comes from northeast Vietnam.

Daphne

Daphne's Oldfield Mouse *Thomasomys daphne* **Thomas,** 1917

Daphne, when pursued by Apollo, prayed to Gaea (Mother Earth) for aid and was changed into a laurel tree—possibly not the sort of aid she had in mind, though it succeeded in keeping her chaste. This is another example of Thomas using a name from mythology without any obvious logic for his choice. The mouse is found in the Andes from southern Peru to central Bolivia.

Darling

Mashona Mole-Rat *Fukomys darlingi* **Thomas,** 1895 [formerly *Cryptomys darlingi*]

Darling's Horseshoe Bat *Rhinolophus darlingi* **K. Andersen,** 1905

James ffolliott Darling (dates not found) was an Irish collector of zoological specimens in Rhodesia (now Zimbabwe) around the end of the 19th century and the beginning of the 20th (certainly between 1894 and 1897). In 1896 De Winton wrote *On Collections of Rodents Made by Mr. J. ffolliott Darling in Mashunaland and Mr. F. C. Selous in Matabeleland,* and in 1902 G. A. Boulenger wrote "A List of the Fishes, Batrachians and Reptiles Collected by Mr. J. ffolliott Darling in Mashonaland, with Descriptions of New Species," published in the *Proceedings of the Zoological Society of London.* There are records of collections of insects and arachnids being donated to the South Africa Museum in 1896–1897. In 1882 and 1883 Darling reported on interesting ornithological sightings in Ireland. He is also commemorated in the scientific name of an arachnid, the African Horned Tarantula *Ceratogyrus darlingi.* The mole-rat is found in northeast Zimbabwe and adjacent Mozambique. The bat occurs in Zimbabwe, Namibia, Angola, Botswana, Swaziland, Malawi, Mozambique, and Tanzania.

Darlington

Large Forest Bat *Vespadelus darlingtoni* **G. Allen,** 1933

Professor Dr. Philip Jackson Darlington Jr. (1904–1983) was an American evolutionary biologist, zoogeographer, and beetle taxonomist. His early career was as a taxonomist and collector, and he traveled to Colombia, Puerto Rico, Haiti, Cuba, New Guinea, Australia, and, in later years, Tierra del Fuego. During one collecting expedition to Cape York in northeast Australia he found the Rufous Spiny Bandicoot *Echymipera rufescens*, which had previously been thought to be confined to New Guinea. Between that year and 1940 he was Assistant Curator of Insects at the Museum of Comparative Zoology, Harvard. He went on to become Curator, and eventually Professor of Zoology, there—although he did take a year off from 1956 to 1957 to live out of the back of a truck with his family in Australia. He wrote many papers and articles, but his two major works were *Zoogeography: The Geographical Distribution of Animals* (1957) and *Biogeography of the Southern End of the World* (1965). The bat is found in eastern Australia, including Tasmania.

Dartmouth

Dartmouth's Vlei Rat *Otomys dartmouthi* **Thomas,** 1906 [Alt. Ruwenzori Vlei Rat]

William Henage Legge, 6th Earl of Dartmouth (1851–1936), who succeeded to the title in 1891, sponsored scientific exploration. The animal was named after him with the words, "to whose generosity this splendid exploration of Mount Ruwenzori is primarily due." The rat is found in the Ruwenzori Mountains on the Uganda DRC (Zaire) border.

Darwin

Darwin's Fox *Pseudalopex fulvipes* **Martin,** 1837 [Syn. *Lycalopex fulvipes*]
Darwin's Leaf-eared Mouse *Phyllotis darwini* **Waterhouse,** 1837
Darwin's Sheep *Ovis ammon darwini* **Przewalski,** 1883 [Alt. Gobi Argali]
Darwin's Galápagos Mouse *Nesoryzomys darwini* **Osgood,** 1929 extinct

Charles Robert Darwin (1809–1882) was the prime advocate (together with Wallace) of natural selection as the way in which speciation takes place. To quote from his most famous work, *On the Origin of Species by Means of Natural Selection*, first published in 1859, "I have called this principle, by which each slight variation, if useful, is preserved, by the term Natural Selection." From 1831 to 1836 Darwin was the naturalist on HMS *Beagle* on her scientific expedition around the world. In South America he found fossils of extinct animals that were similar to extant species. On the Galápagos Islands he noticed many variations among plants and animals of the same general type as those in South America. Darwin collected specimens for further study everywhere he went. On his return to London he conducted a thorough review of his notes and specimens. Out of this study grew several related theories: evolution did occur; evolutionary change was gradual, taking thousands or even millions of years; the primary mechanism for evolution was a process he called natural selection; and the millions of species alive today arose from a single original life form through a branching process called speciation. It is often said that Darwin held back on publication for many years through not wanting to offend Christians, in particular his wife. However, this view has been challenged by scholars who put the apparent "long delay" down to the fact that Darwin was working on other books and also suffered periods of ill-health. He is also commemorated in the names of other taxa, such as Darwin's Frog *Rhinoderma darwinii* and many birds. The fox is found in Chile, with the main population on Chiloé Island and a smaller number in Nahuelbuta National Park on the mainland. The leaf-eared mouse lives in northern and central Chile. The sheep, as its alternative common name suggests, dwells in the Gobi Desert of southern Mongolia and adjacent China. The Galápagos Mouse was found only on Santa Cruz Island, Galapagos, but has not been recorded since 1930 and is believed extinct.

Daubenton

Aye-aye *Daubentonia madagascariensis* **Gmelin,** 1788

Daubenton's Bat *Myotis daubentoni* **Kuhl,** 1817

Daubenton's Free-tailed Bat *Myopterus daubentonii* **Desmarest,** 1820

Dr. Louis Jean-Marie d'Aubenton, or Daubenton, as his name is more commonly spelled (1716–1800; some sources say 1799, as he died on 1 January 1800), was a French naturalist. His work covered many fields including comparative anatomy, plant physiology, paleontology, mineralogy, and experimental agriculture. He was Professor of Mineralogy at the Jardin des Plantes and of Natural History at the School of Medicine in Paris. In 1793 he became the first Director of the Museum of Natural History in Paris. He wrote *Les planches elumineez d'histoire naturelle* in 1765 and collaborated with Buffon on the many volumes constituting *Histoire naturelle des oiseaux.* He completed a great number of zoological descriptions (including 182 species of quadrupeds) for the first section of Buffon's *Histoire naturelle générale et particulière,* which was published between 1794 and 1804. Daubenton was a strange man. He was unusual for his day in being a vegetarian, once saying, "It is to be presumed that man, while he lives in a natural state and a [temperate] climate, where the earth spontaneously produces every type of fruit, . . . feeds himself with these and does not eat animals." Less unusual for his time, he was a racist, believing in the superiority of Europeans. He once said that Europeans were "the model for beauty," and that African babies were not born with flat noses and thick lips but that "African parents, after judging their children to be lacking in beauty, would crush their noses and squeeze their lips so that they swell and thus believe they have beautified nature while disfiguring it." He is also credited with introducing Merino sheep into France. He is commemorated in the common names of two birds, Daubenton's (or Yellow-knobbed) Curassow *Crax daubentoni* and Daubenton's Parakeet *Psittacula eques* (extinct). The Aye-aye is a nocturnal lemur endemic to Madagascar. Daubenton's bat is widespread across Eurasia, from Ireland and Portugal eastward to Kamchatka, Japan, and China. The free-tailed bat is known from Senegal, Ivory Coast, Central African Republic, and northern DRC (Zaire).

David

David's Echymipera *Echymipera davidi* **Flannery,** 1990 [Alt. Kiriwina Bandicoot, Trobriand Bandicoot]

David Flannery (b. 1984) was born in 1984, and his father named the bandicoot after him with these words: "For my son, David, whose birth, and my ejection from the neonatal ward due to termination of visiting hours, allowed me to wander into the mammal collections of the Australian Museum. Here I first recognised the holotype of *E. davidi* as a different kind of bandicoot." The bandicoot is found on Kiriwina Island (Trobriand Islands), Papua New Guinea.

Davies

Davies' Big-eared Bat *Glyphonycteris daviesi* **Hill,** 1964 [Alt. Bartica Bat; Syn. *Barticonycteris daviesi*]

J. N. Davies was a member of the University College Bangor expedition to British Guiana (now Guyana) and collected the type specimen of the bat in 1963. The bat's range is now known to extend from Costa Rica south to Amazonian Brazil, and on the island of Trinidad. It was formerly placed in its own genus, *Barticonycteris,* but this is no longer generally recognized.

Davis, D. D.

Davis' Maroon Langur *Presbytis rubicunda chrysea* Davis, 1962

Dr. D. Dwight Davis (1908–1965) was Curator of Vertebrate Anatomy at the Chicago Natural History Museum. In 1964 he wrote a monograph, *The Giant Panda,* that showed that the Giant Panda is a bear. This was controversial at the time, but today it is almost universally accepted that Davis was right. He was co-author, with Karl Patterson Schmidt, of *Field Book of Snakes of the United States and Canada* (1941). The langur is found is a small area of eastern Sabah, Malaysia.

Davis, W. B.

Davis' Long-tongued Bat *Glossophaga alticola* Davis, 1944 [Syn. *G. leachii* Gray, 1844]
Davis' Round-eared Bat *Tonatia evotis* Davis and Carter, 1978
Davis' Pocket Gopher *Geomys personatus davisi* Williams and **Genoways,** 1981

Professor Dr. William B. Davis (1902–1995) was Head of the Department, and latterly Professor Emeritus, of the Department of Wildlife and Fisheries Sciences at Texas A&M University. He studied the mammals of Texas for over 50 years and is regarded as the father of Texan mammalogy. He took his doctorate in zoology at the University of California, Berkeley. He wrote the early editions of *The Mammals of Texas.* In 1938 he established the Texas Cooperative Wildlife Collection and was founder of the Department of Wildlife Management at Texas A&M University. He was President of the American Society of Mammalogists from 1955 to 1957. His interest in bats is evidenced by the papers he wrote in the early 1950s on the subject of banding bats. The long-tongued bat was named as a new taxon in 1944 but is now believed to be the same species as one named by Gray a hundred years earlier. It is found from southern Mexico to Costa Rica. The round-eared bat occurs in southeast Mexico, Belize, Guatemala, and Honduras. The gopher has a small range in the extreme south of Texas (Zapata County).

Davy

Davy's Naked-backed Bat *Pteronotus davyi* **Gray,** 1838

Dr. John Davy F.R.S. (1790–1868) was a physician who was "well known for his physiological papers." He was the brother of Sir Humphry Davy, who, as all British schoolchildren know, invented the miner's safety lamp. He studied medicine at Edinburgh, then was in the Ceylon Medical Service from 1816 to 1820 and accompanied the Governor of Ceylon (now Sri Lanka), Sir Robert Brownrigg, on his tour of the Central and Uva provinces, visiting Kataragama on 5 April 1819. He later joined the British army as a medical officer and saw service in Malta from 1828 to 1835 at the General Military Hospital at Valletta (during his stay there he treated Sir Walter Scott). He published his impressions as *An Account of the Interior of Ceylon and of Its Inhabitants with Travels in the Island* in 1821 and also wrote on the natural history, geology, agriculture, climate, and medical organization of Malta in 1842. In 1854 he published *The West Indies before and since Slave Emancipation Comprising the Windward and Leeward Islands Military Command Founded on Notes and Observations Collected during a Three Years Residence.* The type specimen of the bat was collected in Trinidad, but it is also found from Mexico south to northwest Peru and northern Venezuela, and on the southern islands of the Lesser Antilles.

Dawson

Dawson's Caribou *Rangifer tarandus dawsoni* Seton-Thompson, 1900 [Alt. Queen Charlotte Caribou]

Dr. George Mercer Dawson (1849–1901) was a Canadian geologist, botanist, and ethnographer. He fractured his spine as an infant and lived the rest of his life with a deformity but did not allow it to hamper him in any way. He was educated at McGill University and at the Royal School of Mines in London, from where he graduated in 1872. He returned to Canada and in 1873 was appointed Geologist and Botanist of the British North American Boundary Commission. In 1875 he started working for the Canadian Geological Survey, which meant that he explored large areas of northwestern North America. In 1878 he visited the Queen Charlotte Islands, off northern British Columbia, and it was only after he had examined a skull and a hide of the local deer that he realized that what he had taken for elk were in fact caribou. At the time of his death from acute bronchitis, he was still the Director of the Geological Survey of Canada as well as an editor of the *American Anthropologist*. The deer, which was originally described by Seton-Thompson as a full species *(Rangifer dawsoni)*, became extinct in the 20th century, possibly as late as 1935, through the introduction to the area of firearms for sport as well as hunting for food. Fans of westerns may perhaps know the name of Dawson City in the Yukon Territory; it is named after him.

Day, F.

Day's Shrew *Suncus dayi* **Dobson,** 1888

Dr. Francis Day (1829–1889) was a prominent ichthyologist and zoologist and a Fellow of both the Zoological Society and the Linnean Society. He qualified in medicine in 1851, joining the Madras Medical Service the following year and serving in the second Burmese war. He became Inspector General of Fisheries in India and Assistant Surgeon General, from which posts he retired in 1876. In 1883 Day sold a large collection of fish specimens to the Australian Museum for £200. Normally such an important collection would have gone to the British Museum, but Day couldn't abide Dr. Albert Günther, who was Keeper of Zoology there. The shrew comes from southern India.

Day, L. G.

Day's Grass Mouse *Akodon dayi* **Osgood,** 1916

Colonel Lee Garnett Day (fl. 1890–1960) was a New York businessman. He was also interested in zoology and was involved in the financing of the Collins-Day expedition to South America in 1915 (the mammals collected on this expedition were described by Osgood in a paper published in 1916, "Mammals of the Collins-Day South American Expedition") and the Mount Roraima expedition in 1927. He graduated from Yale in 1911. In 1914 he spent five months in Brazil exploring and hunting. He is also remembered in the scientific name of a bird, the Great Elaenia *Elaenia dayi*. The grass mouse is endemic to Bolivia.

De Balsac

De Balsac's Mouse *Heimyscus fumosus* **Brosset, Dubost,** and Heim de Balsac, 1965 [Alt. African Smoky Mouse]

Henri Heim de Balsac (1899–1979) was a French zoologist, mammalogist, and ornithologist. He was a member of the French Academy of Sciences and published widely on a large number of subjects, particularly the fauna of West Africa. He is commemorated in the name of the genus *Heimyscus*, as well as in the mouse's common name. This rodent is found in Gabon and southern Cameroon.

De Beaux

De Beaux's Grivet *Chlorocebus aethiops zavattarii* De Beaux, 1943

Professor Oscar de Beaux (1879–1955) was an Italian mammalogist and conservationist. He was a scientific assistant at the Carl Hagenbeck Zoo in Hamburg from 1911 to 1913. He was Professor of Zoology at Genoa University and then worked at the Civic Museum of Natural History in Genoa (Director from 1934 to 1947). In 1930 he published *Biological Ethics: An Attempt to Arouse a Naturalistic Conscience.* The monkey is found in Ethiopia. It is not always regarded as a valid subspecies.

De Brazza

De Brazza's Monkey *Cercopithecus neglectus* **Schlegel,** 1876 [Alt. Schlegel's Monkey]

Jacques (or Giacomo) C. Savorgnan de Brazza (1859–1888) was the younger brother of Count Pierre Paul François Camille Savorgnan de Brazza (1852–1905), the distinguished French explorer. They were of Italian descent. Jaques de Brazza accompanied his brother on his explorations of the Congo, and while the elder brother became Governor of the French Congo, Jacques was the Director of the West African Mission and was responsible for the collection of many ethnographical and natural history specimens. This collection was sent to Paris in about 1885. The monkey is found in central Africa from southern Cameroon and Gabon east to Uganda, western Kenya, and southwest Ethiopia. The species was long known as *Cercopithecus brazzae* (Milne-Edwards, 1886) until it was realized that this represented the same taxon that Schlegel had named 10 years earlier.

Decken

Decken's Horseshoe Bat *Rhinolophus deckenii* **Peters,** 1867
Decken's Sifaka *Propithecus deckenii* Peters, 1870

Baron Carl Claus von der Decken (1833–1865) was a German explorer in East Africa and Madagascar. He was the first European to try to climb Mount Kilimanjaro. Decken explored the region of Lake Nyasa on his first expedition in 1860. In 1861, together with a geologist, he visited the Kilimanjaro massif. In 1862 he ascended Kilimanjaro to 4,200 m (13,780 feet), seeing its permanent snowcap. He also established its height as about 6,100 m (20,000 feet) and mapped the area. His 1863 expedition took him to Madagascar, the Comoro Islands, and the Mascarene Islands. He collected in Somalia in 1865 and sent a considerable quantity of specimens to the Museum in Hamburg. He sailed the river Giuba in Somalia where his ship, the *Welf,* foundered in the rapids above Bardera. There Somalis killed him and three other Europeans. In 1869, after his death, his letters were edited and published in book form under the title *Reisen in Ost-Afrika.* He is also remembered in the scientific name of an African catfish, *Chiloglanis deckenii,* and in the name of a bird, Von der Decken's Hornbill *Tockus deckeni.* The bat is found in Uganda, Kenya, and Tanzania, including the islands of Zanzibar and Pemba. The lemur is from western Madagascar.

De Graaff

De Graaff's Soft-furred Mouse *Praomys degraaffi* Van der Straeten and Kerbis Peterhans, 1999

Dr. Gerrit De Graaff (1931–1996) was a zoologist with the National Parks Board of South Africa. In 1978 he wrote, with A. J. Hall-Martin, "A Note on the Feasibility of Introducing Giraffe to the Kalahari Gemsbok National Park," and in 1981 he wrote *Rodents of Southern Africa.* The mouse occurs in montane forests of the Albertine Rift, in Uganda, Rwanda, and Burundi.

DeKay

De Kay's Shrew *Sorex dekayi* **Bachman,** 1837 [Alt. Northern Short-tailed Shrew; Syn. *Blarina brevicauda* Say, 1823]

James Ellsworth DeKay (1792–1851) was an American zoologist. He was born in Lisbon, but his parents took him to America when he was two years old. He studied at Yale from 1807 to 1812 but did not graduate. Later he studied medicine at the University of Edinburgh, qualifying in 1819. He returned to the USA and married, then traveled with his father-in-law to Turkey as a ship's physician, and later published *Sketches of Turkey in 1831 and 1832*. On returning to America he forsook medicine to study natural history and was involved with the Geological Survey of New York from 1835. He published *The Zoology of New York* in several volumes from 1842 to 1849. He also has a snake named after him, the Northern Brown Snake *Storeria dekayi*. *Sorex dekayi* is a junior synonym of *Blarina brevicauda*—that is, it was first thought to be a new species but had already been described under another name. The shrew is found in the northeast and north-central USA and adjacent southern Canada.

Dekeyser

Dekeyser's Nectar Bat *Lonchophylla dekeyseri* **Taddei** et al., 1983

Dr. Pierre Louis Dekeyser (1914–1984) was a zoologist and ethnologist. In the 1950s he worked at the French Institute of Black Africa in Dakar (Senegal). He published a number of papers on both of his subjects, including "Les mammifères de l'Afrique noire française" in 1955 and, jointly with A. Villiers, "Contribution à l'étude du peuplement de la Mauritanie: Notations écologiques et biogéographiques sur la faune de l'Adrar" in 1956. He also published several articles on the avifauna of Brazil and seems to have worked in São Paulo in the early 1980s, as did Taddei who was a Brazilian bat expert. The bat is endemic to eastern Brazil.

Delacour

Delacour's Marmoset-Rat *Hapalomys delacouri* **Thomas,** 1927
Indochinese Leopard *Panthera pardus delacouri* **Pocock,** 1930
Delacour's Leaf Monkey *Trachypithecus delacouri* **Osgood,** 1932 [Alt. Delacour's Langur]

Dr. Jean Theodore Delacour (1890–1985) was a French-American ornithologist renowned not only for discovering but also for keeping and breeding some of the rarest birds in the world. He was born in Paris and died in Los Angeles. In France, in the years 1919 and 1920, he created the zoological gardens at Clères and donated them to the French Natural History Museum, Paris, in 1967. He wrote *Birds of the Philippines* (with Ernst Mayr) in 1946 and *Birds of Malaysia* in 1947 as well as many other ornithological books, including *The Living Air: The Memoirs of an Ornithologist*. He undertook a number of expeditions to Indochina, in particular Vietnam, and collected specimens there, especially pheasants. He is commemorated in the common name of a pheasant, Delacour's Crested Fireback *Lophura ignita macartneyi*, and in the scientific names of other taxa such as the catfish *Oreoglanis delacouri*. The marmoset-rat is found in Laos, Vietnam, and the island of Hainan (China). The leopard subspecies occurs throughout Indochina, Thailand, and the Malay Peninsula. The monkey is a critically endangered species endemic to Vietnam.

Delany

Delany's Swamp Mouse *Delanymys brooksi* **Hayman,** 1962

Professor Michael James Delany (b. 1928) is a British zoologist, ecologist, and entomologist who has spent over 40 years studying small animals on four continents. After a spell working on a fish-killing microorganism in the Gulf of Mexico, he returned to the UK and switched to the ecology of small mammals. His very varied experiences include one that might easily have proved fatal: he once mistook Idi Amin for a waiter. He has been Professor at Makerere (Uganda), Bradford, and Sultan Qaboos Oman universities. He has written much and published widely, for instance in 1975 *The Rodents of Uganda* and in 1989 *The Zoogeography of the Mammal Fauna of Southern Arabia.* The mouse occurs in southwest Uganda, Rwanda, and adjacent DRC (Zaire).

De la Torre

De la Torre's Yellow-shouldered Bat *Sturnira magna* de la Torre, 1966 [Alt. Greater Yellow-shouldered Bat]

Dr. Luis de la Torre is a zoologist and anatomist and Curator at the Field Museum in Chicago. He is also associated with the Departments of Histology and Physiology, Colleges of Dentistry and of Medicine of the University of Illinois in Chicago. He has made a number of collecting expeditions to Central and South America and has published a number of papers on, mainly, smaller mammals (e.g. in 1961, with Donald F. Hoffmeister as co-author, "Geographic Variation in the Mouse *Peromyscus difficilis*"). The bat is found in Colombia, Ecuador, Peru, and Bolivia.

Demidoff

Demidoff's Galago *Galago demidoff* Fischer, 1806 [Alt. Prince Demidoff's Bushbaby]

Pavel Grigorevich (Paul) Demidoff (1738–1821) was a Russian nobleman, traveler, and scientist. With his brother Grigorij he was a student of Linnaeus in Uppsala between 1760 and 1761. He created a natural history museum in Moscow, which housed his own collections, including his large malacological collection, now housed in the Zoological Museum of Moscow University. Unfortunately many of Demidoff's specimens were lost in a fire, in 1812. The bushbaby is found across most of west and central Africa, from Senegal east to Uganda and western Tanzania, and on the island of Bioko.

Dent

Dent's Horseshoe Bat *Rhinolophus denti* **Thomas,** 1904
Dent's Vlei Rat *Otomys denti* Thomas, 1906
Dent's Monkey *Cercopithecus denti* Thomas, 1907
Dent's Shrew *Crocidura denti* **Dollman,** 1915

Captain R. E. Dent (b. 1882) was a British explorer who collected in tropical Africa in both 1901 and 1906, in which year he was a member of the Ruwenzori expedition. He was a game warden in Africa in 1931 and was one of the very few people to have seen the Marozi or Spotted Lion. This mysterious big cat has been given a scientific name, *Panthera leo maculatus,* but most zoologists dispute its existence as a valid subspecies. A bird, Dent's Short-tailed Warbler *Sylvietta denti,* is also named after him. The bat is found both in southern Africa (South Africa, Namibia, Botswana, Zimbabwe, Mozambique) and in West Africa (Guinea, Ivory Coast, and Ghana). The vlei rat has a sporadic distribution from Mount Ruwenzori, Uganda, through the Virunga volcanoes to the Nyika Plateau of northern Malawi, and to the Usambara and Uluguru mountains of Tanzania. The monkey is found in eastern DRC (Zaire). The shrew was first found in the Ituri Forest of eastern DRC (Zaire) but has also been recorded from Gabon and Cameroon.

Deppe

Deppe's Squirrel *Sciurus deppei* **Peters,** 1863

Ferdinand Deppe (1794–1861) was a German horticulturalist, collector, and artist. He first arrived in Mexico in December 1824 with the

Count von Sack, who had conceived the idea of a scientific expedition to that country. However, the Count seems to have been an irresolute "expedition leader" who soon returned to Germany, while Deppe stayed on in Mexico until January 1827. After a fairly brief visit back to Germany, Deppe returned to Mexico in the company of the botanist Wilhelm Schiede and stayed until 1836. Many of the specimens Deppe collected went to the Zoological Museum of Berlin. He also has a fish, *Herichthys deppii,* named after him, and at least one plant, *Euphorbia deppeana,* which he collected during his travels in Mexico. The squirrel is found from east-central Mexico to northern Costa Rica.

Derby

Banded Palm Civet *Hemigalus derbyanus* **Gray,** 1837
Derby's Pale-eared Woolly Opossum *Caluromys derbianus* **Waterhouse,** 1841 [Alt. Central American Woolly Opossum]

The Hon. Edward Smith Stanley, 13th Earl of Darby (1775–1851), was an avid zoologist and collector. He founded the Derby Museum, which was formed from his specimen collection and from material derived from the live animals he kept at Knowsley Park, near Liverpool. He was President of the Linnean Society and of the Zoological Society of London for over 20 years. He was first elected Member of Parliament for Preston, in Lancashire, in 1796, when he was just 21, and served until 1812. In 1826 he met Audubon, who gave him a few Passenger Pigeons *Ectopistes migratorius.* These started breeding in 1832, but they quickly became a nuisance-sized flock of 70, and Stanley allowed them to fly free. If only he had built up the flock he might have saved the species from extinction. He employed Edward Lear, between 1832 and 1837, to draw the plates for the *Knowsley Menagerie,* which was published in 1846. (Lear's *Book of Nonsense,* published in the same year, was written for Stanley's grandchildren.) Stanley purchased Bartram's original work, some of which is still in the present Earl of Derby's library. His father gave his name to the world-famous horse race, and his son, the 14th Earl, was three times Prime Minister of Great Britain. As both the Earl of Derby and as Stanley he is also commemorated in the names of various birds, such as the Derbyan Parakeet *Psittacula derbiana* and the Stanley Rosella *Platycercus icterotis.* The civet is found in the Malay Peninsula, Borneo, Sumatra, and the Mentawai Islands. The opossum ranges from southern Mexico to northern Ecuador. See also **Lord Derby.**

Deroo

Deroo's Mouse *Praomys derooi* Van der Straeten and **Verheyen,** 1978 [Syn. *Myomys derooi*]

Antoon Emeric Marcel De Roo (1936–1971) was a Belgian naturalist. The original description reads, "One of the main results of this study is the description of a new species of Myomys. We are very pleased to name it after our deeply regretted *[sic]* colleague ANTOON DE ROO who died some years ago and who was one of the most active members of our fieldteam." There is also a frog, described in 1972 by Hulselmans, with the scientific name *Conraua derooi,* and we believe it honors the same person. De Roo was co-author, with Verheyen and De Vree, of *Contribution à l'étude des ciroptères de la République du Togo* (1969). The specimen of the mouse came from Togo, and the species has also been found in Ghana, Benin, and western Nigeria.

Desmarest

Desmarest's Fig-eating Bat *Stenoderma rufum* Desmarest, 1820 [Alt. Red Fig-eating Bat]
Desmarest's Hutia *Capromys pilorides* Say, 1822
Antillean Giant Rice Rat *Megalomys desmarestii* Fischer, 1829 extinct
Desmarest's Spiny Pocket Mouse *Heteromys desmarestianus* **Gray,** 1868 [Alt. Forest Spiny Pocket Mouse]

Professor Anselme Gaëtan Desmarest (1784–1838) was a French paleontologist. In 1805 he wrote *Histoire naturelle des tangaras, des manakins et des todiers.* In 1825 he published *Considérations générales sur la classe des crustacés—et description des espèces de ces animaux, qui vivent dans la mer, sur les côtes, ou dans les eaux douces de la France.* He was also the author, with André-Marie Dumeril, of the *Dictionnaire des sciences naturelles,* published in Paris by F. G. Levrault between 1816 and 1830. He is also commemorated in the name of a bird, Desmarest's Fig Parrot *Psittaculirostris desmarestii.* The bat comes from Puerto Rico and the U.S. Virgin Islands. The hutia is endemic to Cuba. The rice rat used to be found on Martinique in the Lesser Antilles but is now extinct; the devastating volcanic eruption of Mount Pelée on 8 May 1902 may have finished the species off. The pocket mouse is found from southern Mexico to northwest Colombia.

De Vis

De Vis' Bare-backed Fruit Bat *Dobsonia pannietensis* De Vis, 1905 [Alt. Panniet Naked-backed Fruit Bat]
De Vis' Woolly Rat *Mallomys aroaensis* De Vis, 1907

Charles Walter De Vis (1829–1915) was a British-born cleric who gave up the church to concentrate on ornithology. Between 1882 and 1905 he served as the first Director of the Queensland Museum. Before joining the museum De Vis published many popular articles under the pen name of Thickthorn. He was a founding member of the Royal Society of Queensland in 1884 and its President from 1888 until 1889. He was also a founding member of the Australasian Ornithologists' Union in 1901 and its first Vice President. He described 551 new fossil and living species, including the two mammals listed above. He is also commemorated in the name of De Vis' Banded Snake *Denisonia devisi.* The bat is found in the Louisiade Archipelago, the D'Entrecasteaux Islands, and the Trobriand Islands, off Papua New Guinea. The rat comes from the highlands of the Papua New Guinea mainland.

De Vivo

De Vivo's Disk-winged Bat *Thyroptera devivoi* Gregorin et al., 2006
De Vivo's Rice Rat *Cerradomys vivoi* Percequillo, Hingst-Zaher, and Bonvicino, 2008

Professor Dr. Mario de Vivo is a Brazilian zoologist working for the University of São Paulo as Curator of Mammals of the Zoological Museum. The original etymology for the disk-winged bat reads, "*Thyroptera devivoi* is named in honor of Dr. Mario de Vivo, who has contributed greatly to the fostering of mammalian taxonomy among students and who has been responsible for a considerable increase in the understanding of both mammal diversity and systematics in Brazil." The bat is found in northeast Brazil and adjacent southern Guyana. The rice rat is known from the Brazilian states of Minas Gerais, Bahia, and Sergipe.

De Winton

De Winton's Shrew *Chodsigoa hypsibius* de Winton, 1899
De Winton's Tree Squirrel *Funisciurus substriatus* de Winton, 1899 [Alt. Kintampo Rope Squirrel]
De Winton's Long-eared Bat *Laephotis wintoni* **Thomas,** 1901
De Winton's Golden-Mole *Cryptochloris wintoni* Broom, 1907

William Edward de Winton (1856–1922) was at one time Superintendent of the Zoological Gardens in London (London Zoo). He wrote a number of papers such as "On the Moulting of the King Penguin" (*Proceedings of the Zoological Society,* 1900) and "On the Hares of Western Europe and North Africa" (*Annals and Magazine of Natural History,* 1898). Most of his scientific papers were commentaries on collections made by others from all over the world. He completed *Zoology of Egypt—Mammalia,* after Dr. John Anderson's death in 1900 (it was published in 1902). The shrew is from central and southwest China. The squirrel ranges from Ivory Coast to Nigeria. The bat is an apparently rare species, with scattered records from Ethiopia, Kenya, and Tanzania and also from South Africa. The golden-mole has a very small distribution in Namaqualand, western South Africa.

Dian

Dian's Tarsier *Tarsius dianae* Niemitz, Nietsch, Warter, and Rumpler, 1991

Dr. Dian Fossey (1932–1985) was most famed for her work with gorillas, encapsulated in her 1981 book *Gorillas in the Mist,* which was later made into a feature film. She studied occupational therapy at San Jose State University. However, in 1963 she took a trip to Africa, where she met Dr. Louis Leakey and began to study Mountain Gorillas. She started observing them in DRC (Zaire) in 1976 before moving to Rwanda, where she set up the Karisoke Research Center. She lived there among the gorillas for 18 years, eventually becoming totally accepted by them. When one of her favorite subjects (whom she called Digit) was killed by poachers, she started a campaign to save the species. She dedicated the rest of her life to this cause, establishing the Digit Fund. She obtained her Ph.D. from Cambridge in 1980, after which she took a job at Cornell University but soon returned to work with gorillas back in Rwanda. In December of 1985 she was found murdered in her cabin there. The crime was blamed on "poachers," but according to at least one author "she was killed because she knew too much about the illegal trafficking by Rwanda's ruling clique." Her work continues. The describers' citation reads, "Our two reasons for proposing the specific name are: firstly, we wish to bestow the name of Diana, the goddess of hunting, on these fierce little creatures. Secondly, we wish to honor the late primatologist and conservationist Dian Fossey." The tarsier, which is also named after the goddess Diana (see below) is endemic to Sulawesi, Indonesia. The taxonomy of tarsiers is still not fully resolved, and some think that *Tarsius dianae* is actually a junior synonym of *T. dentatus* (Miller and Hollister, 1921).

Diana

Large-eared Sheath-tailed Bat *Emballonura dianae* **Hill,** 1956

Mrs. Diana Mary Bradley (b. 1932) was the collector of the holotype of the bat. She worked for a short time for the bird-banding committee of the British Trust for Ornithology at the Natural History Museum in London. She was Treasurer of the British Ornithologists' Union from 1978 to 1990. In 1953 she, with her husband Dr. J. D. Bradley of the Entomological Department of the Natural History Museum, visited Rennell Island in the Solomon Islands. In 1956 she co-authored, with Torben Wolff, the birds section of *Natural History of Rennell*

Island, British Solomon Islands. The bat was first found on Rennell Island and was later found on other islands in the Solomons and in Papua New Guinea.

Diana

Diana Monkey *Cercopithecus diana* **Linnaeus,** 1758
Dian's Tarsier *Tarsius dianae* Niemitz, Nietsch, Warter, and Rumpler, 1991

In Roman mythology, Diana was goddess of the moon and of hunting. She was often portrayed as a huntress with a stag or a hunting dog. The inspiration for naming the monkey after Diana is thought to lie in the white strip across the monkey's forehead, believed to resemble the crescent moon and/or the bow carried by the goddess. The species is found from Sierra Leone to Ghana (although populations to the east of the Sassandra River, Ivory Coast, are sometimes classed as a separate species, *Cercopithecus roloway*). The tarsier (also named after Dian Fossey; see above) is from Sulawesi.

Diard

Diard's Cat *Neofelis diardi* **Cuvier,** 1823
[Alt. Bornean Clouded Leopard]

Pierre-Medard Diard (1794–1863) was a French explorer. He left Europe in 1817 for India to collect for the Natural History Museum in Paris. In 1818 he was employed as a naturalist by Sir Thomas Stamford Raffles. In 1821 the Honourable East India Company confiscated his entire collection. He then went to Indochina, visiting Annam and becoming one of the first Europeans to visit Angkor Wat. He next visited Malaya, then joined the Dutch East India Company in Batavia (now Jakarta) and collected in the East Indies between 1827 and 1848. Temminck reports that specimens sent from Borneo by Diard arrived at Leiden in 1828. Diard created the Buitenzorg Botanical Gardens in Java, where he was largely responsible for the introduction of sugar cane and the breeding of the silkworm moth. Many other taxa are named after him, including birds such as Diard's Trogon *Harpactes diardii* and the Siamese Fireback Pheasant *Lophura diardi*, the spider *Hyllus diardi*, and the snake *Typhlops diardi*—all from his former haunts in Southeast Asia. The vernacular name Diard's Cat is regarded as archaic, with the taxon long being regarded as a subspecies of the Clouded Leopard *Neofelis nebulosa*. However, in 2006 an article in the journal *Current Biology* argued that differences between the typical (mainland) form of Clouded Leopard and the Indonesian form were great enough for *Neofelis diardi* to be re-recognized as a full species. It is found on Sumatra and Borneo.

Diaz

Volcano Rabbit *Romerolagus diazi* Ferrari-Pérez, 1893

Augustin Diaz (1829–1893) was Director of the Mexican Geographical and Exploring Commission; Professor Fernando Ferrari-Perez was the Chief of its Natural History Section, its leader in the field when it got under way in 1879, and the author of its findings in 1886. Diaz is also remembered in the scientific name of the Mexican Duck *Anas diazi*. This endangered species of rabbit is found around the volcanoes of central Mexico.

Dice

Dice's Rabbit *Sylvilagus dicei* **Harris,** 1932
[Alt. Dice's Cottontail]

Professor Dr. Lee Raymond Dice (1887–1977) was an American ecologist. He was the Director of the Laboratory of Vertebrate Biology at the University of Michigan and later Emeritus Professor of Human Genetics and of Internal Medicine there. He also kept a mouse colony for study there from 1925 until 1975. He wrote numerous articles, including "Interior Alaska in 1911 and 1912: Observations by a Naturalist" and "Life Zones and Mammalian Distribution" (1923). He was President of the American Society of Mammalogists from 1947 to 1949. The rabbit is found in the highlands of southern Costa Rica and northern Panama.

Dickey

San Nicolas Island Fox *Urocyon littoralis dickeyi* Grinnell and Linsdale, 1930
Dickey's Deer Mouse *Peromyscus dickeyi* **Burt,** 1932

Donald Ryder Dickey (1887–1932) was an American ornithologist, collector, and photographer. He was born in Iowa but lived most of his life in California. Between 1908 and 1923 he took 7,000 black-and-white photographs and collected 50,000 specimens of birds and mammals. These form the Donald Ryder Dickey Collection, which his widow presented to University College, Los Angeles. His publications were many, the most notable on birds being *The Birds of El Salvador* (1938), written jointly with A. J. van Rossem. He is remembered in the common names of two birds: Dickey's Egret *Egretta rufescens dickeyi* and Dickey's Jay *Cyanocorax dickeyi*. The fox is one of six races of the Island Fox, each endemic to one of the Channel Islands of California. The mouse is found only on Isla Tortuga, Gulf of California, Mexico.

Diehl

Cross River Gorilla *Gorilla gorilla diehli* **Matschie,** 1904

The German zoologist Paul Matschie originally described this gorilla as a full species, based on skulls sent to him by Mr. Diehl, an employee of the German Northwestern Cameroon Company. However, as Matschie tended to name every individual specimen that came his way as a new species, his taxonomy was long ignored. Recently, however, this form of gorilla—found on the Nigeria-Cameroon border—has been reinstated as a valid subspecies of the Western Lowland Gorilla. It may be the most endangered form of primate on Earth, with only about 150 surviving individuals. Little more is known of Mr. Diehl, but a man of this name acted as temporary Governor of Cameroon for a few months in 1900; it may well be the same person.

Dieter

Dieter's Myotis *Myotis dieteri* Happold, 2005

Dr. Dieter Kock is a German research zoologist based at the Senckenberg Museum in Frankfurt. The original etymology reads, "I have much pleasure in naming this bat in honour of Dr. Dieter Kock of the Senckenberg Museum, Frankfurt, Germany, in recognition of the great contribution he has made to the knowledge of African mammals, in recognition of the kindness and generosity with which he has helped so many others with their studies, and in gratitude for our long friendship." The only known specimen of this bat was collected sometime in the 1960s by Jean-Paul Adam, at Loudima in the Congo Republic. It has only recently been recognized as a distinct species.

Dieterlen

Johnston's Pygmy Shrew *Sylvisorex johnstoni dieterleni* **Hutterer,** 1986
Dieterlen's Brush-furred Rat *Lophuromys dieterleni* **Verheyen** and Hulselmans and Colyn and Hutterer, 1997

Dr. Fritz Dieterlen (b. 1929) is a German zoologist. From 1963 until 1967 he headed the Mammal Section of the IRSAC (Institute for Scientific Research in Central Africa) center at Lwiro-Bukavu, East Congo, working on ecology and taxonomy of small mammals occurring in savannas and rainforests of central

Africa. From 1967 to 1969 he was Curator of Mammals at the Museum Alexander Koenig, Bonn. From 1969 until 1994 he was head of the Mammals Section at Staatliches Museum für Naturkunde, Stuttgart. He is still a volunteer there since retiring and continues to research the taxonomy and ecology of African rodents, as well as mammals of Baden-Württemberg. The rat is endemic to Mount Oku, Cameroon. The shrew named after Dieterlen is the eastern subspecies of Johnston's Pygmy Shrew; it comes from northwest Tanzania, Rwanda, and Uganda.

Dingan

Dingaan's Yellow Bat *Scotophilus dinganii* **A. Smith,** 1833 [Alt. African Yellow Bat]

This bat is almost certainly named after Dingaan (or Dingan) (ca. 1795–1840), who was King of the Zulus. He killed his brother Chaka, the previous King, in 1828, succeeded him, and ruled for 12 years. History records that he treated Cape Dutch settlers with great treachery, massacring a large number of men, women, and children near the Drakensberg Mountains. This caused the Boers and the British to make common cause against him, and the Zulus were comprehensively defeated at the Battle of Blood River in 1838. His half-brother, Panda, tried to dethrone him and eventually, with the help of Boer settlers, killed him. The bat was first found in South Africa and named when Dingaan was at the height of his power. The species is widespread in sub-Saharan Africa.

Dobson

Dobson's Long-tongued Fruit Bat *Eonycteris spelaea* Dobson, 1871 [Alt. Lesser Dawn Bat, Cave Fruit Bat]

Dobson's Horseshoe Bat *Rhinolophus yunanensis* Dobson, 1872

Dobson's Large-eared Bat *Micronycteris brachyotis* Dobson, 1878 [Alt. Yellow-throated Big-eared Bat]

Dobson's Painted Bat *Kerivoula africana* Dobson, 1878 [Alt. Tanzanian Woolly Bat]

Dobson's Shrew-Tenrec *Microgale dobsoni* **Thomas,** 1884

Dobson's Epauletted Bat *Epomops dobsoni* **Bocage,** 1889

Dobson's Fruit Bat *Dobsonia chapmani* **Rabor,** 1952 [Alt. Negros Bare-backed Fruit Bat]

The fruit bat genus *Dobsonia* **Palmer,** 1898 [ca. 14 species]

George Edward Dobson F.R.S. (1848–1895) was an Irish zoologist. He was an army surgeon, having joined in 1868, serving in India from that year and in the Andaman Islands from 1872. He retired in 1888 at the rank of Surgeon Major. Among other things he was an expert on the Chiroptera and Insectivora. He published several articles and papers, including two articles on the Andamanese in 1875 and 1877 in the *Journal of the Royal Anthropological Institute.* An early aficionado of photography, he also took pictures of the local people. In 1876 he published a monograph on Asian bats. He also published a collection of medical hints for travelers. In 1878 he was appointed Curator of the Royal Victoria Hospital's museum. The long-tongued fruit bat is widespread in Southeast Asia, being found from India east to the Philippines, and south through the Malay Peninsula to Indonesia (east to Sulawesi and Timor). The horseshoe bat is known from northeast India, Yunnan

(China), northern Myanmar, and Thailand. The large-eared bat is found from southern Mexico to northern Brazil. The painted bat is an endangered species limited to coastal forest in Tanzania. The shrew-tenrec is a Madagascan endemic. The epauletted bat is found in Angola, southern DRC (Zaire), Zambia, Malawi, northern Botswana, Rwanda, and Tanzania. The *Dobsonia* fruit bat is endemic to the Philippines.

Doggett

Doggett's Blue Monkey *Cercopithecus doggetti* **Pocock,** 1907 [Alt. Silver Monkey; Syn. *C. mitis doggetti*]

Walter Grimwood Doggett (ca. 1876–1905) was a collector and naturalist in East Africa from 1899 to 1903. In 1901 he was working for Sir Harry H. Johnston, Special Commissioner for Uganda, who records being accompanied by him on a mission to collect the skin of an okapi and on another occasion the pair of them shooting a "five-horned giraffe." Ogilvie-Grant published, in 1905, *On the Birds Collected by the Late W. G. Doggett on the Anglo German Frontier of Uganda.* He was drowned in the Kagera River in Uganda. The monkey is found in southern Uganda, Rwanda, northwest Tanzania, and eastern DRC (Zaire).

Dogramaci

Dogramaci's Vole *Microtus dogramacii* Kefelioglu and Krystufek 1999

Professor Dr. Salih Dogramaci (d. 1998) of Ondokuz Mayýs University in Samsun was a Turkish mammalogist. The vole was named "in honour of his contribution to our knowledge of Turkish mammals." He published a book in 1969 on Turkish fauna. The vole is found only in central Anatolia, Turkey.

Dollman

Dollman's Tree Mouse *Prionomys batesi* Dollman, 1910

Dollman's Vlei Rat *Otomys dollmani* **Heller,** 1912

Dollman's Rock Rat *Aethomys chrysophilus dollmani* Hatt, 1934 [Syn. *A. nyikae* Thomas, 1897]

Dollman's Mosaic-tailed Rat *Melomys dollmani* Rümmler, 1935

Dollman's Spiny Rat *Maxomys dollmani* **Ellerman,** 1941

Captain John Guy Dollman (1886–1942) was Curator of Mammals in the Zoology Department of the British Museum of Natural History. He was an innovator at creating displays of specimens that presented them in an interesting way. Perhaps this was a skill inherited from his father, who was a noted artist. Dollman even donated several of his father's paintings to the museum. He wrote a number of scientific papers, often with Rothschild, such as "Notes on Records of Tree Kangaroos in Queensland" (*Australian Zoology,* 1936). He edited and revised at least one book, Lydekker's *The Game Animals of India, Burma, Malaya, and Tibet* (1926). He described a number of mammals himself. Robert Hatt wrote of the rock rat, "The naming of this handsome animal for Captain Guy Dollman is but poor acknowledgment of my indebtedness to that gentleman." The tree mouse is known from Cameroon, Congo Republic, and the Central African Republic. The vlei rat is known only from Mount Gargues, central Kenya. The rock rat is from Katanga, southern DRC (Zaire); its taxonomic status is uncertain, as it may be synonymous with *Aethomys nyikae* of Malawi and northeast Zambia. The mosaic-tailed rat occurs in the highlands of Papua New Guinea. The spiny rat is endemic to montane forests of Sulawesi (Indonesia).

Dolores

Dolorous Grass Mouse *Akodon dolores* **Thomas,** 1916

The type specimen of this mouse was collected near the Villa Dolores in Córdoba Province, Argentina, and is not named after a person.

D'Orbigny

D'Orbigny's Round-eared Bat *Tonatia silvicola* D'Orbigny, 1836
D'Orbigny's Tuco-tuco *Ctenomys dorbignyi* **Contreras** and Contreras, 1984

Alcide Dessalines d'Orbigny (1802–1857) was a French traveler, collector, illustrator, and naturalist. He was the author of *Dictionnaire universel d'histoire naturelle.* His father, Charles-Marie Dessalines d'Orbigny (1770–1856), was a ship's surgeon. He and Alcide studied shells. Alcide went to the Academy of Science in Paris to pursue his methodical paintings and classification of natural history specimens. The Natural History Museum in Paris sent him to South America in July 1826. There the Spanish briefly imprisoned him, mistaking his compass and barometer for "instruments of espionage." After his release he lived for a year with the Guarani Indians, learning their language. He spent five years in Argentina and then traveled north along the Chilean and Peruvian coasts before moving into Bolivia. He returned to France in 1834 and donated thousands of specimens of animals, as well as plants, rocks, fossils, land surveys, and pre-Columbian pottery, to the Natural History Museum. His fossil collection led him to conclude that the Earth was composed of many geological layers, laid down over millions of years. This was the first time such an idea was put forward. He is remembered in the scientific names of several birds, such as the Grey-breasted Seedsnipe *Thinocorus orbignyianus* and the Andean Parakeet *Bolborhynchus orbygnesius.* The round-eared bat is found from Honduras south to Bolivia and northeast Argentina. The tuco-tuco comes only from northeast Argentina.

Dorcas

Dorcas Gazelle *Gazella dorcas* **Linnaeus,** 1758
Bontebok *Damaliscus dorcas* **Pallas,** 1766 [Syn. *D. pygargus* Pallas, 1767]

According to the Bible, Dorcas was a Christian woman who lived in Joppa and was very charitable toward the poor. The apostle Peter was persuaded to raise her from the dead (Acts 9.36–42). As Dorcas means "gazelle" in Greek, the name Dorcas Gazelle means "Gazelle gazelle," and the scientific name means that, too. The gazelle is found across North Africa. The Bontebok is a South African antelope that Pallas named twice.

Doria

False Serotine Bat *Hesperoptenus doriae* **Peters,** 1868
Borneo Roundleaf Bat *Hipposideros doriae* Peters, 1871
Doria's Tree Kangaroo *Dendrolagus dorianus* Ramsay, 1883
Red-bellied Marsupial-Shrew *Phascolosorex doriae* **Thomas,** 1886
Brazilian Big-eyed Bat *Chiroderma doriae* Thomas, 1891
Sumatran Mastiff Bat *Mormopterus doriae* **K. Andersen,** 1907

Marchese Giacomo Doria (1840–1913) was an Italian zoologist who collected in Persia (now Iran) with de Filippi between 1862 and 1863, and in Borneo with Beccari between 1865 and 1866. He was the founder and first Director of the Civic Museum of Natural History in Turin, serving in that position from 1867 until his death. Doria's Hawk *Megatriorchis doriae* is also named after him. The false serotine is a poorly known species recorded from Sarawak and the Malay Peninsula. The roundleaf bat is endemic to Borneo. The tree kangaroo and the marsupial-shrew are both from New Guinea. The big-eyed bat is endemic to southeast Brazil, and the mastiff bat to Sumatra.

Dormer

Dormer's Pipistrelle *Pipistrellus dormeri* **Dobson,** 1875

Major General the Hon. Sir James Charlemagne Dormer C.B. (1834–1893) was the Commander in Chief of the Madras army. Dobson served in the Indian Medical Service at the same time, reaching the rank of Surgeon Major. Dormer is noted for having been mauled by a tiger while hunting; we speculate that Dobson may even have treated his injuries. Dormer is also noteworthy as being nicknamed "Cesspool," because he hid in one to escape from rebels during the Indian Mutiny. The bat is found in India and Pakistan. The type specimen was "a specimen preserved in alcohol, which had been obtained in the Bellary Hills, Southern India, by the Hon. J. Dormer, and presented by him to the British Museum."

Dorothy

Dorothy's Slender Mouse-Opossum *Marmosops dorothea* **Thomas,** 1911 [or *M. ocellatus;* see below]

Thomas makes no mention in his original description of a particular Dorothy, so who, if anyone, he was referring to is lost in the mists of time. Thomas seems to have sometimes given small mammals a feminine-sounding scientific name for no obvious reason. The opossum is found in Bolivia. Its nomenclature is somewhat confused: *Marmosops ocellatus* was named by Tate in 1931 but was long regarded as a synonym of *M. dorothea.* It has now been suggested that *ocellatus* is a valid species, whereas *dorothea* itself should be relegated to a synonym of *M. noctivagus.* The vernacular name Dorothy's Slender Mouse-Opossum is now sometimes used for *M. ocellatus.*

Doucet

Doucet's Musk Shrew *Crocidura douceti* **Heim de Balsac,** 1958

Dr. Jean Doucet (dates not found) was a research scientist working for ORSTOM at Adjopodoumé in the Ivory Coast. In 1958 he was Assistant Director of the local ORSTOM center. (ORSTOM, a French organization now called the Institute of Research for Development, concentrates on scientific projects involving the interrelationship between man and the environment in tropical regions.) In 1972 Doucet was Professor of Parasitology at the Faculty of Medicine at the Pasteur Institute in Abidjan. He described a number of animals from the Ivory Coast, as did Heim de Balsac. He sent the type specimen, which had been collected at Adjopodoumé, to Heim de Balsac. The shrew is known from Guinea, Ivory Coast, and Nigeria.

Douglas, A. M. and M.

Yellow-lipped Bat *Vespadelus douglasorum* Kitchener, 1976

Julia Creek Dunnart *Sminthopsis douglasi* **Archer,** 1979

Athol M. Douglas (dates not found) and Marion Douglas (dates not found) collected bats extensively in Western Australia, mainly for the Western Australian Museum, Perth, where Athol worked in the 1950s and 1960s. The bat is named after both, but the dunnart is named solely after Athol. The bat was originally described under the name *Eptesicus douglasi,* but because the etymology makes it clear that *two* persons are being commemorated in the scientific name, the latter was corrected to the plural *douglasorum.* Athol Douglas wrote an article in 1986 entitled "Tigers in Western Australia?" concerning the putative survival of the Thylacine (Tasmanian "Tiger") on the Australian mainland. The bat is found in northern Western Australia. The dunnart inhabits a small range in northwest Queensland.

Douglas, D.

Douglas' Squirrel *Tamiasciurus douglasii* **Bachman,** 1839

David Douglas (1799–1834) was a Scottish botanist and traveler who collected in North America from 1823 to 1834 and in Hawaii in 1834. Suffering from severe problems with his eyes, he fell into a pit trap while in Hawaii and was gored to death by a feral bull that had been similarly caught. The Douglas Fir is named after him, and he also introduced the Sitka Spruce and the Lodgepole Pine to the UK, along with many other plant species. He is commemorated in the name of a bird, Douglas' Quail *Callipepla douglasii.* The squirrel is found from southwest British Columbia, Canada, to central California.

Drouhard

Drouhard's Shrew-Tenrec *Microgale drouhardi* **Grandidier,** 1934 [Alt. Striped Shrew-Tenrec]

Monsieur E. Drouhard (dates not found) collected the type specimen of the tenrec while he was an Inspector of Forests in Madagascar, a post he held from 1929. He is also remembered in the trinomial of the Western Vasa Parrot *Coracopsis vasa drouhardi.* The shew-tenrec occurs in the forests of north and east Madagascar.

Dubost

Dubost's Bristly Mouse *Neacomys dubosti* **Voss,** Lundi, and Simmons, 2001

Professor Gerard Dubost is a French zoologist and Director of the Laboratory for the Conservation of Animal Species at the Museum of Natural History in Paris. He was Director of the Paris Zoo from 1996 to 2001. He has written a great many scientific papers, such as "Le comportament de Cephalophus monticola Thunberg et C. dorsalis Gray et la place des céphalophes au sein des ruminants" (1983) and "Saisons de reproduction des petits ruminants dans le nord-est du Gabon, en fonction des variations des ressources alimentaires" (1992). The mouse is found in French Guiana, southeast Suriname, and northeast Brazil.

Dudu

Dudu's Brush-furred Rat *Lophuromys dudui* **Verheyen** et al., 2002

Professor Dr. Benjamin Dudu Akaibe Migumiru (b. 1949) is Head of the Laboratory of Ecology and Animal Resource Management in the Democratic Republic of Congo and Vice Dean of Science at the University of Kisangani. His early education was in schools and universities in the Congo, after which he studied at the University of Antwerp, where he earned his doctorate in 1991. In 1992, on his return from Belgium, he became Professor of Zoology and Ecology in the Faculty of Sciences at the University of Kisangani. He holds a number of other posts, such as being, since 1992, Director of the Ecology Laboratory and Secretary of the Faculty of Sciences at Kisangani. He has, with Van der Straeten, described a number of new rodent taxa. The rat is found in northeast DRC (Zaire).

Duke of Abruzzi

Duke of Abruzzi's Free-tailed Bat *Chaerephon aloysiisabaudiae* Festa, 1907

HRH Prince Luigi Amedeo of Savoy, Duke of Abruzzi (1873–1933), was an Italian explorer. He led a polar expedition in 1900, and in 1906 he led an expedition to explore and climb the mountains of the Ruwenzori Range in Uganda, otherwise known as the Mountains of the Moon. The major peaks had already been named after early explorers such as Stanley and Speke. The Duke's expedition made the first detailed maps of the area and climbed a number of the peaks for the first time, including Mount Stanley. They named most of the peaks and features, giving the names Margherita and Alexandra to the twin summits of Mount Stanley in honor of members of the Italian and British royal families. The Duke chose to name one of the smaller peaks in the range Luigi de Savoia after himself. In 1909 he returned to Africa to climb Mount Kenya. He went on to serve as Commander in Chief of the Italian navy from 1915 until 1917. Not only did he climb his eponym, he even managed to die eponymously too, in Abruzzi City, Somalia. The binomial, *aloysiisabaudiae*, is just a Latinized way of saying Luigi of Sabaudia, which was one of the territories of the ancestors of the House of Savoy. The bat has been recorded in Ghana, Gabon, DRC (Zaire), and Uganda.

Dunn

Dunn's Gerbil *Gerbillus dunni* **Thomas,** 1904

Colonel Henry Nason Dunn (1864–1952) was a British army surgeon who became a big-game hunter. He left his diaries relating to his time (1897–1906) in Sudan and Somaliland to the National Army Museum. He is also commemorated in the name of a bird, Dunn's Lark *Eremalauda dunni,* which he collected. The gerbil is found in Ethiopia, Somalia, and Djibouti.

Dupras

Fat-tailed Gerbil *Pachyuromys duprasi* **Lataste,** 1880

Monsieur Dupras (dates not found) was a friend of Lataste, who in the original description does not give more than this surname but does remark that Dupras was a "graveur sur pierres fines" (an engraver of semiprecious stones). The gerbil is found in the northern Sahara, from Morocco to Egypt.

Dupre

Madagascar Straw-colored Fruit Bat *Eidolon dupreanum* **Schlegel,** 1867

Admiral Marie-Jules Dupré (1813–1881) was a French naval officer. He visited Madagascar in 1862 on an official mission from Napoleon III to negotiate and sign a treaty with the local ruler. He returned to France and in 1864 was made Governor of Réunion. He was Governor of French Cochin China (modern southern Vietnam) from 1871 to 1874. He wrote *Trois mois de séjour à Madagascar* in 1863. The description says the bat is named "en honneur de M. le Gouverneur de l'ile de la Réunion, M. Jules Dupré, bien connu des voyageurs de Madagascar, qui nous a comblé sans cesse de ses bontés en facilitant nos recherches scientifiques de tous les moyens don't il était à même de disposer" (in honor of the Governor of the Island of Réunion, Mr. Jules Dupré, well known to travelers from Madagascar, who has ceaselessly been kind to us by facilitating our scientific researches by any means that he had at his disposal). The bat is endemic to Madagascar.

Durga Das

Durga Das' Leaf-nosed Bat *Hipposideros durgadasi* **Khajuria,** 1970 [Alt. Khajuria's Leaf-nosed Bat]

Durga Das Dogra (dates not found) was an Indian teacher of sciences. He was also the father-in-law of the describer, Dr. H. Khajuria, who described him as "a source of great inspiration and encouragement during my research work." The bat is found in central India.

Dussumier

Southern Plains Grey Langur *Semnopithecus (entellus) dussumieri* **I. Geoffroy,** 1843

Jean-Jacques Dussumier (1792–1883) was a French collector, traveler, and trader, and a ship owner in the French merchant navy. He was also interested in cetaceans and reported on sightings he had made while at sea and on harpooned specimens of dolphins and porpoises. He corresponded on the subject with Cuvier, who wrote a number of the formal scientific descriptions. Otherwise he seems to have collected mainly molluscs and fish, a number of which are named after him (e.g. Dussumier's Halfbeak *Hyporhamphus dussumieri*). He is also commemorated in the scientific names of other taxa such as the Seychelles Sunbird *Cinnyris dussumieri* and the Round Island Keel-scaled Boa *Casarea dussumieri*. The langur is from west-central and southwest India.

Duthie

Duthie's Golden-Mole *Chlorotalpa duthieae* Broom, 1907

Dr. Augusta Vera Duthie (1881–1963) was a South African botanist. In 1902 she established the Stellenbosch Herbarium. She also lectured on botany from 1902 to 1920 at what later became Stellenbosch University. In the same period Broom, who described the golden-mole, was Professor of Geology and Zoology at the same institution. On her death, part of the proceeds from the sale of her homestead Belvidere in the Knysna district were bequeathed to St. Andrew's College, where she had taught, for the purpose of funding scholarships. She is also commemorated in the scientific names of a number of cacti: *Ornithogalum duthiae, Psilocaulon duthieae, Ruschia duthiae,* and *Stomatium duthieae.* The golden-mole is found in the extreme south of South Africa.

Duvaucel

Swamp Deer *Cervus duvaucelii* **Cuvier,** 1823 [Alt. Barasingha; Syn. *Rucervus duvaucelii*]

Alfred Duvaucel (1792–1824) was a French naturalist who explored India. He was the son, from her first marriage, of Madame Cuvier. In 1807 he became Naturalist to the King. In 1818 he was sent to India by his stepfather to collect for the Natural History Museum in Paris. With Diard, whom Cuvier had also sent along, he established a botanical garden in Chandernagor in 1818. In 1819 Sir Thomas Stamford Bingley Raffles hired them to collect natural history objects in Sumatra. However, when Raffles discovered that they had sent most of the material they had collected to the museum in Paris, they were summarily dismissed. A lizard, Duvaucel's Gecko *Hoplodactylus duvaucelii,* was erroneously named after him when the museum specimens taken to Europe were credited to him. Only later was it discovered that these specimens actually came from New Zealand and must have been collected by someone else. Duvaucel is also celebrated in the scientific names of a squid, *Loligo duvauceli,* and birds such as the Scarlet-rumped Trogon *Harpactes duvaucelii.* He died in Madras, India. The deer is found in southern Nepal and north and central India.

Dwyer

Large-eared Pied Bat *Chalinolobus dwyeri* Ryan, 1966

Dr. Peter David Dwyer (b. 1937) is a New Zealand–born zoologist and anthropologist who is a Research Fellow at the University of Melbourne. He has undertaken research on bats, rats, rock wallabies, birds, ants, and people in New Zealand, Australia, and Papua New Guinea. He has written a book, *The Pigs That Ate the Garden: A Human Ecology from Papua New Guinea,* and over a hundred articles, including a great many papers on bats published in the 1960s and 1970s. Peter told us that this brief biography was very dry and didn't reveal that the *only* reason he sent a specimen of the pied bat off to Ryan was that when a friend and he were coming out of the mine tunnel in which the bats lived, his friend accidentally sat on one and killed it. That unfortunate beast became the type specimen. Serendipity and systematics! Prior to that time, before he learned that the bats represented a new species, he had always captured and released the animals. The type locality is now under many meters of water. Copeton Dam flooded the area, and with it the mines in which the population of bats he studied had lived. *Chalinolobus dwyeri* is found only in eastern Australia.

Dybowski, B.

Dybowski's Sika Deer *Cervus nippon hortulorum* **Swinhoe,** 1864 [Alt. Pekin Sika Deer]

Benedykt Dybowski (1833–1930) was a Polish biologist born in what is now known as Belarus. He was an ardent proponent of Darwin's theory of evolution. In 1862 he was appointed Adjunct Professor of Zoology in Warsaw, but after the failure of the 1863 revolt he was banished and spent time as a political exile in Siberia. There, support from the Zoological Cabinet at Warsaw allowed him to undertake investigations into the natural history of Lake Baikal and other parts of the Soviet Far East. He was pardoned in 1877 and in 1878 went to Kamchatka as a physician. In 1883 he was appointed to the Chair of Zoology at the University of Lemburg (in those days belonging to Poland, but since 1945 known as Lwow in Ukraine) and held that post until his retirement in 1906. He is also commemorated in the name of Dybowski's Frog *Rana dybowskii* and the scientific names of several fish species, as well as the trinomial of the eastern race of Great Bustard *Otis tarda dybowskii*. The deer is found in northeast China and Ussuriland (Russian Far East).

Dybowski, J.

African Groove-toothed Rat *Mylomys dybowskii* **Pousargues,** 1893
Pousargues' Mongoose *Dologale dybowskii* Pousargues, 1893

Jan (Jean-Thadée) Dybowski (1856–1928) was a botanist and an explorer of Africa, especially the equatorial regions. He led a Congo expedition in 1891 and wrote accounts of his travels, *La route du Tchad* (1893) and *Le Congo meconnu* (1912). Dybowski established new gardens and plantations in Tunisia and organized schools of agriculture. Later (ca. 1908) he became French Inspector General of Colonial Agriculture. The Dybowski family included many scientists, including an outstanding arachnologist, so it is difficult to track down quite what is named after whom. Dybowski was largely responsible for the isolation and introduction of the psychotropic drug ibogaïne. The first botanical description of the iboga plant dates from 1889 in the Congo. In 1901 Dybowski and Landrin isolated the alkaloid, which they named ibogaïne from the bark of the root, and showed it to have the same psychoactive properties as the root itself. He is also remembered in the name of a bird, Dybowski's Dusky Twinspot *Euschistospiza dybowskii*. The rat is found from Ivory Coast eastward to Kenya, Uganda, and Rwanda. The mongoose is found in Central African Republic, southern Sudan, Uganda, and northeast DRC (Zaire).

Ealey

See **Tim Ealey.**

Ebi

Ebian Palm Squirrel *Epixerus ebii* **Temminck,** 1853 [Alt. Western Palm Squirrel]

It is hard to know what or whom Temminck had in mind when he named this squirrel. The type specimen came from what is now Ghana, and there is an Ebi River in that country as well as a town of the same name. Another theory that has been postulated is that the name comes from the Ebo (Igbo) people of southern Nigeria, but the latter area is outside the squirrel's distribution. As well as Ghana, the species is found in the Ivory Coast, Liberia, and Sierra Leone.

Echidna

Echidna **Cuvier,** 1797 [now *Tachyglossus* Illiger]

Cuvier must have viewed the Australian Spiny Anteater as something of a monstrosity, to give its genus the name *Echidna.* This was the name of a mythical female monster, half woman and half serpent. She was also the mother of many other monsters: the Dragon that guarded the Golden Apples of the Hesperides, the Hydra, the Chimaera, the three-headed dog Cerberus, the Crommyonian Sow, the Caucasus Eagle (which had a daily diet of Prometheus' liver), the Sphinx, and the Nemean Lion. As an egg-laying mammal, the spiny anteater could perhaps be crudely considered "half mammal and half reptile." Unfortunately for Cuvier, Johan Reinhold Forster had used the name *Echidna* twenty years earlier, for a genus of moray eels. Cuvier's use is therefore only a junior homonym, but the word has survived as an alternative vernacular name for these animals. In 1811 Johann Karl Wilhelm Illiger came up with the name *Tachyglossus,* which is now regarded as the valid replacement name for the genus.

Eden

Eden's Whale *Balaenoptera edeni* **Anderson,** 1879 [Alt. Pygmy Bryde's Whale]

Sir Ashley Eden (1831–1887) was a British career diplomat. He was appointed as the first civilian Governor of Burma (now Myanmar) under British colonial rule in 1871. He was the third son of Robert John Eden, 3rd Lord Auckland and Bishop of Bath and Wells. He was educated at Rugby, Winchester, and the East India Company's college at Haileybury, entering the Indian civil service in 1852. In 1855 he gained distinction as assistant to the special commissioner for the suppression of the Santal rising, and in 1860 he was appointed Secretary to the Bengal government with an ex officio seat on the legislative council, a position he held for 11 years. In 1861 he negotiated a treaty with the Raja of Sikkim, which led to his being sent on a similar mission to Bhutan in 1863; but being unaccompanied by any armed force, his demands were rejected and he was forced to come to a highly unfavorable treaty. The result was the repudiation of the treaty by the Indian government and the declaration of war against Bhutan. After Burma, in 1877, he was appointed Lieutenant Governor of Bengal. In 1878 he returned to England upon his appointment to the council of the Secretary of State for India, of which he remained a member until his death. The success of his administration of Bengal was attested by the statue erected in his honor at Calcutta after his retirement. It was he who made the type specimen of the whale (which was found stranded on the Burma coast) available for scientific description. The species occurs in coastal waters of the eastern Indian Ocean and western Pacific Ocean.

Edith

Puerto Rican Nesophontes *Nesophontes edithae* **Anthony,** 1916 extinct

Edith I. Anthony (dates not found) was married to Dr. Harold E. Anthony (q.v.). The citation reads, "In honor of the author's wife, Edith I. Anthony, who found the first skull and directed attention to the presence of the animal." The nesophontes was endemic to Puerto Rico. Other species of these shrewlike insectivores inhabited Cuba and Hispaniola. All are now believed to be extinct, though some certainly survived until the coming of Europeans, as their remains are found in conjunction with those of introduced rat *(Rattus)* species.

Edith

Edith's Leaf-eared Mouse *Graomys edithae* **Thomas,** 1919

Thomas is uncharacteristically reticent in his description of this mouse and does not indicate which Edith he is naming the species after. As he uses the genitive, we can assume he had a real person in mind and was not simply making a casual use of a female name in the nominative (see the entry for **Eva** below for an example of that). This omission on Thomas' part is highly unusual for him—and thus presumably deliberate. The mouse is found in Catamarca Province, northwest Argentina.

Edward

Edward's Long-clawed Mouse *Notiomys edwardsii* **Thomas,** 1890 [Alt Milne-Edwards' Long-clawed Mouse]

See the entry for **Milne-Edwards** for biographical details. The species comes from southern Argentina.

Edward

Southern Long-finned Pilot Whale *Globicephala melas edwardii* **A. Smith,** 1834

Cape Rock Elephant-Shrew *Elephantulus edwardii* A. Smith, 1839

Jean Baptiste Edouard Verreaux (1810–1868) was a remarkable man with a remarkable family. Smith wrote of the Pilot Whale, "For the description and a drawing of this species I am indebted to Mr. E. Verreaux, who some time ago had a good opportunity of examining a specimen which had been cast on the shore near Slang-Kop." The whale has a circumglobal distribution in the temperate and subpolar waters of the Southern Hemisphere. The elephant-shrew is found in Cape Province, South Africa. See **Verreaux** for biographical details.

Edwards

Indian Grey Mongoose *Herpestes edwardsii* **E. Geoffroy,** 1818

George Edwards (1697–1773) was an English illustrator, naturalist, and ornithologist. As a young man he traveled widely in Europe. In 1733 he was appointed Librarian of the Royal College of Physicians in London. He wrote *History of Birds,* which appeared in four volumes from 1743 to 1751, and three subsequent volumes called *Gleanings of Natural History,* which appeared between 1758 and 1764, in which year he retired from his position as Librarian. He corresponded frequently with Linnaeus, who supplied him with Linnaean names to go alongside the English and French in a general index that he added to his work. Geoffroy used an illustration of a mongoose by Edwards as the equivalent of a "type specimen" (the rules on this sort of thing were much laxer then than they are now). The mongoose is found in the Arabian Peninsula, Afghanistan, Pakistan, India, Nepal, and Sri Lanka. Human introductions have extended its range to Malaysia, the Ryukyu Islands, Mauritius, and Réunion.

Eha

Smoke-bellied Rat *Niviventer eha* **Wroughton,** 1916

Edward Hamilton Aitken (1851–1909) was a writer, humorist, and naturalist. Wroughton wrote that he named the rat "in memory of my old friend the late E. H. Aitken. . . . I have used his nom-de-plume 'Eha,' under which he is known to such a wide circle by his books on the Field Natural History of the Bombay Presidency." Aitken was born near Bombay of a Free Church family and was educated by his father. He took his degrees at Bombay and Poona, where he read Latin and Greek. In 1876 he entered government service in Customs, serving in various parts of India and becoming Chief Collector of Customs in Karachi in 1905. He co-founded, with H. M. Phipson, the Bombay Natural History Society in 1883 and was for a long time the editor of its journal, in which he published many papers. He retired to Edinburgh in 1906. He wrote *The Naturalist on the Prowl* (1894), about his travels through the local jungle and hills, and *Behind the Bungalow* (1889), about his love for India and its people. He also wrote *The Tribes of My Frontier: An Indian Naturalist's Foreign Policy* (1883) and *The Common Birds of Bombay* (1900). In 1902 he was sent to Goa to investigate the prevalence of malaria there. During this trip he discovered a new species of anopheline mosquito that was named after him as *Anopheles aitkeni*. Leading a mostly lonely and solitary life, he urged others not to despair of this condition but to acquire a diverting hobby, saying, "I am only an exile endeavouring to work a successful existence in Dustypore, and not to let my environment shape me as a pudding takes the shape of its mould, but to make it tributary to my own happiness." The type specimen of this rat was collected in Sikkim by Crump (q.v.). Its range extends from Nepal and northeast India to Yunnan (China) and northern Myanmar.

Ehrenberg

Palestine Mole-Rat *Nannospalax ehrenbergi* **Nehring,** 1898

Professor Christian Gottfried Ehrenberg (1795–1876) was a German natural scientist. He started studying theology at Leipzig in 1815, but in 1817 he changed direction and went to Berlin to study medicine. In 1820 he was working on fungi (publishing a work entitled *De Mycetogenesi*) and was a lecturer at the University of Berlin, where he became a Professor of Medicine in 1827. Between 1820 and 1825 he traveled extensively, mainly in the company of his friend Hemprich. They covered a lot of ground in northeast Africa and the Middle East, from Lebanon to Sinai and from the Nile to Abyssinia (now Ethiopia). In 1829 he traveled with Humboldt to Asia. In 1847 he visited England in order to meet Darwin at Oxford. He is regarded as the founder of micropaleontology. On his death his collection of specimens was deposited at the Natural History Museum in Berlin. He published a great many articles and books, especially on fungi and corals. He was the first person to establish that phosphorescence in the sea is caused by the presence of planktonlike microorganisms. He is also remembered in the vernacular name of a bird, Ehrenberg's Redstart *Phoenicurus phoenicurus samamisicus*. The type specimen of the mole-rat was collected at Jaffa in what is now Israel. The species' range covers Syria, Lebanon, Jordan, and Israel through to northern Egypt and Libya.

Eisentraut

Eisentraut's Shrew *Crocidura eisentrauti* **Heim de Balsac,** 1957
Eisentraut's Mouse-Shrew *Myosorex eisentrauti* Heim de Balsac, 1968
Eisentraut's Pipistrelle *Hypsugo eisentrauti* **Hill,** 1968 [formerly *Pipistrellus eisentrauti*]
Eisentraut's Striped Mouse *Hybomys eisentrauti* Van der Straeten and **Hutterer,** 1986

Professor Dr. Martin Eisentraut (1902–1994) was a German zoologist and collector. He was on the staff of the Berlin Zoological Museum, working on bat migration and on the physiology of hibernation, when he went on his first overseas trip, to West Africa in 1938. He left Berlin in 1950 to become Curator of Mammals at the Stuttgart Museum, remaining there until 1957. He then became Director of the Alexander Koenig Museum in Bonn, where he lived for the rest of his life. Between 1954 and 1973 he made six trips to Bioko and Cameroon. Much of the material collected on these trips is still being studied. Eisentraut published many scientific papers and three books, including *Notes on the Birds of Fernando Pó Island, Spanish Equatorial Africa* (1968). He also published a slim volume of poems. As well as the above mammals, he is commemorated in the name of a bird, Eisentraut's Honeyguide *Melignomon eisentrauti*. The *Crocidura* shrew is found on Mount Cameroon, while the mouse-shrew is endemic to Bioko (formerly Fernando Pó). The pipistrelle was first taken in Cameroon, but it is now known to occur from Liberia east to Kenya and Somalia. The striped mouse is known from only two mountains, Lefo and Oku, in western Cameroon.

Eld

Eld's Deer *Cervus eldii* **MacClelland,** 1842 [Alt. Brow-antlered Deer, Thamin; Syn. *Rucervus eldii*]

Colonel Lionel Percy Denham Eld (1808–1863), of the 9th Native Infantry of the Bengal Army of the Honourable East India Company, was Assistant to the Commissioner of Assam. He discovered the deer in the Manipur Valley in 1838 while serving in the British colonial administration of India. He published a letter about it in the *Calcutta Journal of Natural History*, saying, "Its favourite haunts are the low grass and swamps round the edge of the Logta [lake] at the western edge of the valley, and the marshy ground at the foot of the hills. It is gregarious in its habits, and after the annual grass burning, I have frequently seen herds of two and three hundred." In 1857, during the Indian Mutiny, he was wounded, presumably quite seriously as six years later "he died of his wounds" at Weymouth, Dorset, UK. The deer is now a rare species, occurring in Manipur (India), Myanmar, Cambodia, and Laos and on Hainan Island (China). It is believed to no longer occur in Thailand or Vietnam.

Electra

Melon-headed Whale *Peponocephala electra* **Gray,** 1846

Electra was the name of one of the Oceanids, sea-nymphs who were daughters of Oceanus and Tethys. Gray certainly knew his nymphs, for in 1850 he named a dolphin after another Oceanid (see Clymene). The whale is found worldwide in tropical and warm-temperate waters.

Elery

Bristle-faced Freetail Bat *Mormopterus eleryi* Reardon et al., 2008

Professor Dr. Elery Hamilton-Smith is an Australian sociologist who is alive and well and gave us the following information about himself. He has enjoyed a wide-ranging career as a consultant and academic. He sees sociology as a licence to work on almost everything, and so has adopted a holistic and transdisciplinary perspective. As a central life interest he has been a nature conservation activist for 60 years (as of 2008)and in particular has pursued a continuing interest in speleology for 55 of those years. He has spent the last 10 years as Chairman of the IUCN/WCPA (International Union for Conservation of Nature/World Commission on Protected Areas) Task Force on Cave and Karst Protection. His underground interests have always included studies of subterranean fauna, and in 1972 he published an overview paper on "The Bat Population of the Naracoorte Caves Area." The freetail bat is found in central Australia.

Elias

Rio de Janeiro Spiny Rat *Trinomys eliasi* Pessôa and Reis, 1993

Professor Elias Pacheco Coelho (1950–1987) was a pioneering biologist whose work "made possible the training of young zoology students at the Universidade Federal do Rio de Janeiro, Rio de Janeiro, Brazil." He was a marine biologist who converted to ornithology, specializing in the study of sea birds. He died when he fell from a sea cliff on the island of Cabo Frio while climbing to investigate some nests. The spiny rat is found in Rio de Janeiro State, southeast Brazil.

Ellerman

One-toothed Shrew-Mouse *Mayermys ellermani* Laurie and Hill, 1954
Ellerman's Tufted-tailed Rat *Eliurus ellermani* Carleton, 1994

Sir John Reeves Ellerman (1910–1973) was a wealthy recluse. He was described in 1948 as one of England's richest men, having inherited a shipping fortune, at the time estimated to be £37 million, which he then wisely invested. He was also obsessed with privacy, being photographed only three times in 30 years. Despite his wealth, his real interest was the world of rodents, and he devoted his life to their study. He wrote *The Families and Genera of Living Rodents*, begun in 1930 and published in two volumes from 1940 to 1941. He went on to write other books such as *The Fauna of India Including Pakistan, Burma and Ceylon: Mammalia* and, later, *Checklist of Palaearctic and Indian Mammals 1758–1946*. The shrew-mouse is endemic to Papua New Guinea, and the rat to Madagascar.

Elliot, D.

Elliot's Short-tailed Shrew *Blarina hylophaga* Elliot, 1899
Elliot's Red Colobus *Piliocolobus foai ellioti* **Dollman,** 1909

Dr. Daniel Giraud Elliot (1835–1915) was Curator of Zoology at the Field Museum in Chicago and was one of the founders of the American Ornithologists' Union. He had a great interest in mammalogy and ornithology and, being independently wealthy, was able to travel widely. He was also able to produce a series of books illustrated with magnificent color plates that included his own excellent work, long after most publishers had come to employ smaller formats and cheaper techniques. Elliot could afford to commission the best artists of the day, including Josef Wolf and Josef Smit, both formerly employed by John Gould. The lithograph series includes works on big cats, pittas, pheasants, hornbills, and birds of prey. Elliot is also commemorated in the names of at least nine birds, including Elliot's Pheasant *Syrmaticus ellioti* and Elliot's Laughing-thrush *Garrulax elliotii*. The short-tailed shrew is found in the USA from southern Nebraska and southwest Iowa to Oklahoma and eastern Texas. The colobus monkey comes from the Ituri Forest region of eastern DRC (Zaire).

Elliot, W.

Indian Bush Rat *Golunda ellioti* **Gray,** 1837
Madras Tree-Shrew *Anathana ellioti* **Waterhouse,** 1850

Sir Walter Elliot (1803–1887) was a career civil servant. He was born in Edinburgh and joined the Indian Civil Service of the Honourable East India Company in Madras in 1821, working there until 1860. He was Commissioner for the administration of the Northern Circars from 1845 to 1854 and a member of the council of the Governor of Madras from 1854 until 1860. He corresponded regularly with Charles Darwin and, at his request, sent him skins of domestic pigeons and poultry. He was a distinguished orientalist, and his interests included botany, zoology, Indian languages, numismatics, and archeology. He saved the Amaravathi Marbles, which are now in the British Museum along with his collection of coins and other artifacts. He was a great collector of Indian plants, and his Indian herbarium was given to the Royal Botanic Garden of Edinburgh. After his retirement he went back to his home in Roxburghshire, where despite blindness and poor health he worked on local natural history and other projects until his death in 1887. He was elected a Fellow of the Royal Society in 1878. The rat is found in southeast Iran, Pakistan, Nepal, and south through the Indian Peninsula to Sri Lanka. The tree-shrew is endemic to India.

Emilia

Lesser Pygmy Flying Squirrel *Petaurillus emiliae* **Thomas,** 1908

Emilia Hose (dates not found) was the wife of Dr. Charles Hose, who collected the type specimen (see **Hose** for biographical details). The flying squirrel is found in Sarawak (East Malaysia).

Emilia

Emilia's Gracile Mouse-Opossum *Gracilinanus emiliae* **Thomas,** 1909
Emilia's Short-tailed Opossum *Monodelphis emiliae* Thomas, 1912
Eastern Amazon Climbing Mouse *Rhipidomys emiliae* **J. A. Allen,** 1916
Snethlage's Marmoset *Callithrix emiliae* Thomas, 1920
Tuft-tailed Spiny Tree Rat *Lonchothrix emiliae* Thomas, 1920

Dr. Maria Elizabeth Emilia Snethlage (1868–1929) was a German ornithologist, a Doctor of Natural Philosophy, and an Assistant in Zoology at the Berlin Museum specializing in ornithology. She collected in the Amazon forests from 1905 until her death, having been recommended by Reichenow to the Goeldi Museum. She succeeded Goeldi as head of the Zoological Section of the museum in 1914 but was suspended in 1917 when Brazil entered WW1 against Germany. However, she was reinstated after the Armistice in 1918. She wrote *Catalogo das Aves Amazonicas* in 1914. She also wrote on local languages. Snethlage was the first woman scientist to direct a Brazilian museum and to work in Amazonia. She is also commemorated in the common name of a bird, Snethlage's Tody-Tyrant *Hemitriccus minor.* All the mammals named after her are found in tropical South America. The mouse-opossum occurs from Colombia east to French Guiana and northeast Brazil. The short-tailed opossum is found in the Amazon Basin of Brazil, eastern Peru, and northern Bolivia. The climbing mouse was described as a new species in 1916, then synonymized with *Rhipidomys mastacalis,* and now has been re-elevated to a full species. The marmoset, from central Brazil, is also plagued by taxonomic uncertainty and may be a local form of the Silvery Marmoset *Callithrix argentata.* The tree rat is found south of the Amazon in the area of the Río Tapajoz and Río Madeira.

Emily

Emily's Tuco-tuco *Ctenomys emilianus* **Thomas** and St. Leger, 1926 [Alt. Emilio's Tuco-tuco]

Though sometimes referred to as Emily's Tuco-tuco, this rodent should more properly be called Emilio's Tuco-tuco. The original description states that the species was named after Señor Budin's Christian name. The tuco-tuco is endemic to west-central Argentina. See **Budin** for biographical details.

Emin

Emin's Gerbil *Taterillus emini* **Thomas,** 1892

Emin's Giant Pouched Rat *Cricetomys emini* **Wroughton,** 1910 [Alt. Forest Giant Pouched Rat]

Emin Pasha (1840–1892) was the name by which Eduard Schnitzer became known. He was a German explorer and administrator in Africa. He made important contributions to the geographical knowledge of the Sudan and central Africa. Having studied medicine, he became a physician in Albania, which was then a part of the Ottoman Empire. The people there called him Emin, meaning "faithful one." In 1876 he was appointed as Medical Officer to the staff of General Charles George Gordon, British Governor General and Administrator of the Sudan. Gordon, who became world famous by being killed by the Mahdi at Khartoum in 1885, appointed Emin, with the title of Bey, to be the Pasha (governor) of the southern Sudanese province of Equatoria in 1878. Emin then began his explorations and activities as a naturalist and collector. As a ruler, Emin's claim to fame was that he abolished slavery in the territories he commanded. A Sudanese uprising in 1885 forced him to retreat into what is now Uganda. In 1888 a search party led by Henry Morton Stanley, on what was to be his last African expedition, reached Emin only to find that he didn't want to be rescued. In 1890 Emin joined the German East Africa Company, which controlled what is now Tanzania. He led an expedition to the upper Congo River region but was beheaded by slave traders in the region of Lake Tanganyika. Mount Emin, which was named after him, is one of the six mountains of the Ruwenzori range. He is also commemorated in the common name of Emin's Shrike *Lanius gubernator* and the scientific name of the Chestnut Sparrow *Passer eminibey*, among others. The gerbil occurs in Sudan, northeast DRC (Zaire), southwest Ethiopia, Uganda, and northwest Kenya. The rat is found from Sierra Leone to Gabon and Uganda, and on the island of Bioko.

Emma

Emma's Giant Rat *Uromys emmae* Groves and **Flannery,** 1994

Emma Flannery (b. 1986).The rat is named after T. F. Flannery's daughter. The etymology reads simply, "For the junior author's daughter, Emma." The rat is known only from the island of Owi, off northwestern New Guinea.

Emmons, G. T.

Glacier Bear *Ursus americanus emmonsi* **Dall,** 1895 [Alt. Blue Bear]

Lieutenant George Thornton Emmons (1852–1945) was in the U.S. Navy. He served on board a ship called *Pinta* in Alaska from 1882 until he retired in 1899. During this service he developed an interest in the local inhabitants and became an expert on their ethnography. After his retirement he took on special projects in Alaska for the U.S. government and thus came into contact with the American Museum of Natural History, which purchased his collections of native artifacts. He published many articles and reports, especially on the Tlingit and Tahltan Indians and other tribes in Alaska and British Columbia. This race of the American Black Bear is found in southeast Alaska from Yakutat Bay to Glacier Bay.

Emmons, L. H.

Emmons' Rice Rat *Euryoryzomys emmonsae* **Musser** et al., 1998

Dr. Louise Hickok Emmons (b. 1943) is a noted tropical mammalogist. She has a Ph.D. from Cornell. She is an independent researcher based at the Smithsonian and has worked with the Amazon Conservation Association (Washington, DC). Her principal interest is the ecology of tropical mammals, and she has worked in Gabon, Madagascar, Borneo, and Papua New Guinea as well as in 10 countries of the Neotropics. She has authored numerous scientific papers, such as "Two New Species of *Juscelinomys* (Rodentia: Muridae) from Bolivia" (*American Museum Novitates*, 1999) and "The Identity of Winge's *Lasiuromys villosus* and the Description of a New Genus of Echimyid Rodent (Rodentia: Echimyidae)" (1998). She is the author of a book with illustrations by F. Feer, *Neotropical Rainforest Mammals: A Field Guide* (1997), and of *Tupai: A Field Study of Bornean Treeshrews* (2000), as well as audio-guides and various biological assessments with Conservation International's Rapid Assessment Program. One of Emmons' fascinating discoveries occurred in 1997 during a rapid inventory undertaken in collaboration with the Field Museum, Chicago. While on a mountain trail in Vilcabamba in south-central Peru, Emmons came upon a huge tree rat new to science. *Cuscomys ashaninka* was a rare find—not only a new species, but a new genus. Emmons later discovered that her specimen was a relative of the large rats discovered in 1916 buried alongside humans in the Inca tombs of Machu Picchu. The rice rat is found south of the Amazon in Brazil, between the lower Xingu and Tocantins rivers.

Enders

Enders' Small-eared Shrew *Cryptotis endersi* Setzer, 1950

Professor Dr. Robert Kendall Enders (1899–1988) was a mammalogist and endocrinologist. He was a Professor in the Biology Department of Swarthmore College from 1932 to 1970 and took students with him on many field expeditions to the Colorado Rockies and to Central America. He wrote a number of scientific papers, such as "Changes Observed in the Mammal Fauna of Barro Colorado Island, 1929–1937" (1939) and "Observations on Syntheosciurus: Taxonomy and Behavior" (*Journal of Mammalogy*, 1980), as well as larger works like *Reproduction in the Mink* (1952). In 1955 he co-wrote *The Nature of Living Things*. He was Secretary of the American Society of Mammalogists from 1932 until 1937. He also worked for the Philadelphia Academy of Natural Science and was first sent by that organization to Panama in 1935. Enders collected the holotype of the shrew in 1941 on another foray into Panama; it was a subadult of indeterminate sex. Setzer described it from the holotype in 1950, and it was not until 1980 that a second specimen was found. The shrew is restricted to cloud forests of western Panama.

Endo

Endo's Pipistrelle *Pipistrellus endoi* **Imaizumi,** 1959

Kimio Endo (dates not found). Imaizumi wrote, "This new species is named in honour of Mr. Kimio Endo, who recently collected several important specimens of bats in Iwate Pref., including the type specimen of this species." The bat is endemic to Honshu, Japan.

Entellus

Entellus Langur *Semnopithecus entellus* Dufresne, 1797 [Alt. Hanuman Langur, Grey Langur]

The presumably mythical Entellus was an elderly Sicilian who boxed ferociously and packed a punch that any heavyweight would envy. As a demonstration, he killed a bull with one punch—literally a killer blow. He appears as a character in Book 5 of Virgil's *Aeneid*. The langur is found in Sri Lanka, India, Nepal, Pakistan, and Bangladesh. Some primatologists now split this taxon into seven different species, and the reader will find references to some of these under entries for **Ajax, Dussumier, Hector,** and **Priam.**

Erlanger

Erlanger's Dik-dik *Madoqua saltiana erlangeri* **Neumann,** 1905 [Alt. Salt's Dik-dik]
Erlanger's Gazelle *Gazella gazella erlangeri* Neumann, 1906

Baron Carlo von Erlanger (1872–1904) was a German collector from Ingelheim, in the Rhineland. He traveled in the Tunisian Sahara in 1893 and 1897 and wrote two trip reports. He visited Abyssinia (now Ethiopia) and Somaliland in 1900–1901, accompanied for part of the time by the describer of both the above mammals, Oskar Rudolph Neumann. Erlanger also named 40 new avian taxa as well as having birds named after him, including one in the vernacular: Erlanger's Lark *Calandrella erlangeri.* He is commemorated in other taxa too, such as the frog *Ptychadena erlangeri* and the plant bug *Phytocoris erlangeri.* He died in a car accident in Salzburg. The dik-dik comes from the east Arussi/west Ogaden region of Ethiopia. The gazelle comes from the western Arabian Peninsula.

Ernst Mayr

Ernst Mayr's Water Rat *Leptomys ernstmayri* **Rümmler,** 1933
Long-fingered Triok *Dactylopsila palpator ernstmayri* **Stein,** 1932
Wondiwoi Tree Kangaroo *Dendrolagus dorianus mayri* **Rothschild** and **Dollman,** 1933

Dr. Ernst Mayr (1904–2005) was a German ornithologist and zoologist who began his serious studies of birds and other taxa in the South Pacific. He is best known as a writer on evolution. It has even been said of him, "He is, without a doubt the most influential evolution theoretician of the twentieth century." As a 10-year-old boy he could recognize all the local birds on sight and by their song. In 1928 he led ornithological expeditions to New Guinea, an experience that he said "fulfilled the greatest ambition of my youth." He collected 7,000 skins in two and a half years. He later joined an expedition to the Solomon Islands, then returned to his academic career at the Berlin museum. In 1931 Mayr was employed by the Department of Ornithology of the American Museum of Natural History, at first on a one-year appointment as a Visiting Curator, to catalogue the collection of South Sea birds obtained by the Whitney expedition, in which Mayr had participated. He wrote 12 research papers, describing 12 new species and 68 new subspecies, during his first year there. He went on to become the Alexander Agassiz Emeritus Professor of Zoology at Harvard University. He was the originator of the "founder effect" idea of speciation and was a leading proponent of the biological species concept. His many important books include *Systemics and the Origin of Species* (1942), *Animal Species and Evolution* (1963), and, with Diamond, *The Birds of Melanesia: Speciation, Ecology and Biogeography* (2001). He is also commemorated in the names of at least eight birds, including Mayr's Forest

Rail *Rallina mayri*. The water rat is found in the mountains of New Guinea. The triok (a striped possum) comes from central and eastern New Guinea, with the subspecies *ernstmayri* being from the Huon Peninsula. The tree kangaroo is one of the world's least-known mammals, known from a single specimen found on the Wondiwoi Peninsula, western New Guinea.

Erxleben

Erxleben's Guenon *Cercopithecus pogonias grayi* **Fraser,** 1850 [Alt. Gray's Crowned Guenon]

Johann Christian Polycarp Erxleben (1744–1777) was a German naturalist who was Professor of Physics and Veterinary Medicine at the Georg-August University in Göttingen. He wrote *Anfangsgründe der Naturlehre* and *Systema regni animalis* (1777). He founded the first academic veterinary school in Germany, the Institute of Veterinary Medicine, in 1771 and is credited with having invented the generic name "guenon." He was the son of Dorothea Christiane Erxleben, the first woman in Germany to become a qualified physician. In 1856 a specimen of monkey from an uncertain location in Africa was given the name *Cercopithecus erxlebenii* in his honor. However, this is now regarded as a junior synonym of *C. pogonias grayi*, and the common name of Erxleben's Guenon has fallen into disuse.

Eschricht

Grey Whale *Eschrichtius robustus* Lilljeborg, 1861

Professor Daniel Frederik Eschricht (1798–1863) was a Danish physiologist and naturalist. He studied medicine and surgery at Frederiks Hospital in Copenhagen, and in 1822, after having qualified, he became a general practitioner on the island of Bornholm. He was appointed as Reader in Physiology and Obstetrics in 1829, Assistant Professor in 1830, and Professor of Anatomy and Physiology at the University of Copenhagen in 1836. He was a famous authority on whales, on which subject he wrote a number of papers. Lilljeborg originally described the Grey Whale under the name *Balaenoptera robusta*. The name of the genus was changed to *Eschrichtius* by Gray in 1864, presumably as a memorial to Eschricht, who had died the previous year. The whale is found in the North Pacific.

Eugene

Tammar Wallaby *Macropus eugenii* **Desmarest,** 1817 [Alt. Dama Wallaby]

The type specimen was collected on St. Peter's Isle in the Nuyts Archipelago, southern Australia. Desmarest, being a Frenchman, used the French name for this island: L'île Eugène. Therefore this animal is not named directly after a person but after an eponymous island. The wallaby is found in southern Australia (a feral population on Kawau Island, New Zealand, appears to represent a subspecies originally from the mainland of South Australia that became extinct in its natural range).

Euryale

Mediterranean Horseshoe Bat *Rhinolophus euryale* **Blasius,** 1853

Euryale and Stheno were the sisters of Medusa and the daughters of Phorcys and Ceto. All three were Gorgons—horrendous creatures from Greek mythology, with snakes for hair and a gaze that could turn living things to stone. Euryale and Stheno were immortal, but Medusa could be killed—and was, by Perseus. The bat is found in North Africa, southern Europe as far north as southern France and Slovakia, and east to Iran and Turkmenistan.

Eva

Eva's Deer Mouse *Peromyscus eva* **Thomas,** 1898

Eva's Red-backed Vole *Caryomys eva* Thomas, 1911 [Alt. Gansu Vole; Syn. *Eothenomys eva*]

Thomas makes no mention in his descriptions of a particular Eva, so who, if anyone, he was referring to is lost in the mists of time. The names may not be eponymous. The mouse is confined to southern Baja California, Mexico. The vole is found in the mountains of central China.

Everett

Philippine Forest Rat *Rattus everetti* **Günther,** 1879

Bornean Mountain Ground Squirrel *Dremomys everetti* **Thomas,** 1890

Everett's Grizzled Langur *Presbytis hosei everetti* Thomas, 1892

Mindanao Tree-Shrew *Urogale everetti* Thomas, 1892

Everett's Ferret-Badger *Melogale everetti* Thomas, 1895 [Alt. Kinabalu Ferret-Badger, Bornean Ferret-Badger]

Alfred Hart Everett (1848–1898) was a British civil servant who worked as an administrator in the East Indies. He collected widely, and it is believed that the jawbone of an orangutan *(Pongo pygmaeus)* that he found in a cave may have been used in the "Piltdown Man" hoax. He seems to have been interested in all aspects of natural history and anthropology. He not only collected mammals, birds, and reptiles in Borneo but also published reports on the island's caves and volcanic phenomena. His death made the front page of the *Sarawak Gazette.* He is also commemorated in the names of at least nine birds, including Everett's White-eye *Zosterops everetti* and the Sumba Hornbill *Aceros everetti,* and in the names of other taxa such as the frog *Hydrophylax everetti.* The forest rat occurs on many islands of the Philippines, the tree-shrew only on Mindanao and the small islands of Dinagat and Siargao. The ground squirrel, langur, and ferret-badger are all endemic to Borneo.

Eversmann

Steppe Polecat *Mustela eversmannii* **Lesson,** 1827

Eversmann's Hamster *Allocricetulus eversmanni* **Brandt,** 1859

Professor Dr. Alexander Eduard Friedrich Eversmann (1794–1860), or, in the Russian style, Eduard Aleksandrovich Eversmann, was a pioneer Russian physician and entomologist of German extraction. After being educated in Germany he worked for two years as a physician at an arms factory in Zlatoust. He became disenchanted with medicine and followed his fascination for zoology, eventually becoming Professor of Zoology and Botany at the University in Kazan in Russia. He traveled in, and published on, remote areas of the Russian Empire in Asia, visiting Bukhara in 1820, the Kirghiz steppes in 1825, the Urals in 1827, Orenburg and Astrakhan in 1829, and the Caucasus in 1830. He became recognized as the greatest expert on the fauna of southern Russia. He seems to have concentrated on Lepidoptera, but during his travels he collected widely, giving detailed scientific descriptions of many mammals, birds, and insects. Many butterflies are named after him, such as Evermann's Parnassian *Parnassius eversmanni.* He is also commemorated in the common names of five birds, including Eversmann's Ptarmigan *Lagopus mutus eversmanni* and Eversmann's Redstart *Phoenicurus erythronotus.* The polecat is found from eastern Austria and the Czech Republic east through the steppes and subdeserts of the former USSR to Mongolia and China. The hamster is found on the steppes of northern Kazakhstan.

Fagan

Ethiopian Hare *Lepus fagani* **Thomas,** 1903

Charles Edward Fagan (1855–1921) was a British naturalist who was Secretary of the British Museum. He collected in tropical Africa in 1905–1906 and in New Guinea from 1910 to 1912. An obituary described him as "singularly sweet-tempered, always kind, while his consideration for others was one of the marked characteristics of his nature." He is also commemorated in the scientific name of a bird, the Yemen Accentor *Prunella fagani*. The hare is found in western Ethiopia, southeast Sudan, and the northwest corner of Kenya.

Falconer

Markhor *Capra falconeri* **Wagner,** 1839

Dr. Hugh Falconer (1808–1865) was one of the leading botanists and geologists of his day. He studied fossil mammals in the Siwalik Hills in India. Among his friends were Charles Darwin and the botanist Joseph Dalton Hooker, who named *Rhododendron falconeri* after him. Due to ill health, Falconer returned to England in 1842 bringing with him 70 large chests of dried plants and 48 cases of fossils, bones, and geological specimens. In 1843 he was made a member of a special commission to study the possibility of growing tea in India. Although he wrote widely on natural history and taught botany, his real delight was fossils, and he visited almost every collection in England, France, Italy, and Germany. In 1865 he was made Vice President of the Royal Society, having been a Fellow for 20 years. His brother Alexander (his elder by nearly 20 years) was a very successful businessman in Calcutta. He left a large sum of money, £1,000, to set up the Falconer Museum at Forres in Scotland to commemorate his own and his brother's work and achievements. When Hugh died he left £500 and a collection of his Siwalik fossils to the museum. The markhor is found in Afghanistan, Pakistan, Kashmir, Uzbekistan, and Tadzhikistan.

Fardoulis

Fardoulis' Blossom Bat *Melonycteris fardoulisi* **Flannery,** 1993

Emmanuel Fardoulis is, in Tim Flannery's words in response to our enquiries, "a young man who works in the airline business. He has a great interest in mammals, and made a . . . donation to the Mammal Section at the Australian Museum to support our work in the southwest Pacific. We named the bat for him in recognition of his generosity and interest in our work." The original citation reads, "For Mr. Emmanuel Fardoulis, whose generosity in supporting mammal research at the Australian Museum has allowed important studies to be carried out into the biology of the mammals of the Solomon Islands, which otherwise could never have been done." The bat is endemic to the Solomon Islands.

Father Basilio

Father Basilio's Striped Mouse *Hybomys basilii* **Eisentraut,** 1965

See **Basilio** for biographical details. The mouse is endemic to the island of Bioko (Equatorial Guinea).

Fea

Fea's Muntjac *Muntiacus feae* **Thomas** and **Doria,** 1889
Fea's Tree Rat *Chiromyscus chiropus* Thomas, 1891

Leonardo Fea (1852–1903) was an Italian explorer, zoologist, painter, and naturalist. Fea, who was an assistant at the Natural History Museum in Genoa, liked exploring in far-off and little-known countries. He visited the Cape Verde Islands in 1898, but these islands, west of Senegal, were not as challenging to visit as some of his other destinations; he had already made an expedition to Burma (now Myanmar). He is also commemorated in the names of various other taxa including two birds, Fea's Petrel *Pterodroma feae* and Fea's Thrush *Turdus feae,* and a snake, Fea's Viper *Azemiops feae.* The muntjac is found in Tenasserim (Myanmar), Thailand, and southern Yunnan (China). The rat ranges from eastern Myanmar to Laos and Vietnam.

Felipe

Don Felipe's Weasel *Mustela felipei* Izor and **de la Torre,** 1978 [Alt. Water Weasel, Colombian Weasel]

The weasel was named after "Don Felipe" Hershkovitz with the words "in recognition of his numerous contributions to mammalogy, specifically to our knowledge of the South American fauna." The species is found in the highlands of Colombia and northern Ecuador. See **Hershkovitz** for biographical details.

Felix

Spiny Seram Rat *Rattus feliceus* **Thomas,** 1920

Felix Pratt was a collector, and *feliceus* is here used by Thomas as a version of Felix. Pratt found the type specimen of this rat on an expedition to Seram (Indonesia) in 1920. The rat is endemic to that island and appears to be a rare species. See **Pratt** for biographical details.

Fellowesgordon

Sri Lanka Shrew *Suncus fellowesgordoni* **Phillips,** 1932

Mrs. Marjory Tutein-Nolthenius, née Fellowes-Gordon (dates not found), was the wife of an estate owner. The family owned an estate at Horton Plains, Central Province, Ceylon (now Sri Lanka), and it was Mr. A C Tutein-Nolthenius, after whom Nolthenius' Long-tailed Climbing Mouse *Vandeleuria nolthenii* is named, who collected the type specimen. Phillips says that he received specimens of this shrew "from my friend Mr. A. C. Tutein-Nolthenius" and that he is naming the species "in honour of Mrs. A.C. Tutein-Nolthenius, née Fellowes-Gordon."The shrew is endemic to Sri Lanka.

Fellows

Red-bellied Melomys *Protochromys fellowsi* Hinton, 1943 [Alt. Red-bellied Mosaic-tailed Rat; formerly *Melomys fellowsi*]

The attribution to Fellows for this animal is a complete mystery. The type specimen was collected by Fred Shaw-Mayer and found its way to the Natural History Museum in London, where Hinton was in the Zoology Department. He made no mention in the original, brief description as to why *fellowsi* was chosen as the binomial. Both Shaw-Mayer and Hinton are no longer living, so there is no one we can ask. The type specimen of this rat was collected in June 1940, in the Bismarck Range, New Guinea, and was presumably dispatched to London before the period of intense fighting in New Guinea in 1942–1943. The species is endemic to the highlands of Papua New Guinea.

Felten

Felten's Vole *Microtus felteni* Malec and Storch, 1963 [Alt. Balkan Pine Vole]

Dr. Heinz Felten (dates not found) was a German zoologist who was Curator of Mammals at the Senckenberg Museum, Frankfurt. From 1952 until 1954 he collected mammals in El Salvador, recording 73 species, among which were two new subspecies that he described in papers published between 1955 and 1958. In 1959 he was studying and writing on the flight of bats, having published *Fledermaüse fressen Skorpione* in 1956. In 1980, jointly with D. Kock, he published *Zwei Fledermäuse neu für Pakistan* (Two Bats New for Pakistan). In 1986 was editor of *Contributions to the Knowledge of the Bats of Thailand.* He clearly had wide zoological interests, as he is mentioned in an article on toothed whales. The pine vole is found in Macedonia and Greece.

Fernandez

Fernandez's Sword-nosed Bat *Lonchorhina fernandezi* Ochoa and Ibanez, 1982

Professor Dr. Alberto Fernandez Badillo is a Venezuelan zoologist who has made a particular study of vampire bats. He has twice served three-year terms as President of the Venezuelan Agricultural Zoology Institute, serving first in 1993–1996 and then reelected for 1996–1999. Typical of his many scientific papers is, with Ochoa and others, "Mamiferos de Venezuela: Lista y Claves para Su Identificacion" (1988). He is now retired from his post at the Central University of Venezuela. The bat is known only from southern Venezuela and was discovered when technicians with the Venezuelan Agriculture and Cattle Ministry were implementing a vampire control program. With ignorant abandon they applied the vampiricide poison, an anticoagulant paste, to all captured bats regardless of species. *Lonchorhina fernandezi* was found among these hapless victims.

Fernandina

Fernandina Galápagos Mouse *Nesoryzomys fernandinae* **Hutterer** and Hirsch, 1979

As the name implies, this mouse is named not after a person but after an island. Fernandina is the largest of the Galápagos group.

Fernando

Sri Lankan Spiny Mouse *Mus fernandoni* **Phillips,** 1932

E. C. Fernando (fl. 1900–1966) was a collector working at the Colombo Museum, Ceylon (now Sri Lanka), in 1937. He appears to have also worked as a collector for the Field Museum in Chicago, for which he supplied a number of specimens of the Slender Loris. His official job was as taxidermist at the museum; he tanned the pelt of the largest leopard—2.7 m (8 feet 10 inches) long—ever shot in Sri Lanka. His son, Dr. Henry Fernando, was an entomologist with the Department of Agriculture in Colombo. The mouse is endemic to Sri Lanka.

Ferreira

Ferreira's Fish-eating Rat *Neusticomys ferreirai* Percequillo, Carmignotto, and de J. Silva, 2005

Alexandre Rodrigues Ferreira (1756–1815) was the first Brazilian naturalist to explore the Amazon and Pantanal biomes in the states of Pará and Mato Grosso, which he did in the late 18th century. He was educated at the University of Coimbra in Portugal and taught natural history there until 1778, when he went to Lisbon to work at the Ajuda Museum. He spent five years there cataloguing specimens and writing scientific papers, as a result of which he became a corresponding member of the Lisbon Academy of Science. The Portuguese government sponsored him to explore in his native Brazil, which he did from 1783 until 1792. He followed the course of the Amazon and its tributaries, studying the indigenous people, their languages and customs, and the fauna and flora of the region, becoming known as the Brazilian Humboldt. He returned to Lisbon in 1793 and remained there until his death, working as the Director of the Natural History Museum and Botanical Gardens. The rat was discovered in Mato Grosso.

Field

Field's Mouse *Pseudomys fieldi* Waite, 1896 [Alt. Alice Springs Mouse, Shark Bay Mouse, Djoongari]

J. Field (dates not found) may have been a resident of Alice Springs, as he supplied Professor Spencer with specimens of rodents from that area, and it was Spencer who asked Waite to name this mouse after Field "in recognition of his valuable services in the collection of certain of the material dealt with in this article." The mouse seems to have become extinct on the Australian mainland by the end of the 19th century and survived only on Bernier Island in Shark Bay, Western Australia. Conservationists are trying to establish additional viable populations by translocating animals from Bernier Island.

Findley

Findley's Myotis *Myotis findleyi* Bogan, 1978

Dr. James Smith Findley (b. 1926) is a mammalogist who, from 1955, was Curator of Mammals at the University of New Mexico Museum of Southwestern Biology. Between 1955 and 1978 over 36,000 specimens were added to the collection there. In 1978 Findley was appointed Chairman of the Biology Department and Director of the Museum of Southwestern Biology, and has continued his research program at the museum as Curator Emeritus. The bat is found on the Tres Marias Islands, western Mexico.

Finlayson, G.

Finlayson's Squirrel *Callosciurus finlaysonii* **Horsfield,** 1824

George Finlayson (1790–1823) was a Scottish surgeon and naturalist. In 1821–1822 he accompanied the East India Company mission to Siam (now Thailand) and Cochin China (now part of Vietnam) as surgeon and naturalist, returning with the mission to Calcutta in 1823. Finlayson's health had been broken by the trip, and he died soon after returning to Britain. He wrote an account of the trip, *The Mission to Siam and Hue 1821–1822*, which was published in 1826 with a memoir of the author written by Raffles. He is also commemorated in the scientific name of a bird, the Stripe-throated Bulbul *Pycnonotus finlaysoni*. The squirrel is found from Myanmar east to Vietnam.

Finlayson, H. H.

Finlayson's Cave Bat *Vespadelus finlaysoni* Kitchener, Jones, and Caputi, 1987 [Alt. Inland Cave Bat, Finlayson's Forest Bat]

Professor Hedley Herbert Finlayson (1895–1991) held the Chair of Chemistry at the South Australian Museum in Adelaide, but at heart he was a mammalogist and explorer. As a young chemistry student during WW1, he experimented with explosives and succeeded in losing an eye and blowing off his right hand and part of his left hand. That did not prevent him from taking up the post of Assistant Lecturer and Demonstrator in Chemistry. Becoming interested in mammals, he published his first paper as early as 1927. In 1931 his article "Flying Opossum" was published in a newspaper, and he was still publishing well into his 60s. He traveled whenever he could and did his exploring by camel or on horseback. In roughly 40 years of traveling, Finlayson amassed a collection of approximately 3,000 specimens and about 5,000 photographic negatives. He particularly enjoyed visiting and exploring the center of Australia and wrote about it in 1935 in a book called *The Red Centre: Man and Beast in the Heart of Australia*. The book has been reprinted nine times, and its title has passed into the language as shorthand for the central Australian deserts. The bat is endemic to western and central Australia.

Finsch

Finsch's Tree Kangaroo *Dendrolagus inustus finschi* **Matschie,** 1916

Friedrich Hermann Otto Finsch (1839–1917) was a German ethnographer, naturalist, and traveler. He visited the Balkans, North America, Lapland, Turkestan, and northwest China with Alfred Brehm, and also the South Seas, spending nearly a year on the Marshall Islands from 1879 to 1880. In 1884 Bismarck appointed him Imperial Commissioner for the German Colony of Kaiser-Wilhelm-Land, in what is now Papua New Guinea. He founded the town of Finschhafen there in 1885, which remained the seat of German administration until 1918. He was the Director of a number of museums at various times, including Bremen (where he succeeded Hartlaub as Curator in 1884) and Brunswick. Among many other publications he co-wrote *Die Vogel Ost Afrika* with Hartlaub. He is also remembered in the names of many birds, including Finsch's Duck *Euryanas finschii,* a New Zealand endemic that became extinct at some time between the Polynesian and European colonizations. The tree kangaroo comes from northern New Guinea.

Fischer, A. F.

Philippine Pygmy Fruit Bat *Haplonycteris fischeri* **Lawrence,** 1939

Dr. Arthur Frederick Fischer (b. 1888) was the Chief Director, Investigation Bureau of Forestry in Manila, Philippines. The original citation reads, "It has given me great pleasure to name this interesting new bat for Mr. Arthur Fischer, retired Director of the Bureau of Forestry in Manila, through whose kind and interested assistance I was able to obtain the help and cooperation of members of the Bureau in many of the outlying districts of the Philippines." Fischer followed an academic career, being Assistant Professor at Los Baños, Luzon, in 1912 and Professor of Tropical Forestry and Dean of the Forest School at the University of the Philippines in 1917. He was the author of many scientific papers, such as "Philippine Mangrove Swamps" published in the *Philippine Forestry Bulletin.* He served as a Colonel in the U.S. Intelligence Service in WW2. In 1946 he became Director of the Museum of Natural History in San Diego, California, but had to take several leaves of absence due to recurring bouts of malaria during his tenure, which ended in 1955. He had contracted malaria while serving in the Philippines and had pioneered the introduction of cinchona to the Philippines in 1924, to be grown for the extraction of quinine. In 1942 he was evacuated to Australia by air, taking with him cinchona seeds to establish plantations in the Western Hemisphere (Central and South America). He is also commemorated in the shrub *Eugenia fischeri.* The fruit bat is a Philippine endemic.

Fischer, G.

Fischer's Shrew *Crocidura fischeri* Pagenstecher, 1885

Dr. Gustav Adolf Fischer (1848–1886) was a German explorer of east and central Africa on behalf of the Geographical Society of Hamburg. He was originally an army surgeon and in 1876 went to Africa with the Denhardt brothers, exploring with them until 1878. From 1878 to 1882 he worked in Zanzibar as a physician. He undertook an expedition at the end of 1882 on the mainland, where he explored the Masai and Tana regions and discovered Lake Naivasha in 1883. At the end of 1883 he returned to Germany. In 1885 he was back in Africa with the aim of finding Emin Pasha (q.v.). He got to Lake Victoria but could not reach the Upper Nile and returned to Zanzibar in the middle of 1886. He then returned to Germany, where he died of a tropical fever at the end of that year. He was the first person to demonstrate that the Great Rift Valley extends through equatorial Africa. He published *Das Masai-Land* in 1885. He is commemorated in the common names of at least six birds, including Fischer's Whydah *Vidua fischeri* and Fischer's Lovebird *Agapornis fischeri*. The shrew is found in Kenya and Tanzania.

Fischer, S.

Fischer's Little Fruit Bat *Rhinophylla fischerae* **Carter,** 1966

Miss Sigrid Fischer was honored in the name of this bat for contributing "significantly to the success of our collecting trip on the rivers of Ucayali and Tamaya." So wrote Dilford C. Carter of the Department of Wildlife Science, Texas A&M University. The first specimens of this species were collected in the province of Loreto, Peru, during the summer of 1964, but it has also been recorded in Ecuador, Amazonian Brazil, and southeast Colombia.

Fitzgibbon

Baluchistan Pygmy Jerboa *Salpingotus michaelis* Fitzgibbon, 1966

Michael Fitzgibbon was an animal importer. The species was discovered when "a consignment of mammals from Baluchistan, received in June 1965 by the animal importer, Mr. Michael Fitzgibbon of Hornchurch, Essex, included fifteen specimens of *Salpingotus* now preserved in the British Museum (Natural History)." These proved to be a new species, which J. Fitzgibbon named after Michael Fitzgibbon. Presumably the two men were related in some way or the describer would more likely have used the surname in the binomial. As the common name suggests, the jerboa comes from Baluchistan (west Pakistan).

Fitzroy

Fitzroy's Dolphin *Lagenorhynchus obscurus* **Gray,** 1828 [Alt. Dusky Dolphin]

Admiral Robert Fitzroy (1805–1865) was a hydrographer and meteorologist who invented the Fitzroy Barometer. He is probably best remembered for having been in command of HMS *Beagle* between 1828 and 1836, during the last five years of which Charles Darwin was on board as naturalist. In 1841 he became Member of Parliament for Durham, and between 1843 and 1845 he was Governor of New Zealand; he was dismissed for taking the view that Maori land claims were just as valid as those of the settlers. In 1854 he founded and became first Director of the Meteorological Office in London. He was a creationist who felt guilty that the voyage of the *Beagle* had been used to undermine the scriptures, and he appealed to Darwin to recant. Fitzroy committed suicide during a bout of severe depression. The dolphin is found in cold-temperate waters of the Southern Hemisphere.

Flamarion

Flamarion's Tuco-tuco *Ctenomys flamarioni* Travi, 1981

Dr. Luiz Flamarion Barbosa de Oliveira is a Brazilian zoologist at the Department of Vertebrates at the National Museum in Rio de Janeiro. He graduated in 1980 from the Federal University of Rio Grande do Sul and took his M.Sc. degree at the same university in 1985. His doctorate was awarded in 1990 by Saarland University in Germany. He has published quite widely, often as a co-author with others including Travi, who described the tuco-tuco. This species is found in coastal Rio Grande do Sul, southern Brazil.

Flannery

Greater Monkey-faced Bat *Pteralopex flanneryi* Helgen, 2005

Professor Dr. Timothy "Tim" Fridtjof Flannery (b. 1956) is an Australian zoologist famed for his academic achievements, conservation work, and the discovery of four species of kangaroo (including a black-and-white tree kangaroo) and many extinct mammals—as well as for having a tapeworm named after him. He had what he calls "a Catholic-school upbringing in Melbourne" before taking a degree in English literature. But disaffection with teaching, and the mining boom in 1970s Australia, led him into geology, which in turn led him into paleontology and a love for animals in general and mammals in particular. He went on to complete his Ph.D. in biological sciences at the University of New South Wales in Sydney. He spent many years in New Guinea, an area in which he continues to have a deep interest. During his first stay in New Guinea he several times saw a local man eating a tapeworm extracted from the stomach of a possum. He collected one and sent it to a parasitologist in Melbourne who found it to be new to science and named it *Burtiella flanneryi*. His discovery of many extinct mammals in New Guinea led the press to dub him "the Indiana Jones of Australian Science." He has published at least 11 books, including an account of his time in New Guinea entitled *Throwim Way Leg*. Perhaps his most influential work was a book about the impact of man on Australia entitled *The Future Eaters*. He is currently Principal Research Scientist (Mammalogy) at the Australian Museum. The bat is found in the Solomon Islands.

Fleurete

Fleurete's Sportive Lemur *Lepilemur fleuretae* Louis et al., 2006

Madame Fleurete Andriantsilavo was formerly the General Secretary of MINENVEF (the Malagasy Ministry for the Environment, Water, and Forests). The lemur is found in southeast Madagascar.

Flower

Horn-skinned Bat *Eptesicus floweri* **de Winton,** 1901
Flower's Shrew *Crocidura floweri* **Dollman,** 1915
Flower's Gerbil *Gerbillus floweri* **Thomas,** 1919

Captain Stanley Smyth Flower (1871–1946) was Director of the Cairo Zoological Gardens in Giza, Egypt, from 1898 until 1924. He had previously spent two years as scientific adviser to the government of Siam (now Thailand). Flower visited the zoo at Madras (now Chennai) as an adviser in 1913 and described many zoos of the time. Among other works he wrote *Zoological Gardens of the World*, published in six parts from 1908 to 1914. The bat is found in Sudan and Mali. The shrew is endemic to Egypt, and the gerbil is only known from Sinai. Flower collected the type specimens of all of these mammals.

Foa

Foa's Red Colobus *Piliocolobus foai* **Pousargues,** 1899 [Alt. Central African Red Colobus]

Édouard Foà (1862–1901) was a French explorer and big-game hunter in Africa. In a crossing of that continent, Foà is said to have shot about 500 animals, mainly to supply specimens to the Paris Natural History Museum. He traveled over 11,600 km (7,200 miles), mostly on foot, from the Zambezi delta in the east to the Congo River mouth in the west. He became a Fellow of the Royal Geographical Society in 1894. He wrote an account of his exploits, *After Big Game in Central Africa*, published in 1899. His death at such a young age was blamed on "germs . . . contracted during his African journeys." The colobus is found in the Congo Republic and DRC (Zaire) north and east of the Congo River.

Foch

Foch's Tuco-tuco *Ctenomys fochi* **Thomas,** 1919

Marshall Ferdinand Foch (1851–1929) was a French soldier who became the Supreme Commander of the Allied armies in 1918 and was still in command when Germany asked for an armistice in November of that year. He disapproved of the peace terms in the Treaty of Versailles, stating that they represented not peace but an armistice for 20 years. He was proved right in 1939. He is remembered in monuments and statues in many parts of the world. This rodent comes from Catamarca Province, northwest Argentina.

Fontainer

Fontainer's Cat *Catopuma temmincki tristis* **Milne-Edwards,** 1872

This is a mistranscription for Fontanier; see below

Fontanier

Chinese Zokor *Eospalax fontanierii* **Milne-Edwards,** 1867
Fontanier's Spotted Cat *Catopuma temminckii tristis* Milne-Edwards, 1872 [Alt. Fontainer's Cat (above), Asian Golden Cat]

Henri Victor Fontanier (1830–1870) was in Colombia around the year 1850. He became the French Consul at Tientsin in China in 1870 and combined that job with being a collector for the Paris Museum. On 21 June 1870 a crowd of locally prominent representatives at Tianjin (Tientsin) marched on a Roman Catholic orphanage run by French and Belgian nuns (the French Sisters of Charity) whom they accused of kidnapping children and mutilating them by taking their eyes to make medicine. The crowd demanded a search to reveal the truth. Fontanier lost his temper and fired into the crowd, narrowly missing the District Magistrate but killing his servant. The already xenophobic Chinese mob attacked, killing 24 foreigners, including Fontanier and the nuns, and mutilating their bodies. Fontanier is also remembered in the trinomial of the Tiny Hawk *Accipiter superciliosus fontanieri,* a specimen he sent from Colombia to Bonaparte in 1853. He was possibly the son of another French naturalist, Victor Fontanier (1796–1857), a pharmacist who was sent out by the French government as an envoy to the Persian Gulf in 1834 and who wrote *Voyage in the Indian Archipelago,* published in 1852. The zokor (a kind of mole-rat) is found in central China. The cat, from southwest China, is now regarded as a heavily patterned subspecies of the Asian Golden Cat.

Forbes

Painted Ringtail Possum *Pseudochirulus forbesi* **Thomas,** 1887 [Alt. Moss-forest Ringtail]
Forbes' Tree Mouse *Chiruromys forbesi* Thomas, 1888

Henry Ogg Forbes (1851–1932) was a Scottish explorer and natural history collector. He retraced Wallace's footsteps in the Moluccas on one of his expeditions and published *A Naturalist's Wanderings in the Eastern Archipelago* in 1885. He had originally planned on a medical career, but this was abandoned after he lost an eye in an accident shortly before he was due to qualify. Forbes led two expeditions to New Guinea with the aim of exploring Mount Owen Stanley, but neither expedition succeeded. He was appointed Meteorological Observer in Port Moresby and was later Director of the Canterbury Museum, New Zealand, serving from 1890 until 1893. In 1894 he was back in the UK as Director of the Liverpool Museums. Forbes is also commemorated in the names of three birds: Forbes' Forest Rail *Rallina forbesi,* Forbes' Mannikin *Lonchura hunsteini,* and Forbes's Parakeet *Cyanoramphus (auriceps) forbesi.* The ringtail possum and tree mouse are both endemic to New Guinea.

Fornes

Fornes' Pygmy Rice Rat *Oligoryzomys fornesi* Massoia, 1973

Abel Fornes (d. 1974) was an Argentine zoologist who was described as being Massoia's "inseparable friend and colleague." The citation states that Fornes is particularly remembered as such and that his accidental death was a great tragedy. He and Massoia wrote many scientific papers as co-authors. At the time of his death, we believe, he was working with the Pan American Zoonoses Center in Buenos Aires. The species of rice rat was first found in Formosa Province, northern Argentina, but is also now known from Paraguay and south-central Brazil.

Forrest, G.

Forrest's Rock Squirrel *Sciurotamias forresti* **Thomas,** 1922
Forrest's Mountain Vole *Neodon forresti* Hinton, 1923
Forrest's Pika *Ochotona forresti* Thomas, 1923

George Forrest (1873–1932) was a British collector, primarily of botanical specimens, but he would collect anything from butterflies to birds. He started work in a chemist's shop, where he learned about the medicinal qualities of plants and how to dry and preserve them. He went to Australia at the height of the 1891 gold rush and spent 10 years panning for gold. He returned to Britain in 1902. In 1903 he was employed as a clerk at the Royal Botanic Gardens in Edinburgh. He first went to Asia in 1904, when he arrived at Talifu in China, and in the middle of 1905 he set out for northwest Yunnan, close to the border with Tibet. At Tzekou his party was based at a French mission, and after a collecting foray, Forrest and his team of 17 collectors returned to the mission to find that the locals had turned on the foreigners. Of his group, only Forrest escaped with his life. Continuing to collect despite this tragedy, and despite being struck down by malaria, he and his new team of collectors amassed a very considerable weight of specimens that he took back to Britain in 1906. During his life he made six further expeditions to Yunnan, Sichuan Province, Tibet, and Upper Burma. He never seemed to have time to write, but he discovered more than 1,200 plant species that were new to science as well as many birds and mammals. He died from a heart attack in China while still collecting on what proved to be his last expedition. The squirrel is found in Yunnan and Sichuan provinces, China. The vole is also from Yunnan. The pika occurs in southeast Tibet, Assam and Sikkim (India), Bhutan, and northern Myanmar.

Forrest, J.

Forrest's Mouse *Leggadina forresti* **Thomas,** 1906

The Rt. Hon. Sir John Forrest (1847–1918) was an Australian explorer and statesman who was the first Prime Minister of Western Australia, serving from 1890 to 1901. In 1869 he led an expedition in search of Leichhardt, and in 1870 he led another from Perth (Western Australia) to Adelaide (South Australia) around the Great Australian Bight. In 1874 he crossed the Great Victoria Desert from Champion Bay, Western Australia, to the Overland Telegraph Line in Northern Territory, collecting botanical specimens along the route. He went on to become Deputy Surveyor General for Western Australia before becoming Secretary to the Treasury and Prime Minister. After 1901 he held other ministerial posts. He was noted for having no sense of humor but a love of pomp and ceremony. He died at sea, on 3 September 1918, while en route to England. Two towns, Forrestdale and Glen Forrest, both in Western Australia, are named after him. He is also commemorated in the name of a bird, Forrest's Honeyeater *Lichenostomus virescens forresti*. The mouse is found in inland regions of Australia.

Forster, J. G. A.

Forster's Fur Seal *Arctocephalus forsteri* **Lesson,** 1828 [Alt. New Zealand Fur Seal]

Johann Georg Adam Forster F.R.S. (1754–1794), known as George, was the artist who accompanied Captain James Cook on the *Resolution* on Cook's second expedition (1772–1775). His father, Johann Reinhold Forster (see below), was the naturalist aboard. George was elected a Fellow of the Royal Society in 1777 for his share in the description of the flora, fauna, and ethnology of the South Seas. Later in his life he became Professor of Natural History at the Collegium Carolinum in Kassel and Head Librarian at the University of Mainz. He became a strong supporter of the French Revolution and ended his days as a political exile in Paris. He is also commemorated in the name of an extinct bird, Forster's Dove of Tanna, *Gallicolumba ferruginea*. The fur seal is found around New Zealand and nearby sub-Antarctic islands, as well as on the south Australian coast.

Forster, J. R.

Forster's Shrew *Sorex forsteri* **Richardson,** 1828 [Alt. Cinereous Shrew; Syn. *S. cinereus* Kerr, 1792]

Johann Reinhold Forster (1729–1798) was originally a clergyman in Danzig. He became a naturalist and accompanied James Cook on his second voyage around the world (1772–1775). This voyage extended further into Antarctic waters than anyone had previously reached. Forster gained a reputation as a constant complainer and troublemaker. His complaints about Cook continued after his return and became public, destroying Forster's career in England. He went to Germany and became a Professor of History and Mineralogy. Unpleasant and troublesome to the end, Forster refused to relinquish his notes of the voyage. They were not found and published until almost 50 years after his death. The Emperor Penguin *Aptenodytes forsteri* is named after him in the binomial. In 1772 Forster published an article entitled "Account of Several Quadrupeds from Hudson's Bay," which included references to shrews. It was probably this article that later inspired Richardson to name a shrew from the "Hudsons Bay Countries" as *Sorex forsteri*. Unfortunately for the cantankerous naturalist, this proved to be a junior synonym of the Cinereous (or Masked) Shrew *Sorex cinereus*, named earlier by Kerr, and the name Forster's Shrew is now found only in obsolete references. The shrew is found in Alaska, Canada, and the northern USA.

Forsyth Major

Forsyth Major's Sifaka *Propithecus verreauxi majori* **Rothschild,** 1894 [Alt. Major's Sifaka Lemur]

See **Major** for biographical details.

Foster

Foster's Shrew

We believe that occasional references to this species are merely transcription errors and should be Forster's Shrew; see above.

Fox

Nigerian Mole-Rat *Fukomys foxi* **Thomas,** 1911 [formerly *Cryptomys foxi*]
Fox's Shaggy Rat *Dasymys foxi* Thomas, 1912
Fox's Shrew *Crocidura foxi* **Dollman,** 1915

The Rev. George T. Fox (d. 1912) worked at the Cambridge University Mission in Northern Nigeria. In 1912, when describing the shaggy rat,

Thomas wrote that Fox "died of fever early in this year," and he seemed rather put out by the loss of supply of specimens this entailed. It appears that Fox regularly sent collections of animal specimens back to Britain, as evidenced by Thomas's paper of 1911 in the *Annals and Magazine of Natural History*, "On Mammals Collected by the Rev. G. T. Fox in Northern Nigeria." See the entry for **Johan** for Dr. John C. Fox, who was his brother. The mole-rat and shaggy rat are known from the Jos Plateau of central Nigeria. The shrew was first collected in the same area, but its range probably extends from Senegal to southern Sudan.

François

François' Leaf Monkey *Trachypithecus francoisi* **Pousargues,** 1898 [Alt. François' Langur]

Auguste François (1857–1935) was the French Consul at Lungchow in southern China, where he was the first person to bring this monkey to the attention of Western scientists. He presumably supplied the type specimen, since that originates from Lungchow. He was a professional diplomat who, apart from one posting as Consul in Paraguay, spent his career in Asia: first in Tonkin (now in Vietnam) and then in China from 1896 to 1905. He was a very early film enthusiast who was lent a movie camera by the Lumière brothers. In 1901 he made films of ordinary Chinese life, as well as of the violent and dramatic days of the Boxer Uprising. He traveled widely, including visiting Tibet where he found himself caught up in a violent revolt by the indigenous Yi people against the Chinese; this he also caught on film. Using his material and photographs from other sources, a film called *Through the Consul's Eye* has been released, showing China as it was and on the brink of a completely new age. The monkey is found in southeast China and northern Vietnam.

François Moutou

Réunion Little Mastiff Bat *Mormopterus francoismoutoui* **Goodman** et al., 2008

Dr. François Moutou is a French veterinary surgeon who is employed by the Epidemiological Unit, French Food Safety Agency, Maison Alfort, France, as a specialist in epidemiology and zoology. Goodman named the bat after him for his important contributions to the knowledge of the vertebrate fauna of the Mascarene Islands. In 1989 he wrote *Observing British and European Mammals*, with Christian Bouchardy, and in 2007 *La vengeance de la civette masquée*. The bat comes from the island of Réunion. It was formerly considered conspecific with *Mormopterus acetabulosus* of Mauritius.

Franklin

Franklin's Ground Squirrel *Spermophilus franklinii* Sabine, 1822

Sir John Franklin (1786–1847) was an officer in the Royal Navy and is best known as an explorer of the Northwest Passage. The youngest of 12 boys, at the age of 14 Franklin joined the navy and spent the rest of his life in its service. He undertook his first Arctic voyage in 1818, commanding a vessel trying to reach the North Pole. In 1819 he led his first attempt to find the Northwest Passage, the sea route across the Arctic to the Pacific Ocean. He returned empty-handed after two years amid rumors of starvation, murder, and cannibalism. In 1837 he was appointed as Governor of Van Diemen's Land (now Tasmania), and the following year he played host to John Gould for three months—and to Eliza Gould even longer, as she was pregnant. Lady Franklin even accompanied John Gould on one of his collecting trips. Franklin was involved in several more voyages to the north before disappearing in 1845 in another attempt to cross the Arctic by sea. Several searches were undertaken, including one under the command of Sir Clements Robert

Markham, and it was well over 100 years before the site of his death was discovered. He is remembered in the common names of birds such as Franklin's Grouse *Falcipennis canadensis franklini*, Franklin's Gull *Larus pipixcan*, and Franklin's Nightjar *Caprimulgus monticolus*. The ground squirrel is found in south-central Canada and north-central USA, from Alberta and Saskatchewan south to Kansas, Illinois, and Indiana.

Franquet

Franquet's Epauletted Bat *Epomops franqueti* **Tomes,** 1860

Dr. Franquet (dates not found) was an employee of the French Imperial Navy. The original citation does not give any other names, but possibly this is the same person as François Xavier Franquet who in 1860 published a book called *La vaisseau patron: Solution du problème de l'organisation du personnel matelot de la marine française* (proposing a solution to the manning problems of the French navy). The bat is found from Ivory Coast east to Sudan and Uganda, and south to northern Zambia and Angola. The first known specimen was forwarded by Dr. Franquet to Geoffroy St. Hilaire at the French National Collection.

Fraser, F.

Fraser's Dolphin *Lagenodelphis hosei* F. Fraser, 1956

Dr. Francis Charles Fraser (1903–1978) identified the dolphin as a new species in 1956, from a skull discovered in Sarawak 60 years earlier by Hose, who is remembered in the binomial. Fraser took part in the *Discovery* investigations of 1925–1933 and was a member of the Danish *Atlantide* expedition to West Africa from 1945 to 1946. He was Keeper of the Department of Zoology at the Natural History Museum in London from 1957 to 1964. In 1976 he published *British Whales, Dolphins and Porpoises*. The dolphin has a worldwide distribution in warm-temperate and tropical waters.

Fraser, L.

Fraser's Musk Shrew *Crocidura poensis* L. Fraser, 1843

Louis Fraser (1810–1866) was a British zoologist and collector. He was also variously a curator, explorer, zookeeper, consul, author, dealer, and taxidermist. He was employed first as Office Boy, then as Clerk, Assistant Curator, and finally Curator to the Museum of the Zoological Society of London, serving in the latter position from 1832 until 1841 and again from 1842 until 1846. He collected in western Africa between 1841 and 1842 as the official naturalist on the Niger expedition; in North Africa in 1847 (self-financed); in the Bights of Benin (Nigeria) from 1851 until 1853; and in Ecuador, Guatemala, and California for P. L. Sclater from 1857 to 1860. He wrote *Zoologica typica* (published in 14 parts between 1845 and 1849) and *Catalogue of the Knowsley Collection* (1850), as well as around 40 papers appearing in the *Proceedings of the Zoological Society of London* between 1839 and 1866. He took charge of Lord Derby's zoological collections at Knowsley from 1848 until 1850, after which Fraser was Vice Consul at Whydah (Ouida) to the Kingdom of Dahomey (1850–1853). According to Sharpe he tried to establish himself as a natural history dealer, opening a shop in London to sell exotic birds. This venture appears not to have been a success, as he went to the USA again and spent his last years there. He is commemorated in the common names of at least seven birds, including Fraser's Eagle Owl *Bubo poensis* and Fraser's Warbler *Basileuterus fraseri*. The shrew is found on the islands of Bioko and Principe, and on the African mainland from Liberia to Cameroon.

Frémont

Fremont's Squirrel *Tamiasciurus hudsonicus fremonti* **Audubon** and **Bachman,** 1853

Major General John Charles Frémont (1813–1890) was an American army officer who, as a Lieutenant, led a number of expeditions across the Rockies and played a big part in the campaign to take California from Mexico. He explored between 1843 and 1848. In 1861 Abraham Lincoln made him a Major General and sent him to St. Louis, Missouri, to command the Western Department of the Union army. He was an ardent abolitionist but not a success as a commander and was quickly removed from his command. He became Commander in 1862 of the newly created Mountain Department but was defeated in battle by General "Stonewall" Jackson, was sidelined, and then resigned. In 1864 he tried, unsuccessfully, to form a third political party to run in that year's presidential election. He first owned then lost properties in California, and after 1864 he became gradually poorer. In the early 1880s he was Governor of the Territory of Arizona. In his last years he depended heavily on what income he could make as a writer. The squirrel was originally described as a full species but is now regarded as a subspecies of the American Pine Squirrel. It is found in the Rocky Mountains of Colorado and Utah.

Frith

Frith's Tailless Bat *Coelops frithi* **Blyth,** 1848 [Alt. East Asian Tailless Leaf-nosed Bat]

Long-beaked Common Dolphin *Delphinus frithii* Blyth, 1859 [Syn. *D. capensis* **Gray,** 1828]

Robert W. G. Frith (dates not found) owned an indigo factory at Khulna, in what is now Bangladesh, and was also a natural history collector. He was a great friend of Edward Blyth, who mentions him in his 1875 *Catalogue of Mammals and Birds of Burma.* Blyth named the dolphin from a specimen "procured during the voyage from England to India" that was presented to the Calcutta Museum by Frith. However, this specimen seems to have been lost. Based on published information, it was probably synonymous with the species *Delphinus capensis.* The bat is found from northeast India to southern China, Taiwan, and Vietnam, south through the Malay Peninsula to Java and Bali.

Frost

Small-toothed Fruit Bat *Neopteryx frosti* **Hayman,** 1946

Wilfred J. C. Frost (1875–1958) was mainly a bird collector. He appears to have been a scenery shifter in a theater, possibly the Shepherd's Bush Empire, as he lived in Shepherd's Bush at the end of the 19th century. He had a sideline in trapping birds, at a time when bird-trapping in the UK was still legal. He published just two articles, plus a couple of letters, in the *Avicultural Magazine.* He is known to have been collecting birds on Salawati Island, west of New Guinea, in 1906 and appears to have spent the rest of his life in Java, Singapore, and New Guinea, and to have developed a wider interest including mammals. He left England, after a brief visit, in November 1957, on what was to be his 54th and last collecting expedition. He died sometime early in 1958, probably in Singapore. Between 1942 and 1945 he was interned by the Japanese in Changi Camp as a civilian prisoner. While there he was in charge of the rice store and trapped the rats and snakes that wandered in. He is known to have roasted the rats and shared the meat with fellow prisoners—who were extremely grateful until they discovered what they had been eating. Hayman recounts that in 1939 the British Museum received a large collection of mammal specimens from the Dutch East Indies, sent by Mr. Frost. However, there were no bats among this collection. Then, "after four years in a prison camp at Singapore Mr. Frost returned to England and found a few more small mammals . . . which had been overlooked earlier." One of these was the type specimen of the Small-toothed Fruit Bat. The bat is endemic to Sulawesi, where Frost collected it in 1938 or 1939 at Tamalanti; this remained the only known specimen until three others were collected in 1985.

Furness

Amami Rabbit *Pentalagus furnessi* W. Stone, 1900

Dr. William Henry Furness III M.D., F.R.G.S. (1866–1920) was an American anthropologist and explorer most famed for a paper he wrote about time spent with head-hunters in Borneo, "Home Life of Borneo Head Hunters" (1901). He also wrote *The Island of Stone Money: Uap of the Carolines* (1919) and various articles such as "Observations on the Mentality of Chimpanzee and Orangutan" (1916). He was, with H. M. Miller, the first to describe the rabbit, which they collected in 1896. It is an endangered species confined to the islands of Amami-Oshima and Tokuno-Shima, in the Ryukyu Islands, Japan.

Gabb

Bushy-tailed Olingo *Bassaricyon gabbii* **J. A. Allen** 1876

Professor William More Gabb (1839–1878) was an American invertebrate paleontologist. He acquired his knowledge of geology in the Academy of Natural Sciences in Philadelphia. From 1862 until 1865 he was a paleontologist on the geological survey of California, under Professor Whitney, during which all Cretaceous and Tertiary fossils were classified by him. He wrote the relevant chapter in the 1864 *Geological Survey of California* as well as the entire second volume. In 1868 he surveyed Santo Domingo for the Santo Domingo Land and Mining Company, remaining on there until 1872. In 1873 he wrote a memoir of this time entitled "On the Topography and Geology of Santo Domingo," published in the *Transactions of the American Philosophical Society.* He went to Costa Rica on government contract and undertook a topographical and geological survey of that country. It is here that he made extensive ethnological and natural history collections for the Smithsonian Institution. He wrote "On the Topography of Costa Rica" and "Ethnology of Costa Rica," published in the *Transactions of the American Philosophical Society,* and also contributed papers to scientific journals and proceedings of societies. He was considered to have greater knowledge of American invertebrate paleontology of the Cretaceous and Tertiary periods than any other scientist of his time. He is also commemorated in the names of various molluscs: the genus *Gabbia* and species such as *Crepitacella gabbi, Glyphostoma gabbii,* and *Turbonilla gabbiana.* The olingo is found from Nicaragua south to western Colombia and Ecuador.

Gabriella

Gabriella's Crested Gibbon *Nomascus gabriellae* **Thomas,** 1909 [Alt. Buff-cheeked Gibbon, Yellow-cheeked Gibbon; Syn. *Hylobates gabriellae*]

Madame Gabrielle M. Vassal (1880–1959) was a British lady whose husband, Dr. J. J. Vassal, was a French army doctor. She accompanied him on his postings, most notably to Vietnam between 1907 and 1910. He was stationed at Langbian in Annam, where the type specimen of the gibbon was collected. She wrote *Three Years in Vietnam (1907–1910): Medicine, Chams and Tribesmen in Nhatrang and Surroundings,* which was originally published in 1910 under the title of *On and Off Duty in Annam.* Her husband was later appointed as Director of Public Health for the French Congo, and in 1925 she published *Life in French Congo.* She was a first-class shot and greatly enjoyed hunting tiger in Annam and buffalo in Africa. The White-cheeked Laughing Thrush *Garrulax vassali* is named after her husband. The gibbon is found in eastern Cambodia, southernmost Laos, and southern Vietnam.

Gaimard

Tasmanian Bettong *Bettongia gaimardi* **Desmarest,** 1822 [Alt. Eastern Bettong, Tasmanian Rat-Kangaroo]

Joseph (or Jean, according to some sources) Paul Gaimard (1796–1858) was a French naval surgeon, explorer, and naturalist. He made a voyage to Australia and the Pacific from 1817 to 1819 aboard the *Uranie,* during which time he kept a journal, *Journal du voyage de circumnavigation, tenu par Mr Gaimard, chirurgien à bord de la corvette l'Uranie.* Though he continued with his journal, further entries were lost when the ship was wrecked off the Falklands and he had to continue his journey on board *Physicienne*, the ship that had rescued the expedition and then been purchased as a replacement. He was aboard the *Astrolabe*, under the command of Dumont d'Urville, when it visited New Zealand in 1826. From 1838 until 1840 he led an expedition aboard the *Récherche* to northern Europe, visiting Iceland, the Faeroe Islands, northern Norway, Archangel, and Spitsbergen. His contemporary, the zoologist Henrik Krøyer, who went with Gaimard to Spitsbergen in 1838, described him thus: "He was of medium build, with curly black hair and a rather unattractive face, but with a charming and agreeable manner." He was something of a dandy and, when visiting Iceland, handed out sketches of himself. He is also commemorated in the scientific name of the Red-legged Cormorant *Phalacrocorax gaimardi* and of a fish species, the Yellowtail Coris *Coris gaimard*. The bettong used to occur in southeast Australia but is now extinct on the mainland. It is still found on Tasmania.

Gairdner

Gairdner's Shrew-Mouse *Mus pahari* **Thomas,** 1916

Kenneth G. Gairdner (fl. 1900–1950) was a collector employed by the Raffles Museum in Singapore. In 1914 he produced *Notes on the Fauna and Flora of Ratburi and Petchaburi Districts*. The species was first collected by C. A. Crump in Sikkim and named by Thomas, so one could legitimately ask, Where and why does Gairdner come into its name? Actually the common name Gairdner's Shrew-Mouse should strictly apply only to the eastern subspecies, *Mus pahari gairdneri* (named by Kloss). It seems as if Gairdner's name became attached to the full species because there was no preexisting widely accepted common name for *Mus pahari*. The mouse is found from northeast India through Myanmar, Yunnan (China), and east to Vietnam.

Gaisler

Gaisler's Long-eared Bat *Plecotus teneriffae gaisleri* Benda, Kiefer, **Hanak,** and Veith, 2004

Professor Dr. Jiri Gaisler (b. 1934) is a Czech zoologist at the Institute of Botany and Zoology, Faculty of Science and Ecology at Masaryk University in Brno. He has been at Brno since 1957, and his speciality is the biology of bats. Additionally, he has studied small rodents and insectivores. He has been on expeditions to several countries, notably Afghanistan, Egypt, and Algeria. In addition to his duties in the Czech Republic, he has taught for several years as a visiting professor at Shippensburg University, Pennsylvania, USA, and at the University Centre, Setif, Algeria and has also lectured at universities in Vienna, Brussels, Cairo, and Jalallabad. In 2005 he was invited to give a plenary lecture on "The Importance of Long-term Monitoring of Bat Populations," at the 10th European Bat Research Symposium in Galway, Ireland. His publications include "Bats of Northern Algeria and Their Winter Activity" and, as co-author with J. Zejda, *Enzyklopädie der Säugetiere* (1997). The bat is found in northern Libya. It may be a subspecies of *Plecotus kolombatovici* rather than of *P. teneriffae*.

Gallagher

Gallagher's Free-tailed Bat *Chaerephon gallagheri* **Harrison,** 1975

Major Michael Desmond Gallagher (b. 1921) is a zoologist with a strong interest in herpetology and small mammals. He was Curator of the Oman Natural History Museum from 1985 until his retirement in 1998. He made four visits to Oman between 1970 and 1976 before moving there to live in 1977. He published *The Amphibians and Reptiles of Bahrain* in 1971 and *Snakes of the Arabian Gulf and Oman* in 1993. He is a member of the Royal Geographical Society in London and collected the type specimen of the free-tailed bat on the Zaire River expedition of 1974–1975. During this expedition he made a collection of over 200 bats. The species is only known from one locality in Kivu, DRC (Zaire), and is regarded as critically endangered.

Gapper

Gapper's Red-backed Vole *Myodes gapperi* Vigors, 1830 [Alt. Southern Red-backed Vole; Syn. *Clethrionomys gapperi*]

Dr. Anthony Gapper (1799–1883) was trained as a doctor and practiced medicine between 1823 and 1826. In 1826 he emigrated from Bristol, England, to northern Canada, where he joined up with a family called Southby. He spent his time cataloguing fauna and flora and wandered the country in search of rare specimens. He returned to England in 1829 and inherited the Southby family home, Bulford House, near Stonehenge, in 1835. He then changed his name to Southby. In the census of 1861 Anthony is listed as aged 61, occupation landholder and papermaster, and the employer of five men, one boy and three women. He published an article in the *Zoological Journal* entitled "Observations on the Quadrupeds Found in the District of Upper Canada Extending between York and Lake Simcoe, with the View of Illustrating Their Geographical Distribution, as Well as of Describing Some Species Hitherto Unnoticed." The vole is found over most of Canada and the northern USA, extending south in the Rocky Mountains to New Mexico.

Gardner

Gardner's Spiny Rat *Proechimys gardneri* da Silva, 1998

Gardner's Climbing Mouse *Rhipidomys gardneri* **Patton,** da Silva and Malcolm, 2000

Dr. Alfred L. Gardner (b. 1937) is an American biologist whose main area of interest is mammalian systematics and nomenclature, primarily with regard to mammals of the Western Hemisphere. He took his bachelor's degree in wildlife management and his master's in zoology at the University of Arizona in 1962 and 1965, respectively, and his doctorate in vertebrate zoology, with a minor in paleontology, at Louisiana State University in 1970. He held a postdoctoral appointment at the University of Texas M. D. Anderson Hospital and Tumor Institute, Houston, from 1970 to 1971. From 1958 to 1960 and 1961 to 1962, he was a field collector in Mexico for the Sheffler Collection and the Western Foundation of Vertebrate Zoology in Los Angeles. In 1972 he was an Assistant Professor of Zoology at Louisiana State University, Baton Rouge, and from 1972 to 1973 he was Assistant Professor of Biology at Tulane University in New Orleans. In 1973 he joined the U.S. Fish and Wildlife Service as a research biologist stationed at the National Museum of Natural History, Smithsonian Institution. Transferred to the U.S. Geological Survey in 1996, he remains at the museum as the Curator of the National Collection of North American Mammals and National Collection of Mammal Type Specimens. He has published widely, writing with others or alone. So far in his career he has authored or co-authored descriptions of 2 new genera and 14 species of marsupials, bats, and rodents from Mexico and South America. The two rodents named after him are found in western Amazonia (Brazil and Peru).

Garlepp

Garlepp's Mouse *Galenomys garleppi* **Thomas,** 1898 [Alt. Garlepp's Pericote]

Gustav Garlepp (1862–1907) was a German who, sometimes with his brother Otto Garlepp (1864–1959), collected in Latin America between 1883 and 1897. He was a bank employee who first went to South America to collect insects for the Natural History Museum in Dresden. He also collected birds for Graf

Hans von Berlepsch, including 40 new species, and eggs for Maximilan Küschel. He landed in Pará, Brazil, and crossed the continent to Peru, where he stayed for four years. After a short break in Germany he went back, this time to Bolivia. When he returned to Germany in 1892, he had amassed 1,530 bird specimens. In 1893 he returned to Bolivia together with his wife and his brother Otto. He visited Germany for the last time in 1900, presenting 3,000 specimens of 600 species of birds at the Annual Meeting of the German Ornithologists' Society, Leipzig. He was murdered during a collecting expedition in Paraguay. His collection is now inthe Senckenberg Museum, Frankfurt. He is also commemorated in the common names of two birds, Garlepp's Tinamou *Crypturellus (atrocapillus) garleppi* and Gustav's Parakeet *Brotogeris cyanoptera gustavi*. The mouse is found on the high Altiplano of southern Peru, southwest Bolivia, and northern Chile.

Garnett

Garnett's Galago *Otolemur garnettii* **Ogilby,** 1838 [Alt. Northern Greater Galago, Garnett's Greater Bushbaby, Small-eared Greater Galago]

Thomas Garnett (1799–1878) was a British businessman, cloth manufacturer, and naturalist who was one of Darwin's correspondents. He started life as an independent weaver but around 1811 got a job with Garnett and Horsfall of Clitheroe, a town of which he was to become Mayor several times. The company had been founded by his uncle, Jeremiah Garnett. Thomas obviously showed promise, for he became manager, then partner, andfinally head of the firm, a position he held for many years until his death. He had an enquiring turn of mind and ceaselessly experimented in natural history, agriculture, and medicine. He was one of the first to experiment with using guano as a fertilizer and was the first to see the economic value of alpaca wool, though he could not convince his partners. In 1832, in an article in the *Magazine of Natural History*, he proposed the artificial propagation of fish. After his death a privately produced collection was printed under the title *Essays in Natural History and Agriculture, by Thomas Garnett*. He wrote many letters to the press on natural history matters. This was greatly helped by the fact that his brother was the editor, and one of the founders, of the *Manchester Guardian*. The original citation reads, "Mr. Ogilby likewise called the attention of the Society to certain peculiarities in the structure of the hand, in a living specimen of a new species of Galago, which he proposes to call *Otolicnus Garnettii*, after the gentleman to whom he was indebted for the opportunity of describing it, and who has already conferred many advantages upon science by the introduction of numerous rare and new animals." The galago is found in East Africa from southern Somalia to southern Tanzania (including the islands of Zanzibar and Pemba).

Garrido

Garrido's Hutia *Mysateles garridoi* **Varona,** 1970

Varona and Garrido's Hutia *Mesocapromys sanfelipensis* Varona and Garrido, 1970 [Alt. San Felipe Hutia, Little Earth Hutia]

Orlando H. Garrido is a Curator in the Zoology Department of the National Museum of Natural History in Cuba. Primarily an ornithologist and herpetologist, he has produced a number of books on various classes of birds found there. These books are mainly aimed at awakening the interest of young people. He has also researched and published other guides, such as, jointly with Arturo Kirkconnell, who works with him at the same museum, *Field Guide to* the Birds of Cuba (2000). He is commemorated in the names of five reptiles, including Garrido's Anole *Anolis pumilis* and Garrido's False Chameleon *Chamaeleolis barbatus*. Garrido's Hutia is found on small islets off southern Cuba, east of the Isla de la Juventud (formerly Isle of Pines). The San Felipe Hutia is also found on small islets, but west of Juventud. It may now be extinct, as a survey in 2003–2004 failed to find any.

Gaskell

Gaskell's False Serotine *Hesperoptenus gaskelli* **Hill,** 1983

B. H. Gaskell is a zoologist who has written fairly extensively about bats, including the bats of Sulawesi. In 1984 he contributed a paper to the Leeds Symposium, "Flying Fruit-bat Faunas of the Upper Canopy in Two Palaeotropical Rain Forests." In 1986, jointly with P. G. Kevan, he published "The Awkward Seeds of *Gonystylus macrophyllus* (Thymelaeaceae) and Their Dispersal by the Bat, *Rousettus celebensis*, in Sulawesi, Indonesia." The bat is endemic to Sulawesi.

Gaumer

Gaumer's Spiny Pocket Mouse *Heteromys gaumeri* **J. A. Allen** and **Chapman,** 1897

Dr. George F. Gaumer (1850–1929) collected in Mexico between 1885 and 1893. He acted as a collector of botanical specimens, in particular for the Field Museum in Chicago, as well as supplying a collection of birds from the Yucatan Peninsula to the Smithsonian Institution in Washington, DC. Among other taxa, he collected the type specimen of the species of agouti *Dasyprocta ruatanica*, which is endemic to the island of Roatan in the Bay of Honduras. He is also commemorated in the scientific name of a subspecies of the Caribbean Dove *Leptotila jamaicensis gaumeri*. The pocket mouse is found in the Yucatan Peninsula of Mexico and in northern Guatemala.

Gee

Golden Leaf Monkey *Trachypithecus geei* **Khajuria,** 1956

Edward Prichard Gee (ca. 1910–1966) was a British tea planter who spent most of his life in India, where he developed a great interest in natural history. He was one of the first and most persistent advocates of the establishment of national parks. In 1964 he published *The Wild Life of India*. The monkey is found in Bhutan and Assam (India).

Genoways

Genoways' Yellow Bat *Rhogeessa genowaysi* **Baker,** 1984

Professor Dr. Hugh H. Genoways is a zoologist and mammalogist associated with the University of Nebraska, which he joined in 1986. Now Professor Emeritus at the university and at its museum, he was President of the American Society of Mammalogist from 1984 to 1986. Bats are among his interests, as is evidenced by a paper he published jointly with J. Knox Jones in 1967 called "A New Subspecies of the Fringe-tailed Bat *Myotis thysanodes*, from the Black Hills of South Dakota and Wyoming." In 1994 Professor Genoways became the first recipient of the Hugh H. Genoways annual award, established to recognize outstanding contributions to museums in Nebraska. The yellow bat is found in the lowlands of the Pacific coast of southern Mexico.

Geoffroy, E. G.

Geoffroy's Bat *Myotis emarginatus* E. G. Geoffroy, 1806
Geoffroy's Rayed Bat *Platyrrhinus lineatus* E. G. Geoffroy, 1810 [Alt. White-lined Broad-nosed Bat]
Geoffroy's Rousette *Rousettus amplexicaudatus* E. G. Geoffroy, 1810
Geoffroy's Woolly Monkey *Lagothrix cana cana* E. G. Geoffroy, 1812
Geoffroy's Monk Saki *Pithecia monachus monachus* E. G. Geoffroy, 1812
Geoffroy's Marmoset *Callithrix geoffroyi* **Humboldt,** 1812
Amazon River Dolphin *Inia geoffrensis* **Blainville,** 1817 [Alt. Pink Dolphin, Boto]
Geoffroy's Ground Squirrel *Xerus erythropus* **Desmarest,** 1817
Geoffroy's Spider Monkey *Ateles geoffroyi* **Kuhl,** 1820 [Alt. Central American Spider Monkey]
Lesser Nyctophilus *Nyctophilus geoffroyi* **Leach,** 1821 [Alt. Lesser Long-eared Bat]
Geoffroy's Horseshoe Bat *Rhinolophus clivosus* Cretzschmar, 1828
Geoffroy's Tailless Bat *Anoura geoffroyi* **Gray,** 1838
Western Quoll *Dasyurus geoffroii* **Gould,** 1841 [Alt. Chuditch]
Geoffroy's Cat *Oncifelis geoffroyi* **d'Orbigny** and **Gervais,** 1844 [Syn. *Leopardus geoffroyi*]
Geoffroy's Tamarin *Saguinus geoffroyi* **Pucheran,** 1845

Professor Étienne Geoffroy Saint-Hilaire (1772–1844) was a French naturalist. He originally trained for the Church but abandoned theology to become Professor of Zoology at the age of 21, when Le Jardin du Roi was renamed Le Muséum National d'Histoire Naturelle—the Natural History Museum in Paris. In his *Philosophie anatomique* (published from 1818 to 1822) and other works he expounded the theory that all animals conform to a single plan of structure. This notion was strongly opposed by Cuvier, who had been his friend, and in 1830 a widely publicized debate between the two took place. Despite their differences, the two men did not become enemies; they respected each other's research, and Geoffroy delivered one of the orations at Cuvier's funeral in 1832. Modern developmental biologists have confirmed some of Geoffroy Saint-Hilaire's ideas. He also had several birds named after him, including the Red-cheeked Parrot *Geoffroyus geoffroyi* (where both the genus and the species honor his name). Geoffroy's Bat is found over much of Europe (but not in the UK), northern Africa, and east to Iran and Afghanistan. The rayed bat occurs in South America as far south as Uruguay and northern Argentina. The rousette (a fruit bat) extends from Thailand through Malaysia and Indonesia east to New Guinea and the Solomon Islands. The woolly monkey and saki are from tropical South America, as is the river dolphin. The marmoset is confined to eastern Brazil. The ground squirrel ranges from Senegal east to western Ethiopia and western Kenya. The spider monkey can be found from southern Mexico to Panama. The nyctophilus bat occurs across most of Australia, and the horseshoe bat in Arabia, Israel, and Jordan and in eastern and southern Africa. The tailless bat ranges from Mexico to Bolivia and Brazil and is also found on Trinidad. The quoll is found in southwest Australia. The cat is found in South America from Bolivia and Uruguay to southern Argentina. The tamarin is found in Costa Rica, Panama, and northwest Colombia.

Geoffroy, I.

Geoffroy's Pied Colobus *Colobus vellerosus* I. Geoffroy, 1834 [Alt. Ursine Colobus]
Geoffroy's Saddle-back Tamarin *Saguinus fuscicollis nigrifrons* I. Geoffroy, 1851

Isidore Geoffroy Saint-Hilaire (1805–1861) was a French zoologist, son to Étienne (see above). Having studied medicine and natural history he became Assistant to his father in 1824 at the Natural History Museum in Paris. From 1829 until 1832 he lectured on ornithology and taught zoology there. He was particularly interested in deviant forms rather than the norm. Between 1832 and 1837 he published his work on teratology (i.e. what makes organisms deviate from normal), *Histoire générale et particulière des anomalies de l'organisation chez l'homme et les animaux*. In 1837 he became Deputy to his father at the Faculty of Science in Paris and went to Bordeaux to organize a faculty there. He became Inspector of the Academy of Paris in 1840, Professor of the Museum upon the retirement of his father in 1841, Inspector General of the University of Paris in 1844, and a member of the Royal Council for Public Instruction in 1845. In 1850, upon the death of Henri Marie Ducrotay de Blainville, he succeeded Blainville as Professor of Zoology at the Faculty of Science. In 1854 he founded the Acclimatization Society of Paris and was its President. He published a number of scientific papers, essays, and longer works on zoology, paleontology, anatomy, and the domestication of animals, including *Essais de zoologie generale* (1841) and *Histoire naturelle générale des règnes organiques* (published in three volumes from 1854 to 1862). The colobus is found from the Bandama River (Ivory Coast) east to western Nigeria. The tamarin comes from Amazonian Peru.

Gerbe

Pyrenean Pine Vole *Microtus gerbei* Gerbe, 1879

Dr. Jean-Joseph Zéphirin Gerbe (1810–1890) was a French naturalist and a doctor of science. He completed, and in 1867 published as a jointly authored work, *Ornithologie Européenne*, the work that Côme-Damien Degland was revising at the time of his death. He also published a number of articles mostly on ornithological subjects and a French translation of Alfred Edmund Brehm's works under the title *La vie des animaux illustrée*. The pine vole is found in northern Spain and southwest France.

Germain

Indochinese Leaf Monkey *Trachypithecus germaini* Schlegel, 1876 [Alt. Indochinese Silvered Langur]

(Louis) Rodolphe Germain (1827–1917) was a veterinary surgeon in the French colonial army, serving in Indochina (Vietnam) from 1862 until 1867. He traveled in New Caledonia from 1875 until 1878. He made considerable zoological collections in his spare time, donating them to the Natural History Museum in Paris. He is also commemorated in the name of Germain's Peacock-Pheasant *Polyplectron germaini*. The monkey is found in Thailand, Cambodia, and Vietnam.

German

German's Melomys *Melomys paveli* Helgen, 2003
German's Shrew-Mouse *Mayermys germani* Helgen, 2005

Pavel German (b. 1950) is an Australian wildlife photographer and naturalist. He collected the holotype of the shrew-mouse in 1992 in New Guinea, and is also the collector of many other important mammal specimens from throughout the Melanesian region. He was born in the Soviet Union but relocated to Australia largely because of his fascination with the region's wildlife. In the late 1980s and early 1990s he worked for four years in the Australian Museum's Mammal Section with Tim Flannery, carrying out biodiversity surveys in Papua New Guinea, the South Pacific region, and eastern Indonesia. His award-winning photographs have been published all over the world in hundreds of publications, including books such as *Nature Guide Insects of Australia*. The melomys is from the Indonesian island of Seram; it was originally described as a subspecies of *Melomys rufescens* but is now sometimes treated as a full species. The type specimen was collected by German (along with E. Tasker) in May 1993 but was not described as a new taxon until 10 years later. The shrew-mouse is from southeast New Guinea.

Gervais

Gervais' Funnel-eared Bat *Nyctiellus lepidus* Gervais, 1837
Gervais' Beaked Whale *Mesoplodon europaeus* Gervais, 1855
Gervais' Fruit-eating Bat *Artibeus cinereus* Gervais, 1856
Gervais' Large-eared Bat *Micronycteris minuta* Gervais, 1856 [Alt. Tiny Big-eared Bat]

Professor François Louis Paul Gervais (1816–1879) was a French zoologist, paleontologist, and anatomist. He studied under Blainville, who was Cuvier's successor, and in 1868 followed him to become Professor of Comparative Anatomy at the Natural History Museum, Paris. From 1835 until 1845 he was an Assistant at the museum, and in 1845 he was appointed Professor of Zoology and Comparative Anatomy at the Faculty of Sciences of Montpellier, becoming head of the faculty in 1856. The funnel-eared bat comes from Cuba and the Bahamas. The whale is found in the western North Atlantic (with a few records from the eastern Atlantic). The fruit-eating bat occurs in northern South America, and the large-eared bat is found from Honduras to southern Brazil.

Gibbs

American Shrew-Mole *Neurotrichus gibbsii* **Baird,** 1857

George Gibbs (1815–1873) was a lawyer who practiced in New York. He was extremely interested in both geology and linguistics. He went west to join the California Gold Rush, and in 1853 he was employed as a geologist on the Pacific Railway Survey. He developed a fascination with the northwest of the USA, especially its natural history, its native peoples, and their languages. That proved very useful when in 1857 he was employed as both a geologist and an interpreter during the Northwest Border Survey that settled the final points of the border with Canada. The American Civil War intervened, and the survey's findings and reports were never published. Gibbs was in regular touch with the Smithsonian about his observations, and today the institution houses a collection of his papers. The shrew-mole is found in western North America from southwest British Columbia (Canada) south to west-central California.

Giffard

Giffard's Shrew *Crocidura giffardi* **de Winton,** 1898 [Syn. *C. olivieri giffardi*]

Colonel W. C. Giffard (1859–1921) was in the Gold Coast (now Ghana) from 1897 to 1898. He collected a number of mostly bird specimens in that period that he sent to the museum at Tring, where they were described by Ernst Johann Otto Hartert, who worked for Rothschild. Today the shrew is regarded as a West African subspecies of the widespread African Giant Musk Shrew *Crocidura olivieri*.

Gigas

Steller's Sea-Cow *Hydrodamalis gigas* Zimmermann, 1780 extinct
Australian False Vampire Bat *Macroderma gigas* **Dobson,** 1880
Merriam's Desert Shrew *Megasorex gigas* **Merriam** 1897
Giant Thicket Rat *Grammomys gigas* **Dollman,** 1911

Gigas was a giant in Greek mythology, the child of Uranos and Gaea. The name is applied to taxa that are "giants" of their kind. The sea-cow was found in the Commander Islands, Bering Sea. It is believed to have become extinct by 1770. The bat is from north and central Australia, and the shrew from Mexico. The thicket rat is confined to Mount Kenya.

Gilbert

Gilbert's Potoroo *Potorous gilbertii* **Gould,** 1841
Gilbert's Dunnart *Sminthopsis gilberti* Kitchener, Stoddart, and Henry, 1984

John Gilbert (1812–1845) was a naturalist and explorer who was the principal collector of birds and other animals, although he also collected plants, for John Gould in southwestern Western Australia between 1840 and 1842, taking passage on a ship with the curious name of *Houghton le Skerne*. While in Australia he stayed for a while with a settler, Mrs. Brockman, who wrote of him, "He used to go out after breakfast, provided with some luncheon, and we seldom saw him until late afternoon, when he would come in with several birds and set busily to work to skin and fill them out before dark. . . . He was an enthusiast at his business, never spared himself, and often came in quite tired out from a long day's tramp after some particular bird, but as pleased as a child if he succeeded in shooting it." Despite all his efforts he was poorly served by Gould, who left him with insufficient funds and equipment yet had high expectations of him. Gilbert was born in London, trained as a taxidermist, and was employed as such by the Zoological Society there. He became Curator of the Shropshire and North Wales Natural History Society in Shrewsbury, but as the society was short of funds, his contract was terminated in 1837, and the next year he left for Australia. There he was speared to death on 28 June 1845 by Aboriginal people at the Gulf of Carpentaria. This happened when he was naturalist on Ludwig Leichhardt's expedition to Port Essington. While the expedition was thought heroic at the time, it is likely that they were attacked because two members of the expedition, Aboriginals themselves, treated the people they met very badly and probably provoked the attack, having raped a local Aboriginal woman. Leichhardt's account says Gilbert heard the noise of the attack in the night and rushed from his tent with his gun, only to receive a spear in the chest. He uttered the words "Charlie, take my gun, they have killed me," before dropping lifeless to the ground. Gould barely acknowledged Gilbert's huge contribution of specimens, descriptions, and detailed observations, without which Gould's seminal work on Australia's birds and mammals would be greatly diminished. Gilbert shipped most of his specimens back to Europe on board the *Beagle* and the rest on board the *Napoleon*. The potoroo was thought to have become extinct by 1880 but was rediscovered in 1994 near Albany, southern Western Australia. The dunnart also comes from southwest Western Australia.

Giles

Giles' Planigale *Planigale gilesi* **Aitken,** 1972 [Alt. Paucident Planigale]

William Ernest Powell Giles (1835–1897) was born in England and went to Australia in 1850 to join his family in Adelaide. He was an explorer who was never satisfied and always kept pushing on. He wanted to be the first person to penetrate the central area of Western Australia. Over a period from 1872 to 1876 he led five expeditions into the Outback, two of them on camels. During his travels he discovered Mount Olga and named the Gibson Desert after his companion. They ran out of supplies and had just one horse left. Giles told Gibson to take it and get help while he himself walked. Giles made it out of the desert, but Gibson was never seen again. Giles said of himself that he was "the last of the Australian explorers." He never received any monetary reward for his efforts but was awarded the Gold Medal of the Royal Geographical Society. Giles died of pneumonia, which he developed while working as a clerk in the gold fields at Coolgardie. Aitken said that he named the animal after Giles because, like the planigale, he was "very good at surviving in a desert." The planigale (a small carnivorous marsupial) is native to the deserts of east-central Australia.

Gill

Gill's Bottle-nosed Dolphin *Tursiops gilli* **Dall,** 1873 [Syn. *T. truncatus* Montagu, 1821]

Professor Theodore Nicholas Gill (1837–1914) was an ichthyologist and zoologist at George Washington University and was associated with the Smithsonian Institution for more than half a century. His father wanted him to enter the Church, but this seemed an unattractive calling to Gill, who decided to qualify as a lawyer instead. In around 1857 he was fortunate enough to come to the attention of Spencer Baird, whom he met in Washington while en route to the West Indies, where he made an important collection, especially of freshwater fish from Trinidad. In 1859 he went to Newfoundland, and those two trips—West Indies and Newfoundland—were the only extensive fieldwork Gill ever carried out. In 1862 he was put in charge of the Smithsonian's Library, which was transferred in 1866 to the Library of Congress, where Gill served as Senior Assistant Librarian until 1874. During his career he produced over 500 papers of which 388 were on ichthyology. He also for a time edited the ornithological magazine *The Osprey*. The taxonomy of the genus *Tursiops* is not fully resolved, and some zoologists regard *T. gilli* as a valid species of bottle-nosed dolphin from the eastern North Pacific. However, most authorities believe it to be a form of the widespread *T. truncatus*.

Gilliard

New Britain Flying Fox *Pteropus gilliardorum* **Van Deusen,** 1969

Ernest Thomas Gilliard (1912–1965) was an American ornithologist who was closely associated with the American Museum of Natural History in New York. His wife Margaret (dates not found) often assisted him. He wrote chiefly on birds of New Guinea, his books including *Living Birds of the World* (1959) and *Birds of Paradise and Bower Birds* (1969). His *Handbook of New Guinea Birds*, written with A. L. Rand, was first published in 1968. He also published a book that included ethnographic photographs, which he took himself in 1953 and 1954, of a Kanganam village on the middle Sepik River. He took part in the New Britain expedition and an exploration of the Whiteman Mountains between 1958 and 1959. He is also commemorated in the scientific name of a frog, *Platymantis gilliardi*. The binomial *gilliardorum* is, of course, plural; it was actually Margaret Gilliard who collected the type specimen of the flying fox. Van Deusen's etymology reads, "Archbold Expeditions owes a debt of gratitude to the abilities of Margaret Gilliard who assembled the largest and most important collection of mammals ever made on the island of New Britain. The collecting, as well as the preparation, of specimens was accomplished during an arduous overland trip to the Whiteman Range despite her other duties which often included the running of a base camp, the supplying of advance camps, and assistance to her husband in the study and collection of birds. The late Dr. Gilliard enthusiastically encouraged her in collecting mammals on an expedition in which the study of the avifauna was of paramount importance. This new species of *Pteropus* is named in their honor." The flying fox is confined to New Britain.

Giovanni

Giovanni's Big-eared Bat *Micronycteris giovanniae* **Baker** and Fonseca, 2007

Professor Yolande Cornelia "Nikki" Giovanni (b. 1943) is an American poet, activist, and teacher. She is a survivor of lung cancer and is also noted for having taught Seung-Hui Cho at Virginia Tech University, where she is a Professor of English. She found him "mean" and asked that he be removed from her class. He was not, and is now notorious for the murders he committed on the campus in 2007. The bat, which is found in western Ecuador, was named after her "in recognition of her poetry and writings."

Glass

Glass' Shrew *Crocidura glassi* **Heim de Balsac,** 1966

Professor Dr. Bryan P. Glass (dates not found), a noted ornithologist and zoologist, finished his career as Professor Emeritus at Oklahoma State University. In 1946 he was a member of the Department of Fish and Game at the Agricultural and Mechanical College of Texas (now better known as Texas A&M). From 1957 to 1977 he was the Secretary Treasurer of the American Mammal Society. He wrote *A Key to the Skulls of North American Mammals,* published in 1997. He was an associate editor of the Oklahoma Academy of Science for many years and has written numerous scientific papers, including descriptions of new taxa. The shrew is found in the highlands of Ethiopia.

Gleadow

Indian Hairy-footed Gerbil *Gerbillus gleadowi* Murray, 1886

Sand-colored Soft-furred Rat *Millardia gleadowi* Murray, 1886

F. Gleadow (dates not found) was a naturalist and artist in northern India (including what is now Pakistan) at the end of the 19th century and the beginning of the 20th. He may have had some involvement in the Indian forestry industry, as in 1901 he translated and published, in *The Indian Forester,* a work by the French author A. Melard, "Insuffisance de la production des bois d'oeuvre dans le monde." His translation was entitled "Insufficiency of the World's Timber Supply," with a note on the cover stating that it was "applied to India." He reported in 1887, in the *Journal of the Bombay Natural History Society,* on a new lizard, and in 1901 on a new species of spider. He was also a keen philatelist and was involved in a long correspondence on the dye stuffs used in the production of postage stamps and postmarks, particularly in regard to Kashmir and Jammu. The gerbil is from northwest India and the Indus Valley in Pakistan. The rat has a similar but slightly larger range, extending into Afghanistan.

Glen

Glen's Wattled Bat *Glauconycteris gleni* Peterson and Smith, 1973 [Alt. Glen's Butterfly Bat; Syn. *Chalinolobus gleni*]

Glen's Long-fingered Bat *Miniopterus gleni* Peterson, Eger, and Mitchell, 1995

Glen Holland is a South African now living in New Zealand, where he is Director of the Auckland Zoo. Before being appointed to that position he spent three years running the Mount Bruce Wildlife Centre and managed a bird park in South Africa. As a child he kept and bred birds, and to quote him, "I also had snakes in the bedroom, and tortoises and everything I could get my hands on." The Auckland Zoo supports the Limbe Wildlife Centre in Cameroon, which is one of the countries where the wattled bat has been recorded, the other being Uganda. The long-fingered bat is found in Madagascar.

Glover

Glover's Pika *Ochotona gloveri* **Thomas,** 1922

Professor Dr. Glover Morrill Allen (1879–1942). The pika is found in parts of China (western Sichuan, northwest Yunnan, northeast Tibet and southwest Qinghai). See **Allen, G. M.** for biographical details.

Gloverallen

Barbados Raccoon *Procyon gloveralleni* **Nelson** and **Goldman,** 1930 extinct

Professor Dr. Glover Morrill Allen (1879–1942). As the name implies, the raccoon was found only on Barbados. It has not been seen since the 1960s. However, most zoologists now believe that this "species" was not a native Barbados mammal but that the island population derived from human introductions of the North American Raccoon *Procyon lotor.* See **Allen, G. M.** for biographical details.

Gmelin

Gmelin's Shrew *Crocidura gmelini* **Pallas,** 1811

Mouflon *Ovis gmelini* **Blyth,** 1841

Tarpan *Equus przewalski gmelini* Antonius, 1912 extinct [Alt. European Wild Horse; Syn. *E. ferus*]

Professor Johann Friedrich Gmelin (1748–1804) belonged to a well-known family of German naturalists. He was Professor of Medicine in Göttingen and wrote a 10-volume work on botany as well as several works on chemistry, but he is possibly remembered mainly as the publisher of edition 13 of Linnaeus' *Systema naturae,* published in three volumes between 1788 and 1796. Other members of his family accompanied Pallas on one of his expeditions. The shrew is found in steppes and semidesert from central Iran to central China. The Mouflon occurs in southern and eastern Turkey, Armenia, Azerbaijan, northern Iraq, and western Iran. (Many references give the scientific name of the Mouflon as *Ovis musimon,* but this was applied to populations in Corsica and Sardinia that are feral, deriving from early human introduction. The name is therefore not regarded as valid for the truly wild sheep of southwest Asia.) The Tarpan was found in eastern Europe until the early 19th century and survived longest in the Ukraine; the last known specimen died in 1879. There has been disagreement as to the valid scientific name for this form of wild horse.

Godman

Godman's Long-tailed Bat *Choeroniscus godmani* **Thomas,** 1903

Atherton Antechinus *Antechinus godmani* Thomas, 1923

Godman's Rock Wallaby *Petrogale godmani* Thomas, 1923

Dr. Frederick du Cane Godman F.R.S. (1834–1919) was a British naturalist who, with his friend Osbert Salvin, compiled the massive *Biologia Centrali Americana,* which was issued in parts from 1888 until 1904. Godman and Salvin also presented their joint collection to the British Museum of Natural History over a 15-year period, starting in 1885. Godman qualified as a lawyer but was wealthy and had no need to earn a living, so he devoted his life to natural history, particularly ornithology. He visited Norway, Russia, the Azores, Madeira, the Canary Islands, India, Egypt, South Africa, Guatemala, British Honduras (Belize), and Jamaica. Some of his travels were made with Salvin. In 1870 he published *Natural History of the Azores, or Western Islands,* and between 1907 and 1910 *A Monograph of the Petrels.* After his death the Godman Memorial Exploration Fund was set up by his widow and daughters. This resulted in the antechinus and the rock wallaby having *godmani* as their binomial. To quote the words of Oldfield Thomas when naming the antechinus, "I have named the species, the first-fruits of the Exploration Trust, in honour and affectionate remembrance of the late Mr. F. DuCane Godman, Trustee and life-long benefactor of our National Museum, in whose memory the Godman Collecting Fund has been founded by his widow." The long-tailed bat is found from Mexico to northern South America. The antechinus in restricted to the Atherton Tablelands of northeast Queensland. The rock wallaby is also confined to northeast Queensland, north of Cairns.

Goeldi

Goeldi's Marmoset *Callimico goeldii* **Thomas,** 1904 [Alt. Goeldi's Monkey]
Goeldi's Spiny Rat *Proechimys goeldii* Thomas, 1905

Emil August Goeldi (1859–1917) was a Swiss zoologist who was born in Zurich. He went to Brazil in 1880, where at first he worked at the National Museum. Later he reorganized the Pará Museum of Natural History and Ethnography, which was founded in 1866. The institution today bears his name: Museu Paraense Emílio Goeldi. He became well known for his studies of Brazilian birds and mammals. Goeldi returned to Switzerland in 1907 to teach biology and physical geography at the University of Bern until his death 10 years later. He wrote *Aves do Brasil* (1894) and *Die Vogelwelt des Amazonensstromes* (1901). He is also commemorated in the name of a bird, Goeldi's Antbird *Myrmeciza goeldii*, and in the name of an amphibian, Goeldi's Frog *Flectnotus goeldi*, as well as in the binomial of many other taxa, including 17 different species of South American ants. He was reputed to be a racist who disliked Brazilians and worked in a Swiss enclave. The marmoset is found in the western Amazon basin (western Brazil, northern Bolivia, eastern Peru, and eastern Ecuador). The spiny rat occurs only in Amazonian Brazil.

Goff

Goff's Pocket Gopher *Geomys pinetis goffi* **Sherman,** 1944 extinct

Carlos C. Goff (d. 1939) was a scientist at the University of Florida. He was the first person to seriously investigate the networks of tunnels made by gophers. He was also interested in herpetology, publishing jointly (and posthumously) with C. J. Goin, in 1941, "Notes on the Growth Rate of the Gopher Turtle *Gopherus polyphemus*." The pocket gopher, endemic to just one location in Florida (Pineda Ridge, Brevard County), was last seen in 1955 and is presumed to be extinct.

Goldman

Goldman's Pocket Gopher *Cratogeomys goldmani* **Merriam,** 1895
Goldman's Small-eared Shrew *Cryptotis goldmani* Merriam, 1895
Goldman's Pocket Mouse *Chaetodipus goldmani* **Osgood,** 1900
Goldman's Spiny Pocket Mouse *Heteromys goldmani* Merriam, 1902
Goldman's Woodrat *Neotoma goldmani* Merriam, 1903
Nelson's and Goldman's Woodrat *Nelsonia goldmani* Merriam, 1903
Goldman's Water Mouse *Rheomys raptor* Goldman, 1912
Goldman's Nectar Bat *Lonchophylla concava* Goldman, 1914

Major Edward Alphonso Goldman (1873–1946) was a field naturalist and mammalogist who was born in Mount Carroll, Illinois. Nelson hired him in January 1892 to assist his biological investigations of California and Mexico, and then employed him as Field Naturalist and eventually Senior Biologist with the U.S.

Bureau of Biological Survey. He spent nearly 14 years collecting in every region of Mexico. The biological explorations and collecting expeditions made by Nelson and Goldman in Mexico from 1892 to 1906 are said to have been "among the most important ever achieved by two workers for any single country." They conducted investigations in every state in Mexico, collecting 17,400 mammals and 12,400 birds, as well as amassing an enormous fund of information on the natural history of the country. The best account of their work is Goldman's "Biological Investigations in Mexico" (*Smithsonian Miscellaneous Collections*, vol. 115, 1951). Goldman also held an honorary position with the Smithsonian Institution, as an Associate in Zoology, from 1928 to 1946. In 1911 through 1912 he was part of the Biological Survey of Panama, during the construction of the canal. His results were published in *The Mammals of Panama* in 1920. In 1936 he assisted the U.S. government in negotiating with Mexico to protect migratory birds. Goldman's bibliography includes more than 200 titles. He named over 300 forms of mammals, most of them subspecies. Many other taxa bear his name in their scientific appellations, such as the Russet-crowned Quail-Dove *Geotrygon goldmani* and the Mexican Freshwater Toadfish *Batrachoides goldmani*. Goldman Peak in Baja California was also named in his honor. He was President of the Biological Society of Washington from 1927 until 1929 and of the American Society of Mammalogists in 1946. The shrew is found in the highlands of Mexico and Guatemala. The pocket mouse is confined to northwest Mexico. The spiny pocket mouse occurs in the extreme south of Mexico and adjacent Guatemala. Both woodrats are endemic to Mexico. The water mouse is found in Costa Rica and Panama. The nectar bat occurs from Costa Rica south to Ecuador.

Goliath

Eastern White-eared Giant Rat *Hyomys goliath* **Milne-Edwards,** 1900
Goliath Shrew *Crocidura goliath* **Thomas,** 1906
Harrison's Fruit Bat *Lissonycteris goliath* Bergmans,1997

Goliath of Gath (ca. 1,030 B.C.) was a Philistine warrior of giant size who was killed with a slingshot by David, later King of the Jews (1 Sam. 27.4–50). The giant rat is found in the highlands of Papua New Guinea. The shrew occurs in Cameroon, Gabon, Equatorial Guinea, the Congo Republic, and DRC (Zaire). The fruit bat, originally described as a race of *Lissonycteris angolensis,* is known from the area of the Mozambique-Zimbabwe border.

Gonzales

Mount Isarog Striped Rat *Chrotomys gonzalesi* **Rickart** and **Heaney,** 1997

Dr. Pedro C. Gonzales is a member of the Department of Zoology at the Philippine National Museum. He has specialized in the fauna of Mount Isarog and published on it—in 1990, with S. M. Goodman, "The Birds of Mt. Isarog National Park, Southern Luzon, Philippines, with Particular Reference to Altitudinal Distribution," and in 1999 with four other co-authors, "Mammalian Diversity on Mt. Isarog, a Threatened Center of Endemism on Southern Luzon Island, Philippines." The rat is found only on Mount Isarog in Luzon (Philippines).

Goodfellow

Goodfellow's Tree Kangaroo *Dendrolagus goodfellowi* **Thomas,** 1908
Goodfellow's Tuco-tuco *Ctenomys goodfellowi* Thomas, 1921

Walter Goodfellow (1866–1953) was a British ornithologist and explorer. His first expedition was to Colombia and Ecuador between 1898 and 1899, during which he and Claude

Hamilton collected about 4,000 bird skins, covering 550 species. He also led the British Ornithologists' Union's expedition of 1909 to 1911 to New Guinea. His last expedition was in 1935 to Melville Island, Australia. Goodfellow's collection was later acquired by the British Museum. He was awarded the British Ornithologists' Union Medal in 1912. He is also commemorated in the names of birds such as Goodfellow's White-eye *Lophozosterops goodfellowi*. The tree kangaroo is found in New Guinea. The tuco-tuco is found in Bolivia; it is sometimes treated as a subspecies of *Ctenomys boliviensis*.

Goodman

Goodman's Mouse-Lemur *Microcebus lehilahytsara* Roos and Kappeler, 2005

Dr. Steven "Steve" Michael Goodman (b. 1957) is a mammalogist who has dedicated two decades to the study of land vertebrates in Madagascar for the Field Museum of Chicago and the World Wildlife Fund. He joined the Zoology Department of the Field Museum in 1989. As a rebellious teenager Goodman declared himself uncomfortable living indoors, and at 14 he began sleeping outside. He conducted field research projects in numerous portions of the world until he settled in Madagascar, "a place I could sink my teeth into," where he now works and trains local scientists.In 2005 he was awarded the prestigious MacArthur Foundation Fellowship for his discoveries and research in Madagascar. Kappeler said, "Goodman's field research in all remote parts of Madagascar has contributed enormously to our knowledge about the diversity of Madagascar's unique and threatened fauna and flora." He has been described as "the Michael Jordan of vertebrate field biology." The mouse-lemur is confined to eastern Madagascar. (This is another example of a clever binomial, as *lehilahytsara* is a combination of the Malagasy words meaning "good" and "man").

Goodwin

Goodwin's Small-eared Shrew *Cryptotis goodwini* Jackson, 1933

Goodwin's Mouse-like Hamster *Calomyscus elburzensis* Goodwin, 1938

Goodwin's Spiny Pocket Mouse *Heteromys (desmarestianus) nigricaudatus* Goodwin, 1956

Goodwin's Bat *Chiroderma trinitatum* Goodwin, 1958 [Alt. Little Big-eyed Bat]

Goodwin's Water Mouse *Rheomys mexicanus* Goodwin, 1959

George Gilbert Goodwin (b. 1895) was Assistant Curator of Mammals at the American Museum of Natural History in New York. He took part in various expeditions, such as the Legendre expedition to Iran in 1938. He published over 70 books and articles between 1924 and 1977, including "Mammals from the State of Oaxaca, Mexico, in the American Museum of Natural History" (1969). The shrew is found in southern Mexico, Guatemala, and El Salvador. The hamster occurs in the mountains of northern Iran and southern Turkmenistan. The pocket mouse and the water mouse both have a restricted range in Oaxaca State, southern Mexico. The bat ranges from Panama to Bolivia and Amazonian Brazil, and also occurs on Trinidad.

Gordon

Gordon's Red Colobus *Piliocolobus gordonorum* **Matschie,** 1900 [Alt. Udzungwa Red Colobus, Iringa Red Colobus]

Von Gordon (dates not found). The word *gordonorum* is genitive plural and refers to the brothers Von Gordon, whom we believe to have been resident in Tanganyika (now Tanzania) at the end of the 19th century and the beginning of the 20th. At that time Tanganyika was a German colony. The colobus is an endangered Tanzanian endemic.

Gordon, A. C. G.

Gordon's Wild Cat *Felis silvestris gordoni* **Harrison,** 1968 [Alt. Arabian Wild Cat]

Major A. C. G. Gordon (dates not found) is the person who obtained the type specimen of the cat in Oman. We have not been able to find out more about him. The cat is found in the Arabian Peninsula.

Gorgas

Gorgas' Rice Rat *Oryzomys gorgasi* **Hershkovitz,** 1971

General William Crawford Gorgas (1854–1920) was an epidemiologist and Surgeon General of the U.S. Army. He is known throughout the world as the conqueror of the mosquito and of the malaria and yellow fever it transmits. His pioneering efforts in halting an epidemic of yellow fever by the application of sanitary measures enabled the USA to complete the Panama Canal after earlier attempts had fallen before the onslaught of the insidious insect. The Gorgas Memorial Laboratory was named in his honor. The rice rat was long known from one specimen from northwest Colombia, but in 2001 it was reported rediscovered at a new location in northwest Venezuela.

Gosling

African Water Rat *Colomys goslingi* **Thomas** and **Wroughton,** 1907

Captain George Bennet Gosling (1872–1906) was an explorer and zoologist. He was part of the Boyd Alexander expedition of 1904–1906, which aimed to cross Africa from the Niger to the Nile. He explored the area around Lake Chad but died of blackwater fever while on the Uele River. He is also commemorated in the names of two birds, Gosling's Apalis *Apalis goslingi* and Gosling's Rock Bunting *Emberiza tahapisi goslingi*. The rat has been recorded in Liberia and scattered locations from Cameroon to western Ethiopia, northeast Angola, and northwest Zambia.

Goudot

Malagasy Mouse-eared Bat *Myotis goudoti* **A. Smith,** 1834

Falanouc *Eupleres goudotii* Doyère, 1835 [Alt. Small-toothed Mongoose]

Jules Prosper "Bibikely" Goudot (dates not found) was a French traveler, collector, and entomologist in Madagascar in the first half of the 19th century. His nickname was Bibikely, which is Malagasy for "insect," emphasizing his interest in bugs. The locals thought he was a harmless fool and just ignored him, at a time when Europeans were often regarded as very suspicious characters. He learned Malagasy, adopted local customs, married a local woman, and became so immersed in everything Malagasy that he resigned his position as a collector for the Natural History Museum in Paris. He found the remains of a number of huge eggs and showed them to the museum's Professor Gervais, who at first thought they had been laid by some kind of ostrich. Only later was it discovered that they were eggs of the extinct Elephant Bird *(Aepyornis)*. Goudot is also remembered in the scientific name of the frog *Boophis goudoti*. The Falanouc is endemic to Madagascar, while the bat is found on Anjouan (Comoro Islands) in addition to the Madagascan mainland.

Gould

Gould's Mouse *Pseudomys gouldii* **Waterhouse,** 1839 extinct
Gould's Wattled Bat *Chalinolobus gouldii* **Gray,** 1841
Black-footed Tree Rat *Mesembriomys gouldii* Gray, 1843
Gould's Rat-Kangaroo *Bettongia gouldii* Gray, 1843 [Syn. *B. penicillata* Gray, 1837]
Gould's Nyctophilus *Nyctophilus gouldi* **Tomes,** 1858 [Alt. Gould's Long-eared Bat]

John Gould (1804–1881), the son of a gardener at Windsor Castle, became an illustrious British ornithologist, artist, and taxidermist. Born in Dorset, England, in 1804, Gould went on during his 76-year lifespan to be acknowledged around the world as "the Bird Man." He was employed as a taxidermist by the newly formed Zoological Society of London and traveled widely in Europe, Asia, and Australia. He was arguably the greatest, and certainly the most prolific, publisher and original author of ornithological works in the world. Between 1830 and 1881, in excess of 46 volumes of reference works were produced by him in color. He published 41 works on birds, with 2,999 remarkably accurate illustrations executed by a team of artists that included his wife. His first book, on Himalayan birds, was based on skins shipped to London, but later in his career he traveled to see birds in their natural habitats. In 1838 Gould and his wife, Elizabeth, arrived in Australia to spend 19 months studying and recording the natural history of the continent. By the time they left, Gould had not only recorded most of Australia's known birds, and collected information on nearly 200 new species, but had also gathered data for a major contribution to the study of Australian mammals. His best known works include *The Birds of Europe, The Birds of Great Britain, The Birds of New Guinea,* and *The Birds of Asia.* He also wrote monographs on the Macropodidae, Odontophorinae, Trochilidae, and Pittidae. Gould was a commercially minded man, and Victorian England was fascinated by the exotic, including those exquisite jewels the hummingbirds, a group with which Gould's name is particularly associated. His superb paintings and prints of these and other birds were greatly sought after—so much so that he probably had trouble keeping up with the demand. He is commemorated in the common names of at least 28 birds, including Gould's Bronze Cuckoo *Chrysococcyx russatus* and Gould's Storm-Petrel *Fregetta tropica.* The mouse was found in southern Australia but was last collected in 1857 and is presumed extinct. The wattled bat is found over most of Australia and is also known from Norfolk Island and New Caledonia. The tree rat comes from northern Australia, and the long eared bat from southwest and eastern Australia. Gray named two forms of Rat-Kangaroo *(Bettongia),* but they are now known to be conspecific, and *Bettongia gouldii* is thus only a junior synonym of *B. penicillata.*

Graells

Graells' Tamarin *Saguinus graellsi* Jimenez de la Espada, 1870 [Alt. Rio Napo Tamarin; Syn. *S. nigricollis graellsi*]

Professor Dr. Mariano de la Paz Graells y de la Aguera (1809–1898) of Madrid was a botanist, entomologist, and malacologist. He qualified in medicine and natural sciences at the University of Barcelona, where he was firstly an Associate and later a full Professor of Physics and Chemistry. There was an epidemic in Barcelona in 1835, and he was much to the fore in combating it. In 1837 he moved to Madrid, where he was appointed Professor of Zoology at the Museum of Natural Sciences and Director of the Botanical Gardens. In 1845 he went on a scientific expedition to the Pacific, resulting in the establishment by the Spanish Natural History Society of facilities to allow for the acclimatization of tropical plants. The Phylloxera outbreak virtually wiped out European vineyards in the latter part of the 19th century, and Graells, as a very senior man in the Council of Agriculture, had to deal with its effect on Spanish viticulture. He was one of the founding members of the Spanish Academy of Exact Sciences and was honored not only by the Spanish but also by several foreign governments. Wallace mentions him in connection with plants collected in Spain. In 1897 he published a paper about the Iberian Lynx entitled "Subfamilia felina fauna mastodologica." The tamarin is found in northeast Peru and eastern Ecuador.

Graffman

Graffman Dolphin *Stenella graffmani* Lönnberg, 1934 [Alt. Coastal Spotted Dolphin; Syn. *S. attenuata graffmani*]

J. Holger Graffman (dates not found) was a Swedish civil engineer living in Mexico who sent the type specimen to the Stockholm Natural History Museum. He returned to Sweden sometime before 1943, as at that time he was Managing Director of Transfer AB, Stockholm, an oil company. During WW2 he was a point of contact between American secret agents and Felix Kersten, who as Heinrich Himmler's personal masseur was able to get him to order the release of a number of Jews from concentration camps. The dolphin is found off the Pacific coast of Central America and is usually regarded as a form of the Pantropical Spotted Dolphin *Stenella attenuata*.

Grandidier

Grandidier's Free-tailed Bat *Chaerephon leucogaster* Grandidier, 1869

Grandidier's Mongoose *Galidictis grandidieri* Wozencraft, 1986 [Alt. Giant Striped Mongoose]

Grandidier's Tufted-tailed Rat *Eliurus grandidieri* Carleton and **Goodman,** 1998

Alfred Grandidier (1836–1921) was a French explorer, geographer, herpetologist, and ornithologist who collected in Madagascar in 1865 and contributed his knowledge of Malagasy birds to Sir Alphonse Milne-Edwards' 1876 publication *Histoire naturelle des oiseaux*. Grandidier subsequently edited the second edition of that work, which appeared in 1878. In 1866 he recovered bones of what turned out to be *Aepyornis maximus*—the huge extinct Elephant Bird. The mineral grandidierite, which is found in Madagascar, is named after him, as is Mont Alfred Grandidier, in the French part of the Antarctic. He is further commemorated in the scientific name of a bird, the Madagascar Spine-tailed Swift *Zoonavena grandidieri*, and eight reptiles including Grandidier's Gecko *Geckolepis typica* and Grandidier's Water Snake *Liopholidophis grandidieri*. All three mammals named after him are endemic to Madagascar (records of the bat on mainland Africa are regarded as questionable).

Grant, C. H. B.

Grant's Golden-Mole *Eremitalpa granti* Broom, 1907

Grant's Bushbaby *Galago granti* **Thomas** and **Wroughton,** 1907 [Alt. Mozambique Galago]

Grant's Rock Rat *Aethomys granti* Wroughton, 1908

Captain Claude Henry Baxter Grant (1878–1958) was a British ornithologist and collector. He was editor of the *Bulletin of the British Orni-*

thologists' Club from 1935 to 1940. Grant was co-author of *Birds of Eastern and North Eastern Africa* and *Birds of the Southern Third of Africa*, and he also wrote the *African Handbook of Birds*, published in 1952. He is commemorated in the common names of several birds, including Grant's Bluebill *Spermophaga poliogenys*. The golden-mole is found in western South Africa and the Namib Desert. The bushbaby is found in Mozambique and southern Tanzania. The rock rat has a small range in Cape Province, South Africa.

Grant, J. A.

Grant's Gazelle *Gazella granti* **Brooke,** 1872
Grant's Zebra *Equus quagga boehmi* **Matschie,** 1892 [Alt. Böhm's Zebra]

Colonel James Augustus Grant (1827–1892) was a Scottish naturalist and explorer. After completing his education at Marischal College, Aberdeen, he joined the British army and served in India during the Sikh Wars (1849) and the Indian Mutiny (1857–1858), during which he was wounded. He spent considerable time from 1860 to 1863 in Africa with John Hanning Speke, searching for the source of the Nile. He never saw the source, as he was unable to walk for six months because of debilitating leg ulcers. He kept a record of the journey, however, and published it as *A Walk across Africa* in 1864, in which he described "the ordinary life and pursuits, the habits and feelings of the natives." Primarily a botanist, he nevertheless discovered the gazelle in what is now Tanzania. In 1868 he served as an intelligence officer in the Abyssinian campaign and retired at the rank of Lieutenant Colonel. The gazelle and the zebra have similar distributions in southeast Sudan, southwest Ethiopia, Uganda, Kenya, and Tanzania.

Grant, M.

Grant's Caribou *Rangifer tarandus granti* **J A Allen,** 1902 [Alt. Barren-ground Caribou]

Madison Grant (1865–1937) was an American lawyer. He was noted for his conservation work and philanthropy and was a close friend of such influential people as Theodore Roosevelt. We prefer to remember him for this rather than for his work as a eugenicist. As such he was responsible for one of the most infamous works of scientific racism, a 1916 book entitled *The Passing of the Great Race*, which was greatly admired by Adolf Hitler and was used as part of the "scientific" basis for the Nazi Party's theories on race, justifying its policies of compulsory sterilization and euthanasia. The book also played a role in the crafting of strong immigration restrictions and antimiscegenation polices in the USA. As a conservationist, Grant is credited with saving many species, including the Redwood Tree and the American Bison; founding several environmental and philanthropic organizations; helping to develop the science of wildlife management; and helping to found the Bronx Zoo. From 1925 until his death from nephritis in 1937 he was the Secretary of the New York Zoological Society and used his position to lobby strongly for Ota Benga, a "pygmy" from the Congo, to be exhibited alongside the apes at the Bronx Zoo—a concept hard to believe in our times. It was through the New York Zoological Society that he secured funds for the Andrew J. Stone expedition to Alaska of 1901, during which the type specimen of this subspecies of caribou was collected. The caribou occurs in Alaska and northwest Canada.

Grant, W. R. O.

Luzon Hairy-tailed Rat *Batomys granti* **Thomas,** 1895
Grant's Shrew *Sylvisorex granti* Thomas, 1907

William Robert Ogilvie-Grant (1863–1924) was a Scottish ornithologist. He was Curator of Birds at the British Museum (Natural History) from 1909 until 1918, having started work there at the age of 19. He enlisted with the First Battalion of the County of London Regiment at the beginning of WW1 and suffered a stroke while helping to build fortifications near London in 1916. He is famed for describing a number of well-known species, such as the Philippine Eagle *Pithecophaga jefferyi.* He wrote *A Hand-book to the Game Birds* (1895) and is remembered in the name of Ogilvie-Grant's Warbler *Phylloscopus subaffinis.* As the name implies, the rat is found only on Luzon in the Philippines. The shrew occurs in montane forests of DRC (Zaire), Rwanda, Uganda, Kenya, and Tanzania.

Grasse

Grasse's Shrew *Crocidura grassei* **Brosset, Dubost,** and **Heim de Balsac,** 1965

Professor Pierre-Paul Grassé (1895–1985) was a French zoologist. His interests included cellular structure, animal sociology, and protistology. He was President of the French Academy of Sciences. He is perhaps most famous for his book *Evolution of Living Organisms*, which claims that while mutation is random, the course of evolution is not. He was also editor of the 35-volume *Traité de zoologie.* He is commemorated in the names of a number of gastropods such as *Diplectanum grassei* and *Porpostoma grassei.* The shrew is known from Cameroon, Gabon, Equatorial Guinea, and the Central African Republic.

Grauer

Grauer's Gorilla *Gorilla beringei graueri* **Matschie,** 1914 [Alt. Eastern Lowland Gorilla]
Grauer's Shrew *Paracrocidura graueri* **Hutterer,** 1986

Rudolf Grauer (1870–1927) was an Austrian explorer and zoologist who collected extensively during an expedition to the then Belgian Congo (now Zaire) in 1909 and again between 1910 and 1911 on an expedition paid for by the Austrian Imperial Museum. He suffered from actinomycosis contracted in Africa and eventually succumbed to this bacterial infection. He is also commemorated in the names of other taxa including Grauer's Cuckoo-Shrike *Coracina graueri,* and Grauer's Blind Snake *Rhinotyphlops graueri.* The gorilla comes from the eastern DRC (Zaire). The shrew was collected in 1908 but not described as a new species until 1986 and is known only from the type locality in the Itombwe Mountains of DRC (Zaire).

Gray

Gray's Spinner Dolphin *Stenella longirostris* Gray, 1828 [Alt. Pantropical Spinner Dolphin]
Gray's Dolphin *Stenella coeruleoalba* **Meyen,** 1833 [Alt. Striped Dolphin, Meyen's Dolphin]
Gray's Monk Saki *Pithecia irrorata irrorata* Gray, 1842 [Alt. Gray's Bald-faced Saki]
Gray's Long-tongued Bat *Glossophaga leachii* Gray, 1844
Gray's Spotted Dolphin *Stenella attentuata* Gray, 1846 [Alt. Pantropical Spotted Dolphin]
Gray's Crowned Guenon *Cercopithecus pogonias grayi* **Fraser,** 1850 [Alt. Erxleben's Guenon]
Gray's Four-striped Squirrel *Funisciurus isabella* Gray, 1862 [Alt. Lady Burton's Four-striped Squirrel]
Gray's Beaked Whale *Mesoplodon grayi* von Haast, 1876
Luzon Shrew *Crocidura grayi* **Dobson,** 1890

John Edward Gray (1800–1875) was a British zoologist and entomologist. He started work at the British Museum in 1824, with a temporary appointment at 15 shillings a day, but rose to become Keeper of Zoology. Gray published descriptions of a large number of animal species, including many Australian reptiles and mammals. He was regarded as the leading authority on many reptiles, including turtles. Gray was also an ardent philatelist and claimed that he was the world's first stamp collector. He worked at the museum with his brother George Robert Gray (1808–1872), and together they published a *Catalogue of the Mammalia and Birds of New Guinea in the Collection of the British Museum* (1859). He wrote *Gleanings from the Menagerie and Aviary at Knowsley Hall,* published between 1846 and 1850, which was illustrated by Lear. In 1869 Gray suffered a severe stroke that paralyzed his right side, including his writing hand. Yet he continued to publish to the end of his life by dictating to his wife, Maria Emma, who had always worked with him as an artist and occasional co-author. As shown above, Gray's name has been attached to three different dolphin species of the genus *Stenella*—a recipe for confusion that has led to all of them being generally referred to today by alternative common names. All three occur in tropical and warm-temperate waters worldwide. The saki monkey is from western Amazonia, south of the Amazon itself. The long-tongued bat is found from central Mexico south to Costa Rica. The guenon and the squirrel are both from west-central Africa. The beaked whale is found in cold and temperate waters of the Southern Hemisphere. The shrew is from the islands of Luzon and Mindoro in the Philippines.

Grayson

Tres Marias Cottontail *Sylvilagus graysoni* **J. A. Allen,** 1877

Andrew Jackson Grayson (1819–1869) was an American ornithologist and artist. He was considered to be the most accomplished bird painter in North America of his time and was often referred to as "the Audubon of the West." When Grayson began to paint the birds of western America in 1853, there was no systematic avifaunal record of the region from the Sierra Nevada to the Pacific Ocean. Grayson regarded his *Birds of the Pacific Slope* as a completion of Audubon's *Birds of America,* which did not include all the birds of the West. The Smithsonian Institution recruited Grayson as a field ornithologist, and he became one of its principal collectors for California, Mexico, and Mexico's offshore islands. He discovered many new species, some of which were named after him. He also wrote species accounts, recorded scientific data, and published articles on travel and natural history. However, his greatest achievement remained his paintings of birds, which are virtually color-perfect and depict their subjects in natural settings and activities. Over 16 years he painted more than 175 bird portraits, of which 156 survive, preserved in a single collection. Misfortune often seemed to beset Grayson in his latter years: he was shipwrecked, his son was mysteriously murdered, he was bankrupted, and he contracted yellow fever on a field expedition, from which he died. Birds named after him include Grayson's Parrotlet *Forpus cyanopygius insularis* and Grayson's Thrush *Turdus graysoni.* The endangered cottontail rabbit is confined to the Tres Marias Islands off western Mexico.

Green

Green's Puma *Puma concolor greeni* **Nelson** and **Goldman,** 1931

Edward C. Green (dates not found) was a collector who seems to have been a resident of Rio Grande do Norte in Brazil, where he is recorded in 1921 collecting a new subspecies of bat and in 1930 collecting the type specimen of the puma. Nelson and Goldman wrote, "The new form is named for the collector of the type, Mr. Edward C. Green, a collaborator of the Biological Survey for many years." This race of puma is found only in eastern Brazil. Many authorities do not regard it as a valid subspecies.

Greenhall

Greenhall's Dog-faced Bat *Molossops greenhalli* **Goodwin,** 1958

Arthur Merwin Greenhall (1911–1998) was an American zoologist and field naturalist associated with the Detroit Zoological Park and Portland Zoo. In 1953 he moved to Trinidad and assumed several positions simultaneously: Zoologist, Director of the Emperor Valley Zoo, and Curator of the Royal Victoria Institute, as well as a staff position with the Trinidad Virus Laboratory. His home there came to be known as the "Bat Cave." He returned to the USA in 1963. He published, jointly with George Gilbert Goodwin, "A Review of the Bats of Trinidad and Tobago" (1961). He produced many papers and pamphlets, such as, for the U.S. Fish and Wildlife Service, "House Bat Management" (1982) and "The Natural History of Vampire Bats" (1988). He later published a memoir, *Past and Present Thoughts of a Trinidad Field Naturalist 1934–1990.* His Central American and Caribbean exploits are documented in Ditmars and Bridges' 1935 book, *Snake-Hunter's Holiday.* His priorities might well be revealed by the way in which he proposed to his girlfriend in 1942 while riding on the New York City Subway, saying, "I'm moving to become the Curator of the Portland Zoo, so now we can get married!" The bat was first found on Trinidad but ranges from Mexico to Ecuador and northern Brazil.

Greenwood

Greenwood's Shrew *Crocidura greenwoodi* **Heim de Balsac,** 1966

Mrs. M. Greenwood (dates not found) was a lady who made a study of the fossil insectivores of East Africa. De Balsac expressly refers to her as "Mrs. Greenwood" and so really ought to have made the binomial *greenwoodae* (feminine). The shrew comes from southern Somalia.

Gregory, A. T.

Gregory's Red Wolf *Canis rufus gregoryi* **Goldman,** 1937 [Alt. Swamp Wolf]

Arthur Tappan Gregory (1886–1961) was trained as a lawyer but became a noted mammalogist and photographer. He was an early pioneer of self-portrait flash photography, and Goldman named the wolf after him because he had been successful in 1934 in getting photographs of red wolves through that method. He published *The Camera's Catch of North American Wild Animals* in 1939, and among his other publications was *The Nuremberg Trial* (1946), an event which he attended as an official observer and photographer. This subspecies of red wolf used to be found from southwestern Indiana, southern Missouri, and eastern Oklahoma to southern Mississippi, central Louisiana, and the Big Thicket area of Texas. By the early 1970s it was nearing extinction, and the remaining animals were brought into captivity to begin a breeding program. The program was a success, and red wolves have been reintroduced into the wild, but hybridization with coyotes is a problem in some areas.

Gregory, J. L.

Lesser Cane Rat *Thryonomys gregorianus* **Thomas,** 1894

Professor John Walter Gregory (1864–1932) was a Scottish explorer, stratigrapher, invertebrate paleontologist, and geomorphologist who undertook a number of adventures to India, Spitzbergen, Australia, Africa, and the Himalayas. He was Professor of Geology at the University of Melbourne from 1899 and was Director of the Geological Survey of Victoria from 1901 to 1904. In 1904 he returned home to take up a professorship at the University of Glasgow. He also collected some eyewitness accounts of the 1910 Glasgow earthquake. His final expedition to Peru ended in disaster when his canoe was overturned and he drowned in the Urubamba River. His publications include *The Great Rift Valley* (1896) and *Geography, Structural, Physical and Comparative* (1909). The rat is found from Cameroon east to southern Sudan and Ethiopia and south to Mozambique and Zimbabwe.

Gressitt

Gressitt's Melomys *Paramelomys gressitti* **Menzies,** 1996

Dr. Judson Linsley Gressitt (1914–1982) was an American botanist and entomologist at the Bishop Museum in Hawaii. Using specially equipped aircraft to collect insects as high as the planes could fly, he demonstrated that the winds and jet streams over the Pacific carried an enormous number of insects. In 1984 the J. L. Gressitt Rare Plant Sanctuary at West Maui was founded in his memory. In 1955 he initiated the museum's ongoing faunistic surveys of New Guinea, and he was later seconded as Director of the Wau Ecology Institute in Papua New Guinea, which began in 1961 as a field station of the Bishop Museum, later becoming a fully independent institution. He remained its Director until his death. In 1985 the institute published a book he had co-authored, *Handbook of Common New Guinea Beetles.* The melomys is endemic to Papua New Guinea.

Grevy

Grevy's Zebra *Equus grevyi* **Oustalet,** 1882

François Jules Paul Grevy (1807–1891) was President of the Third Republic of France from 1879 to 1887. In 1882 the Emperor of Abyssinia (now Ethiopia) sent a zebra to France as a gift. The animal died soon after setting hoof on Gallic soil and was promptly stuffed. Émile Oustalet of the Paris Natural History Museum loyally, some might say sycophantically, named it in honor of the President. This species of zebra is found in southern Ethiopia and northern Kenya (and formerly also in Somalia).

Grewcock

Grewcock's Sportive Lemur *Lepilemur grewcocki* Louis et al., 2006

William L. "Bill" Grewcock (b. 1926) is a businessman in Omaha who, together with his wife, Mrs. Berniece E. Grewcock, provides support for fieldwork in Madagascar and facilities and support for Malagasy students who study at the Henry Doorly Zoo's Center for Conservation and Research in Omaha. The binomial ought to be amended to *grewcockorum,* as it is for two people, and some attempts seem to be being made to correct this lapse in grammar. In 2002 "Bill" was appointed Chairman of the Nebraska Game and Parks Commission. He and his wife are well known for philanthropic donations and have set up the Bill and Berniece Grewcock Foundation. The lemur is found in northwest Madagascar.

Grey

Little Broad-nosed Bat *Scotorepens greyii* **Gray,** 1842
Toolache Wallaby *Macropus greyi* **Waterhouse,** 1846 extinct

Sir George Grey (1812–1898) was a British soldier, explorer, colonial governor, premier, and scholar. He explored Western Australia on government-financed expeditions to Hanover Bay and to Shark Bay between 1837 and 1839. On Grey's first expedition he was speared by a native Australian, whom he shot. Nevertheless, he championed the cause of assimilation. Respect for the Aboriginal people was a trait he shared with John Gould, who was a frequent correspondent of his. In 1845 he was appointed Governor of New Zealand, where he faced even greater difficulties than in South Australia. Grey's greatest success was his management of Maori affairs. He scrupulously observed the terms of the Treaty of Waitangi and assured Maoris that their land rights were fully recognized. In 1853 he became Governor of the Cape Colony and High Commissioner for South Africa. Grey's problem there, again, was race relations. He sought to convert the frontier tribes to Christianity to "civilize" them. Grey supported mission schools and built a hospital for African patients. He returned to New Zealand, where he was elected to its Parliament. Though politics left him little time to devote to scholarship, he was a keen naturalist and botanist, and he established extensive collections and important libraries at Cape Town and Auckland. He wrote books on Australian Aboriginal vocabularies and on his Western Australian explorations, as well as taking a scholarly interest in the Maori language and culture. He is also commemorated in the name of Grey's Fruit Dove *Ptilinopus greyii* and in the names of the town Greymouth and the Grey River on South Island, New Zealand. The bat is widespread in Australia except for the far south. The wallaby formerly occurred in southeastern Australia but was apparently extirpated in the wild by 1925, although a few captive specimens lingered until the late 1930s.

Grim

Grey Duiker *Sylvicapra grimmia* **Linnaeus,** 1758 [Alt. Common Duiker, Bush Duiker]

Dr. Herman Nicolaj Grim (1641–1711) was a Swedish scientist who described the duiker as long ago as 1686. Linnaeus subsequently named it after him. Grim studied in Copenhagen and the Netherlands, and between 1664 and 1665 he was a ship's surgeon on a Dutch vessel. In 1668 he was practicing as a doctor in Fredericia (Denmark), and between 1671 and 1681 he was employed by the Dutch East India Company as a doctor in both Ceylon (now Sri Lanka) and Java. After leaving Dutch service he held a number of medical posts in the Netherlands, Germany, and Sweden before finishing his career in Stockholm as a doctor for infectious diseases such as the plague. When he returned to Europe in 1681, he brought with him a collection he had made while abroad, and this included an antelope from the Cape of Good Hope area—"en antilop frün Kaplandet," to quote from a description. In 1682 he wrote *Anatome coralloidis. Miscellanea curiosa medico-physica Academiae naturae curiosorum.* The duiker is very widespread in Africa, from Senegal east to Ethiopia and southern Somalia, south to South Africa.

Griselda

Griselda's Striped Grass Mouse *Lemniscomys griselda* **Thomas,** 1904
Grey Short-tailed Shrew *Blarinella griselda* Thomas, 1912

As Thomas makes no mention of a person named Griselda in his original descriptions, we think these species are not named after a person at all, especially as *griselda* is not in the genitive. The name was probably applied fancifully. It is said to mean "grey fighting maid," and Thomas may have chosen it because of the mammals' grey coloration. The mouse comes from Angola. The shrew is found in central and southern China and has also been recorded in Vietnam.

Grobben

Grobben's Gerbil *Gerbillus grobbeni* Klaptocz, 1909

Dr. Karl Grobben (1854–1945) was an Austrian zoologist. He worked in invertebrate zoology at the University of Vienna, and his main interests were molluscs and crustaceans. Nevertheless, he published a revised edition of the well-known *Lehrbuch der Zoologie,* and he invented the terms Protostomia and Deuterostomia. He is commemorated in the scientific name of a gastropod, *Sphaerophthalmus grobbeni.* He undertook at least one deep-sea expedition. The gerbil is known only from the type locality in Cyrenaica, northeast Libya.

Güldenstädt

Güldenstädt's Shrew *Crocidura gueldenstaedtii* **Pallas,** 1811

Professor Johann Anton Güldenstädt (1745–1781) was a Baltic-German physician, natural scientist, and traveler (born in Riga in Latvia, at that time part of the Russian Empire). He is known to have made expeditions under the auspices of the Imperial Academy of Science to the Caucasus and trans-Caucasus regions between 1768 and 1774. He was a Professor at the Academy of Sciences in St. Petersburg from 1771 and was the author of diaries containing extensive geographical, biological, and ethnographical material on the Caucasus and Ukraine, commissioned by the Empress Catherine II. His *Reisen durch Russland und im caucasischen Gebürge* was published posthumously by Pallas in St. Petersburg between 1787 and 1791. He was the first describer of a number of birds, including, in 1770, the Ferruginous Duck *Aythya nyroca.* He is commemorated in the common name of a bird, Güldenstädt's Redstart *Phoenicurus erythrogaster.* The shrew was first found in Georgia (the country, not the U.S. state), but its distribution and taxonomy are still unclear. It has been treated as a subspecies of *Crocidura russula,* and as conspecific with *C. suaveolens.*

Gundlach

Gundlach's Hutia *Mysateles gundlachi* **Chapman,** 1901 [Alt. Chapman's Prehensile-tailed Hutia; Syn. *M. prehensilis gundlachi*]
Sabana Hutia *Capromys (pilorides) gundlachianus* **Varona,** 1983

Dr. Johannes Christoph (Juan Cristóbal) Gundlach (1810–1896) was a German zoologist and ornithologist. He began to learn the arts of dissection and taxidermy by watching his older brother who was a zoology student. An event that nearly cost him his life ironically allowed him to follow his chosen profession. During a hunting accident he discharged a small gun so close to his nose that he lost his sense of smell. After that he could calmly dissect, macerate, and clean skeletons. He was a Curator at the University of Marburg, which had awarded his doctorate in 1837, and later at the Senckenberg Museum of Frankfurt. He took part in a collecting expedition in 1839 to Cuba and stayed on, collecting there and in Puerto Rico, where he took refuge from the Cuban Civil War of

1868–1878 until his death. He wrote the first major work on Cuba's birds, *Ornitología Cubana*. He met the American explorer Charles Wright and explored the then virgin forest of what is now Alejandro de Humboldt National Park. He was zealous and single-minded, and tended to keep what he collected and describe it for science himself. He discovered several new species of land snails, which were his main zoological interest. He is also commemorated in the common names of two birds, Gundlach's Hawk *Accipiter gundlachi* and Gundlach's Mockingbird *Mimus gundlachii*. Gundlach's Hutia is found on the Isla de la Juventud (formerly Isle of Pines) off southern Cuba. The Sabana Hutia is found in the Sabana archipelago off northern Cuba.

Gunn

Eastern Barred Bandicoot *Perameles gunnii* **Gray** 1838

Ronald Campbell Gunn F.R.S., L.S. (1808–1881) emigrated in 1829 from Scotland to Australia, where he became a superintendent of prisons and a police magistrate. However, it is as a botanist and collector that he is remembered, having corresponded with many of the greats of natural history of his time such as J. E. Gray. Between 1855 and 1860 he was elected to the House of Assembly and subsequently became Deputy Commissioner for Crown Lands in northern Tasmania. He edited the *Tasmanian Journal of Natural Science* for seven years from 1842. He was married twice and had 12 children. When he died he left his herbarium to the Royal Society of Tasmania, from where it went on to the National Herbarium in Sydney. William Jackson Hooker said of him, "Ronald Campbell Gunn . . . to whose labours the Tasmanian Flora is so largely indebted, was the friend and companion of the late Mr. Lawrence, from whom he imbibed his love of botany. Between 1832 and 1850, Mr. Gunn collected indefatigably over a great portion of Tasmania. . . . There are few Tasmanian plants that Mr. Gunn has not seen alive, noted their habits in a living state, and collected large suites of specimens with singular tact and judgement. These have all been transmitted to England in perfect preservation, and are accompanied with notes that display remarkable powers of observation, and a facility for seizing important characters in the physiognomy of plants, such as few botanists possess." Various plants are named after him, including the Cider Gum *Eucalyptus gunnii*. The bandicoot is found in Tasmania, with a small population in Victoria on the Australian mainland.

Gunning

Gunning's Golden-Mole *Neamblysomus gunningi* Broom, 1908

Dr. Jan Willem Bowdewyn Gunning (1860–1913) was a Dutch physician. He went to South Africa in 1884 and was appointed Director at the (now) Transvaal Museum, Pretoria, in 1896, a post which he held until shortly before his death. He founded the Pretoria National Zoo, which he seems to have started in his own garden. Gunning was co-founder of the African Ornithologists' Union. He is also commemorated in the name of a bird, Gunning's Robin *Sheppardia gunningi*. The golden-mole is confined to the eastern Transvaal, South Africa.

Gunnison

Gunnison's Prairie Dog *Cynomys gunnisoni* **Baird,** 1855

Captain John Williams Gunnison (1812–1853) was an American army officer in the Corps of Topographical Engineers. He graduated from West Point Military Academy in 1837 and spent 1838 in Florida in the campaign against the Seminoles. From 1841 to 1849 his main area of exploration was around the Great Lakes in Michigan and Wisconsin. In 1849 he was sent to survey the valley of the Great Salt Lake in what is today the state of Utah. Here he came across the Mormons, and in 1852 he published *The Mormons or Latter-Day Saints, in the Valley of the Great Salt Lake: A History of Their Rise and Progress, Peculiar Doctrines, Present Condition.* It is clear that he admired them, as he called for the Mormons to be allowed to govern themselves. He was again in the Great Lakes in 1852 through 1853, when he was ordered to survey a possible route for a railway to the Pacific. He crossed through what is now Colorado, where both the town called Gunnison and the River Gunnison are named after him. To avoid a canyon, his party turned south into Utah where a war band of Pahvant Utes attacked them. Gunnison and seven of his men were killed and their bodies mutilated. There were rumors at the time that the Pahvant Utes had acted under the instruction of Brigham Young and the Church of the Latter Day Saints. These rumors were later shown to have no basis in fact. The prairie dog is found in southeast Utah, southwest Colorado, northeast Arizona, and northwest New Mexico.

Günther

Günther's Spiny Rat *Trinomys dimidiatus* Günther, 1877 [Alt. Soft-spined Atlantic Spiny Rat]
Günther's Vole *Microtus guentheri* Danford and **Alston,** 1880
Günther's Dik-dik *Madoqua guentheri* **Thomas,** 1894

Dr. Albert Carl Ludwig (Charles Lewis) Gotthilf Günther (1830–1914) was a British zoologist of German extraction. Having studied theology in Berlin and Bonn, and medicine in Tübingen, he went to work as Curator of Zoology at the British Museum of Natural History in 1856. He became a naturalized British subject in 1862 and changed his second two forenames to Charles Lewis. His particular interest was ichthyology, but he also worked on the reptile and amphibian collections. He wrote the eight-volume *Catalogue of the Fishes of the British Museum,* The *Reptiles of British India,* and other books, and a great many scientific papers and catalogues of fish, reptiles, amphibians, and mammals. More than 60 reptiles are named after him in the vernacular or the scientific name or both—as examples, Günther's Philippine Shrub Snake *Oxyrhabdium leporinum,* the Purple Shieldtail *Plectrurus guentheri,* and Günther's Blind Snake *Ramphotyplops guentheri.* The spiny rat comes from southeast Brazil. The vole is found in the southern Balkans and western Turkey. The dik-dik is found in southeast Sudan, northeast Uganda, southern Ethiopia, Somalia, and Kenya.

Gwatkins

Nilgiri Marten *Martes gwatkinsii* **Horsfield,** 1851

This animal was named after a Mr. R. Gwatkins, but it really ought not to have been. It was all Horsfield's fault and can be regarded as a great example of the troubles that can accrue through jumping to conclusions. Walter Elliot collected a specimen of the marten in Madras (now Chennai) around 1850. This came to Horsfield's attention, and he was aware of a similar animal having been shot in the Himalayas by a Mr. Gwatkins. Assuming that the two specimens representedthe samespecies, he gave the Madras specimen the name of *gwatkinsii.* However, the Himalayan specimen would have been the similar Yellow-throated Marten *Martes flavigula,* not a Nilgiri Marten. So the name of Gwatkins has ended up being attached to a species he played no part in collecting, describing, or discovering. The marten is found in southern India.

Hagen

Hagen's Flying Squirrel *Petinomys hageni* **Jentink,** 1888

White-striped Dorcopsis *Dorcopsis hageni* **Heller,** 1897 [Alt. Greater Forest Wallaby]

Dr. Bernhard Hagen (1853–1919) was a German physician and amateur natural historian. After studying medicine at Munich University he was employed by a planting company in Sumatra, during which time he made some collecting expeditions accumulating mostly zoological specimens. In 1893 he was employed by the Astrolabe Company in New Guinea. In 1895 he returned to Germany but is known to have visited New Guinea again in 1905, along with his wife. Between 1897 and 1904 he was a section head at the Senckenberg Museum in Frankfurt, where he founded the Ethnology Department. He published widely on zoology, geography, and ethnography. He is also commemorated in the fern *Asplenium hagenii.* The squirrel occurs in Sumatra and Borneo. The wallaby is found in northern New Guinea.

Hagenbeck

Hagenbeck's Rhinoceros *Dicerorhinus sumatrensis* G. Fischer, 1814 [Alt. Sumatran Rhinoceros]

Hagenbeck's Mangabey *Cercocebus hagenbecki* **Lydekker,** 1900 [Alt. Agile Mangabey; Syn. *C. agilis* **Milne-Edwards,** 1886]

John Hagenbeck (1866–1940) was the half-brother of Carl Hagenbeck (1844–1913), the great animal dealer and zoo owner. John was 20 when he first visited Ceylon (now Sri Lanka), and he settled there in 1891. He started capturing and dealing in animals, and was soon able to expand his interests and buy a number of tea plantations. The outbreak of WW1 in 1914 saw all his property confiscated, and he had to flee to Germany, returning to Colombo after the war was over. In the late 1920s he started a menagerie, and this became the nucleus of the National Zoological Gardens of Sri Lanka. The outbreak of WW2 in 1939 meant that he was interned as an enemy alien in a camp, where he died in 1940. In addition to his activities in Ceylon he had a base in Sumatra from 1898. At some stage, probably in 1899, he came across the tracks of a rhinoceros in the jungle; it was a female with her calf. Unfortunately someone shot the mother, and the local "helpers" rushed to cut off its valuable horn; the fact that only one horn is mentioned has led to suspicions that the animal was a Javan Rhinoceros *Rhinoceros sondaicus* (which also occurred on Sumatra at that time). The female calf was captured and kept in Hagenbeck's camp. After three months he sold her to the Madras Zoological Gardens, where 14 years later she was still alive, as evidenced by Colonel S. Flower (q.v.), who identified her as being a Sumatran Rhinoceros. (Flower certainly had enough expertise to tell the two species apart.) The mangabey, described in 1900 from a specimen brought into captivity, was named in honor of Carl Hagenbeck, but it is now known to be a junior synonym of *Cercocebus agilis,* which comes from Equatorial Guinea and Cameroon east to DRC (Zaire) north of the Congo River.

Haggard, J. G.

Haggard's Oribi *Ourebia ourebi haggardi* **Thomas,** 1895

John George Haggard (1850–1908) was a naval officer who became a career diplomat. In 1883 he was appointed to be the British Vice Consul at Lamu, Kenya, where he collected the type specimen of this antelope subspecies. He subsequently served in British Consulates in Zanzibar, Brest (France), Trieste (Italy), Noumea (New Caledonia), and Malaga (Spain). He was a brother of Sir William Henry Doveton Haggard (see below). The oribi is found in coastal Kenya and southern Somalia.

Haggard, W. H. D.

Haggard's Leaf-eared Mouse *Phyllotis haggardi* **Thomas,** 1898

Sir William Henry Doveton Haggard (1846–1926) was a distinguished British diplomat who served for many years in various South American countries, including as British Minister and Consul General at Quito, Ecuador, and at Rio de Janeiro. His younger brothers were Sir Henry Rider Haggard, the novelist who is famous for stories such as *King Solomon's Mines* and *She*, and John George Haggard (see above). Thomas writes, "At the request of Mr. Söderström I have named it in honour of Mr. W. H. D. Haggard, Her Majesty's Minister at Caracas, to whose kindness he has been at various times indebted." The mouse is endemic to the Andes of central Ecuador and was first collected by Ludovic Söderström.

Hahn

Hahn's Short-tailed Bat *Carollia subrufa* Hahn, 1905 [Alt. Grey Short-tailed Bat]

Dr. Walter Louis Hahn (d. 1911) was an American naturalist and schoolmaster. He appears to have operated mainly in the USA, probably because he could not, as a schoolmaster, disappear for months at a time. Up to 1910 he was Head of the Biology Department of a school in South Dakota. In that year the Bureau of Fisheries wanted a scientific study of the fur-seals on the Pribilof Islands in the Bering Sea, and Hahn was appointed, with the mandate of studying all the fauna and flora of the islands. He died after less than a year as a result of exposure to the freezing waters after a boat capsized. He wrote and published quite widely, including "Notes on the Mammals and Cold-blooded Vertebrates of the Indiana University Farm, Mitchell, Indiana" (1908), "Notes on Mammals of the Kankakee Valley" (1907), and (in the same year) "A Review of the Bats of the Genus Hemiderma (Carollia)." The bat is found from Mexico, where the type specimen was obtained, to northern Nicaragua.

Haig

Haig's Tuco-tuco *Ctenomys haigi* **Thomas,** 1917

Field Marshall Douglas, Earl Haig (1861–1928), took part in the Omdurman campaign of 1897–1898 and the Boer War from 1899 to 1902. He was Inspector General of Cavalry in India from 1903 to 1906, when he became Director of Military Training at the War Office in London. In 1909 he was Chief of Staff of the Indian army, and when WW1 broke out in 1914 he commanded the First Army Corps. In 1915 he was appointed to command the British and Empire forces in Europe. In February 1916 the Germans attacked the French at Verdun. After five months of fighting, 700,000 men had become casualties, and the French were having the greatest difficulty in holding on. Haig knew that the pressure on the French had to be relieved, so the British attacked along the line of the Somme. The Battle of the Somme was the bloodiest battle of the war, when it was over the British had gained only 16 km (10 miles) of land. After the war he was made a Peer. He is sometimes called the Butcher of the Somme. Thomas's citation reads, "Named in honour of General Sir Douglas Haig, Commander-in-Chief of the British armies." The tuco-tuco comes from the provinces of Rio Negro and Chubut, southern Argentina.

Hainald

Hainald's Rat *Rattus hainaldi* Kitchener, **How,** and Maharadatunkamsi, 1991

T. Hainald (dates not found) was the Head of the Bureau of Science and Technology Cooperation (IPTEK), Indonesian Institute of Science (LIPI), South Jakarta. The citation by Kitchener et al. reads, "In memory of the late Mr. Hainald . . . for his untiring and gracious efforts to facilitate the bureaucratic aspects of this series of expeditions to Nusa Tenggara." (Rumor has it that the process is *incredibly* bureaucratic.) The rat is endemic to Flores, Indonesia.

Hall

Patagonian Opossum *Lestodelphys halli* **Thomas,** 1921

T. H. Hall (dates not found) was a collector, but Thomas throws little light on his identity, merely writing in the etymology that the type specimen was among some small mammals "collected by Mr. T. H. Hall at Cape Tres Puntas," Patagonia. The opossum is found in central and southern Argentina.

Hallstrom

New Guinea Singing Dog *Canis hallstromi* **Troughton** 1957 [Syn. *C. lupus hallstromi*]

Sir Edward John Lees Hallstrom (1886–1970) was an Australian industrialist. He was born in Coonamble, New South Wales, and became a pioneer of refrigeration, a philanthropist, and a leading aviculturist. He began work in a furniture factory at the age of 13, but he later opened his own factory to make ice-chests and then wooden cabinets for refrigerators. He eventually designed and manufactured the first popular domestic Australian refrigerator. Hallstrom made generous donations to medical research, children's hospitals, and the Taronga Zoo in Sydney, becoming an Honorary Life Director of the zoo. In 1940 he commissioned the artist Cayley to paint all of the Australian parrots. Twenty-nine large watercolors were produced and presented to the Royal Zoological Society of New South Wales. There is a fine research collection of 1,600 rare books on Asia and the Pacific known as the Hallstrom Pacific Collection. It was purchased with funds given by Hallstrom to the Commonwealth government in 1948, for the purpose of establishing a Library of Pacific Affairs and Colonial Administration. The National Library transferred the collection to the University of New South Wales Library, on permanent loan. Among the rare books in the collection is John Gould's *Birds of New Guinea*. In the 1950s Hallstrom established a center in the southern highlands of Papua New Guinea from which, ostensibly, to introduce local sheep farming—but his real reason was to have a base from which a well-known collector, Fred Shaw Mayer, could devote his time to collecting birds-of-paradise for Hallstrom and the Sydney Zoo. He is also credited with enabling the rapid captive breeding of rare Golden-shouldered Parrots *Psephotus chrysopterygius* at Taronga Zoo in the 1950s. However, there was corruption, which involved trafficking in endangered species, during his time as Director there, and he stands accused of giving in to pressure to appoint the corrupt officials who carried out this trade. He is also remembered in the name of Hallstrom's Bird-of-Paradise *Pteridophora alberti hallstromi*. Although dogs are not native to New Guinea and the singing dog's ancestors must have been brought by early human settlers, today the singing dog avoids humans and lives mainly in mountainous regions. At least one recent study (2003) supports Troughton's original designation of this taxon as a distinct species.

Hamadryas

Hamadryas Baboon *Papio hamadryas* **Linnaeus,** 1758 [Alt. Sacred Baboon]

Hamadryads were wood-nymphs in Greek mythology. Most people would probably not see a resemblance between a baboon and a nymph, but Linnaeus was fond of fanciful names drawn from mythology. The baboon is found in Ethiopia, Eritrea, and northern Somalia, and also in Yemen and western Saudi Arabia.

Hamilton

Hamiton's Tomb Bat *Taphozous hamiltoni* **Thomas,** 1920
Hamilton's Serval *Leptailurus serval hamiltoni* Roberts, 1931

Lieutenant Colonel James Stevenson-Hamilton (1867–1957) was the collector of thetype specimen of the bat in the Sudan and was appointed the first Head Warden of the Kruger National Park after the Boer War in 1902. He wrote *Animal Life in Africa* in 1912 and published a number of maps for parts of southern Africa. In 1917 he was employed in the Sudan civil service. He was known as "Skukuza" by his staff at Kruger National Park, a Shangaan name meaning either "he who sweeps clean" or "he who turns everything upside down." Later in 1936 the main rest camp's name was changed from Sabie Bridge to Skukuza to honor him. He retired in 1946. The bat has been found in Sudan, Kenya, and Chad. The serval subspecies is from the eastern Transvaal.

Hamlyn

Hamlyn's Monkey *Cercopithecus hamlyni* **Pocock,** 1907 [Alt. Owl-faced Monkey, Hamlyn's Owl-faced Guenon]

John D. Hamlyn (dates not found) was an animal dealer whose shop was near the docks in the East End of London. He was notoriously independent and rude to his clients. He was the first person to bring this particular species of monkey to the London Zoo. He and his wife kept chimpanzees as pets and treated them as though they were their own children. The chimpanzees wore clothes, ate at the same table as Mr. and Mrs. Hamlyn, and undressed to go to sleep in their own small bed at night. Visitors to the shop were often greeted by a fully dressed chimpanzee that would then go off to find a human to deal with the customer. The monkey is found in eastern DRC (Zaire) and Rwanda.

Hammond

Hammond's Rice Rat *Mindomys hammondi* **Thomas,** 1913

Gilbert Hammond (dates not found) collected natural history specimens in Ecuador, including the rat named after him, sending them to the British Museum of Natural History. He was certainly still collecting in the 1920s. He also supplied entomological specimens to Ludovic Söderström, who was the Swedish Consul in Ecuador. The rice rat has a limited range in northwest Ecuador.

Hanak

Hanak's Pipistrelle *Pipistrellus hanaki* Benda, Hulva, and **Gaisler,** 2004

Professor Dr. Vladimír Hanák (b. 1931) is a Czech zoologist who specializes in the taxonomy and distribution of bats. In 1949 he began to study natural sciences at Charles University in Prague. He published his first scientific papers while still a student and was appointed as an assistant to Professor Julius Komárek. From the middle of the 1950s, Hanák concentrated on the study of bats. He was a staunch opponent of the Warsaw Pact's invasion of Czechoslovakia in 1968 and spoke his mind openly, with the result that the state prevented him from meeting foreign scientists, put obstacles in his way in regard to his teaching, and slowed his career down for 20 years. In 1989, with the collapse of communism, he was appointed Assistant Professor and was then elected Head of the Department of Zoology at Charles University. His 60th birthday was marked by a *Festshrift* (commemorative publication) presented by his domestic and international colleagues. He retired in 1996. By the time he was 60 he had written 132 books and papers, among them *The Illustrated Encyclopaedia of Mammals*, published in 1979. He tells us that since he retired he has published "about 50 scientific papers in collaboration with my younger colleagues (systematics, distribution, ecology, and conservation of bats in Middle Europe, the Mediterranean, and the Middle East)." The pipistrelle is found in Cyrenaica (Libya).

Handley

Handley's Tailless Bat *Anoura cultrata* Handley, 1960
Handley's Nectar Bat *Lonchophylla handleyi* **Hill,** 1980
Handley's Slender Mouse-Opossum *Marmosops handleyi* **Pine,** 1981
Handley's Red Bat *Lasiurus atratus* Handley, 1996
The Colombian rice rat genus *Handleyomys* **Voss** et al., 2002 [2 species: *fuscatus* and *intectus*]
Handley's Short-tailed Opossum *Monodelphis handleyi* Solari, 2007

Charles O. Handley Jr. (1924–2000) was a mammalogist working as a Curator for the Smithsonian Institution, Department of Vertebrate Zoology, U.S. National Museum of Natural History. After WW2 service in Europe he was taken on as a collector of birds by the Smithsonian and, after completing his Ph.D. in 1955, became Curator—having made four expeditions to the High Arctic, one to Labrador, one to Guatemala, and one to the Kalahari Desert in southern Africa. He made numerous visits to Panama between 1957 and 1967, making an inventory of all the mammals there. He made a similar inventory in Venezuela and organized a bat-trapping program to study their movements. He continued with trips to an island off Panama until 1993. In his time he named a bat, a sloth, an armadillo, and an agouti. He worked at the museum for 53 years. He published as many as 188 scientific papers, such as "Review of Bats" (*Audubon Naturalist*, 1993) and "A New Species of Three-toed Sloth (Mammalia: Xenarthra) from Panama, with a Review of the Genus *Bradypus*" (2001), as well as two books. The museum created in his honor the Handley Memorial Fund, which supports research in mammalogy and tropical biology. The tailless bat is found from Costa Rica to Peru, Bolivia, and Venezuela. The nectar bat is found in southern Colombia, Ecuador, and Peru. The mouse-opossum is known from only one locality in Antioquia, northern Colombia. The red bat occurs in Venezuela and the Guianas. The short-tailed opossum is currently known from the east bank of the Ucayali River, department of Loreto, Peru.

Hanuman

Hanuman Langur *Semnopithecus entellus* Dufresne, 1797 [Alt. Entellus Langur, Grey Langur]

Hanuman is a god in the Hindu religion. He is one of the most important characters in the Ramayana epic and helps Lord Rama rescue his consort, Sita. Hanuman is depicted as a monkey. The langur is found throughout the Indian subcontinent, including Sri Lanka. Some taxonomists now split this form into as many as seven different species.

Hardwicke

Hardwicke's Woolly Bat *Kerivoula hardwickii* **Horsfield,** 1824

Major General Thomas Hardwicke (1756–1835) served with the Bengal Artillery, which in his day was a regiment in the Bengal Army of the Honourable East India Company. He was an amateur naturalist and collector who is credited with being the first to make the Red Panda known to Europeans, through a paper he presented to the Linnean Society of London in 1821: "Description of a New Genus . . . from the Himalaya Chain of Hills between Nepaul *[sic]* and the Snowy Mountains." However, Cuvier stole a march on Hardwicke in giving the Red Panda a scientific name *(Ailurus fulgens)* because Hardwicke's return to England was delayed. Hardwicke collected reptiles in India and published on them in 1827 together with Gray. He is also remembered in the name of a bird, Hardwicke's Leafbird *Chloropsis hardwickii.* Horsfield was appointed Keeper of the Museum of the East India Company in 1920 and would have received Hardwicke's specimens. The bat is found over much of southern Asia, from India and Sri Lanka to the Philippines, Java, the Lesser Sunda Islands, and Sulawesi.

Harlan

Harlan's Musk-ox *Bootherium bombifrons* Harlan, 1825 extinct
Harlan's Gibbon *Nomascus concolor concolor* Harlan, 1826 [Alt. Black Crested Gibbon, Concolor Gibbon; Syn. *Hylobates concolor*]
Harlan's Ground Sloth *Glossotherium harlani* **Owen,** 1840 extinct

Dr. Richard Harlan (1796–1843) was an American physician, naturalist, and writer. When still a medical student in 1816 and 1817 he acted as Ship's Surgeon on an East Indiaman bound for Calcutta. He was a member of the Philadelphia Academy of Natural Sciences and an amateur paleontologist, but not an infallible one. It is due to Harlan that a fossil whale is called *Basilosaurus,* meaning "king lizard," because he thought it was a reptile. He published his major work, *Fauna Americana,* in 1825. He is also commemorated in the name of a bird, Harlan's Hawk *Buteo jamaicaensis harlani.* Two of the species named after him are long extinct: the musk-ox and ground sloth are Pleistocene North American mammals. The gibbon is found in northern Vietnam between the Red and Black rivers and extending into adjacent Yunnan (China).

Harrington

Harrington's Rat *Desmomys harringtoni* **Thomas,** 1902 [Alt. Dega Rat]
Harrington's Gerbil *Taterillus harringtoni* Thomas, 1906

Lieutenant Colonel Sir John Lane Harrington (1865–1927) was the British government's Resident Agent, Consul, and later Minister Plenipotentiary in Addis Ababa (Ethiopia) from 1897 until 1909. In 1903 he personally led an expedition to the border region between Ethiopia and Sudan, and included in it was Charles Singer (1876–1960), a trained zoologist who doubled as the expedition's medical officer. Thomas wrote in the etymology of the rat that he named it "in honour of Col. Harrington, the British Resident at Addis Ababa, to whose assistance all British travellers in Abyssinia are so much indebted." The rat is found in the Ethiopian highlands. The gerbil is found from the eastern Central African Republic east to southern Somalia and south to Kenya and northern Tanzania.

Harris, E.

Harris' Antelope Squirrel *Ammospermophilus harrisii* **Audubon** and **Bachman,** 1854 [Alt. Harris' Marmot Squirrel, Antelope Ground Squirrel]

Edward Harris (1799–1863) was a farmer, landowner, breeder of horses, and amateur ornithologist who lived a life of leisure. Harris helped Audubon financially during the preparation of his *Birds of America* when Audubon, in his own words, had been reduced "to the lowest degree of indigence." He accompanied Audubon on his Missouri River trip of 1843.

Audubon called Harris "one of the best friends I have in the world" and dedicated the Harris' Hawk *Parabuteo unicinctus harrisi* in his honor. He is also commemorated in the common name of another bird, Harris' Sparrow *Zonotricha querula*. The antelope squirrel is found in the deserts of Arizona and southwest New Mexico, and adjacent Sonora (Mexico).

Harris, R.

Tasmanian Tiger *Thylacinus harrisii* **Temminck,** 1824 probably extinct [Alt. Thylacine; Syn. *T. cynocephalus* Harris, 1808]

Tasmanian Devil *Sarcophilus harrisii* Boitard, 1841

George Prideaux Robert Harris (1775–1810), always known as Robert, was Deputy Surveyor on expeditions to map Tasmania. He was the first to describe the Thylacine, using the name *Didelphis cynocephala,* meaning "dog-headed opossum," in 1808. Temminck later coined the name *Thylacinus* for the genus and decided to honor Harris in the binomial. However, under the rules of taxonomy the correct name is *Thylacinus cynocephalus* (first valid name for the genus is *Thylacinus;* first valid specific name is *cynocephalus*). The Tasmanian Tiger is believed to have become extinct by the end of the 1930s, though occasional reports of sightings still surface. Both it and the Tasmanian Devil were confined to Tasmania since before Europeans reached Australia, though both species once also occurred on the Australian mainland.

Harris, W. P.

Harris' Olingo *Bassaricyon lasius* Harris, 1932

Harris' Rice Water Rat *Sigmodontomys aphrastus* Harris, 1932

William P. Harris Jr. (1897–1972) was a zoologist who collected in Central and South America. He was also a life member of the American Ornithologists' Union. In 1932 he published a paper, for the University of Michigan Museum of Zoology, entitled "Four New Mammals from Costa Rica." A number of other papers by him published around the same time all refer to squirrels or similar mammals, which seem to have been his particular interest. The olingo is known only from one locality in central Costa Rica; it may be conspecific with *Bassaricyon gabbii*. The water rat—not a very good vernacular name, as the species doesn't seem to be aquatic—is known from Costa Rica, western Panama, and northwest Ecuador. It may be more widespread than current records indicate.

Harrison, D. L.

Harrison's Fruit Bat *Lissonycteris goliath* Bergmans, 1997

Harrison's Tube-nosed Bat *Murina harrisoni* **Csorba** and Bates, 2005

Dr. David Lakin Harrison is the Chairman of Trustees of the Harrison Institute. This institute, based in Sevenoaks, Kent, UK, is dedicated to the taxonomic study of mammals and birds. The etymology for the tube-nosed bat reads, "Named in honour of Dr. David Lakin Harrison, who, as Chairman of Trustees, has supported, encouraged and actively participated in the extensive researches of the Harrison Institute into the bats of Southern and Southeast Asia." The fruit bat is known from the region of the Mozambique-Zimbabwe border. The tube-nosed bat is known at the time of writing from a single specimen collected in Cambodia.

Harrison, E. N.

Banana Bat *Musonycteris harrisoni* Schaldach and McLaughlin, 1960

Ed N. Harrison (1914–2002) was a naturalist and ornithologist. He supported the Mexican fieldwork of W. J. Schladach, who was one of the first to describe the Banana Bat, and was President of the Cooper Ornithological Society. He established the Western Foundation of

Vertebrate Zoology in 1956. The WFVZ was formed at a time when many natural history museums were unwilling to add eggs to their holdings, and so a respected repository for egg collections was needed. Harrison himself contributed approximately 11,000 egg sets, 2,000 nests, and 1,700 study skins after the founding of the WFVZ. There is an Ed N. Harrison Memorial Scholarship for ornithology students established in his memory. He is also remembered in the trinomial of a subspecies of the Spotted Nightingale-Thrush *Catharus dryas harrisoni*. The bat is endemic to western Mexico.

Hart

Hart's Fruit-eating Bat *Enchisthenes hartii* **Thomas,** 1892 [Alt. Velvety Fruit-eating Bat, Little Fruit-eating Bat; Syn. *Artibeus hartii*]

Dr. John Hinchley Hart (1847–1911) was a British botanist. After serving 11 years in Jamaica, John Hart took up the post of Superintendent of the Trinidad Botanic Gardens in 1887, holding the post until 1908. This was at a time when there was a great role for the gardens to play in the economic life of the island, principally in helping decision makers to decide upon appropriate crops and associated planting. Hart immediately began the process of reorganizing the department, as he found specimens poorly preserved and labeled. He presented the British Museum with a collection of bats, which included the type specimen of the fruit-eating bat, which was collected in the Trinidad Botanic Gardens. In addition to Trinidad, the species' distribution ranges from Mexico south to Bolivia and Venezuela.

Hartmann, A.

Hartmann's [Mountain] Zebra *Equus zebra hartmannae* **Matschie,** 1898

Mrs. Anna Hartmann, née Anna Woermann (d. 1941), was the daughter of the German shipowner Adolph Woermann. There were two ships named *Anna Woermann* after her, the second of which was scuttled as a blockship in 1914, was raised and repaired in 1916 by the British, as the British flag *Polonia,* and was eventually torpedoed and sunk by a German submarine in 1917. The zebra was named after her as the wife of the man who first recorded the Mountain Zebra in southwest Africa—Dr. Georg Hartmann, who sent two skins to the Berlin Museum as a gift. Hartmann was an official in the German administration of German South-West Africa. He and his wife, whom he had married in 1898, lived on in South-West Africa (now Namibia) under South African rule until 1939, when they returned to Germany. He died in a refugee camp in Schleswig-Holstein in 1946. The zebra is found in Namibia.

Hartmann, R.

Hartmann's Water Mouse *Rheomys raptor hartmanni* **Enders,** 1939

Señor Ratibor Hartmann owns Finca Hartmann, a farm in Panama where R. K. Enders and others did much field study. Hartmann is a naturalist and was a scientific lab technician working with zoologists before taking over the family farm, which was founded in 1912. He is intent on developing conservation and ecotourism as well as farming. He has an amazing collection of insects collected from the farm. The mouse is found in the Talamancan highlands of Panama and Costa Rica.

Hartwig

Hartwig's Soft-furred Mouse *Praomys hartwigi* **Eisentraut,** 1968

Wolfgang A. Hartwig is an artist. He illustrated the *Collins Field Guide to the Birds of West Africa* (1988) and Serle, Morel, and Hartwig's *Birds of West Africa* (1977). His work also adorned Eisentraut's book *Im Schatten des Mongo-ma-loba,* a title that can be roughly translated as *In the Shadow of Mongo-ma-loba.*

The mouse is known from the mountains of eastern Nigeria and the Lake Oku area of western Cameroon.

Harvey

Harvey's Duiker *Cephalophus harveyi* **Thomas,** 1893 [Alt. Harvey's Red Duiker]

Sir Robert G. Harvey (1847–1930) was an entrepreneur who seems to have made his fortune mining gold, diamonds, and other minerals in South Africa. He was fond of "sporting trips," judging by an account written in 1889 by his friend Sir John Willoughby, *East Africa and Its Big Game—The Narrative of a Sporting Trip from Zanzibar to the Borders of the Masai,* for Harvey wrote a postscript. The type specimen of the duiker was presented to the British Museum by F. C. Jackson, but Thomas writes, "At Mr. Jackson's suggestion I propose to name the species in honour of Sir Robert Harvey, who was the first of their party to shoot a specimen of this interesting species." The duiker is found in eastern Ethiopia, southern Somalia, eastern Kenya, eastern and southern Tanzania, and northern Malawi.

Harwood

Harwood's Gerbil *Gerbillus harwoodi* **Thomas,** 1901 [Syn. *Dipodillus harwoodi*]

Leonard C. Harwood (dates not found) was an English naturalist and taxidermist who operated his taxidermy business in Hammersmith, London. He accompanied Lord Delamere on an expedition to Kenya at the end of the 19th century. As well as the gerbil he has a bird, Harwood's Francolin *Francolinus harwoodi,* named in his honor. The gerbil is found in the Lake Naivasha area of Kenya. It is sometimes regarded as a race of *Gerbillus bottai.*

Hasselt

Hasselt's Myotis *Myotis hasseltii* **Temminck,** 1840 [Alt. Lesser Large-footed Bat]

Johan Coenraad van Hasselt (1797–1823) was a Dutch biologist. He studied medicine at the University of Groningen, graduating in 1820. However, he was more interested in natural history, like his fellow student and close friend Heinrich Kuhl. They undertook collecting trips together in Europe and visited a number of natural history museums, where they met famous zoologists of the time. They subsequently published several papers. In 1820 they were sent to Java to study its natural history. They started work en route studying pelagic fauna, as well as that of Madeira, the Cape of Good Hope, and the Cocos Islands before they arrived in Java. When Kuhl died after less than a year on the island, van Hasselt continued collecting, until he himself died two years later. The bat is found in Sri Lanka and from northeast India to Vietnam, the Malay Peninsula, Java, and Borneo.

Hatt

Hatt's Vesper Rat *Otonyctomys hatti* **Anthony,** 1932 [Alt. Yucatan Vesper Mouse]

Robert Torrens Hatt (b. 1902) was an American zoologist and paleontologist. He was Corresponding Secretary of the American Society of Mammalogists from 1932 to 1935, when he was working at the American Museum of Natural History. He wrote at least three books: *Island Life: A Study of Land Vertebrates of the Islands of Eastern Lake Michigan* (1948); *Faunal and Archeological Researches in Yucatan Caves* (1953); and *Cranbrook Institute of Science: A History of Its Founding and First Twenty-five Years* (1959). Judging by these and his many published articles, Hatt was as interested in modern mammals as he was in those in the fossil record. In the 1960s he was Director of the Institute of Science at the Cranbrook Institutions. He retired in 1967. The mouse is found in the Yucatan Peninsula of Mexico, Belize, and northeast Guatemala.

Haviside

Haviside's Dolphin *Cephalorhynchus heavisidii* **Gray,** 1828 [Alt. Heaviside's Dolphin]

Captain Haviside (dates not found) was a ship's captain employed by the Honourable East India Company. He took the type specimen of the dolphin and carried it from Cape Town to England in 1827 along with the rest of the Villet collection. Gray mistakenly put an additional *e* in the animal's scientific name, and this led to it being usually called Heaviside's Dolphin (possibly a case of mistaken identity; another captain, this one Captain Heaviside—complete with *e*—was a naval surgeon who collected natural history specimens). The East India Company had two Captain Havisides at the right time, and it is not known if the dolphin is named after Thomas Haviside, who was master of *Elphinstone,* or William Haviside, the master of *Thames.* The dolphin is found in coastal waters from the Cape Town area to southern Angola.

Hawk

Hawk's Sportive Lemur *Lepilemur tymerlachsoni* Louis et al., 2006

Howard Hawk and Mrs. Rhonda Hawk have made contributions and given much support to the activities in Madagascar of the Henry Doorly Zoo's Center for Conservation and Research at Omaha, Nebraska. Because two persons are mentioned in the original etymology, the scientific name may one day be changed to the plural form *tymerlachsonorum.* The lemur comes from the island of Nosy Be, northwestern Madagascar.

Hayden

Hayden's Shrew *Sorex haydeni* **Baird,** 1857 [Alt. Prairie Shrew]

Professor Dr. Ferdinand Vandeveer Hayden (1829–1887) was an American physician, geologist, and explorer. He took his degree in medicine in 1853. From 1854 to 1855 he was part of a geological expedition to the Yellowstone and Missouri rivers, and from 1856 to 1857 he was the geologist on the Warren expedition to Nebraska and Dakota. In 1859 he was a member of an expedition, lasting until 1862, exploring the Rocky Mountains and Yellowstone, and on this expedition he collected vertebrates for the Museum of the Academy of Natural Sciences in Philadelphia. During the American Civil War from 1862 to 1865 he served as a surgeon in the U.S. Army. He was a Professor of Geology at the University of Pennsylvania from 1865 to 1872 and served on the U.S. Geological Survey from 1867 to 1886. His work led to the establishment of Yellowstone National Park. He resigned from the army only in 1886. He is also commemorated in the name of Hayden's Garter Snake *Thamnophis radix haydeni.* The shrew is found in south-central Canada and the north-central USA, as far south as Kansas and east to Minnesota.

Hayman

Hayman's Climbing Mouse *Dendromus haymani* **Hatt,** 1934 [Alt. Banana Climbing Mouse; Syn. *D. messorius* **Thomas,** 1903]

Hayman's Dwarf Epauletted Fruit Bat *Micropteropus intermedius* Hayman, 1963

Robert William Hayman (dates not found) was a British mammalogist at the British Museum of Natural History. He published "Notes on Some African Bats, Mainly from the Belgian Congo" (1954) and translated *A Field Guide to the Mammals of Africa* by Haltenorth and Diller (1980). The mouse was found by Hayman in

1930, in the then Belgian Congo. However, it is now regarded as conspecific with *Dendromus messorius,* which ranges from Benin to Kenya. The bat is found in northern Angola and southern DRC (Zaire).

Heaney

Panay Bushy-tailed Cloud Rat *Crateromys heaneyi* **Gonzales** and Kennedy, 1996

Dr. Lawrence Richard Heaney (b. 1952) is an American ecologist and mammalogist who is Curator of Mammals at the Department of Zoology of the University of Chicago. His current research program focuses on the ecology and evolution of mammals on the islands of Southeast Asia, especially the Philippines. He graduated with a B.S. from the University of Minnesota in 1971, took his M.A. at the University of Kansas in 1975, and was awarded his Ph.D. there in 1979. He was Research Fellow, Smithsonian Institution, 1986–1988; Research Associate, Smithsonian Institution, 1988–present; Research Associate, American Museum of Natural History, 1991–present; Research Associate, Utah Museum of Natural History, 1994–present; Honorary Curator, Department of Zoology, Philippine National Museum, 1990–present; and Science Adviser, Center for Tropical Conservation Studies, Silliman University, Philippines, 1992–present. He has written a great many scientific papers and articles as well as contributing chapters to a variety of books. The cloud rat is from the island of Panay, central Philippines.

Heath

Greater Asiatic Yellow Bat *Scotophilus heathi* **Horsfield,** 1831

Josiah Marshall Heath (d. 1851) was a metallurgist and businessman who had enormous influence on the creation and development of the steel industry in India and, later in his career, at Sheffield in England. He was greatly in favor of India having its own independent sources of supply and production of iron, and in 1830 he opened an iron smelter and established the East Indian Steel and Iron Company at Porto Novo on the Madras coast. Heath later (ca. 1840) produced metallic manganese in England. He was an innovator who was responsible for the general plan of the "open-hearth process "and took out several patents in both the UK and the USA during the 1830s and 1840s. It was in Madras that the bat was first collected. According to Horsfield, the type specimen was presented to the Zoological Society of London "with a numerous and valuable collection of birds formed at Madras by Josiah Marshall Heath, Esq., F.L. and Z.S." The species is widespread from Afghanistan and Pakistan through India to Sri Lanka, and east to southern China and Vietnam.

Heaviside

See **Haviside.**

Heck

Heck's Macaque *Macaca hecki* **Matschie,** 1901

Heck's Wildebeest *Connochaetes taurinus hecki* **Neumann,** 1905

Professor Dr. Ludwig Franz Friedrich Georg Heck (1860–1951) was a German zoologist. At the age of just 26 he became Director of the Cologne Zoo. In 1888 he became Director of the Berlin Zoo and held that position until 1931. Both his sons also became directors of zoos, Lutz Heck (1892–1983) taking over from his father at Berlin. In addition to zoology, Ludwig was also interested in ethnology and racial theory. This may help explain his son Lutz becoming embroiled to a degree in the Nazi movement; it is uncertain whether he was a Nazi Party member or not. A film made in 1926 of an expedition made by Ludwig Heck to Abyssinia still exists. His co-star in the film and on the expedition was Oskar Neumann, who described the wildebeest. Heck is also commemorated in the name of Heck's Finch *Poephila acuticauda hecki*. The macaque is found on Sulawesi (Indonesia). The wildebeest occurs in northwest Tanzania and southern Kenya west of the Rift Valley.

Hector

Tarai Grey Langur *Semnopithecus (entellus) hector* **Pocock,** 1928

Hector, Prince of Troy (ca. 1,200 B.C.), who was killed by Achilles. The best source for the full story is Homer's *Iliad*. There was something of a fashion for naming Indian langurs after characters from Homer and Virgil. This taxon is found in the Himalayan foothills.

Hector, J.

Hector's Beaked Whale *Mesoplodon hectori* **Gray,** 1871
Hector's Dolphin *Cephalorhynchus hectori* **Van Beneden,** 1881 [Alt. Maui Dolphin]

Dr. Sir James Hector (1834–1907) was a Scottish-born Canadian geologist who took his medical degree at Edinburgh and, as both geologist and surgeon, was part of the Palliser expedition to western North America from 1857 to 1860. He discovered and named many landmarks in the Rockies, including Kicking Horse Pass, which was later the route used by the Canadian Pacific Railway. He returned to Scotland via the Pacific Coast, the California goldfields, and Mexico. In 1865 he became the Director of the Geological Survey of New Zealand and eventually became the Curator of the Colonial Museum in Wellington (now the Museum of New Zealand Te Papa), which housed the first known specimen of the beaked whale. In 1886 he published *Outlines of New Zealand Geology*. The whale is found in temperate waters of the Southern Hemisphere (records from California are now known to be of a separate species, *Mesoplodon perrini*). The dolphin is found only in New Zealand's coastal waters.

Heermann

Heermann's Kangaroo-Rat *Dipodomys heermanni* **Le Conte,** 1853

Dr. Adolphus Lewis Heermann (1827–1865) was a U.S. Army physician and naturalist. He was one of many naturalists who came to the attention of Spencer Baird at the Smithsonian Institution, and he was assigned to a surveying party for the Pacific Railroad line. Heermann was especially interested in collecting birds' eggs, and he is credited with coining the term "oology" for the study of eggs. He retired from the army early due to illness and died two years later in a hunting accident; he stumbled and his rifle discharged and killed him. He appears to have looked many years older than he was, the effect of syphilis, among other things. He is also remembered in the name of Heermann's Gull *Larus heermanni*. The rat is found in central California.

Heinrich

Heinrich's Hill Rat *Bunomys heinrichi* **Archbold** and **Tate,** 1935
Montane Long-nosed Squirrel *Hyosciurus heinrichi* Archbold and Tate, 1935 [Alt. Hog Squirrel, Long-snouted Squirrel]

Dr. Gerd Herrmann Heinrich (1896–1984) was a German field biologist and explorer of the East Indies, as was his wife Hildegard. He was the world's leading authority on ichneumonid wasps and described some 1,500 species and subspecies of them. He latterly lived and worked in the USA. He wrote *Der Vogel Schnarch—zwei Jahre rallenfung un urwaldfoeschung in Celebes* (Two Years of *Rallidae* Trapping and Jungle Exploration in the Celebes), published in 1932. He also collected in Angola, South Africa, and Tanzania and discovered various new taxa. He worked at various times in the Balkans, Myanmar, Iran, and Mexico as well as Canada and the USA. He is commemorated in the names of several birds, such as Heinrich's Brush Cuckoo *Cacomantis heinrichi* and Heinrich's Robin-Chat *Cossypha heinrichi*. The rat is found in southwest Sulawesi. The squirrel inhabits the mountains of central Sulawesi.

Heinsohn

Australian Snubfin Dolphin *Orcaella heinsohni* Beasley, Robertson, and Arnold, 2005

Dr. George E. Heinsohn is an Australian biologist who, until his retirement, was a member of the faculty at the James Cook University in Townsville, Queensland. He took his doctorate at Berkeley University in California before returning to Queensland, where he concentrated on marine biology, in particular studying the Dugong *Dugong dugon*. In the 1960s and 1970s he intensively studied what was believed to be the Australian population of the Irrawaddy Dolphin *Orcaella brevirostris*. It was only much later that DNA techniques confirmed Heinsohn's suspicions that these animals were of a distinct species. Currently known from the coasts of northern Australia, it may also occur off New Guinea.

Heller, E.

Heller's Pipistrelle *Neoromicia helios* Heller, 1912 [Alt. Samburu Pipistrelle]
Heller's Vervet *Chlorocebus pygerythrus arenarius* Heller, 1913
Heller's Rock Rat *Aethomys helleri* **Hollister,** 1918

Edmund Heller (1875–1939) was an American zoologist and ornithologist. He collected in the Colorado and Mohave deserts in 1896 and 1897 while still a student at Stanford University, from which he graduated in 1901. He interrupted his studies in 1899 to spend seven months in the Galápagos Islands. In 1900 he was with Wilfred Hudson Osgood in Alaska, and in 1907 he was part of the Chicago Field Museum's African expedition. When he returned from Africa he was appointed Curator of Mammals at the Museum of Vertebrate Zoology of the University of California, and in 1908 he was a member of the Alexander Alaskan expedition. Between 1909 and 1912 he was in Africa with the Smithsonian-Roosevelt and Rainey African expeditions. In 1914 he was a member of the Lincoln Ellsworth expedition to British Columbia and later to Alberta. In 1915 an expedition was sent to explore the newly discovered "lost" city of Machu Picchu in Peru. Heller was employed as the expedition's naturalist and oversaw the collecting of 891 mammal specimens, 695 birds, about 200 fishes, and several tanks of reptiles and amphibians. In 1916 he joined Roy Chapman Andrews on the American Museum of Natural History expedition to China. Rainey became official photographer for the Czech army in Siberia, and he invited Heller to Russia with him. From the summer of 1918 until the end of WW1, they traveled by rail across Siberia to the Ural Mountains and back to their starting point. In 1919 Heller was leader of the Smithsonian Cape-to-Cairo expedition and then worked for a short time for the Roosevelt Wild Life Experiment Station, studying large game animals in Yellowstone National Park. Later in 1919 he became Assistant Curator of Mammals at the Field Museum under Osgood. During his six years in that position, Heller visited Peru from 1922 to 1923 and Africa from 1923 to 1926, the latter being his last collecting expedition. In 1928 he became Director of the Milwaukee Zoological Garden, a position that he held until 1935 when he became Director of the Fleishhacker Zoo in San Francisco, filling that post until his death. The pipistrelle bat is found in northern Tanzania, Kenya, Somalia, Uganda, and southernmost Sudan. The vervet monkey is found in northern Kenya. The rock rat occurs from Cameroon east to southern Sudan and south to northeast Tanzania (it is sometimes treated as a race of *Aethomys hindei*).

Heller, K. B.

Heller's Broad-nosed Bat *Platyrrhinus helleri* **Peters,** 1866

Karl Bartholomaus Heller (1824–1880) was a German traveler and naturalist. He visited Mexico and collected the type specimen there in 1850. In 1853 he published *Reisen in Mexiko in den Jahren 1845–1848* (Travels in Mexico in the Years 1845–1848). He appears to have revisited Mexico as a member of the 1864 scientific mission for Mexico instituted by Napoleon III. He mapped the Mexican state of Tabasco. The bat is found from southern Mexico to Peru and Amazonian Brazil, and on Trinidad.

Hellwald

Hellwald's Spiny Rat *Maxomys hellwaldii* **Jentink,** 1878

Friedrich Anton Heller von Hellwald (1842–1892) was an Austrian ethnographer, geographer, anthropologist, and writer on Polynesia and Australia. He joined the Austrian army in 1858 and served in the war of 1866. He left the army in 1866 and became editor of a military magazine in Vienna. In 1872 he moved to Stuttgart to edit a weekly newspaper, but in 1882 he was forced to resign his post as his support for Darwinism was extremely unpopular. He was a leading Darwinian social theorist in the 1870s and in 1875 published *Culturgeschichte* (History of Culture), which a number of German correspondents recommended to Darwin. We think we have identified the right person but cannot be 100 percent sure, as Jentink made no mention in his original description of who Hellwald was and our evidence is circumstantial. The rat is found on Sulawesi.

Hemprich

Hemprich's Long-eared Bat *Otonycteris hemprichii* **Peters,** 1859 [Alt. Desert Long-eared Bat]

Wilhelm Friedrich Hemprich (1796–1825) was a physician, traveler, and collector. In 1828, *Natural Historical Journeys in Egypt and Arabia* was published, a joint work undertaken with Ehrenberg, whom he had met while studying medicine in Berlin. In 1820 they were invited to serve as naturalists on an expedition to Egypt, and they continued to journey and collect in the region, including Lebanon and the Sinai Peninsula, before moving on to Ethiopia. Hemprich died of fever in the Eritrean port of Massawa. Much of his written work was not published for some years, including his species descriptions, and many citations for species attributed to him appeared many years after his death. A ribbon worm *(Baseodiscus hemprichii)* and a stone coral *(Lobophyllia hemprichii)* are both named after him, as are Hemprich's Gull *Larus hemprichii* and Hemprich's Hornbill *Tockus hemprichii.* The bat is found from Morocco and Niger in the west through Egypt and Arabia to Tadzhikistan, Afghanistan, and Kashmir.

Hendee

Hendee's Woolly Monkey *Oreonax flavicauda* **Humboldt,** 1812 [Alt. Yellow-tailed Woolly Monkey]

Hendee's Spiny Rat *Proechimys hendeei* **Thomas,** 1926

Russell W. Hendee (1899–1929) was an American zoologist. In 1921 to 1922 he was in Alaska, working for the Colorado Museum of Natural History. He was a member of the Godman-Thomas expedition to Peru in 1925 to collect specimens for the British Museum, and in 1929 was in Indochina with Theodore Roosevelt. Hendee was struck down by a tropical fever and went into a delirium. He was put into a hospital but committed suicide by throwing himself out of a window. In 1927 Thomas named a woolly monkey as *Lagothrix hendeei*, unaware that Humboldt had described the same species much earlier. The vernacular name Hendee's is, however, still sometimes used. The monkey is a rare inhabitant of the cloud forests of northeast Peru. The spiny rat is also found in northeast Peru and in southern Colombia. It is sometimes regarded as conspecific with *Proechimys simonsi*.

Henley

Henley's Gerbil *Gerbillus henleyi* **de Winton,** 1903 [Alt. Pygmy Gerbil]

The Hon. Francis Robert Henley (1877–1962), who became the 6th Baron Henley upon the death of his father in 1922, also appears to have changed his surname to Eden at that time. He was well known in scientific circles as a biochemist, publishing under the name F. R. Henley or Francis Robert Henley, such articles as "The Function of Phosphates in the Oxidation of Glucose by Hydrogen Peroxide" (1922), with Arthur Harden as co-author. The gerbil was collected in the Natron Valley, Egypt, by Henley and Nathaniel Charles Rothschild, whose daughter was the famous parasitologist, Miriam Rothschild. The gerbil is found from Morocco east to Israel, Jordan, and the Arabian Peninsula. It has also been recorded from Burkina Faso, Niger, and northern Senegal.

Heptner

Heptner's Pygmy Jerboa *Salpingotus heptneri* Vorontsov and Smirnov, 1969

Vladimir Georgievich Heptner (1901–1975) was a Professor of Biology at the Laboratory of Zoological Geography and Taxonomy at the Vertebrates Zoology Department, Moscow State University. He worked on *Mammals in the Soviet Union* with others over a number of years; so far five volumes have been published. He was widely regarded as the leading mammalogist of the Soviet Union of his day. The jerboa comes from southern Kazakhstan and Uzbekistan.

Herbert

Herbert's Rock Wallaby *Petrogale (penicillata) herberti* **Thomas,** 1926

Dr. George Frederick Herbert Smith (1872–1953) was a mineralogist who was Associate Secretary of the British Museum (Natural History). While there he reworked the catalogue to the collection, published in 1937 as *A Guide to the Mineral Gallery*. He devised a refractometer that allows one to discriminate between gemstones. He also has a mineral named after him, herbertsmithite, in honor of the fact that he discovered another mineral, paratacamite. The rock wallaby is found in eastern Queensland.

Herman

Herman's Myotis *Myotis hermani* **Thomas,** 1923

G. Herman (dates not found), according to Thomas, was a man "to whom the Amsterdam Museum owes many valuable accessions." Thomas described the species when he was in Holland working on some small mammals from the East Indies, which Beaufort at the Amsterdam Museum allowed him to examine. This bat is a Sumatran endemic.

Hernández-Camacho

Hernández-Camacho's Black Tamarin *Saguinus nigricollis hernandezi* **Hershkovitz,** 1982
Hernández-Camacho's Short-tailed Bat *Carollia monohernandezi* Muñoz, Cuartas-Calle, and Gonzalez, 2004
Hernández-Camacho's Night Monkey *Aotus jorgehernandezi* Defler and Bueno, 2007

Jorge Ignacio Hernández-Camacho (1935–2001) was a Colombian mammalogist and conservationist. He was particularly noted for his efforts to protect Colombia's biodiversity via a network of protected areas. He was apparently known by the nickname "El Mono" Hernández—hence the form of the bat's scientific name. An animal sanctuary is being established in his memory, specifically to conserve the tamarin named after him. All three mammals are found in Colombia. The night monkey was described from a captive specimen, and the species' precise distribution in the wild is not known at the time of writing.

Hershkovitz

Hershkovitz's Marmoset *Callithrix intermedia* Hershkovitz, 1977 [Alt. Aripuanã Marmoset]
Hershkovitz's Night Monkey *Aotus hershkovitzi* Ramirez-Cerquera, 1983
Hershkovitz's Grass Mouse *Abrothrix hershkovitzi* **Patterson,** Gallardo, and Freas, 1984

Dr. Philip Hershkovitz (1909–1997) was an American zoologist whose particular field of interest was Neotropical primatology, although he shifted his interest in his last few years to rodents and marsupials. He was Research Curator for Mammals at the Field Museum of Natural History in Chicago, having been on the staff since 1947 and retiring in 1971, although continuing his work as Emeritus Curator. He published over 160 scientific papers and another 100 articles, as well as *Living New World Monkeys*, volume 1 (volume 2 was still being written when he died). He is credited with having discovered about 75 species and subspecies of Neotropical mammals. One of his students, R. A. Mitterrneier, said in remembrance of him, "He was a field mammalogist of the old school, with tireless energy and an understanding of the creatures on which he worked that only comes from decades of hands-on work in nature and in the museum." Some consider him the greatest Neotropical mammalogist of the 20th century. The marmoset has a small distribution along the Rio Aripuanã in west-central Brazil. The night monkey is known from a small number of specimens from central Colombia. A study published in 2001 concluded that the species was invalid, and that *Aotus hershkovitzi* is a junior synonym of *A. lemurinus*. The grass mouse is found on small islands off the tip of Tierra del Fuego.

Heude

Heude's Pig *Sus bucculentus* Heude, 1892 [Alt. Vietnamese Warty Pig]

Pierre Marie Heude (1836–1902) was a French conchologist, naturalist, and Jesuit missionary who collected and described specimens while living in China at the end of the 19th century. He spent so much time on his natural history specimens that he did not get around to converting many Chinese to Christianity. While his records were preserved and used by others, his collection was thought lost for almost 100 years. However, in 1997 Colin Groves of the Australian National University rediscovered it in Beijing, still with labels in Heude's handwriting. Heude wrote *Mémoires concernant l'histoire naturelle de l'Empire Chinois, par des pères de la Compagnie de Jésus*, which was published in a series of installments bound in different volumes printed in Shanghai. His best-known work is said to be *Conchyliologie fluviatile de la province de Nanking (et la Chine centrale)*. He was a fantastic "splitter," in the zoological sense, creating new genera and species with apparent abandon. He died at Zi-ka-wei in China. He is also remembered in the name of a bird, Heude's Parrotbill *Paradoxornis heudei*. No scientist has seen a living example of Heude's Pig, but there are indications it may still be extant in remote areas of Indochina.

Heuglin

Heuglin's Gazelle *Gazella rufifrons tilonura* Heuglin, 1868 [Alt. Eritrean Red-fronted Gazelle]
Heuglin's Olive Baboon *Papio anubis heuglini* **Matschie,** 1898

Theodor von Heuglin (1824–1876) was a German mining engineer, traveler, and ornithologist. The son of the local pastor, he was born in Ditzingen, where the local school is now named after him. He is recorded as exploring eastern Africa in 1861 and published an account of its birds in 1869, entitled *Ornithologie Nordost-Afrik.* He first went to Egypt, where he learned Arabic, in 1850 and visited the Sinai Peninsula with Brehm. He then got a job with the Austrian consulate in Khartoum, which allowed him to travel extensively in Abyssinia. Between 1857 and 1858 he was in East Africa as part of an expedition financed by the Archduke Ferdinand Max of Austria. Not only did he catch malaria on this trip but he also survived being speared in the neck by an irate local. He wrote *Reise nach Abessinien, den Galaländern, Ostsudan und Chartum in den Jahren 1861 und 1862.* After visiting Spitzbergen in 1870, he returned to Africa and the Abyssinian Mountains in 1861 and finally to Egypt and Abyssinia from 1875 to 1876. He was a vocal opponent of evolutionary theories. He died of pneumonia and is buried in the Prague Cemetery in Stuttgart. He is also commemorated in the names of several birds, including Heuglin's Robin-Chat *Cossypha heuglini* and Heuglin's Wheatear *Oenanthe heuglini.* The gazelle is found in northern Eritrea and adjacent regions of northern Ethiopia and eastern Sudan. The Olive Baboon occurs from Mauritania east to Sudan and Ethiopia; the name *heuglini* has been used for eastern populations but is not generally now regarded as a valid subspecies.

Heuren

Flores Warty Pig *Sus heureni* Hardjasasmita, 1987

It appears that Hardjasasmita's reference to "Doctor van Heuren" is merely a mistranscription of "van Heurn." The pig is found on Flores (Indonesia), but some mammalogists believe this "species" is actually a feral population of the Sulawesi Warty Pig *Sus celebensis.* See **Van Heurn** for biographical details.

Hewitt

Hewitt's Red Rock Hare *Pronolagus saundersiae* Hewitt, 1927

Dr. John Hewitt (1880–1961) was Director of the Albany Museum, Grahamstown, South Africa, from 1910 until his retirement in 1958. He had earlier been Curator at the Sarawak Museum. He had two fields of study, vertebrate zoology and archeology. The hare comes from South Africa.

Higgins

Long-tailed Mouse *Pseudomys higginsi* **Trouessart,** 1897

E. T. Higgins (ca. 1816–1891) was an Australian biologist and paleontologist. Many of his publications were joint works with W. F. Petterd, such as "Description of a New Cave-inhabiting Spider, Together with Notes on the Mammalian Remains from a Recently Discovered Cave in the Chudleigh District" (1889) and a paper on "Tasmanian Native Rodents" (1883). The mouse is found only in Tasmania.

Hildebrandt

Hildebrandt's Horseshoe Bat *Rhinolophus hildebrandtii* **Peters,** 1878
Hildebrandt's Multimammate Mouse *Mastomys hildebrandtii* Peters, 1878

Johann Maria Hildebrandt (1847–1881) was a German botanist and explorer who collected and traveled in Arabia, East Africa, Madagascar, and the Comoro Islands from 1872 until his death. He was also interested in languages and in 1876 published *Zeitschrift für Ethiopia,* which deals with the vocabularies of dialects in the Johanna Islands. He died in Madagascar of yellow fever. Two birds are named after him: Hildebrandt's Francolin *Francolinus hildebrandti* and Hildebrandt's Starling *Lamprotornis hildebrandti.* A beetle, *Sternocera hildebrandti,* was named after him by his father, who was a painter and entomologist. The bat is found from southern Sudan and Ethiopia south to Transvaal (South Africa). The mouse—sometimes viewed as synonymous with *Mastomys natalensis*—can be found from Senegal east to Somalia and south to Kenya.

Hildegarde

Hildegarde's Broad-headed Mouse *Zelotomys hildegardeae* **Thomas,** 1902
Hildegarde's Shrew *Crocidura hildegardeae* Thomas, 1904
Hildegarde's Tomb Bat *Taphozous hildegardeae* Thomas, 1909

Hildegarde Beatrice Hinde (1871–1959) was the wife of Dr. Sidney Langford Hinde. Dr. and Mrs. Hinde were in the Congo from 1891 to 1894, and in the East African Protectorate (now Kenya) from 1895 to 1915. Hildegarde Hinde was the author of *Some Problems of East Africa* (1926) and, with her husband, *The Last of the Masai* (1901). The mouse is found in the Central African Republic and southern Sudan, south to Angola and Zambia. The shrew is from east and central Africa—DRC (now Zaire), Rwanda, Burundi, Kenya, and Tanzania, with localized populations in Nigeria and Cameroon. The bat is from Kenya, northeast Tanzania, and Zanzibar.

Hilgendorf

Hilgendorf's Tube-nosed Bat *Murina hilgendorfi* **Peters,** 1880 [Alt. Japanese Tube-nosed Bat]

Franz Martin Hilgendorf (1839–1904) was a German naturalist and zoologist. He entered the University of Berlin in 1859 to study philology and moved to Tübingen University in 1861, where he was awarded a doctorate in 1863 for a thesis on a geological subject. From 1863 to 1868 he worked at, and studied in, the Zoological Museum in Berlin. Between 1868 and 1870 he was Director of the Hamburg Zoological Gardens. From the beginning of 1871 until the end of 1872 he was a private lecturer at the Polytechnic Institute in Dresden, and in 1873 he became a lecturer at the Imperial Medical Academy in Tokyo, staying in Japan until 1876. On his return to Germany he became an assistant to Peters and worked in various departments of the Berlin Museum until ill-health forced him to retire in 1903. This bat has often been treated as a race of *Murina leucogaster,* but many authorities now recognize it as a full species. It is found in Japan, Korea, and in the Russian Far East westward to the Altai.

Hill

Hill's Tomb Bat *Taphozous hilli* Kitchener, 1980
John Hill's Roundleaf Bat *Hipposideros edwardshilli* **Flannery** and Colgan, 1993
Hill's Shrew *Crocidura hilliana* **Jenkins** and Smith, 1995

John Edwards Hill (1928–1997) was the senior bat systematist at the British Museum of Natural History, retiring in 1988. In 1984 he published *Bats: A Natural History,* with J. D. Smith. He described 24 new species and 26 subspecies (13 rodents, 37 bats). He also wrote *A World List of Mammalian Species* (1980), with G. B. Corbett, as well as numerous scientific papers including 11 after his retirement and a memoir and bibliography of Oldfield Thomas. The tomb bat is found in Australia (Western Australia and Northern Territory). The roundleaf bat is from Papua New Guinea. The shrew is found in Thailand and Laos. See also **(The) Hills** and **Koopman and Hill.**

Hillier

Hillier's Mulgara *Dasycercus hillieri* **Thomas,** 1905 [Alt. Ampurta; Syn. *D. cristicauda*]

Henry "Harry" James Hillier (1875–1958) was a collector and farmer. He was born and brought up in Kent, UK, and at age 18 was diagnosed with tuberculosis and told he had six months to live. He went to South Australia and from there to Lake Eyre for the dry climate. He was the schoolteacher at Killalpaninna Mission between 1894 and 1905, and his health improved greatly. In 1904 he marked up a map of the Lake Eyre region with 2,468 Aboriginal place names for different language groups. The map had been drawn by J. G. Reuther, a Lutheran missionary, who recorded detailed descriptions and translations for these place names. Hillier appears to have also been interested in entomology and botany, as he produced papers on aspects of them both in relation to the Lake Eyre region, and was skilled as a painter and illustrator, too. In 1905 he returned to England but found that his health started to deteriorate again, and through the offices of the Church Missionary Society he returned to Australia and taught at Hermannsburg from 1906 to 1910. Among his students was the young Albert Namatjira, who became one of Australia's greatest artists. From 1916 until 1927 he worked for as Diocesan Registrar and Secretary to the Anglican Bishop of Willochra. From 1927 to 1931 he lived in Laura and worked as an accountant for the local Ford dealer. He lived in Western Australia from 1931 to about 1951 and then returned to South Australia, where he spent the rest of his life. The taxonomy of the mulgaras (small carnivorous marsupials) has been in flux; generally, *hillieri* is no longer considered a separate species from *Dasycercus cristicauda*. The mulgara inhabits the central Australian desert.

(The) Hills

Hills' Horseshoe Bat *Rhinolophus hillorum* **Koopman,** 1989 [Alt. Upland Horseshoe Bat]

John Eric Hill (1907–1947) and John Edwards Hill (1928–1997). John Eric Hill was Associate Curator of the Department of Mammals and Acting Curator of the Department of Comparative Anatomy at the American Museum of Natural History, New York. He wrote a number of scientific papers including descriptions of new bat species and at least one book : *Morphology of the Pocket Gopher, Mammalian Genus Thomomys* (1937). The unrelated John Edwards Hill (see above) was an English zoologist and author who worked at the British Museum for many years. His all-consuming passion was bats and their taxonomy, a study he continued at his home in Kent even after he retired from the British Museum in 1988. Despite a heart attack and a stroke he continued to work on bats until a few weeks before his death. Koopman's citation reads, "I name this subspecies after the two (often confused) eminent mammalogists named John E. Hill: J. Eric (1907–1947) and J. Edwards (1928–1997). I have known both of them and feel that each, in different ways, has been important in my understanding of African bats, to which both made outstanding contributions." The bat's range is West Africa from Liberia to Cameroon. It was originally described as a subspecies of *Rhinolophus clivosus* but is now elevated to the rank of a full species.

Hinde

Hinde's Lesser House Bat *Scotoecus hindei* **Thomas,** 1901
Hinde's Rock Rat *Aethomys hindei* Thomas, 1902
Hinde's Rat *Beamys hindei* Thomas, 1909 [Alt. Long-tailed Pouched Rat]

Dr. Sidney Langford Hinde (1863–1931) was Medical Officer of the Interior in British East Africa and a Captain in the Congo Free State Forces, as well as a naturalist and collector. He was also a Provincial Commissioner in Kenya and collected there, too. He wrote *The Fall of the Congo Arabs* (1897) and, with his wife Hildegarde, *The Last of the Masai* (1901). A viper, *Montatheris hindii,* which he collected in Kenya, is named in his honor. He also has a bird, Hinde's Pied Babbler *Turdoides hindei,* named after him. The bat is found from Sudan and Somalia south to Mozambique, and has also been found in Nigeria and Cameroon. The rock rat comes from Kenya and southwest Ethiopia. (Some references give a much larger range, because *Aethomys helleri* is sometimes included in the same species.) Hinde's Rat can be found in southern Kenya and northeast Tanzania.

Hobbit

Moss-forest Blossom Bat *Syconycteris hobbit* **Ziegler,** 1982

Ziegler must have been a fan of J. R. R. Tolkien, since he thought of naming this species after the small race of xenophobic rustics who dwell in The Shire. Apparently this bat, like the fictional hobbits, has very hairy feet. The bat is endemic to the mountains of central New Guinea.

Hodgson

Tibetan Antelope *Pantholops hodgsonii* **Abel,** 1826 [Alt. Chiru]
Hodgson's Bat *Myotis formosus* Hodgson, 1835
Hodgson's Giant Flying Squirrel *Petaurista magnificus* Hodgson, 1836
Hodgson's White-bellied Rat *Niviventer niviventer* Hodgson, 1836
Hodgson's Brown-toothed Shrew *Episoriculus caudatus* **Horsfield,** 1851

Brian Houghton Hodgson F.R.S. (1800–1894) was an official of the East India Company and Assistant Resident in Nepal from 1825 until 1843 and in Darjeeling between 1845 and 1859. He was very interested in Buddhism and was among those who introduced the idea to Britain in the 19th century. He also took an interest in the languages of Nepal and northern India. Hodgson wrote books on Indian birds with Hume and others. He amassed a collection of 9,512 specimens of birds, belonging to 672 species, of which 124 had never been described previously. He got the credit for describing 79 of these but failed to describe the rest before others did. A fiercely patriotic man, he once said that Cuvier (who had stolen a march on Hardwicke by naming the Red Panda, because Hardwicke's return to England was delayed) would "prevent England reaping the zoological harvest of her own domains." He is also commemorated in the names of many birds, including Hodgson's Frogmouth *Batrachostomus hodgsoni* and Hodgson's Redstart *Phoenicurus hodgsoni.* The antelope is found on the Tibetan Plateau. The bat is widespread in Asia, from Afghanistan to Korea, Taiwan, the Philippines, and Java. The squirrel is found in Nepal and West Bengal (India), the rat having a similar distribution but extending west to northeast Pakistan. The shrew occurs from Kashmir east to northern Myanmar and southwest China.

Hodson

Hodson's Puma *Puma concolor hudsoni* **Cabrera,** 1957

References to Hodson's Puma almost certainly arise from a misspelling, and the name ought to be Hudson's Puma; note that the trinomial is *hudsoni* and not *hodsoni,* and also that the vernacular Norwegian name for this animal is Hudsonspuma. We believe that this refers to William Henry Hudson (1841–1922) who was born in Buenos Aires of American parents. He spent his childhood on the pampas but developed a heart condition and had to emigrate, leaving Brazil for England in 1870. He is generally regarded as a British author, naturalist, and ornithologist. However, he is best known for his exotic romances, especially *Green Mansions* published in 1904, which is set in a South American jungle. His novel *Far Away and Long Ago* (1918) lovingly recalls his childhood. Hudson was a sensitive observer of nature, particularly birds. His books describe plants and animals in a highly personal manner with great force and beauty. Other works are *The Purple Land* (1885), *Argentine Ornithology* (1888), *The Naturalist in La Plata* (1892), *A Shepherd's Life* (1910), and *A Hind in Richmond Park* (1922). The novelist John Galsworthy said of him, "Hudson is the finest living observer, and the greatest living lover of bird and animal life, and of Nature in her moods." He is remembered in the names of two birds: Hudson's Black Tyrant *Knipolegus hudsoni* and Hudson's Canastero *Asthenes hudsoni.* Cabrera, who described this subspecies of puma, moved from Spain to work in Buenos Aires in 1925, and although he may never have met Hudson, he is sure to have known about him. This subspecies of puma, from Argentina, is now regarded as synonymous with *Puma concolor cabrerae* (**Pocock,** 1940).

Hoffmann, B.

Hoffmann's Rat *Rattus hoffmanni* **Matschie,** 1901

Dr. B. Hoffmann (dates not found) was a German zoologist who was based in Dresden. In 1887 he published *Uber Säugethiere aus den Ostindisch Archipel (mäuse, fledermäuse, büffel)* in which he described this rodent as a subspecies of the Black Rat *Mus rattus celebensis* at a time when all mice and rats were lumped together in the genus *Mus.* Matschie raised it to full species level in 1901, but as J. E. Gray had already named a species as *Mus celebensis* in 1867, Matschie had to think of another name. He called it *Mus hoffmanni* after the original describer. The genus was later changed, and it became *Rattus hoffmanni.* It is found on Sulawesi (formerly Celebes), Indonesia.

Hoffmann, C.

Hoffmann's Two-toed Sloth *Choloepus hoffmanni* **Peters,** 1858

Dr. Carl (or Karl) Hoffmann (1823–1859) was a German physician and naturalist. He went to Costa Rica in 1853 with Alexander von Frantzius and began to explore the country and collect specimens, mainly botanical ones. He was later a physician to the Costa Rican army. He is remembered in the names of three birds, all described by Cabanis after Hoffmann's death: Hoffmann's Ant-thrush *Formicarius analis hoffmanni,* Hoffmann's Conure *Pyrrhura hoffmanni,* and Hoffmann's Woodpecker *Melanerpes hoffmannii.* The two-toed sloth can be found from Honduras south to Bolivia.

Hoffmann, R. S.

Hoffmann's Pika *Ochotona hoffmanni* Formozov, Yakhontov, and Dmitriev, 1996

Professor Dr. Robert "Bob" S. Hoffmann (b. 1929) grew up in Illinois but graduated from Utah State University in 1950 and took both his master's degree in 1954 and his doctorate in 1955 at the University of California, Berkeley. He worked at the Department of Zoology at the University of Montana in Missoula from 1955 until 1968, at which point he became a Professor and Curator of the Zoological Museum. He was a Professor at the University of Kansas, and Curator of Mammals at the university's Museum of Natural History from 1965 to 1988. He became Director of the National Museum of Natural History in 1985, serving until 1988, when he became Assistant Secretary for Research at the Smithsonian Institution in Washington, DC, a post he held until his retirement in 2006. He was a Visiting Assistant Professor at the University of British Columbia's Department of Zoology in Vancouver, Canada. In 1988–1989 he was President of the Society of Systematic Biologists. He is a past President of the American Society of Mammalogists. He specializes in insectivores, rodents, and ungulates, particularly those found in North America, Russia, and China, to which countries he has traveled on many expeditions. He co-authored the section on squirrels (Sciuridae) in *Mammal Species of the World* and was one of the editors of the section on lagomorphs in the third edition of that title, published in 2005, and has many publications in his own right. He has also acted as scientific editor for a number of English translations of Russian publications, such as *Voles* by Gromov and Polyakov, published in English in 1992. The pika is found on the southern edge of the Hentey Mountains in Mongolia.

Hoffmanns

Hoffmanns' Titi *Callicebus hoffmannsi*
Thomas, 1908

Wilhelm Hoffmanns (1865–1909) was a German naturalist who collected in Peru and Brazil between 1903 and 1908. Unfortunately he contracted malaria while in Brazil and then pneumonia and died shortly after returning to Germany. He is also remembered in the name of a bird, Hoffmanns' Woodcreeper *Dendrocolaptes hoffmannsi.* The titi monkey is found in central Brazil, south of the Amazon.

Hollister

Hollister's Lion *Panthera leo hollisteri*
J. A. Allen, 1924 [Alt. Congo Lion]

Dr. Ned Hollister (1876–1924) was an American zoologist, ornithologist, and zoo superintendent. He wrote, jointly with Ludwig Kumlien, "The Birds of Wisconsin," published in 1903. Today there is a Ned Hollister Bird Club in Wisconsin named after him. In 1909 he was a member of the Alexander expedition to Alaska, and in 1912 he went on the Harvard-Smithsonian expedition to the Altai Mountains of Siberia. In 1916 he was appointed as Superintendent of the National Zoo in Washington, DC. He made a particular study of over 100 lion skulls and skins in the National Museum in Washington, where he had been Assistant Curator of Mammals until his appointment to the zoo, and in 1917 he published a study called "Some Effects of Environment and Habit on Captive Lions." He also has a subspecies of the Thirteen-lined Ground Squirrel named after him: *Spermophilus tridecemlineatus hollisteri.* The type specimen of Hollister's Lion was taken on the east side of Lake Victoria. It is not always regarded as a valid race but may be a synonym of *Panthera leo nubicus* (Blainville, 1843).

Home

Home's Wombat *Vombatus ursinus ursinus*
Shaw, 1800

This is not the name of a species but an individual animal, which we list as there are plenty of references to it that may confuse those coming across mentions of "Home's Wombat." Sir Everard Home (1756–1832), an English naturalist and physician, was an early wombat owner. In the early 19th century it was fashionable to own exotic pets from Australia. The wombat was probably captured by George Bass (of Bass Strait fame) on King Island. If so, it would have been of the race we assign it to above, as this is the only one that occurred on the island, although it is now extinct there. It was reported that the animal greatly disliked being captured, and only after it had taken strips off Bass' coat sleeves was it eventually calmed down. Bass looked after it well and sent it to England, and there in Hume's house in London it lived for two years in a domesticated state. It was reported to be attached to people it knew and was particularly good and tolerant with children. Home wrote a paper in 1809 entitled "An Account of Some Peculiarities in the Anatomical Structure of the Wombat."

Homez

Homez's Big-eared Bat *Micronycteris homezi* Pirlot, 1967

Professor Jorge Homéz Chacín (1921–2001) and M. A. Homéz (dates not found) are both mentioned by Pirlot in his description of the type specimen (number 143 in his collection, as he carefully remarks), which was taken on their property. Professor Homez was a Venezuelan physician and parasitologist who went to France to study in 1940 and landed in Bordeaux without being able to speak a word of French. He ended up staying in France until 1947 as WW2 prevented him returning to Venezuela. He studied at the Sorbonne from 1945 to 1947, returning to Venezuela and requalifying as a physician there in 1948. In 1963 he became Professor of Microbiology at the Faculty of Medicine of the University of Zulia, Maracaibo, Venezuela, retiring as an Emeritus Professor in 1979. He wrote and published on a number of topics. His thesis in France was on bilharzia, and among his other works is *Diez casos de cromoblastomicosis de los Estados,* published jointly with Fernandez and Maso in 1954. We do not know who M. A. Homez was or is. Professor Homez's wife was named Simone, and none of his children have M. A. as initials. Perhaps the property mentioned was owned jointly with a relation? Pirlot, in mentioning two people in his dedication, ought to have used the genitive plural, and it would be more correct to call the bat *Micronycteris homezorum.* The bat was discovered in western Venezuela, but the species has also since been found in Guyana, French Guiana, and northern Brazil.

Hoogerwerf

Hoogerwerf's Rat *Rattus hoogerwerfi* Chasen, 1939

Andries Hoogerwerf (1907–1977) wrote much on Indonesian vertebrates, particularly those of Java, from the late 1940s through to the 1970s. His works include *The Birds of Cibodas* (1949) and *Udjung Kulon: The Land of the Last Javan Rhinoceros* (1970). He is believed to have been the only person to photograph the now-extinct Javan Tiger *Panthera tigris sondaica* in the wild, which he did in 1938. Hoogerwerf's Pheasant *Lophura (inornata) hoogerwerfi* is also named after him. The rat has a limited range in the uplands of Sumatra.

Hoogstraal

Busuanga Squirrel *Sundasciurus hoogstraali* **Sanborn,** 1952

Hoogstraal's Gerbil *Gerbillus hoogstraali* Lay, 1975

Hoogstraal's Striped Grass Mouse *Lemniscomys hoogstraali* **Dieterlen,** 1991

Harold "Harry" Hoogstraal (1917–1986) was an American expert in the field of medical zoology, parasitology, entomology, and ecology. He gained two degrees from the University of Illinois in 1938 and 1942, then served in the U.S. Army in WW2 from 1942 to 1946. After the war he took two further degrees at the London School of Hygiene and Tropical Medicine.

He died in Cairo, Egypt, on his 69th birthday, having suffered for some time from lung cancer. He made a number of field trips: to Mexico in 1940 and to the Solomon Islands, New Guinea, and New Hebrides in 1945 while serving at a Military Medical research establishment nearby. His next trip was to Mindanao and Palawan in 1946 through 1947, then Africa in 1948 through 1949, culminating in Madagascar and Egypt. He then moved to Egypt to organize and become Head of the Department of Medical Zoology, U.S. Naval Medical Research, a position he remained in for the rest of his life. He was renowned for working 18-hour days and had over 500 publications to his name, as well as editing many more. He was a member of 30 professional societies and a volunteer in 20 bodies and collected numerous professional honors. He amassed the biggest collection of ticks outside of the British Museum, and these were donated to the Smithsonian. He is commemorated in the names of several insects and other invertebrates, including the tick *Ixodes hoogstraali*. The squirrel is confined to Busuanga Island in the western Philippines. The gerbil is known only from southwest Morocco. The grass mouse is known only from the type locality in central Sudan.

Hook

Hook's Duiker *Cephalophus (nigrifrons) hooki* St. Leger, 1934

Raymond Hook (1892–1964) was a farmer, guide, hunter and naturalist at Nanyuki in Kenya from 1912. Hook was one of the people behind a scheme to introduce cheetah racing, as an addition to greyhound racing, into the UK in the 1930s. It got as far as a number of animals being captured and shipped to England, where they went into quarantine. The scheme was not a success, and we don't know if any cheetah races ever took place. Hook also bred Zebroids, a cross between a male Grevy's Zebra and a pony or donkey mare. Hook and his wife ran a hotel called Silverbeck at Nanyuki, and it was constructed so that the Equator ran through the bar, in order that guests could choose in which hemisphere they would do their drinking. The duiker comes from the region of Mount Kenya.

Hooker

Hooker's Sea-Lion *Phocarctos hookeri* **Gray,** 1844 [Alt. New Zealand Sea-Lion]

Sir Joseph Dalton Hooker M.D. (1817–1911) was a botanist, biogeographer, and naturalist. He graduated with an M.D. from Glasgow University where his father was Professor of Botany. He used to go to his father's lectures from the age of seven. He recalled sitting on his grandfather's knee, looking at the pictures in Captain Cook's *Voyages*, and was particularly struck by one of Cook's sailors killing penguins on Kerguelen's Land. He remembered thinking, "I should be the happiest boy alive if ever I would see that wonderful arched rock, and knock penguins on the head." He was appointed as Assistant Surgeon on HMS *Erebus* under James Ross. He traveled widely, including on expeditions to Australia and New Zealand. His *Himalayan Journals: or, Notes of a Naturalist in Bengal, the Sikkim, and Nepal Himalayas, the Khasia Mountains* was published in two volumes in 1854. He also wrote *The Rhododendrons of Sikkim-Himalaya; Being an Account, Botanical and Geographical, of the Rhododendrons Recently Discovered in the Mountains of Eastern Himalaya, from Drawings Made on the Spot*, published from 1849 to 1851, and he had a number of other publications. He has been called the most important British botanist of the 19th century. He became a Director of the Royal Botanic Gardens at Kew. Darwin invited him to help examine his Galápagos collection, and they became firm friends. The orchid genus *Sirhookera* is also named after Sir Joseph, possibly a unique use of a knighthood in a scientific name. The sea-lion is found around the sub-Antarctic islands of New Zealand, especially the Auckland Islands.

Hoolock

Hoolock's Gibbon *Hylobates hoolock* **Harlan,** 1834 [Alt. Hoolock Gibbon, White-browed Gibbon; Syn. *Hoolock hoolock*]

This animal is not named after a person at all and is more correctly called the Hoolock Gibbon. "Hoolock" derives from the Hindi and Bengali word *ulluck,* which may come from the Assamese word *houlou.* The noise of the call of the Hoolock is very similar to the pronunciation of those two words. The gibbon comes from Assam (India), Bangladesh, Myanmar, and Yunnan (China).

Hooper

Hooper's Deer Mouse *Peromyscus hooperi* Lee and **Schmidly,** 1977

Professor Emmet Thurman Hooper (1911–1992) was a zoologist at the Museum of Zoology of the University of Michigan. He was Corresponding Secretary of the American Society of Mammalogists from 1941 to 1947 and President from 1962 to 1964. He wrote a great many articles on mammals, particularly rodents. The deer mouse is found in northeast Mexico.

Hopkins

Hopkins' Groove-toothed Swamp Rat *Pelomys hopkinsi* **Hayman,** 1955

George Henry Evans Hopkins (1898–1973) was always known as "Harry "to his friends. He was an entomologist who studied fleas, lice, and disease-carrying insects on behalf of the London School of Tropical Medicine, for which organization he lived and worked in Uganda for a number of years. In 1952 he published a paper on "Mosquitoes of the Ethiopian Region." He worked closely with Theresa Clay on the subject of lice, and they produced a number of publications together, of which the first, appearing in 1950, was "The Early Literature on Mallophaga. Part I, 1758–1762." He also worked with Miriam Rothschild, and through her he was asked to catalogue the Rothschild Collection of Fleas at Tring, UK. Hayman, who described the Swamp Rat, was one of Hopkins' co-authors of a paper on the location of type specimens of some African mammals. The rat is found in Rwanda, Uganda, and southwest Kenya.

Horsfield

Horsfield's Tarsier *Tarsius bancanus* Horsfield, 1821 [Alt. Raffles' Tarsier, Western Tarsier, Malaysian Tarsier]

Horsfield's Flying Squirrel *Iomys horsfieldii* **Waterhouse,** 1838 [Alt. Javanese Flying Squirrel]

Horsfield's Bat *Myotis horsfieldii* **Temminck,** 1840

Horsfield's Fruit Bat *Cynopterus horsfieldi* **Gray,** 1843

Horsfield's Shrew *Crocidura horsfieldii* **Tomes,** 1856

Dr. Thomas Horsfield (1773–1859) was an American naturalist. He was trained as a physician but became an explorer and prolific collector of plants and animals. He began his career in Java in 1796 while it was under Dutch rule. When Napoleon Bonaparte annexed Holland, it enabled the British East India Company to take control of the island in 1811. In 1819 Horsfield's poor health made him seek other employment, and the Company moved him to London, to continue his research under its direction as Curator and then Keeper of the East India House Museum. While in Java he became a good friend of Sir Thomas Stamford Raffles. He wrote *Zoological Researches in Java and the Neighbouring Islands* (1824). He is also commemorated in the common names of at least 10 birds, including Horsfield's Babbler *Trichastoma sepiarium* and Horsfield's Thrush *Zoothera horsfieldi*. The tarsier is found on Sumatra, Borneo, and some small nearby islands. The flying squirrel is found on the Malay Peninsula, Sumatra, Java, and Borneo. The fruit bat has a similar distribution to the squirrel but also extends to the Lesser Sunda Islands. The *Myotis* bat can be found from India east to Vietnam and the Philippines and south to Java, Borneo, and Sulawesi. The shrew occurs in Sri Lanka and southern India, and also in Nepal east to southern China, Vietnam, and Taiwan.

Hose

Hose's Leaf Monkey *Presbytis hosei* **Thomas,** 1889 [Alt. Hose's Langur, Grey Leaf Monkey]
Four-striped Ground Squirrel *Lariscus hosei* Thomas, 1892
Hose's Palm Civet *Diplogale hosei* Thomas, 1892
Hose's Shrew *Suncus hosei* Thomas, 1893
Hose's Hill Rat *Bunomys fratrorum* Thomas, 1896 [Alt. Fraternal Hill Rat]
Hose's Pygmy Flying Squirrel *Petaurillus hosei* Thomas, 1900
Bornean Dolphin *Lagenodelphis hosei* **Fraser,** 1956 [Alt. Fraser's Dolphin]

Dr. Charles Hose (1863–1929) was a naturalist who lived in Sarawak and other parts of Malaysia from 1884 to 1907. He was "Resident of Baram "in 1902. Hose successfully investigated the principle cause of the disease beri-beri. He was also a good cartographer who produced the first reliable map of Sarawak. Fort Hose in Sarawak, which is now a museum, was named after him. He wrote *Fifty Years of Romance and Research* (1927) and *The Field Book of a Jungle Wallah* (1929). He is also commemorated in the common name of a bird, Hose's Broadbill *Calyptomena hosii,* and in the scientific name of another, the Black Oriole *Oriolus hosii.* All the terrestrial species listed above are endemic to Borneo, except for the hill rat, which is found in northern Sulawesi. The binomial *fratrorum* means "of the brothers" and here refers to Charles Hose and his brother Ernest. The dolphin was described from a skeleton collected by Ernest Hose in Borneo in 1895; it has a worldwide distribution in warm-temperate and tropical waters.

Hosono

Azumi Shrew *Sorex hosonoi* **Imaizumi,** 1954
Hosono's Myotis *Myotis hosonoi* Imaizumi, 1954

Atsushi Hosono (dates not found) was a teacher of the Tokiwa Middle School in Nagano Prefecture, Japan. He collected several important specimens of mammals from Nagano, including the type specimens of both the species listed above. They are endemic to the island of Honshu, Japan.

Hotson

Hotson's Jerboa *Allactaga hotsoni* **Thomas,** 1920
Hotson's Mouse-like Hamster *Calomyscus hotsoni* Thomas, 1920

Sir John Ernest Buttery Hotson K.C.S.I. (1877–1944) was a Scottish soldier and colonial administrator. When he was a Colonel in the British army he made a collection of mammals in Shiraz, Iran. In 1921 R. E. Cheesman published a "Report on a Collection of Mammals Made by Col. J. E. B. Hotson in Shiraz" in the *Bombay Natural History Society Journal.* He was Acting Governor of Bombay in 1931, when an attempt was made on his life (he was shot at). He must have been a keen collector of stamps as well as mammals, as he was President of the Philatelic Society of India and, between 1923 and 1928, was Editor of the *Philatelic Journal of*

India. He was also a freemason, District Grandmaster of the United Lodge of Scottish Freemasonry in India and Pakistan from 1929 to 1932. The jerboa is found in southeast Iran, southern Afghanistan, and western Pakistan. The hamster is restricted to Baluchistan (Pakistan).

How

How's Melomys *Melomys howi* Kitchener and Suyanto, 1996

Dr. Richard Alfred How (b. 1944) is an Australian zoologist at the Western Australian Museum, Perth. He has collaborated and published with both D. J. Kitchener and A. Suyanto in regard to taxonomy and geographic morphological variation. Additionally, with others, they produced "Population Ecology and Physiology of the Common Rock Rat, *Zyzomys Argurus* (Rodentia, Muridae) in Tropical Northwestern Australia" (1988). He has also written on reptiles and, jointly with Pearson, Desmond, and Maryan, published "Reappraisal of the Reptiles on the Islands of the Houtman Abrolhos, Western Australia" (1998). The melomys is endemic to Riama in the Tanimbar Islands, eastern Indonesia.

Howell

Howell's Shrew *Sylvisorex howelli* **Jenkins,** 1984

Professor Kim M. Howell (b. 1945) is the Professor of Zoology and Wildlife Conservation at the University of Dar es Salaam,Tanzania. His broad interests include small mammals (especially bats) and birds,and most recently he has published on herpetology, collaborating with S. Spawls, R. Drewes, and J. Ashe on *A Field Guide to the Reptiles of East Africa* (2002), with Sprawls and Drewes on *A Pocket Guide to the Reptiles and Amphibians of East Africa* (2006), and with A. Channing on *Amphibians of East Africa* (2006). The shrew is found in the Usambara and Uluguru Mountains of Tanzania.

Hoy

American Pygmy Shrew *Sorex hoyi* **Baird,** 1857

Dr. Philip Romayne Hoy (1816–1892) was an American physician, explorer, and naturalist. He qualified as a physician in Ohio in 1840, practiced medicine there for some years, and then, in 1846, settled in Racine on Lake Michigan. In the 1850s he collected specimens in Wisconsin for Baird and Girard and was Naturalist for the Geological Survey of Wisconsin and Fish Commissioner for Wisconsin. He was so enthusiast a naturalist that whenever he was on his rounds, calling on patients, he always took extra equipment like a botany book and a butterfly net. He wrote "Catalog of the Cold-blooded Vertebrates of Wisconsin" in 1883. His bird collection is housed in the Racine Heritage Museum. He is also remembered in the scientific name of a terrapin, the Missouri Slider *Pseudemys floridana hoyi*, but this is now regarded as a junior synonym of *P. concinna*. The shrew occurs in the taiga zone of Alaska, Canada, and the northern USA, as well as in the Appalachian Mountains.

Hubbard

Hubbard's Sportive Lemur *Lepilemur hubbardi* Louis et al., 2006

This primate is named after the Theodore F. and Claire M. Hubbard Family Foundation rather than directly after any one person. The foundation has given much support to Malagasy graduate students in the field and in the laboratory at Henry Doorly Zoo's Center for Conservation and Research at Omaha, Nebraska. Dr. Theodore F. Hubbard was a cardiologist who died around 1999. The foundation also funded the Hubbard Gorilla Valley exhibit, opened in 2004 at the same zoo. The lemur is found in southwest Madagascar. See **Claire** for biographical details.

Hubbs

Hubbs' Beaked Whale *Mesoplodon carlhubbsi* Moore, 1963

Professor Carl Leavitt Hubbs (1894–1979) was Professor of Biology at the Scripps Institution of Oceanography, California. The beaked whale is named after him in both the vernacular and the scientific form. The Hubbs family were all ichthyologists, so it is no wonder that many aquatic species carry the scientific name *hubbsi*. *Octopus hubbsorum* was named for Carl as well as his wife, Laura Cornelia (Clark) Hubbs (1893–1988), and their son Clark Hubbs. Professor and Mrs. Hubbs had three children, all of whom became ichthyologists. The daughter, Frances, married yet another ichthyologist, Robert Rush Miller. The whale is found in cold-temperate waters of the North Pacific.

Hubert

Hubert's Multimammate Mouse *Mastomys huberti* **Wroughton,** 1908

Captain Hubert G. Cock (dates not found) was a soldier in the Royal Artillery who collected the type specimen. The species is found from northern Nigeria (where the type was taken) westward to Senegal. It was long regarded as synonymous with *Mastomys natalensis*.

Huet

Huet's Bush Squirrel *Paraxerus ochraceus* Huet, 1880 [Alt. Ochre Bush Squirrel]
Huet's Dormouse *Graphiurus hueti* Rochebrune, 1883 [Syn. *G. nagtglasii* **Jentink,** 1888]

Joseph Huet (dates not found) was a French zoologist. In 1883 he published *Note sur les carnassiers du genre Bassaricyon* (Note on the Carnivores of the Genus *Bassaricyon*). He described a number of mammal species, including the Palawan Stink Badger *Mydaus marchei* and the Congo Golden-Mole *Chlorotalpa leucorhina*. The squirrel occurs in Tanzania, Kenya, southern Ethiopia, and southern Sudan. The dormouse's distribution ranges from Sierra Leone east to Gabon and the Central African Republic. Many authorities now prefer to call this species Nagtglas' Dormouse *Graphiurus nagtglasii* (see **Nagtglas**).

Huey

Huey's Kangaroo Rat *Dipodomys antiquarius* Huey, 1962

Laurence Markham Huey (1892–1963) was a field collector employed by the San Diego Natural History Museum, where there is a collection named after him. In 1913 he took a trip to the Coronado Islands and met Donald R. Dickey, with whom he worked for the next 10 years. In 1923 Huey was offered full-time employment at the San Diego Natural History Museum. He was Curator for the Department of Birds and Mammals at that museum from 1923 to 1961. He published a number of papers and articles, among which are "Field Notes: Trip to the Coronado Islands, June 1926" and "Range Extension of Pocket Gophers along a New Road in the Arid Southwest" (1941). This kangaroo rat, from Baja California, Mexico, is now usually regarded as synonymous with *Dipodomys simulans* (Merriam, 1904).

Hugh

Hugh's Hedgehog *Mesechinus hughi* **Thomas,** 1908 [Alt. Central Chinese Hedgehog]

The Rev. Father Hugh Scallan (dates not found), also known as Padre Hugo, was a missionary and botanist in central China at the end of the 19th century. He is recorded as collecting mosses on Mount Thae-pei-san in 1898. He discovered a beautiful yellow rose, now known as Father Hugo's Rose *Rosa hugonis*, which he sent to London in 1899. The hedgehog is found only in Shanxi and Shaanxi provinces, China.

Humboldt

Humboldt's Woolly Monkey *Lagothrix lagothricha* Humboldt, 1812
Humboldt's Hog-nosed Skunk *Conepatus humboldtii* **Gray,** 1837
Humboldt's Squirrel Monkey *Saimiri sciureus cassiquiarensis* **Lesson,** 1840
Orinoco River Dolphin *Inia geoffrensis humboldtiana* Pilleri and Gihr, 1978
Humboldt's Big-eared Brown Bat *Histiotus humboldti* **Handley,** 1996

Baron Friedrich Wilhelm Heinrich Alexander von Humboldt (1769–1859) was a Prussian naturalist, explorer, and politician. After attending universities at Frankfurt an der Oder and Göttingen, in 1791 he enrolled at the Freiberg Mining Academy to learn natural history and earth sciences to help him with his intended future travels. To complete his experience he then worked as an Inspector of Mines in Prussia for five years. After two years of disappointments and delays, he set off in 1799 and explored in South America until 1804, collecting thousands of specimens, mapping, and studying natural phenomena. The trip took in parts of Venezuela, Peru, Ecuador, Colombia, and Mexico, He returned via the USA, where he was entertained by Thomas Jefferson at Monticello. In 1829 he made a journey of similarly epic proportions ranging from the Urals east to Siberia. Humboldt's *Personal Narrative* was inspirational to later travelers in the tropics, notably Darwin and Wallace. His most famous writing was the five-volume work *The Cosmos*, published between 1845 and 1862. Humboldt did research in many other fields, including astronomy, forestry, and mineralogy. The Humboldt Current, running south to north off the Pacific coast of South America, was named after him. He is also remembered in the name of the Humboldt Penguin *Spheniscus humboldti* and of a hummingbird, Humboldt's Sapphire *Hylocharis humboldti*. The woolly monkey is found in northwest Brazil, eastern Ecuador, and southeast Colombia. The skunk is found from Paraguay south to the Straits of Magellan. The squirrel monkey, the dolphin, and the bat are all found mainly in Venezuela.

Hume

Manipur Bush Rat *Hadromys humei* **Thomas,** 1886
Hume's Argali *Ovis ammon humei* **Lydekker,** 1913 [Syn. *O. ammon karelini* Severtzov, 1873]

Allan Octavian Hume C.B. (1829–1912) was a famous theosophist and poet but is probably mainly remembered as an ornithologist. He was born in London, the son of a radical Member of Parliament, Joseph Hume. At 13 years of age Allan Hume was sent to sea as a junior midshipman and served in the Mediterranean on board HMS *Vanguard*. After attending the East India Company's own school, Haileybury, and further training in medicine and surgery at University College Hospital, London, he joined the Bengal Civil Service at the age of 20 and was appointed to the district of Etawah, Uttar Pradesh. Here he introduced free primary education and founded a vernacular newspaper, *Lokmitra* (People's Friend). When the Indian Mutiny broke out in 1857, Hume was heavily involved in fighting. He caught cholera but managed to get over it and returned to Etawah with 50 men and occupied the town. He was again involved in fighting, leading charges against the rebels and, on one occasion, capturing their artillery. He was made a Commander of the Order of the Bath in 1860 as a reward for his services to the crown during the mutiny. When the mutiny was over, Hume was accused of having been too lenient with mutineers, as "only" seven men were hanged in his administrative area on his judgments. He started *Stray Feathers*, a quarterly ornithological journal, in 1872. He traveled widely throughout India and accumulated an enormous collection of skins. He wrote *Agricultural Reform of India* and the three-volume classic, *The Game Birds of India Burma and Ceylon.* After his retirement, he co-founded the Indian National Congress in 1885 and became its General Secretary. He lived in Simla after his retirement and suffered the loss of 25 years' work when, during the winter of 1884–1885 (during which he and all European returned to the plains), his servants sold all his papers and correspondence; it weighed well over 100 kg (220 pounds), so they sold it in the local bazaar as waste paper. This episode killed his interest in ornithology, and he decided to get rid of his collection as well. He was a generous man and gave it to the British Museum (Natural History) in London. No less a person than Richard Bowdler Sharpe was dispatched to Simla to pack up and escort the collection to England. There were 47 crates, each weighing about 500 kg (1,100 pounds), which had to be moved from Hume's museum, at about 2,380 m (7,800 feet) above sea level in the mountains, to the nearest railway station and thence to a port. Hume is remembered in the vernacular names of more than a dozen birds. The bush rat is found in northeast India and western Yunnan (China). The subspecies of Argali occurs in the Tien Shan Mountains of central Asia. It is now usually regarded as a junior synonym of *Ovis ammon karelini.*

Hummelinck

Hummelinck's Vesper Mouse *Calomys hummelincki* **Husson,** 1960

Dr. Pieter Wagenaar Hummelinck (1907–2003) was a Dutch naturalist. He took his first degree at the University of Utrecht in 1935 and started working in the Zoological Laboratory there in 1940. His consuming passion was the study of the fauna of the Netherlands Antilles in the Caribbean, which he frequently visited, as well as other West Indian Islands. There he made significant collections of land and marine animals, which, on his retirement in 1972, he left at the university. When the lab closed down in 1988, the collection was transferred to the Zoological Museum of Amsterdam. He started two journals and a number of societies as well as writing a great many articles and scientific papers. The mouse is found in northeast Colombia, northern Venezuela, and the islands of Curacao and Aruba (Netherlands Antilles).

Hunter

Hunter's Hartebeest *Damaliscus hunteri* **P. L. Sclater,** 1889 [Alt. Hirola; Syn. *Beatragus hunteri*]

Henry Charles Vicars Hunter(1861–1934) was a British big-game hunter and amateur naturalist. In 1888, 240 km (150 miles) up the Tana River in Kenya, he discovered the hartebeest that would later be named after him by Sclater. He is also remembered in the names of two birds, Hunter's Sunbird *Nectarinia hunteri* and Hunter's Cisticola *Cisticola hunteri.* The hirola is found in southern Somalia and northern Kenya. It is now a rare and seriously endangered species.

Husson

Husson's Water Rat *Hydromys hussoni* **Musser** and Piik, 1982 [Alt. Western Water Rat]

Husson's Yellow Bat *Rhogeessa hussoni* **Genoways** and Baker, 1996

Dr. Antonius Marie Husson (1913–1987) worked at the Rijksmuseum van Natuurlijke Historie, now part of the National Museum of Natural History at Leiden, Holland. In 1967 he published *The Bats of Suriname* and in 1978 *The Mammals of Suriname.* The water rat is found in western and central New Guinea. The yellow bat is known from Suriname and eastern Brazil.

Hutterer

Hutterer's Brush-furred Rat *Lophuromys huttereri* **Verheyen,** Colyn, and Huselmans, 1996

Dr. Leonhard Rainer Georg Hutterer (b. 1948) is a German zoologist at the Museum Alexander Koenig in Bonn, where he is Curator of Mammals. He served also as editor of the *Bonner zoologische Beiträge* and *Myotis*. He is a member of the German Mammals Society and of the American Society of Mammalogists. He is a prolific writer and has published a great many scientific papers on mammals and on the history of science. His research interests include systematics and biogeography of tropical small mammals in South America and Africa, the phylogeny of the Soricidae, and faunal turnover on islands, especially in the Canary Islands. The rat is found in DRC (Zaire).

Hutton

Hutton's Tube-nosed Bat *Murina huttoni* **Peters,** 1872

Captain Thomas Hutton (1806–1875) was an officer in the 37th Bengal Native Infantry who served in the Afghan War of 1839–1840. He retired sometime before 1850, as it was around then that he bought a piece of land near Mussoorie, near Dehra Dun, in Uttar Pradesh, and built the Manor House, which he sold in 1853. It appears that he stayed on in the Mussoorie area. He corresponded regularly with Allan Octavian Hume, who quoted him frequently in his writings. He also started experimental sericulture in Mussoorie, and in 1872 wrote an appendix to a book on sericulture and gave the names of all the species of Indian silkworms known to him up to 1871. He was an all-round naturalist and was one of those who introduced the Giant African Snail *Achatina fulica* to India, though the specimens he released at Mussoorie died from cold during the winter. Peters' original description of the tube-nosed bat refers to "Captain Hutton" and says that he trapped the type specimen at Dehra Dun (India). The bat is found from northern India east to Vietnam and the Malay Peninsula.

Ihering

Ihering's Short-tailed Opossum
Monodelphis iheringi **Thomas,** 1888
Ihering's Brucie *Brucepattersonius iheringi* Thomas, 1896 [Alt. Ihering's Hocicudo, Ihering's Akodont]
Ihering's Spiny Rat *Trinomys iheringi* Thomas, 1911 [Alt. Thomas' Spiny Rat]

Hermann Friedirich Ibrecht von Ihering (sometimes spelled Jhering) (1850–1930) was a German-Brazilian zoologist, malacologist, and geologist. He was trained as a doctor and served in the German army. He arrived in Brazil in 1880 and settled in Rio Grande do Sul. He founded the São Paulo museum in 1894 and spent 22 years as its first Director. In 1907 he published *Catálogos da fauna Brasileira* and *Aves do Brazil,* with his son Rudolpho Teodoro Gaspar Wilhelm von Ihering (1883–1939) as his co-author. In 1924 he returned to Germany and remained there until his death. He is also commemorated in the scientific names of other taxa, such as the Narrow-billed Antwren *Formicivora iheringi* and the catfish *Bunocephalus iheringi.* All three mammals are found in southeast Brazil, with the brucie extending into northeast Argentina. (The brucie was formerly placed in the genus *Oxymycterus,* but Hershkovitz moved the species into *Brucepattersonius.*)

Ikonnikov

Ikonnikov's Myotis *Myotis ikonnikovi* **Ognev,** 1912 [Alt. Ikonnikov's Whiskered Bat]

Nikolaus F. Ikonnikov (dates not found) was a Russian nobleman and entomologist. He collected bats in the Russian Far East and always sent what he collected to Ognev for identification. He described a number of grasshoppers and related taxa around 1911–1913, all collected in southern Siberia, such as *Euchorthippus unicolor* in 1913. (The collection is mostly housed in Moscow University Zoological Museum, but there are samples in other museums such as Berlin.) He also amassed a collection of cockroaches from Peru. The type specimen of the bat was taken by Ikonnikov at Primorsk in the Ussuri region. The species ranges from the Altai Mountains to Mongolia, northeast China, Sakhalin Island (Russia), and Hokkaido Island (Japan).

Illiger

Illiger's Saddle-back Tamarin *Saguinus fuscicollis illigeri* **Pucheran,** 1845

"Carl" Johann Karl Wilhelm Illiger (1775–1813) was a German zoologist. He was the first Director of the Zoological Museum of the University of Berlin, holding that post from 1811 until his premature death. He is also remembered in the common name of Illiger's Macaw *Primolius maracana.* Illiger is credited with inventing the word Proboscidean for the order of mammals that includes elephants and their extinct relatives. He wrote *Versuch einer systematischen vollständigen Terminologie für das Thierreich und Pflanzenreich,* published in 1800. The tamarin is found in eastern Peru, between the Ucayali and Huallaga rivers.

Imaizumi

Lesser Japanese Mole *Mogera imaizumii* Kuroda, 1957
Imaizumi's Vole *Myodes imaizumii* Jameson, 1961
Imaizumi's Horseshoe Bat *Rhinolophus imaizumii* **Hill** and **Yoshiyuki,** 1980

Professor Dr. Yoshinori Imaizumi (1914–2007) was Director of Animal Research at the National Science Museum in Tokyo and, after his retirement in 1984, Emeritus Curator of the museum's Department of Zoology. He wrote *The Natural History of Japanese Mammals* in 1949. He was also the author of the first description, in 1967, of the rare Iriomate Cat *Prionailurus bengalensis iriomotensis,* now usually considered to be a race of the Leopard Cat. The mole occurs on the islands of Honshu and Shikoku, Japan. The vole is known only from the Kii Peninsula, Honshu. It has sometimes been regarded as an isolated population of *Myodes andersoni.* The bat is found on Iriomote in the Ryukyu Islands.

Imhaus

Crested Rat *Lophiomys imhausi* **Milne-Edwards,** 1867 [Alt. Maned Rat]

Monsieur Imhaus of Aden (dates not found) purchased a specimen of the rat's skull, which Milne-Edwards obtained. Nothing more seems to be known about him. Although it is assumed that the skull was brought to Aden from the Horn of Africa, it is possible that the species occurs on the Arabian Peninsula. The known distribution of the species is eastern Sudan, Ethiopia, Djibouti, Somalia, Kenya, Uganda, and Tanzania.

Inez

Inez's Red-backed Vole *Caryomys inez* **Thomas,** 1908 [Alt. Kolan Vole; Syn. *Eothenomys inez*]

Thomas gives no indication as to the identity of Inez. He sometimes used female names as binomials without putting them in the genitive, and such cases may not be eponymous. It is, however, tempting to speculate, and of the notable women of his time, Professor Dr. Inez Luanne Wilder, née Whipple (1871–1929), stands out. She went from being a high school teacher to becoming a Professor of Zoology at Smith College in Massachusetts. In 1901 she was one of the first graduate students in the Zoology Department of Smith College, where she was taught by Harris Hawthorne Wilder (1864–1928), whom she married in 1906. In 1914 she was made full Professor of Zoology. In 1904 she published a paperthat is considered a landmark in the field of genetics, "The Ventral Surface of the Mammalian Chiridium—With Special Reference to the Conditions Found in Man," which suggests that the development of the surfaces of the hands and feet (chiridia) of all mammals are to some extent similar.Her work is regarded as seminal in the development of the practice of identifying criminals through fingerprints. She wrote *Laboratory Studies in Mammalian Anatomy* in 1913 and *Morphology of Animal Metamorphosis* in 1925. She is also commemorated in the scientific name of the Blue Ridge Two-lined Salamander *Eurycea wilderae.* The vole is found in central China.

Ingraham

Bahaman Hutia *Geocapromys ingrahami* **J. A. Allen,** 1891

D. P. Ingraham (dates not found) was an American naturalist who traveled widely in the West Indies and collected specimens of this rodent on East Plana Cay in the Bahamas. He sent his specimens to the American Museum of Natural History and wrote a number of papers, including "Observations on the American Flamingo" in 1893 and, in *The Auk* (1897), "Additional Records of the Flammulated Owl *(Megascops flammeola)* in Colorado." He wrote at least one paper with Allen, "Description of a New Species of *Capromys* from the Plana Keys, Bahama." The hutia is endemic to East Plana Cay. To help preserve the species, it has been introduced onto two other small islets. Other subspecies once occurred on some of the other Bahaman islands but are now extinct.

Ingram

Ingram's Squirrel *Sciurus (aestuans) ingrami* **Thomas,** 1901
Ingram's Planigale *Planigale ingrami* Thomas, 1906 [Alt. Flat-skulled Marsupial Mouse, Long-tailed Planigale]

Sir William Ingram (1847–1924) inherited the *Illustrated London News* in 1860 after his father drowned. He was also a cocoa plantation owner in Trinidad and an amateur ornithologist and collector. Oldfield Thomas published a paper in 1906 entitled "On Mammals from Northern Australia Presented to the National Museum by Sir Wm. Ingram, Bt., and the Hon. John Forrest." In 1919 Ingram introduced 50 Greater Birds-of-Paradise to Little Tobago (an island off Tobago that he had purchased) from their native New Guinea, but the population eventually died out there. The island became a bird sanctuary in 1928 willed by the Ingram estate. Clearly in the vein of great English eccentrics, although he already owned one of the finest villas on the Riviera near Monte Carlo, he bought a local castle in 1911 and then, in 1921, gave it to the town (Roquebrune-Cap-Martin) where it is located. He was also one of the sponsors of a collecting expedition to Australia. He is commemorated in the name of one reptile, Ingram's Brown Snake *Pseudonaja ingrami*. The type specimen of the squirrel was collected by Alphonse Robert in Brazil. Thomas writes, "It has been by the generous assistance of Sir William Ingram, Bart., that Mr. Robert has been enabled to undertake a collecting trip to Southern Brazil; and it is therefore with very great pleasure that I have connected his name with this interesting squirrel." The taxon is usually regarded as a subspecies of the Guianan Squirrel *Sciurus aestuans*. The planigale is found in northern Australia.

Io

Great Evening Bat *Ia io* **Thomas,** 1902
Thomas' Yellow Bat *Rhogeessa io* Thomas, 1903
Thomas' Sac-winged Bat *Balantiopteryx io* Thomas, 1904

We assume that Thomas had Greek mythology in mind when describing these bats, but why he developed such a fondness for the name Io during 1902–1904 is harder to explain. Io was a priestess of the goddess Hera in Argos. She was seduced by Zeus, Hera's husband, who then changed her into a heifer in a vain attempt to escape detection. Hera could not prevent her bovine rival getting loose to roam the world, but the spiteful goddess ensured that the cow/priestess was pursued by an infuriating gadfly that would not allow Io to rest. Perhaps the idea of restless wandering appealed to Thomas as a name for "flighty" bats. The evening bat—owner of the shortest recognized scientific name in the animal kingdom—is found in Nepal, northeast India, southern China, northern Thailand, Laos, and Vietnam. The yellow bat occurs from Nicaragua south to central Brazil and northern Bolivia. The sac-winged bat comes from southern Mexico, Belize, and Guatemala.

Irene

Chilean Climbing Mouse *Irenomys tarsalis* **Philippi,** 1900

Originally named by Philippi as *Mus tarsalis*, the generic name *Irenomys* was coined by **Thomas** in 1919. It is not named after an individual called Irene but was inspired by the derivation of that name, a Greek word meaning "peace." Thomas wrote that the rodent is "so named as a memento that its recognition coincided with the arrival of a glorious peace" (referring to the end of WW1). The climbing mouse is found in central and southern Chile and adjacent western Argentina.

Iris

Zulu Golden-Mole *Amblysomus iris* **Thomas** and Schwann, 1905 [Syn. *A. hottentotus iris*]

In Greek mythology, Iris was the goddess of the rainbow and a personal messenger of the goddess Hera. In zoology, *iris* is usually applied to very colorful species, such as the Iris Lorikeet *Psitteuteles iris*, but this is not the case with the golden-mole. Thomas and Schwann's reasons for choosing this scientific name are not recorded. The golden-mole comes from eastern South Africa. A taxonomic revision has demoted *iris* from a full species to a subspecies of *Amblysomus hottentotus*.

Irma

Western Brush Wallaby *Macropus irma* Jourdan, 1837

No reason is known for Jourdan's choice of scientific name, as he does not mention a person by the name of Irma in his description. The name Irma means "whole" or "complete," and it is unclear whether this meaning has any relevance in the current context. The wallaby comes from southwest Australia.

Isabella

Isabella Shrew *Sylvisorex isabellae* **Heim de Balsac,** 1968 [Alt. Bioko Forest Shrew]

The shrew is found on the island of Bioko (Equatorial Guinea), and the type specimen was taken at "Pic Santa Isabel," so we believe this shrew is named after a locality rather than directly after a person.

Isabella

Lady Burton's Rope Squirrel *Funisciurus isabella* **Gray,** 1862 [Alt. Gray's Four-striped Squirrel]

Lady Isabel Burton (1831–1896). See **Lady Burton** for biographical details. The squirrel occurs in Cameroon, Gabon, the Congo Republic, and the Central African Republic.

Isabelle

Isabelle's Ghost Bat *Diclidurus isabellus* **Thomas,** 1920

We are *almost* sure this bat isn't named after a person. When Thomas named an animal after an individual, he virtually always noted that fact in his description but never deigned to explain or translate his scientific names in any other way. In his description of this bat no mention is made of any person. Thomas does, however, comment on the pale brown coloration of the bat, so we think he may have intended to name it an *isabelline* ghost bat. The species is found in Venezuela, Guyana, and northern Brazil.

Ismael

Ismael's Broad-nosed Bat *Platyrrhinus ismaeli* Velazco, 2005

Dr. Ismael Ceballos Bendezú is a mammalogist from Cuzco, Peru. He has written a number of scientific papers such as "Contribución al conocimiento de los Quirópteros del Cuzco (alrededores de la ciudad)" (*Editorial Universidad Nacional del Cuzco,* 1955), "Los mamíferos colectados en el Cusco por Otto Garlepp" (1981), and "Las aves de la familia Tinamidae (Tinamiformes) en el Peru" (1985). The citation reads that the bat is named after him in "recognition of his important contributions to the study of Peruvian bats." The species is known from the Andes of Colombia, Ecuador, and Peru.

Issel

Issel's Grove-toothed Swamp Rat *Pelomys isseli* **de Beaux,** 1924 [Alt. Lake Victoria Pelomys]

Professor Arturo Issel (1842–1922) was an Italian geologist and paleontologist. He was appointed to the Chair of Geology at the University of Genoa in 1866. He conducted marine research along the Eritrean coast in the 1870s. He recorded his experiences of the 1870 expedition to the Red Sea area in *Viaggio nel Mar Rosse e tra in Bogos,* published in 1876. His malacological interests led to molluscan taxa being named after him, such as *Tellina isseli, Pomatias isselianus, Cingulina isseli,* and *Ancistruma isseli.* The swamp rat is endemic to the islands of Kome, Bugala, and Bunyama in the Ugandan section of Lake Victoria.

Jacchus

Common Marmoset *Callithrix jacchus*
Linnaeus, 1758

This is another case of Linnaeus giving a primate a fanciful name from mythology (see also **Diana, Oedipus, Silenus**). Jacchus (or Iacchus) is sometimes regarded as an epithet of the god Dionysus, sometimes as a separate character associated with the Eleusinian mysteries (where he is depicted as a torch-bearer, leading the worshipers). We do not know what led Linnaeus to connect this obscure deity to a small monkey from eastern Brazil.

Jackson, F. J.

Mount Elgon Vlei Rat *Otomys jacksoni*
Thomas, 1891
Jackson's Hartebeest *Alcelaphus buselaphus jacksoni* Thomas, 1892
Jackson's Mongoose *Bdeogale jacksoni* Thomas, 1894
Jackson's Soft-furred Mouse *Praomys jacksoni* **de Winton,** 1897
Jackson's Shrew *Crocidura jacksoni* Thomas, 1904

Sir Frederick John Jackson (1859–1929) was an English administrator, diplomat, and explorer but also a naturalist and a keen ornithologist. He led a British expedition to make contact with Emin Pasha after the latter was isolated by the Mahdi's victory in the Sudan. In 1889 he led another expedition, financed by the British East Africa Company, to explore the new Kenya colony. He later became the colony's first Governor. He was also Governor of Uganda from 1911 until 1918 and described the country as "a hidden Eden, a wonderland for birds." He wrote *The Birds of Kenya Colony and the Uganda Protectorate,* published posthumously in 1938. He is also commemorated in the name of a reptile, Jackson's Chameleon *Chamaeleo jacksonii,* and in the names of at least nine birds, including Jackson's Widowbird *Euplectes jacksoni* and Jackson's Francolin *Francolinus jacksoni.* The vlei rat is confined to Mount Elgon, on the Uganda-Kenya border. The hartebeest is found in the region of Lake Victoria. The mongoose comes from central Kenya and southeast Uganda. The soft-furred mouse ranges from Nigeria through Cameroon and the Central African Republic to southern Sudan and Kenya, and south to northern Angola and Zambia. The shrew occurs in eastern DRC (Zaire), Uganda, Kenya, and northern Tanzania.

Jackson, F. W. F.

Jackson's Fat Mouse *Steatomys jacksoni*
Hayman, 1936

Major Francis W. F. Jackson C.M.G., D.S.O. (1881–1936) served in the Royal Artillery in both the Boer War and WW1. In 1922 he was Commissioner of the Eastern Province of Gold Coast (now Ghana). From 1933 until his retirement in 1935 he was Chief Commissioner of Ashanti. The type specimen of the fat mouse was caught in 1934 by Willoughby P. Lowe, when the latter was traveling around the Gold Coast with his companion, Miss Waldron. Hayman writes, "I have pleasure in associating with this find the name of Major Jackson, C.M.G., D.S.O., Chief Commissioner of Ashanti, to whom Mr. Lowe and Miss Waldron are indebted for much valuable assistance." The mouse is known from Ghana, Togo, and southwest Nigeria.

Jackson, H. H. T.

Lawrence Island Shrew *Sorex jacksoni* Hall and Gilmore, 1932

Dr. Hartley H. T. Jackson (1881–1976) was born in Wisconsin, where he began studying zoology at the age of 11. He knew Ludwig Kumlien, who later became his college teacher, and Ned Hollister, both local naturalists. He graduated in 1904 from Milton College and then taught science in Missouri, Wisconsin, and Illinois before taking a graduate scholarship at the University of Wisconsin, where he received his master's degree in 1909. At Wisconsin he taught laboratory zoology and identified, arranged, and catalogued the department's bird collection. During the summers he worked with the Wisconsin Geological and Natural History Survey. In 1910 he joined the research staff of the Bureau of Biological Survey and was put in charge of its mammal collection. In 1924 he became Chief of the Division of Biological Investigations and in 1936 Head of Wildlife (later Biological) Surveys. He remained in the latter post until 1951, when the section was merged and he became mammalogist in the new Section of Distribution of Birds and Mammals. He was a founder of the American Society of Mammalogists and served as Chairman in 1919; Corresponding Secretary from 1919 until 1925; editor of the *Journal of Mammalogy* from 1925 until 1929; and President from 1938 to 1940. An annual award is given in his name by the society. His primary research interests were the mammalogy of his native state, the life zone concept of Clinton Hart Merriam, and the taxonomy of mammals and mammal distribution. He published extensively from the age of 15, his major work being the *Mammals of Wisconsin* published in 1961. His articles include "A Preliminary List of Wisconsin Mammals" (1908), "New Species and Subspecies of *Sorex* from Western America" (1922), "A Taxonomic Revision of the American Long Tailed Shrews" (1928), and "The Summer Birds of Northwestern Wisconsin" (1943). This shrew comes from St. Lawrence Island in the Bering Sea, west of Alaska. Portenko's Shrew, from northeastern Russia, is sometimes regarded as a subspecies *(Sorex jacksoni portenkoi).*

Jacobita

Andean Mountain Cat *Leopardus jacobita* Cornalia, 1865 [Syn. *Oreailurus jacobita*]

Jacobita Tejeda de Montemajor (b. 1841) was an Argentinian whose father was Senator for the province of Salta. She was 15 when she married Paolo Mantegazza (1831–1910), the prominent Italian neurologist, physiologist, anthropologist, and pioneer sexologist. He was practicing as a physician in Argentina when they married in 1856. The family, which eventually included five children, returned to Italy in 1858, when Paolo became Professor of General Pathology at the University of Pavia. He subsequently became Professor of Anthropology and Ethnology in Florence in 1870. We do not know when Jacobita died, but it was certainly before 1891, as in that year Paolo married his second wife, Countess Maria Fantoni. The cat is found in southern Peru, western Bolivia, northern Chile, and northwest Argentina.

Jagor

Greater Musky Fruit Bat *Ptenochirus jagori* **Peters,** 1861

Peters' Trumpet-eared Bat *Phoniscus jagorii* Peters, 1866

Professor Dr. Fedor Jagor (1817–1900) was a German ethnographer and naturalist who traveled around the Philippines and other parts of Asia in the second half of the 19th century, collecting for the Berlin Museum. In 1873 he published *Reisen in den Philippinen* (Travels in the Philippines). He described the country thus: "Few countries in the world are so little known and so seldom visited as the Philippines, and yet no other land is more pleasant to travel in than this richly endowed island kingdom. Hardly anywhere does the nature lover find a greater fill of boundless treasure." He also wrote about Indonesia and southern Malaya. Peters wrote a scientific paper about the amphibians that Jagor collected in Siam. The fruit bat is endemic to the Philippines. The trumpet-eared bat was first found on the Philippine island of Samar but is mainly known from Indonesia (Borneo, Sulawesi, Java, and the Lesser Sunda Islands). It has also been recorded in Vietnam, Laos, and Thailand.

James

James' Gerbil *Gerbillus jamesi* **Harrison, 1967** [Syn. *Dipodillus jamesi*]

Dr. James Maurice Harrison (1892–1971) was a British ornithologist and a prominent member of many ornithological and zoological societies and organizations worldwide. He was a member of the Harrison shipowning family, J. and C. Harrison—known in the shipping business as "Hungry Harrison." He started to read medicine before WW1, but in 1914 he joined the Royal Navy. After service in destroyers with the Dover Patrol, he returned to St. Thomas' Hospital and qualified. He rejoined the navy and was sent to the eastern Mediterranean, where his ship was sunk by German battle cruisers. He had to swim for it to the island of Thasos. He was awarded the Distinguished Service Cross for his actions in caring for the wounded. In 1920 he became a general practitioner in Sevenoaks, Kent, retiring only in 1970. He made many collecting expeditions both at home and abroad, from North Africa to Lapland. He left a massive collection of more than 35,000 skins and specimens, which formed the basis for the Harrison Zoological Museum Trust. He wrote *The Birds of Kent,* published in 1953. He will always be remembered for his role in raising suspicions about the "Hastings Rarities," a series of very rare birds supposedly collected in Hastings, UK, but later proved to have been collected elsewhere. The gerbil is endemic to Tunisia and was described by James' son, David L. Harrison, the noted zoologist of the Harrison Institute.

James, L., J., and B.

James' Sportive Lemur *Lepilemur jamesi* Louis et al., 2006

This primate is named after Dr. Lawrence "Larry" James, Mrs. Jeannette James, and Barry James, and so the binomial should more correctly be the plural form *jamesorum*—and will probably be so amended. The James family has given generous, long-term support to Malagasy graduate students in the field, and in the laboratory at the Henry Doorly Zoo's Center for Conservation and Research at Omaha, Nebraska. Dr. and Mrs. James have also endowed a Chair of Business Administration at the University of Nebraska. The lemur is endemic to southeast Madagascar.

Jameson

Jameson's Red Rock Hare *Pronolagus randensis* Jameson, 1907

Dr. Henry Lyster Jameson (d. 1922) was an Irish physician and naturalist, and an amateur speleologist. He published several articles, including one in the *Proceedings of the London Zoological Society* entitled "On the Origin of Pearls" (1903); he had spent a year in Papua unsuccessfully experimenting with pearl shell cultivation. There are also some papers in the *Irish Naturalist,* one called "On the Exploration of the Caves of Enniskillen" (1896) and the other "The Bats of Ireland" (1897), and yet another in the *Proceedings of the Royal Irish Academy,* "Notes on Irish Worms: 1. The Irish Nemertines with a List of Those Contained in the Science and Art Museum, Dublin" (1898), which we assume are all from the same man.

Jameson collected the hare in the Transvaal (South Africa). It is also found in eastern Botswana and Zimbabwe, with a disjunct population (subspecies *caucinus*) in Namibia.

Jayakar

Arabian Rock Hyrax *Procavia capensis jayakari* **Thomas,** 1892
Arabian Tahr *Hemitragus jayakari* Thomas, 1894

Colonel Atmaram Sadashiv G. Jayakar (1844–1911) was an Indian surgeon. His nickname was "Muscati," after the town of Muscat where he was sent by the Indian Medical Service in 1878. During his 30 years in the Oman area he studied the local wildlife and collected specimens that he donated to the British Museum (Natural History). He is also commemorated in the scientific names of the Arabian Sand Boa *Eryx jayakari*, the lizard *Lacerta jayakari*, and several fish including the seahorse *Hippocampus jayakari*. The hyrax is found in the southern Arabian Peninsula. The endangered tahr is found in the mountains of Oman, extending just over the border into the United Arab Emirates.

Jeffery

Jeffery's Tamarin

This is a transcription error for Geoffroy. See **Geoffroy, E. G.**

Jelski

Jelski's Altiplano Mouse *Chroeomys jelskii* **Thomas,** 1894 [Syn. *Abrothrix jelskii*]

Professor Constantine (Konstanty) Roman Jelski (1837–1896) was a Polish naturalist who was involved with the Zoological Museum of Warsaw. He collected widely, including in Peru as correspondent for that museum. Between 1874 and 1878 he was employed by the Peruvian government as Curator at the museum in Lima, where many of his specimens are still displayed. He returned to Poland in 1878 to become Curator of the Krakow Museum, a post he held until his death. He is credited with discovering about 60 new species of birds and is commemorated in the name of Jelski's Chat-Tyrant *Ochthoeca jelskii*, among others. The mouse is found in southern Peru, western Bolivia, and northwest Argentina.

Jenkins

Jenkins' Shrew *Crocidura jenkinsi* Chakraborty, 1978
Jenkins' Shrew-Tenrec *Microgale jenkinsae* **Goodman** and Soarimalala, 2004

Paulina (Paula) D. Jenkins is a British zoologist. The shrew-tenrec is found in southwest Madagascar. The shrew is known from only two localities on South Andaman Island (India) in the Bay of Bengal. The binomial should really be in the feminine, *jenkinsae*, as with the shrew-tenrec. See **Paulina** for biographical details.

Jentink

Jentink's Guenon *Cercopithecus signatus* Jentink, 1886
Jentink's Squirrel *Sundasciurus jentinki* **Thomas,** 1887
Jentink's Dormouse *Graphiurus crassicaudatus* Jentink, 1888 [Alt. Thick-tailed African Dormouse]
Jentink's Duiker *Cephalophus jentinki* Thomas, 1892

Dr. Fredericus Anna Jentink (1844–1913) was the Director of the Dutch National Museum of Zoology at Leiden. He was one of five zoologists chosen by the Third International Congress of Zoology (Leiden, 1895) to deliberate and form a "codex" on zoological nomenclature, the precursor of today's process. He described and named many animals, particularly mammals, and wrote numerous papers such as "On Two Mammals from the Calamianes-Islands: Notes from the Leyden Museum" (1895) and "Mammals Collected by the Members of the Humboldt Bay and the Merauke River Expeditions: Nova Guinea" (1907). More significant, however, were his *Catalogue ostéologique des mammifères* (1887) and *Catalogue systématique des mammifères* (1892). The guenon is something of a mystery. It is known from a small number of specimens from imprecise locations in Africa. It may be a valid species or it could be a hybrid; the jury is still out. The squirrel is found only in the mountains of northern Borneo, while the dormouse ranges from Liberia east to Cameroon. The duiker, which Jentink himself collected in 1884, is native to Sierra Leone, Liberia, and western Ivory Coast.

Jerdon

Jerdon's Palm Civet *Paradoxurus jerdoni* **Blanford,** 1885

Thomas Claverhill Jerdon (1811–1872) was a British physician who had both zoological and botanical interests. He was born in Durham and educated at the University of Edinburgh. He studied medicine and became an Assistant Surgeon in the East India Company. Jerdon published *Birds of India* (1862–1864), which according to Darwin was *the* book on Indian birds. He also wrote *Illustrations of Indian Ornithology* and *The Game Birds and Wildfowl of India*, as well as *Mammals of India*, among others. He is also commemorated in the common names of no fewer than 13 birds, including Jerdon's Courser *Rhinoptilus bitorquatus*, Jerdon's Bushchat *Saxicola jerdoni*, and Jerdon's Starling *Sturnus burmannicus*. The civet is restricted to southern India.

Joffre

Joffre's Pipistrelle *Hypsugo joffrei* **Thomas,** 1915 [formerly *Pipistrellus joffrei*]

General Joseph Joffre (1852–1931) was a Marshall of France and Commander in Chief of the French army. At the age of 18 he took command of a battery during the Paris uprising of 1870. He served in Indochina and in Africa, where he distinguished himself when he led a column of men across the desert and captured Timbuktu. Between 1904 and 1906 he proved to have exceptional organizational skills as Director of Engineers, and in 1911 he was appointed as Chief of the General Staff. This meant he was the senior officer in the French army when WW1 started in 1914. He rightly got the credit for stopping the Germans from advancing on and capturing Paris at the Battle of the Marne in 1914. After that the fighting degenerated into the stalemate of trench warfare, and Joffre became associated with that stalemate and with the failure of any leader on either side to come up with a strategy to end it. He was promoted to Marshall of France in December 1916 and was replaced as Commander in Chief. In 1917 he was appointed president of the Allied War Council. His duties were more ceremonial than strategic. He held a number of posts in the Ministry of War between 1918 and 1930, when he retired. The bat is from northern Myanmar and is now regarded as critically endangered.

Johan

Johan's Spiny Mouse *Acomys johannis* **Thomas,** 1912

Dr. John C. Fox (dates not found) was the brother of the Rev. George T. Fox (see **Fox**). He was based at Kabwir in Nigeria and ran the local hospital established by the missionary society. Both brothers sent specimens of the local wildlife to the Zoological Society in London. Dr. Fox was in England when WW1 broke out. When he was returning to West Africa in 1916, the Elder Dempster liner, the *Falaba*, on which he was traveling was torpedoed with the loss of 102 lives. Fortunately, Fox was among the survivors. Thomas used a Latinized version of "John" for the mouse's binomial. We suspect

"Johan" became the vernacular name because somebody somewhere was too lazy to get the facts right. We humbly suggest that Fox's Spiny Mouse would be a better common name. The species is found from Burkina Faso and Ghana east to Nigeria and northern Cameroon. (There is another mammal with the scientific name *johannis*, the tuco-tuco *Ctenomys johannis*, but in this case the word refers to the Argentine province of San Juan.)

John

John's Langur *Trachypithecus johnii* Fischer, 1829 [Alt. Nilgiri Langur, Hooded Leaf Monkey]

The Rev. Dr. Christoph Samuel John (1747–1813) was a medical missionary at the Danish trading station of Tranquebar (now Tharangambadi) in Tamil Nadu, not far from Madras (now Chennai). It was a Danish colony from 1620 until 1845, when Denmark sold all its possessions in India to Great Britain. John was a botanist, herpetologist, and naturalist among whose friends was William Roxburgh, the botanist who lived in Madras and was in charge of the botanical gardens there. John was awarded an honorary doctorate in 1795 for his studies in natural history. He arrived in Tranquebar in 1771 and stayed there until his death. He wrote on the subject of snakes, and the Indian Sand Boa *Eryx johnii* owes its scientific name to him. We have no definite proof that he was the "John" in Fischer's mind when the latter named the langur, but there is a reference, dated 1884, that says, "This monkey was named after a member of the Danish factory at Tranquebar, M. John," so it seems very likely. The langur is endemic to southern India.

John Hill

John Hill's Roundleaf Bat *Hipposideros edwardshilli* **Flannery** and Colgan, 1993

The bat is from Papua New Guinea. See **Hill** for biographical details.

Johnson, D.

Johnson's Hutia *Plagiodontia ipnaeum* Johnson, 1948 extinct

Dr. David H. Johnson (b. 1912) is an American zoologist and a member of the Biological Society of Washington, DC. This society has published a number of his papers, including "The Spiny Rats of the Riu Kiu Islands" (1946). In 1941 he was working at the Smithsonian Institution as an Associate Curator, and he became Curator in 1957. He was in charge of the Mammals Division from 1948 until 1965. The hutia was found on the island of Hispaniola. No precise date of extinction is known, but it may have survived into the 17th century.

Johnson, K.

Johnson's Mouse *Pseudomys johnsoni* Kitchener, 1985 [Alt. Central Pebble-mound Mouse]

Dr. Ken Johnson is head of the Conservation Commission of the Northern Territory, Australia. Kitchener named the species in his honor "in recognition of his contributions to the study of mammals in the Northern Territory." He spent three years as a wildlife research officer with the Tasmanian Parks and Wildlife Service before joining the Conservation Commission of the Northern Territory, based in Alice Springs, in 1978. He was involved with wildlife research and conservation management in central Australia for 14 years, focusing on mammal ecology in relation to the management of endangered desert marsupials, and in flora and fauna survey work. In 1992 he became Regional Director South for the Northern Territory Parks and Wildlife Commission and also led the commission's team responsible for planning, construction, and management of the Alice Springs Desert Park. In 2000 he became Chair of the Desert Knowledge Project Steering Committee and in 2004 was appointed Chief Executive Officer of the subsequent Desert Knowledge Australia statutory corporation. The mouse comes from a small area of central Northern Territory, Australia.

Johnston

Johnston's Shrew *Sylvisorex johnstoni* **Dobson,** 1888
Johnston's Nyassa Wildebeest *Connochaetes taurinus johnstoni* **Sclater,** 1896 [Alt. Nyassaland Gnu]
Johnston's Dormouse *Graphiurus johnstoni* **Thomas,** 1897
Johnston's Grey-cheeked Mangabey *Lophocebus (albigena) johnstoni* **Lydekker,** 1900
Okapi *Okapia johnstoni* Sclater, 1901 [Alt. (archaic) Johnston's Horse, Forest Giraffe]
Johnston's Genet *Genetta johnstoni* **Pocock,** 1907

Sir Harry Hamilton Johnston (1858–1927) was a formidable English explorer and colonial administrator. He was a larger-than-life character, though physically small. Just 152 cm (5 feet) tall, he became known as the "Tiny Giant." Johnston was also an accomplished painter, photographer, cartographer, linguist, naturalist, and writer. He started exploring tropical Africa in 1882 and the following year met up with Henry Morton Stanley in the Congo. In 1884 he was in East Africa and the following year joined the colonial service, serving in a number of posts in various African countries: Cameroon, Nigeria, Liberia, Mozambique, Tunisia, Zanzibar, and Uganda. He also established a British Protectorate in Nyasaland. Johnston was Queen Victoria's first Commissioner and Consul General to British Central Africa. He was a member of the Royal Academy of Art, and his paintings of African wildlife are exceptional. He spoke over 30 African languages, as well as Arabic, Italian, Spanish, French, and Portuguese. He was knighted in 1896 and retired in 1904, after which he continued his pursuit of natural history. He discovered more than 100 new birds, reptiles, mammals, and invertebrates. Perhaps the most notable of all these is the Okapi. Johnston wrote more than 60 books, including *The Story of My Life* (1923), as well as more than 600 shorter works. He made the very first Edison cylinder recordings in Africa, which have passed down his squeaky voice to posterity. He is also commemorated in the scientific name of the Ruwenzori Turaco *Ruwenzorornis johnstoni.* The shrew is found in lowland forest from Cameroon and Gabon to Uganda and Burundi (genetic studies indicate that more than one species might be currently lumped into this one). The wildebeest occurs in southern Tanzania, Mozambique, and, formerly, in Malawi. The dormouse is a little-known species from southern Malawi. The mangabey comes from Rwanda, Burundi, and DRC (Zaire) north and east of the Congo River, while the okapi is endemic to DRC (Zaire). The genet is known from Guinea, Liberia, Ivory Coast, and Ghana.

Johnstone, R. A.

Coppery Brushtail Possum *Trichosurus johnstonii* Ramsay, 1888

Robert Arthur Johnstone (1843–1905) was a Scottish-educated Australian explorer and policeman. Between 1865 and 1871, having returned from Scotland, he managed a property and stock in Queensland. In 1871 he was appointed as Sub-Inspector of Native Police to the Cardwell district; the gold rushes of the times were creating friction with native peoples. He was accused but cleared of having taken overextreme measures against the Aboriginals who killed and ate the captain and crew of the *Maria.* He was put in charge of native police in 1873 and accompanied G. E. Dalrymple on the northeast coast expedition to explore the coastal lands as far as Cooktown. Dalrymple named mounts Annie and Arthur in the Seymour range after members of Johnstone's family and named the Johnstone River after him; he had first found it when investigating the Green Island massacres. In 1876 Johnstone discovered and named the Barron River while searching for a route over the ranges behind Trinity Bay to serve the new goldfields. His explorations led to the founding of such towns as Cairns and Innisfail. He was sometimes known as "Black" because of his dark tanned complexion. Another nickname was "Snake," because he often gave children a surprise by producing snakes from his shirt. He was also a keen amateur naturalist. His reminiscences of North Queensland appeared as a series of articles, "Spinifex and Wattle," published in *Queenslander* from 1903 to 1905. He is also remembered in the name of Johnston's Crocodile *Crocodylus johnstoni.* The possum is found in the rainforests of northeast Queensland and is sometimes treated as a race of the Common Brushtail *Trichosurus vulpecula.*

Johnstone, R. E.

Johnstone's Giant Mastiff Bat *Otomops johnstonei* Kitchener, **How,** and Maryanto, 1992 [Alt. Alor Mastiff Bat]

Dr. Ronald "Ron" Eric Johnstone (b. 1949) is an ornithologist at the West Australian Museum in Perth, where he was a colleague of both Kitchener and How. He has made many trips to Indonesia to make comparisons between the avifauna in the Kimberley Ranges and Indonesian islands. He also studies reptiles and frogs in these two areas. He is also remembered in the scientific name of a lizard, the Rough Brown Rainbow-skink *Carlia johnstonei.* The bat is found on the island of Alor, Lesser Sunda Islands, Indonesia.

Jolly

Jolly's Mouse-Lemur *Microcebus jollyae* Louis et al., 2006

Dr. Lady Alison Jolly (b. 1937) is a primatologist who is Senior Visiting Scientist at Sussex University. She is married to Sir Richard Jolly, recently of UNICEF. She graduated from Cornell and then earned her Ph.D. at Yale. She first began studying lemurs in 1963. She taught part-time at Cambridge and at Sussex and briefly at the Rockefeller University. She has written a number of books, such as *The Evolution of Primate Behavior* (1972), *Lucy's Legacy: Sex and Intelligence in Human Evolution* (1999), and *Lords and Lemurs: Mad Scientists, Kings with Spears, and the Survival of Diversity in Madagascar* (2004). She has also written around 100 scientific papers and made about 15 TV documentaries. She was President of the International Primatological Society from 1992 until 1996. The mouse-lemur comes from eastern Madagascar.

Jones

Jones' Pocket Gopher *Geomys knoxjonesi* **Baker** and **Genoways,** 1975 [Alt. Knox Jones' Pocket Gopher]

The gopher is found only in southeast New Mexico and western Texas, USA. See **Knox-Jones** for biographical details.

Jones, T. S.

Jones' Roundleaf Bat *Hipposideros jonesi* **Hayman,** 1947

Theodore S. Jones (dates not found) was an agronomist who was Deputy Director of Agriculture in Sierra Leone in 1947, in which year he started a collection of the snakes of that country. He was active as a naturalist, collecting both fish and freshwater mussels for the British Museum. He was Sierra Leone's representative on the committee of the Nigerian Field Society in 1953. He published at least one article with Hayman. The type specimen was, according to the original description, part of "a collection of small mammals from Sierra Leone, presented to the British Museum by Mr. T. S. Jones." The bat is found from Sierra Leone and Guinea to Burkina Faso and Nigeria.

José

José's Hocicudo *Oxymycterus josei* Hoffmann, Lessa, and Smith, 2002 [Alt. Cook's Hocicudo]

Professor Dr. Joseph Anthony Cook (b. 1958) is Professor and Curator of Mammals at the Museum of Southwestern Biology, University of New Mexico, the university where he earned his first degree and doctorate. Previously he had been Chair and Professor of Biology, Idaho State University, and Professor of Biology and Chief Curator at the University of Alaska Museum. He has been fortunate to mentor a number of excellent students, publish scientific papers, and conduct fieldwork in the USA, Canada, Mexico, Central America, Bolivia, Paraguay, Uruguay, and Russia. The type description says, "We name this species after Joseph 'José' A. Cook (currently at Idaho State University, Pocatello, Idaho), who has worked hard for the development of Uruguayan mammalogy." The hocicudo (a mouselike rodent) is found in Uruguay. See also **Cook, J. A.**

Jouvenet

Jouvenet's Shrew *Crocidura jouvenetae* **Heim de Balsac,** 1958

Mademoiselle A. Jouvenet (dates not found) was a helper of the describer, who wrote, "Dédiée à notre collaboratrice Mlle. a. jouvenet, qui a pris une grande part à la présentation de ce travail" (Dedicated to our collaborator Miss A. Jouvenet, who has taken a major part in the presentation of this work).Whether she did all the hard research, was the typist, or merely kept Heim de Balsac supplied with coffee is not made clear. This shrew is found in Guinea, Liberia, and the Ivory Coast.

Juldasch

Juldasch's Vole *Neodon juldaschi* **Severtzov,** 1879 [Alt. Juniper Vole, Pamir Vole]

The type specimen was collected in Kirghizia, Karakul Lake basin, near Aksu in Uzbekistan. Juldasch is an Uzbek name—as an example, there was an Uzbek politician called Juldasch Achunbabajew (1885–1943)—but we have been unable to trace who this particular Juldasch might have been. The vole is found from northeast Afghanistan to Tibet.

Julian

Julian's Gerbil *Gerbillus juliani* St. Leger, 1935

Julian Francis Drake-Brockman (1919–1986). The type specimen was part of a collection of mammals from Somaliland, put together by Ralph Drake-Brockman (see **Brockman**). St. Leger writes, "I have much pleasure in naming this little gerbil after the schoolboy son of Dr. Drake-Brockman." When he was grown up he entered the colonial service and was an administrator in Sarawak. He survived being a prisoner-of-war of the Japanese and went on to be District Officer in the Kanowit District from 1950 to 1952 (when there was still a "White Rajah" of Sarawak). He was co-author with several other people of *The Peoples of Sarawak* (1959). The gerbil is endemic to Somalia. It may be conspecific with *Gerbillus watersi.*

Juliana

Juliana's Golden-Mole *Neamblysomus julianae* Meester, 1972

Juliana Meester is the wife of a South African zoologist. In a rather brief citation, the author, J. A. J. Meester, of the Department of Zoology,

University of Natal, and past President of the Zoological Society of Southern Africa, names the golden-mole "for my wife, Juliana." The species is known from only three isolated populations in the Pretoria area, South Africa.

Julio

The mouse genus *Juliomys* González, 2000 [3 species: *ossitenuis*, *pictipes*, and *rimofrons*]

Dr. Julio Rafael Contreras Roque. See **Contreras** for biographical details.

Juscelino

The burrowing-mouse genus *Juscelinomys* **Moojen,** 1965 [4 species]

Juscelino Kubitschek de Oliveira (1902–1976) was President of Brazil from 1956 to 1961. He was a surgeon by profession who went into politics. Before becoming President, he was Mayor of Belo Horizonte in Minas Gerais Province and Governor of that province. He raised enormous sums of money for capital projects, including the construction of the new capital city of Brasília. In 1964 there was a military takeover, and he was deprived of his political rights and, temporarily, went into exile. Moojen first coined the name of this genus when he described the Candango Mouse *Juscelinomys candango,* which has been recorded only on the site of Brasília during that city's construction.

Ka'apori

Ka'apori Capuchin *Cebus (olivaceus) kaapori* Queiroz, 1992

This monkey is named not after an individual but after the Urubu-Ka'apor Indians, who live in the region of Brazil where it was discovered. It is sometimes regarded as a full species but is often treated as a subspecies of the Weeper Capuchin *Cebus olivaceus*. It occurs in eastern Amazonian Brazil.

Kaiser

Kaiser's Rock Rat *Aethomys kaiseri* **Noack,** 1887

Dr. Wilhelm Kaiser (1841–1884) was a German zoologist. We have not been able to trace any further details of him. In his description Noack used the following wording: "Durch Verarbeitung und Kombinirung dieses Materials lässt sich eine genügende Uebersicht über die Säugethier-Fauna in den von den Herren reichard, böhm und kaiser durchzogenen Gebieten gewinnen" (By absorbing and combining these materials one can achieve a decent overview of mammal fauna in the sections covered by Messrs. Reichard, Böhm, and Kaiser). We are of the opinion that the man referred to is Dr. Kaiser and not the German Emperor, Kaiser Wilhelm I (1797–1888). The type specimen was collected at Marungu in Zaire, and the rat is also found in Angola, Zambia, Malawi, Tanzania, Rwanda, and southern Kenya.

Kalinowski

Kalinowski's Mastiff Bat *Mormopterus kalinowskii* **Thomas,** 1893
Kalinowski's Oldfield Mouse *Thomasomys kalinowski* Thomas, 1894
Kalinowski's Agouti *Dasyprocta kalinowskii* Thomas, 1897
Kalinowski's Mouse-Opossum *Hyladelphys kalinowskii* **Hershkovitz,** 1992

Jan Kalinowski (1860–1942) was a Polish zoologist who was employed by the Branicki family as a collector and expedition leader. His expeditions took him to the Ussuri Territory and Lake Chanke between 1883 and 1885, to Korea and Japan from 1885 to 1888, and to Peru from 1889 to 1902. He settled in Peru, married, and had no fewer than 18 children, one of whom is Celestino Kalinowski, a Peruvian ornithologist. The mastiff bat occurs in Peru and northern Chile. The mouse is confined to the Andes of central Peru. The agouti is known only from southeast Peru. The mouse-opossum was first recorded from Peru but has also been found in French Guiana and Guyana.

Kandt

Golden Monkey *Cercopithecus kandti* **Matschie,** 1905

Dr. Richard Kandt (1867–1918) was a German civil servant, physician, explorer, and naturalist. He went to Africa in 1898, partly to explore and search for the source of the Nile and partly to begin the process of establishing the German Residency in Rwanda. He was the first Imperial Resident (Governor) for the German colonial regime in Kigali, serving from 1906 to 1911 and again from 1913 to 1914. His 16-room house has been restored and turned into a museum celebrating the history of Rwanda and its wildlife. He published *Caput-Nili, eine empfindsame Reise zu den Quellen des Nils* in 1904. The monkey is found in the Virunga Mountains (borders of Rwanda, Uganda, and DRC [Zaire]) and the Nyungwe Forest of southern Rwanda.

Kano

Kano's Mole *Mogera kanoana* Kawada et al., 2007

Dr. Tadao Kano (1906–1945) was a Japanese anthropologist who engaged in fieldwork covering almost all aboriginal villages in Taiwan, which in his day was a Japanese possession. He published many articles on ethnological, archeological, and entomological subjects. He made at least eight trips to Botel Tobago, an island between Taiwan and the Philippines with a unique flora and fauna. During WW2, as an officer of the Educational Section of the Japanese Military Administration, he was stationed in the Philippines to protect its museums and was largely responsible for saving a portion of the Beyer Collection from massive looting and destruction during the Japanese occupation (he had known Beyer at Harvard before the war). He disappeared in Borneo in 1945. The etymology for the mole reads, "The specific name '*kanoana*' is dedicated to the late Dr. Tadao Kano in recognition of his comprehensive studies of the nature and folklore of Taiwan and for his foresight regarding the existence of a second species of Taiwanese mole." It had generally been believed that only one species of mole occurred on Taiwan: *Mogera insularis*. In 1940 Tadao Kano named a second species, as *Mogera montana,* but his work was largely ignored and the name *montana* is invalid in this context. Only in 2007 was the existence of a second species of mole on Taiwan confirmed and a valid scientific name bestowed on it. The species is found in central and eastern Taiwan.

Kappler

Kappler's Armadillo *Dasypus kappleri* Krauss, 1862 [Alt. Greater Long-nosed Armadillo]
Greater Dog-like Bat *Peropteryx kappleri* **Peters,** 1867

August Kappler (1815–1887) was a German botanist and collector in Dutch Guiana (now Suriname). He first went to that country on military service in 1836 but had spare time to devote to collecting plant and insect specimens. After leaving the military and making a short visit to relatives in Stuttgart in 1841, Kappler was back in Suriname in 1842 and stayed there until 1879. He wrote a detailed description of the Carib and Arawak Indians. In 1846 he founded the town of Albina (named after his wife) on the River Marowijne, which forms the border with French Guiana. He wrote *Zes Jaren in Suriname (1836–1842),* among other works. The armadillo is found over much of northern South America. The bat is found from southern Mexico to Peru and Brazil.

Karimi

Karimi's Fat-tailed Mouse-Opossum *Thylamys karimii* **Petter,** 1968

Dr. Y. Karimi is an epidemiologist who has made a particular study of the fleas that cause bubonic plague and has studied these in countries as far apart as Iran and Brazil. He has published widely on such matters, often in conjunction with Petter, who described the opossum. Petter wrote, "A l'occasion d'une étude de la peste et des rongeurs qui en sont les vecteurs dans le Nord-Est du Brésil (région d'Exu, Pernambuco), le Dr. Y. Karimi, chef de la mission a capturé une petite sarigue terrestre" (On the occasion of a study of disease and the rodents that are the vectors of it in the northeast of Brazil [region of Exu, Pernambuco], Dr. Y. Karimi, in charge of the project, caught a small terrestrial opossum). This "small terrestrial opossum" turned out to be a new species, which Petter named after Karimi. However, some authorities believe *Thylamys karimii* to be the same species as *Thylamys velutinus* Wagner, 1842.

Kastschenko

Forest-steppe Marmot *Marmota kastschenkoi* Stroganov and Yudin, 1956

Professor Dr. Nikolai Feofanovich Kastschenko (1858–1935) was the founding Professor of Zoology at Tomsk University in 1888 and described a number of mammals, naming them for others. During his time at Tomsk he undertook and encouraged wildlife research in the area and led expeditions into wider Siberia. His actions and studies led to his university being recognized as the center for zoology in Siberia. His background was in medicine, as he qualified originally as a physician. He published a number of papers, including one on the results of his zoological expedition of 1898, and wrote the first manual in Russian on biology for medical students. His collections provided the beginning of the museum at the university, where there are now 120,000 zoological specimens. His speciality was marmots. The marmot named after him was formerly regarded as a race of *Marmota baibacina*. It is found in western Siberia, in the area south of Tomsk and east of Novosibirsk.

Keays

Keays' Rice Rat *Nephelomys keaysi* **J. A. Allen,** 1900
Hairy-legged Myotis *Myotis keaysi* J. A. Allen, 1914

Herbert H. Keays (dates not found) collected the type specimens of both mammals in Peru, for the American Museum of Natural History. J. A. Allen published a paper in the AMNH *Bulletin* in 1900 entitled "On Mammals Collected in Southeastern Peru, by Mr. H. H. Keays: With Descriptions of New Species," and another in 1901, "On a Further Collection. . . ." He is also known to have collected a new species of swift in Peru in 1899. The rice rat is found in montane forests of the eastern Peruvian Andes. The bat is found from Mexico to northern Argentina and on Trinidad.

Keen

Northwestern Deer Mouse *Peromyscus keeni* **Rhoads,** 1894
Keen's Bat *Myotis keeni* **Merriam,** 1895

The Rev. John Henry Keen (1852–1950) was a British Anglican missionary from the Church Missionary Society in London. In 1875 he went to Moose Fort in British Columbia, and in 1890 to Masset on Graham Island (Queen Charlotte Islands), staying there until 1898 when he returned to England on leave. While at Masset he translated the Book of Common Prayer into the local Haida language, and it was published in London by the Missionary Society in 1899. Keen returned to British Columbia in 1899 and was sent to Metlakatla, staying there until 1913. He was clearly an enthusiastic ornithologist and naturalist. The type specimen of the bat was collected at Masset. In 1896 he also collected the type specimen of the local subspecies of Saw-whet Owl. The deer mouse is found in the coastal regions of southeast Alaska, British Columbia, and northwest Washington State. The bat has a similar distribution (populations east of the Rocky Mountains are now regarded as a separate species, *Myotis septentrionalis*).

Kelaart

Kelaart's Leaf-nosed Bat, *Hipposideros lankadiva* Kelaart, 1850 [Alt. Indian Roundleaf Bat]
Kelaart's Long-clawed Shrew *Feroculus feroculus* Kelaart, 1850
Kelaart's Pipistrelle *Pipistrellus ceylonicus* Kelaart, 1852

Lieutenant Colonel Edward Frederick Kelaart (1819–1860) was a British physician and zoologist who was born in Ceylon (now Sri Lanka).

He qualified as a doctor at Edinburgh and also attended classes in medicine in Paris. While at Edinburgh he made his first contribution as a naturalist, delivering a paper in 1839 on "The Timber Trees of Ceylon." He was in the Ceylon medical service and also served in Gibraltar from 1843 to 1845. While stationed there he studied the local plants and wrote *Flora Calpensis: A Contribution to the Botany and Topography of Gibraltar and Its Neighbourhood.* He returned to Ceylon in 1849 and stayed until 1854, during which time he wrote *Prodromus fauna Zeylanica,* published in 1852, which was the first work to give scientific classifications to the mammals of Ceylon. Also in these five years he produced publications on geology, mammals, birds, reptiles, and the cultivation of cotton. He was appointed Naturalist to the Ceylon government, which paid £200 per annum, which was a lot of money in the middle of the 19th century, plus expenses on top of his army pay. One of his tasks as an official naturalist was to investigate why the Ceylon Pearl Fisheries had not produced any profit. He investigated the life history of the pearl oyster and produced no fewer than four reports on the subject that were published between 1858 and 1863, the latter ones being posthumously published. In 1860 the Governor of Ceylon fell ill and went home to England on board the *Nubia.* His health was of such concern that Kelaart, who was accompanied by his wife and five children, was sent along as the Governor's medical attendant. The Governor died two days before *Nubia* arrived at Southampton, and Kelaart died the very next day and was buried in Southampton. He is also remembered in the scientific name of a bird, the Black-throated Munia *Lonchura kelaarti.* The leaf-nosed bat is found in Sri Lanka and in southern and central India. The shrew is endemic to the central highlands of Sri Lanka. The pipistrelle ranges across Sri Lanka, India, Pakistan, and Burma. It has also been recorded in southeast China, Vietnam, and Borneo.

Kellen

Kellen's Dormouse *Graphiurus kelleni* Reuvens, 1890

Pieter J. van der Kellen (dates not found) of the Dutch Natural History Museum at Leiden. He took part in the Dutch Ethnographic Museum's Southwest Africa expedition, which lasted from 1884 to 1885, and stayed on in Angola until 1888. The leaders of the expedition were D. D. Veth and L. D. Godeffroy. Van der Kellen's job was to act as hunter and to search out and collect "interesting" animals. He returned to Africa and is recorded as exploring in 1896 in the area around Mossamedes in Angola, and between 1899 and 1900 he led an expedition for the Mossamedes Company to the Zambezi. Hermann Baum, a botanist, was a member of this expedition and wrote the trip report in 1903, and he and van der Kellen published descriptions of a number of African ants. The dormouse is found in Angola, Zambia, Zimbabwe, and Malawi. (If *kelleni* includes populations formerly assigned to *olga* and *parvus,* as has been suggested, the distribution of this species extends over much of the savannah regions of sub-Saharan Africa.)

Kellogg

Kellogg's Rice Rat *Euryoryzomys kelloggi* **Avila-Pires,** 1959

Arthur Remington Kellogg (1892–1969) was an American mammalogist. He studied mammalogy at the University of Kansas and the University of California, specializing in marine mammal evolution. He joined the Bureau of Biological Survey of the U.S. Department of Agriculture as an Assistant Biologist in 1920. He also held a research appointment at the Carnegie from 1921 to 1943. He joined the U.S. National Museum in 1928 as Assistant Curator of Mammals, becoming Curator in 1941. He became Director of the U.S. National Museum in 1948, retiring in 1962. He was also an Assistant Secretary of the Smithsonian from 1958 to 1962. After retirement he held an emeritus position with the Smithsonian Division of Vertebrate Paleontology and continued to study marine mammalogy until his death. He was involved in the international regulation of whaling from 1930 to 1967, serving as delegate to the League of Nations whaling conference, 1930; State Department representative to the International Conference on Whaling at London, 1937; Chairman of the Washington Conference, 1946; U.S. Commissioner on the International Whaling Commission, 1947–1967; and Chairman of the International Whaling Commission, 1952–1954. The rice rat is found in southeast Brazil. It may be synonymous with *Euryoryzomys russatus.*

Kemp

Kemp's Gerbil *Gerbilliscus kempi* **Wroughton,** 1906 [Alt. Northern Savannah Gerbil]
Kemp's Spiny Mouse *Acomys kempi* **Dollman,** 1911
Kemp's Thicket Rat *Thamnomys kempi* Dollman, 1911
Kemp's Grass Mouse *Deltamys kempi* **Thomas,** 1917

Robin (Robert) Kemp (b. 1871) was originally an accountant who worked for a company building a railway in Sierra Leone. However, he became a naturalist and professional collector and combined his collecting in Sierra Leone, between 1902 and 1904, with his formal employment. He collected in other parts of Africa around 1906 to 1908, in Australia from 1912 to 1914, and in Argentina from 1916 to 1917. He is also commemorated in the names of two birds, Kemp's Longbill *Macrosphenus kempi* and Kemp's Olive-bellied Sunbird *Cinnyris chloropygius kempi.* (Confusingly, there is also a marine turtle called Kemp's Ridley *Lepidochelys kempi,* but this is named after a fisherman, Richard M. Kemp.) The gerbil ranges from Senegal east to southern Sudan and Uganda. The spiny mouse comes from southern Somalia, Kenya, and northeast Tanzania. The rat is found in the region of the Kivu and Virunga volcanoes in eastern DRC (Zaire), and in western Uganda, Rwanda, and Burundi. The grass mouse is found in Uruguay and the adjacent southeast corner of Brazil, and in the delta of the Rio Paraná, Argentina.

Kenneth

Kenneth's White-toothed Rat *Berylmys mackenziei* **Thomas,** 1916

Kenneth G. Gairdner (fl. 1900–1950) was a collector. In 1918 Kloss gave the name *Rattus kennethi* to a specimen collected by Gairdner in Siam (Thailand). This rat was later regarded as belonging to the same species as one that Thomas had named two years earlier *(mackenziei).* Thus the name Kenneth became attached to *Berylmys mackenziei* as a vernacular name. Later studies indicated that *Rattus kennethi* was not a specimen of *B. mackenziei* after all, but of the related species *B. bowersi.* So Kenneth Gairdner's name has become attached to a species he had nothing at all to do with. The white-toothed rat is found in Assam (India) and Myanmar. Other specimens have been taken in Sichuan Province of China and in southern Vietnam, but not in the intervening areas. See **Gairdner** for biographical details.

Kermode

Kermode's Bear *Ursus americanus kermodei* Hornaday, 1905 [Alt. Spirit Bear, White Black Bear]

Francis "Frank" Kermode (dates not found) was a Canadian zoologist who joined the Provincial Museum of British Columbia (now called the Royal British Columbia Museum) as an apprentice taxidermist in 1891. He retired as the museum's Director in 1940, having been appointed to that post in 1904. This bear, which in 2006 was adopted as the mammalian provincial emblem of British Columbia, lives along the central coast of that province. It was originally named from white-colored individuals—hence the alternative common names—but not all Kermode's Bears are white; they are a race of American Black Bear in which a recessive gene gives a white or cream-colored coat to a small percentage of the total population.

Kerr

Kerr's Tree Rat *Phyllomys kerri* **Moojen,** 1950 [Alt. Moojen's Atlantic Tree Rat]

Dr. John Austin Kerr (b. ca. 1900) was a physician who joined the staff of the International Health Division of the Rockefeller Foundation in 1926. He was involved in many campaigns to improve public health in general and to eradicate malaria and yellow fever in particular. His job took him all over the world. He lived and worked in Brazil from 1933 to 1938; in Colombia from 1938 to 1940; again in Brazil as Director of the Serviço de Estudos e Pesquisas sôbre a Febre Amarela (Yellow Fever Research Service) from 1940 to 1943; in Egypt, where he succeeded in eliminating the malarial mosquito, from 1943 to 1946; and in Sardinia as Regional Director in the campaign against malarial mosquitoes from 1946 to 1947. After his retirement from the Rockefeller Foundation he served as a consultant to the World Health Organization. He also joined the U.S. Agency for International Development in the Philippines, where he lived from 1965 to 1966. The tree rat is found in the state of São Paulo, eastern Brazil.

Khajuria

Khajuria's Leaf-nosed Bat *Hipposideros durgadasi* Khajuria, 1970 [Alt. Durga Das' Leaf-nosed Bat]

Dr. H. Khajuria (fl. 1935–1982) was a zoologist who was a member of the Zoological Survey of India. He wrote extensively on Indian bats including, for example, in 1980, "Taxonomical and Ecological Studies on the Bats of Jabalpur Dist. Madhyapradesh, India." The leaf-nosed bat is found in Central India.

Kihaule

Kihaule's Mouse-Shrew *Myosorex kihaulei* **Stanley** and **Hutterer,** 2000

Philip M. Kihaule (b. 1931) had a long career as a medical-entomological technician. He participated in early surveys of anopholine mosquitoes in southeast Tanzania in 1965 and conducted extensive small-mammal trapping as part of research into plague from 1966 to 1968. He was employed at the School of Medicine, Muhimbili Hospital, as Senior Entomological Technician, and in 1959 transferred to what was then the Department of Zoology and Marine Biology (now Zoology and Wildlife Conservation) of the University of Dar es Salaam, where he rose to the level of Chief Technician. Since his retirement in 1993 he has continued to conduct small-mammal surveys as a collaborator of Bill Stanley of the Field Museum of Natural History, Chicago, and to participate in biodiversity baseline surveys as part of the environmental impact process for development projects. The mouse-shrew was named for Kihaule as, according to the original description, he is responsible for the success of many of the surveys documenting the natural history of the small mammals of the Eastern Arc Mountains, and other areas of Tanzania. Kihaule also collected the type specimen of the shrew in the Udzungwa Mountains of Tanzania, where it appears to be endemic.

Kikuchi

Taiwan Vole *Volemys kikuchii* Kuroda, 1920 [Syn. *Microtus kikuchii*]

Yonetaro Kikuchi (1869–1921) was a collector for the Taipei Museum in Formosa (now Taiwan). He also collected birds, particularly examples of the Mikado Pheasant *Syrmaticus mikado*, for Alan Owston, an English collector who lived in Japan. A reptile, the Botel Gecko *Gekko kikuchii*, is also named after him. The vole comes from the highlands of Taiwan.

Kilonzo

Kilonzo's Brush-furred Rat *Lophuromys kilonzoi* Verheyen et al., 2007

Dr. Bukheti Swalehe Kilonzo is a microbiologist at the Pest Management Centre of Sokoine University of Agriculture, Morogoro, Tanzania. His bachelor's degree in microbiology was awarded in 1972 by the University of London and his master's in medical parasitology a year later by London School of Hygiene and Tropical Medicine, University of London, and his doctorate in 1984 by the University of Dar es Salaam in Tanzania. The species is known from Magamba, West Usambara Mountains, Tanzania.

Kinloch

Selangor Pygmy Flying Squirrel *Petaurillus kinlochii* **Robinson** and **Kloss,** 1911

The description of the type specimen of the squirrel states that it was collected by V. Kinloch on the Jeram Estate, Selangor, Malaya (Peninsular Malaysia). We strongly suspect that this Kinloch was Vincent Kinloch (b. 1887), who was born in Bengal and whose father was General Alexander Angus Airlie Kinloch, a man famous for hunting big game all over India including the Himalayas and in parts of Tibet. The squirrel seems to be confined to Selangor State.

Kirk

Kirk's Galago *Otolemur crassicaudatus kirkii* **Gray,** 1863
Kirk's Dik-dik *Madoqua kirkii* **Günther,** 1880
Kirk's Red Colobus *Piliocolobus kirkii* **Matschie,** 1900 [Alt. Zanzibar Red Colobus]

Sir John Kirk (1832–1922) was a Scottish diplomat, explorer, and naturalist. He was David Livingstone's chief assistant, physician, and naturalist during his second Zambesi expedition from 1858 to 1863. He later served as Vice Consul, then Consul General, in Zanzibar from 1866 to 1887. In 1873 he succeeded in obtaining the local sultan's agreement to a treaty abolishing the slave trade and to concede mainland territories to the British East Africa Company in 1887. He served on the Tsetse Fly Commission in 1886. Kirk was a polymath whose interests included botany, geography, history, geology, chemistry, and photography, as well as the study of Swahili, Arabic, Spanish, Portuguese, and French. He was a Fellow of the Royal Botanical Society and sent specimens to the Royal Botanical Gardens at Kew in London. He also collected the first fishes from Lake Nyasa to reach Western science and observed that they were almost all endemic. He received numerous honorary degrees and awards, including the fellowships of the Royal Society of London and the Geographical Society of Marcella. He is also remembered in the name of Kirk's Rock Agama *Agama kirkii*. The galago is found in Transvaal (South Africa), Mozambique, and Malawi. The dik-dik comes from southern Somalia, Kenya, and Tanzania, with an isolated subspecies *(damarensis)* in southwest Angola and Namibia. The colobus is endemic to Zanzibar.

Kitchener

Red-brown Pipistrelle *Hypsugo kitcheneri* **Thomas,** 1916 [formerly *Pipistrellus kitcheneri*]

Field Marshall Horatio Herbert, Viscount Kitchener of Khartoum (1850–1916), was a soldier who was commissioned into the Royal Engineers. In 1886 he was appointed Governor of the British Red Sea territories, and in 1892 he became Commander in Chief of the Egyptian army. He crushed the Mahdi's army at the Battle of Omdurman in 1898 and captured Khartoum, from which victory he derived his title. After the early British disasters in the Boer War, he was appointed Commander in Chief in South Africa, where his ruthless way of waging war resulted in the eventual defeat of the Boers. In 1902 he became Commander in Chief in India, and in 1911 he returned to Egypt as Proconsul and ruled there and in the Sudan until 1914. When WW1 broke out he was on leave in England and was recruited to the Cabinet as Secretary of State for War. Everyone except Kitchener thought the war would be over by Christmas. He believed it would be a very long war and immediately started recruiting a large number of volunteers, known as "Kitchener's Armies." In 1916 he was sent to Russia and was killed when HMS *Hampshire* was sunk by a German mine—the same year in which Thomas named the pipistrelle in his memory. The bat is endemic to Borneo.

Kitti

Kitti's Hog-nosed Bat *Craseonycteris thonglongyai* **Hill,** 1974 [Alt. Bumblebee Bat; Khun Kitti Bat]

Kitti Thonglongya (1928–1974) was a Thai mammal researcher. He first discovered this tiny species of bat, the world's smallest mammal, in 1973. He worked for a while for the Southeast Asia Treaty Organization. He also has the distinction of having discovered, in 1968, one of the world's rarest birds, the White-eyed River-Martin *Pseudochelidon sirintarae.* He co-wrote *Bird Guide of Thailand* in 1968 with Dr. Boonsong Lekalul. The bat is found in western Thailand and southeast Myanmar.

Kloss

Kloss' Squirrel *Callosciurus albescens* **Bonhote,** 1901

Kloss' Gibbon *Hylobates klossii* **Miller,** 1903 [Alt. Dwarf Siamang]

Kloss' Mole *Euroscaptor klossi* **Thomas,** 1929

Cecil Boden Kloss (1877–1949) was an ethnologist and zoologist. He was a member of the staff of the museum in Kuala Lumpur in 1908, for which he traveled extensively as a collector of specimens. He was then Director of the Raffles Museum in Singapore from 1923 to 1932. Several reptiles including Kloss' Sea Snake *Hydrophis klossi* and Kloss' Forest Dragon *Gonocephalus klossi* are named after him. He is also commemorated in the scientific name of a bird, the South Nicobar Serpent-Eagle *Spilornis klossi.* The squirrel is confined to Sumatra. The gibbon is found on the Mentawai Islands west of Sumatra. The mole is found in the highlands of Thailand, Laos, and the Malay Peninsula.

Knight

Knight's Tuco-tuco *Ctenomys knighti* **Thomas,** 1919 [Alt. Catamarca Tuco-tuco]

Colonel Charles Lewis William Morley Knight (1863–1937) was a soldier and entrepreneur. In 1884, while still a Captain, he wrote *Hints on Driving,* referring to the driving of horse-drawn carriages. Thomas wrote that the tuco-tuco "is named in honour of Col. C. Morley Knight, by whom, in conjunction with his partner Col. J. J. Porteous, the explorations of Messrs. Kemp and Budin have been so much facilitated in various ways." As he was resident in Buenos Aires, Knight could evidently facilitate expeditions to Argentina. He was joint owner with

Porteous of a successful horse-breeding operation at a farm called Las Tres Lagunas in Las Rosas, Santa Fé Province. Knight and Porteous first went to Argentina in 1889 to look for horses and mules on behalf of the British army. They liked what they saw and decided to become partners, not only in horse-breeding but also in cattle, and they introduced the Aberdeen Angus breed of cattle into Argentina in 1890. Whereas Porteous returned to the UK and looked after the European end of the business, visiting Argentina only occasionally thereafter, Knight stayed on in Argentina for the rest of his life. The tuco-tuco is found in northwest Argentina.

Knox-Jones

Knox Jones' Pocket Gopher *Geomys knoxjonesi* **Baker** and **Genoways,** 1975 [Alt. Jones' Pocket Gopher]

Professor Dr. J. Knox Jones Jr. (1929–1992) was an academic and curator. He was the Paul Whitfield Horn Professor of Biological Science and Museum Science at Texas Tech University, Curator of the Museum of Texas Tech University, and also Director of Academic Publications there. He graduated in the early 1950s and then served in the U.S. Army in Korea before returning to the University of Kansas to complete his Ph.D. He stayed on at that university as Assistant Professor and then full Professor. He took up the post at Texas Tech in 1971. He wrote *Distribution and Taxonomy of Mammals of Nebraska* in 1964 and *Handbook of Mammals of the North-Central States, Guide to Mammals of the Plains States,* and *Mammals of the Northern Great Plains* in 1983. He edited a number of journals, such as the *Journal of Mammalogy, Evolution,* and the *Texas Journal of Science,* as well as his own university's occasional scientific papers. The gopher is found only in southeast New Mexico and western Texas (USA).

Kobayashi

Kobayashi's Serotine *Eptesicus kobayashii* Mori, 1928

Professor Teiichi Kobayashi (1901–1996), an Honorary Fellow of the Royal Society of New Zealand, was primarily a geologist and paleontologist. He studied at the Imperial University in Tokyo from 1924 to 1927. He remained at the university after he had graduated until his retirement in 1962. In retirement he continued to conduct research and by 1989 had published nearly 800 books and papers. His works include *Sakawa Orogenic Cycle and Its Bearing on the Genesis of the Japanese Island* (1941), which is regarded as a seminal publication. When well into "retirement" in 1970, jointly with H. Abe, K. Maeda, and T. Miyao, he wrote "Faunal Survey of the Mt. Ishizuchi Area, Jibp Main Area—II: Results of the Small Mammal Survey on the Mt. Ishizuchi Area." Mori, who described the bat, worked with Kobayashi: in the 1950s they were studying Jurassic bivalves. The serotine bat comes from Korea, but its taxonomic status as a valid species has been questioned.

Koepke

Koepke's Hairy-nosed Bat *Mimon koepckeae* **Gardner** and **Patton,** 1972

Maria Koepcke (1924–1971) was born Maria Emilia Ana von Mikulicz-Radecki in Leipzig, Germany. She went to Peru in 1950, where she married Hans-Wilhelm Koepcke, an ecologist at the Museum of Natural History in Lima. She too began working for the museum as an ornithologist. She eventually came to be known as "the Mother of Peruvian Ornithology." She wrote *Corte ecológico transversal en los Andes del Peru Central* (1954), *Die Vögel des Waldes von Zarate* (1958), and *Las aves del departmento de Lima* (1964). She was killed in an air crash in the Amazon Forest on Christmas Eve 1971. Miraculously, her daughter Juliane survived the disaster. Koepcke is also remembered in the name of a hummingbird, Koepcke's Hermit *Phaethornis koepckeae*, and together with her husband in the scientific name of a Peruvian lizard, *Microlophus koepckeorum*. The bat is found in the highlands of central Peru.

Koford

Koford's Grass Mouse *Akodon kofordi* **Myers** and **Patton,** 1989
Koford's Puna Mouse *Punomys kofordi* Pacheco and Patton, 1995

Dr. Carl B. Koford (1915–1979) was an American naturalist and conservationist. He was known as an explorer and outdoorsman and was also an authority on the California Condor, as well as on the flora and fauna of South America. He was Research Associate and Associate Research Ecologist at the Museum of Vertebrate Zoology at the University of California, Berkeley. He published many scientific papers. In 1980 a Carl B. Koford Memorial Fund for the study of vertebrates was set up at Berkeley. He served in the U.S. Navy in WW2 and reached the rank of Commander. He worked for eight years in Puerto Rico at the Cayo Santiago Field Research Unit of the National Institutes of Health. He was Director of the Smithsonian's Canal Zone Biological Area for a year in the middle 1960s. He is also remembered in the scientific name of the Coastal Leaf-toed Gecko *Phyllodactylus kofordi*. Both of the mammal species named after Koford are found in southern Peru.

Kolb

Kolb's White-collared Monkey *Cercopithecus albogularis kolbi* **O. R. Neumann,** 1902 [Alt. Kolb's Guenon]

Dr. George Kolb (d. 1899) was a German zoologist and explorer. He spent much time in Kenya and made two unsuccessful attempts, in 1894 and in 1896, to become the first person to climb Mount Kenya. Arthur H. Neumann supplied the type specimen of this monkey and suggested to Oskar R. Neumann that the latter name it after Kolb, who had been killed by a rhinoceros. The monkey is found in the highlands of Kenya.

Kollmannsperger

Kollmannsperger's Multimammate Mouse *Mastomys kollmannspergeri* **Petter,** 1957 [Alt. Verheyen's Multimammate Mouse]

Dr. Franz Kollmannsperger (b. 1907) was a German zoologist, ecologist, and ornithologist at the University of Saarbrücken. He took part in the international Sahara expedition of 1953–1954. He wrote *Von Afrika nach Afrika* in 1965. The mouse was first found in Niger, and its distribution probably extends east to southern Sudan. *Mastomys verheyeni*, named in 1989 from the Lake Chad region, appears to be the same species.

Kolombatavic

Kolombatovic's Long-eared Bat *Plecotus kolombatovici* Dulic, 1980

Professor Juraj Kolombatovic (1843–1908) was a Croatian natural scientist and ichthyologist collecting around the Adriatic, mostly in what used to be Yugoslavia, particularly Dalmatia in Croatia, in the 1880s, 1890s, and 1900s. He wrote notes and scientific papers on mammals, reptiles, and amphibians, and described nine new species of fish. The bat is found in Greece, in the countries of the former Yugoslavia, and in southwest Turkey.

Koopman

Koopman's Fruit Bat *Koopmania concolor* **Peters,** 1865 [Alt. Brown Fruit-eating Bat]
Koopman's Pencil-tailed Tree Mouse *Chiropodomys karlkoopmani* **Musser,** 1979
Koopman's Pipistrelle *Pipistrellus westralis* Koopman, 1984
Koopman's Rat *Rattus koopmani* Musser and Holden, 1991
Koopman's Porcupine *Coendou koopmani* **Handley** and **Pine,** 1992 [Syn. *C. nycthemera* Olfers, 1818]
Malagasy Mountain Mouse *Monticolomys koopmani* Carleton and **Goodman,** 1996

Dr. Karl F. Koopman (1920–1997). He was considered by many to be the world's leading authority on bat distribution and taxonomy and was a founding member of Bat Conservation International. From 1961 until 1997 he was on the staff of the American Museum of Natural History, the last 11 years as Curator Emeritus. He wrote numerous articles over five decades, such as "Land Bridges and Ecology in Bat Distribution on Islands off the Northern Coast of South America" (1958), "Systematics of Indo-Australian Pipistrellus" (1973), and "Distribution and Systematics of Bats in the Lesser Antilles" (1989). He also undertook a number of field trips all over the world. He was noted for his droll sense of humor, an example being his tongue-in-cheek support of global warming on the grounds that it would allow some of his favorite bat species to extend their ranges. The fruit bat genus *Koopmania* was created in 1991 for the species previously known as *Artibeus concolor.* It is found in northern South America. The tree mouse is endemic to the Mentawai Islands, west of Sumatra. The pipistrelle comes from northern Australia. The rat is known only from Peleng Island, near Sulawesi, Indonesia. The porcupine is found in the Amazon basin of Brazil, south of the Amazon itself; although described as a new species in 1992, it appears to be synonymous with one named almost 175 years earlier. The mountain mouse occurs in the highlands of Madagascar. See also **Koopman and Hill** below.

Koopman and Hill

Koopman and Hill's Yellow-shouldered Bat *Sturnira koopmanhilli* McCarthy, Albuja, and Alberico, 2006

Dr. Karl F. Koopman and John Edwards Hill. The bat is found in the western Andes of Colombia and Ecuador. See **Koopman** and **Hill** for biographical details

Kopsch

Kopsch's Deer *Cervus nippon kopschi* **Swinhoe,** 1873 [Alt. South China Sika]

Henry Charles Joseph Kopsch (dates not found) was a British civil servant seconded to the Chinese Customs Service, which he joined in 1862, becoming Deputy Commissioner in 1867 and Commissioner a year later. He served in China for 38 years in a number of posts. From 1891 to 1897 he was Statistical Secretary and then Postal Secretary. In 1903 he wrote a pamphlet on bi-metallism, which attracted attention when it was published. He was a friend of Swinhoe, who asked him to be on the lookout in the market in Kiukiang as there were rumors of antlered deer being brought there for sale. If Kopsch was successful in obtaining a specimen, and it turned out to be a new taxon, Swinhoe said he would dedicate the animal to Kopsch. Kopsch succeeded, and Swinhoe did as he promised. In 1870 Kopsch published "Notes on the Rivers in Northern Formosa" in the *Proceedings of the Royal Geographical Society of London.* The deer is found in east-central China.

Korinch

Korinch's Rat *Rattus korinchi* **Robinson** and **Kloss,** 1916 [Alt. Mount Kerinci Rat, Sumatran Mountain Rat]

As one of the alternative names of this rat indicates, the species is named not after a person but after a locality. The name Korinch's Rat derives from a corruption of Mount Kerinchi (or Kerinci, Kerintji, or Korinci) in western Sumatra. The species has been recorded from only two Sumatran mountains: Kerinchi and Talakmau.

Koseritz

Big-eared Opossum *Didelphis koseritzi* **Ihering,** 1892 [junior synonym of *D. aurita* **Wied,** 1826]

Carlos (originally Karl) von Koseritz (1830–1890) was a German who went to live in the Brazilian province of Rio Grande do Sul in about 1860. He was a journalist and a politician who was a convinced follower of Darwin. He campaigned against slavery and for reasonable policies to allow settlement. He was regarded as leader of the local German community in Porto Allegre. In 1861 he wrote for *Deutsche Zeitung*, a German-language periodical for traders in Porto Allegre, and in 1881 he started, and until his death edited, a newspaper called *Koseritz' Deutsche Zeitung*. Intellectual society in Porto Allegre was based on a literary society called the Parthenon. This club was started in 1868 and lasted until 1885, during which period von Koseritz was a leading member. He was interested in the manifestation of nature inasmuch as he reported in detail on eruptions of rodent populations in Rio Grande do Sul in both 1876 and in 1885. Ihering, who named the opossum, arrived in Rio Grande do Sul in 1880 and undoubtedly would have met Koseritz and read his newspaper. However, the name Ihering gave this opossum is now regarded as a junior synonym of *Didelphis aurita*. It is found in eastern Brazil, eastern Paraguay, and northeast Argentina.

Kozlov

Kozlov's Pika *Ochotona koslowi* Büchner, 1894

Kozlov's Pygmy Jerboa *Salpingotus kozlovi* **Vinogradov,** 1922 [Alt. Three-toed Dwarf Jerboa]

Kozlov's Long-eared Bat *Plecotus kozlovi* Bobrinskoj, 1926

Kozlov's Shrew *Sorex kozlovi* Stroganov, 1952

General Petr Kuzmich Kozlov (1863–1935) was a Russian researcher of central Asia who was one of Prjevalksy's companions on his fourth and last expedition. He wrote *Mongoliya I Kam*. He led the Mongol-Tibetan expedition from 1899 to 1901 and 1923 to 1926 and the Mongol-Sychuan expedition from 1907 to 1909. Kozlov was sent to Tibet to improve relations there but stopped on the Silk Road in 1908 when he discovered Khara-Khoto, the "Black City," which had been described by Marco Polo. Khara-Khoto lies just inside the present-day Chinese border with Mongolia. It was "the city of his dreams," and he made excavations uncovering many scrolls, which he took back for study, and collected geographic and ethnographic materials. He was the husband of a well-known Russian ornithologist Dr. Elizaveta Vladimirovna Kozlova (1892–1975). Kozlov's Accentor *Prunella koslowi* and the Tibetan Bunting *Emberiza koslowi* are both also named after him, as is a reptile, Koslow's Toadhead Agama *Phrynocephalus koslowi*. The pika comes from the Arkatag Range, Kunlun Mountains, western China. The jerboa is found in the Gobi Desert of southern Mongolia and northern China. The shrew is known from just one locality in Tibet, but its taxonomic validity is uncertain (it may be a race of *Sorex thibetanus*).

Kraemer

Admiralty Cuscus *Spilocuscus kraemeri* Schwartz, 1910 [Alt. Manus Spotted Cuscus]

Professor Augustin Friedrich Kramer (1865–1941) was born in Chile to German parents who returned to Germany in 1867. Between 1883 and 1889 he studied medicine and zoology at the universities of Tübingen, Kiel, and Berlin, but 1884 was spent doing his compulsory military service. In 1889 he went to sea as a ship's doctor in the Imperial German Navy and made several voyages, mainly to German colonies in the Pacific. This included a stay of 12 months at Apia in Samoa, where he became interested in Samoan ethnography. He made a collection of ethnographic objects that he sent to the Museum of Ethnography in Stuttgart and the collection at the University of Tübingen. Over a period of years he visited the Marshall Islands, the Gilbert Islands, the Hawaiian Islands, and, in 1897, Peru and Chile, including the Straits of Magellan. In 1907 he explored in the Bismarck Archipelago and the Caroline Islands. In 1909 he was leader of the Hamburg Pacific expedition to the Carolines on board the *Peiho*. In 1911 he returned to Germany, and in that year he was given a three-year contract as Director at the Stuttgart Museum of Ethnography, after which he lectured on ethnology at Tübingen University. He settled down in Stuttgart and stayed there until his death. In addition to the cuscus, which he probably collected in 1908 when he visited Admiralty Island, a plant is named after him: *Actinophloeus kraemerianus*. The cuscus is found only on the Admiralty Islands, Bismarck Archipelago, Papua New Guinea.

Krebs

Krebs' Fat Mouse *Steatomys krebsii* **Peters,** 1852

Georg Ludwig Engelhard Krebs (1792–1844) was a German immigrant to South Africa who became a farmer and a prodigious collector. He was trained as an apothecary and, after working in Hamburg, took a four-year contract in 1817 with a firm of apothecaries in Cape Town, Pallas and Poleman. Here he succeeded Carl Heinrich Bergius, who had lost his job but collected for Lichtenstein. Krebs and Bergius collected together until Bergius' death from tuberculosis in 1818. Krebs is known to have met other well-known naturalists in the Cape, including Brehm, von Chamiso, and Delalande. He wrote to his brother to ask if he could approach Lichtenstein to see if the latter would buy specimens. Lichtenstein agreed, and the first consignment was sent to him late in 1820, followed by a further three large consignments in 1821. Krebs had finished his contract with the apothecaries in 1821 and now worked as a collector. Lichtenstein had him officially authorized by the Prussian government to collect specimens for the Natural History Museum in Berlin. This resulted in an unusual title, and there is a book, published in 1971, about his life that uses it as its theme: *Ludwig Krebs—Cape Naturalist to the King of Prussia 1820–8*. In 1826 Krebs opened his own pharmacy in Grahamstown. In 1828 he was severely stricken with "rheumatism," and this eventually brought much of his active collecting to an end. His last consignment to the Berlin museum was sent off in 1829 and included a complete "Bushman" (San tribesman) pickled in brine in a barrel, plus a rhinoceros, an elephant, and a (now extinct) quagga. At this time Krebs succeeded in gaining permanent residence and the right to own property, so he bought a farm and named it Lichtenstein. He continued to make expeditions into the unknown parts of southern Africa and sent consignments of specimens to Lichtenstein for disposal, normally by public auction. The fat mouse is found in southern Angola, western

Zambia, and parts of Botswana, Namibia, and South Africa. Its somewhat patchy distribution may indicate a "species complex" rather than a single taxon.

Krefft

Northern Hairy-nosed Wombat *Lasiorhinus krefftii* **Owen,** 1873

Johann Ludwig Gerhard Krefft (1830–1881) was a German adventurer, artist, and naturalist who settled in Australia. He went to the USA in 1851 and worked as an artist in New York but in 1852 sailed for Australia to join the gold rush. He was a miner until 1857, when he joined the National Museum in Melbourne as a collector and artist. He seems to have had a bad temper, as he feuded with the museum trustees and was dismissed in 1874. The temper is evidenced by the fact that he refused to accept the dismissal, barricaded himself in his office, and had to be carried out of the building, still sitting on his chair, deposited in the street and the door locked behind him. Feeling that he had been badly treated, he set up a rival "Office of the Curator of the Australian Museum" and successfully sued the trustees for a substantial sum of money. That was the end of his career, and he never worked in a formal position again, but he did write natural history articles for the Sydney press. He wrote *The Snakes of Australia* (published in 1869) and *The Mammals of Australia* (1871). Krefft's River Turtle *Emydura krefftii* is also named after him. His most famous discovery was the Australian Lungfish *Neoceratodus forsteri*. The wombat is highly endangered, with a single population in the Epping Forest National Park, Queensland.

Kuhl

Kuhl's Pipistrelle *Pipistrellus kuhlii* Kuhl, 1817

Kuhl's Night Monkey *Aotus (azarai) infulatus* Kuhl, 1820 [Alt. Feline Night Monkey]

Kuhl's Tree Squirrel *Funisciurus congicus* Kuhl, 1820 [Alt. Congo Rope Squirrel]

Lesser Asiatic Yellow Bat *Scotophilus kuhlii* **Leach,** 1821

Kuhl's Marmoset *Callithrix kuhlii* **Wied**-Neuwied, 1826 [Alt. Wied's Marmoset, Wied's Tufted-ear Marmoset]

Bawean Deer *Axis kuhlii* **Müller,** 1840

Dr. Heinrich Kuhl (1797–1821) was a German zoologist who became an assistant to Conrad Jacob Temminck when the latter was Director of the Dutch National Museum at Leiden. He also worked with Johan Conrad van Hasselt in the Dutch East Indies, collecting for the Netherlands Committee for Natural Science from 1820 to 1821. He wrote *Conspectus psittacorum* in 1820. Kuhl died in Buitenzorg, Java, of a tropical disease. He is also commemorated in the names of other taxa, such as Kuhl's Lorikeet *Vini kuhlii*, the Blue-spotted Stingray *Dasyatis kuhlii*, and a number of reptiles including Kuhl's Gliding Gecko *Ptychozoon kuhli*. The pipistrelle is found in southern Europe as far north as the Channel Islands, east to Kazakhstan and Pakistan, and in southwest Asia, much of Africa, and the Canary Islands. Probably more than one species should be recognized within this range. The night monkey comes from Brazil, south of the Amazon. The squirrel is found in DRC (Zaire) south of the Congo River and southwestward to Angola and northern Namibia. The yellow bat is found from Pakistan, India, and Sri Lanka east to Taiwan and the Philippines and south to Malaysia and Indonesia. The marmoset is restricted to the Atlantic forest of southern Bahia, Brazil. The deer is endemic to Bawean Island, north of Java, Indonesia.

Kuhn

Liberian Mongoose *Liberiictis kuhni* **Hayman,** 1958

Professor Dr. Hans-Jürg Kuhn is a German anatomist and zoologist. He is Professor of Anatomy at the Georg-August University, Göttingen. He has written numerous scientific papers on mammals, including "A Provisional Check-list of the Mammals of Liberia" (1965). The mongoose occurs in Liberia and western Ivory Coast.

Kulinas

Kulinas' Spiny Rat *Proechimys kulinae* da Silva, 1998 [Alt. Javari Spiny Rat]

This rat is named after the people (and their language) of the area in Western Amazonia where it occurs, not after an individual. The rat is found in parts of Peru and western Brazil.

Kuns

The Neotropical rat genus *Kunsia* **Hershkovitz,** 1966 [2 species: *fronto* and *tomentosus*]
Pygmy Short-tailed Opossum *Monodelphis kunsi* **Pine,** 1975

Dr. Merle L. Kuns (1923–2008) was an American biologist and explorer. His childhood interest in biology was triggered by growing up on a farm. He was a pilot in Europe during WW2 and then completed his education at the University of Wisconsin. He worked for many years for the U.S. Public Health Service National Institutes of Health, during which time he visited most countries in Latin America and investigated epidemics of new and little-known tropical diseases. He lived in Panama, Bolivia, and Argentina but retired to Florida. He has published articles on tropical disease, but also on mammals, and some about both, such as "Chronic Infection of Rodents by Machupo Virus" (1965) and "Isolation of Machupo Virus from Wild Rodent, *Calomys callosus*" (1966). The etymology for the rat genus reads, "Named in honor of Dr. Merle L. Kuns, of the Middle American Research Unit, National Institutes of Health, investigating hemorrhagic fever in Bolivia. Dr. Kuns was particularly concerned with rodent reservoirs of the disease and was responsible for the collection of mammals which included the first known Bolivian representatives of the remarkable genus now bearing his name." The short-tailed opossum is found in Bolivia, northwest Argentina, and southern Brazil.

Kusnoto

Javan Tailless Fruit Bat *Megaerops kusnotoi* **Hill** and **Boedi,** 1978

Professor Kusnoto Setyadiwirja (dates not found) became the first Indonesian Curator of the Indonesian Botanical Gardens, now known as the National Biological Institute. He became its Director in 1959. He also founded the Academy of Biology at Bogor. The bat is found on Java, Bali, and Lombok.

Lacépède

Lacépède's Tamarin *Saguinus midas* **Linnaeus,** 1758 [Alt. Midas Tamarin, Red-handed/Golden-handed Tamarin]

Lacépède's Bottle-nosed Dolphin *Tursiops nesarnack* Lacépède, 1804 [Syn. *T. truncatus* Montagu, 1821]

Bernard Germaine Étienne de la Ville, Comte de Lacépède (1756–1825), was a French naturalist who as a young man came to the attention of Buffon, whose work on the classification of animals he was encouraged to continue. Buffon got him a job at the Jardin du Roi (later the Jardin des Plantes). His published works deal with reptiles, fish, and whales, and are often printed with Buffon's works, which they supplement. Lacépède was active in politics and went into exile during the Reign of Terror to avoid the guillotine. After his return to France he gave up scientific work for a political career and held several offices of state. He has other taxa named after him, such as the fish *Lophotus lacepedei.* The tamarin is found in the Guianas and northeast Brazil. ("Lacépède's Tamarin" is hardly used nowadays as a vernacular name and is regarded as somewhat archaic). The bottle-nosed dolphin is an unusual case, in that the scientific name given by Lacépède predates the one given by Montagu, but the latter is regarded as the valid name. (In taxonomic-technical jargon, Montagu's name is a *nomen conservandum,* a name that would normally be invalid but that the International Code of Zoological Nomenclature has decided to retain—as is sometimes the case when the name has enjoyed long-term common usage.)

Ladew

Ladew's Oldfield Mouse *Thomasomys ladewi* **Anthony,** 1926

Harvey Smith Ladew (1887–1976) was a self-taught American gardener. He was also a wealthy socialite, fox hunter, artist, and traveler. He bought a large estate in Maryland to serve as his private fox-hunting territory but also developed there the Ladew Topiary Gardens, which are famous for their 15 themed "garden rooms" covering 22 acres. Ladew financed the American Museum of Natural History to send a collector, George Tate (q.v.), to accompany Ladew on an expedition to Bolivia in 1926. Tate gathered botanical and zoological specimens, including the type specimen of the oldfield mouse, which is found in the Andes of northwest Bolivia.

Lady Burton

Lady Burton's Rope Squirrel *Funisciurus isabella* **Gray,** 1862 [Alt. Gray's Four-striped Squirrel]

Lady Isabel Burton (1831–1896) was the wife of Sir Richard Burton, the English traveler and author (see **Burton, R.**). She wrote *The Romance of Isabel, Lady Burton,* an unfinished autobiography that was published posthumously in 1897. She is also commemorated in the binomial of a bird, the Mountain Robin-chat *Cossypha isabellae.* The squirrel is found in Cameroon, Gabon, the Congo Republic, and the Central African Republic.

Lamberton

Western Red Forest Rat *Nesomys lambertoni* **Grandidier,** 1928

Charles Lamberton (fl. 1912–1956) was a French paleontologist who wrote about the subfossil fauna of Madagascar, having lived there from 1927 to 1948 and having undertaken a number of paleontological expeditions to the southwest of the country in the 1930s. He was a Professor of the Gallieni College and Secretary of the Malagasy Academy. He published many papers, especially on the fossil fauna of Madagascar and its recent extinctions, such as "On a New Kind of Fossil Lemur the Malagasy Prohapalemur" (1936) and "Examen de quelques hypothéses de Sera concernant les lémuriens fossiles et actuels" (1956). Two reptiles, including the Fito Leaf Chameleon *Brookesia lambertoni*, are named after him. The rat is endemic to western Madagascar.

Lamotte

Nimba Otter-Shrew *Micropotamogale lamottei* **Heim de Balsac,** 1954
Lamotte's Shrew *Crocidura lamottei* Heim de Balsac, 1968
Lamotte's Roundleaf Bat *Hipposideros lamottei* **Brosset,** 1984
Oku Rat *Lamottemys okuensis* **Petter,** 1986

Dr. Maxime Lamotte (1920–2007) was an Honorary Professor of Zoology at the University Paris VI. He collected in West Africa during 1942 and for several months in 1946. He has also trained others to collect and conserve specimens. He became head of the Zoology Department of the École Normale Supérieure in Paris. Lamotte is also honored in the name of the spider genus *Lamottella*. The otter-shrew and the roundleaf bat are both confined to the Mount Nimba area, where Guinea, Liberia, and Ivory Coast come together. The *Crocidura* shrew is found from Senegal to western Cameroon. The rat is endemic to Mount Oku in Cameroon.

Lander

Lander's Horseshoe Bat *Rhinolophus landeri* **Martin,** 1838

Richard Lemon Lander (1804–1834) was born in Cornwall. At the age of nine he left home and walked to London. Between 1825 and 1828 he was in northern Nigeria with Hugh Clapperton, with the intention of traveling down the River Niger. At that time no one knew where it started or finished or much at all about it. Clapperton died in 1827, and it seems that Lander was the only European member of the expedition to survive—and *survive* is very much the operative word, as African tribesman accused him of witchcraft and forced him to drink poison to see if he was a witch or not. Since Lander didn't die, they concluded he was not a witch after all. He returned to England in 1828 and published two books: *Journal of Richard Lander from Kano to the Sea Coast* (1829) and *Records of Captain Clapperton's Last Expedition to Africa, with the Subsequent Adventures of the Author* (1830). In 1830 he returned to Africa, accompanied by his brother, John. They followed the lower Niger River from Bussa to the sea, traveling in canoes. On their way they had a number of adventures, including being captured and kidnapped by the King of the Ibos. They were sold on to another monarch, King Boy of Brass, who held them for ransom. Lander later recounted all this in *Journal of an Expedition to Explore the Course and Termination of the Niger* (1832). The Lander brothers were the first Europeans to discover much about the River Niger, and by doing so they opened up an important trade route into the hinterland of what today is Nigeria. Lander was killed in a skirmish with unfriendly tribesmen while leading the first trade expedition up the River Niger. He was buried on the Island of Fernando Pó (now Bioko), in Equatorial Guinea. The type specimen of the bat was taken on Bioko, but it is found very widely in sub-Saharan Africa.

Langguth

Langguth's Rice Rat *Cerradomys langguthi* Percequillo, Hingst-Zaher, and Bonvicino, 2008

Professor Dr. Alfredo Ricardo Langguth Bonino (b. 1941) is a Uruguayan zoologist. His first degree in biological sciences was awarded in 1964 by the University of the Republic in Montevideo. Subsequently he studied at Johann Wolfgang Goethe University in Frankfurt and was awarded his doctorate by that university in 1968. He was a professor at the University of the Republic and at the Uruguayan National Museum. Now retired and living in Brazil, he is a Volunteer Professor at the Federal University of Paraíba. This rice rat, found in the Brazilian states of Pernambuco, Paraíba, Ceara, and Maranhao, was named after Langguth "for his long-term dedication and commitment to the development of Brazilian mammalogy."

Langheld

Langheld's Baboon *Papio cynocephalus langheldi* **Matschie,** 1892

We believe this to be Lieutenant, later Major, Wilhelm Langheld (1867–1917), who was part of the German colonial service in Africa. He explored with Emin Pasha and Dr. Franz Stuhlmann, and succeeded Stuhlmann at Bukoba, in what is now Tanzania. He joined an artillery regiment in 1885 as a Cadet, was promoted to Lieutenant in 1886, and transferred from the Prussian army in 1889 to the German Colonial Army, the Wissmann Protection Force. In 1891 Germany created a new Imperial Protection Force, and he was again transferred. He is recorded as rescuing women and children fleeing from civil unrest in Uganda in the 1890s, and in 1893 he led an expedition to Lake Victoria Nyanza. He was promoted several times, and from 1894 to 1899 he was the Officer-in-Charge of the Nyanza Province of German East Africa. In 1896 he was promoted to Captain and was posted to Cameroon in 1900, where he was initially Deputy Commander of the security forces. He led a number of expeditions in Cameroon, and in 1908 he returned to Germany and was placed on the retired list. He worked for AEG, a German electrical company, from 1908 to 1914. In 1909 he published *Zwanzig Jahre in Deutschen Kolonien* (Twenty Years in German Colonies). He was recalled to the army in 1914 and served in an infantry regiment during WW1. He died in a field hospital in Galicia of wounds received in action. The type specimen of this baboon came from what is now Tanzania. The subspecies is not usually now regarded as being distinct from the typical form of Yellow Baboon *Papio cynocephalus cynocephalus,* which ranges from Tanzania to Mozambique, Malawi, and eastern Zambia.

Lar

Lar's Gibbon *Hylobates lar* **Linnaeus,** 1771 [Alt. White-handed Gibbon, Lar Gibbon]

References to "Lar's Gibbon" are transcription errors; the correct form is "Lar Gibbon." This is probably another case of Linnaeus using rather obscure names from mythology—as, for example, with his choice of *jacchus* for the Common Marmoset. A Lar (plural Lares) was a kind of guardian deity. Originally gods of cultivated fields, Lares were later regarded as minor gods of the household. The household Lar was often represented as a youthful figure holding a drinking horn and cup. This gibbon is found in southern Yunnan (China), eastern Myanmar, Thailand, the Malay Peninsula, and northern Sumatra.

Lataste

Lataste's Gundi *Massoutiera mzabi* Lataste, 1881 [Alt. Mzab Gundi]

Lataste's Gerbil *Gerbillus latastei* **Thomas** and **Trouessart,** 1903

Professor Fernand Lataste (1847–1934) was a French zoologist. He made a collection of the reptiles and batrachians of Barbary (the area that we know today as Morocco, Algeria, and Tunisia) between 1880 and 1884, and published *Les missions scientifiques de Fernand Lataste en Afrique noire et au Maghreb (1880–1885)*. His *Étude de la faune des vertebres de Barbarie*, published in 1885, was the standard work on animals of North Africa for many years. A few years later he appears to have turned his attention to South America, as he is reported to have written a number of important articles on the birds of Chile. He is also remembered in the name of a snake, Lataste's Viper *Vipera latastei*. The gundi is found in southeast Algeria, southwest Libya, northeast Mali, northern Niger, and northern Chad. The gerbil is found in Tunisia and Libya.

Latona

Latona's Shrew *Crocidura latona* **Hollister,** 1916

Hollister doesn't spell out the reason for his choice of scientific name, but the clue is that he points out that this species is closely related to *Crocidura niobe*. Niobe and Latona are linked in mythology: Niobe had given birth to six daughters and six sons. When she heard people praying to Latona (also called Leto), Niobe cried out "Why do you praise her? Praise me! For where she has only two children, I have twelve." But Latona's children were the gods Apollo and Artemis, and they promptly slew Niobe's offspring in return for this slight against their mother (see also **Niobe**). Latona's Shrew is found in northeast DRC (Zaire) and in the Nyungwe Forest, Rwanda.

La Touche

La Touche's Free-tailed Bat *Tadarida latouchei* **Thomas,** 1920

John David Digues La Touche (1861–1935) was French born and English educated. He was Inspector of Customs in China from 1882 until 1921 and was an amateur naturalist and collector. He retired to Ireland but died at sea in May 1935 on the way home from Majorca, where he had spent the winter. He wrote *A Handbook of the Birds of Eastern China*, published in 1925. He is also commemorated in the common name of a bird, La Touche's Leaf Warbler *Phylloscopus yunnanensis*, and in the scientific name of a snake, the Sichuan Mountain Keelback *Opisthotropis latouchii*. He collected the type specimen of the bat in China. The species has also been recorded from Thailand, Laos, and the Ryukyu Islands (southern Japan).

LaVal

LaVal's Disk-winged Bat *Thyroptera lavali* **Pine,** 1993

Dr. Richard K. LaVal is an American tropical biologist and wildlife photographer. He has a bachelor's degree from Carnegie Mellon University, a master's in Vertebrate Zoology from Louisiana State University, and a doctorate in Wildlife Science from Texas A&M. He has worked mainly in Costa Rica since 1968, and his primary area of research has been bats. He also acts as a tour guide, particularly in the Monteverde Cloud Forest Reserve. He has been based full time at Monteverde since 1980 and is known locally as "Batman." The bat has been recorded in eastern Peru, Ecuador, Venezuela (the Orinoco Delta), and northern Brazil.

Lawrance

Lawrance's Dik-dik *Madoqua saltiana lawrancei* Drake-Brockman, 1926

Major Sir Arthur Salisbury Lawrance (1880–1965) was Governor of Somaliland from 1932 to 1939. The describer, R. E. Drake-Brockman, says that he asked Major Lawrance to provide specimens of dik-dik from the Somaliland region "to fill up the gaps in our knowledge now that the Mad Mullah and his followers had ceased to exist." Among the specimens provided by Major Lawrance were examples of a new taxon, which Drake-Brockman named after him. This small antelope is endemic to Somalia.

Lawrence

Lawrence's Howler Monkey *Alouatta pigra* Lawrence, 1933 [Alt. Guatemalan Howler Monkey, Black Howler Monkey]

Barbara Lawrence Schevill (1909–1997) was an American zoologist. She graduated from Vassar College in 1931, after which she joined the Museum of Comparative Zoology, Harvard University, as Museum Assistant to the Director, Dr. Thomas Barbour, and the Curator of Mammals, Dr. Glover Allen. She mounted an expedition to collect natural history specimens and to study bats in the Philippines and Sumatra in 1936 through 1937. People often assumed that she had further academic qualifications, but when anyone addressed her as *Doctor* Lawrence she would gently insist that she be called *Miss* Lawrence—although in 1838 she had married William Schevill, who was the Librarian of the Museum of Comparative Zoology and Assistant Curator of Invertebrate Paleontology. It was through him that she developed an interest in zoo-archeology. Together they produced the first scientific recordings of whale sounds, and they studied the anatomy and food-locating abilities of a variety of marine species. In 1942 she became Associate and Acting Curator of Mammals, and then Curator of Mammals in 1952, retiring in 1976. From the 1960s through the early 1980s, "Miss Lawrence" published a number of articles in mammalogy and zoo-archeology, including descriptions of new taxa such as the above-listed howler monkey. She joined in the excavation of the Neolithic site of Çayönü in southeastern Turkey in 1962 and continued to travel there even after retirement. The monkey occurs in southern Mexico, Belize, and Guatemala.

Lawrence

Lawrence's Dik-dik *Madoqua saltiana lawrancei* Drake-Brockman, 1926

See **Lawrance**. The spelling "Lawrence" is a common transcription error.

Laxmann

Laxmann's Shrew *Sorex caecutiens* Laxmann, 1785

Erich Laxmann (sometimes Erik Laxman) (1737–1796) was a Swedish Finn who was a clergyman, natural historian, and geologist. He made several scientific expeditions to various parts of Russia and Siberia, and was appointed as Professor at the Academy of Sciences in St. Petersburg and "Mineralogical Traveler" of the Imperial Russian Cabinet. He was also notable for procuring the right for Russian ships to call at the port of Nagasaki. He would certainly have corresponded with Linnaeus. The shrew occurs in Scandinavia and from Poland eastward to the Russian Far East, Mongolia, Korea, and Hokkaido (Japan).

Layard

Layard's Palm Squirrel *Funambulus layardi* **Blyth**, 1849

Layard's Beaked Whale *Mesoplodon layardii* **Gray**, 1865 [Alt. Strap-toothed Whale]

Edgar Leopold Layard (1824–1900) was born in Italy. He spent 10 years in Ceylon (now Sri Lanka) before going to the Cape Colony of South Africa in 1854 as a civil servant on the staff of the Governor, Sir George Grey. In 1855 he became Curator of the South African Museum, attending to the duties of that position in his spare time. Layard later worked in Brazil, Fiji, and New Caledonia. In 1867 he wrote *The Birds of South Africa*, which was later updated by Sharpe. He is commemorated in the names of several birds, including Layard's Tit-Babbler *Parisoma layardi*, and a lizard, Layard's

Nessia *Nessia layardi*. The squirrel is found in southern India and Sri Lanka. The whale is found in temperate waters of the Southern Hemisphere.

Leach

Leach's Single-leaf Bat *Monophyllus redmani* Leach, 1821

Dr. William Elford Leach (1790–1836) was a British zoologist. He originally studied medicine, but he did not practice and instead was employed from 1813 to 1821 at the British Museum, where he became a world-renowned expert on crustaceans. He also worked on insects, mammals, and birds, and wrote *The Zoological Miscellany*, published in 1814. While on a visit to Italy in 1836 he caught cholera, from which he died. He was well known for idiosyncratic nomenclature: for example, in 1818 he named nine genera after Caroline (or various anagrams of that name); Caroline may have been the name of his mistress. He is also commemorated in the common names of such birds as Leach's Storm Petrel *Oceanodroma leucorhoa*, and in the name of a lizard, Leach's Anole *Anolis leachii*. The bat is found in Jamaica, Cuba, Hispaniola, Puerto Rico, and the Bahamas.

Leadbeater

Leadbeater's Possum *Gymnobelideus leadbeateri* McCoy, 1867

John Leadbeater (1831–1888) was Chief Taxidermist at the National Museum, Melbourne, Australia, from 1858. His claim to fame is that he stuffed the first known specimen of this possum, which had been found in forests near Melbourne. The species was thought to have become extinct in the early 20th century but was then rediscovered in 1961. It now appears on the coat of arms of the state of Victoria. Still endangered, the possum is restricted to pockets of mountain ash forests in the highlands of Victoria, Australia.

Leander

Ecuador Fish-eating Rat *Anotomys leander* **Thomas**, 1906

Although Thomas doesn't spell out his reasoning—he never did when using names from mythology—we can be pretty sure that he named this aquatic rat after the Greek mythical figure Leander. Leander, who lived in Abydos, was in love with Hero, a priestess of Aphrodite, who lived on the opposite shore of the Hellespont at Sestos. So that they could be together, Leander swam over to Sestos every night and back home to Abydos in the morning. As in nearly all Greek myths, the ending is sad. One night there was a storm, and Leander lost his way and drowned. Hero was so upset that she threw herself into the sea and also drowned. As a theme it has been immensely popular with artists and writers in many languages, from Ovid onward. The myth is referred to in Christopher Marlowe's poem "Hero and Leander," in three of Shakespeare's plays, and in works by a number of Spanish writers. Lord Byron wrote a poem called "Written after Swimming from Sestos to Abydos." History records that Byron not only found the swim hard going but caught a cold afterward. The rat is found in the Andes of northern Ecuador.

Le Conte

Leconte's Pine Mouse *Microtus pinetorum* Le Conte, 1830 [Alt. Woodland Vole]
Leconte's Four-striped Tree Squirrel *Funisciurus lemniscatus* Le Conte, 1857 [Alt. Ribboned Rope Squirrel]

Dr. John Lawrence Le Conte (1825–1883) was an American entomologist and biologist. He was also a physician during the American Civil War, reaching the rank of Lieutenant Colonel. His father, John Eatton Le Conte (1784–1860), was also a naturalist, as well as a U.S. Army engineer, and some of his writings are addressed to his son, who may also have contributed some illustrations. While the son was still a student he made a number of field trips to the Rocky Mountains and Lake Superior. In 1848 he made a second trip to Lake Superior, accompanied by Louis Agassiz. In 1849 Le Conte went to California and explored the Colorado River until 1851. In 1852 he moved to Philadelphia, which was his base for the rest of his life, though he made a number of overseas expeditions to Panama, Europe, Egypt, Algiers, and Honduras. In 1878 he appears to have made a career change, as in that year he became Chief Clerk of the U.S. Mint in Philadelphia. A bird, Le Conte's Thrasher *Toxostoma lecontei*, is also named after him, as is a reptile, the Western Long-Nosed Snake *Rhinocheilus lecontei*. The vole is found in the eastern USA, from Maine south to northern Florida and west to Wisconsin and eastern Texas. The squirrel is known from Cameroon, Gabon, Equatorial Guinea, the Congo Republic, and western Central African Republic.

Leib

Leib's Myotis *Myotis leibii* **Audubon** and **Bachman,** 1842 [Alt. Eastern Small-footed Myotis]

Dr. George C. Leib (d. 1888) was a physician in Ohio. He had been trained as a physician and graduated from the University of Pennsylvania in 1834. He was a friend of both Audubon and Bachman and in 1842 sent them a specimen of a small brown bat, which was duly described and named after him by them. The myotis is found in southeast Canada and the eastern USA, as far south as Tennessee and northern Georgia.

Leighton

Leighton's Linsang *Poiana leightoni* **Pocock,** 1908 [Alt. West African Linsang]

Leonard Leighton (dates not found) made a small collection of mammals in Liberia in the early years of the 20th century. He passed them on to Pocock, who reported on them in 1908, mentioning the linsang in particular and naming it after Leighton. This relative of the genets was long classified as a subspecies of the (Central) African Linsang *Poiana richardsoni* but is now considered by most authorities to be a full species. The linsang is found in Sierra Leone and the Ivory Coast as well as Liberia.

Leisler

Leisler's Bat *Nyctalus leisleri* **Kuhl,** 1817 [Alt. Lesser Noctule]

Dr. Johann Philipp Achilles Leisler (1771–1813) was a Dutch (some say German) naturalist. He was a friend of Temminck and named a number of birds after him, including Temminck's Stint *Calidris temminckii*. The bat has a wide but local distribution across Europe (except the far north), including Britain and Ireland. It is also found on Madeira, in northwest Africa, and east to the Urals and the western Himalayas.

Lekagul

Large Asian Roundleaf Bat *Hipposideros lekaguli* **Thonglongya** and **Hill,** 1974

Dr. Boonsong Lekagul (1907–1992) was a Thai physician, biologist, and conservationist. He qualified as a physician in 1933, graduating from Chulalongkorn University in Bangkok. He established the Bangkok Bird Club in 1962. His major publication was *Bird Guide of Thailand*, which appeared in 1968. He is credited with being the man who really launched nature conservation in Thailand, and his work in lobbying for legislation resulted in a National Parks Act in 1961. He is remembered in the trinomial of two birds, the Grey-eyed Bulbul *Iole propinqua lekhakuni* and the Hill Blue Flycatcher *Cyornis banyumas lekhakuni*. He collected the type specimen of the bat on his farm in Thailand. It is found in southern Thailand and the Malay Peninsula and has also been recorded on the Philippine islands of Luzon and Mindoro.

Lelwel

Lelwel's Hartebeest *Alcelaphus buselaphus lelwel* **Heuglin,** 1877

This is a transcription error, as the correct form of the name is Lelwel Hartebeest; the name does not refer to a person. This antelope is found in the Central African Republic, southern Sudan, northern DRC (Zaire), southwest Ethiopia, and Uganda.

Lemerle

Lemerle's Hippopotamus *Hippopotamus lemerlei* **Grandidier,** 1868 extinct [Alt. Madagascar Dwarf Hippopotamus]

Grandidier, in his book *Souvenirs de voyages d'Alfred Grandidier,* wrote that he had named this extinct hippopotamus after the odd-job man at Tuléar. He says no more than that, and we have not been able to trace any further details about this person. Tuléar (Toliara) is a town in southwest Madagascar. The species was endemic to Madagascar but died out in the relatively recent past, perhaps no more than 500 years ago.

Leopold

Long-tailed Giant Rat genus *Leopoldamys* **Ellerman,** 1947 [6 species]

Ellerman made no reference in his first printed use of this name as to which Leopold he had in mind. We think that one person of whom Ellerman would certainly have heard, and possibly have corresponded with, was Aldo Leopold (1887–1948), who was a specialist in environmental biology. He worked at the U.S. Forest Service from 1909 to 1933 and made extensive surveys and studies of North American wildlife.

Leschenault

Leschenault's Rousette *Rousettus leschenaulti* **Desmarest,** 1820

Jean Baptiste Louis Claude Theodore Leschenault de la Tour (1773–1826) was a French botanist who served as naturalist to two Kings of France: Louis XVIII (1814–1824) and Charles X (1824–1830). He wrote one of the first descriptions of coconuts and the extraction of their oil. He was botanist on the voyage of the *Casuarina, Le Géographe,* and *Le Naturaliste* between 1801 and 1803, and he collected widely in Australia in 1801–1802. Bunbury in Western Australia was seen and named Port Leschenault after him by the French explorer Captain De Freycinet, from his ship the *Casuarina* in 1803. Leschenault collected in Java between 1803 and 1806 and in India between 1816 and 1822. He also visited the Cape Verde Islands, Cape of Good Hope, Ceylon (now Sri Lanka), Brazil, and British Guiana (now Guyana). He is also commemorated in the scientific name of a bird, the White-crowned Forktail *Enicurus leschenaulti,* and in the names of some reptiles, including Leschenault's Leaf-toed Gecko *Hemidactylus leschenaulti.* The rousette, a fruit bat, is found from India and Sri Lanka east to Vietnam and south China, and south to Sumatra, Java, and Bali.

Lesson

Lesson's Saddle-back Tamarin *Saguinus fuscicollis fuscus* Lesson, 1840

René Primevère Lesson (1794–1849) was a French ornithologist and naturalist of enormous influence and importance, considered one of the giants of natural history. Though best known as a zoologist, he was also a skilled botanist and was Professor of Botany at Rochefort. He was employed on the *Coquille,* in 1822, as a botanist and then on the *Astrolabe,* between 1826 and 1829, as a naturalist and collector. The *Astrolabe*'s surgeon and botanist was his brother, Pierre Adolphe Lesson (1805–1888), who later became a professor at a school of naval medicine. Both voyages were to the Pacific and called at many islands, including New Zealand. Lesson published a considerable number of ornithological texts, including *Manuel d'ornithologie* in 1828 and

Traité d'ornithologie in 1831. The tamarin is found in southern Colombia.

Lesueur

Lesueur's Bettong *Bettongia lesueur* Quoy and **Gaimard,** 1824 [Alt. Lesueur's Rat-Kangaroo, Burrowing Bettong, Boodie]

Lesueur's Hairy Bat *Cistugo lesueuri* **Roberts,** 1919 [Alt. Lesueur's Wing-gland Bat]

Charles Alexandre Lesueur (1778–1846) was a French naturalist, artist, and explorer. At the age of 23 he set sail for Australia and Tasmania aboard *Le Geographe* as an assistant gunner. He was appointed as an official expedition artist by Baudin when the original artists jumped ship in Mauritius. During the next four years he and fellow naturalist François Péron collected more than 100,000 zoological specimens representing 2,500 new species, and Lesueur made 1,500 drawings. From these drawings he produced a series of watercolors on vellum, which were published between 1807 and 1816 in the expedition's official report, *Voyage de découvertes aux terres australes.* On their return he and Peron continued to work on the collections. From 1815 until he returned to France in 1837 he lived in the southern USA and undertook some local travels and collections. During 1824 he met Audubon and so admired his work that he suggested Audubon should try again to get them published in France. In 1845 Lesueur was appointed Curator of the Museum of Natural History in Le Havre, which was created to house his drawings and paintings. He is remembered in the scientific names of a bird, the White-shouldered Triller *Lalage sueurii,* and a lizard, the Eastern Water Dragon *Physignathus lesueurii.* The bettong was extirpated from the Australian mainland and survived only on offshore islands (Barrow, Bernier, and Dorre). Some attempts are being made to reintroduce the species into protected sites on the mainland. The bat is found in South Africa and Lesotho.

Lewis, J. S.

Lewis' Tuco-tuco *Ctenomys lewisi* **Thomas,** 1926

John Spedan Lewis (1885–1963) was a British businessman who took over his father's shops and turned them into the now well-known employee-owned John Lewis Partnership. He was a great sponsor of expeditions, and in 1926 Thomas and St. Leger published *The Spedan Lewis South American Exploration* and listed the mammals that had been "obtained by Señor E. Budin in Neuquen." Lewis owned important bird collections in Britain, and in 1928 he also financed an expedition to Vietnam for his friend Delacour, during which the bird known as Lewis' Silver Pheasant *Lophura nycthemera lewisi* was collected. The tuco-tuco is endemic to Bolivia.

Lewis, M.

Lewis' Marmot *Arctomys lewisii* **Audubon** and **Bachmann,** 1853 BOGUS SPECIES

Captain Meriwether Lewis (1774–1809) was one half of the "Lewis and Clark" duo, famous for their expedition of 1804–1806 across North America. Lewis was chosen to lead the expedition by President Thomas Jefferson; he was the latter's private secretary at the time. Before becoming Jefferson's secretary he had grown up in the country, managed the family plantation, and spent time in the army both as an ordinary soldier and then as an officer. Lewis chose Clark, a friend he had made while in the army, to accompany him. The famous journey of exploration covered more than 6,400 km (4,000 miles) to the Pacific Ocean. Lewis was fascinated with the Native Americans, plants, animals, fossils, geological formations, topography, and other facets of the trip, all of which he recorded in his journal entries. As a reward for his success, he was appointed to the governorship of the Louisiana Territory. In 1809 he undertook a journey to Washington, DC, in order to clear his name after having been publicly accused of misusing public money. En route he met his death from two gunshot wounds to the head and chest. It is still not known whether this was murder or suicide, but as he was a sufferer of bipolar syndrome, suicide appears more likely.

The circumstances behind the naming of the marmot are strange and well worth mentioning. Audubon and Bachman based their description of this species on a preserved specimen in the collections of the Zoological Society of London. The specimen was labeled as having come from "Columbia," which they took to mean the Columbia River. They then named the species after the explorer Captain Meriwether Lewis, even though there was no direct connection between the man and the marmot. No more specimens of this marmot turned up, and it was largely forgotten. Finally, Robert S. Hoffmann of the University of Kansas reexamined the type specimen that was still in the British Museum of Natural History and concluded that it was actually a specimen of the Altai Marmot *Marmota baibacina*. How a Siberian marmot came to London labeled as having come from "Columbia" we will doubtless never know, but "Lewis' Marmot" is the North American mammal-that-never-was. Meriwether Lewis is, however, remembered in the name of a bird, Lewis's Woodpecker *Melanerpes lewis*.

L'Hoest

L'Hoest's Monkey *Cercopithecus lhoesti* **P. Sclater,** 1899

Francois L'Hoest (1839–1904) was Director of Antwerp Zoo from 1888 to 1904. Directing Antwerp Zoo appears to be a hereditary occupation, as Francois L'Hoest was succeeded by his son, known as M. L'Hoest Sr. (1869–1930), who was Director from 1905 to 1930, and then by his grandson, M. L'Hoest Jr., who was Director from 1931 to 1944. The monkey's range covers eastern DRC (Zaire), Rwanda, Burundi, and western Uganda.

Lichtenstein

Lichtenstein's Hartebeest *Sigmoceros lichtensteinii* **Peters,** 1849 [Syn. *Alcelaphus lichtensteinii*]

Lichtenstein's Jerboa *Eremodipus lichtensteini* **Vinogradov,** 1927

Martin Heinrich Carl Lichtenstein (1780–1857) was a German traveler and zoologist who was head of the Berlin Zoological Museum from 1813 and founded the Berlin Zoo in 1844. He was trained as a physician. Between 1802 and 1806 he traveled widely in South Africa and while there became the personal physician to the Dutch Governor of the Cape of Good Hope. He wrote *Reisen in Südlichen Africa* in 1810. Lichtenstein studied many species sent to the Berlin Museum by others, and it has been said that "whilst he gave every species, or what he judged to be a species, a name, this was done without consulting the recent English and French literature. His only aim was to give the specimens in question a distinguishing mark for his personal needs. These names were used in Lichtenstein's registers and reappeared on the labels of the mounted specimens, but only exceptionally were they published by himself in connection with a scientific description." This caused much unnecessary confusion and trouble to others. He was also commemorated in the names of several birds, including Lichtenstein's Sandgrouse *Pterocles lichtensteinii*, and several reptiles, including Lichtenstein's Night Adder *Causus lichtensteini* The hartebeest is found from northeast Angola and southern DRC (Zaire) east to Tanzania and Mozambique. The jerboa is found in Kazakhstan, Turkmenistan, and Uzbekistan.

Lindbergh

Lindbergh's Grass Mouse *Akodon lindberghi* **Hershkovitz,** 1990
Scott's Rice Rat *Cerradomys scotti* **Langguth** and Bonvicino, 2002

Scott Morrow Lindbergh (b. 1943) farms at Vão dos Bois in the Cerrado area of Brazil and makes his land and services readily available to naturalists. He has written scientific articles such as, with others, "Small Non-flying Mammals from Conserved and Altered Areas of Atlantic Forest and Cerrado: Comments on Their Potential Use for Monitoring Environment," published in 2002 in the *Brazilian Journal of Biology*. His father, Colonel Charles Augustus Lindbergh, became one of the most famous men in the world in 1927 after becoming the first person to fly the Atlantic solo, in his plane *The Spirit of St. Louis*. The grass mouse has a restricted range in the Brasília area of Brazil. The rice rat is also endemic to the Brazilian cerrado (tropical savanna) region.

Linnaeus

Linnaeus' False Vampire Bat *Vampyrum spectrum* Linnaeus, 1758
Linnaeus' Mouse-Opossum *Marmosa murina* Linnaeus, 1758 [Alt. Common Mouse-Opossum]

Carolus Linnaeus (1707–1778). Both of the above species come from tropical South America, with the bat extending northward to southern Mexico. See **Linné** for biographical details.

Linné

Linné's Two-toed Sloth *Choloepus didactylus* Linnaeus, 1758

Carl Linné (1707–1778) is much better known by the Latin form of his name, Carolus Linnaeus, or just as Linnaeus. Later in life (in 1761) he was ennobled and so could call himself Carl von Linné. In the natural sciences he was undoubtedly one of the greatest heavyweights of all time, ranking with Darwin and Wallace. He is thought of primarily as a botanist, but he invented the system for naming, ranking, and classifying organisms that is still in use today, albeit with modifications. In 1727 he entered the University of Lund to study medicine and a year later transferred to the University of Uppsala—then, as now, the most prestigious university in Sweden. At that time the study of botany was part of medical training, as doctors had to prepare drugs derived from plants with medicinal qualities. His first expedition was to Lapland in 1732. In 1734 he mounted an expedition to central Sweden. He went to the Netherlands in 1735 and finished his studies as a physician there before enrolling at Leiden. In that same year he published the first edition of his classification of living things, the *Systema naturae*. He returned to Sweden in 1738, lecturing and practicing medicine in Stockholm. In 1742 he became Professor at Uppsala and restored the university's botanical garden. He had a number of famous students, including Daniel Solander, who was a naturalist on Captain James Cook's first round-the-world voyage and brought back the first Australian plant collections to Europe. Linnaeus became the Swedish royal family's personal physician. In 1758 he bought the manor estate of Hammarby, outside Uppsala, where he built a small museum for his extensive personal collections. This house and garden still exist and are now run by the University of Uppsala. His son, also named Carl, succeeded to his professorship at Uppsala but was never noteworthy as a botanist. When Carl the younger died, five years later, with no heirs, his mother and sisters sold the elder Linnaeus's library, manuscripts, and natural history collections to the English natural historian Sir James Edward Smith, who founded the Linnean Society of London to take care of them. He is remembered in the names of a number of reptiles, including Linné's Reed Snake *Calamaria linnaei*. The sloth is found in the Guianas and the Amazonian forests of Brazil, Venezuela, southeast Colombia, eastern Ecuador, and northeast Peru.

Littledale, G.

Littledale's Argali *Ovis ammon littledalei* **Lydekker,** 1902 [Syn. *O. ammon karelini* Severtzov, 1873]

St. George R. Littledale (1855–1921) was a British explorer, traveler, geographer, and hunter. He traveled widely to look for unusual animals to shoot. He took his wife with him on all his major journeys and was sometimes also accompanied by his pet terrier. He attempted to reach Lhasa, in Tibet, but the Tibetans would not at that time countenance unwanted visitors to the city, so Littledale was forced to turn back. He was a member of the Royal Geographical Society and was awarded the Patron's Medal in 1896. As a hunter he was a great friend of such luminaries of the age as Theodore Roosevelt and Frederick Courtney Selous. When he died he left his collection of stuffed animals to the British Museum of Natural History, except for one which he left to King George V. The type specimen of this subspecies of wild sheep was taken in "Chinese Turkestan." However, it is now generally regarded as synonymous with *Ovis ammon karelini.*

Littledale, H.

Littledale's Whistling Rat *Parotomys littledalei* **Thomas,** 1918

Major Harold A. P. Littledale (1853–1930) was an English collector. He is known to have collected orchids in Nepal in 1895. His diary of egg-collecting in Southern Africa between 1906 and 1911 is held by the Natural History Museum. Littledale collected the type specimen of the whistling rat in Cape Province, South Africa. The species occurs in arid regions of western South Africa and Namibia.

Livingstone

Livingstone's Eland *Taurotragus oryx livingstonii* **Sclater,** 1864

Livingstone's Suni *Neotragus moschatus livingstonianus* **Kirk,** 1864

Livingstone's Flying Fox *Pteropus livingstonii* **Gray,** 1866 [Alt. Comoro Black Flying Fox]

David Livingstone (1813–1873) was a Scottish doctor and missionary, and undoubtedly the most famous African explorer of all time. Livingstone is remembered as the first European to have gone into the heart of Africa, and as someone who came to be regarded as a saint in his own lifetime. He worked in a cotton mill from the age of 10, earning extra income by selling tea from farm to farm. He studied Latin and Greek on his own and elected to become a missionary when he was persuaded that science and theology were not in opposition. He trained at the London Missionary Society and, in medicine, in Glasgow. Livingstone left for South Africa in 1840. His many expeditions brought him fame as a surgeon and scientist over the next few years, but his missionary efforts were less successful. He sympathized with the lot of the indigenous people and so made enemies among white settlers. It annoyed some of them that he learned the languages and tribal customs of the peoples he tried to convert. Nevertheless, his indictment of the slave trade did much to make antislavery laws enforced. In 1853 he undertook an expedition into the interior of the continent that lasted three years, during which he discovered the Victoria Falls, a find that sealed his fame upon his return to Britain in 1856. His last expedition, begun in 1866, was to search for the source of the Nile. False reports of his death, and the public's need to know where the lost explorer was, led to Stanley's equally famous mission to find him. Livingstone is also commemorated in the name of a bird, Livingstone's Turaco *Tauraco livingstonii.* The eland is found in Angola, southern DRC (Zaire), Zambia, and Zimbabwe. The suni (a dwarf antelope) comes from eastern Africa, from Kenya to Natal (South Africa). Southern populations are sometimes treated as a separate subspecies, *zuluensis*. The endangered flying fox is found on the islands of Anjouan and Moheli in the Comoro Islands.

Lockwood

Grey Leaf-eared Mouse *Graomys lockwoodi* **Thomas,** 1918 [Syn. *G. griseoflavus* **Waterhouse,** 1837]

Charles Lockwood (1856–1952) was a member of an English family of cutlers in Sheffield. He had the misfortune to be trapped in Paris during the siege of 1870–1871 during the Franco-Prussian War. He nearly starved and was reduced to eating rats. He emigrated from England to Argentina, where he stayed for the rest of his life, except for a visit to England in 1892 to marry. In Argentina he set up a company that was an associate of the parent company in Sheffield. The leaf-eared mouse was collected by Emilio Budin, who was occasionally employed by Lockwood. Budin also collected for Thomas and sent the type specimen of the mouse to him. Thomas originally described it as a full species. He wrote in his etymology, "Named in honour of Mr. Charles Lockwood, of Buenos Ayres, by whose kindness as intermediary all the business arrangements with Messrs. Kemp and Budin have been so greatly facilitated." Later the taxon was thought to be a subspecies *(Graomys domorum lockwoodi)* of the Pale Leaf-eared Mouse. Nowadays *Graomys lockwoodi* is regarded as synonymous with *G. griseoflavus*, but it is likely that the taxonomy of this genus will undergo further revision.

Loder

Loder's Gazelle *Gazella leptoceros loderi* **Thomas,** 1894 [Alt. Slender-horned Gazelle, Rhim]

Sir Edmund Giles Loder, 2nd Baronet (1849–1920), was a wealthy Victorian naturalist, traveler, sportsman, and horticulturalist. He spent a considerable amount of time hunting and exploring in northern and eastern Africa in the last decade of the 19th century. He is probably the man who shot the last Red Gazelle *Gazella rufina* in Algeria sometime in the early 1890s. In 1897 he was on safari in Somaliland. Nowadays he is remembered much more as an important horticulturalist. In 1889 he moved to Leonardslee in Sussex and expanded its already famous garden. He was a great fan of the rhododendron and bred many cultivars. Among his many creations is the rhododendron 'Pretty Polly'. This subspecies of gazelle is from the western and central Sahara Desert, in Algeria, Tunisia, Libya, and probably Mali.

Longman

Longman's Beaked Whale *Indopacetus pacificus* Longman, 1926 [Alt. Tropical Bottlenose Whale]

Heber Albert Longman (1880–1954) was a paleontologist who was born in England and moved to Australia for health reasons in 1902. He was fascinated by Australian natural history, and his work earned him the Mueller Medal in 1952. He spent many years in Queensland and published over 70 papers, in the most part through the Queensland Museum. He described, among other fossils, *Kronosaurus queenslandicus*, which was one of the largest marine reptiles ever to have lived. For a long time this eponymous species of whale was known only from two skulls and jawbones, one of which languished in the Queensland Museum for 44 years before someone (Longman) got around to considering it. The second skull was found in Somalia in 1955. In 2003 an article in *Marine Mammal Science* described further specimens from strandings and finally put on scientific record the physical appearance of the species. Sightings and strandings now suggest that Longman's Beaked Whale is found across the Indian and Pacific oceans, from South Africa to Hawaii and Mexico.

Lord Derby

Lord Derby's Scaly-tailed Squirrel *Anomalurus derbianus* **Gray,** 1842

Lord Derby's Giant Eland *Taurotragus derbianus* Gray, 1847 [Alt. Giant Eland]

The Hon. Edward Smith Stanley, 13th Earl of Derby (1775–1851). The scaly-tail, which is not a true squirrel but an anomalure, is found from Sierra Leone east to Kenya and south to Angola, Zambia, and Mozambique. The giant eland occurs from Senegal to southern Sudan. See **Derby** for biographical details.

Lord Howe

Lord Howe Long-eared Bat *Nyctophilus howensis* McKean, 1975 extinct

This bat is named after Lord Howe Island, east of Australia, and not after Lord Howe personally. The island was discovered in 1788 and named after the First Lord of the Admiralty, Richard Howe, 1st Earl Howe. The bat is known only from a fossil skull. The species may have survived until human settlement of the island.

Lorentz

Lorentz's Melomys *Paramelomys lorentzii* **Jentink,** 1908 [Alt. Lorentz's Mosaic-tailed Rat]

Speckled Dasyure *Neophascogale lorentzi* Jentink, 1911

Long-footed Tree Mouse *Lorentzimys nouhuysi* Jentink, 1911

Dr. Hendrikus Albertus Lorentz (1871–1944) was a Dutch explorer who studied law and biology. In 1901 he participated in the expedition of Professor Wichmann to northern (Dutch) New Guinea (now West Papua), and from 1905 to 1906 and 1909 to 1910 he himself led expeditions in southern New Guinea, which led to the discovery of the Wilhelmina Peak (named after the Dutch Queen Wilhelmina) in the Snow Mountains. Upon his return from New Guinea he entered the Dutch consular services, becoming Ambassador in Pretoria, South Africa, 1929. There is a Lorentz National Park in Indonesian Papua named after him. The mammals named after him are all found in New Guinea.

Loria

Loria's Mastiff Bat *Mormopterus loriae* **Thomas,** 1897 [Alt. Little Northern Free-tailed Bat]

Large Tree Mouse *Pogonomys loriae* Thomas, 1897

Dr. Lamberto Loria (1855–1913) was an Italian ethnologist who collected in New Guinea in 1889 and 1890. He also founded the first Italian Museum of Ethnography, in 1906, in Florence. The museum was subsequently transferred to Rome, after Loria organized the first ethnography exhibition there in 1911. When the 5,000 objects in the Florence museum were transferred to the capital, over 30,000 objects collected by Loria and his assistants were added to the collections. Loria also has a bird named after him, Loria's Bird-of-Paradise *Cnemophilus loriae*. The bat is found in Papua New Guinea and northern and eastern Australia. The tree mouse is found in New Guinea (it is sometimes also listed as occurring in northeast Queensland, Australia, but this population probably represents a separate species).

Loring

Loring's Rat *Thallomys loringi* **Heller,** 1909

John Alden Loring (1871–1947) was an American field collector. He was a field biologist with the Bureau of Biological Survey of the U.S. Department of Agriculture from 1892 to 1897. He joined Theodore Roosevelt and his son Kermit on the Roosevelts' mammoth safari, which lasted from 1909 to 1910, as a professional collector, and wrote about and lectured on it under the title *Through Africa with Roosevelt.* The object of the expedition was to collect large game mammal specimens for the U.S. National Museum in Washington, DC. They observed and collected more than 160 species of mammals in Kenya, Uganda, and Sudan. Heller, who described the rat, was also a member of the expedition. The rat is found in eastern Kenya and Tanzania.

Los Chalchaleros

Chalchalero Viscacha Rat *Salinoctomys loschalchalerosorum* Mares, Braun, Barquez, and Díaz, 2000

The Chalchalero Viscacha Rat is probably the only animal named after a modern musical combo. It was "named for the great Argentine folklore group 'Los Chalchaleros' in honour of their 52 years singing the traditional music of western Argentina, its habitats, and its history." The rat comes from the province of La Rioja, western Argentina.

Louis XV

Louis XV's Rhinoceros

This is not a species but a famous individual. Louis XV (1710–1774) became King of France at the age of five upon the death of his great-grandfather, Louis XIV. The animal referred to was an Indian Rhinoceros *Rhinoceros unicornis* that was kept in the Royal Menagerie in Paris in the 18th century. When it died it was the subject of an extensive anatomical investigation before being stuffed and put on display. It can still be seen today, in the Zoological Museum at the Jardin des Plantes in Paris, and is well worth a visit. You will find it in the Appendices, including a scientific name. This is to show the species of the animal on display, not to imply that all animals of that species are named after a King of France.

Louise

Louise's Spiny Mouse *Acomys louisae* **Thomas,** 1896

Mrs. Louise Lort Phillips (1857–1944) was the wife of Ethelbert Lort Phillips, a British big-game hunter in East Africa between 1884 and 1895. Thomas wrote that he was naming the spiny mouse "in honour of Mrs. Lort Phillips, who took a considerable share in the collecting done by the expedition." She is also remembered in the binomial of a bird, the Somali Grosbeak *Rhynchostruthus louisae.* The mouse is endemic to Somalia.

Lovat

Lovat's Climbing Mouse *Dendromus lovati* **de Winton,** 1900

Simon Joseph Fraser, 14th Lord Lovat (1871–1933). Though legally the 14th Lord, he was referred to as the 16th Lord Lovat. He was an aristocrat and soldier who was commissioned into the Queen's Own Cameron Highlanders. In 1899 he raised the Lovat Scouts, who fought in the South African War. Lovat was awarded the Distinguished Service Order. He led an expedition through Abyssinia (Ethiopia) from Berbera to the Blue Nile in 1899, which collected specimens of mammals and birds. The climbing mouse is found in the highlands of central Ethiopia.

Low

Pen-tailed Tree-Shrew *Ptilocercus lowii* **Gray,** 1848

Low's Squirrel *Sundasciurus lowii* **Thomas,** 1892

Sir Hugh Brooke Low (1824–1905) was a civil servant in the British administration in Malaya and an amateur collector in the Malay Archipelago. He was the first successful British Administrator of Perak, serving from 1877 until 1889. Subsequently his methods became models for British colonial operations in Malaya. Previously he had been an unremarkable Colonial Secretary of Labuan, an island off Borneo, for 30 years from 1848 until 1877. There is a "Historical Trail" to the summit of Mount Kinabalu, Sabah, named after him, as he used to collect specimens from the summit, having been the first person to climb it in 1851. He knew Sir Charles Brooke, the second "White Rajah" of Sarawak, having traveled out to Malaya with him when first appointed by the East India Company. He worked as Brooke's secretary for a few months before returning to England, where he published *Sarawak: Its Inhabitants and Productions.* He is best remembered for the many new orchids he found and that are named in his honor. He was also instrumental in encouraging the planting of rubber trees all over Malaya. Low's Dwarf Snake *Calamaria lowi* is also named after him. The tree-shrew is found on the Malay Peninsula, Sumatra, Borneo, and some small nearby islands. The squirrel has a similar distribution.

Lowe

Lowe's Shrew *Chodsigoa parca* **G. M. Allen,** 1923
Lowe's Gerbil *Gerbillus lowei* **Thomas** and Hinton, 1923
Lowe's Monkey *Cercopithecus lowei* Thomas, 1923
Lowe's Otter-Civet *Cynogale lowei* **Pocock,** 1933 [Alt. Tonkin Otter-Civet]
Lowe's Servaline Genet *Genetta servalina lowei* Kingdon, 1977

Willoughby Prescott Lowe (1872–1949) was a British naturalist, explorer, and collector for the British Museum. He was in the USA in the closing years of the 19th century and in virtually all parts of Asia and Africa between 1907 and 1935. He is reputed to have sent over 10,000 specimens to the Bird Room at the British Museum of Natural History alone. He is also notorious for having shot eight specimens of Miss Waldron's Red Colobus *Piliocolobus badius waldroni* in Ghana in 1933. Miss Waldron also worked at the museum and accompanied him on this trip. The colobus was already rare and is now thought to be extinct. The shrew is found in southwest China, northeast Myanmar, northern Thailand, and northern Vietnam. The gerbil is known only from the highlands of the Jebel Marra area of Sudan. Lowe's Monkey can be found in the Ivory Coast and Ghana. The otter-civet is known from a single specimen taken in northern Vietnam; its taxonomic status is uncertain (it may be a distinct species or be conspecific with *Cynogale bennettii*). The genet is almost equally little known; the type specimen was taken in 1932, after which the subspecies "disappeared" until photographs of live specimens were taken in 2002 by camera-traps in the Udzungwa Mountains of Tanzania.

Lucas

Lucas' Short-nosed Fruit Bat *Penthetor lucasi* **Dobson,** 1880

Frederic Augustus Lucas (1852–1929) was a professional curator. In 1871 he entered Ward's Natural Science Establishment in Rochester, New York, where he remained for 11 years working at taxidermy, osteology, and museum technique. He joined the Smithsonian Institution in 1882 as an "osteological preparatory" and became Assistant Curator to True (q.v.) in 1887. In 1893 he became Curator, a post he held until 1904. From 1904 to 1911 he was the Curator in Chief of the Museum of the Brooklyn Institute of Arts and Sciences, and in 1911 he became the Director of the American Museum of Natural History, New York, a post he held until 1923. From 1924 to 1929 he served as Honorary Director of the museum. He was awarded an honorary D.Sc. degree from the University of Pittsburgh in 1909. Dobson says that the type specimen of the bat was "forwarded to me by Mr. Frederic A. Lucas, who had correctly recognised it as representing a hitherto un-described species." The bat is found in the Malay Peninsula, Sumatra, and Borneo.

Lucifer

Black-and-Red Bush Squirrel *Paraxerus lucifer* **Thomas,** 1897
Lucifer Titi *Callicebus (torquatus) lucifer* Thomas, 1914

Depending upon your religious convictions, Lucifer (another name for Satan) may or may not actually exist. He (it?) is considered by some to be a "fallen angel." The name literally means "light bearer." Thomas seems to have used the name fancifully for two species with black-and-red coloration (suggesting the fires of Hell?). The squirrel is found in northern Malawi and southwest Tanzania. The titi monkey comes from southern Colombia, eastern Ecuador, northeast Peru, and adjacent northwest Brazil.

Lucina

Lucina's Shrew *Crocidura lucina* Dippenaar, 1980 [Alt. Moorland Shrew]

Lucina was the Roman goddess of childbirth. The original etymology wrongly refers to her as a Greek goddess and gives no explanation for why this name was chosen. The shrew is endemic to the montane moorlands of Ethiopia.

Ludia

Ludia's Shrew *Crocidura ludia* **Hollister,** 1916 [Alt. Dramatic Shrew]

The original description makes no suggestion that the name is eponymous, and we assume that the vernacular name was coined via a misunderstanding. The word *ludia* in Latin can mean a female gladiator or an actress, which probably led to someone inventing the alternative name Dramatic Shrew. It has been recorded in the Central African Republic and northern DRC (Zaire).

Ludovic

Highland Yellow-shouldered Bat *Sturnira ludovici* **Anthony,** 1924 [Alt. Anthony's Bat]

Ludovic Söderström (1843–1927) was the Swedish Consul General at Quito in Ecuador from about 1886 until the year of his death. The bat is found from Mexico south to Ecuador and Guyana. See **Söderström** for biographical details.

Lugard

Lugard's Mole-Rat *Fukomys damarensis lugardi* **de Winton,** 1898 [formerly *Cryptomys damarensis lugardi*]

Frederick John Dealtry Lugard, 1st Baron Lugard (1858–1945), was a soldier and colonial administrator in Africa. His parents were missionaries in India, where he was born. After passing out at Sandhurst as an officer, he was sent to India in 1878 and saw action in the Afghan War of 1879–1880. In 1885 he was sent with the Indian regiment to the Sudan in the relief of Khartoum. In 1886 he was sent to Burma. In 1887 he went to England but was unable to resume his commission for medical reasons. He was also in despair over an unhappy love affair and decided to go to East Africa, arriving in Mozambique in 1888. In that same year he took command of a private military expedition organized by the British settlers in Nyasaland (now Malawi) against Arab slave traders on Lake Nyasa and was severely wounded. After his experiences in Nyasaland he took a position with the African Lakes Company in 1889. He went to Uganda in 1890 and was instrumental in securing British dominance there and persuading the British government to declare it a Protectorate in 1894. He became British Commissioner for Nigeria, created the West African Frontier Force in 1897, and "subdued" Nigeria by 1903. He became Governor there from 1912 until 1919, binding its various territories into one administrative whole. He developed the doctrine of "indirect rule" that Britain afterward employed in many of its African colonies, the doctrine being that the colonial administration should exercise its control of the subject population through traditional native institutions. Lugard expounded this theory in *The Dual Mandate in British Tropical Africa*, published in 1922. He was raised to the peerage in 1928. There is also a waterfall in South Africa and a Nile steamer, SS *Lugard*, named after him. The type specimen of the mole-rat was collected in the Kalahari Desert by Lugard.

Luis

Luis' Yellow-shouldered Bat *Sturnira luisi* **Davis,** 1980

Luis de la Torre. The bat is found from Costa Rica to Ecuador and northwest Peru. See **De la Torre** for biographical details.

Luis Manuel

Luis Manuel's Tailless Bat *Anoura luismanueli* Molinari, 1994

Luigi Emanuele Molinari Tani (1927–1991) was the father of a zoologist. The describer, Jesus Molinari (Department of Biology, University of the Andes, Merida, Venezuela), wrote in the formal description, "The specific name, a noun in the genitive case, is a patronymic that I dedicate to the memory of my father, Luigi Emanuele Molinari Tani, 1927–1991, who admired and loved the mountains in which the new species thrives, and who stimulated my curiosity toward animals when I was a child. The Latinized name *luismanueli* is a derivative from Luis Manuel, the Spanish translation of his originally Italian name that my father often employed while living in his native Madrid and in Venezuela." The bat is found in the Andes (Eastern Cordillera) of Colombia and western Venezuela.

Lumholtz

Lumholtz's Tree Kangaroo *Dendrolagus lumholtzi* **Collett,** 1884 [Alt. Boongary]

Dr. Carl Sophus Lumholtz (1851–1922) was a Norwegian naturalist, ethnologist, humanist, and explorer. Growing up in Norway in the 1870s he fell in love with mountainous country. In 1880, having just graduated with a natural science degree, he set off for northeastern Australia, where he spent time living with Aboriginal people until 1884. In 1882 he said, on first hearing of tree kangaroos in Australia, "According to the statement of the blacks, it was a kangaroo which lived in the highest trees on the summit of the Coast Mountains. It had a very long tail, and was as large as a medium-sized dog, climbed the trees in the same manner as the natives themselves, and was called boongary. I was sure that it could be none other than a tree-kangaroo (Dendrolagus)." He organized a number of expeditions, including one to explore the Sierra Madre in Mexico in 1890, for the American Museum of Natural History. He visited Borneo in 1914 and later planned a trip to New Guinea, a country he had wanted to visit for many years, but the outbreak of WW1 prevented this, and he traveled extensively in Borneo instead. Lumholtz National Park in Queensland is named after him. The tree kangaroo inhabits northeast Queensland.

Lund

Lund's Amphibious Rat *Lundomys molitor* Winge, 1887

Lund's Atlantic Tree Rat *Phyllomys lundi* Leite, 2003

Peter Wilhelm Lund (1801–1880) was a Danish physician, botanist, zoologist, and paleontologist who lived and worked in Lagoa Santa, Minas Gerais, Brazil, during the 19th century. He first traveled to Brazil in 1833 but decided to settle there for health reasons. His interest in fossils led him to explore many of the caves of the area. He assembled one of the most important mammal collections from a single locality in the Neotropics and made outstanding contributions toward describing the Pleistocene and recent mammal faunas of Brazil, including the description of the genus *Phyllomys* in 1839. During his life he was a regular correspondent of Darwin's. He is also remembered in the name of a reptile, Lund's Teiid *Heterodactylus lundii*. Voss and Carleton created the genus *Lundomys* in 1993 to accommodate the species originally described by Winge as *Hesperomys molitor*. It is found in Uruguay and adjacent southernmost Brazil. The tree rat comes from southeast Brazil.

Lycaon

African Wild Dog *Lycaon pictus* **Temminck,** 1820 [Alt. Cape Hunting Dog, Painted Dog]

There are two people in Greek mythology called Lycaon. One was a son of Priam, King of Troy, who was killed by Achilles. The other, and a much more likely source for the name, was the mythical first King of Arcadia. According to one legend, he killed and cooked his own son, Nyctimus, and served the flesh to Zeus. The god refused the dish and turned Lycaon into a wolf, so that the King's physical form now matched his brutal nature. The dog was once found widely in sub-Saharan Africa but is now a rare species, largely confined to protected areas in eastern and southern Africa.

Lydekker

Grey Slender Loris *Loris lydekkerianus* **Cabrera,** 1908

Richard Lydekker (1849–1915) was an English naturalist, geologist, and vertebrate paleontologist who worked at the British Museum of Natural History, where he catalogued the mammals, reptiles, and amphibians. In 1874 he joined the Geological Survey of India and studied the fossils there, particularly in Kashmir. His published works include *A Manual of Palaeontology* (1889) and *A Hand-book to the British Mammalia* (1895). He delineated the biogeographical boundary known as Lydekker's Line, which separates the Oriental faunal and floral region on the west from the Australia New Guinea region on the east. This line is further east than the more famous Wallace's Line. The loris is found in southern India and Sri Lanka.

Lyle

Lyle's Flying Fox *Pteropus lylei* **K. Andersen,** 1908

Shield-faced Roundleaf Bat *Hipposideros lylei* **Thomas,** 1913

Sir T. H. Lyle (1878–1927) was a British consular official who served in Siam (now Thailand) for many years. He was appointed Vice Consul at Nan in 1896 and Consul at Chiang Mai in 1907. He was one of the founding members of Chiang Mai Gymkhana Club in 1898. He was only 157 cm (5 feet 2 inches) tall and suffered from extremely bad health but insisted in joining in all activities. He was undoubtedly very brave, as he played a prominent part in the Shan Rebellion by going alone to confront the insurgents. He persuaded many of them to lay down their arms and go home. He left Chiang Mai in 1913 and was subsequently knighted. At the end of WW1 he returned to Bangkok as Consul General and was still serving there when he died. The flying fox is found in Thailand, Cambodia, and Vietnam. The roundleaf bat is from Myanmar, Thailand, and the Malay Peninsula.

Lynn

Lynn's Brown Bat *Eptesicus lynni* **Shamel,** 1945

Professor Dr. William Gardner Lynn (1905–1990) was a naturalist who collected the type specimen of the bat, one of a series of 27 specimens he collected. Although he obtained them in 1932, they were not described as a new species until 1945. Lynn graduated from Johns Hopkins University and also received a doctorate in biology from that university, where he worked as a member of the faculty from 1930 to 1942. In 1942 he became Professor of Biology at Catholic University in Washington, DC. He retired in 1971 and in retirement taught at a number of institutions, including Montgomery College. Lynn wrote a number of papers about reptiles in the 1930s and later, including, with C. Grant, "The Herpetology of Jamaica" (*Bulletin of the Jamaica Institute of Science*, 1940). The bat is found on Jamaica. It is sometimes treated as a race of *Eptesicus fuscus*.

MacArthur, C.

MacArthur's Shrew *Crocidura macarthuri* St. Leger, 1934

C. G. MacArthur (fl. 1915–1945) collected entomological specimens in 1932 at Merifano on the Tana River in Kenya, where the type specimen of the shrew was obtained. He is also recorded as having collected the type specimen of a beetle in 1938. Between 1939 and 1945 he was employed by the Kenya Game Department. The identification is speculative in as much as all Jane St. Leger put in her original description was that the shrew was named after "Mr. MacArthur"—not even supplying an initial to help future researchers. The shrew is found in savannah regions of Kenya and Somalia.

MacArthur, J.

MacArthur's Mouse-Lemur *Microcebus macarthurii* Radespiel et al., 2008

John Donald MacArthur (1897–1978) was the founder of the MacArthur Foundation, along with his wife Catherine. Born into a poor family in Pennsylvania, MacArthur made his fortune in the insurance business. At the time of his death he was said to be worth in excess of $1 billion and was reportedly one of the three richest men in the USA. MacArthur left 92 percent of his estate to begin the John D. and Catherine T. MacArthur Foundation. This foundation supplied funding for the research that led to the discovery of the mouse-lemur, which is found in the Makira region of northeast Madagascar.

MacClelland

Himalayan Striped Squirrel *Tamiops macclellandi* **Horsfield,** 1840

Dr. John MacClelland (1805–1875) was employed by the East India Company. He was one of the persons with botanical knowledge sent by the British government of India to Burma (now Myanmar) to survey the growth timber from Bago Division in 1852. He pointed out that the utilization of timber should not depend upon teak alone but should include other hardwood species as well. He is also commemorated in the name of MacClelland's Coral Snake *Calliophis (Sinomicrurus) macclellandi* and in the scientific name of a fish, the Spotted Codlet *Bregmaceros macclellandi.* The squirrel is found from eastern Nepal and Assam (India) east to Vietnam and south to the Malay Peninsula.

MacConnell

MacConnell's Climbing Mouse *Rhipidomys macconnelli* **de Winton,** 1900
MacConnell's Bat *Mesophylla macconnelli* **Thomas,** 1901
Guyanan Red Howler Monkey *Alouatta (seniculus) macconnelli* **Elliot,** 1910
MacConnell's Rice Rat *Euryoryzomys macconnelli* Thomas, 1910

Frederick Vavasour McConnell (1868–1914) was an English traveler and collector. He made some of his collections between 1894 and 1898 with J. J. Quelch, who wrote *Animal Life in British Guiana* (1901). His collections inspired a book by C. H. Chubb, *The Birds of British Guiana,* written between 1916 and 1921, to which

McConnell's wife wrote the foreword. He presented his specimens to the British Museum of Natural History. They comprised collections of birds and spiders as well as mammals, all obtained in British Guiana (now Guyana). He is also commemorated in the names of two birds, McConnell's Flycatcher *Mionectes macconnelli* and McConnell's Spinetail *Synallaxis macconnelli*. The climbing mouse is found in the highlands of southern Venezuela and adjacent corners of Brazil and Guyana. The bat occurs from Nicaragua south to Peru, Bolivia, Amazonian Brazil, and Trinidad. The howler monkey inhabits Trinidad, the Guianas, and adjacent parts of northern Brazil. The rice rat is widespread in northern South America east of the Andes.

Macdonald

Fijian Blossom Bat *Notopteris macdonaldi* **Gray,** 1859

Professor Dr. Sir John Denis Macdonald (1828–1906) joined the Royal Navy in 1849 and was Assistant Surgeon on board HMS *Herald*. Macdonald was a Fellow of the Royal Society and a correspondent of Charles Darwin. He led an expedition to explore the Fijian island of Viti Levu, where the type specimen of the bat was collected, and wrote of his experiences under the title "Proceedings of the Expedition for the Exploration of the Rewa River and its Tributaries, in Na Viti Levu, Fiji Islands," published in the *Proceedings of the Royal Geographical Society* in 1857. He was also quite an accomplished artist, painting portraits of Fijians, some of which are in Australian galleries and museums. He became Professor of Hygiene at Netley in 1872 and was Inspector General of Hospitals and Fleets from 1880 to 1886. The bat is from Fiji and Vanuatu. Populations on New Caledonia are now often treated as a separate species, *Notopteris neocaledonica*.

MacDonnell

Fat-tailed Pseudantechinus
Pseudantechinus macdonnellensis **Spencer,** 1896

This small marsupial is not named directly after a person but after the MacDonnell Mountains in central Australia, where the first known specimens were collected. These mountains take their name from Sir Richard Graves MacDonnell (1814–1881), who was Governor of South Australia from 1855 to 1862.

Machang'u

Rungwe Brush-furred Rat *Lophuromys machangui* Verheyen et al., 2007

Professor Dr. Robert S. Machang'u is a veterinary surgeon and an Associate Professor at Sokoine University of Agriculture, Morogoro, Tanzania, where he is Director of the Pest Management Center. The rat is found on Mount Rungwe, in the southern highlands of Tanzania.

MacInnes

MacInnes' Mouse-tailed Bat *Rhinopoma macinnesi* **Hayman,** 1937

Donald Gordon MacInnes (dates not found) was a paleontologist who organized the mammal collecting that took place during the 1934 Lake Rudolph Rift Valley expedition. This expedition was primarily concerned with making a geological survey. The original description credits MacInnes, "to whom much of the success of the mammal-collecting was due." He worked in East Africa from the middle 1920s for a number of seasons, with, among others, Dr. Louis Leakey. In the 1950s he taught paleontology at the National Museum in Nairobi. He published a number of books as well as contributing to reports and learned papers, such as "Fossil Mammals of Africa." An African gibbonlike fossil ape was named after him by LeGros Clark and Leakey in 1950 as *Dendrop-*

ithecus macinnesi. The bat is known from Uganda, Kenya, Somalia, and Ethiopia.

Mackenzie

Kenneth's White-toothed Rat *Berylmys mackenziei* **Thomas,** 1916

J. M. Donald Mackenzie (dates not found) was a biologist and ornithologist who is remembered for designing a simple nesting box. He was a divisional forestry officer in Burma (now Myanmar). At times during 1913, 1914, and 1915 he and J. C. Hopwood were in the northern part of the Chin Hills of Burma, where the type specimen of the rat was taken. In 1917, with Hopwood as co-author, he published "A List of Birds from the North Chin Hill." In addition to Myanmar, the rat is found in Assam (India) and has also been recorded in Sichuan (China) and southern Vietnam, but not in the intervening areas.

Mackilligin

Mackilligin's Gerbil *Gerbillus mackilligini* **Thomas,** 1904 [Alt. Mackilligin's Dipodil; Syn. *Dipodillus mackilligini*]

Arthur M. Mackilligin (dates not found) was the person Thomas credits with collecting the type specimen of this gerbil, in the eastern desert of southern Egypt. We do not know anything more about him. (A common misspelling is the addition of an extra *n* into Mr. Mackilligin's name, with the species thus becoming Mackillingin's Gerbil/Dipodil.) The gerbil occurs in southern Egypt and probably also in northern Sudan.

Macklot

Sunda Acerodon *Acerodon mackloti* **Temminck,** 1837 [Alt. Sunda Fruit Bat]

Heinrich Christian Macklot (1799–1832) was a German taxidermist and naturalist who worked at the Natural History Museum at Leiden. He was appointed to assist members of the Dutch Natural Science Commission and traveled to the East Indies. He took part in an expedition to New Guinea and Timor on the *Triton* from 1828 to 1830. In 1832 he was apparently murdered while in Java. Macklot's Python *Liasis mackloti* is named after him, as is Macklot's Pitta *Pitta erythrogaster macklotii*. He is also remembered in the binomial of the Sumatran Trogon *Apalharpactes mackloti*. The bat occurs on the Lesser Sunda Islands, Indonesia, from Lombok to Timor.

Maclaud

Maclaud's Horseshoe Bat *Rhinolophus maclaudi* **Pousargues,** 1897

Dr. Charles Maclaud (1866–1933) was the French Resident at Timbo in French Guinea (now Republic of Guinea) in 1898. He published *Notes anthropologiques sur les Diolas de la Casamance* (1907) and *Gouvernement général de l'Afrique occidentale française. Notes sur les mammifères et les oiseaux de l'Afrique occidentale: Casamance, Fonta Dialon, Guinée française et portugaise* (1906). (Fonta-Dialon, or Fouta-Djallon, is a highland region in west-central Guinea.) The bat is known only from Guinea; specimens from Liberia have now been classified as a separate species, *Rhinolophus ziama*.

Maclear

Maclear's Rat *Rattus macleari* **Thomas,** 1887 extinct

Admiral John Fiot Lee Pearse Maclear (1838–1907) served in the Royal Navy from 1851 to 1891. His first appointment, as a Cadet, was to the frigate HMS *Castor*, in which he served until 1854 when he transferred to HMS *Algiers*, being on board her until 1856. He saw action in both the Baltic and the Black Sea, including the siege of Sebastopol during the Crimean War. In 1857 he was Master's Mate on HMS *Cyclops*, and in 1859 he was appointed to HMS *Sphinx* on the China station. Between 1860 and 1862 he fought in the Opium War, including taking part in the capture of the Taku forts. In 1863 he was appointed Gunnery Lieutenant of HMS *Excellent,* in 1864 he was on HMS Princess *Royal,* the flagship on the China Station, and in 1867 he became First Lieutenant of HMS *Octavia,* flagship of the squadron on the East Indies station. In 1868 he was in action again in the Abyssinian campaign, and in the same year he was promoted to Commander. In 1872 he joined HMS *Challenger,* a ship that was commanded by Captain Sir John Nares for the famous *Challenger* expedition. From 1879 to 1882 he commanded HMS *Alert,* which was employed in surveying work in the Straits of Magellan, and from 1883 to 1887 he was in command of HMS *Flying Fish.* In 1886 he discovered an anchorage in a bay on Christmas Island, which was then uninhabited. Maclear named it Flying Fish Cove and went ashore, where he made a small but interesting collection of the flora and fauna. He was promoted to Rear Admiral and retired in 1891, becoming a Vice Admiral in 1897 and an Admiral in 1903. It is perhaps worth explaining that in those days promotion up the higher rungs of the flag rank ladder was achieved by seniority and continued even in retirement. Two islands are named Maclear Island after him: one in Papua New Guinea and the other off Queensland, where there is also a Mount Maclear. The rat was endemic to Christmas Island but has not been recorded since 1903.

Macleay, W. J.

Macleay's Dorcopsis *Dorcopsulus macleayi* Miklouho-Maclay, 1885 [Alt. Papuan Forest Wallaby]

William John Macleay (1820–1891) was a politician and naturalist who wrote widely on entomology, ichthyology, and general zoology. He took part in several collecting expeditions, including to New Guinea in 1875, for which expedition he bought and fitted out a barque called *Chevert.* The whole of the Macleay family were avid naturalists and collectors—so much so that the Macleay Museum University of Sydney was built in 1887 to house their vast natural history collection. They began collecting insects in the late 18th century. When Alexander Macleay (1767–1848), a diplomat and entomologist, went to Sydney as Colonial Secretary in 1826, he already had one of the largest private insect collections in the world. It was added to by his son, William Sharp Macleay (see below), and expanded to include all aspects of natural history by William's cousin, William John Macleay. He, in turn, donated the collections to Sydney University in 1887. He is also commemorated in the name of a bird, Macleay's Honeyeater *Xanthotis macleayana.* The wallaby is found in southeast New Guinea.

Macleay, W. S.

Macleay's Moustached Bat *Pteronotus macleayi* **Gray,** 1839

William Sharp Macleay (1792–1865) was one of the Macleay family's several collectors of natural history specimens (see above). In 1825 he became British Commissioner of Arbitration to the joint British and Spanish Court of Commission in Havana for the abolition of the slave trade. He remained in Havana until 1836, and in 1839 he arrived in Sydney, Australia, where he spent the rest of his life. He collected the type specimen of the bat while in Cuba but did not have to work hard to obtain it: he found it in his bedroom in Havana. The species occurs on Jamaica as well as Cuba.

Macmillan

Macmillan's Thicket Rat *Grammomys macmillani* **Wroughton,** 1907
Macmillan's Shrew *Crocidura macmillani* **Dollman** 1915

Sir William Northrop Macmillan (1872–1925) was a huge Scot who was raised in St. Louis, Missouri, in a rich family. He arrived in East

Africa to go big-game hunting, fell in love with an area of Kenya, and bought half a mountain, Ol Donyo Sabuk (2,146 m; 7,041 feet). He developed a ranch at its base, and in 1911, while on his famous safari, Theodore Roosevelt stayed at this ranch, Juja Farm, which is now a popular location for film crews. Macmillan and his wife were philanthropists and founded the Macmillan Library in Nairobi. They also bequeathed their ranch "to the nation"; it is now a national park, and both Sir William and his wife, Lady Lucie, are buried there. During WW1 the 25th Royal Fusiliers were in Kenya. This regiment was known officially as "the Legion of Frontiersman" but by its members as "the Old and Bold." Among its members was Macmillan, and we can assume that he was a gentleman of considerable presence and size, as his sword belt measured 1.63 m (64 inches). The rat is known from Sierra Leone, Liberia, Central African Republic, southern Sudan, northern DRC (Zaire), southern Ethiopia, Kenya, Uganda, Tanzania, Malawi, Mozambique, and Zimbabwe. The shrew is endemic to Ethiopia.

MacNeill

See **McNeill.**

Macow

Macow's Shrew *Crocidura macowi* **Dollman** 1915 [Alt. Nyiro Shrew]

We do not know the origin of "Macow." The original description says that the type specimen was collected by Arthur Blayney Percival on Mount Nyiro in Kenya. Perhaps Macow is the name of a settlement in the area, as we can find no eponymous candidate. The shrew is known only from the type locality.

Magdalena

Magdalena Rat *Xenomys nelsoni* **Merriam** 1892

This rodent is named after a farm rather than a person. The rat is found in Colima and southwest Jalisco, western Mexico, where it was first collected at the Hacienda Magdalena.

Maggie Taylor

Maggie Taylor's Roundleaf Bat *Hipposideros maggietaylorae* Smith and **Hill,** 1981 [Alt. Maggie Taylor's Leaf-nosed Bat]

Mrs. Margaret Reese Taylor (1916–1996) was a philanthropist and volunteer at the Los Angeles County Natural History Museum. She ran a team of 215 volunteers at the Los Angeles Zoo and was President of the Greater Los Angeles Zoo Association. She set up the Taylor Science Fund, and as well as sponsoring an expedition to southwest Africa in 1972, she was an important sponsor of the Taylor South Seas expedition of 1979. This set out from the Los Angeles Natural History Museum to study the bats of the Bismarck Archipelago and Solomon Islands; a further expedition was mounted in 1981. The bat is found on New Britain, New Ireland, and the mainland of Papua New Guinea.

Mahomet

Mahomet's Mouse *Mus mahomet* **Rhoads,** 1896

Though the common name has been "personalized," we do not think this mouse is named directly after the prophet of Islam. The locality of the type specimen is given as "Sheikh Mahomet" in south-central Ethiopia, so we conclude that the species is named after the place. The spelling is not as is usually shown today, but is consistent with that in use in earlier times. The mouse is found in Ethiopia, Uganda, and Kenya.

Main

Arnhem Land Rock Rat *Zyzomys maini* Kitchener, 1989

Professor Albert (Bert) Russell Main C.B.E., Ph.D., F.A.A. (b. 1919) was Professor of Zoology at the University of Western Australia and a well-known ecologist. He is now retired as Professor Emeritus. He is married to Barbara York Main, who is also a zoologist, specializing in spiders. Main served in the Royal Australian Air Force during WW2. From 1952 to 1967 he was a researcher in the Department of Zoology at the University of Western Australia, where he completed his doctorate in 1956, becoming Professor of Zoology there in 1967. Main's Frog *Cyclorana maini* is also named after him, as is a gecko *(Diplodactylus maini)* and a skink *(Menetia maini)*. The rock rat is found in the Northern Territory of Australia, in the region of the East and South Alligator rivers.

Major

Major's Sifaka *Propithecus verreauxi majori* **Rothschild,** 1894 [Alt. Forsyth Major's Sifaka; Syn. *P. verreauxi verreauxi*]
Major's Tufted-tailed Rat *Eliurus majori* **Thomas,** 1895
Major's Long-fingered Bat *Miniopterus majori* Thomas 1906
Major's Pine Vole *Microtus majori* Thomas, 1906

Charles Immanuel Forsyth Major (1844–1923) was born in Glasgow but educated in Switzerland, graduating in medicine at Basel in 1868. However, his real interest was fossil mammals, and he is remembered as a vertebrate paleontologist rather than as a physician. He did fieldwork in Italy and Corsica but is most noted for his expedition to Madagascar from 1894 to 1896. This was the first systematic survey of Madagascar's mammalian fauna. The sifaka (lemur) named after him was later found to be only a dark color-morph of *Propithecus verreauxi* and so does not warrant separate taxonomic status. Oddly, this does not prevent "Major's Sikafa" from appearing on some (badly researched) website lists of extinct animals. The lemur, rat, and bat are all endemic to Madagascar. The vole is found in Turkey and the Caucasus.

Makundi

Hanang Brush-furred Rat *Lophuromys makundii* Verheyen et al., 2007

Professor Dr. Rhodes H. Makundi is an entomologist at Sokoine University of Agriculture, Morogoro, Tanzania. His bachelor's degree was awarded in 1979 by the University of Dar es Salaam in Tanzania, and his master's in 1983 and his doctorate in 1996 by the University of Newcastle-upon-Tyne in England. The rat occurs on Mount Hanang, Tanzania.

Mandelli

Mandelli's Mouse-eared Bat *Myotis sicarius* **Thomas,** 1915

Louis Mandelli (1833–1880) was a tea planter in Assam (India) and was the subject of a book, *L. Mandelli, Darjeeling Tea Planter and Ornithologist* by Fred Pinn. This account refers to his excursions into "the snow" seeking birds, and he is remembered in the common name of Mandelli's Snowfinch *Montifringilla taczanowskii*. His collections are now in the Natural History Museum, Tring, UK. The bat is found in Nepal and northeast India.

Marca

Marca's Marmoset *Callithrix (argentata) marcai* Alperin, 1993

Alex Garcia Marca is a zoologist working at the University of Rio de Janeiro, Brazil. He has written many articles from the 1990s onward. The original etymology is one of the shortest we have seen, reading simply, "Em homenagem ao biólogo Alex Garcia Marca." The marmoset is known only from one locality at the mouth of the Rio Castanho (a tributary of the Rio Roosevelt) in western Amazonian Brazil.

Marcano

Marcano's Solenodon *Solenodon marcanoi* Patterson, 1962 EXTINCT

Professor Eugenio de Jesus Marcano Fondeur (1923–2003) was a botanist who wrote some 15 books, including one on poisonous plants and one on edible plants in the Dominican Republic. He was Professor of Entomology, Botany, and Geology at the Autonomous University of Santo Domingo, having been previously a Professor at the Polytechnic Institute Loyola de San Cristobal since 1955. The solenodon is known only from skeletal remains in Haiti and the Dominican Republic, but as some of these are associated with the remains of (Eurasian) rats of the genus *Rattus*, it seems that the solenodon was not extinct when Europeans first arrived in Hispaniola. A number of other taxa have been named after Marcano, including a spider *Selenops marcanoi* and a lizard *Anolis marcanoi*.

Marcgrave

Marcgrave's Capuchin Monkey *Cebus flavius* **Schreber,** 1774 [Alt. Marcgraf's Capuchin Monkey, Blond Capuchin]

Georg Marcgrave (also Marcgraf) (1610–1648) was a German naturalist and astronomer. He studied botany, astronomy, mathematics, and medicine in Germany and Switzerland until 1633. He then practiced medicine in the Netherlands for four years. In 1637 the Dutch West India Company offered him the position of Personal Physician to Count Johan Maurits van Nassau-Siegen, who was at that time in South America in the capacity of Governor of the colony of Dutch Brazil. Marcgrave arrived there in early 1638 and undertook the first zoological, botanical, and astronomical expedition in Brazil, exploring various parts of the colony to study its natural history and geography, staying on until 1644. Cuvier praised him as one of the best and most diligent observers and recorders of the era. He returned to the Netherlands in 1648 and was sent to Angola by the West India Company, but died there shortly after his arrival. He was the co-author, with Willem Piso, of *Historia naturalis Brasiliae*, an eight-volume work on the botany and zoology of Brazil published in the year he died. Many animal species were drawn and painted by artists on the Brazil expedition, and one, for which there was only a description and an illustration but no physical specimen, was a very blond capuchin monkey. However, no similar specimens turned up in later collections, and so this mysterious primate was either lumped with other, better-known, species or ignored by later taxonomists. Then in 2006 a monkey resembling the old drawing was found in the flesh and described under the name of *Cebus queirozi*. It was later realized that this was the same monkey described by Marcgrave and named as *Simia flavia* by Schreber. The capuchin is found in northeast Brazil's Atlantic forests.

Marche

Palawan Stink Badger *Mydaus marchei* **Huet,** 1887 [Alt. Philippine Stink Badger, Calamian Stink Badger; Syn. *Suillotaxus marchei*]

Antoine-Alfred Marche (1844–1898) was a French naturalist, anthropologist, traveler, and explorer who was on the island of Palawan and elsewhere in the Philippines from 1879 to 1886 and again in 1888. It is supposed that he collected the type specimen of the badger. His explorations were halted by the hostility of the locals. There are also records of him having explored in Africa in 1872–1873 and 1875–1877, and in Hawaii around 1889. He wrote *Three Voyages in Western Africa* in 1879, *Luzon and Palawan* around 1883, and *Six Years in the Philippines* in 1887. The badger is endemic to Palawan and the Calamian Islands in the Philippines.

Marco Polo

Marco Polo Sheep *Ovis ammon polii*
Blyth, 1841

Marco Polo (1254–1324) is perhaps the most famous Western traveler of his era, having traveled the Silk Road and through Asia for 24 years. He also became (according to his own account) a confidant of Kublai Khan. During his stay in China he encountered such wonders as paper money and a postal service. Returning to Venice in 1295, Marco Polo found that not everyone was willing to believe his accounts of distant lands. Polo wrote *The Description of the World, or The Travels of Marco Polo,* which was to become one of the most popular books in the whole of Europe, although the original is now lost. In 1995 a book entitled *Did Marco Polo Go to China?* was published. The writer, Dr. Frances Wood, put forward the theory that the famous explorer never set foot in China. One piece of evidence for this is that the Chinese "Annals of the Empire" (Yuan Shih) make no mention of Polo, though they record the names of foreign visitors who were far less important. However, other historians have defended Polo's accounts. Thus the stories of Marco Polo continue to be as contentious today as they were in the early 14th century. The sheep named after him is found in the Pamir Mountains of central Asia.

Margareta

The Sulawesi rat genus *Margaretamys*
Musser, 1981 [3 species: *beccarii, elegans,* and *parvus*]

Miss Margareta Becker was at one time an employee of the American Museum of Natural History. In 1971 she was working on specimens for Musser in connection with the Archbold expeditions. She accompanied Musser during his trip to central Sulawesi in 1976, and, as he says in his description, "The genus is named for Margareta Becker, who shared the adventure of living and working in the primeval forests of Central Sulawesi." She acted as the expedition's photographer.

Margaretta

Greater Ranee Rat *Haeromys margarettae*
Thomas, 1893

Margaret Brooke (1849–1936), Ranee of Sarawak, was the wife of Sir Charles Johnson Brooke, the second "White Rajah." At Kuching, the capital of Sarawak, Brooke built a palace-cum-fortress for his wife in 1878 and named it Fort Margherita after her. It was completed in 1880 and proved to be very useful in defending Kuching against attacks by pirates. The wall around it was stone and about 5 m (16 feet) high. Margaret Brooke published her memoirs under the title *My Life in Sarawak by the Ranee of Sarawak.* The rat is known from Sarawak and Sabah (East Malaysia, Borneo).

Margarita

Sand Cat *Felis margarita* Loche, 1858

General Jean Auguste Margueritte (1823–1870) was a Frenchman whose family moved to Algeria in 1831. He joined a French regiment stationed in Algeria in 1837, and in 1840 was promoted to Sergeant and then commissioned as a Lieutenant. Between 1862 and 1864 he was part of the French force in Mexico to support the ill-fated attempt to put a Frenchman on the Mexican throne. He finally left Algeria as a Colonel in 1870, as his regiment was needed to fight in the Franco-Prussian War. He fought at the Battle of Sedan in 1870 and was promoted to Major General on the battlefield, only to be fatally wounded a few minutes later. His two sons, Paul (1860–1919) and Victor (1866–1942), both born in Algeria, were notable French novelists. The cat was first found in Algeria. Its range covers the Sahara Desert, the Arabian Peninsula, Turkmenistan, Uzbekistan, and Pakistan.

Margarita

Santa Margarita Island Kangaroo Rat *Dipodomys margaritae* **Merriam,** 1907

As the common name implies, this rodent is named after a place rather than a saint. The species is endemic to the island of Santa Margarita, on the west coast of southern Baja California, Mexico.

Marie

Marie's Vole *Volemys musseri* **Lawrence,** 1982

Marie A. Lawrence (1924–1992) was, at the time she described the vole, a Scientific Assistant in the Department of Mammalogy, American Museum of Natural History. She named the vole "for Dr. Guy G. Musser, Archbold Curator in the Department of Mammalogy, AMNH," but her own name has become attached to the species as its common name. In 1990 she was awarded the Hartley H. T. Jackson Award for long and outstanding service to the American Society of Mammalogists. In addition to scientific descriptions she has published other papers on mammals, especially during the 1980s and 1990s. The vole comes from western Sichuan Province, China.

Marinho

Marinho's Rice Rat *Cerradomys marinhus* Bonvicino, 2003

This rodent is not named after one person but a family: Francis R. Marinho, Jose Roberto Marinho, and Theodoro de Hungria Machado, the owners of Fazenda Sertao do Formoso, a plantation where, in the words of the describer, "conservation of the environment is always considered." The rice rat was first discovered on this property and comes from the state of Goias, south-central Brazil.

Marinkelle

Marinkelle's Sword-nosed Bat *Lonchorhina marinkellei* **Hernandez-Camacho** et al., 1978

Dr. Cornelis Johanes Marinkelle (b. 1925) is Emeritus Professor of the Laboratory of Microbiology and Parasitology at the University of the Andes in Bogota. He was born in Vienna, is a Dutch citizen, and lives in Colombia. His doctorate in biology (1963) comes from the University of Utrecht. He worked and taught at many institutions in many places in England, Belgium, Indonesia, El Salvador, Iraq, and Sudan before going to Colombia. The University of Michigan Museum of Zoology holds 454 specimens he collected, and it is clear that he undertook a number of collecting trips during the 1950s and 1960s, as other specimens from that period appear in other collections. He also amassed a collection of 25,000 birds' eggs, which he donated to the Humboldt Institute in 2001. He did not collect most of these personally but accumulated the collections of others, taken from as early as 1871, although some were bought by him or his father or grandfather. He wrote a number of scientific papers, mostly on diseases and epidemiology of both humans and other mammals, from the 1970s until the 1990s. The bat is found in the northern part of South America, from eastern Colombia to French Guiana.

Marisa

Aellen's Roundleaf Bat *Hipposideros marisae* **Aellen,** 1954

Marisa Aellen (dates not found) was the wife of Professor Villy Aellen. Of the naming of this bat he wrote, "Dédiée à ma femme qui m'a accompagné et secondé au cours de ce voyage" (dedicated to my wife who accompanied me and supported me throughout the course of this journey). The bat is found in Ivory Coast, Liberia, and Guinea.

Marjorita

Bunyoro Rabbit *Poelagus marjorita* St. Leger, 1929 [Alt. Uganda Grass-Hare]

Mrs. Marjorie Pitman (dates not found) was the wife of Colonel Charles Robert Senhouse Pitman. She obviously accompanied her husband on his travels as game warden in Uganda, as Jane St. Leger says in her original description, "I have great pleasure in naming this hare in honour of Mrs. Marjorie Pitman, to whom, together with her husband, we are indebted for its discovery." The hare is found in Uganda and neighboring countries (southern Sudan, Rwanda, northeast DRC [Zaire]), with a disjunct population in Angola.

Markham

Markham's Grass Mouse *Akodon markhami* **Pine,** 1973

We believe this to be named after B. J. Markham (fl. 1950–2003), although so far our evidence is circumstantial. In 1970 Markham was a member of the Department of Zoology at the University of Guelph in Ontario and was co-author with S. N. Gatehouse of an article on raptors that was published in *The Auk*. He has published several articles in the *Anales del Instituto de la Patagonia*, including "Catálogo de los anfibios, reptiles, aves y mamiferos de la provincia de Magallanes (Chile)" in 1971, as well as a number of articles on the mammals of the region. In the 1970s both he and Pine wrote articles on the distribution of introduced hares and other taxa in Chile. He also published a number of other articles, mostly on conservation issues in Canada, during the late 1970s and early 1980s when working for the Alberta Department of Environmental Protection (e.g. "Waterfowl Production and Water Level Fluctuation," 1982). In 2003 he was Acting Director of Wildlife for the Fish and Wildlife Division of the Alberta Sustainable Resource Development program. The grass mouse is confined to Isla Wellington, southern Chile, and we believe Markham did some of his work there.

Marley

Marley's Golden-Mole *Amblysomus marleyi* **Roberts,** 1931

Harold Walter Bell-Marley (1872–1945) was a principal fisheries officer at Durban, South Africa, but also a naturalist with a particular interest in entomology. He was born in England and went to South Africa at the time of the Boer War, then decided to stay on. He collected continuously for nearly 50 years in nearly every part of the country south of the Zambesi River. Museums in many parts of the world have specimens that he sent them. He also collected birds' eggs, and his collection, now in the Pretoria Museum, is regarded as one of the most complete ever assembled. He contracted blackwater fever at the end of 1944 while collecting in northern Zululand and died soon after his return to Durban in January 1945. He was a very well known figure, and Roberts does not actually state, in his description of the golden-mole, who he is naming the species after. Perhaps he felt it unnecessary. The golden-mole is found in KwaZulu-Natal, eastern South Africa.

Marshall

Marshall's Horseshoe Bat *Rhinolophus marshalli* **Thonglongya,** 1973

Dr. Joe Truesdell Marshall Jr. (b. 1918) was a zoologist at the American Museum of Natural History. He had a particular interest in rodents and published widely on them. In 1977 he wrote "A Synopsis of Asian Species of *Mus* (Rodentia, Muridae)," then in 1981, jointly with R. D. Sage, "Taxonomy of the House Mouse." He also wrote on the subject of wild mice trapped in Thailand. The bat is found in Thailand, Malaysia (mainland), Laos, and Vietnam.

Martienssen

Large-eared Free-tailed Bat *Otomops martiensseni* **Matschie,** 1897

Herr Martienssen (dates not found) worked for the Königlich Museum für Naturkunde in Berlin and supplied Matschie with the type specimen of this bat, which came from what is now Tanzania. The species is found from Eritrea and the Central African Republic south to Angola and Natal (South Africa). It has also been found in Yemen.

Martin, R. D.

Martin's False Potto *Pseudopotto martini* Schwartz, 1996

Professor Robert Denis Martin (b. 1943) is a zoologist and conservationist with strong interests in paleontology and evolutionary biology. He is Provost for Academic Affairs at the Field Museum in Chicago, having left the University of Zurich in 2001 to be Vice President for Academic Affairs and Curator in Biological Anthropology. He received both his B.A. and his Ph.D. in zoology from Oxford. From 1969 to 1974 he lectured on physical anthropology at University College London. From 1974 to 1978 he was Senior Research Fellow in charge of the Wellcome Laboratories of Comparative Physiology at the Zoological Society of London, and during that period he was a Visiting Professor at Yale. He returned to University College London in 1978 as a Reader in Physical Anthropology, becoming Professor in that subject in 1982, a post he held until 1986 when he became Professor and Director at the Anthropological Institute and Museum of the University of Zürich, where he is still Professor Emeritus. He is regarded as a pioneer in the study of primitive primates. He has made a particular study of lemurs in Madagascar and Barbary Macaques in Algeria, Morocco, and Gibraltar. The false potto was discovered by Jeffrey Schwartz of the American Museum of Natural History when studying skeletons of the (true) Potto *Perodicticus potto.* He noticed that two specimens displayed differences from typical pottos: the dentition was wrong for a Potto, and the tail was longer than normal. The type specimen of *Pseudopotto* had lived its life in a zoo, masquerading as a Potto, so the appearance in life of this species must be very similar to that of a Potto. Believed to originate in Cameroon, no living specimens of *Pseudopotto* have surfaced since the species was named. Some zoologists have questioned whether the species is truly valid. Certainly the creation of a whole new genus for this primate may prove to have been somewhat premature.

Martin, W.

Martin's Guenon *Cercopithecus nictitans martini* **Waterhouse,** 1838 [Alt. Martin's Putty-nosed Monkey]

William Charles Linnaeus Martin (1798–1864) was a fellow curator of Waterhouse's at the London Zoo from 1830 to 1838. In the latter year the Zoological Society of London had to cut costs, and Martin was made redundant. He became a natural history writer and produced more than 1,000 books and articles including, in 1841, *A Natural History of Quadrupeds and other Mammiferous Animals,* and in 1845 both *The History of the Dog* and *The History of the Horse.* The monkey comes from the island of Bioko (Equatorial Guinea), Benin, southern Nigeria, and western Cameroon.

Martino

Martino's Snow Vole *Dinaromys bogdanovi* Martino, 1922

Dr. Vladimir Martino (1883–1965) was a Yugoslav zoologist and ornithologist who was an Assistant Professor of Zoology at the University of Beograd (Belgrade) in 1927. He later went to work in Bulgaria and gave real impetus to the study of mammalogy there. Vladimir Martino worked closely with E. Martino, for whom we have not been able to establish any identity; certainly they jointly described the Balkan Mole *Talpa stankovici* in 1931, and together they wrote a number of papers, such as "Preliminary Notes on Five New Mammals from Yugoslavia" (1940). It is clear that Martino's Snow Vole *Dinaromys bogdanovi* was described by only one of them, and we think it was Vladimir; therefore, it was probably his eponym that was adopted. The vole is recorded only in mountainous regions of the former Yugoslavia but may also occur in Albania.

Martins

Martins' Bare-faced Tamarin *Saguinus martinsi* **Thomas,** 1912

Oscar Martins (d. 1924) was a Brazilian who collected the type specimen of this tamarin. He was a technical assistant and taxidermist at the Emilio Goeldi Museum in Pará from 1908 to 1910, and in 1910 he was exploring and collecting as an assistant to Maria Emilia Snethlage. The tamarin is found north of the Amazon River in Brazil, on either side of the Rio Nhamundá.

Matambuai

Manus Melomys *Melomys matambuai* **Flannery,** Colgan, and Trimble, 1994

Karol Matambuai Kisokau (b. 1949) was trained as an animal ecologist, graduating in 1971, and is first Permanent Secretary of the Papua New Guinea Department of Environment and Conservation. The melomys (mosaic-tailed rat) is found on the island of Manus, where Karol Matambuai was born, off the north coast of Papua New Guinea.

Matschie

Matschie's Guereza *Colobus guereza matschiei* **Neumann,** 1899

Matschie's Grivet *Chlorocebus aethiops matschiei* Neumann, 1902

Matschie's Tree Kangaroo *Dendrolagus matschiei* Forster and **Rothschild,** 1907 [Alt. Huon Tree Kangaroo]

Matschie's Galago *Galago matschiei* Lorenz, 1917 [Alt. Spectacled Galago, Dusky Bushbaby]

Professor Dr. Georg Friedrich Paul Matschie (1861–1926) was a German zoologist who worked at the Humboldt University Zoological Museum in Berlin as Curator of Mammals. He studied natural sciences and mathematics at Halle and Berlin universities, but he never passed an examination or took a degree. He joined the Berlin Museum in 1886 as an unpaid volunteer in the bird room under Cabanis. He was put in charge of the Department of Mammals in 1892, becoming Curator in 1895 and Professor in 1902—quite an achievement for a man with no formal qualifications. In 1924 he became the museum's second Director, but two years later he died of cancer of the rectum. He named a great many species of mammals, mainly because any small variation was enough to inspire him to give the "new" taxon a scientific name. According to his obituary in the *Journal of Mammalogy* in 1927, his best work was done between 1895 and 1905, and much of his early work was of extremely variable quality. He never accepted Darwinian theory and believed firmly in creationism, a prejudice that clouded his scientific judgment in his later years. The guereza (colobus monkey) is found in Kenya west of the Rift Valley and in adjacent northern Tanzania. The grivet monkey occurs in western Ethiopia but may not be a valid subspecies. The tree kangaroo is restricted to the Huon peninsula of Papua New

Guinea. The galago is found in eastern DRC (Zaire) and Uganda.

Matses

Matses' Big-eared Bat *Micronycteris matses* Simmons, **Voss,** and Fleck 2002

This bat is named after the Matses indigenous tribe of the Amazon basin (northeast Peru and adjacent western Brazil) rather than an individual. The bat was discovered in the Peruvian part of their territory.

Matthey

Matthey's Mouse *Mus mattheyi* **Petter,** 1969

Professor Robert Matthey (1902–1982) was a Swiss biologist at the University of Lausanne. He was very interested in rodents and wrote a great deal about them, publishing, in 1965, "Études cytogenetiques sur des *Murinae* Africains appartenant aux genres *Arvicanthis, Praomys, Acomys* et *Mastomys* (Rodentia)"; and in 1972, jointly with M. Jotterand, "L'analyse du caryotype permet de reconnaître deux espèces cryptiques confondue sous le nom *Taterillus gracilis Th.* (Rongeurs—Gerbillidae)." The mouse is found in Ghana. Supposed records from other West African countries may refer to *Mus haussa.*

Max

Bartels' Rat *Sundamys maxi* **Sody,** 1932
Max's Shrew *Crocidura maxi* Sody, 1936 [Alt. Max Bartels' Shrew, Javanese Shrew]

Max Eduard Gottlieb Bartels (1871–1936). The rat comes from western Java. The shrew is found in Java and the Lesser Sunda Islands (Indonesia). It also occurs on the island of Ambon (Moluccas), where it was presumably introduced by human agency. See **Bartels** for biographical details.

Maximilian

Shaggy Bat *Centronycteris maximiliani* J. Fischer, 1829

Maximilian Alexander Philip, Prince of Wied-Neuwied (1782–1867), was an aristocratic German explorer who collected in Brazil between 1815 and 1817, in Guyana in 1921, and in North America from 1832 to 1834. In 1833 he made his famous journey of some 8,000 km (5,000 miles), principally up the Missouri River, on the second voyage of the steamer *Yellowstone.* Artists and scientists from Europe went with him, gathering specimens and painting scenes of the mostly uninhabited countryside. He wrote *Reise nach Brasilien in den Jahren 1815 bis 1817* (Journey to Brazil in the Years 1815–1817) in 1820, *Beiträge zur Naturalgeschichte von Brasilien* (Contributions to the Natural History of Brazil) in 1825, and *Reise in das innere Nord-Amerika in den Jahren 1832 bis 1834* (Journey to the Interior of North America in the Years 1832–1834) in 1840, on his return to Europe. The bat is found from southern Mexico to Peru and Brazil. See **Wied** for further species attributed to him.

Maximowicz

Maximowicz's Vole *Microtus maximowiczii* Schrenck, 1859

Carl Johann (Karl Ivanovich) Maximowicz (1827–1891) was a Russian of German extraction who worked as a taxonomist and botanist. He was a member of the expedition of 1854–1856 to Eastern Siberia. Schrenck, who described the vole, led this expedition. Maximowicz was the first person to consider cultivating the soybean in Russia. He was also Director of the Botanical Garden in St. Petersburg. He is commemorated in a number of plant scientific names, such as that of the Monarch Birch *Betula maximowicziana.* He traveled widely, especially in Japan and Mongolia. The vole is found in eastern Siberia, eastern Mongolia, and northeast China.

Maxwell, C. W.

Maxwell's Duiker *Cephalophus maxwellii* H. Smith 1827 [Syn. *Philantomba maxwellii*]

Lieutenant General Sir Charles William Maxwell (1776–1848) joined the army as a Lieutenant in 1796 and served in the Royal Africa Corps. He was Governor of Senegal in 1809 with the rank of Lieutenant Colonel and was appointed Governor of Sierra Leone (where the type specimen originated) in 1811. He had been instructed to enquire into the fate of the explorer Mungo Park (1771–1806) and was able to report to London in 1811 that Park was indeed dead. Maxwell resigned on account of ill health in 1814. On his return to England he was sued by an American, George Cook, whose factories Maxwell had burned because they were owned by Europeans involved in slavery. As it was determined these were outside Maxwell's jurisdiction, the trader was awarded £20,000, paid by the British government. In 1815 Maxwell became Lieutenant Colonel in the 21st Regiment of Foot and served in Gibraltar and Malta. In 1819 he became Governor of a number of Caribbean islands. He was promoted to Major General in 1830, knighted in 1836, and finally promoted to Lieutenant General in 1841. The duiker occurs from Senegal east to southwest Nigeria.

Maxwell, G.

Maxwell's Otter *Lutrogale perspicillata maxwelli* **Hayman,** 1956

Gavin Maxwell (1914–1968) was a naturalist, shark-hunter, traveler, writer, and explorer. In *Ring of Bright Water,* published in 1960, he tells how he brought an otter from Iraq back to his native Scotland. The animal turned out to belong to a hitherto-undescribed subspecies of the Smooth-coated Otter *Lutrogale perspicillata.* It was named in Maxwell's honor. He was in many ways disorganized but very adventurous. He started a commercial shark-fishing company in Skye in 1945, but the enterprise was undercapitalized and failed in 1948. It did, however, lead to Maxwell's first book, *Harpoon at a Venture,* published in 1952. He visited Iraq with Wilfred Thesiger, and that trip produced two of the great travel books: Thesiger's *The Marsh Arabs* (1964) and Maxwell's *A Reed Shaken by the Wind* (1959). Found in the Iraqi marshes, this race of otter may well now be extinct because of the massive drainage of its native habitat.

Mayer

See **Shaw-Mayer.**

Maynard

Bahaman Raccoon *Procyon maynardi* **Bangs,** 1898

Professor Charles Johnson Maynard (1845–1929) was an American zoologist, mainly publishing on birds and butterflies. He was often described as "Newtonville's [Massachusetts] enigmatic naturalist." He was a well-known observer of birds, particularly in Florida and the Bahamas. He was a member of the Nuttall Ornithological Club, contemporary with Outram Bangs. He wrote *The Naturalist's Guide in Collecting and Preserving Objects of Natural History* (1873), *Birds of Eastern North America* (1881), *The Butterflies of New England* (1886), *A Manual of North American Butterflies* (1891), and *Birds of Washington and Vicinity* (1898). His own company published many of these books, and he competently illustrated them himself. He is also commemorated in the name of a bird, Maynard's Cuckoo *Coccyzus minor maynardi.* The raccoon is known only from New Providence Island in the Bahamas. It is almost certainly not a valid species but represents an introduced population of the North American Raccoon *Procyon lotor.*

Mayor

Mayor's Mouse *Mus mayori* **Thomas,** 1915

Major E. W. Mayor (dates not found) collected the type specimen of this mouse in 1914. We do not know any further details about him. Thomas congratulated him on "his discovery of so striking an addition to the mammal fauna of Ceylon, and I have much pleasure in connecting his name with it." The mouse is endemic to forested regions of Sri Lanka.

Mayr

Wondiwoi Tree Kangaroo *Dendrolagus dorianus mayri* **Rothschild** and Dillon, 1933

See Ernst Mayr.

McGregor

Jamaican Monkey *Xenothrix mcgregori* Williams and **Koopman,** 1952 extinct

Professor Dr. James Howard McGregor (1872–1954) was an American zoologist who took his doctorate in science from Columbia University and became Professor of Anatomy there. H. A. Anthony originally discovered this monkey in 1919, finding a lower jaw and femur at Long Mile Cave, Jamaica. These remains languished in the American Museum of Natural History until a formal description was finally published in 1952. At first it was assumed to be a long-extinct species, but further remains have apparently been discovered in association with those of the Black Rat *Rattus rattus,* which arrived in the Caribbean only when Europeans did. The monkey may have survived until as late as the 18th century.

McIlhenny

McIlhenny's Four-eyed Opossum *Philander mcilhennyi* **Gardner** and **Patton,** 1972

John Stauffer McIlhenny (1910–1997) was an American sponsor of expeditions. He was extremely wealthy, his family having invented the original recipe for Tabasco sauce in the 1860s and still being in the business of producing it. His father, Edward, who invented the sauce, was also a keen ornithologist and took part in the disastrous *Miranda* voyage to Greenland in 1894; the *Miranda* sank, taking all the specimens that had been collected to the bottom. John is also commemorated in the scientific name of a bird, the Black-faced Cotinga *Conioptilon mcilhennyi.* The four-eyed opossum inhabits the rainforests of western Amazonian Brazil and northeast Peru.

McKenzie

McKenzie's False Pipistrelle *Falsistrellus mackenziei* Kitchener, Caputi, and Jones, 1986 [Alt. Western False Pipistrelle]

Norman Leslie (Norm) McKenzie works for the Department of Conservation and Land Management, Western Australia (formerly the Department of Fisheries and Wildlife). He graduated with a zoology degree (B.Sc. Hons., M.Sc.) at Monash and was employed from 1970 by the Western Australian government as a Research Scientist. Currently he is a Principle Research Scientist for the state's ecological survey program. He has published more than 190 scientific papers on a wide range of topics such as conservation, biogeography, landscape ecology, bat aerodynamics, and foraging ecology. McKenzie told us that the describers had misspelled his surname in the description. The bat is restricted to southwest Australia.

McNeill

McNeill's (Red) Deer *Cervus (canadensis) macneilli* **Lydekker,** 1909 [Alt. Sichuan Deer, Gansu Red Deer]

Captain Malcolm McNeill (dates not found) brought the type specimen of this deer from Sichuan, China, and Lydekker named it after him. Although McNeill spelled his own name without an *a,* the animal's name is variously spelled

MacNeill's and McNeill's Deer. It is found in Sichuan and nearby provinces of China. The taxonomy of red deer/wapiti is complex and disputed. Recently this form has been treated by some zoologists as a full species. It can thus be found in the literature as *Cervus elaphus macneilli, C. canadensis macneilli,* or *C. macneilli,* depending upon which authority one consults.

Mearns

Mearns' Grasshopper Mouse *Onychomys arenicola* Mearns, 1896
Mearns' Squirrel *Tamiasciurus mearnsi* **Townsend,** 1897
Mearns' Luzon Rat *Tryphomys adustus* **Miller,** 1910 [Alt. Luzon Short-nosed Rat]
Mearns' Pouched Mouse *Saccostomus mearnsi* **Heller,** 1910
Mearns' Flying Fox *Pteropus (speciosus) mearnsi* **Hollister,** 1913
Western White-bearded Wildebeest *Connochaetes taurinus mearnsi* Heller, 1913
Mearns' Pocket Gopher *Thomomys bottae mearnsi* **Bailey,** 1914

Lieutenant Colonel Edgar Alexander Mearns (1856–1916) was a surgeon in the U.S. Army. He was stationed first in Mexico, from 1892 until 1894, then in the Philippines, from 1903 to 1907, and was later in Africa, between 1909 and 1911. He published a great deal on natural history during the last decade of his life. Mearns was a friend of Theodore Roosevelt and accompanied him on his trip to East Africa in 1909. In 1911 Childs Frick approached the Smithsonian Institution, looking for a scientist to accompany him on his collecting trip to Africa, and Mearns was chosen. Frick agreed to pay Mearns' salary and expenses and to donate all collections to the U.S. National Museum. This was to be Mearns' last expedition. His life was beset with illness, including a "nervous breakdown complicated by malaria" and various parasitic conditions. Eventually he developed diabetes, and as this was before the development of insulin treatment, little could be done for him. Mearns is also commemorated in the names of several birds, such as Mearns' Gilded Flicker *Colaptes chrysoides mearnsi.* The grasshopper mouse is found in the Chihuahuan Desert of New Mexico, western Texas, and northern Mexico. The squirrel is confined to Baja California (Mexico). The rat, as its name suggests, is restricted to Luzon in the Philippines. The pouched mouse can be found from southern Somalia to Tanzania. The flying fox is from the Philippine islands of Mindanao and Basilan. The wildebeest inhabits the Serengeti-Mara region of East Africa. The gopher—one of a huge number of subspecies of *Thomomys bottae*—is found in the southwest corner of New Mexico, USA.

Mechow

Mechow's Mole-Rat *Fukomys mechowi* **Peters,** 1881 [Alt. Giant Mole-Rat; formerly *Cryptomys mechowi*]

Major Alexander von Mechow (1831–1904) was a German explorer and naturalist who led several expeditions to, and collected in, Angola between 1878 and 1900. He explored the middle Kwango River in Angola in 1880 and there collected reptiles, such as the snake *Xenocalamus mechowi,* and amphibians. He is also commemorated in the name of a bird, Mechow's Long-tailed Cuckoo *Cercococcyx mechowi,* and a number of insects such as the butterflies *Hypolymnas mechowi* and *Papilio mechowi.* The mole-rat is found in Angola, Zambia, and southern DRC (Zaire).

Medem

Medem's Titi *Callicebus medemi* **Hershkovitz,** 1963 [Alt. Black-handed Titi]

Professor Federico Medem (1912–1984) was born in Riga as Friedrich Johann Comte von Medem. He was of noble German origin but thought of himself as a Latvian. His family left Latvia after the Russian Revolution in 1917 and moved to Germany. Medem studied at the Humboldt University in Berlin and later at Tübingen. He then worked for his doctorate at the marine biology station in Naples, run by Gustav Kramer. He served in the Wehrmacht in WW2 and fought on the Russian front. After the war he worked in Germany and Switzerland until 1950, then emigrated to Colombia. Here he changed his name and became a herpetologist and ardent conservationist, working at the research station at Villavicencio and at the National University in Bogota. The national plant specimen collection at the Alexander von Humboldt Biological Resources Research Institute is named in his honor. He wrote numerous scientific papers from the 1950s through the 1980s, mostly on Colombian reptiles, and was a founding member of the Crocodile Study Group. In 1981 and 1983 he published the two volumes that make up his *Los crocodylia de Sur America.* The crocodilians he collected between 1955 and 1966 are in the Field Museum, Chicago. The titi monkey is found only in a small area of southern Colombia.

Mehely

Mehely's Horseshoe Bat *Rhinolophus mehelyi* **Matschie,** 1901

Professor Lajos V. Méhely (1861–ca.1952) was a Hungarian biologist, herpetologist, and anthropologist. He may have died as early as 1946, but the date is unknown as he was held as a political prisoner after WW2. In 1913 he became head of the Zoological Department of the Museum and Professor of Zoology and Anatomy at the University Pázmány Péter in Budapest. He was extremely racist, and when he was head of the Anthropology Department at the Hungarian Natural History Museum in Budapest he insisted that only his views and theories should be taught. He wrote a seminal monograph called "Herpetologia Hungarica," but for some reason it was never published. The excellent collection of amphibians and reptiles was almost entirely destroyed in the Budapest uprising of 1956. The bat is distributed throughout the Mediterranean regions of southern Europe and North Africa and east as far as Iran and Afghanistan. The type specimen was collected at Bucharest, Rumania.

Meinertzhagen

Giant Forest Hog *Hylochoerus meinertzhageni* **Thomas,** 1904

Richard Meinertzhagen (1878–1967) was a soldier, a hunter, an ornithologist, a writer, a spy, an advocate of Zionism, and, at least according to the preface to a second edition of his *Kenya Diary*, a killer. He was an empire builder in the "Boy's Own" mold, although he lived in an era when the British Empire was in its final decline. He is truly one of the most remarkable characters to appear in this list. He was born to wealth and position. Although he was one of ten children he was still pampered, even to the extent of being bought an elephant by an eccentric uncle as a christening gift. The family had several homes and was well connected; family friends included Florence Nightingale. As a soldier, Meinertzhagen served in India, in East Africa before and during WW1, and in Palestine and in France toward the end of the war. His primary role was intelligence gathering, but he was also known as a hardened killer. Meinertzhagen's most famous "exploit" happened during WW1, when the British army targeted Beersheba in the buildup to attacking Turkish-occupied Jerusalem. The word "exploit" is in quotation marks, as Meinertzhagen almost certainly took the credit for someone else's idea and actions. His task was to confuse the Turks about the real target. He rode out toward the enemy hoping to be discovered by a patrol, then, putting on an act, he convinced the Turkish patrol that the papers he had dropped for them to find were authentic. His ruse worked, and the focus of the Turks' defense shifted to Gaza and away from Beersheba. This subterfuge was portrayed in the Australian film *The Light-Horsemen.* Meinertzhagen's tactics included a plan to air-drop cigarettes laced with opium to the Turkish troops the night before the attack to impair their defensive capability. This plan was never put into action.

Meinertzhagen retired from the army in 1925, after serving as Britain's chief political officer in Palestine and Syria. He indulged his twin hobbies of ornithology and promoting a Jewish state. He traveled widely, gathering material for books that became standard references, but also frequently acting in support of British intelligence. A recent study of him and his voluminous diaries show him as a "Walter Mitty" character who almost certainly did not kill anything like the number of people, or to have had most of the adventures, that he claimed. He was at one and the same time xenophobic and a passionate advocate of Zionism—so much so that he met Hitler twice in the late 1930s to plead for the Jews. There is a tale, apocryphal or otherwise—and we have to assume that many of the deeds attributed to him were figments of his imagination—which is too irresistible to leave out. He reports that on one of these occasions, when Hitler saluted with "Heil Hitler," he responded with "Heil Meinertzhagen" and was then treated to a 40-minute rant from Hitler and Ribbentrop. He told people that he had a gun in his pocket at the time and forever afterward regretted not having shot the pair of them on the spot.

Meinertzhagen's Zionism was traceable to an incident that occurred in 1910 when he was visiting Odessa, Russia, where he rescued a young Jewish girl from one of the pogroms. He became a proponent of a Jewish state in Palestine since he was, according to his diary, impressed by the biblical promise that "the Holy Land forever remains Israel's inheritance." He vowed that he would help the Jews whenever and however he could. He was still prepared to help even in 1948 when his ship docked in Haifa, as he was returning to England from a field trip to Saudi Arabia. This was in the middle of the Israeli War for Independence, and there were still three weeks to go before the British Mandate expired. Meinertzhagen, although he was 70 years old, saw the opportunity to help the Jewish cause in an intensely personal way. He borrowed the uniform and equipment of one of the Coldstream Guards who were assigned to protecting government stores. Alone, he found some of the Haganah in a firefight with some Arabs, so he joined in. After an hour of fighting he was caught by one of the Coldstream officers, who ordered him back to the ship. "It mattered little," he wrote in his diary, "as by then I had fired all my 200 rounds."

Meinertzhagen wrote *Nicoll's Birds of Egypt* (1930), *The Birds of Arabia* (1954), *Kenya Diary* (1957), *Pirates and Predators* (1959), *Middle East Diary* (1959), *Army Diary* (1960), and *The Diary of a Black Sheep* (1964). In his will he left his collection of over 25,000 bird specimens, recognized as one of the best of its kind in the world, to the British Museum (Natural History). However, some suspicions about his ornithological materials were expressed, although it was not until 1993 that a serious investigation of his collection began. The outcome is revealed in Alan Knox's article "Richard Meinertzhagen—A Case of Fraud Examined," published in the journal *The Ibis.* According to Pamela Rasmussen, an ornithologist at the Smithsonian Institution and something of a scientific Sherlock Holmes, this respected soldier, war hero, and expert ornithologist is alleged to have systematically engaged in the theft of bird specimens from a variety of museums, restuffed them, and added them to his own collection. Three subspecies of birds have been removed from the British List as a result of the investigation reported in the article. The forest hog is found in western Africa from

Guinea to Ghana, and in central and eastern Africa from eastern Nigeria to southwest Ethiopia, Kenya, and northern Tanzania.

Melck

Melck's House Bat *Neoromicia melckorum* **Roberts,** 1919 [Alt. Melck's Serotine]

The type specimen of this bat was taken at Kersfontein in South Africa. Martin Melck settled there in 1770, and his family is still in possession eight generations later. *Melckorum* means "of the Melcks," so the bat is named after the Melck family as a whole rather than just one person. A taxonomic problem has arisen in that the type material for this bat actually represents a different species, *Neoromicia capensis.* This means that populations of "*melckorum*" will need to be given another name, as the latter name becomes a junior synonym of *capensis.* As currently understood, the species is found from South Africa north to Kenya and has also been recorded on Madagascar.

Melissa

Melissa's Yellow-eared Bat *Vampyressa melissa* **Thomas,** 1926

Melissa was a nymph in Greek mythology. Russell W. Hendee (q.v.) collected the type specimen of this bat in Peru. Thomas makes no mention of anyone by the name of Melissa and does not explain his choice of scientific name. However, as Thomas quite often made use of names from mythology, this offers a possible explanation. Melissa was a nymph who discovered and taught the use of honey, and the Melissae were nymphs who nursed the god Zeus when he was an infant. There is also another species of bat in the same genus, *Vampyressa nymphaea,* which Thomas named in 1909. Perhaps he decided to give a "nymph" name to a related species. Most specimens of this bat come from Peru. It has also been recorded in Colombia, but a record from French Guiana is now believed to be a misidentification.

Meller

Meller's Mongoose *Rhynchogale melleri* **Gray,** 1865

Charles James Meller (1836–1869) was a botanist who worked in Nyasaland (now Malawi) in 1861 and on Mauritius in 1865 where he was Superintendent of the Botanical Gardens. He is also remembered in the name of Meller's Duck *Anas melleri.* The mongoose is found in Tanzania, Malawi, Zambia, Zimbabwe, Mozambique, Swaziland, and northeast South Africa.

Menelik

Menelik's Bushbuck *Tragelaphus scriptus meneliki* **Neumann,** 1902

Menelik II (1844–1913). The mammal is almost certainly named after Menelik II who, after considerable struggles against his opponents, became King of Shewa at the age of 11 and, having been imprisoned by usurpers, succeeded in becoming Emperor of Abyssinia (Ethiopia) in 1889. However, it is possible that Menelik I is the person remembered: he was the legendary son of King Solomon of Israel and the Queen of Sheba. Menelik II claimed to be directly descended from him. This bushbuck is an Ethiopian subspecies of a widely distributed African antelope.

Menzbier

Menzbier's Marmot *Marmota menzbieri* Kashkarov, 1925

Mikhail Aleksandrovich Menzbier (1855–1935) was a Russian zoologist from Moscow. He described species across Russia, the eastern Soviet Union, and northern China. Menzbier was one of the founding members of Russia's first ornithological society, and he has a number of such societies named after him. He was among the first to posit that birds are related to reptiles. His major work was on the taxonomy of birds of prey, and he was responsible for much of their modern classification.

He wrote *Oritnologicheskaya geografiia evropeiskoi Rossii* (1882) and *Ornithologie du Turkestan et des pays adjacents* (1888). He also wrote *Ptitsy Rossii* (Birds of Russia), published from 1893 to 1895, the first critical review of the systematics and biology of birds of Russia. Other works were on the zoogeography of the Palaearctic, comparative anatomy, and Darwinism. He was a Professor at Moscow University from 1886, but in 1911 he left in protest against the oppression of students. In 1917 he became Rector of the university. The marmot is found in the Tien Shan Mountains of Kyrgyzstan and southeast Kazakhstan.

Menzies

Menzies' Mouse *Pogonomelomys sevia* **Tate** and **Archbold,** 1935 [Alt. Highland Brush Mouse; Syn. *Abeomelomys sevia*]
Menzies' Echymipera *Echymipera echinista* Menzies, 1990 [Alt. Fly River Bandicoot]

Dr. James I. Menzies was a Professor in the Biology Department of the University of Papua New Guinea, which awarded him a honorary doctorate in science in 2003. He is now an Emeritus Professor of that university and editor of *Science in New Guinea*. He lived in Papua New Guinea from 1967 to 1977, returning in 1980 and having lived there ever since. He is a Visiting Research Fellow at the University of Adelaide in South Australia. From 1983 to 1987 he was Curator of Natural History at the National Museum in Port Moresby. He has published widely, including a book co-written with Helen Fortune Hopkins in 1995, *The Flora of Motupure Island—Papua New Guinea*, and numerous scientific papers. Although the mouse was described decades ago, Menzies' name seems to have become attached to it only around 1990, when he proposed splitting the genus *Pogonomelomys* and creating a new genus, *Abeomelomys*, solely for the above species. However, this proposal has not received wide support from other taxonomists, who continue to use the name *Pogonomelomys sevia*. Both mammals are endemic to New Guinea.

Mephistophiles

Northern Pudu *Pudu mephistophiles* **de Winton,** 1896

Mephistophiles is the name of a demon, or "fallen angel." He is most well-known for the role he plays in the story of Faust, as in, for example, Christopher Marlowe's *The Tragical History of Doctor Faustus* (1604). The pudu is a species of small deer, resembling Bambi's cuter kid brother. What the deer and the demon have in common is hard to say; maybe a pair of short horns? This is certainly one of the more fanciful scientific names in our list. The pudu comes from the Andean forests of Colombia, Ecuador, and Peru.

Merriam

Merriam's Chipmunk *Tamias merriami* **J. A. Allen,** 1889
Merriam's Kangaroo Rat *Dipodomys merriami* **Mearns,** 1890
Merriam's Shrew *Sorex merriami* **Dobson,** 1890
Merriam's Pocket Mouse *Perognathus merriami* J. A. Allen, 1892
Merriam's Pocket Gopher *Cratogeomys merriami* Thomas, 1893
Merriam's Deer Mouse *Peromyscus merriami* Mearns, 1896 [Alt. Mesquite Mouse]
Merriam's Desert Shrew *Megasorex gigas* Merriam 1897
Merriam's Ground Squirrel *Spermophilus canus* Merriam, 1898
Merriam's Small-eared Shrew *Cryptotis merriami* Thomas, 1898
Merriam's Wapiti (Elk) *Cervus canadensis merriami* **Nelson,** 1902 extinct

Dr. Clinton Hart Merriam (1855–1942) was an American naturalist and physician. His father was a congressman, and through him

Merriam met Baird of the Smithsonian, in 1871, which led to his being invited to work as a naturalist the following year in Yellowstone as a member of the Hayden Geological Survey. This experience sparked his interest and guided his choice of further education. He studied biology and anatomy at Yale and finally graduated as a physician in 1879. He continued to pursue natural history as a hobby while he practiced medicine. In 1883 he forsook his profession for full-time scientific work. He became Chairman of the Bird Migration Committee of the American Ornithologists' Union. Under his chairmanship the AOU applied to Congress for funding to study birds, on the grounds that such work would benefit farmers. With the help of Senator Warner Miller of New York, Merriam's cousin and family friend, the application was successful and Congress granted the AOU sufficient funds. Merriam became the first Chief of the U.S. Biological Survey's Division of Economic Ornithology and Mammalogy. He is most famed for his "life zone" theory, which hypothesized that "temperature extremes were the principal desiderata in determining the geographic distribution of organisms." He is rather notorious in taxonomic history as a "splitter" and creator of species on the flimsiest of evidence. In 1918 he wrote a paper identifying 86 species of Brown Bear in North America; today, the Brown Bear is considered a single species. The chipmunk is found in California. The kangaroo rat occurs in the southern USA from California to Texas and in northern Mexico. The *Sorex* shrew is found in much of the western half of the USA, except in coastal regions. The pocket mouse inhabits southeast New Mexico, Texas, and northeast Mexico. The gopher is restricted to east-central Mexico. The deer mouse occurs in southern Arizona and northwest Mexico. The desert shrew is endemic to Mexico, while the ground squirrel inhabits an area in the western USA (eastern Oregon, western Idaho, and northwest Nevada). The small-eared shrew occurs in highlands from southern Mexico to northern Costa Rica. The wapiti was found in Arizona and New Mexico, but hunting and competition with livestock wiped this subspecies out by 1907.

Mertens

Flores Shrew *Suncus mertensi* Kock, 1974

Robert Friedrich Wilhelm Mertens (1894–1975) was a German zoologist and herpetologist who was born in St. Petersburg. He left Russia in 1912 to study medicine and natural history and obtained his doctorate from the University of Leipzig in 1915. After serving in the German army during WW1, he worked at the Senckenberg Museum in Frankfurt as an assistant from 1919 to 1920. He was then appointed to take Robert Sternfeld's place in charge of Herpetology (Sternfeld had been fired), and he became Curator in 1925 and Director in 1947. He was a man of prodigious energy. Except for his wife, he had no assistants from 1920 to 1943, and he was additionally in charge of Mammals from 1919 to 1953, of Birds from 1923 to 1947, and of Fish from 1920 to 1954. He was Chairman of the Zoology Section from 1934 until 1955. Finally, in 1960, he retired as Director Emeritus. He was a lecturer at the University of Robert Friedrich Frankfurt-am-Main from 1932 to 1939 and Professor from 1939 onward. Despite these incredible responsibilities and huge workload, he still found time to publish close to 800 scientific papers and 13 books. His first collecting trip was in 1913 to Tunisia, and during his time he visited 30 countries in search of specimens, including Indonesia in the early 1920s. In 1924 he published *Über einige Reptilien aus Borneo*. During WW2 he had most of the collection evacuated to small towns, where they were set up in locations like dance halls for use and study. Mertens also encouraged German soldiers fighting outside Germany to collect specimens for him, and a regular supply of reptiles and other creatures reached him courtesy of the German Field Post Office system. In Australia the Mertens Falls are named after him. He is remembered in the name of several reptiles such as Mertens' Monitor *Varanus mertensi*. Mertens died at the age of 81 as the result of a bite from a specimen of *Thelotornis kirtlandi*, a South African rear-fanged snake that he had long kept at home as a pet. At that time no antivenin existed for this species. He took 18 very painful days to die and kept a diary of each day's events, remarking in it, with true gallows humor, "für einen Herpetologen einzig angemessenes Ende" (a singularly appropriate end for a herpetologist). The shrew is endemic to Flores, Indonesia.

Meyen

Meyen's Dolphin *Stenella coeruleoalba* Meyen, 1833 [Alt. Striped Dolphin, **Gray**'s Dolphin]

Franz Julius Ferdinand Meyen (1804–1840) was a German surgeon who was also a botanist and a collector. It was he who first suggested that new cells were created through cell division rather than through the creation of new free cells. He published this theory in his 1830 work *Phytotomie*. He was a member of an expedition that was based on board *Prinzess Luise* and visited Peru and Bolivia between 1830 and 1832. Two birds are named after him in the binomial, the Lowland White-eye *Zosterops meyeni* and the Chilean Swallow *Tachycineta meyeni*. The dolphin has a worldwide distribution in temperate and tropical waters.

Meyer

Trefoil-toothed Giant Rat *Lenomys meyeri* **Jentink,** 1879

Dr. Adolf Bernard Meyer (1840–1911) was a German anthropologist and ornithologist who collected in the East Indies, New Guinea, and the Philippines. He was a Professor at the Anthropological and Ethnographic Museum of Dresden and became Director of the Natural History Museum there in 1872. He wrote *The Birds of the Celebes and Neighbouring Islands* (1898) and is cited as having made the first description of a number of bird species from the East Indies. It was he who first recognized that the red male and green female of the sexually dimorphic Australian King Parrot *Alisterus scapularis* were members of one species, not two. He was interested in the evolution debate and corresponded with Wallace. Several birds are named after him, including Meyer's Friarbird *Philemon meyeri*. The rat is found only in Sulawesi.

Mhorr

Mhorr's Gazelle *Gazella dama mhorr* **Bennett,** 1833 [Alt. Mhorr Gazelle]

"Mhorr's" or "Mohor's" (as this gazelle is also sometimes called) are incorrect terms; this antelope is not named after a person. "Mhorr" is a local North African name for the gazelle. The subspecies was declared to be extinct in the wild in 1968 but has been saved as a result of captive breeding programs. Reintroduction has started into its original habitat, in Senegal and southwest Morocco.

Michie

Michie's Tufted Deer *Elaphodus cephalophus michianus* **Swinhoe,** 1874

Alexander Michie (1833–1902) was a businessman, a Fellow of the Royal Geographical Society, and writer on China. He was Chairman of the Chamber of Commerce in Shanghai and in that capacity—accompanied by Swinhoe, the British Consul and describer of this mammal—conducted a mission into western China, and in particular into the province of Sichuan. He was the editor of the *Chinese Times*, an English-language newspaper published in the 1880s. He first went to China in 1854 and stayed for about 40 years, with occasional visits to the UK. In 1864 he published *The Siberian Overland Route*, an account of his adventures while tracing an old route from Peking (now Beijing) to St. Petersburg. He retired to England in 1895. He published much else, including *Missionaries in China* (1891). This subspecies of deer is from eastern China.

Micklem

Kataba Mole-Rat *Fukomys micklemi* Chubb, 1909 [formerly *Cryptomys damarensis micklemi*]

Thomas Nathaniel Micklem (dates not found). All we have managed to discover about him is that he fought in the Boer War as a trooper in the Rhodesia Regiment and in WW1 with the rank of Lieutenant in a South African Battalion of the Tank Corps. The etymology for the mole-rat reads, "The Rhodesia Museum is indebted to Mr. T. N. Micklem for a collection of small mammals made by him on the Upper Zambezi River between Sesheke and the junction of the Kabompo River with the Zambezi." The species (long regarded as a form of *Cryptomys damarensis*) is found in western Zambia.

Midas

Midas Tamarin *Saguinus midas* **Linnaeus,** 1758 [Alt. Lacepede's Tamarin, Red/Golden-handed Tamarin]

Midas Free-tailed Bat *Mops midas* **Sundevall,** 1843

King Midas of Phrygia is a mythological figure, but the legends may have been based on a real king of that area (now western Turkey) who lived in the eighth century B.C. Midas was reputed to be very fond of gold and was granted the gift of turning everything he touched into gold—often with unfortunate results. The golden-colored hands of the tamarin doubtless inspired Linnaeus to give the species this name. It is found in the Guianas and adjacent northern Brazil. There are occasional references to "Mida's Bat," but we are convinced these are examples of a wrongly placed apostrophe. The bat's name may arise from another legend about King Midas: after he offended Apollo, the god turned Midas's ears into those of a donkey (the bat too has very large ears). It is widespread in savannah and woodland regions of sub-Saharan Africa, and it also occurs in the Arabian Peninsula and Madagascar.

Middendorff

Middendorff's Vole *Microtus middendorffi* Poljakov, 1881

Kodiak Bear *Ursus arctos middendorffi* **Merriam,** 1896

Dr. Alexander Theodor (Aleksander Fedorovich) von Middendorf (1815–1894) was an Estonian of German extraction. A traveler and a naturalist, he was a member of the Imperial Academy of Sciences at St. Petersburg from 1845. He qualified as a physician in 1837. In 1840, together with Karl Maximilian von Baer, he traveled through the Kola Peninsula in far-northern Russia. From 1842 until 1845 he journeyed throughout Siberia and the surrounding regions on an Imperial Academy expedition. His accounts of the Amur River and other remote regions were the fullest by a naturalist and anthropologist of that time. He wrote on the spread of permafrost and how this affected the distribution of plants and animals. In 1870 he explored Barabinsk forest-steppe, and in 1878 the Fergana valley. Apart from these mammals and some birds, plants, and invertebrates, there is a cape on the island of Novaya Zemlya and a bay on the Taimyr Peninsula named after him. Among his observations of the people of the area, he wrote that in northern Siberia no one would remove anything from a sleigh left unattended, even if it contained much-needed food: "It is well known that the inhabitants of the far North are frequently on the verge of starvation, but to use any of the supplies left behind would be what we call a crime, and such a crime might bring all sorts of evil upon the tribe." He was also famed for his selection work on horse breeding and cattle farming. He is further commemorated in the name of Middendorff's Bean Goose *Anser fabalis middendorffi*. The vole is found in Siberia, from the Urals east to the Lena River. The bear comes from Kodiak Island, off the south coast of Alaska.

Millard

The soft-furred rat genus *Millardia* **Thomas,** 1911 [4 species]
Millard's Rat *Dacnomys millardi* Thomas, 1916

Dr. Walter Samuel Millard (1864–1952) was a British naturalist who was the Honorary Secretary of the Bombay Natural History Society and spent 35 years in that city. He was also a keen gardener and botanist. In 1937, with E. Blatter, he wrote *Some Beautiful Indian Trees.* Around 1906 he showed the young Salim Ali, India's most famed ornithologist, the Bombay Natural History Society's bird collection, and he often helped him with identification in his early years. The etymology in the original description of Millard's Rat says, "I have connected with this remarkable animal the name of Mr. W. S. Millard, to whose keenness, energy and generosity the Bombay Society's Survey so largely owes the great success it has attained." The rat is known from northeast India, eastern Nepal, southern Yunnan (China), and northern Laos. It probably has a wider range but is a little-known species.

Miller, G. S.

Miller's Long-tongued Bat *Glossophaga longirostris* G. S. Miller, 1898
Miller's Striped Mouse *Hybomys planifrons* G. S. Miller, 1900
Miller's Andaman Spiny Shrew *Crocidura andamanensis* G. S. Miller, 1902
Miller's Mastiff Bat *Molossus pretiosus* G. S. Miller, 1902
Miller's Myotis *Myotis milleri* **Elliot,** 1903
Miller's Water Shrew *Neomys anomalus* **Cabrera,** 1907 [Alt. Southern Water Shrew]
Miller's Hutia *Isolobodon levir* G. S. Miller, 1922 extinct
Miller's Nesophontes *Nesophontes hypomicrus* G. S. Miller, 1929 extinct
Miller's Grizzled Langur *Presbytis hosei canicrus* G. S. Miller, 1934
Carmen Mountain Shrew *Sorex milleri* Jackson, 1947

Gerrit Smith Miller Jr. (1869–1956) was Curator, Division of Mammals, U.S. National Museum, Washington, DC. He grew up in relative isolation on a country estate, and through the influence of his great uncle, who was an ornithologist, he developed an early interest in natural history. After graduating from Harvard in 1894 he joined the Biological Survey in the Department of Agriculture and worked under Clinton Hart Merriam. In 1898 he joined the U.S. National Museum as Assistant Curator of Mammals, becoming Curator in 1909, a position he held until 1940 when he retired. He remained an Associate in Biology at the Smithsonian Institution until his death. Among his major contributions to mammalogy were a series of checklists of North American mammals published in 1901, 1912, and 1924, as well as *The Families and Genera of Bats* (1907). He also wrote the *Catalogue of the Mammals of Western Europe in the Collection of the British Museum* (1912). He was an early critic of the claimed discovery of "Piltdown Man" in England. Miller was a regular correspondent with Oldfield Thomas. The long-tongued bat is from northern South America, including the islands of the Netherlands Antilles, Trinidad and Tobago, Grenada, and St. Vincent. The striped mouse comes from West Africa (Guinea to the Ivory Coast). The shrew is endemic to the Andaman Islands in the Bay of Bengal. The mastiff bat occurs from southern Mexico south to Colombia, Venezuela, and Guyana. The myotis, restricted to northern Baja California, Mexico, was once feared to be extinct but was then rediscovered; this taxon may be a subspecies of the Long-eared Myotis *Myotis evotis*. The water shrew is found from Portugal east to Ukraine and Turkey. The hutia and the nesophontes were both found on the Caribbean island of Hispaniola; it is now thought that *Isolobodon levir* was conspecific with *I. portoricensis* (Allen, 1916). The langur is found in eastern Borneo; the subspecies is critically endangered, and possibly even extinct, due to habitat loss. The shrew is endemic to the Sierra Madre Oriental Mountains of northeast Mexico.

Miller, L. E.

Miller's Monk Saki *Pithecia monachus milleri* **J. A. Allen**, 1914

Leo Edward Miller (1887–1952) was an American ornithologist, mammalogist, and explorer. He spent the years from 1910 to 1917 in South America, employed continuously by the American Museum of Natural History as a collector. He was with Frank Michler Chapman in Colombia in 1911 and 1912. He was a member of the Roosevelt Brazilian expedition of 1913–1914, and in 1916, jointly with J. A. Allen, he wrote "Mammals Collected on the Roosevelt Brazilian Expedition, with Field Notes by Leo E. Miller." He wrote of his South American experience in the book *In the Wilds of South America*, published in 1918, and in a number of other books including *The Hidden People: The Story of a Search for Incan Treasure* (1920) and *The Jungle Pirates* (1925). This subspecies of saki monkey was first collected by Miller in 1912, in southern Colombia where it is endemic.

Millet, C.

Millet's Shrew *Sorex coronatus* C. Millet, 1828 [Alt. Crowned Shrew]

C. Millet (dates not found) was a French naturalist whose main field of study was freshwater fauna. In 1831 he named a freshwater shrimp, *Atyaephyra desmarestii*, after Anselme Desmarest. He published at least two books on freshwater subjects: *Les poisons* (The Fishes) and *Les merveilles des fleuves et des ruisseaux* (Wonders of the Rivers and Brooks). The shrew is found in western Europe, from Germany and Switzerland south to northern Spain. It is also present on Jersey (Channel Islands). Millet may be the same person as the Charles Millet who is known to have been in Canton (now called Guangzhou) in China in 1830. A plant genus, *Milletia*, is named after this man. He was employed by the local branch of the Honourable East India Company and was actively engaged in collecting botanical specimens. He was a major correspondent for China for William Hooker at Kew.

Millet, F.

Millet's Giant Long-tailed Rat *Leopoldamys milleti* **Robinson** and **Kloss**, 1922

Fernand Millet (dates not found) was the Superintendent of Forests in Annam, in French colonial Vietnam. A professional hunter since 1902, he published in 1930 *Les grands animaux sauvages de l'Annam, leurs moeurs, leurs chasse, le tir*, a title that has been quaintly rendered as *The Large Savage Animals of Annam, Their Manners, Their Hunting, and Their Shooting.* He appears to have been responsible for quite considerable slaughter. He acted as a professional guide for important visitors, such as the French journalist Albert Londres, who wrote that by the time he met him in Indochina, Millet had already shot 47 tigers. Robinson and Kloss also named a bird after him, Millet's Laughingthrush *Garrulax milleti*. Kloss wrote that Millet had given him the (then) only known specimen of the long-tailed rat, and that "I am greatly indebted to Monsieur Millet for assistance and hospitality during my visit to the Langbian Plateau." The species appears to be endemic to Vietnam.

Milne-Edwards

Chinese Serow *Naemorhedus (sumatraensis) milneedwardsii* **David,** 1869
Milne-Edwards' Sifaka *Propithecus edwardsi* **Grandidier,** 1871
Milne-Edwards' Potto *Perodicticus potto edwardsi* **Bouvier,** 1879 [Alt. Cameroon Potto]
Milne-Edwards' Long-tailed Giant Rat *Leopoldamys edwardsi* **Thomas,** 1882
Milne-Edwards' Swamp Rat *Malacomys edwardsi* Rochebrune, 1885
Milne-Edwards' Long-clawed Mouse *Notiomys edwardsii* Thomas, 1890 [Alt Edwards' Long-clawed Mouse]
Milne-Edwards' Sportive Lemur *Lepilemur edwardsi* **Forbes,** 1894

Sir Alphonse Milne-Edwards (1835–1900) was a noted paleontologist who wrote *Histoire naturelle des oiseaux* in 1876. Alfred Grandidier contributed his knowledge of Malagasy birds to this book and edited a later edition in 1878. Milne-Edwards had a close working relationship with Prince Albert I of Monaco and may have been influential in encouraging the Prince to establish the Oceanographic Museum in Monte Carlo. He was sent specimens from China by Père David, the most notable of which was the Giant Panda. David named the panda *Ursus melanoleucus*, meaning "black-and-white bear." Milne-Edwards thought it wasn't a bear and changed the genus to *Ailuropoda* (cat-foot). It is now generally accepted that the Giant Panda is a bear after all. The Prix Alphonse Milne-Edwards was created in 1903 in his memory. The serow comes from central and southern China, Thailand, Laos, Cambodia, and Vietnam. The sifaka is found in the rainforests of eastern Madagascar. The potto subspecies inhabits west-central Africa. The giant rat is found in northeast India, northern Myanmar, southern and central China, and northern Thailand. The swamp rat is known from Liberia, Sierra Leone, Guinea, Ivory Coast, Ghana, and Nigeria. The long-clawed mouse comes from southern Argentina. The sportive lemur is confined to a small area of western Madagascar.

Miner

Miner's Cat *Bassariscus astutus* **Lichtenstein,** 1830 [Alt. Ringtail, Ringtail "Cat," Northern Cacomistle]

This mammal is not named after a person but after the occupation of miner. It is a member of the raccoon family, not a cat, despite some of its vernacular names. It is said to be easily tamed and to make an affectionate pet and excellent mouser, which is why it was a favorite in the mining camps of North America. It is found in the southwest USA and much of Mexico (except the far south).

Minna

Ethiopian Thicket Rat *Grammomys minnae* **Hutterer** and **Dieterlen,** 1984

Marlies Minna Raguschat is a German from East Prussia. Dr. Fritz Dieterlen, who kindly passed this information on to us, made her acquaintance when he was traveling in Sudan. She and her companion, Gerhard Nikolaus, whom she later married, captured the type specimen of the thicket rat in 1976. The species is endemic to southern Ethiopia.

Misonne

Misonne's Soft-furred Mouse *Praomys misonnei* Van der Straeten and **Dieterlen,** 1987

Xavier Misonne (d. 2007) was a Belgian zoologist and anthropologist. He has had a long and varied career that included working for World Health Organization, exploring mountains in Uganda, and staying with tribal peoples in central Africa and central Asia. He was in Nigeria in 1967 when he was head of the Department of Vertebrates at the Royal Institute of Natural Sciences of Belgium, of which he was Director from 1978 to 1988. He had also been a Professor at the University of Louvain and organized scientific expeditions. He wrote several books, including one on his travels through the deserts of the world, and a great many papers from the 1950s through to the 1990s, including *Les rongeurs du Ruwenzori et des régions voisines* (1963) and, with Hayman and Verheyen, *The Bats of the Congo and of Rwanda and Burundi* (1966). He retired to the south of France in 2005. The mouse comes from northern and eastern DRC (Zaire).

Miss Ryley

Miss Ryley's Soft-furred Rat *Millardia kathleenae* **Thomas,** 1914

Miss Kathleen Ryley (dates not found) was a naturalist. In 1913 we find her commenting in the *Journal of the Bombay Natural History Society* on the coloring of the hairs of the Slender Loris *Loris tardigradus*. Thomas writes in his original description of the soft-furred rat, "I have named the species in honour of Miss Kathleen Ryley, to whom the Survey is much indebted for the work she has done on its collections during the temporary absence of Mr. Wroughton." The "survey" Thomas mentions was the Mammal Survey of India, Burma, and Ceylon (now Sri Lanka), carried out by the Bombay Natural History Society between 1911 and 1914. The rat is found in Burma (now Myanmar).

Miss Waldron

Miss Waldron's Red Colobus *Piliocolobus badius waldroni* **Hayman,** 1936 extinct

Miss F. Waldron (dates not found) was an employee of the Natural History Museum in London. She accompanied Willoughby P. Lowe on his expedition to the Gold Coast (now Ghana) in 1934–1935. Lowe collected the type specimen of this monkey. Although the original spelling of the scientific name is *waldroni*, this is sometimes amended to *waldronae*, which is the correct form given that the taxon is named after a woman. The monkey was found in Ghana and eastern Ivory Coast, but it is now believed to be extinct.

Mitchell

Mitchell's Hopping Mouse *Notomys mitchelli* **Ogilby,** 1838

Lieutenant Colonel Sir Thomas Livingstone Mitchell (1792–1855) was a Scottish army surveyor and explorer. He was the Surveyor General of New South Wales from 1828 until 1855, and led various expeditions into eastern Australia between 1831 and 1836, and to tropical Australia from 1845 to 1846. A cockatoo, Major Mitchell's Cockatoo *Cacatua leadbeateri*, is named after him, and a very lifelike colored plate of it appears in Mitchell's *Three Expeditions into the Interior of Eastern Australia*, published in 1838. A town in Queensland is also named after him. The mouse is found in southern Australia, as far east as western Victoria.

Mittendorf

Mittendorf's Striped Grass Mouse *Lemniscomys mittendorfi* **Eisentraut,** 1968

Heiner Mittendorf is a German herpetologist who now lives in Namibia. He has made a special study of puff adders and keeps them in his house as "pets." He accompanied Eisentraut on a collecting expedition to Cameroon and Fernando Pó (now Bioko) between 1966 and 1967. The grass mouse is found in the region of Lake Oku, Cameroon.

Mittermeier

Mittermeier's Mouse-Lemur *Microcebus mittermeieri* Louis et al., 2006
Mittermeier's Sportive Lemur *Lepilemur mittermeieri* Rabarivola et al., 2006

Dr. Russell A. Mittermeier (b. 1949) is a primatologist, herpetologist, and conservationist who has also been President of Conservation International since 1989. Previously he worked, for 11 years, for the World Wildlife Fund. He has undertaken fieldwork for over 30 years in Asia, Africa, and South and Central America and has discovered several new primate species. His fieldwork has focused on protected areas and other conservation issues in Brazil, Suriname, Madagascar, and more than 20 other countries. His areas of expertise include biological diversity and its value to humanity, ecosystem conservation, tropical biology, and species conservation. He has been an Adjunct Professor at the State University of New York at Stony Brook since 1978. He received his B.A. in 1971 at Dartmouth and his Ph.D. from Harvard in 1977. He has published 10 books, including *Lemurs of Madagascar,* and more than 300 papers and popular articles on primates, reptiles, tropical forests, and biodiversity. Both lemurs are known only from small areas of northern Madagascar, the mouse-lemur in the northeast and the sportive lemur in the northwest.

Mo

Mo's Spiny Rat *Maxomys moi* **Robinson** and **Kloss,** 1922

There is no mention in the original description of this rat of any person named Mo—or anyone else. No reason for the scientific name is given. However, we believe it may be named after the Moi people who inhabit the mountains in the area where it was found. Today the term "Moi" is not often used, as it is actually a pejorative; it basically means "savages" and was used for any and all of the "primitive" highland tribes of Vietnam. The type specimen was taken at Arbre Broyé, in the Lang Bian Mountains of Vietnam, and it is also known from southern Laos. This is a species in need of a better common name.

Moellendorff

Calamian Tree-Shrew *Tupaia moellendorffi* **Matschie,** 1898
Culion Tree Squirrel *Sundasciurus moellendorffi* Matschie, 1898

Otto Franz von Möllendorff (1848–1903) was a German expert on living and fossil molluscs. He started his career in China, where he had gone in 1873 to learn to be an interpreter. His elder brother, Paul Georg von Möllendorff, had been in China since 1869, and the brothers appear not to have left China until 1882. In 1876 they co-authored *Manual of Chinese Bibliography,* which was published in Shanghai. Otto wrote a number of articles, including "On the Supposed New Zealand Species of Leptopoma" (1893) and "Nachrichtsblatt der Deutschen Malakozoologischen Gesellschaft" (*Journal of Malacology,* 1903). He is also commemorated in the name of a snake, Moellendorff's Ratsnake *Elaphe moellendorffi.* Both mammals come from the Calamian Islands, western Philippines.

Molina, A.

Molina's Hog-nosed Skunk *Conepatus chinga* A. Molina, 1782

Professor Abbot Juan Giovanni Ignazio (Ignacio) Molina (1740–1829) was a Chilean naturalist who wrote *Saggio sulla storia naturale del Chili,* which was first published in Bologna in 1782. He studied languages and natural history at a Jesuit college and was appointed Librarian at the college after becoming a member of that order. When Jesuits were banned in Chile in 1768 he left for Italy, where in 1774 he was appointed Professor of Natural History in Bologna. All his natural history notes were lost during his move to Italy, and he later rewrote what he could from memory. The grass genus *Molinia* is also named for him, and he is commemorated in the scientific name of a bird, the Green-cheeked Conure *Pyrrhura molinae*, and a number of plants such as the Chilean Guava *Ugni molinae*. The skunk occurs from Peru and southern Brazil to northern Argentina and Chile.

Molina, O.

Molina's Grass Mouse *Akodon molinae* **Contreras**, 1968

We have no direct confirmation, but we believe this mouse was named after Omar José Molina (dates not found). He worked as the Chief Technician in the laboratory for the preparation of specimens at the La Plata Museum of Natural Sciences in Argentina, where his overall boss was Rosendo Pascual. Contreras, who described the grass mouse, also named a tuco-tuco after Pascual. In 1971 Molina was one of a quartet of authors (the others being N. O. Bianchi, O. A. Reig, and F. N. Dulout) of a paper entitled "Cytogenetics of the South American Akodont Rodents." He has also been a member of the Scientific Research Commission of the Province of Buenos Aires, Argentina. The grass mouse is found in east-central Argentina.

Moloch

Moloch Gibbon *Hylobates moloch* Audebert, 1797 [Alt. Silvery Gibbon, Javan Gibbon]

Red-bellied Titi *Callicebus moloch* Hoffmannsegg, 1807

Moloch was the Canaanite sun-god, to whom children were sacrificed. Later Christians viewed him as a demon, or "fallen angel," and he appears as such in Milton's *Paradise Lost*. Some early zoologists seem to have thought that apes and monkeys looked demonic. The gibbon is endemic to Java, and the titi monkey is found in Brazil south of the Amazon.

Moloney, C.

Moloney's Flat-headed Bat *Mimetillus moloneyi* **Thomas**, 1891

Sir Cornelius Alfred Moloney (1848–1913) was a British civil servant. He was Administrator of the Gambia from 1884 until 1886 and Governor of Lagos, Nigeria, from 1886 until 1889 and from 1890 until 1891. From 1891 to 1897 he was Lieutenant Governor of British Honduras (now Belize). From 1900 to 1904 he was Governor of Trinidad and Tobago. He wrote *Forest in West Africa*, which initiated a forestry policy in Nigeria. He is also commemorated in the name of a bird, Moloney's Illadopsis *Illadopsis fulvescens moloneyana*. The bat is found from Sierra Leone east to Ethiopia and south to Angola and Zambia.

Moloney, J.

Moloney's Monkey *Cercopithecus albogularis moloneyi* **Sclater**, 1893

Dr. Joseph Augustus Moloney (1857–1896) was, as Sclater put it, "one of the surviving members of Stairs's Expedition to Katanga." He was a valiant Irish-born explorer and pioneer who, it was said, saw more of Africa than any living man but Stanley. In 1893 he published an account of the expedition, *With Captain Stairs to Katanga*. Their object was to secure the country for the Congo Free State (i.e. for Belgium) and to "free" it from, as expressed at the time, the tyranny of Chief Msiri. The expedition was also staged before the British South Africa Company could claim the same territory. Dr. Moloney took command of the expedition when all senior officers were dead or dying, building a fort that enabled Captain Bea afterward to firmly establish control. In his last year Moloney commanded an English expedition to negotiate treaties with local chiefs west of Lake Nyassa, accompanied by Lieutenants

Elwes and Biscoe, Royal Navy, and seven other Englishmen. Not a shot was fired during the whole journey, and four powerful chiefs were placed under the British flag. The notorious Chief Mpseni, due to Portuguese influence, declined to accept the British flag but showed a remarkably friendly disposition to Dr. Moloney, who stayed two days with the much-dreaded leader. Moloney died at the age of 38 due to the after effects of privation suffered during the African expeditions. The monkey occurs in the Poroto Mountains of southern Tanzania, in northernmost Malawi, and in Zambia east of the Luangwa River.

Monard

Monard's Climbing Mouse *Dendromus leucostomus* Monard, 1933
Monard's Dormouse *Graphiurus monardi* St. Leger, 1936

Dr. Albert Monard (1886–1952) was a Swiss zoologist. He made a number of collecting expeditions to Africa for the Natural History Museum of La Chaux-de-Fonds: to Angola in 1928–1929 and 1932–1933, to Portuguese Guinea (Guinea Bissau) in 1937–1938, and to Cameroon in 1946–1947. He wrote a number of papers in the 1930s, including descriptions of new species. The climbing mouse is known only from specimens collected by Monard in Angola. The dormouse is found in northeast Angola, western Zambia, and southern DRC (Zaire).

Monckton

Monckton's Melomys *Paramelomys moncktoni* **Thomas,** 1904 [Alt. Monckton's Mosaic-tailed Rat]
Earless Water Rat *Crossomys moncktoni* Thomas, 1907

Captain Charles Arthur Whitmore Monckton (1872–1936) was probably a New Zealander, as we find him at Wanganui Collegiate School in 1886. He was a Resident Magistrate in Papua New Guinea from around or before 1897 to 1907, when he resigned in protest at having been officially criticized for his extremely harsh and brutal methods in keeping order. In 1904 he founded the town of Kokoda, the start of the Kokoda Trail over which Australian and Japanese forces fought long and hard in the campaign in New Guinea during WW2. Besides being a magistrate, he was also an explorer, making an ascent of Mount Albert Edward in 1906. He was very interested in natural history and was a Fellow of both the Zoological Society of London and the Royal Geographical Society. He wrote a number of books about his time and experiences in Papua New Guinea, including *Some Experiences of a New Guinea Resident Magistrate* (1922). In 1904 Oldfield Thomas published an article "On Some Mammals from British New Guinea Presented to the National Museum by Mr. C. A. W. Monckton, with Descriptions of Other Species from the Same Region." The melomys and water rat are both endemic to Papua New Guinea.

Mondolfi

Mondolfi's Four-eyed Opossum *Philander mondolfii* Law, Perez-Hernandez, and Ventura, 2006

Professor Dr. Edgardo Mondolfi (1918–1999) was a Venezuelan mammalogist and all-round ecologist and biologist. He was a Professor at Metropolitan University, Venezuela. The rather fulsome etymology given in the original description reads, "The specific epithet of this new form is dedicated to the memory of Edgardo Mondolfi (1918–1999), outstanding Venezuelan mammalogist who devoted his professional life to the study of wildlife, pioneered conservationist activism in Venezuela, and who was a living source of respect, solidarity, and joy for all who knew him." He wrote many articles and books including, with Rafael Hoogesteijn, *The Jaguar*, published in 1993. He had a lifelong interest in manatees and made studies of their distribution. He was on the board of several conservation bodies and, in later life, was Venezuela's Ambassador to Kenya. He died from dengue fever. The opossum is found in Venezuela and northeast Colombia.

Montebello

Japanese Grass Vole *Microtus montebelli* **Milne-Edwards,** 1872

Louis Gustave Lannes, Marquis de Montebello (1838–1907), was a French diplomat. His grandfather was Jean Lannes, a Marshall of France whom Emperor Napoleon I made Duc de Montebello. In 1886 he became French Ambassador in Constantinople, and in 1891 he was transferred as Ambassador to the Imperial Court at St. Petersburg. He became a great friend of Tsar Nicholas II and did much to organize the Franco-Russian alliance. He retired in 1902. Milne-Edwards says in his description that de Montebello collected the type specimen on Mount Fuji and so he named the vole after him. The vole is found on the Japanese islands of Honshu, Sado, and Kyushu.

Monteiro

Silvery Greater Galago *Otolemur monteiri* Bartlett, 1863

L. A. Monteiro (dates not found). Bartlett says that the type specimen was a living animal kept at Monteiro's home and sent to England from Angola by his son, Joachim John Monteiro (1833–1878). The latter was a mining engineer and naturalist who has several birds named after him, including Monteiro's Hornbill *Tockus monteiri*. The galago is found from Angola east to Malawi and Tanzania, extending north to Rwanda and western Kenya.

Moojen

Moojen's Atlantic Tree Rat *Phyllomys kerri* Moojen, 1950 [Alt. Kerr's Tree Rat]
Moojen's Spiny Rat *Trinomys moojeni* Pessôa, Oliveira, and Reis, 1992
Moojen's Pygmy Rice Rat *Oligoryzomys moojeni* Weksler and Bonvicino, 2005 [Alt. Moojen's Colilargo]

Dr. João Moojen de Oliveira (1904–1985) was a Brazilian zoologist who collected extensively in the 1930s, 1940s, and 1950s. He was responsible for assembling a large part of the collection of mammals held by the University of Rio de Janeiro and the National Museum there, where he was Curator of Mammals. In 1959 he designed and oversaw the construction of the new zoo at Brasília, when the new capital was under construction. He described a number of new mammals and was an expert on *Phyllomys* species. He wrote what is regarded as the classic work on Brazilian rodents in 1952, *Brazilian Rodents: Their Habitats and Habits*. He is also commemorated in the binomial of a venomous snake, *Bothrops moojeni*. All three rodents are found in eastern Brazil.

Moore

Moore's Woolly Lemur *Avahi mooreorum* Lei et al., 2008

This primate's name honors members of the Moore family—Gordon and Betty Moore, Ken and Kris Moore, and Steve and Kathleen Moore—in recognition of their long-term commitment to biodiversity and conservation. The support they have provided through the Gordon and Betty Moore Family Foundation has been critical to advancing conservation in some of the world's most important biodiversity hotspots, including Madagascar. The lemur is known only from the Masoala National Park, eastern Madagascar.

Moreno, A.

Western Long-tongued Bat *Glossophaga morenoi* Martinez and **Villa,** 1938

Alfredo Moreno (dates not found) collected the type specimen of this bat. We have been unable to find out anything more about him. The bat is found in southwest Mexico.

Moreno, F. J. P.

Monte Gerbil-Mouse *Eligmodontia moreni* **Thomas,** 1896

Dr. Francisco Josue Pascasio Moreno (1852–1919) was an Argentine zoologist and paleontologist. He traveled widely, exploring little-known parts of Argentina, from 1875 to 1879. He was Director of the Museum at La Plata, which he founded in 1894 and ran until 1905. In 1908 he founded the Argentine Boy Scouts as well as a number of schools and children's homes. Thomas wrote that the type specimen of the gerbil-mouse was "collected and presented by Dr. F. P. Moreno, the distinguished head of the La Plata Museum, in whose honour I have named this very pretty little mouse." The species is found in northwest Argentina.

Morgan

Morgan's Gerbil-Mouse *Eligmodontia morgani* **J. A. Allen,** 1901

John Pierpont Morgan (1837–1913) was born into a wealthy family. He was educated at Boston and at the University of Göttingen in Germany, from where he graduated in 1857. He started work as an accountant in New York, working for a number of banks until he joined his father's firm on the outbreak of the American Civil War. He eventually reorganized his family's holdings and through mergers and acquisitions created the firm of J. P. Morgan and Co., one of the most powerful banking companies in the world. He gathered in ailing railway companies and in 1900 owned 8,000 km (5,000 miles) of railway track. By the early years of the 20th century he was the major force behind the trusts that controlled virtually all of America's basic industries. He invested abroad and in 1902 bought the White Star Line—and so was the real owner of the ill-fated *Titanic.* It is said that he was scheduled to travel on its maiden voyage but canceled at the last minute. He was immensely interested in art and amassed a huge collection, most of which he left in his will to the Metropolitan Museum of Art in New York. His personal wealth was very great, but he was a generous man in that he gave money to charities, schools, hospitals, churches, and, as it turns out, zoological expeditions, as the wording in the original description of the gerbil-mouse shows: "This species is based on a large series of specimens collected at or near Cape Fairweather, Patagonia, by Mr. A. E. Colburn, for the Princeton Patagonian Expedition, generously supported by Mr. J. Pierrepont Morgan, after whom the species is named." (The spelling "Pierrepont" is apparently an error). The gerbil-mouse is found in Patagonian Argentina and adjacent parts of Chile.

Morris, P. A.

Morris' Bat *Myotis morrisi* **Hill,** 1971

Patrick A. Morris was a senior lecturer at the School of Biological Sciences, Royal Holloway College, University of London, and an acknowledged expert in wildlife ecology. In 1968 he was one of the zoologists on the Great Abbai expedition, and after it he wrote, with J. E. Hill, who described the bat, *Bats from Ethiopia Collected by the Great Abbai Expedition, 1968.* Another zoologist on that expedition was D. W. Yalden, and Morris has written a number of articles with him, including "The Analysis of Owl Pellets" and, in 1975, "Lives of Bats." Morris helped in the preparation of the "Catalogue of the Mammals of Ethiopia" and made a number of collecting expeditions to Ethiopia in connection with it. The bat was first found in Ethiopia and has also been recorded from Nigeria; full details of its distribution remain unknown.

Morris, R. C.

Morris' Flying Squirrel *Olisthomys morrisi* Carter, 1942 [Syn. *Petinomys setosus* **Temminck,** 1844]

Colonel Randolph Camroux Morris (1894–1977) was a Scot born in India who became a British army officer and a hunter-naturalist. He grew up on a coffee plantation created by his father, Randolph Hayton Morris, who was gored by a Gaur (Indian wild ox) and died of complications following the injury. He was an authority on big game in India and a member of the Vernay-Hopwood Chindwin expedition of the American Museum of Natural History during 1934 and 1935. Carter, who named the squirrel, also took part in this expedition. Morris published on natural history prolifically, especially in the *Journal of the Bombay Natural History Society,* ranging from an article on "Panthers and Artificial Light" in 1922 to "Rat Snake Mating" in 1958. He also entertained all the great Indian naturalists of his day, including Salim Ali, as well as the great and good such as the Maharaja of Mysore. The squirrel was discovered in northern Burma (now Myanmar), but most authorities do not recognize it as a distinct species, believing it to be synonymous with Temminck's Flying Squirrel *Petinomys setosus.*

Mrs. Gray

Mrs. Gray's Lechwe *Kobus megaceros* Fitzinger, 1855 [Alt. Nile Lechwe]

Maria E. Gray (1787–1876) was the wife of the British Museum Curator J. E. Gray. She was a good artist and spent much time making scientific drawings, particularly of molluscs. Gray originally named the lechwe in 1859 as *Kobus maria,* but Fitzinger's earlier name has precedence. This antelope is found in southern Sudan and a small area of western Ethiopia.

Mrs. Millard

Mrs. Millard's Flying Squirrel *Petaurista sybilla* **Thomas** and **Wroughton,** 1916

Mrs. Millard (we presume Sybil to be her forename, as acknowledged in the binomial) was the wife of Dr. Walter **Millard** (see above). This species is sometimes regarded as a race of *Petaurista elegans,* but other authorities rank it as a full species. The original description reads, "This beautiful Flying Squirrel is named in honour of Mrs. Millard, wife of the Society's Honorary Secretary" (i.e. the Bombay Natural History Society). The squirrel is found in the mountains of Myanmar and adjacent western Yunnan (China).

Muennink

Muennink's Spiny Rat *Tokudaia muenninki* **Johnson,** 1946

Odis Alfred Muennink (b. 1920) was working as a collector for the U.S. Fish Commission in Japan in 1945, at which time David H. Johnson, who described the spiny rat, collected the specimen with him. The original etymology reads, "named for Odis A. Muennink of Hondo, Texas, who during the recent war collected more than a thousand specimens of animals for Naval Medical Research Unit No. 2 in various parts of the Pacific area." During WW2 he was a Pharmacist's Mate in the U.S. Navy and is especially thanked by Lieutenant Rollin H. Baker for his help in preparing "A Study of Rodent Populations on Guam, Mariana Islands," published in 1946. The species is confined to Okinawa Island.

Müller

Brown Dorcopsis *Dorcopsis muelleri* **Lesson,** 1827
Müller's Gibbon *Hylobates muelleri* **Martin,** 1841 [Alt. Borneo Grey Gibbon]
Müller's Giant Sunda Rat *Sundamys muelleri* **Jentink,** 1879

Dr. Salomon Müller (1804–1864) was a Dutch naturalist who collected in Indonesia in 1826, where he worked under Schlegel as a taxidermist assisting members of the Netherlands Natural Sciences Commission. He went on to New Guinea, and in 1829 he explored the interior of Timor. In 1831 he collected in Java, and between 1833 and 1835 he explored western Sumatra. He is also commemorated in the common names of several birds, including Müller's Barbet *Megalaima oorti*. The dorcopsis (wallaby) is found in western New Guinea, including some offshore islands such as Misool, Salawatti, and Japen. The gibbon is endemic to Borneo. The rat occurs on the Malay Peninsula, Sumatra, Borneo, and the western Philippines (Palawan, Balabac, Culion, and Busuanga).

Musschenbroek

Sulawesi Palm Civet *Macrogalidia musschenbroekii* **Schlegel,** 1877
Musschenbroek's Spiny Rat *Maxomys musschenbroekii* **Jentink,** 1878

Samuel Cornelius Jan Willem van Musschenbroek (1827–1883) graduated as a lawyer and also trained as a seaman, passing his First Mate's examination. He then served as a Dutch colonial administrator in the East Indies from 1855 to 1876. He was Resident of Ternate in 1873 and of Menado in 1875. He traveled extensively in the Moluccas with Beccari. In 1879 he was in Leiden and was appointed to be the first Director of the Museum der Koloniale Vereniging (Colonial Museum). Many bird skins that he had collected in Indonesia were exhibited here before they were transferred by Voous to the Zoological Museum of Amsterdam. Musschenbroek was regarded as an expert on large parts of the Dutch East Indies, as is shown by his pioneering maps of Minahassa and the northern parts of the Moluccas. He also has a bird, Musschenbroek's Lorikeet *Neopsittacus musschenbroekii*, named for him. The Artis Library, University of Amsterdam, advised us that he wrote *Iets over de fauna van Noord-Celebes* (1867) and *Kaart van de bocht van Tomini* (1880). The civet and the spiny rat are both found only on Sulawesi (Indonesia).

Musser

Marie's Vole *Volemys musseri* **Lawrence,** 1982
Musser's Shrew-Mouse *Microhydromys musseri* **Flannery,** 1989 [Alt. Torricelli Shrew-Mouse]
Mossy Forest Shrew *Crocidura musseri* Ruedi and Vogel, 1995
Sierra Madre Shrew-Mouse *Archboldomys musseri* **Rickart** et al., 1999
Musser's Bristly Mouse *Neacomys musseri* **Patton,** da Silva, and Malcolm, 2000 [Alt. Manu Bristly Mouse]

Professor Dr. Guy G. Musser works as Archbold Curator for the Department of Mammalogy, American Museum of Natural History, New York. According to the museum, "Dr. Musser researches the geographic origins of murine rodents in Asia by examining the phylogenetic relationships between the rats and mice found on that continent and their Indo-Australian cousins." Musser took his first degree in 1959 and his master's in 1961, both at the University of Utah. Having obtained his doctorate at the University of Michigan in 1967, he began his studies of the American Museum of Natural History collection of *Muridae*. He then spent three years living in the rainforests of Sulawesi, where he collected various samples and specimens including 10 species previously undescribed. He contributed the Muridae section of *Mammal Species of the World*. He has written at least one book on *Philippine Rodents*, published in 1992, and numerous scientific papers and contributions to larger works. He is editor of the *Journal of Mammalogy* and an Adjunct Professor to graduate students at the City University of New York. The vole comes from Sichuan (China). The *Microhydromys* mouse is found in the Torricelli Mountains of Papua New Guinea. The shrew occurs in the mountains of Sulawesi. The *Archboldomys* mouse is endemic to northern Luzon (Philippines). The bristly mouse is found in the Manu National Park (Cuzco, Peru) and adjacent westernmost Brazil.

Musso

Musso's Fish-eating Rat *Neusticomys mussoi* Ochoa and **Soriano,** 1991

Andres Musso (d. before 1991) was a pioneering Venezuelan mammalogist. He spent much of his life teaching and encouraging many young and aspiring Venezuelan zoologists. He published *Lista de los mamiferos conocidos de la isle de Margarita* in 1962. This little-known rat comes from western Venezuela.

Muton

Muton's Soft-furred Mouse *Praomys mutoni* Van der Straeten and **Dudu** 1990 [Alt. Riverine Soft-furred Mouse]

This rodent is not named after a person. The original description says, "Mutoni is a Swahili word for riverbank or wet area (mto or muto = river). It refers to the typical biotope where the specimens were trapped." It should thus more correctly be called the Riverine Soft-furred Mouse. The species was discovered in northern DRC (Zaire).

Myers

Myers' Grass Mouse *Akodon philipmyersi* Pardiñas et al., 2005

Dr. Philip Myers (b. 1947) is Associate Professor and Associate Curator of Mammals at the Museum of Zoology at the University of Michigan. He is also the founder and Director of the Animal Diversity Web. He has published a number of papers on the systematics of *Akodon* and other genera of South American rodents. He has undertaken fieldwork in Nicaragua, Costa Rica, Mexico, Ecuador, Suriname, Peru, Paraguay, Nepal, and Pakistan. In 2007 he was conducting research on the effect of climate change on the structure of small-mammal communities in the northern Great Lakes region. The grass mouse was named after Myers for his contributions to the understanding of *Akodon* systematics and ecology. The species is found in Misiones Province, northern Argentina.

Nagtglas

Nagtglas' Dormouse *Graphiurus nagtglasii* **Jentink,** 1888

Colonel Cornelis Johannes Marius Nagtglas (1814–1897) was twice—from 1857 to 1862 and from 1869 to 1871—Governor of the Dutch settlement at Elmina on the coast of "Guinea" (now Ghana), which traded with the Ashanti in the interior of the Gold Coast (also now Ghana). The Dutch settlement was sold to Great Britain in 1871 and incorporated into the Gold Coast Colony. Nagtglas sent a number of natural history specimens to the museum at Leiden. The dormouse is found in West Africa, from Sierra Leone to Gabon and the Central African Republic. Until recently this species was known as Huet's Dormouse *Graphiurus hueti,* which is an earlier name and would normally have priority. However, for various reasons—including the lack of a type specimen for *Graphiurus hueti*—many authorities now prefer to use *G. nagtglasii* as the valid name for this dormouse. See also **Huet.**

Nancy Ma

Nancy Ma's Night Monkey *Aotus nancymaae* **Hershkovitz,** 1983 [Alt. Ma's Owl Monkey, Nancy Ma's Douroucouli]

Dr. Nancy Shui-Fong Ma was Assistant Professor of Comparative Pathology at the New England Regional Primate Research Center, Harvard Medical School. She became involved in mapping genetics of nonhuman primates and worked principally on the genus *Aotus* (owl or night monkeys) in association with the Southwest Foundation for Biomedical Research in San Antonio, Texas, during the 1970s and 1980s. The monkey is found in a small region of the Peru-Brazil border, mainly south of the Amazon.

Nasarov

Nasarov's Pine Vole *Microtus nasarovi* Shidlovsky, 1938 [Alt. Nazarov's Vole; Syn. *Terricola nasarovi*]

Pavel Stepanovich Nazarov (or Nazaroff) (dates not found) was a Russian zoologist, naturalist, and geologist. He graduated from Moscow University but was living in Tashkent at the time of the Russian Revolution. A Tsarist, he fought against the Bolsheviks and fled from Stalin's secret police, first into central Asia, where he lived among the Kazakhs, then to China and finally to the West. He had been accused of spying for the British government and had been sentenced to death *in absentia.* In 1886 he published an article on marmots called "Recherches zoologiques dans les steppes des Kirghizes." He wrote at least two volumes of memoirs that appeared in English translations as *Hunted through Central Asia* (1930) and *Moved On! From Kashgar to Kashmir* (1935). He also wrote a book on the Saiga Antelope *Saiga tatarica* in 1932. The vole is found in the northeast Caucasus. It is sometimes regarded as being conspecific with the Caucasus Pine Vole *Microtus daghestanicus.*

Nasolo

Nasolo's Shrew-Tenrec *Microgale nasoloi* **Jenkins** and **Goodman,** 1999

Nasolo H. N. Rakotoarison (d. 1996) was a Malagasy zoologist who published extensively on lemurs. He was Curator of Mammals at the Botanical and Zoological Park of Tsimbazaza, Antananarivo, Madagascar. He was killed in a car accident. The shrew-tenrec is endemic to a small area of southwest Madagascar.

Nathusius

Nathusius' Pipistrelle *Pipistrellus nathusii* Keyserling and **Blasius,** 1839

Professor Hermann Engelhard von Nathusius (1809–1879) was a German zoologist and agriculturist. He studied in Berlin under Johannes Müller between 1827 and 1829. He became a civil servant and from 1869 was the senior adviser in the Prussian Ministry of Agriculture. He wrote a number of books on zoological and botanical subjects, most notably a book on pigs, *Über die Rassen des Schweines* (1860). The German Society for Animal Science awards the Hermann von Nathusius Medal, named in his honor. The bat is found from Western Europe, including Britain, east to the Urals and the Caucasus.

Nation

Andean Hairy Armadillo *Chaetophractus nationi* **Thomas,** 1894

Professor William Nation (1826–1907) started work at the age of 14 at Kew, where he received botanical training. He went to Peru as a collector in 1862 and stayed and taught in that country until 1880. Strangely, he was a Professor of languages, not of botany or any other scientific discipline. He was, like so many of his contemporaries, one of Charles Darwin's correspondents. The armadillo is found in Bolivia.

Natterer

Natterer's Bat *Myotis nattereri* **Kuhl,** 1817
Natterer's Tuco-tuco *Ctenomys nattereri* **Wagner,** 1848

Dr. Johann Natterer (1787–1843) was an Austrian naturalist and collector. He studied botany, zoology, mineralogy, chemistry, and anatomy and was appointed as a taxidermist to what is now the Natural History Museum in Vienna. In 1817 he, Spix, and others took part in the expedition to Brazil that started on the occasion of Archduchess Leopoldina's wedding to Dom Pedro, the Brazilian Crown Prince. The entire suite traveled in two Austrian frigates, *Austria* and *Principesse Augusta.* During 1818 and 1819 Natterer explored a potential river route to Paraguay. He continued to explore from 1821 to 1835, during which time he went on five expeditions, exploring the Mato Grosso and the Amazon basin. He returned to Vienna with an accumulated collection of specimens in 37 crates, which he deposited with the Museum of Vienna—a staggering collection of 12,293 birds and about 24,000 insects. They can still be seen there today. He lost the majority of his possessions in the civil war then being waged in Brazil, so one wonders how big his total collection actually was. He ended his career at the Austrian Imperial Museum of Natural History. He died of a lung ailment. He did not publish an account of his travels, and unfortunately for posterity, his notebooks and diary were destroyed by fire in 1848. The result of this was that he never received the credit in Austria that he deserved, but abroad he was held in high esteem, for instance being made an honorary Doctor of Philosophy at Heidelberg University. He is also commemorated in the scientific names of other taxa, including birds such as the Blue Cotinga *Cotinga nattererii* and fish like the Red-bellied Piranha *Pygocentrus nattereri.* The bat is found in Europe (except Scandinavia), northwest Africa, and western Asia east to the Caucasus, Iraq, and Turkmenistan. The tuco-tuco is found only in Mato Grosso, Brazil.

Neave

Neave's Mouse *Mus neavei* **Thomas,** 1910

Dr. Sheffield Airey Neave (1879–1961) was a British entomologist who wrote *The History of the Entomological Society of London, 1833–1933.* He was Secretary of the Zoological Society of

London, for which he edited *Nomenclator zoologicus,* published in 1939, and was employed by the Natural History Museum in London. He traveled extensively in central Africa and collected invertebrates, reptiles, batrachians, and fishes in Northern Rhodesia (now Zambia) in 1911. (His son, Airey Middleton Sheffield Neave, 1916–1979, was a British Conservative Member of Parliament, killed when a car bomb placed by Irish Nationalists exploded under his vehicle as he drove out of the Palace of Westminster parking lot. During WW2 he was the first British officer to successfully escape from the German POW camp at Colditz.) The mouse occurs from southern DRC (Zaire) and southern Tanzania to Zimbabwe and Transvaal (South Africa).

Nehring

Nehring's Blind Mole-Rat *Nannospalax nehringi* **Satunin,** 1898

Carl Wilhelm Alfred Nehring (1845–1904) was a German zoologist who wrote extensively on paleontology. He specialized in the Middle East region, naming various mammals of that area including a subspecies of Stone Marten *Martes foina syriaca,* a subspecies of Fat Dormouse *Glis glis orientalis,* and the Palestine Mole-Rat *Nannospalax ehrenbergi* (after Christian Gottfried Ehrenberg). The *nehringi* mole-rat is found in Turkey, Armenia, and Georgia.

Neill

Neill's Long-tailed Giant Rat *Leopoldamys neilli* **Marshall,** 1976

William A. Neill collected the type specimen of this rat in 1973, while studying bats at Kaengkhoi Cave in Thailand. Seeking to understand the total ecology of the cave, he trapped specimens of other fauna near the cave entrance. This led to the discovery of not one but two new rodent species, the above species and *Rattus hinpoon.* The giant rat is known only from the provinces of Saraburi and Kanchanaburi, Thailand.

Nelson

Magdalena Rat *Xenomys nelsoni* **Merriam,** 1892
Nelson's Antelope Squirrel *Ammospermophilus nelsoni* Merriam, 1893
Nelson's Pocket Mouse *Chaetodipus nelsoni* Merriam, 1894
Nelson's Small-eared Shrew *Cryptotis nelsoni* Merriam, 1895
Desert Bighorn Sheep *Ovis canadensis nelsoni* Merriam, 1897
Nelson's Giant Deer Mouse *Megadontomys nelsoni* Merriam, 1898
Nelson's Rice Rat *Oryzomys nelsoni* Merriam, 1898 extinct
Nelson's Collared Lemming *Dicrostonyx (groenlandicus) nelsoni* Merriam, 1900
Nelson's Coati *Nasua (narica) nelsoni* Merriam, 1901 [Alt. Cozumel Island Coati]
Nelson's Spiny Pocket Mouse *Heteromys nelsoni* Merriam, 1902
Nelson and Goldman's Woodrat *Nelsonia goldmani* Merriam, 1903
Nelson's Woodrat *Neotoma nelsoni* Goldman, 1905
Nelson's Kangaroo Rat *Dipodomys nelsoni* Merriam, 1907
Mexican Grizzly Bear *Ursus arctos nelsoni* Merriam, 1914 extinct
Nelson's Ocelot *Leopardus pardalis nelsoni* Goldman, 1925

Dr. Edward William Nelson (1855–1934) was the Chief of the U.S. Biological Survey and the founding President of the American Ornithologists' Union. Perhaps his greatest achievement was the creation of the Migratory Bird Treaty, which is still in force today. In June 1881 he was on board the *Corwin* during its search for the missing Arctic exploration vessel *Jeanette;* this expedition was the first to reach and explore Wrangel Island. In 1890 Nelson became a Special Field Agent with the Death Valley expedition under Clinton Hart Merriam (as the list above shows, most of the mammals named after Nelson were named by Merriam). After the expedition concluded, Nelson did prolonged fieldwork in Mexico. He is also commemorated in the scientific names of other taxa, such as the Spotted Box Turtle *Terrapene nelsoni* and the Dwarf Vireo *Vireo nelsoni.* Most of the mammals listed above come from Mexico, though the antelope squirrel is found in the San Joaquin Valley of California, and the lemming comes from western Alaska. The bighorn sheep occurs in Nevada, Arizona, southern California, and northern Baja California (Mexico). Of the Mexican species, the rice rat occurred only on Maria Madre Island off the west coast. Introduced Black Rats *Rattus rattus* may have caused its extinction not long after the species was discovered. The Mexican race of grizzly bear was wiped out by direct human persecution; the last few seem to have been killed during the 1960s.

Nereid

Nereid Horseshoe Bat *Rhinolophus nereis* **K. Andersen,** 1905 [Alt. Anamban Horseshoe Bat]

The Nereids are figures from Greek mythology. They were sea-nymphs, the 50 daughters of the sea-god Nereus and his wife, Doris. The best known of them are Amphitrite, Galatea, and Thetis, but the curious will find the names of all 50 in Edmund Spenser's *The Faerie Queen,* 4.11.48–51. The bat is found on the Anamba and North Natuna Islands (Indonesia), east of peninsular Malaysia. Presumably the name *nereis* was applied fancifully because of the bat's habitat on small islands surrounded by the sea.

Netscher

Sumatran Rabbit *Nesolagus netscheri* **Schlegel,** 1880 [Alt. Sumatran Striped Rabbit]

Eliza Netscher (1825–1880) was the Secretary General in Batavia (now Jakarta) for the Dutch authority and a member of the Council of the Dutch East Indies, as well as being an amateur naturalist. He first went to Batavia as a clerk in 1848. In 1861 he became Resident at Riouw and stayed in that post until 1870. He clearly had respect for the locals and was instrumental in ensuring that their knowledge was studied by Europeans. On the other hand, reports indicate that he was not good at getting on with local rulers. He appears to have sent the type specimen of the rabbit to the Royal Dutch Museum, which would explain why Schlegel honored him when naming it. The rabbit is a rare Sumatran endemic.

Neumann, A. H.

Neumann's Hartebeest *Alcelaphus buselaphus neumanni* **Rothschild,** 1897

Arthur Henry Neumann (1850–1907) was a famous big-game hunter. He wrote *Elephant Hunting in East Equatorial Africa,* published in 1898. He spent the greater part of 1893 shooting elephants in the Loroghi Mountains. In 1898 he recounted shooting a black rhino cow with a half-grown calf as a "diversion" during a delay while waiting to go after elephant, because "we could see she had a very fine horn." He later measured the horn at a meter (39 inches). He doesn't say what happened to the calf he orphaned. The hartebeest is now generally regarded as a hybrid between two other subspecies, *lelwel* and *swaynei.* Neumann shot the type specimen to the northeast of Lake Rudolph (Lake Turkana).

Neumann, O. R.

Neumann's Grass Rat *Arvicanthis neumanni* **Matschie,** 1894
Neumann's Olive Baboon *Papio anubis neumanni* Matschie, 1897
Neumann's Black-and-White Colobus *Colobus guereza gallarum* Neumann, 1902

Professor Oskar Rudolph Neumann (1867–1946) was a German ornithologist who collected in East Africa between 1892 and 1894. In 1899 he began a two-year-long expedition to Somaliland and Ethiopia with Carlo von Erlanger. In the early 1900s he studied the birds and mammals of the Rothschild collection. Later in his life he moved to Chicago to escape Nazi persecution and to work at the Field Museum. Several birds are named after him, including Neumann's Short-tailed Warbler *Hemitesia neumanni*. The grass rat is found from Ethiopia and Somalia south to Tanzania. The baboon was described from specimens taken at Lake Natron, northern Tanzania, but is not generally accepted nowadays as a valid subspecies. The colobus monkey is found in the Ethiopian highlands east of the Rift Valley.

Newton

Romanian Hamster *Mesocricetus newtoni* **Nehring,** 1898

Professor Alfred Newton F.R.S. (1829–1907) was a British zoologist who was born in Geneva. He was a co-founder, in 1858, of the British Ornithologists' Union. Between 1854 and 1865 he studied ornithology in Lapland, Iceland, the West Indies, and North America. He became Professor of Zoology and Comparative Anatomy at Cambridge from 1866 until his death. He was awarded both the Royal Medal of the Royal Society and the Gold Medal of the Linnean Society. He edited *The Ibis* from 1865 until 1870. He wrote *Zoology of Ancient Europe* in 1862 and *A Dictionary of Birds,* with Hans Gadow, in 1893. Although it is seldom acknowledged, he was instrumental in launching the bird protection movement in England and the rest of the world. For example, in 1886 he said, "Fair and innocent as the snowy plumes may appear on a lady's hat, I must tell the wearer the truth—she bears the murderer's brand on her forehead." He studied the vanishing birds of the Mascarene Islands, a task made all the easier by the fact that his brother was appointed Assistant Colonial Secretary on Mauritius in 1859. Newton is also commemorated in the scientific names of birds such as the Golden Bowerbird *Prionodura newtoniana*. The hamster is found in eastern Romania and Bulgaria.

Niceforo

Niceforo's Big-eared Bat *Micronycteris nicefori* **Sanborn,** 1949 [Syn. *Trinycteris nicefori*]

Brother Niceforo Maria (1888–1980) was a Frenchman originally called Antoine Rouhaire. He became a missionary in Colombia under his monastic name, Niceforo Maria. He went to Medellin in 1908 and in 1913 was given the task of forming a natural history museum. Primarily a herpetologist, he was also an excellent taxidermist. Niceforo's Marsupial Frog *Gastrotheca nicefori* is also named after him, as are two birds: the extinct Niceforo's Pintail *Anas georgica niceforoi* and Niceforo's Wren *Thryothorus nicefori*. The bat inhabits a wide range from Belize south to the Guianas, Trinidad, Amazonian Brazil, and northeast Peru.

Nicholls

Tree Bat *Ardops nichollsi* **Thomas,** 1891

Dr. Sir Henry Alfred Alford Nicholls (1851–1926) took medical degrees at the University of Aberdeen and at St. Bartholomew's Hospital in London. In 1875, as soon as he had qualified, he left for Dominica where he remained for the rest of his life. There he was Assistant Medical Officer, being promoted to Chief Medical Officer in 1904 and holding that post until he retired in 1925. He was interested in botany and agriculture, corresponded regularly with Kew, and wrote a textbook on tropical agriculture. He discovered the Boiling Lake in Dominica. He found that the temperature at the edges was between 82 and 91°C and that the center of the lake was actively boiling. In 1922 he became a member of the Executive Council of the Leeward Islands. The bat is found in the Lesser Antilles: Dominica, Guadeloupe, Martinique, Montserrat, St. Lucia, and St. Vincent.

Niethammer

Niethammer's Dormouse *Dryomys niethammeri* Holden, 1996 [Alt. Baluchistan Forest Dormouse]

Jochen Niethammer is a zoologist at the Museum Alexander Koening in Bonn who has published widely, especially on rodents. In 1978 he and Franz Krapp published *Handbuch der Säugetiere Europas* (Handbook of European Mammals). The dedication in the original description of the dormouse states that Niethammer's "extensive mammalian research encompasses forest dormice and biogeography of the Middle East and adjacent countries." The dormouse is apparently restricted to juniper forest in Baluchistan Province, Pakistan.

Nikolaus

Nikolaus' Mouse *Megadendromus nikolausi* **Deiterlen** and **Rupp,** 1978 [Alt. Nikolaus' African Climbing Mouse]

Gerhard Nikolaus is a German zoologist with interests in both mammals and birds. He has published numerous scientific papers such as, with Hutterer and Dieterlen, "Small Mammals from Forest Islands of Eastern Nigeria and Adjacent Cameroon, with Systematical and Biogeographical Notes" (*Bonner Zoologische Beiträge,* 1992). He has also collected in a number of African countries. His wife is the "Minna" remembered in the scientific name of the Ethiopian Thicket Rat *Grammomys minnae.* Nikolaus' Mouse is found in scrub moorland of the Bale Mountains, Ethiopia.

Nilsson

Northern Bat *Eptesicus nilssonii* Keyserling and **Blasius,** 1839

Professor Dr. Sven Nilsson (1787–1883) was a Swedish naturalist, primarily a zoologist, and archeologist. In 1806 he began studying for the priesthood at Lund University but was persuaded by his Professor of Zoology, Anders Jahan Retzius, to take up natural history instead. In 1822 he became Director of the Swedish Natural History Museum in Stockholm but resigned in 1831 when he became Professor of Zoology and Curator of the Museum of Lund University, where he had studied and achieved his doctorate. He held those posts until 1856. While at the museum in Stockholm he endeavored to assemble a complete collection of the vertebrates of Sweden, and between 1832 and 1840 he published *Illuminerade figurer till Skandinaviens fauna* (Colored Pictures of Scandinavian Fauna). The Blasius who, with Keyserling, described the bat is not Wilhelm August Heinrich Blasius, who has an entry in this book, but his father Johann Heinrich Blasius (1809–1870), who on at least one occasion visited Lund and would have met Nilsson there. The bat is widely distributed over most of Europe and northern Asia, east to Sakhalin Island (Russia) and Hokkaido (Japan). It occurs as far south as Bulgaria, Iraq, and the Elburz Mountains (northern Iran).

Niobe

Niobe Ground Squirrel *Lariscus niobe* **Thomas,** 1898 [Alt. Montane Three-striped Ground Squirrel]
Moss-Forest Rat *Stenomys niobe* Thomas, 1906 [Alt. Eastern New Guinea Mountain Rat; Syn. *Rattus niobe*]
Niobe's Shrew *Crocidura niobe* Thomas, 1906 [Alt. Stony Shrew]

In Greek mythology, Niobe was the daughter of Tantalus and wife of Amphion, King of Thebes, by whom she had 14 children. She made the mistake of teasing the goddess Latona for only having two. However, Latona's children were Artemis and Apollo, who did not take kindly to anyone insulting their mother. They killed all of Niobe's sons and daughters; it does not pay to mock the gods. (See also **Latona**.) Niobe, who is the personification of maternal grief, wept and wept until she died and was turned into a stone, from which ran water. William Shakespeare put it rather well: "Like Niobe, all tears" (*Hamlet* 1.2.149). Thomas was fond of using names from mythology, sometimes with no obvious reason for his choice of binomials. In the case of the shrew, he described its appearance as "dark blackish grey, with indistinct silvery mottling"; perhaps this mottling made him think of tears. The ground squirrel is found in the mountains of Sumatra and Java. The rat occurs in the mountains of Papua New Guinea. The shrew can be found in the highlands of eastern DRC (Zaire), Burundi, and Uganda.

Noack

Noack's Roundleaf Bat *Hipposideros ruber* Noack, 1893

Professor Dr. Theophil Noack (1840–1918) was a German zoologist. Between 1858 and 1865 he studied natural history and geography at the universities of Halle and Leipzig, being eventually awarded his doctorate by Leipzig in 1865. He taught at the Grammar School in Stettin (now Szczecin in Poland) from 1860 to 1874, and from 1874 to 1911 he was a Professor at the Ducal Reform Gymnasium in Braunschweig. He is best known for describing the Pygmy African Elephant *Loxodonta pumilio* as a separate species in 1906. He based his description on a specimen obtained by the great animal dealer Hagenbeck and bought by the Bronx Zoo in New York. For much of the 20th century, authorities argued as to whether the Pygmy Elephant really existed as a valid species, or whether it was merely based on small specimens of African Forest Elephant. The current belief is that *Loxodonta pumilio* is *not* a separate species. Noack published extensively on African mammals, in such works as *Beiträge zur Kenntnis der Säugetier-Fauna von Ost Afrika* (1891). The bat is found from Senegal to Ethiopia and south to Angola, Zambia, and Mozambique.

Nolthenius

Nolthenius' Long-tailed Climbing Mouse *Vandeleuria nolthenii* Phillips, 1929

Adriaan Constant Tutein-Nolthenius (1892–1954) was a tea-estate owner near the Horton Plains in Ceylon (now Sri Lanka). He was born in the Netherlands, emigrating from there to Ceylon and taking British nationality in 1921. In 1937 he found two examples of the Ceylon Mountain Slender Loris *Loris lydekkerianus nycticeboides* and kept them in captivity, where they bred—a notable achievement at that time. Tutein-Nolthenius was Vice Chairman of the Ceylon Game and Fauna Protection Society from 1928 to 1940. In 1948 he was a Director of Boustead Brothers (plantation owners) and a Member of Parliament in Colombo. The mouse is found in the highlands of Sri Lanka.

Nora

Aberdare Shrew *Surdisorex norae*
Thomas, 1906

Mrs. Nora Holms-Tarn (dates not found) made a small collection of mammals in Kenya and presented it to the Natural History Museum in London. Included in it was the type specimens of this shrew and another that Thomas wrote a paper on. Thomas found that it was so different from other known shrews that it required the creation of a new genus. We have been unable to find out anything about Mrs. Holms-Tarn. "Aberdare" is after the Aberdare Mountains in Kenya where this species is endemic.

Nouhuys

Long-footed Tree Mouse *Lorentzimys nouhuysi* **Jentink,** 1911 [Alt. New Guinea Jumping Mouse]

Captain Jan Willem van Nouhuys (1869–1962) was a Dutch ship's captain. In 1888 he was sent by the Royal Dutch Navy to the East Indies. He took part in the Lorentz expeditions to New Guinea, in 1907 and in 1909–1910. After returning to Holland, Nouhuys was appointed as Director of the Geography and Ethnography Division of the Rotterdam Maritime Museum. As well as the mouse, other taxa are named after him in their binomials, including the Large Scrubwren *Sericornis nouhuysi* and a fish, the Mountain Hardyhead *Craterocephalus nouhuysi*. These, like the tree mouse, are found in New Guinea.

Novaes

Novaes' Bald-headed Uacari *Cacajao calvus novaesi* **Hershkovitz,** 1987

Fernando da Costa Novaes (1927–2004) was a Brazilian zoologist and ornithologist. He worked for many years at the Goeldi Museum in Pará, where he was head of the Zoology Department. In 1952 he went on an expedition to Amazonia with José C. M. Carvalho, with whom he was later to work at the National Museum in Brazil. In 1954 he went to study at Berkeley, California, and was later in Washington, DC, where he met John Todd Zimmer. He published over 60 books and papers, mainly on ornithology, including *Aves da Grande Belém—Município de Belém e Ananindeua*. He is regarded as the founder of modern Brazilian ornithology. The uacari (also spelled uakari) monkey is found in the far west of Amazonian Brazil, in the region of the Jurua River.

Nuttall

Golden Mouse *Ochrotomys nuttalli*
Harlan, 1832
Nuttall's Cottontail *Sylvilagus nuttallii*
Bachman, 1837 [Alt. Mountain Cottontail]

Thomas Nuttall (1786–1859) was an English botanist and zoologist who collected for the University of Pennsylvania. He originally went to America in 1808. He was the basis for the character "Old Curious," the naturalist in Richard Henry Dana's novel *Two Years before the Mast*. Primarily an ornithologist, he was the first to publish a small, inexpensive field guide to U.S. and Canadian birds, and he also wrote *Manual of the Ornithology of the United States and Canada* (1832). He is commemorated in the names of several birds, such as Nuttall's Woodpecker *Picoides nuttallii*. Audubon wrote, when naming a bird after Nuttall, "I hope, kind reader, you will approve of the liberty which I have taken in prefixing the name of my friend Nuttall to the present species, which was discovered by his indefatigable and enthusiastic devotion to science"—words that seem to sum up well his contemporaries' opinion of him. The mouse is found in the eastern USA, from southeast Missouri east to southern Virginia and south to eastern Texas and central Florida. The cottontail is found in much of the western USA except for the southwest corner and coastal regions.

O

O'Connell

O'Connell's Spiny Rat *Proechimys oconnelli* **J. A. Allen,** 1913

Geoffroy M. O'Connell (dates not found) was a naturalist, primarily an ornithologist, who was associated with the American Museum of Natural History. He was a co-author with such luminaries as Louis Agassiz Fuertes, Frank M. Chapman, George K. Cherrie, Robert Cushman Murphy, and others in the production of two books on South American birds: *The Distribution of Bird-life in Colombia: A Contribution to a Biological Survey of South America* (1917) and *The Distribution of Bird-life in Ecuador: A Contribution to a Study of the Origin of Andean Bird-life* (1926). The rat is found in central Colombia, where the type specimen was collected in 1913 during one of the American Museum of Natural History's expeditions. O'Connell was on this expedition with the task of collecting mammals, and Allen, who described the rodent, wrote the expedition report.

Oedipus

Cotton-top Tamarin *Saguinus oedipus* **Linnaeus,** 1758

The literal meaning of "Oedipus" is "swollen foot." However, this species of tamarin doesn't have enlarged feet. We also know that Linnaeus had a penchant for giving primates names derived from mythology, sometimes with little obvious rationale. So he may have named this one after the mythical tragic King of Thebes, who unknowingly married his own mother. What inspired Linnaeus in this choice of name we will probably never know. The tamarin's range is confined to a small area of northern Colombia.

Oersted

Central American Squirrel Monkey *Saimiri oerstedii* Reinhardt, 1872 [Alt. Red-backed Squirrel Monkey]

Professor Anders Sandoe Oersted (1816–1872) was a Danish naturalist. He was brought up in the household of his uncle, Hans Christian Oersted, who discovered electromagnetism. Anders Oersted became Professor of Natural History in Copenhagen in 1837 and was awarded a gold medal and a fellowship by the university in 1844. In 1845 he left Copenhagen and went to study the geography of Central America. He spent the years 1845–1848 in Nicaragua and also spent some time in Costa Rica, where in 1863 he was the first naturalist to visit Poas Volcano. The squirrel monkey is found in the Pacific lowlands of Costa Rica and western Panama.

Ogilby

Ogilby's Duiker *Cephalophus ogilbyi* **Waterhouse,** 1838 [Alt. Fernando-Pó Duiker]

William Ogilby (1804–1873; many authorities say 1808–1873) was a Cambridge-educated Irish barrister and zoologist. He practiced at the bar in London from 1832 to 1846, when he returned to Ireland. He wrote such papers as "Descriptions of Mammalia and Birds from the Gambia" (1835), "Exhibition of the Skins of Two Species of the Genus Kemas" (1838), and "Observations on the History and Classification of the Marsupial Quadrupeds of New Holland" (1839), all published in the *Proceedings of the Zoological Society of London.* He was Honorary Secretary of the Zoological Society from 1839 to 1846 and crossed swords a number of times with Gray of the British Museum. He started building a castle (Altnachree Castle) in Ireland in the 1840s, just as the Great Famine struck. The castle was completed in the 1860s. Unlike many landlords, he kept on all his workers and fed them by importing grain. His son, James Douglas Ogilby (1853–1925), was a notable ichthyologist. The duiker was first found on the island of Fernando Pó (now Bioko). It also occurs in southeast Nigeria, Cameroon, Gabon and the Congo Republic. Another race, *brookei*, found from Sierra Leone east to Ghana, is now often treated as a separate species (see **Brooke, V. A.**).

Ognev

Ognev's Dormouse *Myomimus personatus* Ognev, 1914 [Alt. Masked Mouse-tailed Dormouse]

Ognev's Long-eared Bat *Plecotus ognevi* Kishida, 1927

Professor Sergei Ivanovich Ognev (1886–1951) was a Russian zoologist who specialized in mammals and their taxonomy. He was awarded his doctorate at Moscow University in 1910 and stayed on as a member of the staff, becoming a Professor in 1928. Under the Soviet regime he was awarded the degree of Doctor of Science in 1935 without having to submit and defend a thesis, but he was probably eminent enough by then to deserve it. He was ranked as a Scientist of Merit, was twice awarded the Stalin Prize, and was awarded the Order of Lenin. Based on his own expeditions and field studies he wrote a number of books, including a seven-volume work, *The Mammals of the USSR and Adjacent Countries*, which appeared between 1928 and 1950. The dormouse occurs in Turkmenistan and northeast Iran. The bat, which was long regarded as conspecific with *Plecotus auritus*, occurs from the Altai and central Siberian highlands east to Korea and the island of Sakhalin.

Olalla

Olalla's Titi *Callicebus olallae* Lönnberg, 1939 [Alt. Beni Titi, Olalla Brothers' Titi]

The soft-furred spiny rat genus *Olallamys* **Emmons,** 1988 [2 species: *albicauda* and *edax*]

Alfonso Manuel Olalla (dates not found) was the most prominent member of a family of animal collectors, which also included his brothers Manuel, Ramón, and Rosalino. Their father, Carlos Olalla, an Ecuadorian, began this family business, and he and his sons collected birds and mammals in the Amazon basin from 1922 until the late 1960s. Alfonso later moved to Brazil and made large collections on behalf of the American Museum of Natural History (New York), the Field Museum (Chicago), and the Royal Natural History Museum (Stockholm, Sweden). He has a bird named after him, Olalla's (Blue-winged) Parrotlet *Forpus xanthopterygius olallae*. In Emmons' proposal for the use of *Olallamys*, written in 1987, she lists Carlos and all four sons as the persons honored in this name and comments that Manuel was at that time living in Quito, Ecuador, but that his father and three brothers were deceased. The titi monkey is known only from the province of Beni, northern Bolivia. The spiny rats are found in Colombia and western Venezuela.

Olga

Olga's Dormouse *Graphiurus olga* **Thomas,** 1925

Mrs. Olga Buchanan (dates not found) was the wife of Captain Angus Buchanan. Captain Buchanan collected the type specimen of this dormouse. In 1925 Thomas wrote an article entitled "On Mammals (Other Than Ruminants) Collected by Captain Angus Buchanan during His Second Saharan Expedition." The dormouse was "named after Mrs. Buchanan." It is known only from three localities in Niger, northern Nigeria, and northeast Cameroon. However, this taxon may be conspecific with the more widespread species *Graphiurus kelleni*.

Oliver

Mindoro Warty Pig *Sus (philippensis) oliveri* Groves, 1997

William L. R. Oliver is, at the time of writing, Chairman of the Pigs and Peccaries Specialist Group of the Species Survival Commission of the International Union for Conservation of Nature, which group he has run since its inception in 1981. He worked briefly at Marwell Zoo before joining the Jersey Wildlife Preservation Trust in 1974. He now spends much of his time in the Philippines as Director of the Philippines Biodiversity Conservation Program. The dedication in the description of the warty pig says of him that he "has worked tirelessly for the worldwide conservation of the *Suidae* and other animals." The pig is found on the Philippine island of Mindoro.

Olivier

Olivier's Shrew *Crocidura olivieri* **Lesson,** 1827 [Alt. Giant Musk Shrew, African Giant Shrew]

Guillaume-Antoine Olivier (1756–1814) was a French entomologist and malacologist. He was trained as a physician but became one of the great French naturalists. He collected extensively in Europe and was employed as a naturalist for six years on a major expedition to Persia (now Iran). He returned to France in 1798 with a significant natural history collection from Turkey, Asia Minor, Persia, Egypt, and some Mediterranean islands. In 1800 he was appointed Professor of Zoology at the Veterinary School at Alfort. He was acquainted with Lesson, and his collection is mostly at the National Museum of Natural History in Paris, where Lesson would in all likelihood have catalogued it. The original description of this shrew was based on an ancient mummified specimen from Egypt. The species still occurs in Egypt and is also widely distributed in sub-Saharan Africa except for the far south.

Olrog

Olrog's Chaco Mouse *Andalgalomys olrogi* Williams and Mares, 1978

Olrog's Four-eyed Opossum *Philander olrogi* Flores, Barquez, and Diaz, 2008

Dr. Claes Christian Olrog (1912–1985) was a Swedish ornithologist and zoologist who lived in Argentina and wrote widely on South American birds. As a young man he joined the Stockholm Natural Science Museum banding expeditions to northern Scandinavia, the Danube delta, Iceland, and Greenland. During his doctoral studies at Stockholm he went to South America for the first time, on a two-year expedition to Tierra del Fuego from 1939 to 1941. From 1946 to 1947 he went to Paraguay. In 1948 he started work at the Miguel Lillo Institute of the National University, Tucuman, in northwest Argentina. He published *Las aves Argentinas* in 1959, as well as an annotated Peruvian checklist and over 100 articles on distribution and other topics. The Institute for the Administration of Protected Areas in Buenos Aires is named after "Dr. Claes C Olrog." He also wrote *Destination Eldslandet* (Destination Tierra del Fuego), published in 1943. At his death he left an unfinished manuscript for a book on the birds of Bolivia. He is commemorated in the common names of two birds, Olrog's Cinclodes *Cinclodes olrogi* and Olrog's Gull *Larus atlanticus*. The mouse is found in northwest Argentina. The opossum has been recorded in Amazonian areas of Bolivia and Peru.

Omura

Omura's Whale *Balaenoptera omurai* Wada, Oishi, and Yamada, 2003

Professor Dr. Hideo Omura (d. 2002) was a Japanese cetologist. The National Research Institute of Fisheries Science discovered that this was a new species of whale in Japan using DNA sampling. They named it to honor the recently deceased Dr. Omura, who had been the leading light in cetacean research in Japan in the 1960s and 1970s. Dr. Omura did much of his work at the Whale Research Institute, where Dr. Wada had been one of his students. Among his writings is, with R. L. Brownell, *Whale Meat in the Japanese Diet*, published in 1980. The distribution of Omura's Whale is not yet fully known. Specimens have been found in the Sea of Japan, the Solomon Sea, and the eastern Indian Ocean.

Opdenbosch

Opdenbosch's Mangabey *Lophocebus opdenboschi* **Schouteden,** 1944 [Syn. *L. aterrimus opdenboschi*]

Armand Opdenbosch (dates not found) was a Belgian who was the Chief Technician (Taxidermist) of the Africa Museum at Tervuren between 1930 and 1975. He was a close friend of Max Poll, whom he accompanied on an expedition to the Belgian Congo in 1956 to collect good-quality mammal specimens to be exhibited in the Belgian Congo's pavilion at the Brussels Exposition of 1958. He is also commemorated in the scientific name of an African freshwater fish, *Ivindomyrus opdenboschi*. This monkey species (or subspecies; taxonomists are not of one mind) is found in southwest DRC (Zaire) and adjacent northern Angola.

Orces

Orces' Chibchan Water Mouse *Chibchanomys orcesi* **Jenkins** and Barnett, 1997 [Alt. Las Cajas Water Mouse]

Orces' Nectar Bat *Lonchophylla orcesi* Albuja and **Gardner,** 2005

Professor Gustavo Orces (1902–1999) was an Ecuadorian zoologist who worked at the Polytechnic in Quito. His principal interest was herpetology. The Gustavo Orces Herpetological Foundation in Quito was founded in 1989 and holds his collection. He is also remembered in the scientific names of other Ecuadorian fauna such as the El Oro Parakeet *Pyrrhura orcesi*, the fish *Hemibrycon orcesi*, and a subspecies of coral snake, *Micrurus steindachneri orcesi*. One zoologist who visited Orces, Dr. Janis Roze, made a point of noting that he "made available to me his large coral snake collection and provided other valuable assistance." The mouse and the bat are both known only from Ecuador.

Ord

Ord's Kangaroo Rat *Dipodomys ordii* **Woodhouse,** 1853

George Ord (1781–1866) was an American philologist, collector, and naturalist. He was originally a ships' chandler but became one of the earliest members of the active Philadelphia natural history community. This brought him to the attention of President Thomas Jefferson, who sent him many specimens from the Lewis and Clark expedition, and Ord named many of western North America's familiar birds. In 1824 Bonaparte tried to get Audubon accepted by the Academy of Natural Sciences and was opposed by Ord, who tried to prevent it. He apparently detested Audubon, referring to him as an "impudent pretender" whose illustrations of birds were "vile." Their enmity was something of a cause célèbre in scientific circles. Ord described a number of taxa himself, in the 1810s and 1820s, sometimes with Thomas Say. The Brown-banded Puffbird *Notharcus ordii* is named after him in the binomial. The kangaroo rat ranges over much of the western and central USA and northern Mexico.

Orestes

Borneo Black-banded Squirrel *Callosciurus orestes* **Thomas,** 1895
Afroalpine Vlei Rat *Otomys orestes* Thomas, 1900

Orestes was the son of King Agamemnon and his wife Clytemnestra. Clytemnestra murdered Agamemnon (he was having a bath at the time), and she in turn was murdered by Orestes, helped by his sister Electra, to avenge their father. At least seven plays by three of the great Athenian classical dramatists have survived and tell this story. Thomas was fond of using names from classical mythology, and in these cases he was probably inspired by the meaning of the Greek name: Orestes means "he who stands/dwells on the mountain," and the type specimens of both mammals were taken on mountains. The squirrel came from Mount Dulit in Sarawak, the vlei rat from Mount Kenya. Both species are now known to have wider distributions: the squirrel in Sarawak and Sabah (East Malaysia), the rat in the highlands of west and central Kenya.

Orii

Hokkaido Flying Squirrel *Pteromys volans orii* Kuroda, 1921
Orii's Shrew *Crocidura orii* Kuroda, 1924 [Alt. Amami Shrew, Ryukyu Shrew]

Hyojiro Orii (1883–ca. 1957) was a Japanese collector. His main task was to collect ornithological specimens for Dr. Nagamichi Kuroda, who described the above mammals and named them after Orii. He also worked for Marquis Yamashina, who founded the Yamashina Institute for Ornithology at Chiba in Japan and who published a small work called *On Korean Birds Collected by Mr. H. Orii.* Orii wrote an appreciation of Marquis Yoshimaro Yamashina in 1948. He is also remembered in the trinomial of a number of subspecies of birds, such as the Daito Varied Tit *Parus varius orii,* which probably became extinct about 1940, and the Japanese Light-vented Bulbul *Pycnonotus sinensis orii.* The squirrel, from the Japanese island of Hokkaido, is a subspecies of the Siberian Flying Squirrel. The shrew is endemic to the Amami group of the Ryukyu Islands, southern Japan.

Orion

Orion Broad-nosed Bat *Scotorepens orion* **Troughton,** 1937 [Alt. Eastern Broad-nosed Bat]

In Greek mythology, Orion was a mighty hunter. The name is also linked to "Orient" (i.e. from the east). The bat is a hunter of insects, found in coastal eastern Australia.

Osborn

Osborn's Caribou *Rangifer tarandus osborni* **J. A. Allen,** 1902
Aquatic Genet *Osbornictis piscivora* J. A. Allen, 1919 [Alt. Water Civet; Syn. *Genetta piscivora*]
Osborn's Key Mouse *Clidomys osborni* **Anthony,** 1920 extinct [Alt. Jamaican Giant Hutia]

Professor Dr. Henry Fairfield Osborn (1857–1935) was an American zoologist, paleontologist, humanist, and evolutionist. From 1881 until 1891 he was Professor of Natural Sciences at Princeton, where he had graduated in 1880. From 1891 to 1907 he was Professor of Biology and Zoology at Columbia University, and from 1908 until 1933 he worked at the American Museum of Natural History, where he was to become a member of the Board of Trustees. Both describers, Allen and Anthony, were at different times Curator of Mammals there. Osborn was a leading proponent of the "theory of evolution" and was called as an expert witness at the famous "monkey trial." However, his beliefs clearly led him down some dark avenues, as he was a confirmed racist, once saying, "The Negroid stock is even more ancient than the Caucasian and Mongolians as may be proved by an examination not only of the brain, of the hair, of the bodily characteristics, but of the instincts, the intelligence. The standard intelligence of the average adult Negro is similar to that of the eleven-year-old youth of the species Homo sapiens." The caribou is found in the Cassier Mountains of northern British Columbia. Some authorities include this population within the Woodland Caribou subspecies, *Rangifer tarandus caribou*. The aquatic genet is from DRC (Zaire) north and east of the Congo River. Genetic analysis suggests that the genus *Osbornictis* is invalid and that the species should be placed with other genets in *Genetta*. The hutia was confined to Jamaica. The date of its extinction is unknown, but it seems to have died out well before Europeans reached the Caribbean.

Osgood

Michoacan Deer Mouse *Osgoodomys banderanus* **J. A. Allen,** 1897
Osgood's Mouse *Peromyscus gratus* **Merriam,** 1898 [Alt. Saxicoline Deer Mouse]
Osgood's Aztec Mouse *Peromyscus aztecus evides* Osgood, 1904
Osgood's Short-tailed Opossum *Monodelphis osgoodi* Doutt, 1938
Osgood's Horseshoe Bat *Rhinolophus osgoodi* **Sanborn,** 1939
Osgood's Leaf-eared Mouse *Phyllotis osgoodi* Mann, 1945
Osgood's Rat *Rattus osgoodi* **Musser** and Newcomb, 1985

Wilfred Hudson Osgood (1875–1947) was an ornithologist and mammalogist who began his career in 1897 working as a biologist in the U. S. Department of Agriculture. Here he was in charge of the U.S. biological investigation in Canada. In 1909 he moved to the Field Museum of Natural History in Chicago, holding the position of Assistant Curator of Mammalogy and Ornithology until 1921, then the position of Curator of Zoology until he retired in 1940. He conducted biological explorations and surveys of many areas of North and South America, Ethiopia, and Indochina. He spent three separate years—1906, 1910, and 1930—studying in European museums, and in 1914 he was a special investigator with a fur-seal study. He led the Field Museum Abyssinian expedition of 1926–1927 and the Magellanic expedition of 1939–1940. Osgood was a fellow of the American Academy for the Advancement of Science and of the American Ornithologists' Union. He founded, and became the first President of, the Cooper Ornithology Club of California. In addition, he was Secretary of the Biological Society of Washington from 1900 to 1909; a corresponding member of the Zoological Society of London and the British Ornithologists' Union; a member, and President from 1924 to 1926, of the American Society of Mammalogists; and a member and Trustee of the Chicago Zoological Society. His publications included "Revision of Pocket Mice of the Genus *Perognathus*" in 1900 through to *Mammals of Chile* in 1943. He also wrote around 180 shorter papers on classification, anatomy, and habits of mammals and birds, as well as contributing zoological definitions to *Webster's New International Dictionary*. The deer mouse and subspecies of Aztec Mouse are both endemic to western Mexico. Osgood's Mouse is found in southwest New Mexico and southward through interior Mexico to Oaxaca State. The opossum inhabits southeast Peru and central Bolivia. The bat is found in Yunnan, southern China. The leaf-eared mouse occurs in northeast Chile. The rat comes from the highlands of southern Vietnam.

Osman-Hill

Osman-Hill's Grey-cheeked Mangabey *Lophocebus (albigena) osmani* Groves, 1978

Professor William Charles Osman Hill (1901–1975) was a physician, an anthropologist, and a primatologist at London University. His collection of skeletons and tissue is held by the Royal College of Surgeons in London. In 1949 he published *The Primates—Comparative Anatomy and Taxonomy*. The Primate Society of Great Britain awards an Osman-Hill Medal named in his honor. He is also remembered in the scientific name of the Colombo Wolf Snake *Lycodon osmanhilli*. This monkey is found in Cameroon. Traditionally it was viewed as a race of *Lophocebus albigena*, but in 2007 the taxonomist Colin Groves of the Australian National University suggested that the various subspecies of the Grey-cheeked Mangabey should be elevated to full-species rank.

Osvaldo Reig

Osvaldo Reig's Tuco-tuco *Ctenomys osvaldoreigi* **Contreras,** 1995

Dr. Osvaldo A. Reig (1929–1992). The tuco-tuco is found in Córdoba Province, north-central Argentina. See **Reig** for biographical details.

Otto

Otto's Sportive Lemur *Lepilemur otto* Craul et al., 2007

Dr. Michael Otto (b. 1943) is a German businessman who was Chairman of the Executive Board and Chief Executive Officer, Otto Group, which is the world's largest mail-order company; it consists of 123 main companies and operates in 19 countries in Europe, America, and Asia. He retired from active management at the end of 2007 and is now head of the group's Supervisory Board. He has been Chairman of the World Wildlife Fund as well as heading a number of other committees dedicated to the environmental cause, and has sponsored research projects and environmental education. He is Chairman of the Board of Trustees of the Michael Otto Foundation for Environmental Protection. Craul's etymology states, "The name *Lepilemur otto* was chosen to acknowledge the donation of Dr. Michael Otto for the purpose of research and conservation of Malagasy lemurs." The lemur is found in western Madagascar, between the Mahajamba and Sofia rivers.

Oustalet

Oustalet's Red Colobus *Piliocolobus foai oustaleti* **Trouessart,** 1906

Dr. Jean-Frédéric Emile Oustalet (1844–1905) was a French zoologist. He wrote *Les oiseaux de la Chine* in 1877, with Père Armand David as co-author, and *Les oiseaux du Cambodge* in 1899. In 1873 he succeeded Jules Verreaux as Assistant Naturalist at the Paris Natural History Museum. In 1900 he succeeded Alphonse Milne-Edwards, with whom he had co-authored a number of works, as Professor of Mammalogy. He is remembered in the names of a number of birds, such as Oustalet's Sunbird *Nectarinia oustaleti,* and other taxa such as Oustalet's Chameleon *Furcifer oustaleti*. The colobus is found in central Africa north of the Congo River.

Ouwens

Mouse sp. *Mus ouwensi* **Kloss,** 1921 [Syn. *M. caroli ouwensi*]

Peter A. Ouwens (dates not found) was a Dutch scientist and Director of the Java Zoological Museum and Botanical Gardens in Bogor. He is particularly remembered as the person who first scientifically described the Komodo Dragon, the world's largest lizard, in 1912. A fish, Ouwen's Goby *Sicyopterus ouwensi*, is also named after him. The mouse, which is found on Java, is now generally considered to represent a population of *Mus caroli* introduced by accidental human agency.

Owen, G. F.

Gansu Mole *Scapanulus oweni* **Thomas,** 1912

Captain George Fenwick-Owen (dates not found) was a big-game hunter and collector. In 1912 he undertook a big-game hunting expedition through the interior of China with the object of obtaining specimens of the Golden Takin *Budorcas taxicolor bedfordi* as well as other mammals for the British Museum. Clearly he was successful, as the museum has such an exhibit dated 1913. He also collected the type of a peony in Gansu Province. He often donated antiquities he had collected to the museum and undertook paleontological excavations. He traveled from Shanghai to Omsk through the Gobi Desert. Thomas writes in his description of the mole that he names it "in honour of Mr. Fenwick Owen, to whose interest and kindness the Museum owes this valuable accession to its collections." In 1912 Thomas wrote a paper entitled "On a Collection of Small Mammals from the Tsin–ling Montains, Central China, Presented by Mr. G. Fenwick Owen to the National Museum." The mole is found in the montane forests of central China.

Owen, R.

Owen's Pygmy Sperm Whale *Kogia sima* Owen, 1866 [Alt. Dwarf Sperm Whale; scientific name often given as *K. simus*]
Owen's Marsupial "Lion" *Thylacoleo oweni* McCoy, 1876 extinct

Professor Sir Richard Owen (1804–1892) was a British anatomist and paleontologist. He studied medicine at the University of Edinburgh and St. Bartholomew's Hospital, London, before becoming Assistant Conservator of the museum of the Royal College of Surgeons. In 1836 he was appointed Hunterian Lecturer, charged with giving public lectures on anatomy. These lectures were attended by many important figures of the time. His growing fame as a scientist led to his appointment to teach natural history to Queen Victoria's children; he astounded the court with the fact that tadpoles metamorphosize into frogs. He was largely responsible for the creation of the Natural History Museum in London, having successfully separated that department from the British Museum. However, his career was tainted by his unwillingness to give due credit to others. He was also an opponent of Darwin's theories on evolution, arousing such bitterness that Darwin commented, "I used to be ashamed of hating him so much, but now I will carefully cherish my hatred and contempt to the last days of my life." Owen earned a similar opinion from his fellow paleontologist Gideon Mantell, who said of Owen that it was "a pity a man so talented should be so dastardly and envious." Posterity may remember Owen foremost as the man who coined the term *Dinosauria* ("terrible lizards") for large fossil reptiles, thus creating the present day dinosaur industry. A bird, Owen's Kiwi *Apteryx owenii*, is named after him, as is a lizard, Owen's Galliwasp *Diploglossus oweni*. The pygmy sperm whale is found worldwide in tropical and warm-temperate seas. The *Thylacoleo* is a long-extinct Australian marsupial carnivore. *Thylacoleo oweni* is regarded by some authorities as a synonym of *T. carnifex* (Owen, 1858).

Owl

Owl's Spiny Rat *Carterodon sulcidens* **Lund,** 1841

This rodent is not named after a person called Owl. The type specimen was collected at Lagoa Santa in Minas Gerais Province, Brazil. In 1835 Peter Wilhelm Lund (1801–1880), a Danish physician, zoologist, and botanist, settled there and spent time exploring the extensive caves in the area. One of the caverns is called Gruta da Coruja, which translates as Owl's Grotto. Lund found remains of the spiny rat in the pellets disgorged by owls. Humbly we opine that a better vernacular name should be found for this species, which is endemic to eastern Brazil.

Owston

Owston's Palm Civet *Chrotogale owstoni* **Thomas,** 1912

Alan Owston (1853–1915) was an English collector of Asian wildlife, as well as a businessman and yachtsman. He founded the Yokohama (Japan) Yacht Club, and his business appears to have involved dealing in wildlife specimens. He left for the Orient when still quite young. He married Shimada Rei Jkao in Japan, around 1880, and they had one child, Susie. He later married Kame (Edith) Miyahara, in about 1893, and had eight children by that marriage. Owston's most active collecting period seems to have been in the early 20th century. He is remembered in the scientific names of birds such as the Guam Rail *Gallirallus owstoni* and of several species of fish, including the Goblin Shark *Mitsukurina owstoni.* He died of lung cancer in Yokohama. In 1911 a collector employed by Owston in Vietnam discovered the palm civet that would be named after him. The civet is found in northern Laos, Vietnam, and adjacent southeast China.

Oyapock

Oyapock's Fish-eating Rat *Neusticomys oyapocki* **Dubost** and **Petter,** 1978

The Oyapock River forms the boundary between French Guiana and Brazil. This mammal is named after that river, not after a person, as one might suppose from the form of the common name. The rat has been found in French Guiana and adjacent northeast Brazil.

Pallas

Pallas' Long-tongued Bat *Glossophaga soricina* Pallas, 1766
Pallas' Mastiff Bat *Molossus molossus* Pallas, 1766
Pallas' Tube-nosed Fruit Bat *Nyctimene cephalotes* Pallas, 1767
Pallas' Cat *Felis manul* Pallas, 1776 [Alt. Manul; Syn. *Otocolobus manul*]
Pallas' Squirrel *Callosciurus erythraeus* Pallas, 1779 [Alt. Belly-banded Squirrel, Red-bellied Squirrel]
Pallas' Tarsier *Tarsius spectrum* Pallas, 1779 [Alt. Spectral Tarsier, Eastern Tarsier; Syn. *T. tarsier* **Erxleben,** 1777]
Pallas' Pika *Ochotona pallasi* **Gray,** 1867

Peter Simon Pallas (1741–1811) was a Russian of German extraction, having arrived in Russia in 1767. He was a zoologist and was considered one of the greatest 18th-century naturalists. He was obviously very bright, as he earned his doctorate from the University of Leiden at the age of 19. In 1761 he went to London to study the English hospital system and appears to have been enchanted by the Sussex coast and the countryside in Oxfordshire. The Empress Catherine II summoned him to Russia in 1767 to become the Professor of Natural History at the St. Petersburg Academy of Sciences and to investigate Russia's natural environment. He was also a geographer and traveler and explored widely in lesser-known areas of Russia. Between 1768 and 1774 he headed an Academy of Sciences expedition that studied many regions of Russia, including southern Siberia, Altai, and the Lake Baikal region. He described many new species of mammals, birds, fish, insects, and fossils, including, in some cases, their long-dead bodies preserved in the ice. His works include *A Journey through Various Provinces of the Russian State* (1771), *Flora of Russia* (1774), and *A History of the Mongolian People and Asian-Russian Fauna* (1811), as well as other works on zoology, paleontology, botany, and ethnography. A volcano on the Kurile Islands and a reef off New Guinea were also named in his honor. In 1772 he found a mass of iron weighing 700 kg (1,543 pounds). This turned out to be a meteorite of a new kind and it was named pallasite after him. Pallas is also commemorated in at least 14 bird names, including Pallas' Leaf Warbler *Phylloscopus proregulus* and Pallas' Sandgrouse *Syrrhaptes paradoxus.* The long-tongued bat is found from Mexico south to Peru and Paraguay, with a subspecies on Jamaica. The mastiff bat has a similar distribution but occurs also on many islands of the Greater and Lesser Antilles. The fruit bat is found in eastern Indonesia (Sulawesi, the Moluccas) and southern New Guinea. The cat is found in steppe and semidesert regions from the Caspian Sea to Kashmir, Mongolia, and central China. The squirrel ranges from northeast India to southern China, Taiwan, Indochina, and the Malay Peninsula. The tarsier is from Sulawesi (formerly Celebes), Indonesia. (Though long known by Pallas' name *spectrum,* the species should more correctly be known by Erxleben's earlier name *tarsier.*) The pika has a discontinuous distribution in Kazakhstan, the Altai Mountains, Mongolia, and northern China.

Palmer

Palmer's Chipmunk *Tamias palmeri* **Merriam,** 1897

Dr. Theodore Sherman Palmer (1868–1958) was an American botanist. He started work in 1889 for the U.S. Biological Survey, which

Merriam headed, and worked there for 15 years as Assistant Chief. From 1900 to 1916 he was the Law Enforcement Officer of the U.S. Fish and Wildlife Service. After his retirement in 1933 he devoted himself to wildlife conservation and to ornithology—he was once Secretary of the American Ornithologists' Union—as well as to biographic and bibliographic interests. In 1891 he led an expedition to study the flora and fauna of Death Valley. His most important work, written over 20 years and published in 1904, was *Index generum mammalium*. He was confined to his house for the last 21 years of his life following an accident in which he badly broke a hip. The chipmunk is found only in the Charleston Mountains, southern Nevada, USA.

Pan

The chimpanzee genus *Pan* Oken, 1816 [2 species: *paniscus* and *troglodytes*]

This is another case of primates being named after mythological characters. Pan was, of course, the Greek god of nature. He had the hindquarters and legs of a goat, thus mixing human and animal characteristics. Probably early naturalists viewed chimpanzees and other apes as looking part-human, part-animal. The scientific name of the Bonobo (Pygmy Chimpanzee) is *Pan paniscus*, which can be translated as "Pan, the little Pan."

Pandora

Yucatan Brown Brocket *Mazama pandora* **Merriam,** 1901

In Greek mythology Pandora was the first woman. Zeus ordered the god Hephaestus to make her, as part of his punishment of mankind for the sins of Prometheus, who had stolen fire from the gods and given it to mortals. Pandora was given a box and told that on no account should she open it. Curiosity got the better of her, and she opened the box, letting loose all the evils of the world. She closed it too late but heard that there was something still in it, opened it again, and Hope sprang out. Merriam does not explain why he chose *pandora* as the scientific name for this deer but says that the discovery of a new mammal in an area well studied by naturalists came as a great surprise. Perhaps the implication is that zoological curiosity can still uncover new wonders out of Nature's bottomless box. The brocket is found in the Yucatan Peninsula of southeast Mexico.

Pardelluch

Pardelluch's Lynx *Lynx pardinus* **Temminck,** 1827 [Alt. Iberian Lynx, Spanish Lynx, Pardel Lynx]

Luchs is the German for "lynx," *Pardelluchs* is the German for "Iberian lynx," so occasional references to "Pardelluch's Lynx" are transcription errors (and tautology). This species of lynx is now extremely rare and found mainly in Spain's Coto Doñana National Park.

Parnell

Parnell's Moustached Bat *Pteronotus parnellii* **Gray,** 1843 [Alt. Common Moustached Bat]

Richard Parnell (1810–1882) was a British ichthyologist who collected in Jamaica from 1839 to 1840 and, on the same trip, visited many collections in museums in the USA. He wrote *Prize Essay on the Natural and Economical History of the Fishes Marine, Fluviatile, and Lacustrine, of the River District of the Firth of Forth* (1838) and *The Grasses of Britain* (published in two volumes between 1842 and 1845). However, he is more remembered for the notebooks he kept on the Jamaica expedition. They contain little by way of narrative but are packed with sketches and notes, particularly on his two loves, fishes and grasses. They also contain one detailed sketch of the head of the moustached bat. The fish and other fauna he collected are now in the National Museum in Edinburgh and the British Museum of Natural History, although the grasses are in many other museums. While his publishing ceased in 1845, his collecting went on for many years. The type specimen of the bat was taken in Jamaica, but it is widely distributed from Mexico south to Peru and Brazil and on Cuba, Puerto Rico, Hispaniola, and Trinidad and Tobago.

Pariente

Pariente's Fork-marked Lemur *Phaner (furcifer) parienti* Groves and **Tattersall,** 1991 [Alt. Sambirano Fork-crowned Lemur]

George F. Pariente (1937–1976) was a zoologist who studied lemurs and wrote about them over a number of years. As an example, he wrote "Influence of Light on the Activity Rhythms of Two Malagasy Lemurs: *Phaner furcifer* and *Lepilemur mustelinus leucopus*," published posthumously in 1979. The lemur is found in the Sambirano region of northwest Madagascar.

Parry

Parry's Marmot Squirrel *Spermophilus parryii* **Richardson,** 1825 [Alt. Barrow Ground Squirrel, Arctic Ground Squirrel]

Parry's Wallaby *Macropus parryi* **Bennett,** 1835 [Alt. Whiptail Wallaby, Pretty-faced Wallaby]

Sir William Edward Parry (1790–1855) was an English explorer. At the age of just 13 he joined the flagship of the Channel fleet as a first-class volunteer, and at 16 he became a Midshipman, moving through the ranks to become Commander in 1821. He took advantage of every opportunity to study and practice astronomical observation in northern latitudes, and afterward published the results of his studies in a small volume, *Nautical Astronomy by Night* (1816). In 1818 he received his first command, the brig *Alexander* in the Arctic expedition under Captain (afterward Sir John) Ross. This expedition returned to England without having made any new discoveries, but Parry said "that attempts at Polar discovery had been hitherto relinquished just at a time when there was the greatest chance of succeeding." He was given the chief command of a new Arctic expedition, consisting of the two ships HMS *Griper* and HMS *Hecla*. This expedition returned to England in November 1820 after a voyage of almost unprecedented Arctic success, having accomplished more than half the journey from Greenland to the Bering Strait, the completion of which solved the ancient problem of a Northwest Passage. A narrative of the expedition, entitled *Journal of a Voyage to Discover a Northwest Passage*, appeared in 1821. In 1823 Parry was appointed Acting Hydrographer to the navy. His *Journal of a Second Voyage, &c.*, appeared in 1824. With the same ships Parry undertook a third expedition on the same quest in 1824, but again unsuccessfully, and following the wreck of the *Fury* he returned home in October 1825 with a double ship's company. He published an account of this voyage in 1826. Parry also pioneered the use of canning techniques for food preservation on his Arctic voyages. In the following year Parry attempted to reach the North Pole from the northern shores of Spitzbergen. He published an account of this journey, *Narrative of the Attempt to Reach the North Pole, &c.*, in 1827. In April 1829 he was knighted. Parry served as Commissioner of the Australian Agricultural Company from 1829 to 1834. The first example of the Whiptail Wallaby ever to be identified was one he kept as a pet at his home in 1834. Parry was subsequently selected for the post of Comptroller of the newly created Department of Steam Machinery of the Navy, and he held this office until his retirement from active service in 1846, when he was appointed Captain Superintendent of Haslar Hospital. He attained the rank of Rear Admiral in 1852 and in the following year became a Governor of Greenwich Hospital, retaining this post until his death. Sir Edward Parry had a strong religious side: besides the journals of his different voyages he also wrote a *Lecture to Seamen, and Thoughts on the Parental Character of God*. The ground squirrel is found in Alaska and northern Canada, as well as in the northeast extremity of Asia (Russia). The wallaby comes from eastern Australia (coastal Queensland and northeast New South Wales).

Pascual

Pascual's Tuco-tuco *Ctenomys rosendopascuali* **Contreras,** 1995

Professor Rosendo Pascual (b. 1925) is a paleontologist. He graduated from the Faculty of Natural Sciences and Museum, National University of La Plata, in 1949 as a Doctor of Science. From 1959 he was Professor of Vertebrate Paleontology at the university's Paleontology Department and at its museum until 1990, when he became Professor Emeritus. From 1975 to 2000 he was Chief Investigator for CONICET (the National Scientific and Technical Research Council, Argentina). Since 1982 he has been an honorary member of the National Academy of Exact, Physical, and Natural Sciences of Argentina (Buenos Aires), and since 1992 an honorary life member of the Society of Vertebrate Paleontology (Lincoln, Nebraska, USA). He has traveled widely and worked in the USA for 18 months during 1963 through 1964, and had shorter appointments in Spain and France in 1965, in France again in 1971, and at the University of New South Wales in 1981. He has published more than 180 books and papers on a great number of subjects related to his speciality such as, with B. Patterson, *The Fossil Mammal Fauna of South America* (1972). Most of his papers are on South American fossil mammals, and many extinct animals are named after him. The tuco-tuco occurs in Córdoba Province, north-central Argentina.

Patrizi

Patrizi's Trident Leaf-nosed Bat *Asellia patrizii* **De Beaux,** 1931

Marquis Don Saverio Patrizi Naro Montoro (1902–1957) was an Italian explorer, zoologist, collector, and speleologist. He was the Italian signatory of a 1930s League of Nations treaty on African mammal preservation. Almost all the initial mammal and bird specimens now on display at Ethiopia's Zoological Natural History Museum were collected and prepared by Patrizi during the Italian occupation of Ethiopia. They were donated to the then University College of Addis Ababa following their discovery in storage at Akaki in 1955. Patrizi collected during the 1920s in central Africa as well as in Ethiopia during the 1930s and 1940s. He also collected in Kenya during 1946 through 1947, having been interned there during WW2. The bat is found in Ethiopia and Eritrea and has also been recorded in Saudi Arabia.

Patterson, B. D.

The rodent genus ("brucies") *Brucepattersonius* **Hershkovitz,** 1998 [7 species]

Professor Bruce D. Patterson (b. 1952) is MacArthur Curator, Department of Zoology (Mammals), the Field Museum, Chicago, and also Associate Professor, Museum of Natural History, University of San Marcos, Lima, Peru. He is a former President of both the American Society of Mammalogists and the Society for the Study of Mammalian Evolution. His current three main research areas are, in his own words, "studies of mammalian systematics and biogeography in the Neotropics, mammalian ectoparasites and using them to reconstruct the evolutionary radiations of parasite groups and host-parasite coevolution, and thirdly, ecology, behavior and conservation of lions, especially those living in Tsavo, Kenya." He has published widely around these interests. The "brucie" genus *(Brucepattersonius)* and four new species were described in a paper by Hershkovitz in 1998. All were found in the Atlantic rainforests of southeast Brazil. Subsequently, in 2000, Mares and Braun assigned three further species, found during the 1990s in the neighboring Misiones Province of Argentina, to the same genus. The genus was named in honor of Bruce Patterson, and the common name "brucie" also honors him. Professor Patterson told us that Musser and Carleton gave these rodents common names in 2005 but used "akodont" instead of "brucie"—such as Guarani Akodont, Soricine Akodont, and so on. He thinks that the alternative Guarani Brucie or Misiones Brucie comes nicely off the tongue, and we are inclined to agree with him.

Patterson, J. H.

Colonel Patterson's Eland *Taurotragus oryx pattersonianus* **Lydekker,** 1906

Colonel John Henry Patterson (1867–1947) was an Irishman who joined the British army in 1884, quickly rising through the ranks. He served in India, where experience in shooting tigers proved useful later in his career in Africa. In 1898 he was commissioned by the British East Africa Company to build a bridge over the River Tsavo. Trouble ensued as a couple of lions became man-eaters and preyed on the workers. Patterson hunted the lions down, becoming a hero to his workmen. In 1907 he published a description of his experiences, *The Man-Eaters of Tsavo and Other East African Adventures*. He served in the Boer War with distinction and afterward stayed on in Africa, becoming Chief Game Warden in the East Africa Protectorate from 1907 to 1909 and publishing *In the Grip of Nyika* in 1909 about his time in that occupation. This autobiographical account covered a curious incident that dogged him for many years. He was on safari with James Audley Blyth and Blyth's wife, Ethel. Blyth died of a gunshot wound, and there were nasty rumors about foul play, but he is thought to have committed suicide, as he believed his wife was having an affair with Patterson. The strange thing was that Patterson had Blyth buried in a deep grave and continued on the safari with the widowed Mrs. Blyth for several weeks before returning to report the incident. He was never charged or censured officially, but the scandal forced his resignation, and he and Mrs. Blyth shortly afterward left for England. Ernest Hemingway based his story "The Short Happy Life of Francis Macomber" on Patterson. During WW1 he became a major figure in Zionism, although he himself was a Protestant, as he commanded the Zion Mule Corps and the 38th Battalion of the Royal Fusiliers (also known as the Jewish Legion of the British Army, a unit that eventually served as the foundation for the Israeli Defence Force). His last two books—*With the Zionists at Gallipoli* (1916) and *With the Judeans in Palestine* (1922)—were about his experience in those times. He retired from the army as a full Colonel in 1920. He lived in California for a number of years before his death. During that period he continued to be a strong supporter of Zionism, and after the establishment of the state of Israel in 1948 his ashes were taken from the USA and interred in Israel. The subspecies of eland named after him is found in East Africa, from southern Ethiopia south to the Zambezi River, Mozambique.

Patton

Patton's Spiny Rat *Proechimys pattoni* da Silva, 1998
Rusty-sided Atlantic Tree Rat *Phyllomys pattoni* **Emmons,** Leite, Kock, and Costa, 2002
The spiny rat genus *Pattonomys* Emmons, 2005 [5 species]
Patton's Nectar-feeding Bat *Lonchophylla pattoni* Woodman and Timm, 2006

Professor Dr. James L. Patton (b. 1941) is a zoologist who works at the Museum of Vertebrate Zoology, University of California, Berkeley, USA. He is known as "Jim" to his many colleagues and friends. He was educated at the University of Arizona, completing his doctorate there in 1969. He was appointed as an Assistant Curator at the Museum of Vertebrate Zoology in that year and rose to full Professor and Curator of the Zoology Department in 1979, a post from which he retired in 2001. He is now Emeritus Professor of Integrative Biology. He is also Research Associate, Department of Mammalogy, American Museum of Natural History; Research Associate, Museum of Southwestern Biology, University of New Mexico; and Research Associate, the Museum, Texas Tech University, Lubbock, Texas. He was elected a Fellow of the American Association for the Advancement of Science in 1985, and the American Society of Mammalogists has honored him with the Merriam Award for Excellence in Research in 1893, with the Grinnell Award for Excellence in Teaching in 1998, and with election to honorary membership in 2001. He has also served on various academic boards and as an officer of several professional societies and has edited a number of professional journals. The tree rat was named in his honor because of his "remarkable contributions to knowledge of the evolution and systematics of Neo-tropical mammals, especially toward understanding the diversification of echimyid rodents." His field of interest remains "evolutionary theory; evolution, genetics, ecology, systematics, and biogeography of mammals, with emphasis on Geomyoid and Neotropical rodents and marsupials." So far he has published over 160 scientific papers or contributions to larger works, including descriptions of a number of new species. A fossil porcupine, *Neosteiromys pattoni,* from Argentina and a gopher louse, *Geomydecus pattoni,* have both been named after him. The spiny rat is found in western Brazil and eastern Peru, whereas the tree rat is confined to eastern Brazil. The bat is so far known only from eastern Peru.

Paul

Chiriqui Olingo *Bassaricyon pauli* **Enders,** 1936

Anthony Joseph Drexel Paul Jr. (1915–1999) was an amateur explorer. He was also a Harvard classmate and friend of Franklin Delano Roosevelt Jr. and a member of a wealthy Philadelphia banking family. He graduated from Harvard University in 1937 and gained a law degree from the University of Virginia in 1941. He was a partner with the Philadelphia law firm of Hepburn Willcox Hamilton and Putnam from 1956 until 1985. He was Board Chairman for Drexel from 1970 to 1974. Dr. Robert K. Enders went to Panama for the Philadelphia Academy of Natural Science in 1935, and his party, of which Paul was a member, collected the type specimen of the olingo there. The species was named after its discoverer in the academy's *Proceedings* upon his return. The Paul family had been generous patrons of the academy. This member of the Raccoon family is known only from the type locality in Panama.

Paula

Paula's Long-nosed Rat *Paulamys naso* **Musser,** 1981 [Alt. Flores Long-nosed Rat]

Paula Hamerlinck is the wife of Dr. Theodor Verhoeven. In 1980 Dr. Verhoeven and his wife visited Musser at the American Museum of Natural History in New York, and the doctor asked Musser if he would be willing to name a new taxon after Paula. The genus, which has just this one species, is confined to the island of Flores, Indonesia. It may be the world's commonest extinct mammal species. Though originally described from subfossil fragments and usually listed as extinct, the rat is reported to be alive and well and living in western Flores.

Paulina

Paulina's Rock Rat *Saxatilomys paulinae* **Musser,** Smith, Robinson, and Lunde, 2005

Paulina "Paula" D. Jenkins is a zoologist who is Collections Manager, Mammal Curation Group, at the Natural History Museum in London. She has described new small-mammal species and written numerous papers. She has been involved in the descriptions of several shrews, often in conjunction with Hutterer and others, such as, in 1983, "Species-limits of *Crocidura somalica* Thomas, 1895 and *Crocidura yankariensis* Hutterer and Jenkins, 1980." She was honored in the scientific name of the rock rat because she had given two of the describers access to the museum's rodent collection over several years. The rat comes from central Laos (Lao PDR) and was first described from two whole specimens and fragments found in owl pellets. The species doesn't seem to have acquired a generally recognized common name, so the one suggested above is our invention. See **Jenkins** for other eponymous species.

Peale

Peale's Dolphin *Lagenorhynchus australis* Peale, 1848

Peale's Free-tailed Bat *Nyctinomops aurispinosus* Peale, 1848

Peale's Meadow Mouse *Microtus montanus* Peale, 1848 [Alt. Montane Vole]

Titian Ramsay Peale (1799–1885) was an American naturalist and artist who collected in the South Pacific, particularly Tahiti, Fiji, and Samoa, between 1838 and 1842. This was while he was on the Wilkes expedition, aboard the *Peacock*. This expedition visited Madeira, Cape Verde Islands; recrossed the Atlantic to Rio de Janeiro; worked its way down the east coast of South America, around Cape Horn, and up the west coast; and explored extensively the South Pacific islands, Australia and New Zealand, the Hawaiian Islands, the Philippines, Singapore, St. Helena, the northwest coast of the USA, and California. Peale was also noted for his pen-and-ink sketches of North American Indians and buffalo, made during an extensive expedition in 1820, under Major Long, which followed the Lewis and Clark route. Peale was Assistant Naturalist on that journey. Many of the birds and mammals that he obtained from around the world contributed to the original collections of the Smithsonian Institution. Peale Islet, near Wake Island in the Pacific Ocean, is named after him. He is also commemorated in the common names of at least five birds, including Peale's Parrotfinch *Erythrura pealii* and Peale's Petrel *Pterodroma inexpectata*. The dolphin is found in cold-temperate waters off Chile, southern Argentina, and the Falkland Islands. The bat occurs from Mexico to Peru, Bolivia, and Brazil. The vole occurs in the Cascade, Sierra Nevada, and Rocky Mountain ranges of western North America. (The name "meadow mouse" is viewed as archaic and not found in modern references.)

Pearson, C. A.

Pearson's Puma *Puma concolor pearsoni* **Thomas,** 1901

Sir Cyril Arthur Pearson (1866–1921) was a journalist who launched a British newspaper, the *Daily Express,* in 1900. He went blind and sold out to Max Aitken, the future Lord Beaverbrook, in 1916. He founded the St. Dunstan's Home for soldiers blinded in WW1 and became President of the National Institution for the Blind. There were rumors of a "prehistoric" giant ground sloth, the Mylodon, being still extant in the mountains of Patagonia. Pearson sent his star reporter, Hesketh Pritchard, to find it. Despite pushing further into the Andes than any European before him, Pritchard failed to find any trace of giant sloths. But he did find a lake and the subspecies of puma, and named both after his employer. It is now thought that *pearsoni* is not a valid subspecies of puma, and that it should be included within *Puma concolor puma* along with other populations from southern South America.

Pearson, J.

Pearson's Long-clawed Shrew *Solisorex pearsoni* **Thomas,** 1924

Dr. Joseph Pearson (1881–1971) was primarily a marine biologist who spent 23 years in Ceylon (now Sri Lanka). He was Director of the Colombo Museum in the 1920s and 1930s and Marine Biologist to the government of Ceylon. By 1940 he was in Tasmania as Director of the Tasmanian Museum and as a member of the Tasmanian Fisheries Board. Pearson apparently spent New Year's Day 1924 collecting specimens, as the type of this shrew was taken on that day. It is endemic to the central highlands of Sri Lanka.

Pearson, J. T.

Hairy-footed Flying Squirrel *Belomys pearsonii* **Gray,** 1842

Pearson's Horseshoe Bat *Rhinolophus pearsonii* **Horsfield,** 1851

Dr. John Thomas Pearson (1801–1851) was one of Gray's companions during the time they were studying medicine. He is also recorded, from 1841, as a donor of specimens to the Museum of the East India Company, and in the same year he was Curator of the Museum of the Asiatic Society of Calcutta. The flying squirrel is found from northeast India and Bhutan east to southern China, Indochina, and Taiwan. The bat occurs from northern India east to Vietnam and south through the Malay Peninsula.

Pearson, O. P.

Pearson's Chaco Mouse *Andalgalomys pearsoni* **Myers,** 1977 [Alt. Pearson's Pericote]

Pearson's Tuco-tuco *Ctenomys pearsoni* Lessa and **Langguth,** 1983

Pearson's Long-clawed Mouse *Pearsonomys annectens* **Patterson,** 1992

Professor Dr. Oliver Payne Pearson (1915–2003), known to his friends as "Paynie," was a zoologist with a particular interest in mice. He was Professor Emeritus at the University of California, Berkeley, and a former Director of the Museum of Vertebrate Zoology. He made fundamental contributions in many fields, from systematics, field natural history, and biogeography to animal behavior and energetics. He undertook field trips all over Central and South America for over half a century. His first trip was to Panama in 1938, during his master's work at Harvard. He also claimed to be the original hippie: in the early 1950s he outfitted a bus and had it shipped to Peru, where he embarked on a six-month trip to study small-animal populations in Peru and Bolivia, taking his wife and children with him. He retired in 1971 and continued to study in his chosen field right up until his death, still doing fieldwork in Argentina at the age of 85. He served as a Director of the American Society of Mammalogists for a total of 17 years between 1952 and 1990. He published widely, including many papers such as "Mammals in the Highlands of Southern Peru" (1951) and the series of books, *Rare Mammals of the World.* He is also commemorated in the names of other taxa, such as the fossil rodent *Cholomys pearsoni.* The chaco mouse comes from southeast Bolivia and western Paraguay. The tuco-tuco is endemic to Uruguay, and the long-clawed mouse to Valdivia Province, Chile.

Peary

Peary Caribou *Rangifer tarandus pearyi* **J. A. Allen,** 1902

Rear Admiral Robert Edwin Peary (1856–1920) was an American naval officer and Arctic explorer. He made his first expedition to the Polar regions in 1886 and made at least half a dozen more before his expedition of 1908–1909, during which he reached the geographical North Pole. Doubts have been cast upon his achievement on the grounds that he could not have traveled so fast, but in recent years a modern expedition, using the same equipment as Peary, has shown that it *was* possible. He received many honors for his exploits and discoveries, and at least three ships, including the Liberty ship *Robert E. Peary* and a U.S. Navy destroyer *Peary,* have been named after him. Peary brought the first known specimens of this caribou subspecies from Ellesmere Land and gave them to the American Museum of Natural History. It occurs only on islands of Canada's high Arctic.

Peel

Somali Pygmy Gerbil *Microdillus peeli* **de Winton,** 1898

Charles Victor Alexander Peel (dates not found) was a British explorer, big-game hunter, and naturalist who made a series of forays into Somalia in the 1890s. He wrote *Somaliland: Being an Account of Two Expeditions into the Far Interior Together with a Complete List of Every Animal and Bird Known to That Country, and a List of the Reptiles Collected by the Author.* He also wrote *Through the Length of Africa* (1928), which was "an account of a journey from Cape Town to Alexandria and sport in Kenya Colony." He published some scientific papers, such as "In Somaliland with Descriptions of New Species" and "On a Collection of Insects and Arachnids Made in 1895 and 1897 by Mr. C. V. A. Peel FZS, in Somaliland with Descriptions of New Species," both in the *Proceedings of the Zoological Society of London.* In 1912 he was in Franz Josef Land in the Arctic, where in 10 days his "bag" included 21 polar bears. He wrote a book about this in 1928—in those days one boasted about such things rather than being reviled for them—called *The Polar Bear Hunt.* He was a Fellow of the Zoological Society and wrote a book about the zoos he had visited around Europe in the 1920s. He also wrote a dystopian novel, *An Ideal Island,* in 1927. He is commemorated in the names of other taxa, such as the scorpion *Pandinus (Pandinops) peeli.* The gerbil is endemic to Somalia.

Pel

Pel's Scaly-tailed Squirrel *Anomalurus pelii* **Schlegel** and **Müller,** 1845 [Alt. Pel's Anomalure]

Pel's Pouched Bat *Saccolaimus peli* **Temminck,** 1853

Hendrik Severinus Pel (1818–1876) was a Dutch colonial Governor of the Gold Coast (now Ghana) from 1840 to 1850. He took a keen interest in the fauna of the region and collected specimens for the Dutch State Museum of Natural History at Leiden. He had trained as a taxidermist and acted as such for the museum. In 1851 he published his account of 10 years on the Gold Coast, *Over de jagt aan de Goudkust, volgens eene tienjarige eigene onder-*

vinding. He is also commemorated in the name of Pel's Fishing Owl *Scotopelia peli* and in the scientific names of the Bristle-nosed Barbet *Gymnobucco peli* as well as two fish found off the West African coast, the Pebbletooth Moray *Echidna peli* and the Boe Drum *Pteroscion peli*. The anomalure is found in Liberia, Ivory Coast, and Ghana. The bat occurs from Liberia eastward to western Kenya and south to northern Angola.

Pelzeln

Pelzeln's Gazelle *Gazella (dorcas) pelzelni* Kohl, 1886 [Alt. Somali Dorcas Gazelle]

August Pelzel von Pelzeln (1825–1891) was an Austrian ornithologist. For 40 years he was in charge of the mammal and bird collections at the Imperial Museum of Vienna, where he worked on the 343 species of birds that Natterer, a fellow Austrian, collected in Brazil in 1822. He wrote a pamphlet in 1883, *Brasilische Säugethiere. Resultate von Johann Nattherer's Reisen in den Jahren 1817 bis 1835*, which deals with Brazilian mammals and birds that Johan Natterer gathered between 1817 and 1835 when he was on board the *Novara*. Pelzeln also wrote *Ornithologie Brasileiras* in 1871 and *Beitrage zur Ornithologie Sud Afrikas* in 1882. He is commemorated in the names of birds such as Pelzeln's Grebe *Tachybaptus pelzelnii* and Pelzeln's Tody-Tyrant *Hemitriccus inornatus*. Most authorities currently view this gazelle as a subspecies of *Gazella dorcas*. It is found in Djibouti and northern Somalia.

Pemberton

Pemberton's Deer Mouse *Peromyscus pembertoni* **Burt,** 1932 extinct

John Roy Pemberton (1884–1968) was always known, from childhood, at his own instigation, as "Bill." His father first aroused his interest in natural history by taking him on frequent outings. By the age of 11 he was a keen ornithologist. Until paralyzed by a stroke in 1960, he collected bird skins, nests, and eggs wherever he traveled. He was undecided as to whether to study ornithology or geology but decided on the latter on the grounds that there were more career opportunities, and that being a geologist was likely to mean he would travel and could thus study and collect local fauna in his spare time. Before taking his degree he spent some time working in the petroleum industry. He graduated in 1909 and, as a student, was a renowned athlete and could have become a professional boxer. He took his first appointment as a geologist in Argentina in 1910 and was promoted in 1913, not returning to the USA until 1915. He then worked for several petroleum companies in the USA, all the while continuing to collect and publishing papers on ornithology. Among other things, he taught himself to play the piano and invented a new type of boomerang. In his spare time he used his own yacht to explore the islands off the Californian and Mexican coasts. From 1940 to 1960 he set up in private practice. He continued to publish and to take volunteer positions with the Cooper Bird Club and became recognized as the leading expert on the California Condor. In recognition of his enthusiasm and patronage of scientific field exploration, the following taxa (among others) were named in his honor: *Inoceramus pembertoni* (an enormous Cretaceous mollusc), *Pinnixa pembertoni* (a commensal crab from the Gulf of California), *Lophortyx gambelii pembertoni* (a race of quail from Tiburon Island, Gulf of California), and, of course, the deer mouse. This rodent is thought to be extinct, but it used to live on San Pedro Nolasco Island, Gulf of California, Mexico.

Pennant

Pennant's Marten *Martes pennanti* **Erxleben,** 1777 [Alt. Fisher]
Pennant's Red Colobus *Piliocolobus pennantii* **Waterhouse,** 1838
Northern Palm Squirrel *Funambulus pennantii* **Wroughton,** 1905

Dr. Thomas Pennant F.R.S. (1726–1798) was an extremely highly regarded British naturalist and traveler. His early work *Tour in Scotland,* published in 1771, was instrumental in encouraging tourism in the Highlands. Gilbert White published his *Natural History of Selbourne* in the form of letters to Thomas Pennant and Daines Barrington. Pennant published on the Arctic, Britain, and India and wrote about birds as well as quadrupeds. His other works include *British Zoology* (published between 1761 and 1766), *Genera of Birds* (1777), and *Arctic Zoology* (1785). He was said to make "dry and technical material interesting." He is also commemorated in the common names of several birds such as Pennant's Parakeet *Platycercus elegans.* The fisher is found in much of Canada, except the far north, and in parts of the northern USA. The colobus monkey has a curious, disjunct distribution, in three separate subspecies: one on the island of Bioko (Equatorial Guinea), one in the Niger Delta, and one in the region of the Likouala River, Congo Republic. Its absence from intervening areas of suitable habitat is a zoological mystery. The palm squirrel occurs from southeast Iran to northern and central India.

Percival

Percival's Trident Bat *Cloeotis percivali* **Thomas,** 1901
Percival's Spiny Mouse *Acomys percivali* **Dollman,** 1911
Percival's Gerbil *Gerbillus percivali* Dollman, 1914

Arthur Blayney Percival (1874–1940) was a British game warden in East Africa from 1901 until 1928 and retired there, staying in Kenya for the rest of his life. With the taxidermist W. Dodson he took part in a Royal Society expedition to Arabia in 1899 that collected many specimens. In 1900 he was appointed Assistant Collector, which was purely an administrative job, and one year later he was made Ranger for Game Preservation. He stayed in that job until his retirement in 1923. He was instrumental in the establishment of two big-game reserves and was largely the author of the codified game laws in the East African Game Ordinance of 1906. In 1909 he was a founding member of the East Africa and Uganda Natural History Society. He was known as one of the most knowledgeable wildlife experts and hunters in eastern Africa. Percival wrote *A Game Ranger's Notebook* in 1924 and *A Game Ranger on Safari* in 1928. He is also commemorated in the name of a bird, Percival's (Montane) Oriole *Oriolus percivali.* The bat is found from Kenya south to Swaziland and Transvaal (South Africa). The spiny mouse comes from southern Sudan, Uganda, Kenya, Ethiopia, and southern Somalia. The gerbil is endemic to Kenya, but this taxon may be conspecific with *Gerbillus pusillus.*

Père David

Père David's Deer *Elaphurus davidianus* **Milne-Edwards,** 1866 [Alt. Milu]
Père David's Rock Squirrel *Sciurotamias davidianus* Milne-Edwards, 1867
Père David's Macaque *Macaca thibetana* Milne-Edwards, 1870 [Alt. Tibetan Macaque]
Père David's Vole *Eothenomys melanogaster* Milne-Edwards, 1871
Père David's Mole *Talpa davidiana* Milne-Edwards, 1884

Père Jean Pierre Armand David (1826–1900) was a French Lazarist priest as well as a fine zoologist who taught in Savona, where both Doria and d'Albertis were among his pupils. He went as a missionary to China and was the first Westerner to observe many of the region's animals, including the deer later named after him and the Giant Panda *Ailuropoda melanoleuca*. He did not arrive in China until 1862 and started collecting only a year later. The French naturalist Alphonse Milne-Edwards classified many specimens collected by Père David. This remarkable priest collected thousands of specimens and had many plants and animals named after him. He co-authored, with Oustalet, *Les oiseaux de Chine*, which they completed in 1877. By the time David saw the deer, it was confined to the Imperial Hunting Park near Peking (Beijing), having already died out in the wild. Specimens smuggled to Europe became the founders of a captive population—which was fortunate, as the Chinese population was later eradicated. Zoo-bred specimens have been reintroduced into Chinese reserves. David is commemorated in the common names of at least 11 birds. The squirrel, macaque, and vole are reasonably widespread in China, with the vole's range extending to parts of Myanmar and northern Thailand as well as Taiwan. The mole is known from southeast Turkey and northwest Iran (the latter population having been described as a separate species, *Talpa streeti*; see **Street**).

Pernetty

Pernetty's Spotted Dolphin *Stenella pernettensis* **Desmarest,** 1817 [Alt. Blainville's Spotted Dolphin; probable syn. of *S. frontalis*]

Dom Antoine-Joseph Pernetty (1716–1801) was a Benedictine monk who went as one of the original settlers to Les Îles Malouines, which is the French name for the Falkland Islands. He was the resident priest and doubled up as botanist and chronicler of the expedition that was led by Antoine Louis de Bougainville. He published his account in 1770 under the title *Journal historique d'un voyage fait aux Îles Malouines en 1763 et 1764 pour reconnaître, et y former un établissement; et deux voyages au détroit de Magellan avec une relation sur les Patagons*. In 1763 he described a spotted "porpoise" captured during the voyage, which he painted a likeness of. Desmarest described the species on the basis of Pernetty's drawing and description. Desmarest later amended the scientific name to the more correct version, *pernettyi*. Pernetty also wrote about the Warrah (Falklands Fox) *Dusicyon australis*, which was the only endemic Falklands land mammal; the last one was killed in 1876. Pernetty was a member of the Academy of Sciences in Paris and for several years was the Librarian to Frederick the Great at Berlin. About 1770 he is said to have founded a Rosicrucian order called the Illuminati of Avignon; other members were said to include Cagliostro and Mesmer. By this time he had been unfrocked for being a Cabalist and alchemist. A botanical genus, *Pernettya*, is also named after him. The names *pernettensis* and *pernettyi* are not considered valid, because of uncertainty as to exactly which species Pernetty's illustration depicts. It was probably an individual of the species now known as *Stenella frontalis*.

Perny

Perny's Squirrel *Dremomys pernyi* **Milne-Edwards,** 1867 [Alt. Perny's Long-nosed Squirrel]

Abbé Paul Hubert Perny (1818–1907) was a parish priest at Besançon before becoming a missionary and botanical explorer. He was imprisoned under the Paris Commune until his friend Jean-Pierre Guillaume Pauthier persuaded the authorities that he was harmless. In 1869 he published a 459-page French Mandarin Chinese dictionary; the volume contained not only the dictionary but also notes on the geography and administration of China and on its plants. Perny later published translations of Chinese proverbs and other material. He is also commemorated in the names of other taxa such as the Chinese Oak Silk-moth *Antheraea pernyi*, Perny Holly *Ilex pernyia*, which he discovered in 1858, and the popular garden plant Solomon's Seal *Disporopsis pernyi*. The squirrel is found in Tibet, southern China, Taiwan, northeast India, and northern Myanmar.

Peron

Southern Right-Whale Dolphin *Lissodelphis peronii* **Lacépède,** 1804
Western Bare-backed Fruit Bat *Dobsonia peroni* **Geoffroy,** 1810

François Péron (1775–1810) was a French voyager and naturalist. He was a member of the Nicolas Baudin scientific expedition with the ships *Geographe* and *Naturaliste* to southern and Western Australia from 1800 until 1804. The expedition visited New Holland, Maria Island and Van Diemen's Land (Tasmania), and Timor in Indonesia. Péron observed the dolphins for the first time on 11 January 1802 off south Tasmania. He died of tuberculosis. There is also a national park in Western Australia named after him, at the north end of the Peron Peninsula, and he is also commemorated in the scientific name of the Orange-banded Thrush *Zoothera peronii*. The dolphin is found in cold-temperate waters of the Southern Hemisphere. The bat occurs on the Lesser Sunda Islands, Indonesia.

Perrens

Goya Tuco-tuco *Ctenomys perrensi* **Thomas,** 1896 [scientific name sometimes given as *perrensis*, in error]

Richard Perrens (dates not found) was a collector in Argentina who sent specimens to the British Museum of Natural History in the last decade of the 19th century. Perrens operated around Goya in Corrientes Province, which is the area where he collected the type specimen. The tuco-tuco comes from northeast Argentina.

Perrier

Perrier's Sifaka *Propithecus perrieri* Lavauden, 1931 [Alt. Perrier de la Bathie's Sifaka, Perrier's Simpona]

Joseph Marie Henri Alfred Perrier de la Bâthie (1873–1958) was a French botanist who has been called "the father of Malagasy floristics." He wrote several books on plants as well as many scientific papers. In 1936 he wrote the massive *Biogéographie des plantes de Madagascar* cataloguing virtually all plants native to the island, although many more have been found since. He considered 5,820 of his 7,370 species (about 80%) as endemic, and 238 of the 1,289 native genera (about 20%) as endemic. Naturally there are a great number of plant names that commemorate him, including the orchid genus *Neobathiea*. The sifaka is found only in the extreme north of Madagascar and is severely endangered.

Perrin

Perrin's Beaked Whale *Mesoplodon perrini* Dalebout, Mead, Baker C. S., Baker, A., and van Helden, 2002

Dr. William F. Perrin (b. 1938) is an American cetologist and biologist who helped to collect specimens of this whale. From 1957 to 1961 he was in the U.S. Air Force as a language specialist. He gained his B.S. in biology from the University of San Diego in 1966. He works at the National Oceanic and Atmospheric Administration (Fisheries). Previously he worked at the Scripps Institute of Oceanography of the University of California, where he gained his doctorate in 1972. He was the first person to bring to public attention the slaughter of dolphins by fishing boats using purse-seine nets and was partly responsible for inclusion of a clause in the Marine Mammal Protection Act of 1972 to prevent it. He has published widely and was one of the editors and contributors to the *Encyclopedia of Marine Mammals* (2002). The beaked whale is known from specimens stranded on the Californian coast.

Peters

Peters' Gerbil *Gerbilliscus leucogaster* Peters, 1852 [Alt. Bushveld Gerbil]
Peters' Disc-winged Bat *Thyroptera discifera* **Lichtenstein** and Peters, 1855
Peters' Squirrel *Sciurus oculatus* Peters, 1863
Peters' Ghost-faced Bat *Mormoops megalophylla* Peters, 1864
Peters' Wrinkle-lipped Bat *Mormopterus jugularis* Peters, 1865
Peters' Climbing Rat *Tylomys nudicaudus* Peters, 1866
Peters' Tent-making Bat *Uroderma bilobatum* Peters, 1866
Peters' Trumpet-eared Bat *Phoniscus jagorii* Peters, 1866
Peters' Bat *Balantiopteryx plicata* Peters, 1867 [Alt. Grey Sac-winged Bat]
Peters' Dwarf Epauletted Fruit Bat *Micropteropus pusillus* Peters, 1867
Peters' Angolan Colobus *Colobus angolensis palliatus* Peters, 1868
Peters' False Serotine *Hesperoptenus doriae* Peters, 1868
Peters' Musk Shrew *Crocidura gracilipes* Peters, 1870
Peters' Tube-nosed Bat *Murina grisea* Peters, 1872
Peters' Mastiff Bat *Eumops bonariensis* Peters, 1874 [Alt. Dwarf Bonneted Bat]
Peters' Spiny Pocket Mouse *Liomys adspersus* Peters, 1874
Peters' Duiker *Cephalophus callipygus* Peters, 1876
Peters' Mouse *Mus setulosus* Peters, 1876
Peters' Striped Mouse *Hybomys univittatus* Peters, 1876
Peters' Flat-headed Bat *Platymops setiger* Peters, 1878
Black-and-Rufous Elephant-Shrew *Rhynchocyon petersi* **Bocage,** 1880
Peters' Gazelle *Gazella granti petersi* **Günther,** 1884
Peters' Pipistrelle *Falsistrellus petersi* **Meyer,** 1899

Dr. Wilhelm Karl Hartwich Peters (1815–1883) was a German naturalist and explorer who worked for many years as Director of the Berlin Zoological Museum and was one of the great men of German zoology. He started his career as the assistant of Johannes Müller, the great anatomist. Most bats were named in the 19th century by just two people, Oldfield Thomas and Wilhelm Peters. It is said of Peters that in the mid–1800s he was about 100 years ahead of his time in terms of systematics; his descriptions easily could have been written in the late 20th century. Between 1845 and 1885 he named many mammal species, although his first love was herpetology. He wrote *Naturwissenschaftliche Reise nach Mossambique,* published in nine volumes from 1852 to 1882, following a trip there that started in neighboring Angola in September 1842, when he was among the first scientists to visit the region. He took an enormous collection of natural history specimens back to the museum, and his studies of these led to his being described as the foremost expert on southern African zoology. The gerbil is found in much of southern Africa. The disc-winged bat occurs from Nicaragua south to Amazonian Brazil and Peru. The squirrel is confined to eastern Mexico. The ghost-faced bat is found from southern Texas and Arizona south to northwest Peru, northern Venezuela, and Trinidad. The wrinkle-lipped bat is endemic to Madagascar. The climbing rat is from Central America, and the tent-making bat from Central and tropical South America. The trumpet-eared bat occurs in Malaysia and Indonesia (also recorded from the Philippine island of Samar and recently from Thailand and Vietnam). Peters' Bat is found in Central America. The fruit bat's distribution ranges from Gambia east to Ethiopia and south to Angola and Zambia. The colobus monkey subspecies is found in Tanzania and southeast Kenya, and the false serotine in the Malay Peninsula and Borneo. The musk shrew is known only from the type specimen, believed to have come from the environs of Mount Kilimanjaro. The tube-nosed bat occurs in the foothills of the western Himalayas, but no specimens have been found in the last 100 years, and it may be extinct. The mastiff bat is widespread in Central and South America. The pocket mouse is endemic to Panama. The duiker occurs in west-central Africa. Peters' Mouse can be found from Sierra Leone east to Ethiopia and northern Uganda, and the striped mouse in central Africa. The flat-headed bat is known from southern Sudan, Ethiopia, and Kenya. The elephant-shrew comes from southeast Kenya and eastern Tanzania. The gazelle can be found in eastern Kenya. The pipistrelle has been recorded on Borneo, Sulawesi and the Molucca Islands (Indonesia), and Luzon and Mindanao (Philippines).

Peterson, O. A.

Peterson's Chinchilla-mouse *Euneomys petersoni* **J. A. Allen,** 1903

Olaf A. Peterson (1865–1933) was an American paleontologist. Between 1896 and 1899 he undertook three expeditions to Patagonia for Princeton University. In 1905, as Curator and Field Collector for the Carnegie Museum of Natural History, he searched deposits of Miocene mammals in western Nebraska, USA. The mouse is from southern Argentina and adjacent parts of Chile.

Peterson, R. L.

Peterson's Free-tailed Bat *Mops petersoni* El Rayah, 1981

Peterson's Long-fingered Bat *Miniopterus petersoni* **Goodman** et al., 2008

Dr. Randolph Lee Peterson (1920–1989) was a Canadian zoologist. He was Curator of Mammals at the Royal Ontario Museum of Zoology and Palaeontology in Toronto. He wrote a number of scientific papers on bats, particularly in the 1960s and 1970s, and his books include *North American Moose* and the two-volume *Mammals in Profile*. He co-wrote *Mammal Collectors' Manual* in 1988. He became the Secretary of the American Mammalogist Society in 1950 and was its President from 1966 to 1968. The free-tailed bat has been found in Cameroon and Ghana. The long-fingered bat comes from southeast Madagascar.

Petra

Petra Fruit Bat *Lissonycteris petraea* Bergmans, 1997

Dr. Peter J. H. van Bree is, at the time of writing, Curator Emeritus of Mammalogy at the Institute for Systematics and Population Biology at the Zoology Museum of the University of Amsterdam. He is interested in marine mammals and published an article in 2000, with E. J. O. Kompanje and H. Güclüsoy as co-authors, on "Osteoporosis in an Adult Female Monk Seal *Monachus monachus* from Cesme, Turkey." This fruit bat is found in Ethiopia and was originally regarded as a race of *Lissonycteris angolensis*. The vernacular name Petra appears to have been coined without reference to the fact that the scientific name *petraea* relates to a person by the name of Peter. A better name might be Van Bree's Fruit Bat.

Petter, F.

Petter's Gerbil *Taterillus petteri* Gautun, **Tranier,** and Sicard, 1985

Petter's Tufted-tailed Rat *Eliurus petteri* Carleton, 1994

Petter's Soft-furred Mouse *Praomys petteri* Van der Straeten et al., 2002

Petter's Big-footed Mouse *Macrotarsomys petteri* **Goodman** and Soarimalala, 2005

Professor Dr. Francis Petter works in the Laboratory of Zoology, Mammals and Birds, National Museum of Natural History, Paris. He has written a number of papers and articles, and in 1987 he was elected an honorary member of the American Society of Mammalogists. He has been on collecting trips to a number of countries in Africa, including the Ivory Coast, where he collected a number of specimens of *Pachybolus laminaria*, which is a millipede; microbiologists found parasites on them that appear to be the source of Ebola fever. On this trip he was accompanied by his wife, Dominique, who is a microbiologist. They co-wrote *Les Félins* in 1993. The gerbil is found in southern Mali, Burkina Faso, and western Niger. The tufted-tailed rat and big-footed mouse are both endemic to Madagascar. The soft-furred mouse occurs in Cameroon, the Congo Republic, and southwest Central African Republic.

Petter, J.

Petter's Sportive Lemur *Lepilemur petteri* Louis et al., 2006

Professor Jean-Jacques Petter (1927–2002) was a French zoologist, primatologist, and population biologist who first qualified as a physician. He was Professor of Ecology and Ethnology at the Natural History Museum of Paris and Director of the Zoological Gardens at Vincennes Zoo. He wrote extensively on Madagascan mammals in general and lemurs in particular, and was awarded the World Wildlife Fund Gold Medal in 1980 for his conservation work in Madagascar. The lemur occurs in southwest Madagascar.

Peyrieras

Peyrieras' Woolly Lemur *Avahi peyrierasi* Zaramody et al., 2007

Dr. André Peyrieras is a French biologist and one of Madagascar's most respected naturalists. He has discovered over 3,000 insects new to science. He runs Mandraka Nature Farm, which contains his private collection and where a wide variety of Madagascar's rare reptiles, frogs, and insects are bred in captivity. He has authored numerous scientific papers. He also has a reptile named after him, Peyrieras' Dwarf Chameleon *Brookesia peyrierasi*. The lemur is known from a small area of eastern Madagascar.

Pfeiffer

Pfeiffer's Hairy-tailed Bat *Lasiurus pfeifferi* **Gundlach,** 1861 [Alt. Pfeiffer's Red Bat]

Dr. Ludwig Georg Karl Pfeiffer (1805–1877) was a physician by training and a botanist by preference. He was one of the greatest experts of his day on cacti. He was also a conchologist and one of Charles Darwin's many correspondents. The bat is found on Cuba.

Phayre

Phayre's Leaf Monkey *Trachypithecus phayrei* **Blyth,** 1847 [Alt. Phayre's Langur]
Phayre's Squirrel *Callosciurus phayrei* Blyth, 1856
Indochinese Flying Squirrel *Hylopetes phayrei* Blyth, 1859

Lieutenant General Sir Arthur Purves Phayre (1812–1885) was Commissioner in Burma (now Myanmar) between 1862 and 1867, and Governor of Mauritius from 1871 until 1878. He wrote a *History of Burmah,* published in 1883. He is also commemorated in the name of a bird, Phayre's (or Eared) Pitta *Pitta phayrei*, and in the trinomial of a tortoise, *Manouria emys phayrei*. All the species named after him are found in Myanmar. The monkey's range extends from southern Assam (India) east through Myanmar, southern Yunnan (China), and north and central Thailand to parts of Laos and Vietnam. Phayre's Squirrel is found in Myanmar and Yunnan. The flying squirrel occurs from Myanmar eastward to Vietnam and Hainan (China).

Philippi

Juan Fernandez Fur Seal *Arctocephalus philippii* **Peters,** 1866

Dr. Rodolfo Amando (Rudolph Amandus) Philippi (1808–1904) was a Chilean zoologist of German extraction. He was educated in Berlin, taking a medical degree followed by a dissertation in zoology. He was then Professor of Natural History and Geography at the Polytechnic of Kassel. He went to southern Italy, believing himself to be gravely ill and wanting to end his days in a mild climate. He kept himself busy publishing articles on the geology, paleontology, and molluscs of the area. However, he recovered. In 1851 he left Europe with a letter of introduction from Alexander von Humboldt and settled in Chile. There he became Professor of Botany and Zoology and was then appointed Director of the Santiago Museum, a post he occupied for 30 years (1853–1883). He organized no fewer than 64 expeditions within Chile. He published a staggering 456 papers and described an even more astonishing 6,000 species of plant—he was a giant of Chilean natural history. He finally retired in 1896. He is commemorated in many plant names, including *Austrocactus philippii* and *Maihuenia philippii*. His son Federico was also a botanist and collaborated with him in later life. The fur seal breeds on the Juan Fernandez Islands off the coast of Chile.

Phillips

Phillips' Congo-Shrew *Congosorex phillipsorum* Stanley, Rogers, and **Hutterer,** 2004

This shrew is not named after an individual but after the Phillips family (James, Rachel, Rebecca, Richard, and Victoria) of Iringa, Tanzania. The describers state that this family has been very supportive of scientific and conservation efforts within Tanzania. The shrew occurs in the Udzungwa Mountains of Tanzania.

Phillips, A.

Phillips' Short-eared Shrew *Cryptotis phillipsii* Schaldach, 1966

Allan Robert Phillips (1914–1996) was an American ornithologist and taxonomist whose special area of study was the southwest USA and Mexico. Born in New York, he moved to Mexico in 1957 and lived there for the rest of his life. He co-authored *Birds of Arizona*, published in 1964. He sometimes collected with his friend William Schaldach, who named the shrew. The species is found in Oaxaca, southern Mexico.

Phillips, E.

Phillips' Gerbil *Gerbilliscus phillipsi* **de Winton,** 1898

Phillips' Dik-dik *Madoqua saltiana phillipsi* **Thomas,** 1894

Ethelbert Edward Lort Phillips (b. 1857) was a British traveler and hunter who shot big game in many parts of the world. He was also a collector of natural history specimens, particularly mammals and birds. He was in East Africa between 1884 and 1895. In 1895 he, with a party of friends, explored parts of Somaliland (now Somalia). This must have been a grand Victorian odyssey, as the party included both Mrs. Phillips and a Miss Gillett, with her brother Frederick Alfred Gillett; we surmise that Gillett was related to Phillips, as he changed his name to Frederick Alfred Lort Phillips in 1926. Ethelbert is also remembered in Norway as the man who developed an estate called Vangshaugen on Lake Storvatnet, where, toward the end of the 19th century, he planted a rhododendron garden (the plant was virtually unknown in Norway at that time). He became Vice President of the Zoological Society of London. He is also commemorated in the name of a bird, Phillips' (Somali) Wheatear *Oenanthe phillipsi*. The gerbil is found in Somalia, Ethiopia, and Kenya. The dik-dik is restricted to northern Somalia.

Phillips, J.

Phillips' Kangaroo-Rat *Dipodomys phillipsii* **Gray,** 1841 [Alt. Southern Banner-tailed Kangaroo-Rat]

John Phillips (dates not found) went to Mexico in 1840. He was Secretary to the Board of a company called Company of Adventurers in the Mines of Real del Monte and was also an artist who painted landscapes in Mexico, including a view of Real del Monte. The mine produced silver, and in 1846 Phillips wrote a paper called "Descriptive Notice of the Silver Mines and Amalgamation Process of Mexico." Gray wrote in his original description of the kangaroo-rat, "Mr. John Phillips, who has lately returned from Real del Monte, Mexico, has, at the recommendation of Mr. John Taylor, sent to the British Museum the skins of some very rare and interesting birds." And along with the birds came the type specimen of the kangaroo-rat. (John Taylor was the Treasurer of the Geological Society and also one of the owners of the mining company at Real del Monte.) The rat is endemic to Mexico.

Phillips, J. C.

Blesbok *Damaliscus pygargus phillipsi* Harper, 1939

Dr. John Charles Phillips (1876–1938) was an American physician and traveler. He began his travels at the age of 20, when he went to Greenland, and over the next five decades his journeys took him to Japan and Korea, Mexico, up the Nile to Khartoum, through the Canadian Rockies, and Anglo-Egyptian Sudan, Mount Sinai, and Arabian Petra (Jordan). He commanded a field hospital during WW1. After the war he traveled widely through Kenya, Uganda, and the eastern Belgian Congo, making zoological collections for Harvard's Museum of Comparative Zoology, of which he was a lifelong benefactor. Something of an all-rounder, he made studies of genetics, wrote the four-volume *Natural History of Ducks*, and took a great interest in conservation in the USA and Africa in particular. He often represented the USA in international conservation meetings. He was also largely responsible for the Migratory Bird Treaty, which involved the USA, Canada, and Mexico. In 1935 he became aware that a scientific overview of existing knowledge was needed to identify which species of birds and mammals might be saved from extinction. He raised the money for this and enlisted the services of Glover Morrill Allen and Francis Harper, who described the Blesbok, and they prepared the basic volumes of *Extinct and Vanishing Mammals of the Old and New World*, which appeared in 1942 and 1945. The Blesbok is found in parts of central and eastern South Africa. It has been reintroduced into Swaziland and Lesotho.

Phillips, R. M.

Phillips' Mouse *Mus phillipsi* **Wroughton,** 1912

R. M. Phillips (dates not found) was a Deputy Superintendent of Police in the Imperial Indian Police Service. Between 1905 and 1909 he was Deputy Police Commissioner in Bombay, and in that capacity he attended the ceremonial opening in 1907 of the first electric tramcar service in that city. He was stationed at Dharwar, and Wroughton mentions in his description of the mouse that Phillips had "given . . . whole-hearted assistance to our Collector" (the collector being **C. A. Crump**, q.v.) and that he had much pleasure in naming the mouse after him. The mouse is endemic to peninsular India.

Piacentini

Piacentini's Dik-dik *Madoqua piacentinii* Drake-Brockman, 1911 [Alt. Silver Dik-dik]

Renato Piacentini (dates not found) was an Italian diplomat. In 1909 he was Acting Consul in Eritrea and in 1911 was Consul. He was also Acting Consul General in Aden for Italy and was able to help Drake-Brockman to "send my collector to the Mijertain country" part of Italian Somaliland (now Somalia). Piacentini was a member of the colonial administration of Italian Somaliland after WW1, and because he was experienced in East African affairs, he was appointed Ambassador to Abyssinia (now Ethiopia). He became a friend of the Emperor Haile Selassie during the period of the Italian occupation. He was Ambassador in Riga (Latvia) from 1926 to 1929 and was Minister Plenipotentiary in Sofia (Bulgaria) in 1930. He appears to have been captured by British troops and was repatriated to Italy in 1942. The dik-dik is found only in eastern Somalia.

Pinheiro

Pinheiro's Slender Mouse-Opossum *Marmosops pinheiroi* **Pine,** 1981

Dr. Francisco de Paula Pinheiro is a Brazilian epidemiologist. He worked as Chief of the Virus Section at the Evandro Chagas Institute in Belém, Brazil. Part of his work in looking for potential disease vectors involved trapping animal specimens, and this seems to be how the type specimen of the mouse-opossum came to light. This small marsupial inhabits eastern Venezuela, the Guianas, and northeast Brazil.

Pitman

Pitman's Shrew *Crocidura pitmani* Barclay, 1932

Colonel Charles Robert Senhouse Pitman (1890–1975) was a British colonial administrator who went to Uganda in 1925 and stayed until 1951. He was an all-round naturalist with particular interests in malacalogy, conchology, and ornithology. He became a game warden and wrote about his work in *A Game Warden among His Charges,* published in 1931. He also wrote *A Guide to the Snakes of Uganda* (1938) as well as scientific papers such as "The Breeding of the Standard-wing Night-jar" (1929) and "The Gorillas of the Kayonsa Region" (1935). In 1942 he recounted that there were stories of a flying "monster" that dwelt in swamps on the borders of Angola and the Belgian Congo. Sadly, no specimens of this winged wonder have ever been found. Pitman also laid the foundations of the Kafue National Park in Northern Rhodesia (now Zambia), and he was seconded to Northern Rhodesia to undertake a game survey between 1931 and 1932, which is when the type specimen of the shrew was taken. The Bunyoro Rabbit *Poelagus marjorita* is named after his wife (see **Marjorie**). The shrew is known only from Zambia.

Pittier

Pittier's Crab-eating Rat *Ichthyomys pittieri* **Handley** and **Mondolfi,** 1963

Henri François Pittier, or Henri François Pittier Dormond (1857–1950), was a Swiss botanist, geographer, civil engineer, and all-round naturalist. He published over 300 papers, monographs, and books in various languages on three continents covering a wide variety of subjects including geography, botany, forestry, archeology, ethnography, linguistics, geology, and climatology. He moved to Costa Rica at the age of 30 in 1887 but traveled over much of Latin America. In 1953 a Venezuelan nature reserve was designated the Henri Pittier National Park. It is the oldest nature reserve in Venezuela, created in 1937, and the area was well known to Pittier himself, who visited often over the course of 30 years. The rat is found along streams in northern Venezuela.

Plain

Plain's Gazelle *Oryx gazella* **Linnaeus,** 1758 [Alt. Sub-Saharan Oryx]

Plain's Zebra *Equus quagga burchelli* **Gray,** 1823 [Alt. Burchell's Zebra]

These names are merely common transcription errors and should be spelled "Plains"; they refer to the habitat of these African mammals and not to a person.

Plancy

Chinese Noctule *Nyctalus plancyi* **Gerbe,** 1880

Victor Collin de Plancy (1853–1922) was a diplomat and traveler in China, Korea, and Siam (now Thailand). The type specimen of this bat was sent from Beijing by him. While other European nations' diplomats were interfering in the politics of Korea, de Plancy, who was the head of the French mission there for 15 years, was interested in the language and culture. He made studies of both, catalogued as many Korean books as he could, and also donated a number of Buddhist paintings to the Guimet Museum. The Bibliothèque Nationale de France purchased his impressive collection of books, which included xylographic editions of Korean books and maps. The Chinese Noctule was formerly regarded as a race of the Common Noctule *Nyctalus noctula*. It is found in eastern China and Taiwan.

Pluto

Pluto Monkey *Cercopithecus mitis mitis* Wolf, 1822 [Alt. Angolan Blue Monkey]
Philippine Pouched Bat *Saccolaimus pluto* **Miller,** 1910
Pluto Tamarin *Saguinus mystax pluto* Lönnberg, 1926 [Alt. White-rumped Moustached Tamarin]

Pluto was the Roman god of the underworld and judge of the dead. In keeping with the idea that the underworld is a dark and gloomy place, the name is usually applied to animals of a dark appearance. Pluto was regarded with trepidation, and people were afraid to say his real name for fear they might attract his attention. Black sheep were sacrificed to him. The monkey is found in northern Angola. The bat occurs in the Philippines (it is sometimes treated as a subspecies of *Saccolaimus saccolaimus*). The tamarin comes from western Amazonian Brazil.

Pocock

Pocock's Zebra *Equus burchelli pococki* Brasil and Pennetier, 1909
Pocock's New Guinea Rat *Rattus (Stenomys) pococki* **Ellerman,** 1941
Pocock's White-throated Guenon *Cercopithecus erythrogaster pococki* Grubb, Lernould, and Oates, 1999

Reginald Innes Pocock (1863–1947) was a zoologist who became an assistant at the British Museum Natural History in 1885. He resigned in 1904 to become the Superintendent of the Zoological Garden in London, a post he held until his retirement in 1923. In retirement he worked as a volunteer assistant in the Mammals Department of the British Museum. He was also a keen ornithologist. He published widely, particularly on Arachnida and Myriapoda; he had been in charge of these collections at the Natural History Museum. Though a well-known zoologist, he did not do well in having mammal species named after him. The zebra description was based on a specimen from an unknown locality, and the subspecies is no longer regarded as valid. The rat is found in the highlands of New Guinea, but its taxonomic status is also uncertain; it may be a subspecies of *Stenomys niobe*. The monkey was named long after Pocock's death, when Grubb et al. resurrected an old, unofficial manuscript name of John Guy Dollman's. The existence of this guenon had been known for a long time, but its status was unclear; it was generally regarded as being a color-morph of *Cercopithecus erythrogaster* rather than a subspecies. It is found in southwest Nigeria.

Poeppig

Poeppig's Woolly Monkey *Lagothrix poeppigii* Schinz, 1844 [Alt. Silvery Woolly Monkey]

Professor Eduard Friedrich Poeppig (1798–1868) was a German naturalist and collector. He studied medicine and natural science at

Leipzig University and left to undertake an expedition to Cuba and the USA. In 1826 he went from there to Chile. He then took part in an expedition to Brazil and Peru from 1829 to 1832. In 1835 he published *Reise nach Chili, Peru, und auf dem Amazonen-Flusse* (two volumes), then, in 1845, *Nova genera ac species plantarum quas in regno, Chiliensi, Peruviano, ac Terra Amazonica, anni 1827–1832 lectarum* (three volumes). When he returned to Germany he was appointed as Professor of Zoology at Leipzig. He is commemorated in the scientific names of other taxa such as the tree *Maihuenia poeppigii*. The monkey is found in eastern Ecuador, northeast Peru, and adjacent western Brazil. He is also commemorated in the name of the Basin Ground Snake *Atractus poeppigi*.

Poey

Poey's Pallid Flower Bat *Phyllonycteris poeyi* **Gundlach,** 1860 [Alt. Cuban Flower Bat]

Professor Felipe Poey y Aloy (1799–1891) was a Cuban zoologist, naturalist, and artist. He was brought up in France from 1804 to 1807 and later Spain. He qualified as a lawyer in Madrid, but his ideas were too liberal for the age, and he was forced to return to Cuba in 1823. He concentrated on natural history, describing 85 species of Cuban fish. His *Memorias sobre la historia natural de la isla de Cuba, acompañadas de sumarios latinos y extractos en frances*, published in 1858, depicts mainly fish and snails but also some mammals, Hymenoptera, and Lepidoptera; the drawings are all done by Poey. The Felipe Poey Natural History Museum in Havana—created in 1842, merged in 1849 with the University of Havana, and the first such museum in Cuba—was named in his honor, and he was its first Director. He also became the first Professor of Zoology and Comparative Anatomy at the University of Havana. Many of the museum's exhibits, especially of fish, were prepared by Poey, who also supplied Cuvier in Paris with Cuban fish specimens. He is commemorated in the scientific names of other taxa, especially fish such as Poey's Scabbardfish *Evoxymetopon poeyi* but also moths such as *Nannoparce poeyi*. The bat occurs on Cuba and Hispaniola.

Pohle

Geelvink Bay Flying Fox *Pteropus pohlei* **Stein,** 1933

Pohle's Fruit Bat *Scotonycteris ophiodon* Pohle, 1943

Hermann Pohle (1892–1977) was a German zoologist who was the Curator of Zoology at the Berlin Museum. In 1926 he co-founded the German Mammalogists' Society and became its first Secretary. He wrote "Die Unterfamilie der Lutrinae" (about otters) in 1919 and "Uber die Fledertiere von Bougainville" in 1953. The flying fox is found on Japen Island, Geelvink Bay, northwest New Guinea. The fruit bat is found from Liberia east to the Congo Republic.

Polia

Polia's Shrew *Crocidura polia* **Hollister,** 1916 [Alt. Fuscous Shrew]

This shrew is not named after a person at all, and the vernacular name Polia's Shrew appears to be the result of a misunderstanding. The alternative name, Fuscous Shrew, gives the clue, as *polia* derives from a Greek word for "greyhaired." The species is known only from the type locality in DRC (Zaire).

Poll

Poll's Shrew *Congosorex polli* **Heim de Balsac** and **Lamotte,** 1956 [Alt. Greater Congo Shrew]

Professor Dr. Max Poll (1908–1991) was a Belgian zoologist, ichthyologist, and "connoisseur of the fish fauna." Naturally he is commemorated in the scientific names of numerous fish, in such diverse species as the shark

Galeus polli and the cichlid *Tropheus polli.* He collected and studied in the Congo but also worked at the Congo Museum in Belgium. In 1953 he led an expedition to carry out a survey of the Congo River. The shrew is known only from the type locality in Kasai, southern DRC (Zaire). It has not been found again since its discovery in 1955.

Poncelet

Poncelet's Naked-tailed Rat *Solomys ponceleti* **Troughton,** 1935 [Alt. Poncelet's Giant Rat]

The Rev. Jean-Baptiste Poncelet S.M. (1884–1958) headed the Catholic Mission at Buin, on the island of Bougainville, in the 1930s. He also collected natural history specimens for a person by the name of V. Danis, who worked at the Paris Museum. This endangered species is found on the Solomon Islands of Bougainville and Choiseul.

Portenko

Portenko's Shrew *Sorex portenkoi* Stroganov, 1956 [Alt. Chukotka Shrew]
Wrangel Island Lemming *Lemmus portenkoi* Tchernyavsky, 1967

Dr. Leonid Aleksandrowitsch Portenko (1896–1972) was a Russian zoologist, primarily an ornithologist. He completed his Ph.D. at the St. Petersburg Zoological Institute in 1929. He spent the next 10 years at the Arctic Institute but returned to St. Petersburg in 1940 and spent the rest of his working life there. He made a number of expeditions to little-studied and remote areas of the USSR, including the northern Urals, Novaya Zemlya, Wrangel Island, Kamchatka, and the Kuril Islands. He wrote a great many papers and books and edited and contributed several chapters to the multivolume work *The Birds of the USSR,* published from 1951 to 1954. He is also remembered for having described an important zoogeographical concept on the unity of the fauna of circumpolar tundra; he was an advocate of the single Holarctic zoogeographic region. The shrew is found in northeast Siberia. The lemming is endemic to Wrangel Island off the Siberian coast.

Porteous

Porteous' Tuco-tuco *Ctenomys porteousi* **Thomas,** 1916

Don Cecil John Montague Porteous (1884–1953) and Lieutenant Colonel John James Porteous (1857–1948) are both cited by the describer. The etymology by Thomas reads, "Thanks to the hospitality of the owners of the ranche, and especially to the kind assistance of Don Cecil Porteous, Mr. Robin Kemp has been enabled to make a collection of small mammals at La Maria Luisa Ranche, on the Pampas of Buenos Ayres Province." And later: "I have named this distinct Tuco-tuco in honor of Don Cecil Porteous, but in doing so I may also recall the help Mr. Kemp has received from Col. J. J. Porteous in various matters connected with his trip." Colonel Porteous was a career officer in the Royal Artillery before becoming a livestock breeder and businessman. His partner was another retired officer, Colonel Charles Lewis William Morley Knight. Colonel Porteous and Knight first went to Argentina in 1889 to look for horses and mules on behalf of the British army. They liked what they saw and decided to become partners, not only in horse breeding but also in cattle. Knight stayed on in Argentina and looked after the farm they jointly owned, Las Tres Lagunas, in Las Rosas, Santa Fé Province, while Colonel Porteous returned to the UK. He visited Argentina only occasionally and looked after the European end of the business. They not only bred excellent horses but also introduced the Aberdeen Angus breed of cattle into Argentina in 1890. During WW1 Colonel Porteous was in the Remount Service, as was appropriate for a horse breeder. In 1918 he was awarded the C.M.G. (Companion of the Order of St. Michael and St. George) and was also "Mentioned in Despatches." Colonel Porteous' and Knight's help facilitated other peoples' expeditions, which resulted in the discovery of a number of new mammals. As yet we have found little about Don Cecil Porteous, beyond that he appears to have been Colonel Porteous' nephew, and that he continued to live at La Maria Luisa until 1945, having been given the property by his uncle in 1938. The tuco-tuco occurs in eastern Argentina (Buenos Aires and La Pampa provinces).

Porter

Porter's Rock Rat *Aconaemys porteri*
Thomas, 1917

Professor Dr. Carlos Emilio Mosso Porter (1867–1942) was a Chilean zoologist and entomologist. He founded the *Revista Chilena de Historia Natural* in 1897 and the *Anales De Zoologia Aplicada* in 1914, merging the two publications in 1924. He founded, and was the first Director of, the Natural History Museum in Valparaíso, where he also served as Professor of Physiology and Hygiene in the College of Engineering of the Chilean navy. The museum was destroyed by an earthquake in 1906, after which Porter moved to Santiago before spending a short period (1910–1912) in Europe. After his return to Chile he filled, at one time, no fewer than four professorial chairs simultaneously and directed the Zoological Museum. In 1928 the Institute of Zoology in Santiago was founded at Porter's instigation. During his career he published over 400 papers, articles, and books. He also has other taxa named after him, including several spiders such as *Melanophora porteri*. The rock rat is found in southern Chile and adjacent parts of western Argentina, at altitudes of 900–2,000 m (3,000–6,500 feet) above sea level.

Potenziani

Mentawai Langur *Presbytis potenziani*
Bonaparte, 1856

Marquis Ludovico Potenziani (dates not found) was a Roman aristocrat who was nominated to the Council of State in Italy in 1811 and was Sub-Prefect at Chiavari in 1813. These appointments were made during the period when the Napoleonic Empire included much of Italy. In later life he was a friend of the French writer Stendhal and of the naturalist Prince Bonaparte, who dedicated this monkey to his late friend's memory with the words "Dédié à mon ami le marquis L. Potenziani, puisse-t-il contribuer à perpétuer sa mémoire sur la terre" (Dedicated to my friend the Marquis L. Potenziani, may it contribute to perpetuating his memory on Earth). The langur is found on the Mentawai Islands off the west coast of Sumatra.

Pousargues

Pousargues' Mongoose *Dologale dybowskii*
Pousargues, 1893
Pousargues' Fat Mouse *Steatomys opimus*
Pousargues, 1894
Pousargues' White-collared Monkey
Cercopithecus albogularis albotorquatus
Pousargues, 1896

Eugène de Pousargues (d. 1901) was a zoologist and anatomist. No one seems to have recorded when or where he was born, but it is known that he died at a young age. In 1894 he wrote *Description d'une nouvelle espèce de mammifère du genre Crossarchus,* and in 1895 *Note sur l'appareil génital male des orang-outans.* He became Curator of Mammals and Birds at the Museum of Natural History in Paris in 1900. His major interest seems to have been primates and their sexual anatomy. The mongoose is found in the Central African Republic, northeast DRC (Zaire), southern Sudan, and Uganda. The mouse occurs in forest margins and moist savannah from Cameroon east to the southwest corner of Sudan. The monkey is from southern Somalia and the Tana River area of Kenya.

Powell-Cotton

See **Cotton.**

Prater

Prater's Cat *Felis chaus prateri* **Pocock,** 1941 [Alt. Jungle Cat]

Stanley Henry Prater (1890–1960) was the Curator of the Bombay Natural History Society from 1923 to 1947. He wrote *The Book of Indian Animals,* which first appeared in 1948 and has since been updated and republished several times. (We first thought that a reference to "Prater's cat" was merely a reference to a fictional and individual cat, "Mr. Prater" being a fictional housemaster at a public school in a short story by P. G. Wodehouse entitled "The Tabby Terror".) This race of Jungle Cat comes from western India.

Pratt, A. E.

Pratt's Roundleaf Bat *Hipposideros pratti* **Thomas,** 1891
Pratt's Vole *Eothenomys chinensis* Thomas, 1891 [Alt. Sichuan Red-backed Vole]
Green Acouchi *Myoprocta pratti* **Pocock,** 1913

Antwerp Edgar Pratt (ca. 1850–ca. 1920) was an English explorer and naturalist who was a corresponding member of the Royal Geographical Society. In 1891, while traveling to Tibet, he visited Tatsienlu (China) and there met two famous naturalists, both of whom were Lazarite missionaries: Bishop Felix Biet and Père Jean André Soulie. Pratt wrote *To the Snows of Tibet through China* (1892) and, with his son, *Two Years among New Guinea Cannibals* (1906). Whether or not the following observation of his is true, it is certainly fascinating: "One of the greatest curiosities that I noted during my stay in New Guinea was the spiders' web fishing net near Waley. In the forest at this point huge spiders' webs, six feet in diameter, abounded. These are woven in a large mesh, varying from one inch square at the outside of the web to about one-eighth inch at the center. The web was most substantial, and had great resisting power, a fact of which the natives were not slow to avail themselves, for they have pressed into the service of man this spider, which is about the size of a small hazelnut, with hairy, dark brown legs spreading to about two inches. This diligent creature they have beguiled into weaving their fishing nets. At the place where the webs are thickest they set up long bamboos, bent over into a loop at the end. In a very short time the spider weaves a web on this most convenient frame, and the Papuan has his fishing net ready to his hand. He goes down to the stream and uses it with great dexterity to catch fish of about one pound weight, neither the water nor the fish sufficing to break the mesh. The usual practice is to stand on a rock in a backwater where there is an eddy. There they watch for a fish, and then dexterously dip it up and throw it onto the bank." The bat comes from southern China, Myanmar, Thailand, West (mainland) Malaysia, and Indochina. The vole is recorded from the highlands of Sichuan (China). The acouchi is found in the Amazon basin of Brazil, southern Colombia, eastern parts of Ecuador and Peru, and northern Bolivia. (The "correct" scientific name for the Green Acouchi is now sometimes said to be *Myoprocta acouchy,* but *pratti* can still commonly be found in the literature, and the taxonomy of acouchis remains unresolved.)

Pratt, C., F., and J.

Seram (or Ceram) Bandicoot *Rhynchomeles prattorum* **Thomas,** 1920

The bandicoot was named in honor of Charles, Felix, and Joseph Pratt. Charles and Felix (see **Felix**) were the sons of A. E. Pratt (see above). The brothers explored together and collected on the islands of the East Indies and New Guinea. Joseph was also related, but we are not certain quite how. The bandicoot is found on the Indonesian island of Seram. It has not been recorded since its original discovery, and some zoologists fear it may now be extinct.

Preble

Preble's Kangaroo-Rat *Dipodomys microps preblei* **Goldman,** 1921
Preble's Shrew *Sorex preblei* **Jackson,** 1922
Preble's Meadow Jumping Mouse *Zapus hudsonius preblei* Krutzsch, 1954

Edward Alexander Preble (1871–1957) was an American naturalist and conservationist who undertook significant field exploration of the northwestern USA and Canada, collecting birds and mammals in particular. He was appointed a field naturalist with the U.S. Bureau of Biological Survey in 1892 under Clinton Hart Merriam. In 1900 he went to the Hudson Bay area with his brother, Alfred Emerson Preble. In 1908 Edward published *A Biological Investigation of the Athabaska-Mackenzie Region.* Between 1907 and 1915 he continued to travel and collect, and he compiled *A Biological Investigation of the Pribilof Islands* in 1923. His work with the Biological Survey resulted in faunal surveys and wildlife management reports, but few systematic or taxonomic studies. He meticulously recorded his observations of the geography, local people, weather, and so forth, as well as the local fauna. He was Chairman of the Editorial Committee for the American Society of Mammalogists' *Journal of Mammalogy* from 1930 to 1935 and was made a fellow of the American Ornithologists' Union in 1935. Toward the end of his government career he became Senior Biologist for the Biological Survey and developed an interest in conservation. In 1925 he became Consulting Naturalist for *Nature* magazine, and in 1935 he became its Associate Editor. He is credited with 239 published items amounting to 1,500 pages in the form of articles, books, reports, and the like. The kangaroo-rat occurs in southeast Oregon and northwest Nevada (USA). The shrew is found in the western USA (eastern Washington State south to northeast California and east to Montana, western Wyoming, and northern Utah). The jumping mouse comes from Colorado and southeast Wyoming, but DNA research suggests that it is not a valid subspecies.

Preuss

Preuss' Mouse-Shrew *Myosorex preussi* **Matschie,** 1893 BOGUS SPECIES
Preuss' Monkey *Cercopithecus preussi* Matschie, 1898 [Alt. Cross' Guenon]
Preuss' Red Colobus *Piliocolobus preussi* Matschie, 1900

Professor Paul Preuss (1861–1926) was a Polish-born German naturalist, botanist, and horticulturist. He was known to be collecting in West Africa between 1886 and 1898, in New Guinea around 1903, and again in West Africa in 1910. He was a member of Zintgraff's 1888–1891 military expedition to explore the hinterland of Cameroon, then a German colony. While storming a native village, the troop commander was killed and the second-in-command severely wounded. Preuss took command and led the remaining troops back to the coast. He constructed the botanical gardens of Victoria (Limbe) in Cameroon in 1901 and was in the employ of the colonial government there. He is also commemorated in the names of such birds as Preuss' Cliff Swallow *Hirundo (Petrochelidon) preussi* and Preuss' (Golden) Weaver *Ploceus preussi.* The mouse-shrew is now considered to be bogus, as reexamination of the type series showed that they are based on mismatched parts of three different shrew genera *(Crocidura, Sorex, Sylvisorex).* Both species of monkey occur in the region of the Nigeria-Cameroon border, with Preuss'

Monkey being additionally found on the island of Bioko (Equatorial Guinea).

Prevost

Prevost's Squirrel *Callosciurus prevostii* **Desmarest,** 1822

Florent Prévost (or Prevot) (d. 1870) was a French artist and writer who worked on museum collections. He was Assistant Naturalist at the Paris Museum. His primary interest was birds, and he wrote *Iconographie ornithologique* in 1845 and *Histoire naturelle des oiseaux d'Europe* in 1864. He also illustrated works by Temminck, Bonaparte, and Buffon. He is commemorated in the scientific name of a hummingbird, the Green-breasted Mango *Anthracothorax prevostii.* The squirrel inhabits the Malay Peninsula, Sumatra, Borneo, and some small nearby islands (but not Java).

Priam

Tufted Grey Langur *Semnopithecus (entellus) priam* **Blyth,** 1844

Priam (ca. 1,200 B.C.) was King of Troy. His eldest son was Hector (q.v.), killed by Achilles during the Trojan War. Priam himself was killed by Achilles' son, Neoptolemus, even though he clung for sanctuary to an altar to Zeus. The best source for the full story is Homer's *Iliad.* This langur is found in southern India and Sri Lanka.

Prigogine

Prigogine's Angolan Colobus *Colobus angolensis prigoginei* **Verheyen,** 1959

Alexandre Prigogine (1913–1991) was a Russian-born Belgian naturalist, whose brother Ilya (1917–2003) won the Nobel Prize for Chemistry in 1977. The family left Moscow after the Revolution and settled in Belgium. In his Nobel Prize acceptance speech Ilya mentioned Alexandre by name as also having studied chemistry. Alexandre has written extensively on birds and sponsored a number of expeditions to central Africa, especially the former Belgian Congo (now Zaire). He wrote *Les oiseaux de l'Itombwe et de son hinterland,* published in three volumes between 1971 and 1984. He is remembered in the names of eight birds, the most well-known being the Congo Bay Owl *Tyto (Phodilus) prigoginei.* The subspecies of colobus monkey is known only from Mount Kabobo, eastern DRC (Zaire).

Prince Alfred

Prince Alfred's Deer *Cervus alfredi* **Sclater,** 1870 [Alt. Philippine Spotted Deer, Visayan Spotted Deer; Syn. *Rusa alfredi*]

Prince Alfred Ernest Albert, Duke of Edinburgh and Saxe-Coburg Gotha (1844–1900), was the second son of Queen Victoria and Prince Albert and husband of Her Imperial Highness Maria Alexandrovna, Grand Duchess of Russia. He followed a career in the Royal Navy until 1893, when he succeeded his uncle, Ernst, as the reigning Duke of Saxe-Coburg and Gotha. He was always interested in the sea and geography and traveled widely in his youth. He collected and dried flowers, which he sent to his aunt (a collector of such), and he also had an interest in agriculture and animal husbandry. In 1868 he was shot in the back by an Irish nationalist while visiting Australia but survived the attack. The deer is found on the Philippine islands of Negros and Panay (formerly also on Cebu and Masbate).

Prince Bernhard

Prince Bernhard's Titi *Callicebus bernhardi* M. G. M. **van Roosmalen,** T. van Roosmalen, and Mittermeier, 2002

HRH Prince Bernhard of the Netherlands (1911–2004) was also a noted naturalist. He created the Order of the Golden Ark to honor conservationists internationally. This award has been bestowed on two of those who described the titi monkey: Marc van Roosmalen and Russell Mittermeier. Van Roosmalen presented his discovery of the titi to the Prince at Soestdijk Palace in Holland on 25 June 2002, four days before the Prince's 91st birthday. Prince Bernhard also received a special portrait of "his" monkey by Stephen Nash. The titi is found in Brazil, between the Madeira River and the lower Aripuana River.

Pringle

Pringle's Gerbil *Gerbilliscus pringlei* Hubbard 1970

Dr. Gordon Pringle (b. 1916) is a British parasitologist and malacologist. From 1946 until 1958 he was the senior malacologist at the Ministry of Health in Baghdad. In 1958 he moved to the Malaria Institute at Amani in Tanganyika (Tanzania), where he worked first as Deputy Director until 1960 and then as Director up to 1966, when he returned to the UK. He worked as a parasitologist for the Pfizer Group from 1966 to 1970, in which year he became Senior Lecturer in Tropical Hygiene at the Liverpool School of Tropical Medicine. In 1967 he published *Malaria in the Pare area of Tanzania*. Hubbard used the following words in his description of the gerbil: "This gerbil bears the name of Dr. Gordon Pringle, Director of the Malaria Institute, Amani, Tanzania, through whose kind offices the writer's three years' study in East Africa was made possible." The gerbil is found only in one location in northeast Tanzania, but this "species" is often regarded as being a population of the Gorongoza Gerbil *Gerbilliscus inclusus*.

Prometheus

The vole genus *Prometheomys* **Satunin,** 1901 [1 species: *schaposchnikowi*]

Prometheus, in Greek mythology, was a Titan who stole fire from Zeus and gave it to mortals for their use. Zeus, naturally enough, was not pleased at this, so he had Prometheus chained to a rock. Every day an eagle came to devour his liver. Prometheus was, of course, immortal, so his liver grew back every night, and thus he was forced to suffer agony every day until the hero Heracles eventually freed him. The rock Zeus chose was on the top of the Caucasus Mountains, which is the region where the Long-clawed Mole-Vole *Prometheomys schaposchnikowi* can be found.

Przewalski

Przewalski's Horse *Equus (ferus) przewalskii* Poliakov, 1881 [Alt. Mongolian Wild Horse]
Przewalski's Gerbil *Brachiones przewalskii* Büchner, 1889
Przewalski's Steppe Lemming *Eolagurus przewalskii* Büchner, 1889
Przewalski's Gazelle *Procapra przewalskii* Büchner, 1891
Przewalski's Hare *Lepus oiostolus przewalskii* **Satunin,** 1907

General Nikolai Mikhailovitch Przewalski (1839–1888) was a Russian Cossack naturalist who explored central Asia. There are at least half a dozen different spellings of his name, but he signed himself as Prjevalsky (pronounced "she-val-ski"). He was undoubtedly one of the greatest explorers the world has ever seen and made four major expeditions to parts of central Asia. He wrote *Mongolia and the Tangut Country* in 1875 and *From Kulja, across the Tian Shan to Lob-Nor* in 1879. He died of typhus at the age of 49 while preparing for a fifth expedition. Tsar Alexander III decreed that the town where he died, Karakol, should immediately have its name changed to Prjevalsk. In 1946 the Russian Academy of Sciences instituted the Prjevalsky Gold Medal. All the mammals named after him are found in the steppes and deserts of Mongolia and/or western China. Przewalski's Horse died out in the wild, probably in the late 1960s, but from a small number taken into captivity enough have been bred to allow reintroductions into Mongolia.

Pucheran

Andean Squirrel *Sciurus pucheranii*
Fitzinger, 1867

Jacques Pucheran (1817–1894) was a French zoologist who went on the expedition of the *Astrolabe* with Dumont d'Urville, Gaimard, and Jacquinot. On his return he contributed the ornithological section of *Voyage au Pole Sud et dans l'Oceanie sur les corvettes L'Astrolabe et La Zelée*, in 1842. He is also commemorated in the binomial of the Black-cheeked Woodpecker *Melanerpes pucherani*. The squirrel is found in the Andes of Colombia.

Pundt

Pundt's Tuco-tuco *Ctenomys pundti*
Nehring, 1900

Moritz Pundt (dates not found) was the proprietor of a substantial agricultural property in Argentina. In the original etymology Nehring wrote, "Some time ago I received from Argentina the skull and skin of a type of *Ctenomys*, which seems to be new. It was sent to me by Mr. Moritz Pundt (one of my former students)." The species is found in central Argentina.

Quarles

Mountain Anoa *Bubalus quarlesi* **Ouwens,** 1910 [Syn. *Anoa quarlesi*]

This small buffalo is endemic to the mountains of Sulawesi, Indonesia. One of these ranges is known as the Quarles Mountains, and we are satisfied that the anoa is not named after a person but after these mountains.

Queen Charlotte

Queen Charlotte Caribou *Rangifer tarandus dawsoni* Seton-Thompson, 1900 [Alt. Dawson's Caribou]

This deer is named after the Queen Charlotte Islands, British Columbia, Canada, not directly after Queen Charlotte. Indeed, the islands were named by Captain George Dixon after one of his ships, the *Queen Charlotte*, which in turn was named after the wife of King George III. So the eponym is thrice removed. **See also Dawson.**

Queen of Sheba

Queen of Sheba's Gazelle *Gazella bilkis* Groves and Lay, 1985 extinct

The Queen of Sheba was a biblical character who visited King Solomon in Jerusalem (see 1 Kings 10.1–13). According to the Koran her name was Bilkis, which is used in the scientific name of the gazelle; the name was also used by A. Kopff when he named an asteroid after her. The land of Sheba is believed by many researchers to have been located in modern-day Yemen. The gazelle is known from a handful of specimens collected near the city of Ta'izz, Yemen. It has not been recorded since 1951 and is now feared to be extinct.

Queen Victoria

Queen Victoria's Ibex *Capra pyrenaica victoriae* **Cabrera**, 1911 [Alt. Gredos Ibex]

Queen Victoria Eugenie, formerly Princess Ena of Battenberg (1887–1969), was the goddaughter of Queen Victoria of Great Britain. She married King Alfonso XIII of Spain in 1906. The ibex is found in the mountains of central Spain, in particular the Gredos Mountains.

R

Rabor

Palawan Montane Squirrel *Sundasciurus rabori* **Heaney,** 1979
Leyte Shrew-Rat *Crunomys rabori* **Musser,** 1982
Rabor's Tube-nosed Bat *Nyctimene rabori* Heaney and Peterson, 1984 [Alt. Philippine Tube-nosed Fruit Bat]
Visayan Leopard Cat *Prionailurus bengalensis rabori* Groves, 1997

Dr. Dioscoro S. (Joe) Rabor (1911–1996) was the preeminent Philippine zoologist and conservationist of the 20th century. He graduated from the University of the Philippines in 1934 and completed his Ph.D. at Yale under S. Dillon Ripley in 1958. He is best known for his work on birds and mammals, but his expertise also extended to herpetology and ichthyology. His career spanned 30 positions, mostly in academic and research posts. He led more than 50 expeditions between 1935 and 1977, his wife and, later, all six children usually going along, too. His love for nature was reflected in every aspect of his life, including the fact that all four of his daughters were named after the scientific names of animals (fish for Alectis Cyrene, birds for Iole Irena, Nectarinia Julia, and Ardea Ardeola). His scientific bent clearly rubbed off, as all four became physicians. Rabor was responsible for the most thorough documentation of the mammals and birds of the Philippines ever. In 1977 he published *Philippine Birds and Mammals: A Project of the U.P. Science Education Center.* He found many new mammal taxa. He published nearly 90 books and articles and was also awarded a number of fellowships and other honors. From his collections, 61 subspecies and 8 full species of birds have been described; one was named after him in its binomial, *Napothera rabori,* and another for his wife Lina, *Aethopyga linaraborae.* All the mammals named after him are found in the Philippines. The squirrel comes from the island of Palawan and the shrew-rat from Leyte (though the latter is now regarded as conspecific with *Crunomys melanius* of Mindanao). The bat occurs on Negros, Cebu, and Sibuyan. The subspecies of Leopard Cat is found on Negros, Cebu, and Panay.

Racey

Racey's Pipistrelle *Pipistrellus raceyi* Bates et al., 2006

Professor Paul Adrian Racey (b. 1944) is the Regius Professor of Natural History at the University of Aberdeen, where he was Professor of Zoology from 1985 to 1993, having originally joined the department in 1973. His initial degree is from Cambridge, and after qualifying he worked in a number of posts, including at the Zoological Society of London from 1966 to 1970 and Liverpool University from 1970 to 1973. He is an expert in the biology of European and tropical bats, particularly those of Madagascar, and of Mexico and Trinidad, where he has had projects. The pipistrelle is a Madagascan endemic.

Radde

Ciscaucasian Hamster *Mesocricetus raddei* **Nehring,** 1894
Radde's Shrew *Sorex raddei* **Satunin,** 1895

Gustav Ferdinand Richard Radde (1831–1903) was originally trained as an apothecary. Born in Danzig in Prussia (now Gdansk, Poland), he settled in Russia in 1852. He then participated in numerous expeditions through Siberia, Crimea, the Caucasus, the Trans-Caucasus, and other regions of Russia, and also through Iran and Turkey. During these trips he gathered an extensive zoological, botanical, and ethnographic collection. In 1863 he settled in Georgia and founded the Caucasian Museum in Tbilisi in 1867. He published *Die Vogelwelt des Kaukasus, systematisch und biologisch-geographisch beschrieben* in 1888 and, in two volumes released in 1862 and 1863, *Reisen im Süden von Ost-Sibirien in den Jahren 1855–1859.* He is recognized as being the person to have first given a detailed description of the flora of the Caucasus. He wrote *Ornis Caucasica* in 1884. Toward the end of the 19th century he made two further journeys, both as part of the suite of members of the Russian Imperial family: to India and Japan in 1895 with the Grand Duke Michael, and to North Africa in 1897 with other members of the family. A species of viper, *Vipera raddei,* is also named after him, as are two birds, Radde's Warbler *Phylloscopus schwarzi* and Radde's Accentor *Prunella ocularis.* The hamster is found in the steppes of southern Russia, north of the Caucasus. The shrew comes from northeast Turkey, Georgia, Armenia, and Azerbaijan.

Raffles

Raffles' Tarsier *Tarsius bancanus* **Horsfield,** 1821 [Alt. Horsfield's Tarsier, Western Tarsier, Malaysian Tarsier]

Raffles' Banded Langur *Presbytis femoralis femoralis* **Martin,** 1838

Sir Thomas Stamford Bingley Raffles (1781–1826) was Lieutenant Governor of Java from 1811 until 1815 and Lieutenant Governor of an area of Sumatra from 1818 to 1824. In 1819 he founded what has become the city-state of Singapore. He was noted for his "liberal attitude toward peoples under colonial rule, his rigorous suppression of the slave trade, and his zeal in collecting historical and scientific information." Raffles was also the first President of the Zoological Society of London and the author of a *History of Java,* published in 1817. He employed zoologists and botanists to collect specimens, paying them out of his own pocket. On his return journey to England in 1824 on HMS *Fame,* he lost a huge collection of specimens, notes, and drawings in a fire. His wife, Lady Sophia Raffles, mentioned his zoological collection in her memoirs of him. A bird, Raffles' Malkoha *Phaenicophaeus chlorophaeus,* is also named after him. The tarsier is found on Sumatra, Borneo, and some neighboring smaller islands. The langur comes from Singapore and is now very rare; populations in Johor (southern mainland Malaysia) are usually now included in this subspecies (formerly separated as *Presbytis femoralis australis*).

Raffray

Raffray's Bandicoot *Peroryctes raffrayana* **Milne-Edwards,** 1878

Raffray's Sheath-tailed Bat *Emballonura raffrayana* **Dobson,** 1879

Marie Jacques Achille Raffray (b. 1844) was a French traveler, civil servant, and active collector. He published *Tour du monde* in 1879, which included information on his time in New Guinea in a chapter called "Voyage en Nouvelle Guinée (1876–1877)." He also visited Abyssinia (now Ethiopia) and wrote *Les églises monolithes de Lalibela* in 1882. The bandicoot is found in New Guinea. The bat occurs from the Molucca Islands east through New Guinea to the Solomon Islands.

Rafinesque

Rafinesque's Big-eared Bat *Corynorhinus rafinesquii* **Lesson,** 1827

Professor Constantine Samuel Rafinesque (1783–1840) was born in Constantinople of a French father and a German mother. He was sent to live in Tuscany to get him away from the turmoil of the French Revolution. His father, a merchant, died in Philadelphia in 1793, leaving the family very badly off. Despite being unable to attend a university, Rafinesque was a highly gifted individual, and his accomplishments include being a botanist, a geologist, a historian, a poet, a philosopher, a philologist, an economist, a merchant, a manufacturer, a professor, a surveyor, an architect, an author, and an editor. In 1802 he was apprenticed to a merchant house in Philadelphia, and for the next two years he roamed the fields and woods and made collections of plants and animals. In 1805 he went to Sicily as Secretary to the U.S. Consul and carried on a lucrative trade in commodities. He scoured the whole island for plants and collected previously unrecorded fishes from the stalls of the Palermo market. He stayed in Sicily until 1815, when he sailed for New York, but he was shipwrecked in Long Island Sound, losing all his unpublished manuscripts and collections. In 1818 he sailed down the Ohio River and conducted a comprehensive survey of the fishes there that was published in 1820 as *Ichthyologia Ohiensis*. He visited Henderson, Kentucky, and stayed for eight days with John James Audubon. In 1819 he was appointed to be Professor of Botany and Natural Science at the University of Transylvania (in Lexington, Kentucky), where he stayed until 1826. In 1826 he returned to Philadelphia with 40 crates of specimens. He had a remarkable gift for inventing scientific names, some 6,700 in botany alone. He died in poverty but was later reinterred in Lexington. The bat is found in the eastern USA, from southern Indiana, Kentucky, and Virginia south to Florida and eastern Texas.

Rahm

Rahm's Brush-furred Rat *Lophuromys rahmi* **Verheyen,** 1964

Mrs. Ursula Rahm is a Swiss zoologist. So in spite of *rahmi* being a masculine form, this animal is named after a woman. She first went to Africa at the end of 1951, when she and her husband, Urs, went to the Ivory Coast as the first Directors of the Swiss Research Center there. She went to the Congo (now Zaire) in 1958 when her husband took up his appointment as Director of the Mammalogy Department at Lwiro. She became particularly interested in chimpanzees. In 1966 she wrote, with A. Christiaensen, "Les mammifères de l'île Idjwi (lac Kivu, Congo)." In 1970 she wrote "Ecology, Zoogeography and Systematics of Some African Forest Monkeys." She is also involved in the breeding of Arabian horses and was the Swiss member of the disciplinary committee at the World Arabian Horse Championship in Paris in 2000. The rat is found in montane forests around Lake Kivu, in western Rwanda and adjacent eastern DRC (Zaire).

Rainey

Rainey's Shrew *Crocidura raineyi* **Heller,** 1912

Rainey's Gazelle *Gazella granti raineyi* Heller, 1913

Paul J. Rainey (1877–1923) was an American multimillionaire, hunter, and playboy. His family money came from coal and coke production. He was reckless and wild, the odd man out in a family that was sober and conservative (his brother, Roy, never even owned a car—he thought a bicycle was quite enough). Paul Rainey bought land in the Deep South, in an area where other families had neither running water nor electricity, and built himself an enormous estate that had an indoor heated swimming pool. He had a trophy room the size of a small house to accommodate the trophies he had acquired from hunting all over the world. He also owned a large plantation near Nairobi, racing stables in England and America, and a 23,000-acre duck preserve in Vermilion, Louisiana, that was given to the National Audubon Society after his death. He was a member of the American Geographical Society, the American Museum of Natural History, the Zoological Society of New York, the Smithsonian Institution, and the National Institute of Social Sciences. He was the first person to take a cameraman with him on safari and so provided the first films of African big game. In 1923 he went to England to buy a new pack of hounds with which he intended to sail to India to hunt tigers on horseback, as though they were foxes. Sadly for him, but fortunately for the tigers, he never made it but died at sea before reaching Cape Town. The shrew is known only from the type locality, at Mount Garguez in Kenya. The gazelle (not always regarded as a valid subspecies) comes from northern Kenya.

Ramon

Ramon's Shrew *Crocidura ramona* Ivanitskaya, Shenbrot, and Nevo, 1996 [Alt. Negev Shrew, Ramon White-toothed Shrew]

This shrew is not named after a person but after the locality where the type specimen was taken: Makhtesh Ramon (Ramon Crater) in the Ramon Nature Reserve, Negev, Israel.

Randrianasolo

Randrianasolo's Sportive Lemur *Lepilemur randrianasoli* Andriaholinirina et al., 2006

Georges Randrianasolo (d. 1989) was a Malagasy zoologist with an interest in both mammalogy and ornithology. He was an expert on lemur distribution who worked at the Botanical and Zoological Park of Tsimbazaza, latterly as Director. He co-wrote at least one book, *Faune de Madagascar*, published in 1973. The etymology in the original description of the sportive lemur reads, "*L. randrianasoli* is named in honour of our late colleague, Georges Randrianasolo, who worked from 1970 to 1977 to sample the sportive lemurs necessary for the first taxonomic revision based on cytogenetic studies, and who walked for two weeks to obtain data and samples from the *L. ruficaudatus* from Antsalova." He also had a bird named after him, the Cryptic Warbler *Cryptosylvicola randrianasoloi*. This lemur is known from a small area of western Madagascar.

Ranjini

Ranjini's Field Rat *Rattus ranjiniae* Agrawal and Ghosal, 1969 [Alt. Kerala Rat]

Dr. (Miss) P. V. Ranjini collected the first known specimens of this rat. She worked at the Cytogenetics Laboratory, Department of Zoology, Banaras Hindu University in India in 1966. The rat is known only from Kerala, southwest India.

Ratanaworabhan

Ratanaworabhan's Fruit Bat *Megaerops niphanae* Yenbutra and Felten, 1983

Dr. Niphan Chanthawanich Ratanaworabhan is a Thai scientist whose doctorate is in entomology. Her chief interest seems to be parasites. In 1973 she co-wrote *Some Ectoparasites of the Birds of Asia*. In 1990 she published *Endangered Species and Habitats of Thailand*, and she has written many articles on birds and mammals and their parasites. The bat has been recorded from northeast India, Thailand, Laos, Cambodia, and Vietnam.

Rayner

Rayner's Flying Fox *Pteropus rayneri* **Gray,** 1870 [Alt. Solomons Flying Fox]

Dr. Frederick Matthew Rayner (1818–1873) was a naval surgeon. In early 1841 he was an assistant surgeon on board HMS *Queen*, transferring after a few months to HMS *Thunderer*. He was on board HMS *Herald* as surgeon, in Australian and Pacific waters, from 1857 to 1863. In 1863 he was promoted to the rank of Staff Surgeon, and in 1864 he was appointed to be the Staff Surgeon on board HMS *Eagle*. The flying fox is found in the Solomon Islands, and Rayner collected the type specimen on the island of Guadalcanal.

Redman

Jamaican Long-tongued Bat *Monophyllus redmani* **Leach,** 1821 [Alt. Leach's Single-leaf Bat]

Dom R. S. Redman (dates not found) was the person from whom Leach obtained the type specimen, which came from Jamaica; we can find no further biographical information. As well as Jamaica, the bat occurs on Cuba, Hispaniola, and Puerto Rico.

Reeves

Reeves' Muntjac *Muntiacus reevesi* **Ogilby,** 1839 [Alt. Chinese Muntjac]

John Reeves F.R.S. (1774–1856) was an English amateur naturalist and collector who served in China, chiefly Macao, as a civil servant between 1812 and 1831. He was employed by the East India Company as an "Inspector of Tea." He sent specimens of many types of Chinese fauna and flora back to England, including the Wisteria *Wisteria sinensis*. Reeves also commissioned Chinese artists to paint animals and plants in the Western (naturalistic) style. His daughter-in-law donated this huge collection of watercolors to the British Museum in 1877. Reeves is also remembered in the names of other taxa, such as Reeves' Pheasant *Syrmaticus reevesii*, and Reeves' Terrapin *Chinemys reevesii*. The muntjac is found in the southern half of China and on Taiwan.

Reig

Reig's Grass Mouse *Akodon reigi* Gonzalez, **Langguth,** and Oliveira, 1998

Reig's Montane Mouse *Aepeomys reigi* Ochoa, Aguilera, Pacheco, and **Soriano,** 2001

Dr. Osvaldo A. Reig (1929–1992) was an Argentinean paleontologist. In 1961 he was made a Professor at the University of Buenos Aires, despite not having high academic qualifications, for he was regarded as the leading expert in vertebrate paleontology and the evolutionary genetics of vertebrates. From Buenos Aires he went to Venezuela, where he stayed for 15 years before moving to London in 1971. He obtained a doctorate in zoology and paleontology from the University of London in 1973, then went to Chile, where he organized the Institute of Ecology and Evolution at the Austral University in Valdivia. He returned to Buenos Aires and was reinstated as Professor, but was dismissed in 1974 and went in to exile in Venezuela. Reig returned to Argentina in 1983 after the military junta fell and the dictatorship came to an end, and became the first President of the Argentine Society of Mammalogists. Later in his life he received many honors and was greatly revered. The grass mouse is found in Uruguay and southeast Brazil. The montane mouse comes from the Andes of western Venezuela.

Remy

Remy's Shrew *Suncus remyi* **Brosset, Dubost,** and **Heim de Balsac,** 1965 [Alt. Gabon Dwarf Shrew]

Professor Paul A. Rémy (1894–1962) was a scientist at the Biological Mission at Makakou in Gabon, where he died. He seems to have had a special interest in the Pauropoda, an obscure class of arthropods with 8 to 11 pairs of legs. The shrew occurs in Gabon, Cameroon, the Congo Republic, and Central African Republic.

Rendall

Rendall's Serotine *Neoromicia rendalli* **Thomas,** 1889

Dr. Percy Rendall (1861–1948) was an itinerant zoologist who collected over much of Africa and in Trinidad and other Caribbean locations in the last decades of the 19th century. There is a reference to a collection of new fishes he made from the Upper Shiré River, British Central Africa, which was presented to the British Museum by Sir Harry Johnston. In 1892 he published "Notes on the Ornithology of the Gambia." A fish, *Tilapia rendalli,* is also named after him. The serotine is found from Gambia east to Somalia and south to Botswana and Mozambique.

Rennell

Rennell Flying Fox *Pteropus rennelli* **Troughton,** 1929

This bat is named after an island in the Solomon Islands and not after a person.

Revoil

Somali Elephant Shrew *Elephantulus revoili* **Huet,** 1881

Georges E. J. Révoil (1852–1894) was a French naturalist who collected in Somaliland between 1878 and 1880. He wrote *La vallée du Darro: Voyage aux pays Somalis* (1882). The Somali Bee-eater *Merops revoilii* is also named after him. The elephant shrew is found in northern Somalia.

Rhesus

Rhesus Monkey *Macaca mulatta* Zimmermann, 1780

Rhesus was a character in the saga of the Trojan War. He was a King of Thrace who led his forces to the aid of Troy when that city was besieged by the Greeks. Two Greek "heroes," Odysseus and Diomedes, secretly entered the King's encampment and murdered Rhesus in his sleep. In 1798 Audebert gave the name *Simia rhesus* to a monkey he described. It later turned out that this was the same species that Zimmermann had already described 18 years earlier, so Audebert's scientific name was discarded as a junior synonym. However, the name Rhesus stuck as the monkey's common name. It is found from northern and central India east to southern China and northern Vietnam.

Rhoads

Rhoads' Gerbil *Gerbillus pulvinatus* Rhoads, 1896 [Alt. Cushioned Gerbil]
Rhoads' Oldfield Mouse *Thomasomys rhoadsi* Stone, 1914

Samuel Nicholson Rhoads (1862–1952) was a vertebrate taxonomist from Philadelphia, but he was interested in all areas of zoology such as mammalogy, ornithology, malacology, and so on. He collected widely in the USA (it was he who identified the Wood Bison as a subspecies of the American Bison) and also in Ecuador, donating a substantial part of his collections to the Academy of Natural Sciences of Philadelphia. He wrote scientific papers such as "Geographic Variation in *Bassariscus astutus,* with Description of a New Subspecies" (1893) as well as monographs such as *The Mammals of Pennsylvania and New Jersey* (1903). The gerbil is found in Ethiopia, Eritrea, and Djibouti. The mouse is confined to the Andes of north-central Ecuador.

Richardson, J.

Richardson's Ground Squirrel *Spermophilus richardsonii* Sabine, 1822
Barren Ground Grizzly Bear *Ursus arctos richardsoni* **Swainson,** 1838
Richardson's Vole *Microtus richardsoni* **DeKay,** 1842 [Alt. American Water Vole]
African Linsang *Poiana richardsonii* **Thomson,** 1842 [originally named *Genetta richardsonii*]
Richardson's Collared Lemming *Dicrostonyx richardsoni* **Merriam,** 1900

Sir John Richardson (1787–1865) was a Scottish naval surgeon and Arctic explorer, knighted in 1846. He was a friend of Sir John Franklin, to whom he was also related by marriage, and took part in Franklin's expeditions of 1819–1822 and 1825–1827. He also participated from 1847 in the vain search for Franklin and his colleagues; their fate was not discovered until

Rae's expedition of 1853–1854. He is also the man who named the Franklin's Gull after his then-captain. The Richardson Mountains in Canada are also named after him, as are at least six birds. Thomson writes in his description of the linsang, "I have taken the opportunity of naming the *Genetta* after my friend Dr. John Richardson, the Inspector of the Naval Hospital at Haslar, so well known for his highly scientific acquirements, and so much esteemed in the naval service by all his medical brethren." The ground squirrel can be found in south-central Canada and north-central USA. The bear occurs in northern Canada (Northwest Territories and Nunavut). The vole comes from the Cascade and Rocky mountains (southwest Canada and USA south to Utah). The linsang is found in west-central Africa, from Cameroon to eastern DRC (Zaire), and on the island of Bioko. The lemming is found in the Canadian tundra, on the western side of Hudson's Bay.

Richardson, W. B.

Richardson's Shrew-Mouse *Microhydromys richardsoni* **Tate** and **Archbold,** 1941 [Alt. Groove-toothed Shrew-Mouse]
Glacier Rat *Stenomys richardsoni* Tate, 1949 [Syn. *Rattus richardsoni*]

William Blaney Richardson (b. 1868) was an American zoologist and professional collector with a particular interest in rodents. The original citation for the Glacier Rat mentions that it was named after "W. B. Richardson, mammalogist of the 1938–1939 Archbold New Guinea expedition." Richardson kept a detailed journal of that trip, which was by no means his first. For example, he was part of an expedition to Nicaragua in 1908 and another to Ecuador in 1912. In fact, he collected in Nicaragua many times; in 1891 he settled in Matagalpa, Nicaragua, to grow coffee, and stayed on for the rest of his life. He took part in the Roosevelt-Rondon scientific expedition to Brazil of 1913–1914, and on another expedition to Ecuador in 1916 he collected for the American Museum of Natural History and the Academy of Natural Sciences of Philadelphia. At some point he was employed by the American ornithologist Charles Barney Cory to collect bird specimens for him. Both of the rodents named after him are found in New Guinea, the rat being confined to high altitudes in the Snow Mountains.

Richmond

Richmond's Squirrel *Sciurus richmondi* **Nelson,** 1898

Charles Wallace Richmond (1868–1932) was an American ornithologist who specialized in nomenclature and bibliography. His first job, aged about 13, was as a pageboy at the House of Representatives in Washington, DC. He was part of the U.S. Geological Survey of Montana in 1888 and later worked for the U.S. Department of Agriculture as an "ornithological clerk." The next job he was able to get in Washington was with the U.S. National Museum as a night watchman. He was, however, soon promoted, eventually becoming Assistant Curator of Birds in 1894. He was a colleague of Ridgway at the Smithsonian, and when Ridgway died he collected all of Ridgway's notes and filed them for future use. He created *The Richmond Index to the Genera and Species of Bird.* It was his life's work, started in 1889 and lasting 40 years, culminating in 70,000 file cards. The squirrel inhabits forests and plantations in Nicaragua, where Richmond collected the type specimen.

Rickart

Philippine Large-headed Fruit Bat *Dyacopterus rickarti* Helgen et al., 2007

Dr. Eric A. Rickart (b. 1950) has been Curator of Vertebrates at the Utah Museum of Natural History in Salt Lake City since 1985. He graduated with a bachelor's degree in ecology and systematics in 1974 from the University of Kansas, which also awarded him a master's degree two years later. He was awarded his doctorate in biology in 1982 by the University of Utah. He worked temporarily at the University of Texas at El Paso before returning to Utah. He is also a Research Associate at the Field Museum of Natural History in Chicago. He is an authority on Philippine mammals. The fruit bat comes from the islands of Luzon and Mindanao in the Philippines.

Rickett

Rickett's Big-footed Bat *Myotis ricketti* **Thomas,** 1894 [Alt. Rickett's Large-footed Myotis]

Charles Boughey Rickett (1851–1943) was a British banker and an amateur ornithologist. He left records of his observations in Scotland and Cornwall from 1901 through to 1907. He also collected in various parts of Asia, including China, Japan, the Straits Settlements (now Malaysia), Java, and India. Throughout his 33-year career in Asia he worked for the same bank, the Hong Kong and Shanghai Banking Corporation. He is also remembered in the scientific names of two birds, the Yunnan Parrotbill *Paradoxornis (brunneus) ricketti* and the Sulphur-breasted Warbler *Phylloscopus ricketti*. The bat is found in eastern China, Laos, and Vietnam. The name *Myotis ricketti* may be a junior synonym of *M. pilosus* (Peters, 1869).

Ride

Wongai Ningaui *Ningaui ridei* **Archer,** 1975

Professor Dr. William David Lindsay Ride (b. 1926) is a zoologist and former Director of the Western Australia Museum. He worked for the Commonwealth Scientific and Industrial Research Organisation from 1974 to 1980 and was head of the School of Applied Science from 1982 to 1987. In the latter year he was appointed Principal of the Canberra College of Advanced Education. He is currently Professor in the Department of Earth and Marine Sciences, Faculty of Science at the National Museum of Australia. He served on the International Commission on Zoological Nomenclature from 1964 to 2001, twice in that time as its President. He has written numerous articles and books. The ningaui, a small carnivorous marsupial, inhabits the deserts of western and central Australia.

Ridley

Ridley's Myotis *Myotis ridleyi* **Thomas,** 1898 [Alt, Ridley's Bat]

Ridley's Roundleaf Bat *Hipposideros ridleyi* **Robinson** and **Kloss,** 1911

Henry Nicholas Ridley (1855–1956) was a British botanist and collector on the island of Fernando de Noronha (Brazil) in 1887. He was sometimes known as "Mad Ridley" or "Rubber Ridley," as he was keen to ensure that the rubber tree was transplanted to British territory so as to prevent dependency on Brazil for supplies of latex. From 1888 to 1912 he was Superintendent of the Tropical Gardens in Singapore, where early experiments in growing the rubber tree outside Brazil took place, the first successful growth having been achieved at Kew and plants then shipped to Singapore. He wrote *The Natural History of the Island of Fernando de Noronha Based on the Collections Made by the British Museum Expedition* (1887) and "The Habits of Malay Reptiles" (1889). He is also remembered in the scientific name of a bird, the Noronha Elaenia *Elaenia ridleyana*. The myotis is found in the Malay Peninsula, Sumatra, and Borneo. The type specimen of the roundleaf bat was collected in the Singapore Botanical Gardens while Ridley was the Superintendent. It also occurs in western (mainland) Malaysia.

Riggenbach

Riggenbach's Gerbil *Gerbillus riggenbachi* **Thomas,** 1903

F. W. Riggenbach (1864–1944) was a zoologist and collector employed on the German Central African Expeditions of the first two decades of the 20th century, which were led by Herzog Adolf Friedrich zu Mecklenburg (Duke of Mecklenburg). Riggenbach is also commemorated in the scientific names of two African freshwater fish, *Aphyosemion riggenbachi* and *Ctenopoma riggenbachi*. The gerbil is known from the Western Sahara (Morocco) and northern Senegal. This species is sometimes regarded as conspecific with *Gerbillus tarabuli*.

Rimoli

Rimoli's Hutia *Hyperplagiodontia stenocoronalis* Rimoli, 1977 extinct

Renato O. Rimoli is a paleontologist and biologist at the Academy of Sciences in the Dominican Republic. At one time he was Director of Biodiversity. The hutia is known only from skeletal remains found in Haiti. It is now thought to represent the same species as *Plagiodontia araeum*, so Rimoli's scientific name is reduced to a junior synonym. It may have survived until the 17th century.

Risso

Risso's Dolphin *Grampus griseus* **Cuvier,** 1812

Professor Giovanni Antonio Risso (1777–1845), also known as Joseph Antoine Risso. An Italian naturalist, Risso was until 1826 an apothecary in his native town of Nice (not part of France until annexed in 1860). He later became Professor of Botany and Chemistry at the University of Nizza (Nice). He had an outstanding knowledge of ichthyology and published *Ichthyologie de Nice* in 1810. Other works included *Histoire naturelle de l'Europe méridionale* (1826). The dolphin is found worldwide in temperate and tropical waters.

Roach

Roach's Mouse-tailed Dormouse *Myomimus roachi* Bate, 1937

Edward Keith-Roach (1885–1954) originally trained in banking and worked for the Mercantile Bank of India in Bombay. In WW1 he served in Egypt and became a Major. In 1916 he became District Commissioner and was the only European in Eastern Darfur, which he explored and mapped. Based on his experiences there, he published a number of articles and books, including *Adventures among the "Lost Tribes of Islam" in Eastern Darfur: A Personal Narrative of Exploring, Mapping, and Setting up a Government in the Anglo-Egyptian Sudan Borderland* (1924). He then transferred to the British authority responsible for administering the League of Nations Mandate of Palestine. He was District Commissioner in Galilee and subsequently District Governor of Jerusalem before and during WW2, retiring from his post in 1945. Jointly with Sir Harry Luke, he published *Handbook of Palestine* in 1930 and an enlarged version in 1934 called *Handbook of Palestine and Trans-Jordan*. The original etymology for the dormouse is only a brief note and does not mention Roach; we think we have found the right man but cannot be completely sure. The dormouse was first found as a fossil in Israel in a cave on Mount Carmel, a location where Keith-Roach was accustomed to carry out inspections and reviews of the troops. It was later found to be extant in southeast Bulgaria and western Turkey.

Robert

Voalavoanala *Gymnuromys roberti* **Major,** 1896
Robert's Hocicudo *Oxymycterus roberti* **Thomas,** 1901
Robert's Spiny Rat *Proechimys roberti* Thomas, 1901
Robert's Arboreal Rice Rat *Oecomys roberti* Thomas, 1904
Robert's Snow Vole *Chionomys roberti* Thomas, 1906

Alphonse Robert (dates not found) was a French collector, particularly in South America but also in other parts of the world. He acted as Major's assistant on an expedition to Madagascar between 1894 and 1896. He collected in Brazil in 1901 for the British Museum of Natural History, and there is a paper by Thomas acknowledging this in the *Annals and Magazine of Natural History* for 1902 entitled "On Mammals from the Serra Do Mar, Paraná, Collected by M. Alphonse Robert." In 1903 he was again collecting in Brazil and in that year collected the type specimen of Barbara Brown's Titi *Callicebus barbarabrownae,* the skin of which lay in the museum in London until 1990, when it was described as a new taxon by Dr. Philip Hershkovitz. The voalavoanala (a kind of rat) comes from eastern Madagascar. The hocicudo (also a ratlike rodent) and the spiny rat are both endemic to Brazil. The arboreal rice rat dwells in the Amazon basin, from southern Venezuela and the Guianas to eastern Peru and northern Bolivia. The vole can be found in northeast Turkey and the western Caucasus.

Roberts, F. R.

Roberts' Gazelle *Gazella granti robertsi* **Thomas,** 1903
Roberts' Lechwe *Kobus leche robertsi* **Rothschild,** 1907 extinct

F. Russell Roberts (dates not found) sent the type specimen of the gazelle subspecies to the British Museum. He seems to have been a wealthy individual who could afford to travel. He was reported as being in German East Africa around 1903, and in 1920 the *Illustrated London News* carried some photos of giraffes taken in the wild by him. The gazelle is found in northern Tanzania, in the region west of Lake Natron. The lechwe antelope was found in the Luongo and Luena River drainages of northwest Zambia but is now extinct. (Because Rothschild only refers to "Mr. Roberts" in his original description of the lechwe, allocation of this taxon to F. Russell Roberts is uncertain.)

Roberts, J. A.

Roberts' Flat-headed Bat *Sauromys petrophilus* Roberts, 1917 [Syn. *Mormopterus petrophilus*]
Roberts' Serval *Leptailurus serval robertsi* **Ellerman,** Morrison-Scott, and **Hayman,** 1953

Dr. J. Austin Roberts (1883–1948) was a South African zoologist who had little formal zoological training but was awarded an honorary doctorate in 1935 by the University of Pretoria. During the first half of the 20th century he was the most prominent ornithologist in southern Africa. He worked at the Transvaal Museum for 38 years studying mammals as well as birds. Roberts is best remembered for his *Birds of South Africa,* a landmark publication in African ornithology, which first appeared in 1940 and was a bestseller in its numerous subsequent editions. He was working on a similar opus when he died in a car accident in the Transkei region, and *The Mammals of South Africa* was published posthumously in 1951. The bat occurs in South Africa, Namibia, Botswana, Zimbabwe, and Mozambique. The serval subspecies comes from the western Transvaal, South Africa.

Robinson, H.

Queensland Tube-nosed Bat *Nyctimene robinsoni* **Thomas,** 1904
Malayan Tailless Leaf-nosed Bat *Coelops robinsoni* **Bonhote,** 1908
Robinson's Banded Langur *Presbytis femoralis robinsoni* Thomas, 1910
Peninsular Horseshoe Bat *Rhinolophus robinsoni* **Andersen,** 1918

Herbert Christopher Robinson (1874–1929) was a British zoologist and ornithologist. In 1894, after education at Marlborough College, he went to Switzerland because of ill health. In 1896 he made a trip to Queensland, and when he returned to England he worked as an assistant to Dr. H. O. Forbes at the Liverpool Museum, from 1897 until 1900. After 1900 he spent 30 years in the tropics, initially with Dr. Annandale in the Malay Peninsula. In 1903 he became Curator of the Federated Malay States Museum, at Selangor, and held that post until 1926. In 1908 he joined Cecil B. Kloss in exploring the Indo-Malay region. He sent many of the specimens he collected to Liverpool and to the British Museum of Natural History. He wrote and illustrated *The Birds of Singapore* and *The Birds of the Malay Peninsula* as well as many articles, on a variety of topics from gastropods to Siamese amphibians, that were mostly published in the *Journal of the Federated Malay States Museum*. Robinson is honored in the scientific names of other Southeast Asian fauna, including the Malaysian Whistling Thrush *Myophonus robinsoni*, the copepod *Paramphiascella robinsoni*, and the frog *Kalophrynus robinsoni*. The tube-nosed bat occurs in eastern Queensland and northeast New South Wales (Australia). The leaf-nosed bat is found in the Malay Peninsula and Borneo. The langur is also found in the Malay Peninsula, but not in the southernmost portion. The horseshoe bat is found in Thailand and (mainland) Malaysia.

Robinson, W.

Robinson's Mouse-Opossum *Marmosa robinsoni* **Bangs,** 1898

Wirt Robinson (1864–1929) was a U.S. Army General, whose major area of expertise was the chemistry of explosives. From 1906 to 1928 he taught chemistry at West Point Military Academy, and among his pupils was the future President of the USA, Dwight D. Eisenhower. When on holiday, Robinson was an enthusiastic naturalist. He collected in Venezuela for the U.S. National Museum, first in 1895 and then in 1899, for the Department of Mammalogy with Marcus Lyon, an Assistant Curator. He wrote a number of articles, notably "Some Rare Rhode Island Birds" and "Some Rare Virginia Birds," both published in *The Auk* in 1889. Later there was a series of articles on collections made in South America. The mouse-opossum is found from Panama to northwest Peru and northern Venezuela. It also occurs on Trinidad and Tobago, Grenada, and the island of Roatan off the Caribbean coast of Honduras.

Roborovski

Roborovski's Hamster *Phodopus roborovskii* **Satunin,** 1903 [Alt. Desert Hamster]

Captain Vladimir Ivanovich Roborovski (1856–1910) was a Russian explorer of parts of China and Tibet. He accompanied Prjevalsky on his third and fourth expeditions (1879–1880 and 1883–1885). From 1893 until 1896 he was the leader of his own expedition, that of the Imperial Russian Geographical Society in eastern Tien-Shan, Nanshan, and northern Tibet, during which he gathered zoological, botanical, and geological specimens, including the eponymous hamster. During his last expedition he was hit by paralysis, despite which he continued the handling of field material and published the results of the expeditions:

Ekspeditisii v storonu ot pugey Tibetskoy ekspeditsii in 1896 and *Otchet nachalnika ekspeditskii* in 1900. He is also commemorated in the name of a bird, Roborovski's Rosefinch *Kozlowia roborowskii*. The hamster occurs from extreme eastern Kazakhstan to Mongolia and adjacent parts of northern China.

Rodolphe

Rodolphe's Striped Squirrel *Tamiops rodolphii* **Milne-Edwards,** 1867 [Alt. Cambodian Striped Squirrel]

(Louis) Rodolphe Germain (1827–1917) was a veterinary surgeon in the French colonial army, serving in Indochina. The squirrel comes from Cambodia, southern Laos, Vietnam, and eastern Thailand. See **Germain** for biographical details.

Rodriguez

Rodriguez's Harvest Mouse *Reithrodontomys rodriguezi* **Goodwin,** 1943

Professor Juvenal Valerio Rodriguez (dates not found) was Director of the National Museum of Costa Rica, where the herbarium is named after him. Goodwin named the harvest mouse after Rodriguez "in appreciation of his cooperation in getting collections for the American Museum of Natural History." The species is confined to the highlands of central Costa Rica.

Rohu

Rohu's Bat *Philetor brachypterus* **Temminck,** 1840 [Alt. Short-winged Pipistrelle]

Henry Stewart Boventure Rohu (d. before 1928) was an Scottish naturalist and taxidermist who operated in Sydney, Australia, with his wife Ada (ca. 1845–1928), who was also a taxidermist. They had a shop that traded as "Tost and Rohu, Taxidermists, Furriers and Curiosity Shop," which became a curiosity in itself and was known as "the Queerest Shop in Australia." In 1920, in Poole (England), "H. S. Rohu, Naturalist and Furrier and Tanner" was advertising his business and notes in particular that it had been trading for 50 years in "both England and the Colonies." Rohu collected ethnographic items in northern Australia in the 1870s and beyond. His collection of bird skins is housed in the Oxford University Museum. He collected a specimen of bat in New Guinea, which Thomas, in 1902, thought was a new species and named *Philetor rohui*. In 1971 J. Edwards Hills established that *Philetor rohui* was the same species as that collected in Sumatra and described by Temminck in 1840. Even though Temminck's nomenclature had to take precedence, "Rohu's Bat" has stuck as a vernacular name for the species. The New Guinea population is sometimes recognized as a subspecies, which retains Rohu's name in the trinomial as *Philetor brachypterus rohui*. The species has been recorded from Nepal, the Philippines, Malaysia, Sumatra, Borneo, Sulawesi, New Guinea, and the Bismarck Archipelago.

Rohwer

Rohwer's Shrew *Sorex rohweri* Rausch, Feagin, and Rausch, 2007

Professor Sievert A. Rohwer is Professor of Zoology at the University of Washington. He was formerly Director of the Burke Museum and its Division of Zoology and is now Curator of Birds at the museum. He received the Elliot Coues Award in 2006 for outstanding contributions to ornithology. (There are references to "Rohmer's Shrew *Sorex rohmeri*," but these are just mistranscriptions of the correct name.) The shrew is found in Washington State (northwest USA) and the extreme southwest corner of Canada.

Roig

Roig's Tuco-tuco *Ctenomys roigi* **Contreras,** 1989
Roig's Chaco Mouse *Andalgalomys roigi* Mares and Braun, 1996 [Alt. Roig's Pericote]

Virgilio G. Roig is an Argentinean biologist. During the 1980s and 1990s he held various positions in the Argentina Society for the Study of Mammals. He has written extensively on batrachians, often with J. M. Cei as co-author. The tuco-tuco comes from northeast Argentina. The chaco mouse is found, unsurprisingly, in the Gran Chaco region of northwest Argentina.

Romero

Volcano Rabbit *Romerolagus diazi* Ferrari-Pérez, 1893

Don Matias Romero (1837–1898) was a Mexican government minister who afforded a great deal of assistance to the U.S. Biological Survey. He was originally trained as a lawyer. By 1857 he was working for the Mexican Foreign Office, and in that same year he sided with the government, which had been forced out of Mexico City during the revolution and was obliged to migrate around the country. In 1859 he was appointed Secretary of the Mexican Legation in Washington, DC, and was Chargé d'Affaires until 1863, when he returned to Mexico and resigned from the diplomatic service. He joined the army as a Colonel and worked as Chief of Staff for General Porfirio Diaz. Late in 1863 he returned to Washington as Minister to the USA and held that post until 1868. In 1868 he returned to Mexico and was appointed Finance Minister, holding that job until 1873, when he retired due to ill health and became a farmer, though he remained a member of Congress for his region. In 1876 he was reappointed to the Finance Ministry, leaving in 1879, and in 1880 became Postmaster General. He was instrumental in negotiating treaties between Mexico and Guatemala and Mexico and the USA, thus settling boundary disputes. This endangered species of rabbit is found around the volcanoes of central Mexico.

Ronald

Ronald's Short-tailed Opossum *Monodelphis ronaldi* Solari, 2004 [Alt. Peruvian Short-tailed Opossum]

Professor Dr. Ronald H. Pine is an American zoologist who has spent well over 20 years studying the genus *Monodelphis,* for which heroic effort Solari specifically named this species after him. In addition to working as a Curator at the Smithsonian, he has been on many scientific expeditions to five continents and has many publications to his name. He graduated in zoology from the University of Kansas, was awarded a master's degree by the University of Michigan, and earned a doctorate in wildlife science by Texas A&M University. He is a Research Associate at the Field Museum in Chicago and a Permanent Visiting Scholar at the Natural History Museum and Biodiversity Research Center at the University of Kansas. For a long time he has been a leading opponent of the "pseudoscience" of creationism and "intelligent design" and has returned to live in Kansas in order to be at the center of the argument there about how evolution is taught in schools. The opossum was discovered in the Manu Reserve, southeast Peru.

Roosevelt

Roosevelt's Wapiti *Cervus canadensis roosevelti* **Merriam,** 1897 [Alt. Roosevelt Elk, Olympic Elk]
Roosevelt's Shrew *Crocidura roosevelti* **Heller,** 1910
Roosevelt's Sable Antelope *Hippotragus niger roosevelti* Heller, 1912
Roosevelt's Gazelle *Gazella granti roosevelti* Heller, 1913
Roosevelt's Lion *Panthera leo roosevelti* Heller, 1913

Theodore Roosevelt (1858–1919) was the 26th President of the USA, serving from 1901 to 1909. His distinguished career and his list of achievements are too well known to need reiteration here. It is, however, worth noting that he belonged to a generation that believed that there was nothing wrong with the wholesale slaughter of wildlife. Please bear that in mind when considering the following quotation: "On the 21 April 1909, Teddy Roosevelt's safari set off from Mombasa, Kenya. By the time the entourage arrived in Khartoum 8 months later, they had slaughtered 5,013 mammals, 4,453 birds, 2,322 reptiles and amphibians and similar numbers of fish, invertebrates, shells, and plants. The skins, etc. were sent to the Smithsonian; among these were Roosevelt's gazelle and Roosevelt's sable." The wapiti (usually called elk in North America) is the westernmost subspecies of this deer, occurring in coastal British Columbia, Washington, and Oregon. The shrew is found in forest-savanna margin areas of central Africa, being recorded from Angola, Cameroon, Central African Republic, DRC (Zaire), Uganda, Rwanda, and Tanzania. Heller may have allowed his eagerness to honor Roosevelt to get in the way of good taxonomy when naming the three larger African mammals on the above list: probably none of them are valid. The sable antelope is an isolated population from the Shimba Hills of Kenya, but DNA analysis suggests it is not distinct from Tanzanian populations. The gazelle, from the Athi Plains of Kenya, is doubtfully distinct from the typical race of Grant's Gazelle. The type specimen of the lion was presented to the President by the Emperor of Abyssinia and was kept in a zoo until it died in 1906, its body being then exhibited in the Smithsonian. This subspecies is now regarded as a junior synonym of *Panthera leo nubicus.*

Roosevelts

Roosevelts' Muntjac *Muntiacus rooseveltorum* **Osgood,** 1932 [Alt. Laotian Barking Deer]

This deer was named after two sons of Theodore Roosevelt (see above), the eldest two of four: Theodore Roosevelt Jr. (1887–1944), who went on to be a war hero, and Kermit Roosevelt (1889–1943), who was a keen amateur ornithologist and who accompanied his father on a number of his expeditions. The latter is also commemorated in the name of a bird, Kermit's Antwren *Myrmotherula (sclateri) kermiti.* The type specimen of the deer was collected in Laos in 1929, during the Kelley-Roosevelts Asiatic expedition. Osgood named it as a new species in 1932, then the species "disappeared" for over 60 years. No more specimens turned up, and some authorities began to question the validity of the taxon. Then, in 1995, a living specimen of muntjac was seen, in a menagerie in Laos, that seemed to be different from the well-known species. This proved to be the long-lost *rooseveltorum,* and genetic tests have now shown that it is indeed a valid species.

Roosmalen

Roosmalen's Dwarf Marmoset *Callithrix (Callibella) humilis* van Roosmalen et al., 1998

The marmoset is found in the angle of land between the lower Madeira and Aripuana rivers, southern Amazonian Brazil. See **Van Roosmalen** for biographical detail.

Rory

Rory's Pseudantechinus *Pseudantechinus roryi* Cooper, **Aplin,** and Adams, 2000 [Alt. Tan Pseudantechinus, Rory Cooper's False Antechinus]

Rory Cooper is the son of one of the species' describers, Norah Cooper of the Western Australian Museum, Perth. It is so named as "Rory" is Gaelic for "red," and this species is generally a brighter reddish-brown than the other species of *Pseudantechinus*. This carnivorous marsupial inhabits Western Australia.

Rosalinda

Rosalinda's Oldfield Mouse *Thomasomys rosalinda* **Thomas** and St. Leger, 1926
Rosalinda Gerbil *Gerbillus rosalinda* St. Leger, 1929

We have checked the original etymology of both mammal species named *rosalinda*, and they give no indication that these animals were named after a specific person or any clue as to why that scientific name was chosen. We conclude, therefore, that they are probably just a fanciful application of the name, which means "beautiful rose." Perhaps Jane St. Leger thought the small mammals looked pretty and deserved a "beautiful" name. The mouse is found in the Andes of northern Peru, and the gerbil comes from central Sudan.

Rosamond

Little Red Kaluta *Dasykaluta rosamondae* **Ride,** 1964 [Alt. Little Red Marsupial Mouse]

Rosamond Clifford (ca. 1133–ca. 1176) was the mistress of King Henry II. This small carnivorous marsupial has reddish fur and was discovered on an Australian sheep farm called Woodstock Station, living amid prickly spinifex bushes. Rosamond herself was red- haired and was kept locked in the Royal Manor of Woodstock, which was surrounded by a maze of prickly hedges. She eventually died there, supposedly murdered by Eleanor of Aquitaine, wife to the King, either by poison, stabbing, beheading, or being bled to death in her bath. Henry acknowledged her only in 1174 after he had imprisoned his wife (who was in reality innocent of the aforementioned crimes). The kaluta is found in northwest Western Australia.

Rosenberg

Rosenberg's Dwarf Squirrel *Prosciurillus rosenbergii* **Jentink,** 1879 [Alt. Sanghir Squirrel]

Baron Carl (originally Karl) Benjamin Hermann von Rosenberg (1817–1888) was a German naturalist and geographer who collected in the East Indies. He enlisted as a common soldier in the Dutch colonial army in the Malay Archipelago and served for 30 years, the first 16 as a topographic draftsman on the island of Sumatra, then as a civil servant in the Moluccas and around New Guinea. He traveled there in a Dutch warship, the *Etna*, and during this voyage he met Wallace. He took part in mapping outlying districts of the archipelago, all the while pursuing his interest in natural history, especially ornithology. He published an important zoological and ethnographical study of New Guinea, entitled *Reistochten naar de Geelvinkbaai op Nieuw-Guinea in de jaren 1869 en 1870*. He is remembered in the scientific names of several birds, including the Blue-faced Rail *Gymnocrex rosenbergi*. The squirrel is confined to the Sanghir Islands, north of Sulawesi, Indonesia.

Rosevear

Rosevear's Striped Grass Mouse *Lemniscomys roseveari* Van der Straeten, 1980
Rosevear's Brush-furred Rat *Lophuromys roseveari* **Verheyen** et al., 1997 [Alt. Mount Cameroon Brush-furred Rat]

Donovan Reginald Rosevear (1900–1986) was a British resident of Nigeria, where he was a member of the colonial administration. He took a B.A. at Cambridge and a diploma of forestry in 1923. In 1924 he joined the Colonial Forest Service and was posted to Nigeria, where he became Inspector General of Forests in 1951. He was a member, and eventually Vice President, of the Nigerian Field Society, and while in Lagos he published a handbook called *Vegetation, Forestry and Wild Life in Nigeria* (1953). He retired in 1954 and returned to the UK. In London he worked as an Honorary Associate at the Natural History Museum and prepared a series of monographs: *The Bats of West Africa* (1965), *The Rodents of West Africa* (1969), and *The Carnivores of West Africa* (1974). The grass mouse is known from only two locations in southwest Zambia. The brush-furred rat appears to be confined to forests on Mount Cameroon.

Ross

Ross Seal *Ommatophoca rossii* **Gray,** 1844

Rear Admiral Sir James Clark Ross (1800–1862) discovered the Ross Sea and the Ross Ice Shelf. He joined the Royal Navy at the age of 12. In 1818 he joined his uncle, Sir John Ross, on a voyage in search of the Northwest Passage. Between 1819 and 1827 he joined Edward Parry in four more expeditions to the Arctic. Ross commanded *Erebus* and *Terror* during the Antarctic expedition of 1839–1843. It was while close to the magnetic South Pole that he broke through a wide expanse of pack ice into a large and clear sea that later bore his name. He is also remembered in the name of Ross' Gull *Rhodostethia rosea*. The seal dwells amid the pack ice of the Antarctic Ocean.

Rosset

Thick-thumbed Myotis *Myotis rosseti* Oey, 1951

C. W. Rosset (dates not found). This species of bat was originally described as *Glischropus rosseti* in 1951 from two Cambodian specimens that appear to have been lying around in a museum for decades. It seems likely that it is named after C. W. Rosset, who sold or presented a specimen to the Natural History Institute (a commercial institute) in Frankfurt-am-Main in 1889. Rosset probably collected the specimen himself, as he is known to have traveled in the regions where it occurs. He undertook a three-year-long journey in the Far East, including Indochina. On this journey he traveled on a ship called *Ceylon* from Colombo and was able to go ashore in the Maldive Islands in 1885. The myotis is found in Cambodia, Laos, and Thailand.

Rossetti

Rosetti's Wombat

This is not a species but an individual animal, which we list as there are numerous references to it that may confuse those coming across mentions of "Rosetti's Wombat." The artist Dante Gabriel Rossetti (1828–1882) was a leading member of the Pre-Raphaelite Brotherhood. He bought his first pet wombat, which he named Top, in 1869. After the marsupial's demise, Rosetti painted its portrait. At this time Rossetti was infatuated with Jane Morris, the wife of his friend William Morris, and in the British Museum there is a drawing of Jane Morris and Top, illustrating the degree to which lover and pet merged in Rossetti's mind as objects of sanctification; each of them wears a halo.

Rothschild, L. W.

Rothschild's Cuscus *Phalanger rothschildi* **Thomas,** 1898 [Alt. Obi Cuscus]
Rothschild's Woolly Rat *Mallomys rothschildi* Thomas, 1898 [Alt. Smooth-tailed Giant Rat]
Rothschild's Porcupine *Coendou rothschildi* Thomas, 1902
Rothschild's Giraffe *Giraffa camelopardalis rothschildi* **Lydekker,** 1903 [Alt. Baringo Giraffe]
Rothschild's Rock Wallaby *Petrogale rothschildi* Thomas, 1904
Broad-striped Dasyure *Paramurexia rothschildi* **Tate,** 1938

Lord Lionel Walter Rothschild F.R.S. (1868–1937) was the founder of the Zoological Museum at Tring in 1889. It is now known as the Walter Rothschild Zoological Museum and comprises the Natural History Museum's Ornithology Section. According to the history of the museum, "As a child Walter Rothschild knew exactly what he was going to do when he grew up, announcing at the age of seven, 'Mama, Papa, I am going to make a museum.' He had already started collecting insects and stuffed animals by then, and a year later started setting his own collection of butterflies. By the time he was 10, Walter had enough natural history objects to start his first museum—in a garden shed! Before long Walter's insect and bird collections were so large that they had to be stored in rented rooms and sheds around Tring. Then in 1889, when Walter Rothschild was 21, his father gave him some land on the outskirts of Tring Park. Two small cottages were built, one to house his books and insect collection, the other for a caretaker. Behind these was a much larger building, which would contain Lord Rothschild's collection of mounted specimens. This was the beginning of the zoological museum which opened to the public in 1892 and the beginning of Lord Rothschild's life long passion for natural history." He amassed the largest bird collection in the world: 300,000 skins, 200,000 eggs, and 30,000 books. However, in 1932 he was forced to sell almost all his collection due to blackmail by an unknown party. Rothschild was Member of Parliament for Aylesbury, a Major in the Buckinghamshire Yeomanry, a Justice of the Peace, and a Deputy Lieutenant for the County of Buckinghamshire. Rothschild's own pictures of extinct birds have been used in posters. He had many bird species named after him, as well as a butterfly, Rothschild's Birdwing *Ornithoptera rothschildi.* The cuscus comes from the islands of Obi and Bisa in the Moluccas, Indonesia. The woolly rat is found in the highlands of New Guinea. The porcupine seems to be restricted to Panama. The giraffe subspecies, now very rare in the wild, comes from Uganda and western Kenya. The rock wallaby inhabits Western Australia (Pilbara region and islands of the Dampier Archipelago). The dasyure is from southeast New Guinea.

Rothschild, N. C.

Rothschild's Zokor *Eospalax rothschildi* **Thomas,** 1911

The Hon. Nathaniel Charles Rothschild (1877–1923). He presented the type specimen of the zokor (mole-rat) to the British Museum. The species is found in central China. See **Carol** for biographical details.

Rousselot

Velvet Climbing Mouse *Dendroprionomys rousseloti* **Petter,** 1966

Dr. R. Rousselot (dates not found) was Director of the Brazzaville Zoological Gardens, Congo Republic. He wrote several articles on African birds, including "Notes sur la faune ornithologique du cercle de Mopti, Soudan Français" (1939) and "Notes sur la faune ornithologique des cercles de Maradi et Tanout (Niger Français)" (1947), as well as a paper on mammal parasites, "Ixodes de l'Afrique noire"

(1951). He found the type specimen of the climbing mouse on the grounds of the Brazzaville Zoo and sent it to Petter at the Natural History Museum in Paris, as Petter recounted in his original description: "L'espèce est dédiée au Dr. Vre Rousselot qui était Directeur du Jardin Zoologique de Brazzaville lorsqu'il a envoyé ces spécimens au Muséum après qu'ils aient été capturés dans l'enceinte même du zoo à l'occasion de défrichements." The mouse is known only from the type locality.

Roux

Rufous Horseshoe Bat *Rhinolophus rouxii* **Temminck,** 1835

(Jean Louis Florent) Polydore Roux (1792–1833) was a French naturalist and painter. Even as a child he was interested in natural history and accumulated a large collection of insects. He studied under Cuvier in Paris. In 1819 he was appointed Curator of the Natural History Collection of the city of Marseille. He accompanied Carl Alexander Anselm Freiherr von Hügel (1796–1870) on part of his expedition to India, Ceylon (now Sri Lanka), Australia, New Zealand, and the Himalayas. Roux accompanied Hügel during the first part of the expedition, meeting him in Egypt and traveling on to India, but died in 1833 in Bombay, after they had argued and parted company. His death is something of a mystery, although some sources say it was from plague. An unpublished manuscript of his is held by the Natural History Museum in Vienna. He wrote and illustrated *Ornithologie provençale,* published in two volumes between 1825 and 1830 and considered a major contribution to the ornithology of France at the time. He also wrote and illustrated *Crustacés de la Méditerranée et de son littoral,* published from 1828 to 1830. The copepod *Pandarus rouxi* is also named after him, as are several other marine organisms. The bat is found from India and Sri Lanka east to southern China and Vietnam.

Roxellana

Golden Snub-nosed Monkey *Rhinopithecus roxellana* **Milne-Edwards,** 1870

Roxellane (ca. 1505–1558) was Consort to the Turkish Sultan Sulayman the Magnificent (1494–1566). She was a Russian lady of "doubtful repute" with long golden hair and a turned-up nose, characteristics shared by the monkey. Some accounts spell her name Roxelane and say she was the Sultan's third wife and bore him a son. She stands accused by history of being the cruel influence behind the Sultan. She rose from imprisonment and slavery to become the most powerful woman in the Ottoman Empire. She eliminated rivals for the Sultan's affection and threats to her son's succession. She is said to have been indirectly responsible for the building of the great mosque in Istanbul. The Sultan had planned a palace, but Roxellane did not want to be confined to a harem and persuaded him to build a mosque instead. The monkey is endemic to central China.

Royle

Royle's Pika *Ochotona roylei* **Ogilby,** 1838
Royle's Mountain Vole *Alticola roylei* **Gray,** 1842

Dr. John Forbes Royle F.R.S. (1799–1858) was an Indian-born British botanist. He started his professional life in the service of the East India Company as an assistant surgeon, then devoted himself to studying botany and geology and made large collections in the Himalayas. Combining his two loves, botany and medicine, he made a particular study of Hindu herbalism. From 1823 until 1831 he was Superintendent of the Botanical Garden of Saharanpur in northwest India and collected widely during that time. In 1837 he was appointed Professor of Materia Medica in Kings College, London, a post he held until 1856. He was in charge of the Indian exhibits at the Great Exhibition of 1851. In the last two years of his life he created a technical museum at East India House in London. He wrote several books, including *Antiquity of Hindoo Medicine* (1837) and *Illustrations of the Botany and Other Branches of the Natural History of the Himalayan Mountains and of the Flora of Cashmere* (1839). He also wrote a number of articles and essays, such as "Productive Resources of India" (1840) and "Observations on the Vegetation and Products of Afghanistan, Kashmir and Tibet" (1848). He was also a frequent correspondent with Darwin. The pika is found in the Himalayas, from northwest Pakistan east to Nepal and adjacent parts of Tibet. The vole is known only from the western Himalayas of India (Kumaon and Himachal Pradesh).

Rozendaal

Gilded Tube-nosed Bat *Murina rozendaali* **Hill** and Francis, 1984

Frank G. Rozendaal is a Dutch ornithologist. He has published extensively on bats as well as birds—for example, "Notes on Macroglossine Bats from Sulawesi and the Moluccas, Indonesia with the Description of a New Species of *Syconycteris* matschie, 1899 from Halmahera (Mammalia: Megachiroptera)" in 1984 and, with W. Bergmans, "Notes on Collections of Fruit Bats from Sulawesi and Some Offshore Islands (Mammalia, Megachiroptera)" in 1988. The tube-nosed bat comes from Malaysia (recorded from both peninsular Malaysia and Sabah).

Rozet

North African Elephant Shrew *Elephantulus rozeti* Duvernoy, 1833

Claude-Antoine Rozet (1798–1858) was a French army officer who took part in the invasion of Algeria in 1830. He was also a talented painter, and in 1833 he published a number of images of Algerian women and an account, *Voyage dans la régence d'Alger, ou Description du pays occupé par l'armée française en Afrique*. He was also interested in geology and in 1830 published *Traité élémentaire de géologie*. The elephant shrew occurs from Morocco east to western Libya.

Rudd

Rudd's Mole-Rat *Tachyoryctes ruddi* **Thomas,** 1909

Rudd's Mouse *Uranomys ruddi* **Dollman,** 1909 [Alt. White-bellied Brush-furred Rat, Rudd's Bristle-furred Rat]

Charles Dunell Rudd (1844–1916) was an associate of Cecil Rhodes and attended to their mining business while Rhodes got himself into politics. Rudd obtained the concession, in 1883, for Rhodes to go into Mashonaland to establish mining. In 1888 he co-founded the De Beers Mining Company with Rhodes. Rudd financed Captain C. H. B. Grant to collect zoological specimens in southern Africa. Publications of the Zoological Society of London in the first decade of the 20th century carry many references to "the Rudd Exploration of South Africa" and descriptions of new species unearthed by Captain Grant. There is also one mammal named after Rudd's wife (see **Corrie**). He is also commemorated in the names of two birds, Rudd's Apalis *Apalis ruddi* and Rudd's (Long-clawed) Lark *Mirafra ruddi*. The mole-rat is found in southwest Kenya and southeast Uganda. The mouse has a wide but discontinuous distribution in tropical Africa: from Senegal to northern Cameroon; in northeast DRC (Zaire), Uganda, and Kenya; and in Malawi, Mozambique, and eastern Zimbabwe.

Rudolphi

Rudolphi's Whale *Balaenoptera borealis* **Lesson,** 1828 [Alt. Sei Whale]

Professor Karl Asmund Rudolphi (1771–1832) was a Swedish-born physiologist, botanist, zoologist, and pathologist. In 1810 he became

Professor of Anatomy at Berlin University. He was an early proponent of the notion that the basic structural unit of plants is the cell, and he was also the first person to detail the life cycle of parasitic worms. He was a pioneer in stressing the importance of chemistry to the study of both biology and medicine. The whale is found worldwide, from subpolar to subtropical waters.

Rümmler

Rümmler's Mouse *Coccymys ruemmleri* **Tate** and **Archbold,** 1941 [Alt. Rümmler's Brush Mouse]

Hans Rümmler (dates not found) was a German mammalogist with an interest in rodents. He described at least one genus, *Pseudohydromys,* in 1934 and a number of species such as *Rattus jobiensis* in 1935. In 1938 he published *Die Systematik und Verbreitung der Muriden Neuguineas.* The mouse is found in the highlands of New Guinea.

Rupp

Rupp's Mouse *Stenocephalemys ruppi* Van der Straeten and **Dieterlen,** 1983

Hans Rupp is a zoologist who wrote several papers with Dieterlen in the 1970s on rodents, some of them published in the *Annals* of the Royal Museum for Central Africa. The mouse is found only in the highlands of southwest Ethiopia, where Rupp collected between 1972 and 1976.

Rüppell

Rüppell's Fox *Vulpes rueppelli* Schinz, 1825

Rüppell's Pipistrelle *Pipistrellus rueppelli* J. Fischer, 1829

Rüppell's Guereza *Colobus guereza guereza* Rüppell, 1835

Rüppell's Horseshoe Bat *Rhinolophus fumigatus* Rüppell, 1842

Rüppell's Broad-nosed Bat *Scoteanax rueppellii* **Peters,** 1866 [Alt. Greater Broad-nosed Bat]

Wilhelm Peter Eduard Simon Rüppell (1794–1884) was a German collector. He first went to Egypt and ascended the Nile as far as Aswan in 1817 and later made two extended expeditions to eastern Africa, the first from 1821 until 1827 to Sudan, and the second between 1830 and 1834 to Ethiopia. Although he brought back large zoological and ethnographical collections, his expeditions impoverished him. He wrote *Reisen in Nubien, Kordofan und dem Petraischen Arabien* (1829), *Systematische Ubersicht der Vogel Nord ost Afrikas* (1845), and *Reise in Abyssynien* (1838–1840). He was also a collector in the broadest sense of the word and presented his collection of coins and rare manuscripts to the Historical Museum in Frankfurt (his home town). He is remembered in the names of eight birds, including Rüppell's Griffon Vulture *Gyps rueppellii* and Rüppell's Warbler *Sylvia rueppelli,* and two reptiles, including Rüppell's Desert Chameleon *Chamaeleo affinis.* The fox is found from Morocco east to Arabia and Israel, and in parts of Iran, Afghanistan, and Pakistan. The pipistrelle occurs throughout Africa from Algeria and Egypt south to Botswana and Transvaal. It has also been recorded in Israel and Iraq. The guereza (colobus monkey) is found in the Ethiopian highlands west of the Rift Valley and along the Awash River. The horseshoe bat is widespread throughout sub-Saharan Africa. The broad-nosed bat is confined to eastern Australia.

Ruschi

Ruschi's Rat *Abrawayaomys ruschii* Cunha and Cruz, 1979

Augusto Ruschi (1915–1986) was a Brazilian naturalist who dedicated his life to ecological conservation. He was considered to be a world authority on hummingbirds but is probably even more famous for his second love, orchids. He identified and described 5 species and 11 subspecies of hummingbirds and catalogued about 50 new orchids. He wrote over 400 scientific articles and books, among which his *Birds of Brazil* is perhaps the best known. "Gutti," as he was known, virtually grew up in the forest and was fascinated by plants and animals even as a youngster. Nevertheless, he studied law and agronomy, so was largely self-taught in natural history. Moreover, he learned English, French, Latin, and German in order to read the standard texts on botany. Ruschi lived on the Santa Lucia reserve in Santa Teresa, in a house built by his grandfather in the 19th century. He transformed his house into the "Professor Mello Leitão Biology Museum," and before he died he requested permission from the Pro-Memória Foundation to be buried on the reserve. It was thought by some that the liver disease he died from was caused by touching poisonous frogs during years of research, but the doctor who treated him at the end said his liver problems were caused by an overdose of antimalarial drugs. He also has a museum named after him. The rat has a limited range in eastern Brazil and in extreme northeast Argentina (Misiones Province).

Sabuni

Ufipa Brush-furred Rat *Lophuromys sabunii* Verheyen et al., 2007

Christopher Andrew Sabuni (b. 1961) is a postgraduate student at Sokoine University of Agriculture, Morogoro, Tanzania. This species is known from the Ufipa Plateau, southwest Tanzania.

Sage

Sage's Rock Rat *Aconaemys sagei* **Pearson,** 1984

Dr. Richard Sage is a zoologist at the Museum of Vertebrate Zoology, University of California, Berkeley. He worked with Pearson in his studies of Patagonian mice. Among his writings is a paper he wrote with Marshall in 1981 on "Taxonomy of the House Mouse." A species of tree iguana is named *Liolaemus sagei* after him. The rock rat is known only from Neuquén Province, west-central Argentina, though it may also occur in adjacent parts of Chile.

Salenski

Salenski's Shrew *Chodsigoa salenskii* **Kastschenko,** 1907

Dr. Vladimir Vladimirovich Salenski (1847–1918) was a Russian embryologist, anatomist, and zoologist at the Zoological Museum of the Imperial Academy of Sciences in St. Petersburg. His name is spelled variously Salenski and Zalenski, but we have used the former as being the more familiar Anglicized version. He made studies of the embryology of neural systems in invertebrates and fish. Between 1897 and 1906 he was Director of the Zoological Institute. In 1902 he published a study *Equus Przewalski*, which compared this wild horse with various extinct forms of horses and asses. His interests included ichthyology, as in 1878 he published *Life History of the Sterlet Acipenser ruthenus*. He was an Academician of the St. Petersburg Academy of Sciences (renamed the Russian Academy of Sciences in 1917). Salenski and Kastschenko appear to have been studying the same zoological and geographical area at about the same time, possibly together. This shrew is known only from the holotype's locality in northern Sichuan, China.

Salim Ali

Salim Ali's Fruit Bat *Latidens salimalii* **Thonglongya,** 1972

Dr. Salim Moizuddin Abdul Ali, known as Dr. Salim Ali (1896–1987), was an Indian ornithologist and conservationist. At the age of 10 he shot a sparrow that had a yellow streak below its neck. Plucking up his courage, he took it to the Bombay Natural History Society's offices and was seen by W. S. Millard, who told him it was a Yellow-throated Sparrow and showed the young Salim Ali the society's collection of stuffed birds. That experience persuaded him to become an ornithologist. In 1919 he moved to Burma (now Myanmar) to work in his family's timber business. When he returned to India he found he could not get a job at the Zoological Survey of India because he was only a Bachelor of Science and not a Doctor of Philosophy. He decided that more education was needed and went to Germany, where he studied under Stresemann. He then spent 20 years wandering over India and becoming the foremost expert on the birds of the subcontinent. Unlikely as it seems in view of Richard Meinertzhagen's well-known disdain for "colonials," he and Salim Ali hit it off particularly well, became lifelong friends, and went on a number of expeditions together. On India becoming independent, Salim Ali took over the Bombay Natural History Society and made sure that funds were found to prevent it from closing down. (The society was founded in 1883 but represented nearly 200 years of history, as it had acquired the Bombay records of the Honourable East India Company going back to the early 1750s.) His intervention was responsible for saving the Bharatpur Bird Sanctuary and the Silent Valley National Park. He founded the Salim Ali Centre for Ornithology and Natural History at Chorao (Goa). The fruit bat is found in southern India.

Salomonsen

Mindanao Hairy-tailed Rat *Batomys salomonseni* **Sanborn,** 1953

Finn Salomonsen (1909–1983) was a Danish ornithologist and artist. He led a natural history expedition to northwest Greenland in 1936 and assembled much of the ornithology collection of the Zoological Museum at the University of Copenhagen, which covers Denmark and the North Atlantic Dependencies. In 1952 he co-wrote and illustrated *Birds of Greenland.* He is remembered in the common names of two subspecies of parrots that he described: Salomonsen's Blue-naped Parrot *Tanygnathus lucionensis hybridus* and Salomonsen's Racket-tailed Parrot *Prioniturus discurus whiteheadi.* Salomonsen was in the Philippines from 1951 to 1952 and collected the type specimen of the rat on Mindanao. It also occurs on the islands of Biliran, Dinagat, and Leyte.

Salt

Salt's Dik-dik *Madoqua saltiana* **Desmarest,** 1816

Sir Henry Salt (1780–1827) was an explorer and diplomat who had originally been trained as a painter. He visited Egypt and India from 1802 to 1806 and in 1809 returned to Africa to attempt to establish contact with the King of Abyssinia on behalf of the British government; this task took him two years. He was British Consul General in Alexandria from 1815 until 1827. During his time there he accumulated antiquities that after his death were sent to the British Museum. He carried out major excavations at Giza investigating the Great Pyramid. He employed Caviglia to excavate the Sphinx, but apparently the two men had a falling out because Caviglia kept looking for mummy pits. The typical form of this dik-dik is found in Eritrea. Other subspecies, some of which may prove to be distinct species, occur in eastern Ethiopia and Somalia.

Salvin

Salvin's Big-eyed Bat *Chiroderma salvini* **Dobson,** 1878 [Alt. Salvin's White-lined Bat]

Salvin's Spiny Pocket Mouse *Liomys salvini* **Thomas,** 1893

Osbert Salvin (1835–1898) was an English naturalist who became a Fellow of the Royal Society. He was a lifelong friend of Frederick Godman, whom he met when both were at Cambridge University, where Salvin studied mathematics. In 1861 he wrote that he was "determined, rain or no rain, to be off to the mountain forests in search of quetzals, to see and shoot, which has been a daydream for me ever since I set foot in Central America." He was the first European to record observing a Resplendent Quetzal *Pharomachrus mocinno.* He pronounced it "unequalled for splendour among the birds of the New World"—and promptly shot it. Salvin redeemed himself by co-authoring with Godman the incredible 40-volume *Biologia Centrali Americana* (1879), which provided a near-complete catalogue of Central American species. The Godman-Salvin Medal, a prestigious award of the British Ornithologists' Union, is named after him and Godman. He is remembered in the names of nearly

20 birds. The bat is found from Mexico south to Bolivia and Venezuela. The mouse occurs from southern Mexico to Costa Rica.

Sambirano

Sambirano's Lemur *Microcebus sambiranensis* Rasoloarison, **Goodman,** and Ganzhorn 2000 [Alt. Sambirano Mouse-Lemur]

This lemur is not named after a person but after the Sambirano Forest region in northwest Madagascar where it was discovered. References to "Sambirano's Lemur" are therefore mistranscriptions or misunderstandings.

Sanborn

Sanborn's Flying Fox *Pteropus mahaganus* Sanborn, 1931
Sanborn's Bonneted Bat *Eumops hansae* Sanborn, 1932 [Alt. Hansa Mastiff Bat]
Northern Broad-nosed Bat *Scotorepens sanborni* **Troughton,** 1937
Sanborn's Grass Mouse *Abrothrix sanborni* **Osgood,** 1943
Sanborn's Squirrel *Sciurus sanborni* Osgood, 1944
Sanborn's Big-eared Bat *Micronycteris sanborni* Simmons, 1996

Dr. Colin Campbell Sanborn (1897–1962) was a biologist with an interest in birds as well as mammals. He wrote a number of works, including *Land Mammals of Uruguay* (1929), *Birds of the Chicago Region* (1934), *New Mammals from Guatemala and Honduras* (1935), *Philippine Zoological Expedition 1946–1947: Mammals* (1949), and, with Nicholson, *Bats from New Caledonia, the Solomon Islands, and New Hebrides* (1950). He also compiled the *Catalogue of Type Specimens of Mammals in Chicago Natural History Museum* (1947). He was Assistant Curator in the Ornithology Section of the Chicago museum when Osgood was overall Curator of Zoology. He later became Curator of Mammals there. The flying fox is from the islands of Bougainville (Papua New Guinea), Ysabel, and Choiseul (Solomon Islands). The bonneted bat can be found from southernmost Mexico south to Peru and Brazil. The broad-nosed bat occurs in northern Australia, southeast New Guinea, and Timor. The grass mouse comes from southern Chile and the squirrel from southeast Peru. The big-eared bat was first discovered in northeast Brazil but has also been recorded in eastern Bolivia.

Sanford

Sanford's Lemur *Eulemur (fulvus) sanfordi* **Archbold,** 1932

Dr. Leonard Cutler Sanford (1868–1950) was an American zoologist and Trustee of the American Museum of Natural History. It was Sanford who, in 1928, invited Archbold to participate in the French-British-American expedition of 1929–1931 to Madagascar, which was led by Delacour. Later he was instrumental in persuading Archbold to consider New Guinea for his next expedition. Sanford wrote *The Waterfowl Family* in 1924, with L. B. Bishop and T. S. van Dyke. The scientific names of many species of animals as diverse as dinosaurs and butterflies include *sanfordi* in their binomials, for example the Northern Royal Albatross *Diomedea (epomophora) sanfordi*. The lemur comes from northern Madagascar.

Sarasin

Sulawesi Free-tailed Bat *Mops sarasinorum* **A. Meyer,** 1899 [Alt. Sulawesi Mastiff Bat]

Paul B. (1856–1929) and K. Fritz (1859–1942) Sarasin were cousins. They were Swiss zoologists, explorers, and collectors who collaborated on works such as *Reisen in Celebes* (1905) and *Die Vögel Neu-Caledoniens und der Loyalty Inseln* (1913). They are also remembered in the scientific names of two birds, the Greater Streaked Honeyeater *Myza sarasinorum* and the Sulawesi Leaf-Warbler *Phylloscopus sarasi-*

norum, and in the scientific names of a number of reptiles including Roux's Giant Gecko *Rhacodactylus sarasinorum*. The bat is found on Sulawesi and Peleng (Indonesia) and also in the Philippines.

Sartori

Central American Red Brocket *Mazama sartorii* **Saussure,** 1860 [Syn. *Mazama temama* Kerr, 1792]

Dr. Christian Carl Wilhem Sartorius (1796–1872) was a German naturalist who lived and collected in Mexico from 1826 to 1872. He had a hacienda called El Mirador, which was a magnet for German scientists, especially botanists. Saussure probably visited El Mirador during his visit to Mexico in 1856. Sartorius himself collected everything and anything and is mentioned in the literature in connection with, among other topics, botany, herpetology, and ornithology. He also has a snake named after him, Sartorius' Snail-sucker *Sibon sartorii*. Although the name *Mazama sartorii* is still found in the literature, most authorities regard the correct scientific name for this brocket deer to be *M. temama*. Found from southern Mexico to Panama, it has traditionally been treated as a form of the South American Red Brocket *Mazama americana*, but recent studies suggest that it deserves species status.

Satere-Maues

Sateré-Maués' Marmoset *Callithrix saterei* Silva Jr. and Noronha, 1998 [Alt. Satere Marmoset]

The Sateré-Maués are a tribe of indigenous people living in the Amazon rainforest of Brazil, so this marmoset is named after a people rather than an individual. It is found in the region between the Canuma and Abacaxis rivers, south of the Amazon in Brazil.

Satunin

Caucasian Shrew *Sorex satunini* **Ognev,** 1922

Konstantin Alexeevitsch Satunin (1863–1915) was an eminent Russian zoologist who studied the fauna of the Caucasus region. Like his predecessor Radde, he was initially most interested in birds, writing such works as *A Systematic Catalogue of the Birds of the Caucasian Region*, published in 1912. However, he was an all-rounder, collecting fish, insects, and mammals and writing on an extensive range of topics. He wrote *New Mammals from Transcaucasia* (1914) and also *Mammals of the Caucasian Land* (not published until 1920). He is commemorated in the names of a number of other taxa such as the fish *Leucalburnus satunini* and the katydid *Uvarovistia satunini*. The shrew is found in northern Turkey and the Caucasus.

Saunders

Hewitt's Red Rock Hare *Pronolagus saundersiae* **Hewitt,** 1927

Saunders' Vlei Rat *Otomys saundersiae* **Roberts,** 1929

Miss Enid Saunders (dates not found) lived in the Grahamstown district of South Africa. Hewitt's etymology for the hare reads, "I have pleasure in associating them with the name of Miss Enid Saunders, M.Sc., whose recent studies have added much to our knowledge of the rodents of the district." Roberts made similar remarks when naming the rat. It seems that Miss Saunders never achieved more than local fame for her studies. Both species named after her are found in South Africa.

Saussure

Saussure's Shrew *Sorex saussurei* **Merriam,** 1892

Dr. Henri Louis Frédéric de Saussure (1829–1905) was a Swiss mineralogist, zoologist, and entomologist. After graduating from the University of Giessen, he spent the period 1854–1856 in the West Indies, the USA, and Mexico, and made considerable collections of specimens which he took back to Geneva. He was accompanied on his journey by Adrien Jean Louis de Sumichrast (1828–1882). He wrote *Mémoire sur divers crustacés nouveaux des Antilles et du Mexique.* He was interested in geology and geography, and in 1858 he founded the Geographical Society of Geneva and was its President from 1888 to 1889. The mineral saussurite is also named after him. The shrew is found in Mexico and Guatemala.

Sauvel

Kouprey *Bos sauveli* Urbain, 1937 [Alt. Cambodian Grey Ox]

Dr. René Sauvel (dates not found) was a French veterinary surgeon who practiced in Cambodia. It is said that Achille Urbain first came across the Kouprey as a set of horns, mounted as a hunting trophy, on the walls of Dr. Sauvel's house. Sauvel also sent a live male calf to France in 1936; it survived until at least 1941 in the Paris Zoo. R. Vittoz, who had been Sauvel's predecessor as Veterinary Inspector for Cochin, wrote a description of the Kouprey in 1933 without realizing that the species was new to science. In turn, scientists seem to have taken no notice of Vittoz's writings. The Kouprey was originally found in eastern Thailand, Cambodia, southern Laos, and western Vietnam. Today it is on the verge of extinction, if it is still extant. It has recently been claimed that the Kouprey may not be a true species at all but arose via hybridization between domestic zebu cattle and the Banteng *Bos javanicus.* This theory is not universally accepted.

Savi

Savi's Pygmy Shrew *Suncus etruscus* Savi, 1822 [Alt. White-toothed Pygmy Shrew, Etruscan Shrew]

Savi's Pipistrelle *Hypsugo savii* **Bonaparte,** 1837 [formerly *Pipistrellus savii*]

Savi's Pine Vole *Microtus savii* de Selys-Longchamps, 1838

Paolo Savi (1798–1871) was an Italian naturalist, zoologist, paleontologist, and geologist. He studied physics and natural science at Pisa University, becoming Professor of Natural History there and also Director of the museum. He became an Italian senator in 1862. His greatest work was *Ornitologia Italiana,* published posthumously between 1873 and 1876. His is also commemorated in the common name of a bird, Savi's Warbler *Locustella luscinioides.* The shrew is found in southern Europe, North Africa, the Middle East, and eastward to the Indian subcontinent and Indochina (eastern populations might represent distinct species). The pipistrelle is widely distributed from Morocco, the Canary Islands, and the Iberian Peninsula east through southern Europe and western Asia to Iran, Afghanistan, and northern India. The pine vole occurs in Italy (including Sicily) and southeast France.

Savile

Savile's Bandicoot Rat *Bandicota savilei* **Thomas,** 1916

Sir Leopold Halliday Savile (1870–1953) was a distinguished British civil engineer and was the President of the Institute of Civil Engineers from 1940 to 1941. He went to India in 1896 to work on the construction of the Hoogly River tunnel. There he became a member of the Bombay Natural History Society. According to the original etymology, "The species is named after Mr. L. H. Savile, who was Honorary Treasurer of the Bombay Natural History Society for many years, and has taken great interest in the success of the Mammal Survey." He was interested in the excavation of the docks at Lothal, a city of the Harappan civilization in the Indus Valley, and in 1941 he published a pamphlet entitled *Ancient Harbours*. The rat is found from central Myanmar east to Vietnam.

Say

Say's Least Shrew *Cryptotis parva* Say, 1823 [Alt. North American Least Shrew]

Thomas Say (1787–1834) was a self-taught American naturalist whose primary interest was entomology. He described over 1,000 new species of beetles and over 400 new insects of other orders. Say was appointed Chief Zoologist with Major S. H. Long's expeditions and explored the Rocky Mountains with him. Later, Say lived at the utopian settlement of New Harmony in Indiana (1826–1834). His most famous works include *American Entomology, or Descriptions of the Insects of North America*, published in three volumes, and *American Conchology*. A bird, Say's Phoebe *Sayornis saya*, is also named after him. The shrew is found in the eastern half of the USA and in Mexico.

Scaglia

Scaglia's Tuco-tuco *Ctenomys scagliai* **Contreras,** 2000

Galileo Juan Scaglia (1915–1989) was an Argentine naturalist and paleontologist. He was Director of the Municipal Museum of Natural Sciences at Mar del Plata. This museum is named after his father, Lorenzo Scaglia, who was an avid collector. A dinosaur, *Sarmientichnus scagliai*, has also been named in his honor. The tuco-tuco comes from Tucumán Province in northern Argentina.

Schadenberg

Luzon Bushy-tailed Cloud Rat *Crateromys schadenbergi* **Meyer,** 1895

Alexander Schadenberg (1851–1896) was a German chemist who had gone to the Philippines to join a wholesale drug company. At the end of 1881 he went on a scientific expedition to the south of Mindanao with Otto Koch. They stayed in the area for six months (the so-called Schadenberg-Kock expedition of 1881–1882) and climbed the volcano Mount Apo in February 1882. They made a collection of botanical and zoological specimens, including thousands of butterflies. North of Mount Apo they discovered a giant parasitic plant, *Rafflesia schadenbergiana*, the open flower of which has a diameter of 80 cm (31 inches). Schadenberg sent the first known specimen of the cloud rat to the Dresden Museum in 1894. It is only found in northern Luzon, in the Philippines.

Schäfer

Dwarf Bharal *Pseudois schaeferi* Haltenorth, 1963 [Alt. Dwarf Blue Sheep]

Ernst Schäfer (1910–1992) was a German hunter, biologist, and ornithologist who led the German expedition to Tibet in 1938. He and all the other scientists on the expedition were members of the SS and confirmed Nazis. He had joined the SS in 1933 and was a member of Himmler's inner circle. Schäfer had first visited Tibet in 1930 and returned there in 1931–1932, and again in 1934–1936 as a member of the American Brooke-Dolan expeditions, which also visited China and Siberia. He spent part of 1932 and 1933 studying the collections of the Natural History Museum in London. He appears to have had different aims than those approved by Himmler, who later lost patience with him and had him posted to fight on the Eastern Front from 1943 onward. He was arrested in 1945 and charged with war crimes, but was released at the end of 1947, having been found not guilty. From 1949 to 1954 he lectured on zoology and biology in Caracas and researched Venezuelan fauna, with particular reference to birds. In 1955 he met King Baudouin of Belgium, who invited him to Brussels to advise on an intended expedition to the Belgian Congo. From 1956 to 1959 he accompanied Baudouin and his companions on their travels there. In 1960 he was appointed head of the Zoological Department of the Lower Saxony State Museum in Hanover. He retired in 1970, and between his retirement and 1984 he visited Alaska, Tanzania, Kenya, Uganda, Namibia, Zimbabwe, and Mozambique, paying a final visit to Venezuela in 1984. He suffered badly from arthritis from 1986 until his death. His collection is in the Natural History Museum in Berlin. This endangered species of sheep is found in the upper Yangtze gorge of China.

Schaller

Schaller's Mouse-Shrew *Myosorex schalleri* **Heim de Balsac,** 1966

Dr. George Beals Schaller (b. 1933) is an American zoologist who was born in Berlin and moved to Missouri as a teenager. He was a research assistant at Johns Hopkins University and then a research zoologist for the New York Zoologicqal Society. He is now the Science Director of International Programs for the Wildlife Conservation Society in New York, which is based at the Bronx Zoo. In 1980 he was awarded the World Wildlife Fund Gold Medal. In 1963 he published *The Mountain Gorilla: Ecology and Behavior.* He was followed in his research by Dian Fossey. The shrew is known only from the type locality in the Itombwe Mountains of eastern DRC (Zaire).

Schaposchnikow

See **Shaposhnikov.**

Schaub

Schaub's Myotis *Myotis schaubi* Kormos, 1934

Professor Samuel Schaub (1882–1955) was a Swiss paleontologist who worked at the University of Basel and the Natural History Museum there. In 1929 he published *Über eocäne Ratitenreste in der osteologischen Sammlung des Basler Museums,* with further publications through to the 1950s. He was elected an honorary member of the Society of Vertebrate Paleontology in 1958. This bat was first named from Pliocene fossils found in Hungary. Later it was found to be extant in Armenia and western Iran.

Schauinsland

Hawaiian Monk Seal *Monachus schauinslandi* **Matschie,** 1905

Professor Dr. Hugo Hermann Schauinsland (1857–1937) was a German zoologist and explorer. He studied natural sciences at Geneva in 1879 and zoology from 1880 to 1883 at Königsberg (now Kaliningrad and in Russia), where he was awarded his doctorate. Between 1883 and 1885 he was involved in research in Naples and Munich and in the latter year became a Professor at the University of Munich. In 1887 he became the Superintendent of the Municipal Collections of Natural History and Ethnology in Bremen. In 1896 he established Bremen's Municipal Museum of Natural History, Ethnology, and Trading (today it is known as the Überseemuseum), becoming its first Director and holding that position until 1933, when he retired as Professor Emeritus. He traveled in East Asia and the Pacific between 1896 and 1897, returning home via New Zealand, Australia, Ceylon (now Sri Lanka), and Egypt. It was on that trip, in 1896, that he visited Laysan Island, staying there for three months, during which time he collected specimens of the monk seal. He described his time on Laysan in *Drei Monate auf einer Koralleninsel*, published in 1899. Between 1905 and 1906 he visited the Bismarck Archipelago, China, Korea, Japan, Borneo, and the Celebes. In 1907 and 1908 he was again in the Pacific and East Asia, having traveled via the USA and Hawaii. His expedition of 1913–1914 covered Asia and the South Seas again. His final expedition was to Egypt in 1926. Perhaps there is good reason to consider him to be "ein Bismarck der deutschen Museumlandschaft" (a Bismarck of the German museum world), as he has been described. The seal is found around Laysan and other islands of the northwestern Hawaiian chain.

Scheffel

Scheffel's Sand Cat *Felis margarita scheffeli* Hemmer, 1974 [Alt. Pakistan Sand Cat]

Walter Scheffel is a German citizen who has a particular interest in breeding nondomestic cats. According to our sources, he was the consul in Germany for the Comoro Islands, but he got into trouble with the law for tax evasion. He spent some months in prison, during which time his private collection of exotic *Felidae* was disbanded—a great pity, as it held many rarities, including Sand Cats and Javan Leopards. In 1975, jointly with H. Hemmer, who described the sand cat, he published an article entitled "Breeding Geoffroy's Cat *Leopardus geoffroyi salinarum* in Captivity." This race of Sand Cat occurs in Pakistan (other subspecies are found in the Arabian Peninsula and North Africa).

Schelkovnikov

See **Shelkovnikov.**

Schlegel

Schlegel's Guenon *Cercopithecus neglectus* Schlegel, 1876 [Alt. De Brazza's Monkey]

Arfak Ringtail Possum *Pseudochirulus schlegeli* **Jentink,** 1884 [Alt. Vogelkop Ringtail]

Professor Hermann Schlegel (1804–1884) was a German zoologist and the first person to use trinomials to describe separate races, a practice he began in 1844. In his youth he was tutored by C. L. Brehm, the father of Alfred Brehm. Between 1824 and 1825 he studied in Vienna, and in 1825 he was recruited by Temminck as a "preparator" for the museum at Leiden. He studied at Leiden under Reinwardt in 1831. He was primarily an ornithologist but also wrote extensively on herpetology, and most of his early work was on reptiles. In 1858 he became Director of the Rijksmuseum van Natuurlijke Historie in Leiden in succession to Temminck. He wrote *Fauna Japonica—Aves* and *Kritische Ubersicht der Europaischen Vogel*, both published in 1844, and *De Vogels van Nederland*, published between 1854 and 1858. He is also commemorated in the scientific names of many other taxa, including birds such as the Karoo Chat *Cercomela schlegelii*, reptiles such as the Eyelash Viper *Bothriechis schlegelii*, and fish such as the Yellow Guitarfish *Rhinobatos schlegelii*. The monkey, more commonly known nowadays as De Brazza's Monkey, is found from southern Cameroon east to Uganda and western Kenya. The possum is a little-known species from the Arfak Mountains of western New Guinea.

Schlieffen

Schlieffen's Bat *Nycticeinops schlieffeni* **Peters,** 1859 [Alt. Schlieffen's Twilight Bat]

Graf Wilhelm Schlieffen von Schlieffenberg (b. 1833) was the youngest son of a German family that suffered hereditary tuberculosis. Several of his brothers had died of that disease in Europe, and so in 1851 or 1852 his mother took him to Egypt in order to try a hot dry climate as a possible cure. They traveled as far as the province of Dongala, north of Khartoum, and stayed there several months during which time Schlieffen von Schlieffenberg discovered a historically important stone inscription from the period of the Ethiopian Empire. This stone was of enormous size but was successfully transported down the Nile, despite the cataracts in Nubia, and was presented to the Berlin Museum where it was highly regarded. Heinrich Brugsch, the famous German Egyptologist, reported in his memoirs that he had met Schlieffen von Schlieffenberg in Cairo in 1854. The bat is found in the southwest Arabian Peninsula, Egypt (where the type specimen was taken), and much of sub-Saharan Africa except the far south.

Schliemann

Western Sucker-footed Bat *Myzopoda schliemanni* **Goodman,** Rakotondraparany, and Kofoky, 2006

Professor Dr. Harald Schliemann (b. 1936) is a zoologist and biologist at the University of Hamburg, Germany, from whence he graduated in 1965. He was a guest lecturer at the University of Nairobi (Kenya) between 1968 and 1969. In 1971 he was appointed as Lecturer at the University of Hamburg and became Professor of Zoology there in the following year. He acted as Director of the university's Zoological Institute and Zoological Museum from 1989 to 1992. Since 2002 he has been Chairman of the Scientific Association in Hamburg. He was one of the co-editors, with Martin S. Fischer and Jochen Niethammer, of *Handbuch der Zoologie: Eine Naturgeschichte der Stämme des Tierreiches* (Handbook of Zoology: A Natural History of the Phyla of the Animal Kingdom), published in 2005. The bat is found in the dry forests of western Madagascar.

Schmid

Giant Mole-Shrew *Anourosorex schmidi* **Petter,** 1963

Dr. Fernand Schmid (d. 1998) was an entomologist attached to the Zoological Museum at Lausanne. As the world opened up after the end of WW2, he set himself the project of traveling the whole length of the Himalayas from Afghanistan to Assam, and this he did through a series of small expeditions, walking alone or occasionally with one companion, taking only what he could carry on his back. After that project was completed, he spent the years 1958–1961 in India, but in 1961 he became ill while in a closed military area of Bengal during a period of great tension between India and China. After a lot of trouble with getting permissions, he was extracted and returned to Switzerland. He recovered his health but was unable to find a job at any university or museum in Switzerland and so applied to a Canadian institute, eventually emigrating from Switzerland to Canada and taking with him his unique collection of Himalayan insects. At the time of his death he was an Honorary Research Associate of the Eastern Cereal and Oilseed Research Centre (Agriculture and Agri-Food, Canada). Petter had accompanied Schmid on a collecting trip to Iran, and in his original description says he is naming the shrew "en l'honneur de M. F. Schmid, entomologiste attaché au Musée de Lausanne, et en souvenir de nos chasses communes aux Rongeurs en Iran" (in honor of Mr. F. Schmid, an entomologist attached to the Museum of Lausanne, and in remembrance of our common hunt for rodents in

Iran). A genus of insect, *Fernandoschmidia*, was named after him in 2007. The mole-shrew is found in Bhutan and northeast India.

Schmidly

Schmidly's Deer Mouse *Peromyscus schmidlyi* Bradley et al., 2004

Professor Dr. David James Schmidly (b. 1944) trained as a biologist with a special interest in the taxonomy and natural history of mammals. His bachelor's and master's degrees were awarded by Texas Tech University and his doctorate, in zoology, by the University of Illinois. He has taught at Texas A&M University, Oklahoma State University, and Texas Tech University. He served as President of both Texas Tech and Oklahoma State University and is now, since October 2007, President of the University of New Mexico. In 2002 he published *Texas Natural History: A Century of Change*, an annotated reprinting and updating of *A Biological Survey of Texas*, the landmark 1905 study of the state by Vernon Bailey and a team of 12 other federal biologists. In 2004 he published *The Mammals of Texas* documenting the natural history of the state's extant mammals. He is also a noted conservationist. The etymology reads, "This species is named in honor of Professor David J. Schmidly for his many contributions to the systematics of the genus *Peromyscus* and devotion to mammalian taxonomy." The mouse is found in the Sierra Madre Occidental mountains of western Mexico.

Schmidt, K. P. and F.

Schmidt's Big-eared Bat *Micronycteris schmidtorum* **Sanborn,** 1935

Karl Patterson Schmidt (1890–1957) was an American herpetologist, and his brother Frank Schmidt (dates not found) was a collector. Karl worked under Noble and Dickinson at the American Museum of Natural History from 1916 to 1922. From 1922 to 1940 he was Curator of the newly founded Department of Amphibians and Reptiles, Field Museum, Chicago, becoming Curator of Zoology in 1941 and Emeritus Curator in 1955. He undertook many expeditions, including one to Guatemala in 1933, where his brother collected the holotype of the bat. He was an avid collector of herpetological specimens and wrote such books as, with W. L. Necker, *Amphibians and Reptiles of the Chicago Region* (1935) and, with D. D. Davis *Field Book of Snakes of the US and Canada* (1941). In 1957 he was bitten by a Boomslang *Dispholidus typus*. Incorrectly believing that the juvenile snake could not inject a fatal dose of venom, he went home to his wife and received no medical treatment. He kept a careful note of the development of the symptoms he experienced until he died. The original description of the bat reads, "I take great pleasure in naming this species after the Schmidt brothers, Karl and Frank, who have, by active interest and encouragement on one side and by so perseveringly searching for specimens on the other, added so much to the knowledge of the bats of Guatemala." The bat is found from southern Mexico to northern Peru and northeast Brazil.

Schmidt, R.

Schmidt's Monkey *Cercopithecus ascanius schmidti* **Matschie,** 1892 [Alt. Schmidt's Guenon, Uganda Red-tailed Monkey]

Dr. Rochus Schmidt (1860–1938) was a physician who was a member of the administration and armed forces in German East Africa. In February 1889 he commanded a portion of the troops assigned to Hermann von Wissmann to create a colonial army in Africa and to combat the Arab rebellion on the east coast. He took part in fighting at Bagamayo and Pangani in 1889, and late in that year he was ordered to go to the coast and escort Emin Pasha and Henry Morton Stanley away from it. On his way to meet them he encountered a caravan moving inland, and from it he purchased a young monkey, which he kept alive and eventually, in 1892, delivered as a living specimen to the Berlin Zoo. In 1891 Schmidt was the Chief Administrator in the Bagamayo district. He returned to Germany in 1892 and was promoted to the rank of Major General. He published a number of books about his time in Africa, including *Deutschlands Kolonien* (1894). The monkey is found in eastern DRC (Zaire), Uganda, and western Kenya.

Schmitz

Schmitz's Caracal *Caracal caracal schmitzi* **Matschie,** 1912

Father Ernst Schmitz (1845–1922) was an ornithologist and naturalist as well as a Lazarite priest, ordained in 1869. Between 1875 and 1879 he was in Portugal. In 1880 he was transferred to Madeira, where he remained until 1898. Between 1898 and 1902 he moved to Belgium but then returned to Madeira, where he stayed until being transferred to Jerusalem in 1908. He spent the rest of his life in the Middle East. His job in Jerusalem was to run the Lazarite seminary, which included a small natural history museum. He entered enthusiastically into fieldwork and accumulated many varieties of ant, of which 10 were new to science, such as *Hagioxenus schmitzi.* He also discovered the subspecies of Caracal that is named after him. A subspecies of Barn Owl, *Tyto alba schmitzi,* is also named after him. This race of Caracal is found from Israel and the Arabian Peninsula east to India.

Schnabl

Smoky Thumbless Bat *Amorphochilus schnablii* **Peters,** 1877

Dr. Johann Andreas Schnabl (1838–1912) was originally called Jan Sznabla but reverted to the old spelling of his family name that had been in use when his ancestors moved from Dresden to Warsaw at the end of the 18th century, which was at that time the administrative capital of South Prussia. The change certainly made his name much more accessible to non-Polish speakers. He was born a Pole with Russian nationality. He qualified and practiced as a physician but was also an anatomist, a teacher of natural history at the Classic Academy in Warsaw, and an entomologist. He traveled widely, visiting the Caucasus, the Ural Mountains, Lapland, the Pyrenees, Corsica, Hungary, and Peru, where he collected insect specimens that he sent to the Berlin Museum. He also wrote several descriptions of new Diptera species and is commemorated in the scientific names of a fungus gnat, *Mycetophila schnablii,* and a dipteran, *Cheilosia schnabli.* He traveled with Wladyslaw Taczanowski when the latter was collecting for the Branicki Museum. He published widely in German when writing on entomology and in Polish for his medical publications. The bat is found in Ecuador, Peru, and northern Chile.

Schneider

Schneider's Roundleaf Bat *Hipposideros speoris* Schneider, 1800 [Alt. Schneider's Leaf-nosed Bat]

Johannes Gottlob Theaenus Schneider (1750–1822) was a German scholar in the days when scholars were expected to be polymaths and scholarship covered everything from the natural sciences to dead languages. Jointly with Heinrich Keil he wrote a book demonstrating that a lost manuscript by Marcian was the source of all the extant agricultural writings of Cato and Varro. In 1801, with M. E. Bloch, he published a book on fishes written in Latin, *Systema ichthyologiae iconibus CX illustratum.* He also described a number of reptiles and amphibians, and is commemorated in the names of at least four reptiles, including Schneider's Skink *Novoeumeces schneideri.* The bat is found in southern India and Sri Lanka.

Schomburgk

Schomburgk's Deer *Cervus schomburgki* **Blyth,** 1863 extinct [Syn. *Rucervus schomburgki*]

Sir Robert Hermann Schomburgk (1804–1865) was a German-born English traveler and an explorer for the Royal Geographical Society. He made a botanical and geographical exploration of British Guiana (now Guyana) in 1835. Later, from 1841 to 1843, he surveyed the colony for the British government. During the survey he outlined the "Schomburgk Line," a boundary that played a prominent part in subsequent border disputes with Venezuela. He was knighted in 1844 and was appointed British Consul at Santo Domingo in 1848 and at Bangkok in 1857. He wrote books on British Guiana and Barbados and edited Walter Raleigh's journal of his second voyage to Guiana. He is also remembered in the common name of a bird, Schomburgk's Parrotlet *Forpus sclateri eidos*. The deer was found in south-central Thailand but died out during the 1930s. It was long believed to be extinct, but in 1991 a pair of antlers apparently belonging to this species was seen in a medicine shop in Laos. However, since then there has been no further evidence that the deer is still extant.

Schouteden

Schouteden's Blue Monkey *Cercopithecus mitis schoutedeni* Schwartz, 1928

Schouteden's Shrew *Paracrocidura schoutedeni* **Heim de Balsac,** 1956 [Alt. Lesser Large-headed Shrew]

Henri Eugene Alphonse Hubert Schouteden (1881–1972) was a Belgian zoologist who undertook many expeditions to the Congo (now Zaire). He published on both ornithology and entomology and wrote *De Vogels van Belgisch-Congo en van Ruanda-Urundi*. He is also remembered in the name of a bird, Schouteden's Swift *Schoutedenapus schoutedeni*. The race of Blue Monkey is found on the islands of Idjwi and Shushu in Lake Kivu; populations on the neighboring mainland show degrees of intergradation with other subspecies. The shrew is found in southern Cameroon, Gabon, Central African Republic, Congo Republic, and DRC (Zaire).

Schreber

Schreber's Yellow Bat *Scotophilus nigrita* Schreber, 1774 [Alt. Giant House Bat]

Professor Dr. Johann Christian Daniel von Schreber (1739–1810) was a German naturalist. He studied theology, natural science, and medicine at the universities of Halle and Upsala. He practiced medicine in Mecklenburg and was appointed Professor of Medicine and Botany at Erlangen in 1770, Director of the Erlangen Botanical Gardens in 1773, and Professor of Natural History in 1776. He was made a member of the Royal Swedish Academy of Sciences in 1787 and was knighted in 1791. In 1771 he published *Spicilegium florae lipsicae*, a work that discussed 80 species of fungi, and in 1774 started work on *Die Säugethiere in Abbildungen nach der Natur mit Beschreibungen*, a work in many volumes focusing on the mammals of the world. Many of the animals included received scientific names for the first time, as Schreber followed Linnaeus' binomial system. The bat is found from Senegal east to Sudan and Kenya and south to Malawi and Mozambique.

Schreibers

Schreibers' Bat *Miniopterus schreibersii* **Kuhl,** 1817 [Alt. Common Bentwing Bat, Schreibers' Long-fingered Bat]

Dr. Karl Franz Anton Ritter von Schreibers (1775–1852) was a botanist and all-round naturalist who in 1806 became "Keeper of the Imperial-Royal Repository of Natural Specimens" at the Museum in Vienna—or, less colorfully, Director of the Austrian Imperial Museum. He transformed what had been a repository for curios into a formidable scientific institution. He gathered talented people around him, such as Joseph Natterer, the father of Johann Natterer, and also organized scientific expeditions such as one to Brazil from 1817 to 1835. Schreibers wrote many scientific papers and one longer work on meteorites. When a meteor shower was reported in Moravia in 1808, he immediately traveled there and collected a large number of examples. He became known as the foremost authority on meteorites, and people sent him samples from all over Europe. During his time as Director of the museum he amassed a collection of more than 30,000 scientific books plus a large number of unpublished manuscripts. All this was destroyed in 1848 when the Imperial Army was bombarding the revolutionaries and unfortunately hit the museum, setting it on fire. Virtually all his life's work was lost, and Schreibers found that he could not bear the loss. He retired in 1851 and died the next year. The bat has an enormous distribution extending from southern Europe and Morocco east to Japan and the Philippines, much of the Indo-Malayan region east to the Solomon Islands and Australia, and also Africa and Madagascar.

Schulz

Schulz's Round-eared Bat *Lophostoma schulzi* **Genoways** and Williams, 1980

Dr. Johan P. Schulz (1921–1999), known as "Joop" to his friends, was a Dutch ecologist who was originally trained in tropical agriculture and forestry but became interested in natural history. In particular, he studied birds and turtles. He was employed by the Suriname Forest Service from 1954. In 1965 he set up, and was appointed the first Director of, the Suriname Nature Conservation Department, and he also established the Foundation for Nature Preservation in Suriname. Between 1963 and 1964 he organized and led six expeditions to establish where sea turtles were nesting, not only in Suriname but also in Guyana and French Guiana. Between 1969 and 1973 he designed and operated an extensive tagging program for Green, Leatherback, and Olive Ridley Turtles, and in 1976 he published *Sea Turtles Nesting in Suriname.* He retired from the forestry business in 1980 to devote more time to sea turtle conservation, and between 1984 and 1993 he conducted six turtle surveys in Indonesia from the South China Sea to the Arafura Sea. The bat is found in the Guianas and northeast Brazil.

Schwartz

Schwartz's Myotis *Myotis martiniquensis* **LaVal,** 1973
Schwartz's Fruit-eating Bat *Artibeus schwartzi* Jones, 1978

Professor Albert Schwartz (1923–1992) was an American biologist and entomologist who was Professor Emeritus of Biology at Miami-Dade Community College and was associated with the Florida Museum of Natural History. He was also a Research Associate at the Carnegie Museum of Natural History in Pittsburgh. A specialist in the fauna of the West Indies, he wrote extensively on amphibians, reptiles, and Lepidoptera, including *The Butterflies of Hispaniola*. Both of these bats occur on the islands of the Lesser Antilles, the myotis being restricted to Martinique and Barbados. The fruit-eating bat was formerly regarded as a subspecies of *Artibeus jamaicensis* but has been found to be genetically distinct.

Schweinfurth

Schweinfurth's Chimpanzee *Pan troglodytes schweinfurthii* Giglioli, 1872
[Alt. Eastern Chimpanzee]

Dr. Georg August Schweinfurth (1836–1925) was a German botanist, traveler, and ethnologist who was born in Latvia, then part of the Russian Empire. He studied between 1856 and 1862 at the universities of Heidelberg, Munich, and Berlin. In 1863 he traveled around the Red

Sea. In 1868 the Humboldt Institute in Berlin sent him back to the region to explore in East Africa. In 1869 he left Khartoum and traveled south and eventually, in 1870, discovered the westward-flowing River Uele. He described in detail the cannibal practices of the Mangettu people and proved decisively the presence of very short races in Africa with his discovery of the Akka "pygmy" people. He had made important collections, but a fire in his camp destroyed nearly all of them. He was back in Khartoum in 1871 and in 1874 published an account of his adventures, *Im Herzen von Afrika* (The Heart of Africa). Between 1873 and 1874 he explored the Libyan Desert, accompanied by Gerhard Rohlfs. He settled in Cairo in 1875 and, with the encouragement of Khedive Ismail, founded a geographical society. He himself concentrated on African historical and ethnographical studies. In 1876 he was with Paul Gussfeldt when they penetrated the Arabian Desert, and for the next 12 years he continued to explore there and to make botanical and geographical studies in the Nile valley. He returned to live in Berlin in 1889 but made three further expeditions to Eritrea (at that period an Italian colony), in 1891, 1892, and 1894. This race of chimpanzee is found in eastern parts of the DRC (Zaire), Uganda, Rwanda, and western Tanzania.

Schweitzer

Schweitzer's Shrew *Crocidura schweitzeri* **Peters,** 1877 [Alt. Fraser's Musk Shrew; Syn. *C. poensis* L. Fraser, 1843]

Schweitzer (forenames and dates not found) was a German resident in West Africa. He collected the shrew and sent it to Peters for identification. Peters gives no further details beyond the surname that would help us to identify the particular Schweitzer he had in mind. This shrew is now considered to be conspecific with Fraser's Musk Shrew *Crocidura poensis,* first described from the island of Bioko. It is found from Liberia to Cameroon.

Sclater, P. L.

Sclater's Angolan Colobus *Colobus angolensis angolensis* Sclater, 1860

Sclater's Black Lemur *Eulemur macaco flavifrons* **Gray,** 1867 [Alt. Blue-eyed Black Lemur]

Sclater's Dog *Atelocynus microtis* Sclater, 1883 [Alt. Short-eared Dog, Small-eared Zorro]

Somali Hedgehog *Atelerix sclateri* **Anderson,** 1895

Sclater's Shrew *Sorex sclateri* **Merriam,** 1897

Sclater's Monkey *Cercopithecus sclateri* **Pocock,** 1904

Dr. Philip Lutley Sclater (1829–1913) was a graduate of Oxford and practiced law for many years. He was the founding editor of *The Ibis,* the journal of the British Ornithologists' Union. He edited it from 1858 to 1865 and again from 1877 until 1912. He was also Secretary of the Zoological Society of London from 1860 until 1903. Sclater's study of bird distribution resulted in the classification of the biogeographical regions of the world into six major categories. He later adapted his scheme for mammals, and it is still the basis for work in biogeography. He wrote widely on birds, including *Exotic Ornithology* (1866). He also wrote a monograph on the African monkey genus *Cercopithecus,* published in 1893. The colobus monkey is found in northern Angola and DRC (Zaire) south of the Congo River. The lemur occurs in a small area of northwest Madagascar. The dog comes from western parts of the Amazon basin, from southern Colombia south to northwest Bolivia. The hedgehog is endemic to Somalia. The shrew is known only from the state of Chiapas, southern Mexico, and the monkey only from southern Nigeria.

Sclater, W. L.

Sclater's Forest Shrew *Myosorex sclateri* **Thomas** and Schwann, 1905 [Alt. Sclater's Mouse-Shrew]
Sclater's Golden Mole *Chlorotalpa sclateri* Broom, 1907

William Lutley Sclater (1863–1944) was the son of Philip Lutley Sclater (see above). Like his father, he was educated at Oxford, obtaining a first-class honors degree in natural science in 1885. For a few years he was Deputy Superintendent of the Indian Museum in Calcutta, and in 1896 he was appointed as the first Director of the South African Museum in Cape Town. He was also a onetime President of the South African Ornithologists' Union. Sclater resigned from the South African Museum in 1906 and for the following 30 years worked at the British Museum of Natural History. He succeeded his father as editor of *The Ibis* from 1913 until 1930 and was President of the British Ornithologists' Union from 1928 to 1933. He wrote *Systema avium Aethiopicarum* in 1924. In July 1944 he was killed by a V1 flying bomb in London. The shrew is found in KwaZulu-Natal (South Africa). The golden mole occurs more widely in the southern and eastern parts of South Africa and in Lesotho.

Scott

Scott's Rice Rat *Cerradomys scotti* **Langguth** and Bonvicino, 2002

This rice rat comes from the Brazilian cerrado (tropical savanna). For biographical details, see (Scott Morrow) **Lindbergh.**

Scott, H. H.

Scott's Mouse-eared Bat *Myotis scotti* **Thomas,** 1927

Dr. H. H. Scott of Cambridge acquired the type specimen of this bat near Addis Ababa—so says Thomas's original etymology. We assume this to be Dr. Sir Henry Harold Scott (1874–1956) of the London School of Hygiene and Tropical Medicine, after whom the medical condition Strachan-Scott syndrome was in part named, but Thomas gave no further clues. Henry Scott was also, for a time, the pathologist of the Zoological Society of London and so would have been well known by Thomas. The bat comes from the Ethiopian highlands.

Scott, S. and W.

Scott's Sportive Lemur *Lepilemur scottorum* Lei et al., 2008

The name *scottorum* honors the Suzanne and Walter Scott Jr. Family Foundation. Suzanne and Walter Scott Jr. are prominent supporters of in situ and ex situ conservation throughout the world and have volunteered extensively in conservation programs at the Henry Doorly Zoo (Omaha, Nebraska). Walter Scott Jr. is a member of the Board of Directors of the Omaha Zoological Society. He is a philanthropist and community activist on the board of many charitable organizations and businesses. Suzanne M. Scott is a longtime zoo supporter and became the founding Executive Director of the Omaha Zoo Foundation, responsible for establishing the zoo's endowment funds and raising money for special projects. The lemur is known from the Masoala National Park, eastern Madagascar.

Scott, W. V.

Scott's Tree Kangaroo *Dendrolagus scottae* **Flannery** and **Seri,** 1990 [Alt. Tenkile Tree Kangaroo]

Lady Winifred Violet Scott (d. 1985) was a philanthropist who left a bequest, known as the Scott Foundation, to fund conservation of endangered species. The tree kangaroo is an endangered species found only in the Torricelli Mountains of northern Papua New Guinea. It was so named by Tim Flannery because the Scott Foundation funded his research in the region during the 1980s.

Scully

Scully's Tube-nosed Bat *Murina tubinaris* Scully, 1881

Dr. John Scully (1846–1912) was Nepal's Resident Surgeon between 1876 and 1877. He made a collection of nearly 300 birds and was the first person to describe the status of birds in the Kathmandu valley. He was in the Indian army and was commissioned in Bengal in 1872 as an Assistant Surgeon, eventually being promoted to the rank of Lieutenant Colonel. The bat is found from Pakistan east through northern India to Laos and Vietnam.

Seabra

Angolan Hairy Bat *Cistugo seabrai* **Thomas,** 1912 [Alt. Angolan Wing-gland Bat]

Dr. Anthero Frederico de Seabra (1874–1952) of the Lisbon Museum presumably sent the type specimen to Thomas, as the latter wrote in his description of the bat that it was named in honor of "Senhor A. F. de Seabra, C.M.Z.S., of the Lisbon Museum." Seabra wrote many articles, such as "Nota sobre a existencia de *Diomedia immutabilis* nas costas occidentaes de Africa" (*Jornal de Sciencias Mathematicas*, 1906). The bat comes from southwest Angola, Namibia, and Northern Cape Province (South Africa).

Seal

Seal's Sportive Lemur *Lepilemur seali* Louis, 2005

Ulysses "Ulie" S. Seal III (1929–2003) was a former official with the International Union for Conservation of Nature. With a love of animals, a boundless curiosity, and considerable charm, he helped persuade zoos worldwide to participate in animal conservation. He also helped found the Minnesota Zoo and is credited with creating "computer dating" for zoo animals to help save endangered species. Seal's training was in psychology and biochemistry. His career was spent researching the human endocrine system and its influence on prostate cancer at what is now the Minneapolis Veterans Medical Center. He also held adjunct professorships at the University of Minnesota. In 1969 he found that no one knew what normal blood values were for many exotic animals, so he volunteered to undertake the laboratory work at his own expense. From this, in 1975, he co-developed what is now the International Species Information System (ISIS)—a computer program that can find matches for animals that don't have mates. This system now involves more than 600 zoos and aquaria worldwide. Seal retired in 1990 but continued his conservation work until his death. The lemur lives in the rainforests of northeast Madagascar and was discovered by a team led by Edward Louis from the Henry Doorly Zoo, Nebraska.

Seba

Seba's Short-tailed Bat *Carollia perspicillata* **Linnaeus,** 1758

Albert Seba (1665–1736) was an extremely wealthy Dutch collector and apothecary who lived in Amsterdam and formed what was regarded as the richest museum of his time. In 1717 he sold a huge collection to the Russian Tsar, Peter the Great, and then started collecting all over again. Linnaeus visited him in 1735. Seba's broad collection-cataloguing systems influenced Linnaeus in the shaping of his own system, and many of Seba's animals finished up as type specimens for Linnaeus' descriptions. The bat is found from southern Mexico to Bolivia, Paraguay, and southern Brazil, and also on the islands of Trinidad and Tobago and Grenada.

Selborne

Selborne's Hartebeest *Alcelaphus buselaphus selbornei* **Lydekker,** 1913

William Waldegrave Palmer, 2nd Earl of Selborne (1859–1942), was a British Liberal politician who rose to Cabinet level, serving as First Lord of the Admiralty. He was appointed as High Commissioner for South Africa and Governor of the Transvaal and Orange River colonies from 1905 to 1907, doing much to reconcile the Dutch and British communities. The type specimen of the hartebeest was presented to the British Museum of Natural History by the De Beers Mining Company in 1912, at the prompting of Lord Selborne. The animal was shot on the Kimberley Game Farm, where the stock was originally imported from the Transvaal. Recognition of this subspecies had a "shelf -life" of about 50 years, until it was shown that the type specimen of *selbornei* was an unusually colored individual of the Cape (Red) Hartebeest *Alcelaphus (buselaphus) caama.* So unfortunately for Lord Selborne's zoological claim to fame, *selbornei* is now just regarded as a junior synonym of *caama.*

Selevin

Desert Dormouse *Selevinia betpakdalaensis* Belosludov and Bazhanov, 1939

V. A. Selevin (d. 1938) was a Russian zoologist and ornithologist. He started the bird-banding program at the Chohpak Ornithological Station in 1926. Selevin collected the type specimen of the dormouse in 1938 during an expedition to the Betpak-Dala Desert in Kazakhstan. He died before the taxonomists could work on his collections, so this genus (with just the one species) was named as a memorial to him after his death. The dormouse is confined to the deserts of eastern Kazakhstan.

Selous

Selous' Mongoose *Paracynictis selousi* **de Winton,** 1896
Selous' Zebra *Equus quagga selousi* **Pocock,** 1897
Selous' Sitatunga *Tragelaphus spekii selousi* **Rothschild,** 1898 [Alt. Zambesi Sitatunga]

Frederick Courtney Selous (1851–1917) was a Rhodesian (Zimbabwean) explorer and hunter. He is famous for telling the Headmaster of Rugby School in England, when he was eight years old and being interviewed for a place at the school, that he would one day become a great explorer and hunter in Africa. He was a hunter for 30 years but also collected natural history specimens and was concerned that wildlife should be conserved. As one of very few Europeans in the interior of Africa at the time, Selous was instrumental in opening up south-central Africa for Cecil Rhodes and the British Empire, negotiating with the great chiefs such as Lobegula at his royal "boma" in Bulawayo. Selous first arrived in South Africa in 1871. He published many books on his travels and his hunting exploits; he seems to have shot an awful lot of elephants. He was the model for Sir Henry Rider Haggard's fictional character Allan Quatermain. When WW1 began he was already in his 60s, but he begged the War Office for a commission and headed out to East Africa. There he helped to defend the Uganda railway from the Germans, who at the time were in control of German East Africa (now Tanzania) and who had ambitions across the border in British Kenya. Selous also took part in the successful invasion of northern Tanzania and in driving the German army south, eventually expelling it into the bush. However, while pursuing the Germans Selous was killed by machine-gun fire and was buried where he lay. His grave is marked now, as it is in the eponymous Selous Game Reserve in Tanzania. Bigger than the whole of Switzerland, this game park is a World Heritage Site set aside as a reserve in 1905 and named after Selous in 1922. A Southern Rhodesian regiment, the Selous Scouts, was named after him too. The mongoose is found in Angola, Zambia, Zimbabwe, Malawi, western Mozambique, northern Botswana, and northeast South Africa. Regarding the zebra subspecies, different authorities recognize different numbers of races and disagree on where to draw the boundaries between them. Selous' Zebra was first described from the Manyami Valley of Zimbabwe; some would say zebras from this area are *Equus quagga chapmanni*, while those inclined to "lumping" would say they are *E. quagga burchelli*. Happily, our role is not to make taxonomic judgments. Equally happily, there is no such dispute over recognition of the sitatunga subspecies: this antelope is found in southeast Angola, northern Botswana, Zambia, and southwest Tanzania.

Semon

Semon's Roundleaf Bat *Hipposideros semoni* **Matschie,** 1903

Dr. Richard Wolfgang Semon (1859–1918) was a German embryologist, evolutionary biologist, zoologist, and physiologist. He was particularly interested in memory and whether it could be inherited. He took his doctorate in zoology in 1883 and in medicine in 1886, both at Jena University. In 1892 he became an Associate Professor and led an expedition to Australia to investigate, among other matters, monotreme reproduction; science had only recently been shaken by the revelation that some apparent mammals were egg-layers. The expedition, which lasted until 1893, discovered 207 new species and 24 new genera. Semon traveled in Queensland to study such wonders as the platypus and the Australian lungfish, using native Australians as trappers and collectors. Returning to Germany late in 1893, he stayed at Jena University until 1897 but was forced to resign as a result of becoming involved in an affair with the wife of the Professor of Pathology. He moved to Munich, where he worked as a private scholar. He married the lady with whom he had been involved in Jena, but in 1918 she died of cancer, and this, coupled with his distress over Germany's role in WW1, led him to commit suicide in December of that year. The bat is found in northern Queensland and eastern New Guinea.

Seri

Seri's Tree Kangaroo *Dendrolagus (dorianus) stellarum* Flannery and Seri, 1990

Seri's Sheath-tailed Bat *Emballonura serii* **Flannery,** 1994

Lester Seri is a mammalogist and an official of the Papua New Guinea Department of Environment and Conservation. He has a particular interest in fruit bats, and from the 1970s onward he traveled with, and collaborated closely with, Tim Flannery in research into New Guinea bats. He was one of the people who supplied illustrations for *Bats of Papua New Guinea* by Frank Bonaccorso, published in 2000. The tree kangaroo is found in the central mountain chain of New Guinea. The bat comes from the islands of New Ireland and Manus in the Bismarck Archipelago, Papua New Guinea.

Setzer

Setzer's Hairy-footed Gerbil *Gerbillurus setzeri* Schlitter, 1973

Setzer's Mouse-tailed Dormouse *Myomimus setzeri* Rossolimo, 1976

Setzer's Pygmy Mouse *Mus setzeri* **Petter,** 1978

Dr. Henry W. Setzer (1916–1992) was an American zoologist and mammalogist who was an Assistant Curator at the Smithsonian Institution from 1949 to 1969 and Curator from 1969 to 1979. He ran the Smithsonian's African Mammal Project, and his teams of field collectors acquired in excess of 63,000 specimens from many of the countries of Africa. With J. Meester he was co-author of *The Mammals of Africa—An Identification Manual,* published in 1971. He also wrote *National Geographic Book of Mammals* (1998). The gerbil is found in the Namib Desert of southwest Africa. The dormouse comes from northwest Iran and eastern Turkey. The mouse occurs in western Zambia, parts of Botswana, and northeast Namibia.

Seuánez

Seuánez's Rice Rat *Hylaeamys seuanezi* Weksler et al., 1999

Professor Héctor Nicolás Seuánez Abreu (b. 1947) was born in Montevideo, Uruguay, but is now a Brazilian citizen and a Professor in the Genetics Department of the Federal University of Rio de Janeiro. He trained originally in Montevideo and qualified both as a physician and as a biologist. In 1974 he was awarded a scholarship from the University of Edinburgh that enabled him to work at the Medical Research Council in the UK, where he took his doctorate in 1977. He returned to South America at the end of 1978, and in 1979 he joined his present employer, initially as a visiting Associate Professor but since 1980 as a full member of the teaching staff. So far in his career he has published over 100 articles in learned international journals. The rice rat is found in the Atlantic Forest of eastern Brazil. This species may be conspecific with *Hylaeamys laticeps.*

Severtzov

Severtzov's Ibex *Capra (ibex) severtzovi* **Menzbier,** 1888 [Alt. West Caucasian Tur; Syn. *C. caucasica* **Güldenstaedt** and **Pallas,** 1783]

Severtzov's Argali *Ovis ammon severtzovi* Nasonov, 1914

Severtzov's Jerboa *Allactaga severtzovi* **Vinogradov,** 1925

Severtzov's Birch Mouse *Sicista severtzovi* **Ognev,** 1935

Professor Nikolai Alekseevich Severtsov, sometimes spelled Severzov or Severtzow (1827–1885), was a Russian zoologist who explored in central Asia. He is considered to be one of the pioneers of ecology and evolutionary science in Russia. He wrote works on the zoogeographical division of the regions of the Palaearctic, and on the birds of Russia and Turkestan, including mapping migration routes. After becoming acquainted with Darwin's theory of natural selection, he tested it against his own observations and became a fervent supporter and propagandist for Darwinism in Russia. He made extensive collections on his travels, including 12,000 bird skins. He wrote *Ornithology and Ornithological Geography of European and Asian Russia* (1867), *Journeys through Turkestan Territory and Investigation of the Tien-san Mountain Country* (1873), and *On Zoological (Mainly Ornithological) Regions of Our Mainland* (1877). His *Orographic Sketch of the Pamir Mountain System* was published posthumously in 1886. He is also commemorated in the scientific names of the Chinese Grouse *Bonasa sewerzowi* and of two freshwater fish. A mountain peak in Pamiro-Alai and glaciers in Pamir and Zailijaskoe also bear his name. The ibex inhabits the western Caucasus. (Most authorities use *Capra caucasica* as the scientific name of this taxon, but some have used *severtzovi* for the name of the species or have treated *severtzovi* as a subspecies of *caucasica.* Taxonomy can give you headaches.)

The argali is found in Uzbekistan, and the jerboa in Kazakhstan, Uzbekistan, and northern Turkmenistan. The birch mouse is found in eastern Ukraine and the Voronezh region of southern European Russia.

Sezekorn

Buffy Flower Bat *Erophylla sezekorni* **Gundlach,** 1860

Eduard Sezekorn (dates not found) was a German naturalist and ornithologist. He recorded the birds encountered in North Hesse in the years 1836 to 1839 under the title *Die Vogellisten des Eduard Sezekorn. Eine erste Bestandsaufnahme der Avifauna Nordhessens aus den Jahren 1836–1839.* We think that Gundlach, who came to Cuba in 1839 from Germany, named the bat after Sezekorn, as he would probably have known him through the Society of Natural History of Kassel, of which he was a corresponding member, and it is clear that the two men stayed in contact after Gundlach left Germany. In 1855 Sezekorn published *Dr. J. Gundlach's Beiträge zur Ornithologie Cuba's. Nach Mittheilungen des Reisenden an Hr. Bez.-Dir. Sezekorn in Cassel; von Letzterem zusammengestellt* (Dr. J. Gundlach's Contribution to Cuba's Ornithology, According to the Reports Sent by the Traveler to District Supervisor Sezekorn in Cassel, Put Together by the Latter). The flower bat is found in Cuba, the Cayman Islands, Jamaica, and the Bahamas.

Shamel

Shamel's Horseshoe Bat *Rhinolophus shameli* **Tate,** 1943

Henry Harold Shamel (1885–1963) was an American zoologist who worked at the Smithsonian National Museum of Natural History in Washington, DC, until at least 1954. Tate was his contemporary at Harvard. In the 1920s, 1930s, and 1940s Shamel described several bats and other small mammals. He also published such scientific papers as "Notes on the American Bats of the Genus *Tadarida*." The horseshoe bat is found from Myanmar east to Vietnam and south to peninsular Malaysia.

Shaposchnikow

Long-clawed Mole-Vole *Prometheomys schaposchnikowi* **Satunin,** 1901

Christophor Georgievich Shaposhnikov (1872–1938) was primarily an entomologist and an expert in forestry. In 1904, jointly with C. Deegener, he published a book on ghost-moths, and he is also commemorated in the scientific name of a dipterid insect, *Beris schaposchnikowi*. He was also one of the founders, and the first Director, of the Caucasian Nature Reserve. He died in prison, to which he had been undeservedly sentenced. The vole is found in the Caucasus Mountains and extreme northeast Turkey.

Sharman

Sharman's Rock Wallaby *Petrogale sharmani* Eldridge and Close, 1992 [Alt. Mount Claro Rock Wallaby]

Dr. Geoffrey Bruce Sharman F.A.A. (b. 1925) was Professor of Biology, School of Biological Sciences at Macquarie University, Sydney, until 1985 and Emeritus Professor from 1985. He was a leading Australian research biologist, an expert in both marsupial reproduction and mammalian evolutionary biology. He was born and educated in Tasmania and, after service in the Royal Australian Navy from 1943 to 1946, had many academic and research posts in universities across Australia, and at the Medical Research Council at Harwell in the UK in 1955. The University of Western Australia awarded his doctorate in science in 1961. He was elected a Fellow of the Australian Academy of Science in 1980. He was also a Chairman of the Australian

Biological Resources Study Advisory Committee. The rock wallaby has a small distribution in northeast Queensland, Australia.

Sharpe

Sharpe's Grysbok *Raphicerus sharpei* **Thomas,** 1897
Sharpe's Colobus *Colobus angolensis sharpei* Thomas, 1902

Sir Alfred Sharpe K.C.M.G., C.B., LL.D. (1853–1935) was His Majesty's Commissioner and Consul General for the British Central Africa Protectorate, Governor of Nyasaland (now Malawi), and an amateur naturalist. He is also remembered in the names of two birds, Sharpe's Greenbul *Phyllastrephus alfredi* and Sharpe's Pied Babbler *Turdoides sharpei*. The grysbok is found from Tanzania south to northeast South Africa. The colobus subspecies comes from northern Malawi and southern Tanzania. It is often lumped with *Colobus angolensis palliatus* (the latter name having precedence).

Shaw, F. W.

Shaw's Melomys *Melomys shawi* **Tate** and **Archbold,** 1935 [Syn. *Paramelomys rubex* **Thomas,** 1922]

See **Shaw-Mayer.**

Shaw, G. K.

Shaw's Mastiff Bat *Eumops auripendulus* Shaw, 1800 [Alt. Black Bonneted Bat]

Dr. George Kearsley Shaw (1751–1813) was a British physician, botanist, and zoologist. He lectured on botany at Oxford from 1786 to 1791, when he became Assistant Keeper to the Natural History Section of the British Museum. In 1807 he became Keeper, a position he occupied until his death. He was a co-founder of the Linnean Society. His main works were the *Zoology of New Holland* (today Australia), published in 1794; the *Museum Leverianum* (1792–1796); the *General Zoology* (1800–1812); and the *Naturalists' Miscellany* (1789–1813). The bat is found from southern Mexico to northern Argentina.

Shaw, T.

Shaw's Jird *Meriones shawi* Duvernoy, 1842

The Rev. Dr. Thomas Shaw (1694–1751) was a great traveler in North Africa and the Near East in the period from 1720 to 1732, when he was Chaplain at the British consulate in Algiers. He visited the Sinai Peninsula and Cyprus in 1721, Jerusalem, Jordan, and Mount Carmel in 1722, and Tunis and the ruins of Carthage in 1727. He published a number of books, including *Travels, or Observations Relating to Several Parts of Barbary and the Levant* (1738), which included copious details of everything he observed, including Roman ruins, agriculture, geography, people, flora, and fauna. This book was translated into French as early as 1743 and appears to have been very influential and helpful to later French geographers, among whom was Rozet, who traveled in Algeria from 1830 to 1832 during the French conquest. Shaw's book appears to have been reissued in 1830 under the title *Travel in the Regency of Algiers*. We were unable to consult the original reference, as the library of the Natural History Museum in Paris does not have a copy of the *Mémoires de la Société des Sciences, Lettres, et Arts de Nancy* for the year in question. The jird is found in coastal North Africa, from Morocco to Egypt.

Shaw-Mayer

Pygmy Ringtail Possum *Pseudochirulus mayeri* **Rothschild** and **Dollman,** 1932
Shaw-Mayer's Pogonomelomys *Pogonomelomys mayeri* Rothschild and Dollman, 1932 [Alt. Shaw Mayer's Brush Mouse]
Shaw's Melomys *Melomys shawi* **Tate** and **Archbold,** 1935 [Syn. *Paramelomys rubex* **Thomas,** 1922]
Shawmayer's Ornate Tree Kangaroo *Dendrolagus goodfellowi shawmayeri* Rothschild and Dollman, 1936
Shaw-Mayer's Water Rat *Hydromys shawmayeri* Hinton, 1943 [Syn. *Baiyankamys shawmayeri*]
Fergusson Island Tree Mouse *Chiruromys forbesi shawmayeri* Laurie, 1952
Shaw-Mayer's Shrew Mouse *Mayermys ellermani* Laurie and Hill, 1954 [Alt. One-toothed Shrew Mouse; Syn. *Pseudohydromys ellermani*]

Frederick "Fred" William Shaw Mayer (1899–1989), a native of Australia, has been described as the "last of the great collectors." He collected mammals, birds, and other animals in the Moluccas and western New Guinea from 1928 to 1931, and in many New Guinea locations between 1931 and 1952. He provided bird skins for Lord Rothschild's museum (now the British Museum of Natural History) and live birds for the London Zoo and a number of prominent aviculturists. In 1953 he took charge of Sir Edward Hallstrom's aviaries in the Wahgi Valley, Papua New Guinea. Around 1970, failing health led him to return to Australia. All the mammals named after him are found in New Guinea. (One taxon, the melomys, is no longer considered valid and has been "demoted" to a junior synonym.)

Sheila

Red Brocket *Mazama sheila* **Thomas,** 1913 [now *M. americana sheila*]

Thomas named so many animals in his career that one wonders if he was occasionally at a loss for a suitable name and just grabbed whatever came first to mind. This is one such case, as he does not give any clue in his original description as to why he chose this scientific name. This deer, now regarded as a race of the Red Brocket, comes from northern Venezuela.

Shelkovnikov

Schelkovnikov's Pine Vole *Microtus schelkovnikovi* **Satunin,** 1907
Schelkovnikov's Water Shrew *Neomys schelkovnikovi* Satunin, 1913 [Alt. Transcaucasian Water Shrew; Syn. *N. teres*]

Alexander Bebutovich Shelkovnikov (1870–1933) was a Russian zoologist who in 1922 founded the Herbarium of Armenia, which later became part of the Institute of Botany of the Armenian National Academy of Sciences. He worked for a time as an official in the Armenian agricultural administration. His collections, mainly of zoological and botanical specimens from Georgia, Armenia, and Azerbaijan, are held across parts of the old Soviet Union, particularly in Yerevan, St. Petersburg, and Moscow. He was a graduate of a military school in St. Petersburg, so had no formal zoological training, but was a gifted amateur who contributed much to the knowledge of the natural history of the Caucasus and Persia (now Iran). He spent part of his career as the unofficial Curator of Scientific Collections at the Moscow State University. More than 30 plant species are named after him, as are some 20 invertebrates. The pine vole is found in the mountains of Azerbaijan and northwest Iran. The shrew is found in Georgia, Armenia, Azerbaijan, and northeast Turkey. (The scientific name *Neomys teres,* coined by Miller in 1908, appears to be the oldest valid name for this shrew.)

Shepherd

Shepherd's Beaked Whale *Tasmacetus shepherdi* Oliver, 1937

George Shepherd (1908–1992) was a Curator at the Wanganui Alexander Museum in New Zealand. In 1933 he found the type specimen of the whale cast up on a beach on the North Island of New Zealand and passed it on to his friend, W. R. B. Oliver, who described and named it. The skeleton can be seen in the Wanganui Regional Museum. The whale is found in cold-temperate waters of the Southern Hemisphere.

Sherman

Sherman's Pocket Gopher *Geomys pinetis fontanelus* Sherman, 1940
Sherman's Short-tailed Shrew *Blarina shermani* Hamilton, 1955
Sherman's Fox Squirrel *Sciurus niger shermani* Moore, 1956

Dr. Harley B. Sherman (dates not found) was an instructor at New York University in the 1920s. At some stage he moved to Florida to take up a position as Professor at the University of Florida, and he also established a business making small-animal traps, including specialized models for catching gophers and squirrels. The company he founded, H. B. Sherman Traps Inc. of Tallahassee, Florida, is still in business. He made a collection of Florida mammals, and his unpublished notes were used by other zoologists in the course of their studies. The pocket gopher, from eastern Georgia, USA, was last seen in 1950 and may be extinct. The shrew, which has a very restricted distribution in the region of Fort Myers, western Florida, has been elevated to species level, having previously been regarded as a race of *Blarina carolinensis.* The squirrel is found in Florida and Georgia.

Sherrin, T. V.

Giant White-tailed Rat *Uromys sherrini* Thomas, 1923

Thomas Vaughan Sherrin (b. 1875) was a collector and taxidermist and brother to W. R. Sherrin (see below). He was part of an expedition by the Godman Exploration Fund of 1922 in south Queensland which collected birds, reptiles, and plants, as well as mammals. He also collected in China and Malaya just before WW1. He collected the holotype in Queensland; it is now found in northeast Queensland and may be synonymous with *U. caudimaculatus.*

Sherrin, W. R.

Tasmanian Long-eared Bat *Nyctophilus (timoriensis) sherrini* **Thomas,** 1915

William Robert Sherrin (1871–1955) was a "preparator" at the British Museum of Natural History and also a collector. He was the son of a watercolor artist and had a taxidermist's shop in Ramsgate in Kent until 1895, when he joined the British Museum in London. In 1919 he became Curator at the South London Botanical Institute, from which he organized and led botanical meetings. After 1919 he continued to work part-time for the British Museum, until 1928 for the Department of Zoology and thereafter until retirement, in 1947, for the Department of Botany. Thomas wrote in his original description of the bat, "It is named in honour of Mr. W. R. Sherrin, to whom every mammalogist who has visited the Museum is indebted for assistance, and whose admirable preparation of tiny skulls and tinier bacula has so immensely helped in the mammalian work done both by staff and visitors." As the common name implies, the bat is from Tasmania. While usually regarded as a subspecies of *Nyctophilus timoriensis*, it is sometimes listed as a full species.

Shipton

Shipton's Mountain Cavy *Microcavia shiptoni* **Thomas,** 1925

Stewart Shipton (1869–1939) was an Englishman who spent most of his life in Argentina. He was employed as an accountant and then as General Manager of a sugar mill at Concepción, a town of which he was the first Intendant (administrator). He was by inclination a naturalist and ornithologist. Lillo and Shipton were very close friends, and there is a Shipton Collection of birds, reptiles, mammals, and fish that has been housed since its acquisition in 1938 in the Zoology Department of the Miguel Lillo Foundation, Tucumán, Argentina. The cavy is found in the mountains of northwest Argentina.

Shiras

Shiras' Moose *Alces alces shirasi* **Nelson,** 1914 [Alt. Yellowstone Moose]

Congressman George Shiras III (1859–1942) was the Representative for the state of Pennsylvania and a noted conservationist. He graduated from law school at Yale in 1883 and was called to the bar in the same year. Until 1889 he was an avid hunter, but he gave it up to become a pioneer conservationist. Between 1889 and 1890 he served in his local legislature. He was elected as an independent Republican Congressman and served from 1903 to 1905 but did not seek renomination. While in Congress he served on the Public Lands Committee and prepared and introduced in the House the Federal Migratory Bird Law. He also helped to write legislation creating Olympic National Park. After his political career he undertook biological research and became known as "the father of wildlife photography." He wrote *Hunting Wild Life with Camera and Flashlight,* which was published in two volumes. The moose occurs in Canada (southern Alberta and southeast British Columbia) and the northern USA (Montana, Idaho, Wyoming, Utah, and Colorado).

Shitkov

See **Zhitkov.**

Shortridge

Heath Rat *Pseudomys shortridgei* **Thomas,** 1907

Shortridge's Rousette *Rousettus leschenaulti shortridgei* Thomas and **Wroughton,** 1909

Shortridge's Mouse *Mus shortridgei* Thomas, 1914

Shortridge's Leaf Monkey *Trachypithecus shortridgei* Wroughton, 1915

Shortridge's Horseshoe Bat *Rhinolophus shortridgei* Andersen, 1918

Shortridge's Rat *Thallomys shortridgei* Thomas and Hinton, 1923

Shortridge's Free-tailed Bat *Chaerephon shortridgei* Thomas, 1926

Shortridge's Rock Mouse *Petromyscus shortridgei* Thomas, 1926

Shortridge's Leopard *Panthera pardus shortridgei* Pocock, 1932

Shortridge's Multimammate Mouse *Mastomys shortridgei* St. Leger, 1933

Shortridge's Chacma Baboon *Papio ursinus ruacana* Shortridge, 1942 [Alt. Kalahari Baboon]

Captain Guy Chester Shortridge (1880–1949) was originally trained as a geologist but worked as a taxidermist for the South African Museum from 1902 to 1903. He was an indefatigable zoological and entomological collector for the British Museum and for the New York Museum of Natural History. He worked for these clients in South Asia when he was attached to the Raffles Museum in Singapore, in Australia between 1904 and 1907, and in Africa. It is worth recalling his comment of dismay at the effect of the Europeans on Australia: "Animals are dying out here as fast as Aboriginals—both are fading before the products of a tougher civilisation." He was very far-sighted and appealed to Western Australia to set aside reserves in which the fauna of the state could be preserved. Between 1921 and 1949 he was the Director of the Kaffrarian Museum in King William's Town; the Shortridge Mammal Collection is housed there. He led at least five Percy Sladen and Kaffrarian Museum expeditions. He wrote *The Mammals of South West Africa* in 1934. In 1906 Oldfield Thomas published a paper entitled "List of Further Collections of Mammals from Western Australia, Including a Series from Bernier Island, Obtained for Mr. W. E. Balston; with Field-notes by the Collector, Mr. G. C. Shortridge." The Heath Rat is found in southern Australia. The rousette (a fruit bat) comes from Java and Bali. The mouse is found from Myanmar east to Cambodia and Vietnam. The leaf monkey occurs in Myanmar and the Gongshan district of Yunnan (China). The horseshoe bat is known from Myanmar and northern India (it was long considered to be a race of *Rhinolophus lepidus*). Shortridge's Rat is known only from northwest Cape Province, South Africa. The free-tailed bat is found in Angola, Namibia, Botswana, Zimbabwe, Zambia, and southern DRC (Zaire). The rock mouse occurs in Angola and northern Namibia. The leopard subspecies, from southern Africa, is no longer considered valid; all African leopards are now usually treated as a single form, *Panthera pardus pardus*. The multimammate mouse inhabits northeast Namibia, extreme northwest Botswana, and eastern Angola. The baboon comes from southwest Angola and northern Namibia.

Sibbald

Sibbald's Whale *Balaenoptera musculus* **Linnaeus,** 1758 [Alt. Blue Whale, Sibbald's Rorqual]

Sir Robert Sibbald (1641–1722) was Professor of Medicine at Edinburgh. He was born in Edinburgh, but the family fled to Linlithgow in 1645 to avoid the plague. With Andrew Balfour he established the first Botanical Garden in Edinburgh in 1671. He also founded the Royal College of Physicians in Edinburgh in 1681 and was appointed Geographer Royal to King Charles II (and Physician in Ordinary to His Majesty) in 1682. He was commissioned, as Cartographer Royal for Scotland, to produce a natural history of Scotland and a geographical description combining historical data with contemporary survey results. He eventually published only a natural history: *Scotia Illustrata*. He did manage to collect together all extant maps and manuscripts of relevance. He is also commemorated in the name of a plant genus, *Sibbaldia*. Sibbald was the first person to scientifically describe the Blue Whale, after a specimen was washed up on a Scottish beach. The species has a worldwide distribution but is rare following the period of commercial whaling (it has been protected since 1967).

Sibree

Sibree's Dwarf Lemur *Cheirogaleus sibreei* **Forsyth Major,** 1896

The Rev. James Sibree (1836–1929) originally trained to be a civil engineer and worked for the Hull Board of Health from 1859 to 1863. He joined the London Missionary Society and was sent to Madagascar as an architect, where he designed and built four large stone churches over the period 1863–1867. He returned to England, where he studied and was ordained as a Congregationalist minister. He returned to Madagascar as a missionary in 1870 and worked as such for seven years. He ran into trouble with the government in 1877 and was forced to leave the country, returning only in 1883 after a spell in England and two years in India. He remained in Madagascar until 1915. He resigned as a missionary in 1916 but continued to work for the London Missionary Society into the 1920s. He was an outspoken supporter of Malagasy independence in the year leading up to the French invasion in 1895. He was made a Fellow of the Royal Geographical Society and received an honorary degree of Doctor of Divinity from St. Andrews University. He published extensively in both English and the Malagasy language on a diversity of subjects, including *Madagascar and Its People* (1870), *A Naturalist in Madagascar* (1915), and his autobiography, *Fifty Years in Madagascar* (1923). Although the lemur was named in 1896, it was largely ignored by taxonomists for a century. Only in 2000 was it re-recognized as a distinct species. The precise limits of its distribution in eastern Madagascar are still unclear.

Siebers

Great Kai Island Giant Rat *Uromys siebersi* **Thomas,** 1923

Hendrik Cornelis Siebers (1890–1949) was a Dutch ornithologist who worked for the Amsterdam University Zoological Museum from 1920 to 1947. He took part in the museum's central-east Borneo expedition of 1925 and wrote a number of articles including, in 1930, "Fauna Buruana, aves." The rat is from Great Kai (Kai Besar) Island, eastern Indonesia.

Silenus

Lion-tailed Macaque *Macaca silenus* **Linnaeus,** 1758

Linnaeus was rather fond of giving primates fanciful names from mythology. In this case he named a monkey after Silenus: the elderly, jovial, and usually heavily inebriated companion and adviser of Dionysus (god of wine and madness). The macaque is found in southern India.

Simmons

Simmons' Mouse-Lemur *Microcebus simmonsi* Louis et al., 2006

Dr. Lee G. Simmons D.V.M. is Chairman of the Omaha Zoo Foundation and has been an active supporter of conservation programs in Madagascar. He has also turned a small regional zoo into a leading institution specializing in the captive breeding of a number of endangered mammals. The lemur was named after him by Dr. Edward Louis, head of the Genetics Department of the Grewcock Center for Conservation and Research at the Henry Doorly Zoo. The lemur comes from eastern Madagascar.

Simon

Simon's Dipodil *Gerbillus simoni* **Lataste,** 1881 [Alt. Lesser Short-tailed Gerbil]

Eugene Louis Simon (1848–1924) was a French zoologist. Lataste wrote in his original etymology for the gerbil, "Je suis heureux de dédier cette espèce à M. Eugène Simon, qui, par le don d'un commencement de collection mammalogique, point de départ de ma proper collection, m'a décidé à joindre à l'étude des reptiles celle des petits mammifères" (I am happy to dedicate this species to Mr. Eugene Simon, who, through the gift of the beginnings of a mammalogical collection, the starting point of my own collection, convinced me to join to the study of reptiles that of small mammals). He was an arachnologist and ornithologist who was an expert on hummingbirds. He wrote *Histoire naturelle des Trochilidae* in 1921. He is commemorated in the trinomial of a race of the Swallow-tailed Hummingbird, *Eupetomena macroura simoni*. The gerbil is found in Algeria, Tunisia, Libya, and Egypt west of the Nile Delta.

Simons

Simons' Spiny Rat *Proechimys simonsi* **Thomas,** 1900

Perry O. Simons (1869–1901) was an American citizen who collected in the Neotropics. Around 1900 he was collecting for herpetologists in Peru. He collected birds in Bolivia in 1901, and while crossing the Andes after that expedition he was murdered by his lone guide. His name is remembered in a number of species, such as Simons' Brush Finch *Atlapetes seebohmi simonsi* and the Andean Tapaculo *Scytalopus (magellanicus) simonsi.* There are more than a dozen type specimens in the British Museum of Natural History that Simons collected. A certain F. A. Simons was collecting in Colombia a decade earlier, and we would very much like to know if the two were related. The spiny rat is found in southern Colombia, eastern Ecuador, and northeast Peru.

Simpson

Simpson's Duiker *Cephalophus monticola simpsoni* **Thomas,** 1910 [Syn. *Philantomba monticola simpsoni*]

Captain Melville William Hilton-Simpson (1881–1938) was a British anthropologist who had a particular interest in medicine. He published a number of papers on Berber and Arab medical and surgical methods in use in North Africa. Between 1907 and 1909 he explored the Kasai region of the Congo with Emil Torday, who was collecting for the Pitt-Rivers Museum in Oxford. In 1911 Hilton-Simpson published *Land and Peoples of the Kasai.* The type specimen of the duiker was collected by Torday, but Thomas writes, "I have had much pleasure in naming it in honour of Mr. M.W. Hilton Simpson, Mr. Torday's companion during his expedition and himself a donor to the National Museum of several interesting mammals." This is a subspecies of the Blue Duiker, found in southern DRC (Zaire).

Sir David

Sir David's Long-beaked Echidna *Zaglossus attenboroughi* **Flannery** and Groves, 1998 [Alt. Attenborough's Echidna, Cyclops Long-beaked Echidna]

See Sir David **Attenborough** for biographical details.

Sladen

Sladen's Rat *Rattus tanezumi* **Temminck,** 1844

Lieutenant Colonel Sir Edward Bosc Sladen (1827–1890) was British Chief Commissioner at the Court of Mandalay. In 1868 he led a political mission sent to the Chinese frontier to inquire into the cause of the cessation of overland trade between Burma (now Myanmar) and China. From 1876 to 1885 he was Commissioner of the Arakan division. A specimen of rat from Yunnan was named *Mus sladeni* by Anderson in 1879 at a time when all mice and rats were lumped together in the genus *Mus.* Later the taxon was relegated to being a subspecies of the Black Rat, as *Rattus rattus sladeni.* Later still, as the tides of taxonomy swung toward splitting, the Oriental House Rat was separated from the Black Rat, and the scientific name *Rattus tanezumi* was revived. The vernacular name Sladen's Rat is now used somewhat loosely for human-commensal rats in the Far East—not only for populations of *Rattus tanezumi* but sometimes for other species (e.g. in Hong Kong, the name has been used for *R. sikkimensis*).

Slevin

Slevin's Deer Mouse *Peromyscus slevini* Mailliard, 1924

Dr. Joseph Richard Slevin (1881–1957) was in a tradition of natural history aficionados, as his father was an ornithological collector and member of the California Academy of Sciences.

He served in the merchant navy and made some 20 voyages with the Oceanic Steamship Company. In 1904 he was employed by the California Academy of Sciences under Van Denburgh, then Curator of Herpetology, and together they prepared for an expedition to the Galápagos that took place over 17 months between 1905 and 1906. While they were away the academy's museum was destroyed in the great San Francisco earthquake of 1906. On their return they established the foremost collection of flora and fauna from the Galápagos. Slevin continued working with Van Denburgh until 1928 (apart from WWI service as a submarine officer). They co-wrote *List of Amphibians and Reptiles of Arizona*. Slevin became Curator in 1928, holding that post until 1957. He wrote *The Amphibians of Western North America* in 1934. During all that time he continued to collect in California and beyond, such as trips to Australia and Central America, and to write papers such as "A Contribution to Our Knowledge of the Nesting Habits of the Golden Eagle" (1929). He is also commemorated in the names of at least five reptiles, including Slevin's Skink *Emoia slevini*. The critically endangered deer mouse is found only on Catalina Island in the Gulf of California, Mexico.

Sloggett

Sloggett's Vlei Rat *Otomys sloggetti* **Thomas,** 1902 [Alt. Rock Karroo Rat; Syn. *Myotomys sloggetti*]

Lieutenant General Sir Arthur Thomas Sloggett R.A., M.C. (1857–1929) was a Colonel in the Royal Army Medical Corps and was in charge of No. 21 General Hospital at Deelfontein in Cape Colony during the Boer War. He was on duty at the hospital from March 1900 until the cessation of hostilities. He eventually became Surgeon General and Director General of Medical Services in the British army, and in that capacity in 1916 he received a report on the introduction of steel helmets. He must have had an interest in natural history, as he presented a collection of mammals from Deelfontein to the British Museum. The type specimen of the vlei rat was part of this collection. The rodent is found in the Drakensberg and Maluti mountains of South Africa and Lesotho.

Smith, A.

Smith's Rock Elephant Shrew *Elephantulus rupestris* A. Smith, 1831 [Alt. Western Rock Elephant Shrew]

Smith's Red Rock Hare *Pronolagus rupestris* A. Smith, 1834

Smith's Bush Squirrel *Paraxerus cepapi* A. Smith, 1836

Ruddy Mongoose *Herpestes smithii* **Gray,** 1837

Dr. Sir Andrew Smith (1797–1872) was a Scotsman who started his professional life as a ship's surgeon. He was a zoologist whose first love was reptiles, and he was famed for his scrupulous accuracy. Smith was Director General of the British Medical Services during the Crimean War. He was in the Cape Colony, South Africa, from 1820 until 1837, and led the first scientific expedition into the South African interior, from 1834 until 1836. He wrote *Illustrations of the Zoology of South Africa*. However, Smith stopped his natural history collecting and study after returning to Britain. Later in life he became a British Member of Parliament. Much of his private collection was given to Edinburgh University and is now in the Royal Museum of Scotland. He is also commemorated in the scientific name of a duck, the Cape Shoveler *Anas smithii*. The elephant shrew is found in Namibia and western South Africa. The rock hare has two separate areas of distribution, one in South Africa and one in eastern Africa (from southwest Kenya to eastern Zambia and northern Malawi). The squirrel ranges from southern Angola and northern Namibia east to Mozambique and the Transvaal (South Africa). The mongoose comes from

India and Sri Lanka. (Unfortunately we cannot be certain that the mongoose was named after Sir Andrew Smith. However, he was known to be a friend of Gray, who also named an Indian turtle *Kachuga smithii* in his honor, so we feel he is the likeliest candidate.)

Smith, A. D.

Desert Musk Shrew *Crocidura smithii* **Thomas,** 1895

Dr. Arthur Donaldson-Smith (1864–1939) was a physician, traveler, naturalist, and big-game hunter of American birth who seems to have spent a great deal of time in East Africa. He visited Lake Rudolph (now Lake Turkana) in both 1895 and 1899. He was in Ethiopia in 1896 and 1897 and may have been present at the Ethiopian victory over the Italians at the Battle of Adwa. He published *Through Unknown African Countries* in 1897. He was elected a Fellow of the Royal Geographical Society and is remembered in the names of three birds: Donaldson-Smith's Nightjar *Caprimulgus donaldsoni*, Donaldson-Smith's Sparrow-weaver *Plocepasser donaldsoni*, and Donaldson's Turaco *Tauraco leucotis donaldsoni*. The shrew has a strange distribution. It was first found in Ethiopia, but a subspecies has been described from Senegal on the opposite side of Africa. Whether it occurs in intervening areas is unknown.

Smith, F. C.

Smith's Fruit Bat *Lissonycteris smithii* **Thomas,** 1908

Canon F. C. Smith (dates not found) was presumably a missionary in Sierra Leone from where he sent the type specimen of this bat to the British Museum. He was a member of the Royal African Society and is recorded as being at a number of the society's dinners in London in the years before WW1. The bat, formerly regarded as a race of *Lissonycteris angolensis*, is found in West Africa from Senegal to Togo.

Smith, J. A.

Smith's Woolly Bat *Kerivoula smithii* **Thomas,** 1880

Dr. J. A. Smith (dates not found) was a physician by training but an archeologist and antiquarian by preference. He was associated with the National Museum of Antiquities in Edinburgh. He took part in a number of excavations, including several during the construction of the North British Railway in 1846, during which a number of ancient human remains were uncovered. He was Secretary of the Society of Antiquaries of Scotland, before which he read a paper in 1869 entitled "On the Remains of the Rein-deer in Scotland." According to Thomas, the type specimen of the woolly bat turned up when Smith sent the British Museum "a small collection of bats obtained at Old Calabar by Dr. A. Robb, of the United Presbyterian Mission at that place." The species is found from Nigeria east to Kenya.

Smith, J. A. C.

Smith's Shrew *Chodsigoa smithii* **Thomas,** 1911

Smith's Zokor *Eospalax smithii* Thomas, 1911

Dr. J. A. C. Smith (dates not found) joined Malcolm P. Anderson (q.v.) when the latter was conducting the Duke of Bedford's Exploration of Eastern Asia for the Zoological Society of London. Dr. Smith accompanied Anderson into Sichuan and other parts of the Chinese interior. Both the shrew and the zokor (mole-rat) come from central China.

Smith, R. G.

Smith's Vole *Myodes smithii* **Thomas,** 1905 [Alt. Smith's Red-backed Vole; Syn. *Phaulomys smithii*]

Richard Gordon Smith (1858–1919) was an English animal hunter who spent time in France, Canada, Norway, Sri Lanka, Burma, New Guinea, Fiji, China, Singapore, and Japan—largely as the result of falling out with his wife and not being able to contemplate the shame of divorce. Throughout his travels he kept diaries and embellished them with drawings and keepsakes. He wrote *Ancient Tales and Folklore of Japan*, published in 1918. From 1915 onward he made no further diary entries, being ravaged by beriberi and malaria. He presented to the British Museum a collection of small mammals obtained by him in Japan, including the type specimen of the vole, which is endemic to Japan.

Snethlage

Snethlage's Marmoset *Callithrix emiliae* **Thomas,** 1920

See **Emilia.**

Söderström

Ecuadorian Puma *Puma concolor soderstromii* Lönnberg, 1913

Ludovic Söderström (1843–1927) was the Swedish Consul General in Ecuador from about 1886 until the year of his death. He was also an avid collector of natural history specimens. There are a number of records of his collecting small mammals in 1913. As Lönnberg was also collecting in the same place that year, we believe the two men knew each other and collected together. This form of puma is no longer regarded as a valid subspecies, being included in *Puma concolor concolor* along with other populations from northern South America. See **Ludovic** for biographical details.

Sody

Sody's Tree Rat *Kadarsanomys sodyi* **Bartels,** 1937

Henri Jacob Victor Sody (1892–1959) was an active zoologist and collector in southeast Asia and a well-known expert on systematics. He published quite widely, with articles such as "Six New Mammals from Sumatra, Java, Bali and Borneo" (1931), "Seventeen New Generic, Specific, and Sub-specific Names for Dutch Indian Mammals" (1936), and "Notes on Some Primates, Carnivora, and the Babirusa from the Indo-Malayan and Indo-Australian Regions (with Descriptions of 10 New Species and Subspecies)" (1949). Sody described a species of rat in 1932, naming it after Bartels—so the favor seems to have been returned by Bartels in 1937. The tree rat is found only in western Java.

Soemmerring

Soemmerring's Gazelle *Gazella soemmerringii* Cretzschmar, 1828

Samuel Thomas von Soemmerring (1755–1830) was a German anatomist, physician, paleontologist, and notable freemason. He was the author of a large body of work on anatomy. He was interested in many scientific and philosophical fields and corresponded with Goethe, Kant, Blumenbach, Forster, and von Humboldt. He was Professor of Anatomy at Kassel and also developed the first German hot-air balloon. He was opposed to the use of the guillotine and to the wearing of corsets—two views which, so far as we know, were not connected. The Copper Pheasant *Syrmaticus soemmerringii* and a number of parts of the human body (e.g. Soemmerring's ganglion) are also named after him. Although he was not noted for contributions to zoology, his preeminence in other fields led naturalists to honor him when naming animal species. The gazelle is found in eastern Sudan, Ethiopia, and northern Somalia.

Sokolov

Sokolov's Dwarf Hamster *Cricetulus sokolovi* Orlov and Malygin, 1988

Sokolov's White-toothed Shrew *Crocidura sokolovi* **Jenkins,** Abramov, Rozhnov, and Makarova, 2007

Vladimir Evgenevich Sokolov (1928–1998) was a Russian zoologist, mammalogist, and member of the Russian Academy of Sciences. He cowrote *Guide to the Mammals of Mongolia* in 1980 as well as at least a dozen other books about mammals. The hamster is found in western and southern Mongolia and adjacent parts of northern China. The shrew is known only from Ngoc Linh Mountain, Vietnam.

Someren

Someren's Girder-backed Shrew *Scutisorex somereni* **Thomas,** 1910 [Alt. Armored Shrew, Hero Shrew]

Dr. Robert A. Logan van Someren (1880–1955) was a naturalist who lived in Uganda for many years from 1905. In 1926 he co-wrote a paper with his brother, Dr. Victor Gurner Logan van Someren (1886–1976), about the life cycle of certain butterflies. In 1906 he and his brother started a survey of the birds of Kenya and Uganda, and their collection ultimately exceeded 25,000 specimens. These were deposited in various museums as far away as Los Angeles. The shrew is found in DRC (Zaire), Uganda, Rwanda, and Burundi.

Sommer

Long-nosed Shrew-Mouse *Sommeromys macrorhinos* **Musser** and Durden, 2002

Helmut G. Sommer is an American zoologist who was a Scientific Technician in the Department of Mammalogy at the American Museum of Natural History. He often writes jointly with Guy G. Musser, one of the describers of the genus, which has only this one known species. He has also, with Sydney Andersen as co-author, written a paper entitled "Cleaning Skeletons with Dermestid Beetles: Two Refinements in the Method" (1974). The shrew-mouse comes from Sulawesi, Indonesia.

Sorensen

Sorensen's Leaf-nosed Bat *Hipposideros sorenseni* Kitchener and Maryanto, 1993 [Alt. Pangandaran Roundleaf Bat]

Dr. Kurt Sorensen is a specialist in tropical medicine and was the Officer-in-Charge of the U.S. Naval Medical Research Unit No. 2, Jakarta Detachment in Indonesia. He was based in Djakarta in 1975, and he also worked at San Lazaro Hospital in Manila, the Philippines, in 1979. The dedication states that the bat is named after him "in recognition of his support and encouragement of staff of the Western Australian Museum/Museum Zoologicum Bogoriense with their research programme in Indonesia." The bat is known only from Java.

Soriano

Soriano's Yellow-shouldered Bat *Sturnira sorianoi* Sanchez-Hernandez et al., 2005

Dr. Pascual José Soriano of the Institute of Environmental Science and Ecology at the University of the Andes, Merida, Venezuela, is a Venezuelan zoologist. He collected the type specimen of the bat near Merida in 1980. He has since described a number of new mammal species himself. The bat has been recorded in Bolivia as well as Venezuela. The full extent of its range is unclear.

Sowell

Sowell's Short-tailed Bat *Carollia sowelli* **Baker** et al., 2002

James E. Sowell is an American philanthropist who provides funds and sponsors expeditions for students, particularly from the Texas Tech University, where he graduated himself in 1970 with a degree in finance. He sponsored the James E. Sowell Expeditions to Honduras and Ecuador in 2001. These two expeditions made collections of 75 species, including two that were possibly previously unknown to science. The Library of Texas Tech University also holds the James Sowell Family Collection in Literature, Community, and the Natural World and includes the personal papers of a number of prominent 20th-century writers on nature. Mr. Sowell formed Jim Sowell Construction Co. Inc. in 1972, which is one of the largest subdivision developers in the state of Texas. He served in the U.S. Army in 1970–1971. He has served on the Board of Directors of several New York Stock Exchange companies, including NL Industries, Todd Shipyards Corporation, Lomas and Nettleton Corporation, and Ketchum Drug Company. He has served as Chairman of the Board of Regents for Texas Tech University and Texas Tech University Medical School. The bat is found from southern Mexico to Panama.

Sowerby

Sowerby's Beaked Whale *Mesoplodon bidens* Sowerby, 1804 [Alt. North Sea Beaked Whale]

James Sowerby (1757–1822) was an English naturalist and watercolor artist who described the first known individual of this whale species. It was stranded in the Moray Firth in Scotland in 1800, and Sowerby's description was published in 1804. He was a graduate of the Royal Academy. In 1809 he wrote a book on the theory of color entitled *A New Elucidation of Colours, Original Prismatic and Material: Showing Their Concordance in the Three Primitives, Yellow, Red and Blue: and the Means of Producing, Measuring and Mixing Them: with Some Observations on the Accuracy of Sir Isaac Newton.* He also wrote and illustrated *English Botany, or Coloured Figures of British Plants, with Their Essential Characters, Synonyms, and Places of Growth,* published between 1790 and 1814. In addition, he wrote a work on what we would today call invertebrate paleontology, *Mineral Conchology of Great Britain, or Coloured Figures and Descriptions of Those Remains of Testaceous Animals, or Shells Which Have Been Preserved at Various Times, and Depths in the Earth,* published between 1812 and 1846. Sowerby contributed illustrations to many other works and also hand-colored many of the illustrations in William Curtis' *Flora Londinensis.* He collected both plant and mineral specimens, which were eventually purchased by the British Museum. He came from a talented family: no fewer than 14 members of his family published, wrote, or illustrated natural history works between 1780 and 1954. The whale occurs in the temperate and sub-Arctic waters of the North Atlantic.

Spartacus

Bronze Quoll *Dasyurus spartacus* Van Dyck, 1988

Spartacus (ca. 120–69 B.C.) was about 51 when he either died in battle or was one of those crucified along the Via Appia—quite an old man for his era and occupation. Life for a slave-gladiator must normally have been short and brutal. Plutarch, the Roman historian, reported Spartacus as being cultured and intelligent and more like a Greek than a Thracian, as tradition states he was. For reasons unknown he was enslaved and trained as a gladiator at Capua, and in 73 B.C. he and some companions escaped and started a slave revolt that spread widely until he had a very large army of ex-slaves, plus women and children—perhaps as many as 120,000 people at the height of his revolt. It caused the Roman authorities immense trouble, as he routed a number of armies and crushed several legions. Finally he was defeated by Crassus at the battle of Silarus, and the survivors of that battle were caught and defeated by Pompey the Great. Tim Flannery writes that the quoll, a carnivorous marsupial, was named because of "the many similarities it shares with the notorious Thracian gladiator." It must be said that these "similarities" appear less than obvious to us. In modern times Spartacus has become idealized as a fighter for liberty, even a great Hollywood hero. The species is found in southern New Guinea.

Spegazzini

Spegazzini's Grass Mouse *Akodon spegazzinii* **Thomas,** 1897

Carlos Luigi Spegazzini (1858–1926) was an Italian-born Argentinean mycologist and naturalist. He was trained in oenology but from the outset his main interest was fungi. He traveled from Italy to Brazil in 1879 and swiftly moved from there to Argentina to escape an epidemic of yellow fever. In 1881 he was a member of the Italo-Argentine expedition to Patagonia and Tierra del Fuego, but they were shipwrecked and Spegazzini had to swim for it, bearing all his notes on his shoulder to keep them from the sea. In 1884 he took up permanent residence in Argentina. He was a Professor at the University of La Plata from 1887 to 1912. In the same period he was also Curator of the National Department of Agriculture Herbarium, first head of the Herbarium of the Museo de La Plata, and founder of an arboretum and an Institute of Mycology in La Plata city. He is most remembered for his study of mycological and vascular plants, but he traveled widely and collected natural history specimens wherever he went. He published about 100 papers on vascular plants, mostly in Argentinean journals, and described around 1,000 new taxa. The grass mouse is found in northwest Argentina.

Speke

Speke's Pectinator *Pectinator spekei* **Blyth,** 1856 [Alt. Bushy-tailed Gundi]
Speke's Gazelle *Gazella spekei* Blyth, 1863
Sitatunga *Tragelaphus spekii* **Sclater,** 1863 [Alt. Marshbuck]

Captain John Hanning Speke (1827–1864) was a British explorer. He was the first European to see Lake Victoria (now Lake Nyanza), and it was he who proved it to be the source of the Nile. Speke joined Burton's expedition to discover the Nile's source, not because he was particularly interested in finding it but more because he wanted the chance to hunt big game. By the time he parted from Burton, who went on to Lake Tanganyika, Speke too had caught the source-location obsession. Speke hunted to supply the expedition, but he also observed the behavior and ecology of birds, one of which, Speke's Weaver *Ploceus spekei*, was named after him (as was a Nile steamer, SS *Speke*). His own shotgun killed him when he stumbled over a stile while out shooting in England, although some believe that he committed suicide. The pectinator (a rodent looking somewhat like a guinea-pig, though unrelated) is found in Eritrea, Djibouti, and northern Somalia. The gazelle is from Somalia. The Sitatunga has a wider range: there is a population in Senegal and Gambia, with the main distribution being from Nigeria east to southern Sudan and western Kenya, south to Angola and northern Botswana.

Spencer, E. L.

Long-snouted Phalanger *Tarsipes spencerae* **Gray,** 1842 [Alt. Honey Possum; Syn. *T. rostratus* Gervais and Verreaux, 1842]

Lady Eliza Lucy Spencer (1819–1898) was the wife of Captain Sir George Grey (q.v.). Her maiden name was Spencer. Originally the binomial was spelled variously as *spencerae* or *spenserae*, but in 1970 Ride introduced *spencerae* as the definitive name, and the 1972 Monaco International Conference of Zoology concurred. Lady Grey was the daughter of the government Resident in Albany, Western Australia. She and Grey married in 1839, and when their only son died in Adelaide at the age of five months, Grey blamed her. She hated life in Adelaide and, later, in New Zealand. In 1858 she suffered a nervous breakdown and returned to England, where Grey followed her the next year. While on board ship, she committed what was described as "an indiscretion involving a flirtatious letter." As the ship's doctor opined that Grey would kill either his wife or himself, the ship put about, returned to Rio de Janeiro, and Lady Grey was put ashore. It was 37 years before she finally returned to England in 1896 and met her husband again—a reconciliation that both found extremely difficult. They died in the same month in 1898. John Gray, who named this marsupial, commented that Lady Grey had shown "a great taste for, and paid great attention to, natural history." This marsupial is found in southwest Australia.

Spencer, W. B.

Pouched Jerboa *Antechinomys spenceri* **Thomas,** 1906 [Alt. Marsupial Jerboa, Wuhl-wuhl; Syn. *A. laniger spenceri*]

Professor Sir Walter Baldwin Spencer K.C.M.G., M.A., D.Sc., Litt.D., C.M.G., F.R.S. (1860–1929) was an explorer, biologist, anthropologist, zoologist, and patron of the arts. He went to Australia in 1887 to become the first Professor of Biology at the University of Melbourne and Director of the Natural History Museum in Melbourne, holding the posts until he retired in 1919. He was the photographer and zoologist on the 1904 Horn expedition, which was the first major scientific expedition to the center of Australia. In 1899 he became Honorary Director of the Natural Museum of Victoria and in 1904 was elected President of the Royal Society of Victoria. He donated his personal collection to the museum in 1917 and resigned as Director in 1928 in order to return to England. He died at Navarin Island in Tierra del Fuego, while on his way home. Most of his published works were on ethnographical subjects, such as *The Native Tribes of Central Australia* (1899), *The Northern Tribes of Central Australia* (1904), *Native Tribes of the Northern Territory of Australia* (1914), and *The Arunta: A Study of a Stone Age People* (1927). The marsupial jerboa lives in the deserts of Australia. There has been taxonomic uncertainty as to whether *spenceri* is a valid species or a race (or just a clinal variant) of *Antechinomys laniger*. Nowadays it is not generally regarded as a separate species.

Sphinx

Mandrill *Mandrillus sphinx* **Linnaeus,** 1758

Greater Short-nosed Fruit Bat *Cynopterus sphinx* Vahl, 1797

Linnaeus must have thought that the Mandrill looked like the mythical Sphinx, shown in ancient Egyptian art and sculpture as a human-headed lion. Linnaeus was fond of using fanciful names from mythology for primates, and several other cases can be found in this book. Vahl's logic in choosing this scientific name for a bat is even harder to understand. The Mandrill is found in Cameroon, Gabon, Equatorial Guinea, and the Congo Republic. The bat occurs from India and Sri Lanka east to Vietnam and south to Sumatra and Borneo.

Spix

Spix's Bearded Saki *Chiropotes israelita* Spix, 1823 [Alt. Brown-backed Bearded Saki]

Spix's Black-headed Uacari *Cacajao melanocephalus ouakary* Spix, 1823

Spix's Black-mantled Tamarin *Saguinus nigricollis nigricollis* Spix 1823

Spix's Disk-winged Bat *Thyroptera tricolor* Spix, 1823

Spix's Owl Monkey *Aotus vociferans* Spix, 1823 [Alt. Noisy Night Monkey]

Spix's Round-eared Bat *Tonatia bidens* Spix, 1823 [Alt. Greater Round-eared Bat]

Spix's Saddle-back Tamarin *Saguinus fuscicollis fuscicollis* Spix, 1823

Spix's Yellow-toothed Cavy *Galea spixii* **Wagler,** 1831

Johann Baptist Ritter von Spix (1781–1826) was a German naturalist working in Brazil from 1817 until 1820. He obtained his Ph.D. at the age of 19. He studied theology for three years in Würzburg, then medicine and the natural sciences, qualifying as a medical doctor in 1806. In 1808 he was awarded a scholarship by the King of Bavaria and went to Paris to study zoology. At that time Paris was *the* center for the natural sciences, with renowned scientists such as Cuvier, Buffon, Lamarck, and Étienne Geoffroy Saint-Hilaire at the height of their reputations. In 1810 the King appointed Spix Assistant to the Bavarian Royal Academy of Sciences, with special responsibility for the natural history exhibits. In 1816 a group of academicians was invited to travel to Brazil, and King Maximilian I agreed that two members of the Bavarian Academy of Sciences should accompany them. In 1817 Spix went to South America, along with Natterer. They were part of the suite on board *Austria* and *Principesse Augusta,* in attendance on the Archduchess Maria Leopoldina of Austria, who was on her way to marry the Brazilian Crown Prince, Dom Pedro. Spix returned in 1820 with specimens of 85 mammal species, 350 birds, 130 amphibians, 116 fish, and 2,700 insects as well as 6,500 botanical items. He also brought back 57 species of living animals, mainly monkeys, parrots, and curassows. This was to form the basis for the Natural History Museum in Munich. The King awarded him a knighthood and a pension for life. When he returned, Spix catalogued and published his findings despite extremely poor health caused by his stay in Brazil. The report on the expedition was published in three volumes in 1823, 1828, and 1831. In 1824 Spix published *Avium Brasiliensium species novae* and is himself remembered in the names of a number of birds, notably Spix's Macaw *Cyanopsitta spixii,* and about half a dozen reptiles, including Spix's Whiptail Lizard *Cnemidophorus ocellifer.* The saki was illustrated and named by Spix, but no formal description was written. It was rediscovered in the Rio Negro area of the Amazon basin in 1993 and formally described as a full species in 2003 by Boniviceno et al. All the species named after Spix are found in tropical South America. The disk-winged bat's range extends as far north as southern Mexico.

Spurrell

Spurrell's Free-tailed Bat *Mops spurrelli* **Dollman,** 1911
Spurrell's Woolly Bat *Kerivoula phalaena* **Thomas,** 1912
Chestnut Long-tongued Bat *Lionycteris spurrelli* Thomas, 1913

Professor Dr. Henry George Flaxman Spurrell (1882–1919) was a British zoologist, a Fellow of the Zoological Society who collected in both Ghana and Colombia. In 1917 he published *Modern Man and His Forerunners: A Short Study of the Human Species Living and Extinct.* The free-tailed bat is found from Liberia east to DRC (Zaire). The woolly bat has a similar distribution. The long-tongued bat occurs from eastern Panama to the Amazon basin.

St. Aignan

St. Aignan's Trumpet-eared Bat *Kerivoula agnella* **Thomas,** 1908

Saint Aignan is the former name for Misima Island, off Papua New Guinea, where the type specimen of the bat was taken. Thus the species is named after the island, not directly after a person. The bat is found in the Louisiade Archipelago and D'Entrecasteaux Islands, Papua New Guinea.

Stalker

Kai Myotis *Myotis stalkeri* **Thomas,** 1910
[Alt. Moluccan Mouse-eared Bat]

William Stalker (1879–1910) was an Australian collector and ornithologist. He is reported as having made a collection of mammals in the Northern Territory. The original etymology for the myotis states it is named "for its discoverer Mr. W. Stalker, whose sad death by drowning shortly after the landing of the expedition in New Guinea deprives the Museum of one of its best and most enthusiastic collectors." The bat occurs on the Kai and Gebe Islands, eastern Indonesia.

Stampfli

Stampfli's Putty-nosed Monkey *Cercopithecus nictitans stampflii* **Jentink,** 1888

Franz Xavier Stampfli (1847–1903) was a German (some sources say Swiss) naturalist who was working in Liberia between 1879 and 1887. Büttikofer, who was Stampfli's companion on at least one expedition, wrote a paper in 1886 entitled "Zoological Researches in Liberia: A List of Birds, Collected by Mr. F. X. Stampfli Near Monrovia, on the Messurado River, and on the Junk River with Its Tributaries." This western race of Putty-nosed Monkey occurs in Liberia and Ivory Coast.

Standing

Standing's Hippopotamus *Hippopotamus amphibius standingi* Monnier and Lamberton, 1922 extinct

Dr. Herbert F. Standing (dates not found) was a British medical missionary in Madagascar, where he was a friend of Charles Lamberton. He was interested in paleontology and described a subfossil lemur, *Hapalemur (Prohapalemur) gallieni,* in 1905. He is also remembered in the name of a Madagascan lizard, Standing's Day Gecko *Phelsuma standingi.* In 1887 he published *The Children of Madagascar.* He was also for a time Headmaster of the Boys' High School of the Friends' Foreign Mission Association, Antananarivo. *Hippopotamus amphibius standingi* is now considered an invalid synonym for *H. lemerlei.*

Stanger

Forest Giant Squirrel *Protoxerus stangeri* **Waterhouse,** 1842 [Alt. African Giant Squirrel, Oil-palm Squirrel]

Dr. William Stanger M.D. (1811–1854) was a British geologist and explorer. He took part in the Niger expedition of 1841, which used three ships to sail to and then up the Niger River. He suffered from fever (presumably malaria) intermittently on his return to England. He was the author of the geological report of the expedition. The genus of Natal grass, which he discovered in 1839, was named *Stangeria* in his honor. In 1845 he became the first Surveyor General of Natal, where a town was named after him, serving in that post until 1854. In 1848 he was instrumental in the establishment of the Durban Botanical Gardens. A Durban writer said of him, "Stanger who had come to Natal fresh from exploring the Niger River and was well pickled with tropical diseases died in Durban on the 14th March 1854." He was buried in England in 1857! He is commemorated in the trinomial of a subspecies of the Green-throated Sunbird, *Chalcomitra rubescens stangerii.* The squirrel is found from Sierra Leone east to southern Sudan and western Kenya, south to northern Angola.

Stankovic

Stankovic's Mole *Talpa stankovici* V. and E. **Martino,** 1931 [Alt. Balkan Mole]

Professor Dr. Sinisa Stankovic (1892–1974) was a Serb at a time when his nationality would have been described as Yugoslav. He founded the Belgrade School of Ecology and in 1933 published the first textbook on Yugoslav ecology, called *The Frame of Life.* He established the Hydrobiological Institute at Lake Ohrid (in Montenegro) in 1935. That lake contains a number of endemic species, and in 1960 Stankovic gathered all his research and conclusions together and published *The Balkan Lake Ohrid and Its Living World.* The mole can be found in Greece, countries of the former Yugoslavia, and probably in Albania.

Stanley

Stanley's Brush-furred Rat *Lophuromys stanleyi* Verheyen et al., 2007

William "Bill" Stanley is a zoologist who was brought up in Kenya. His bachelor's degree in 1986 and his master's in 1991 were both awarded by Humboldt State University in California. In 1989 he joined the Field Museum in Chicago as Collection Manager in the Mammals Division. He visits East Africa yearly to collect in cooperation with Tanzanian biologists. He collected specimens of this rat in December 1991, on Mount Ruwenzori-Bujuku.

Starck

Starck's Highland Hare *Lepus starcki* **Petter,** 1963 [Alt. Ethiopian Highland Hare]

Professor Dr. Dietrich Starck (1908–2001) was a renowned German vertebrate morphologist and anatomist who studied medicine, chemistry, botany, and zoology at the Friedrich Schilller University in Jena. He continued his studies in Vienna and then in Frankfurt-am-Main, where he later taught, retiring from his professorships in 1976. He went on expeditions between 1950 and 1960 to Ethiopia, South Africa, and Madagascar. As its common name implies, the hare's distribution is restricted to the Ethiopian highlands.

Starrett

Starrett's Tailless Bat *Anoura werckleae* Starrett, 1969

Emeritus Professor Dr. Andrew Starrett (b. 1930) of the University of Southern California is a biologist and zoologist. He has published many articles about bats, such as, with Richard S. Casebeer, "Records of Bats from Costa Rica" (1968). The tailless bat is found in Costa Rica. It is often regarded as conspecific with *Anoura cultrata.*

Steere

Steere's Squirrel *Sundasciurus steerii* **Günther,** 1877 [Alt. Southern Palawan Tree Squirrel]

Steere's Spiny Rat *Proechimys steerei* **Goldman,** 1911

Professor Dr. Joseph Beal Steere (1842–1940) was an American zoologist, botanist, and ornithologist who earned a bachelor's degree in natural history and a bachelor of laws, both from the University of Michigan. Upon completion of his second degree he took a trip around the world from 1870 to 1875, during which he first went to Brazil and up the Amazon as far as he could by boat, then crossed the Andes to Peru and took ship for China. He was in Formosa (now Taiwan, but then owned by Japan) in 1871 and made two further visits there in 1873 and 1874. He visited China and the Moluccas and collected in the Philippines between 1874 and 1875, and again between 1887 and 1888. *A List of the Birds and Mammals Collected by the Steere Expedition to the Philippines* was published in 1890. On his return to Michigan in 1876 Steere joined the faculty of the University of Michigan, becoming a Full Professor in 1879 and holding that post until 1893. He took a party of students to the Philippines in 1887. His collection from his first round-the-world trip was of enormous proportions. According to the *Representative Men of Michigan* in the Michigan County Histories, it included 3,000 birds, 100,000 seashells, and 12,000 insects plus 300 fishes, 200 reptiles, and 1,000 corals—not to mention collections of Chinese bronzes and such. As the article explains, many items were duplicates and available for swapping with other collections. The number given in another source, which states that he donated 60,000 specimens and artifacts of botanical, anthropological, geological, and zoological interest, is probably more accurate, but the collection was nevertheless so large that to house it the University of Michigan built the first public museum to be constructed by a university in North America. Although he ceased active involvement with the university in 1893 and took up farming and private research, Steere undertook a final expedition, again with students, in 1901 to the Amazon to collect on behalf of the Smithsonian. He is also commemorated in the names of four birds: Steere's Broadbill *Eurylaimus steerii*, Steere's Coucal *Centropus steerii*, Steere's Liocichla *Liocichla steerii*, and Steere's Pitta *Pitta steerii*. His grandson, William Campbell Steere (1907–1989), became a famous botanist and Director of the New York Botanical Garden. The squirrel is confined to Balabac and Palawan islands in the southern Philippines. The spiny rat is found in eastern Peru, western Brazil, and northern Bolivia.

Stein

Stein's Cuscus *Phalanger vestitus* **Milne-Edwards,** 1877
Stein's Melomys *Paramelomys steini* **Rümmler,** 1935
Stein's Rat *Rattus steini* Rümmler, 1935 [Alt. New Guinea Small Spiny Rat]

Georg Hermann Wilhelm Stein (b. 1897) was a German collector and self-educated naturalist with a particular liking for zoology and geology. He was originally a secondary-school teacher. In 1931 he was sent on a collecting expedition to New Guinea by Sterling Rockefeller for the Museum of Comparative Zoology at Harvard. The prime focus was the collection of thousands of bird skins. He made similar expeditions to other parts of Indonesia. While in New Guinea he visited Waigeu and the Weyland Mountains. He did not confine himself to ornithology but also wrote formal descriptions of several nonavian New Guinea species, such as a spiny bandicoot, *Echymipera clara*, named after his wife (see **Clara**), who accompanied him on this expedition. He later wrote *A Contribution to the Biology of Papuan Birds*. In 1933 he named a species of cuscus, *Phalanger interpositus*. This is now believed to be a synonym of *Phalanger vestitus*, but Stein's name has stayed attached as the species' common name, which explains why he is associated with an animal described 20 years before he was born. All three mammals come from New Guinea, the melomys being confined to the Weyland Mountains in the west of the island.

Steinbach

Steinbach's Tuco-tuco *Ctenomys steinbachi* **Thomas,** 1907
Steinbach's Ocelot *Leopardus pardalis steinbachi* **Pocock,** 1941 [sometimes shown as Steinback's Ocelot]

Dr. José Steinbach (1856–1929) was a collector in Argentina and Bolivia for the Chicago Field Museum of Natural History. He collected botanical as well as zoological specimens, and many of the plants are held at the Instituto de Botánica Darwinion, at San Isidro in Argentina. He is also remembered in the name of a bird, Steinbach's Canastero *Asthenes steinbachi*. Both the tuco-tuco and the ocelot subspecies come from Bolivia, east of the Andes.

Stejneger

Stejneger's Beaked Whale *Mesoplodon stejnegeri* **True,** 1885
Kuril Harbor Seal *Phoca vitulina stejnegeri* **J. A. Allen,** 1902 [Alt. Insular Seal]

Dr. Leonhard Hess Stejneger (1851–1943) grew up in Bergen, Norway. He studied philosophy and law at the University of Christiania (now Oslo), where he was awarded his Ph.D. He then became interested in the natural sciences and in his early 20s published several handbooks on the mammals and birds of Norway. In 1881 he left Norway and moved to the USA and by 1884 was Assistant Curator of Birds at the Smithsonian Institution. He wrote the majority of the volumes on birds of the *Standard Natural History*. In 1889 he became Curator of Reptiles and until 1911 was the Head Curator of Biology. He was a life member of the Bergen Museum, a member of the National Academy of Sciences, and a fellow of the American Ornithologists' Union. He also had the honor of "Commander on Nomenclature" and of "Permanent Commander" of the International Zoological Congress. He was an honorary member of the California Academy of Sciences, British Ornithological Union, American Society of Mammalogists, and German Ornithological Society. In 1906 he became a Decorated Knight 1st Class, Royal Norwegian Order of St. Olav. He published a variety of literature, including a biography of Georg Steller (see next entry). Various taxa are named after him, including an Asian bamboo viper, *Trimeresurus stejnegeri*, and a toad, *Bufo stejnegeri*. The whale inhabits cold-temperate waters of the North Pacific. The seal (usually but not universally treated as a race of the Harbor Seal) comes from the western North Pacific.

Steller

Steller's Sea-Lion *Eumetopias jubatus* **Schreber,** 1776
Steller's Sea Cow *Hydrodamalis gigas* Zimmermann, 1780 extinct

Georg Wilhelm Steller (1709–1746) was a German naturalist and explorer in the Russian service. He studied medicine at Halle. He went to Russia in 1731 and between then and 1734 was a physician in the Russian army. In 1734 he became an assistant at the Academy of Sciences in St. Petersburg and in 1737 left for Kamchatka accompanying Vitus Bering on his second expedition to Alaska on board the *St. Peter*, which was accompanied by the *St. Paul*. This expedition ended when the *St. Peter* was wrecked on a desolate island, now called Bering Island, where Bering died and the surviving crew had to spend the winter in crude huts. Steller and the Danish First Lieutenant Waxell proved effective in ensuring their survival. After nine months a boat was constructed from the wreckage of the *St. Peter*, enabling the survivors to leave the island; they arrived in Kamchatka in August 1742. Between 1742 and 1744 Steller worked in Petropavlovsk but died on his return journey from there to St. Petersburg. In 1743 he published *Journal of a Voyage with Bering 1741–1742*. He described the huge sea cow that would later bear his name. Found around Bering Island, it was hunted to extinction by 1770 for its meat, fat, and skin, meaning that Steller was the only scientist ever to see living specimens. The sea-lion is distributed over the North Pacific, from Hokkaido (Japan) to California. He is remembered in the names of a number of birds, including Steller's Eider *Polysticta stelleri* and Steller's Sea Eagle *Haliaeetus pelagicus*.

Steno

Rough-toothed Dolphin *Steno bredanensis* **Lesson,** 1828

The name of this dolphin genus probably originates from the Greek word *stenos*, meaning "narrow." However, at least one author has suggested that it might have been inspired by Dr. Nicholas Steno (1638–1687), a Danish geologist, anatomist, and author. He was born Niels Stensen but, as Linnaeus did later, Latinized his name. In 1660 he went to Leiden to study medicine. In 1665, after a short period in Paris and Montpelier, Steno went to Florence, where he studied anatomy. He was the first person to realize that what looked like sharks' teeth embedded in rocks were in fact fossilized sharks' teeth. From that discovery he was led to formulate his most important contribution to geology, Steno's Law of Superposition. Between 1672 and 1674 he was the Royal Anatomist in Copenhagen. However, in 1667 he converted to Roman Catholicism and abandoned science. In 1675 he was ordained as a priest and in 1677 became a bishop, spending the rest of his life ministering to the minority Catholic populations in Denmark, Norway, and northern Germany. In 1987 he was beatified, the first step on the road to sainthood.

Stephan

San Esteban Deer Mouse *Peromyscus stephani* **Townsend,** 1912

This mouse is named after an island, not a person ("Esteban" is the Spanish form of "Stephen"). The mouse is confined to San Esteban Island, Sonora, northwest Mexico.

Stephen Nash

Stephen Nash's Titi *Callicebus stephennashi* **van Roosmalen,** 2002 [Alt. Nash's Titi Monkey]

Dr. Stephen Nash works for Conservation International as a technical illustrator and is based at the Department of Anatomical Sciences, State University of New York at Stony Brook, where he is a Visiting Research Associate. In 2004 he received the President's Award from the American Society of Primatologists. Local fishermen brought the type specimen of this monkey to Van Roosmalen in Manaus, Brazil, and the species' exact distribution is uncertain. It probably occurs on the east bank of the Rio Purús.

Stephens

Stephens' Woodrat *Neotoma stephensi* **Goldman,** 1905

Stephens' Kangaroo-Rat *Dipodomys stephensi* **Merriam,** 1907

Frank Stephens (1849–1937) was an ornithologist and mammalogist. He was Curator Emeritus of the San Diego Society of Natural History and a member of the Death Valley expedition of 1891. In his early years he collected birds and their nests and eggs for Charles Edward Howard Aiken. His wife, Kate, who lived to be over 100 years old, was a conchologist of note. As her husband had a reputation for being careless as well as deaf, she insisted on traveling with him on all his trips. For example, they were both members of the Alexander Expedition to southeastern Alaska in 1907, and in 1910 they accompanied Dr. Joseph Grinnell on the Colorado River. In that same year Stephens gave 2,000 bird and mammal specimens to the San Diego Society. In 1937 he was knocked down by a tram and died 10 days later; we speculate that he might not have heard the tram coming. The woodrat is found in Arizona and western New Mexico. The kangaroo-rat is endemic to southern California.

Stevenson

Stevenson's Collared Lemming *Dicrostonyx stevensoni* **Nelson,** 1929

Donald H. Stevenson (d. ca. 1925) worked for the U.S. Biological Survey. From 1920 until 1925 he was the Reservation Warden for the Aleutian Islands. He collected specimens of the lemming on Umnak Island, in the Aleutians, between 1920 and 1924. By the time Nelson described this taxon in 1929, he was honoring the "recently deceased" Stevenson. The lemming is usually regarded as a subspecies (or even merely a synonym) of *Dicrostonyx unalascensis* (Merriam, 1900) from Unalaska Island.

Stheno

Lesser Brown Horseshoe Bat *Rhinolophus stheno* **Andersen,** 1905

The bat is found from Myanmar east to Vietnam and south to Sumatra and Java. See **Euryale** for biographical details.

Stirton

Stirton's Deer Mouse *Peromyscus stirtoni* **Dickey,** 1928

Ruben Arthur Stirton (1901–1966) was an American paleontologist, specializing in mammals, and a Professor at the University of California, Berkeley. His earlier work was on extant North American mammals. He conducted fieldwork in South America in the 1940s and turned his attention to Neogene (Miocene and later) mammalian fauna from Australia in the 1950s. Some Australian fossil species, such as Stirton's Thunderbird *Dromornis stirtoni*, are named after him. The mouse has a patchy distribution in Central America, from southern Guatemala to Nicaragua.

Stoliczka

Stoliczka's Trident Bat *Aselliscus stoliczkanus* **Dobson,** 1871
Stoliczka's Mountain Vole *Alticola stoliczkanus* **Blanford,** 1875
Anderson's Shrew *Suncus stoliczkanus* **Anderson,** 1877

Dr. Ferdinand Stoliczka (1838–1874) was an Austrian paleontologist and zoologist who was born in Moravia (Czech Republic). He was educated in Prague and at the University of Vienna, where he obtained his doctorate. He collected during travels throughout India as Assistant Superintendent of the Geological Survey of India from 1864 to 1874. He wrote a number of scientific papers, such as "Contribution towards the Knowledge of Indian Arachnoidea," published in the *Journal of the Asiatic Society*. During the last 10 years of his life he published geological memoirs on the western Himalayas and Tibet and many papers on Indian zoology, from mammals to insects and corals. He took part in the Second Yarkand Mission of 1873–1874 but collapsed and died of spinal meningitis when "returning loaded with the spoils and notes of nearly a year's research in one of the least-known parts of Central Asia," according to his obituary in *Nature*. He is commemorated in the names of species from many phyla, including the common names of three birds: Stoliczka's Bushchat *Saxicola macrorhyncha*, Stoliczka's Tit-Warbler *Leptopoecile sophiae*, and Stoliczka's Treecreeper *Certhia nipalensis*. The bat is found in peninsular Malaysia, Thailand, Laos, Vietnam, Myanmar, and southern China. The vole's distribution extends from Ladakh (northern India) through Nepal and Tibet to central China. The shrew inhabits arid areas of Pakistan, India, and Bangladesh.

Stolzmann

Stolzmann's Crab-eating Rat *Ichthyomys stolzmanni* **Thomas,** 1893

Jean Stanislas Stolzmann, or Szrolcman (1854–1928), was a Polish zoologist who first went to Peru in 1871. He worked as a collector there between 1875 and 1883. He wrote scientific papers such as "On the Ornithological Researches of M. Jean Kalinowski in Central Peru," published in 1896. Oldfield Thomas described him as "one of the best known and most successful of Peruvian collectors; the discoverer of many new mammals." He is also commemorated in the scientific names of five birds, including the Ochre-breasted Tanager *Chlorothraupis stolzmanni* and the Tumbes Swallow *Tachycineta stolzmanni*. The rat is found in Ecuador and Peru.

Stone

Stone's Sheep *Ovis dalli stonei* **J. A. Allen,** 1897
Stone's Caribou *Rangifer tarandus stonei* J. A. Allen, 1901

Andrew Jackson Stone (1859–1918) was an American explorer, photographer, writer, and naturalist with the American Museum of Natural History and a representative for the New York Zoological Society. He was born in Missouri but grew up in Montana and moved, as an adult, to Seattle. For the museum and the Zoological Society he made a series of expeditions from the middle 1890s to the early 1900s to British Columbia and Alaska; his first expedition in 1896 was sponsored by *Recreation* magazine. The sheep was discovered by him on that first expedition, near the headwaters of the Stickeen River in northern British Columbia at an altitude of 1,980 m (6,500 feet). He explored extensively in the Yukon and Northwest Territories. He drowned in heavy seas near Cape Nome, Alaska, when his kayak overturned. The sheep is found in south-central Yukon and north-central British Columbia. The caribou, from the Kenai Peninsula of southern Alaska, appears to have been wiped out by overhunting not long after the taxon was described (caribou have since been reintroduced to the peninsula from other populations).

Storey

Storey's Mole-Rat *Tachyoryctes storeyi* **Thomas,** 1909

Charles B. C. Storey (b. 1868) was a collector in British East Africa, ca. 1906. Thomas says the type specimen of the mole-rat was presented by Storey and came from the vicinity of Lake Elmentaita, Kenya. We have not been able to find out more about him. The species is known only from the type locality.

Strachey

Strachey's Mountain Vole *Alticola stracheyi* **Thomas,** 1880

Lieutenant General Sir Richard Strachey (1817–1908) went to India in 1836, returned to England in 1850, and then in 1855 went back to India. He served in the Bengal Engineers, being a Lieutenant in 1841 and a Lieutenant General by 1875. In 1847 frequent attacks of fever compelled him to go to Nani Tal in the Kumaon Himalayas for his health. There he made the acquaintance of Major E. Madden, under whose guidance he studied botany and geology, making expeditions into the western Himalayas for scientific purposes. He served as an administrator in several capacities from 1862 to 1871 and was a member of the Council of India from 1875 to 1889. He was President of the Royal Geographical Society from 1887 to 1889. Jointly with his brother, Sir John Strachey, who was also a colonial administrator, he published in 1882 a major work, *The Finances and Public Works of India.* He wrote many monographs and papers, such as "A Description of the Glaciers of the Pindur and Kuphinee Rivers in the Kumaon Himalayas" (1847) and "On the Physical Geography of the Provinces of Kumaon and Garhwal, in the Himalaya Mountains, and of the Adjoining Parts of Tibet" (1851). The original citation for the vole reads, "It was collected in Kumaon by Capt. (now Lieut.-Gen.) R. Strachey." The vole is found from Kashmir east through southern Tibet to Sikkim.

Strand

Strand's Birch Mouse *Sicista strandi* Formozov, 1931

Professor Dr. Embrik Strand (1879–1947) was a Norwegian entomologist and arachnologist. He first studied, and then worked as a Curator, at the University of Christiania (now Oslo). Between 1898 and 1903 he made a number of trips inside Norway and gathered a significant collection of insects, including moths and beetles. Most of the collection is still at the University of Oslo's museum. In 1903 Strand went to Germany and worked in a number of different universities and museums there. In 1923 he was appointed Professor of Zoology at the University of Riga (Latvia). He published widely on spiders and insects, describing hundreds of new species. The birch mouse is found in the northern Caucasus and north to the Kursk region of southern Russia.

Street

Persian Mole *Talpa streeti* Lay, 1965 [Syn. *T. davidiana* **Milne-Edwards,** 1884]

William Sherman Street (1905–2000) and Janice Kergan Street (1902–2001) were sponsors of scientific expeditions and may have taken part themselves. They certainly wrote, with Richard Sawyer, *Iranian Adventure—The First Street Expedition.* The mole was discovered during the first Street expedition of the Field

Museum of Natural History (Chicago) to Iran between 1962 and 1963. Douglas M. Lay, who described the mole, wrote up some of the expedition's results under the heading *a Study of the Mammals of Iran Resulting from the Street Expedition of 1962–63.* The article gives the following information: "In 1962 Dr. M. Baltazard, then director of the Institut Pasteur of Iran, generously donated a small but important collection of small mammals primarily from Iranian Kurdistan to the William S. and Janice Kergan Street Expedition of the Chicago Natural History Museum to Iran, 1962–63." The type specimen of the mole (and a series of five others) was among this collection. However, Lay demonstrated his lack of knowledge of Latin grammar by not using the plural form, *Talpa streetorum.* The mole is known only from northwest Iran, but it is now believed to be synonymous with *Talpa davidiana* from southeast Turkey.

Strecker

Strecker's Pocket Gopher *Geomys personatus streckeri* **Davis,** 1943

John Kern Strecker (1875–1933) was an American naturalist whose main interest was herpetology. He was also interested in both conchology and folklore, being President of the American Folklore Society. He began his field surveys of Texas in 1895 and is regarded as the father of Texan herpetology. He was Head Librarian, until his death from heart disease, at Baylor University at Waco, Texas, from 1919 and Curator of the university museum from 1903. The museum was renamed the Strecker Museum in his honor in 1940. In 1964 the John K. Strecker Herpetological Society was formed, only to be disbanded two years later. He published widely on all his interests, such as, in 1915, *Reptiles and Amphibians of Texas* and, in 1928, *Reptiles of the South and Southwest in Folklore.* The gopher is found in a small area of southern Texas.

Strelkov

Strelkov's Long-eared Bat *Plecotus strelkovi* Spitzenberger, 2006

Dr. Petr Petrovich Strelkov is, at the time of writing, Curator of the St. Petersburg Zoological Institute, Russian Academy of Sciences. He was very active from the 1960s through the 1980s, publishing many articles. The citation for the bat reads, "This species is named in honour of Petr Petrovich Strelkov, who produced the first modern revision of the genus *Plecotus* in the former USSR." The bat is found in the Pamir and Tienshan mountains of central Asia. Its range may extend southwest to northern Iran, as there are museum specimens with "Iran" given as their place of origin.

Streltzov

Streltzov's Vole *Alticola strelzowi* **Kastschenko,** 1899 [Alt. Flat-headed Vole; the binomial is sometimes rendered *strelzovi*]

Dr. Zosima Ivanovich Streltzov (1831–1898) was a Russian histologist. He was a physician who worked as a local doctor in the Ekaterina area of Russia from 1854 to 1857. He took an advanced qualification as Doctor of Medicine in 1874. From 1870 he was appointed to teach embryology at the Medical Faculty of Kharkov University (then in Russia, now in the Ukraine). His publications include *Récherches experimentales sur le mecanisme de la production des hydropisies,* published in French in 1864. The vole is found in the Altai Mountains of central Asia.

Stuart

Stuart's Antechinus *Antechinus stuartii* **Macleay,** 1841 [Alt. Brown Antechinus]

John McDouall Stuart (1815–1866) was an Australian explorer and surveyor. He arrived in Australia aboard the *Indus* in 1838 and was described by a fellow passenger as "somewhat delicate, having two rather severe attacks of vomiting blood." And, at 168 cm (5 feet 6 inches) tall and just 57 kg (126 pounds) in weight, he seemed an unlikely hero. He attended the Scottish Naval and Military Academy and graduated as a civil engineer. He got work surveying in the new colony at Adelaide. In 1844 he was engaged as surveyor by Captain Charles Sturt for his expedition to the Australian interior. By the time they returned from the Great Stony Desert, both men were suffering severely from scurvy. While today he is acclaimed as Australia's greatest inland explorer, he left the colony nearly blind and with his constitution thoroughly broken. A history of his explorations, edited from his own notes, was published by William Hardman as *the Journals of John Mcdouall Stuart during the Years 1858, 1859, 1860, 1861, and 1862, When He Fixed the Centre of the Continent and Successfully Crossed It from Sea to Sea*. His name is also commemorated in the Stuart Highway linking Adelaide to Darwin, as well as in many geographical features throughout the lands he explored, including Mount Stuart. The antechinus is found in eastern Australia.

Stuhlmann

Stuhlmann's Monkey *Cercopithecus mitis stuhlmanni* **Matschie,** 1893
Stuhlmann's Golden-Mole *Chrysochloris stuhlmanni* Matschie, 1894

Professor Dr. Franz Stuhlmann (1863–1928) was a German zoologist and naturalist who collected in East Africa from 1888 until 1900. He made his career in the German Colonial Forces and Civil Service. He did not confine himself to zoological specimens, as a number of artifacts he collected in Africa are in anthropological exhibits of various museums. Stuhlmann traveled with Emin Pasha, and after Emin's murder, he and others who had survived an outbreak of smallpox came back from the area of Lake Albert with a large collection, plus cartographic material from which the first comprehensive map of German East Africa was made. In 1891 Stuhlmann became the first European to confirm the existence of snow-covered peaks that fed the Nile. (About 1,800 years earlier, Ptolemy had showed such features on a map and called them the Mountains of the Moon. Today they are known as the Ruwenzori Mountains.) The German government published a monograph by Stuhlmann, *Dr. Franz Stuhlmann: Mit Emin Pasha ins Herz von Africa*, in 1894. He is remembered in the names of three birds: Stuhlmann's (Double-collared) *Sunbird Nectarinia (Cinnyris) stuhlmanni*, Stuhlmann's Starling *Poeoptera stuhlmanni*, and Stuhlmann's Weaver Ploceus baglafecht stuhlmanni. The monkey is found in northeast DRC (Zaire), southern Sudan, northern Uganda, and western Kenya. The golden-mole occurs in the Ruwenzori Mountains of the Uganda DRC (Zaire) border and Rwanda. Other subspecies occur in Kenya and Tanzania, and a very isolated subspecies has been found on Mount Oku, Cameroon.

Sturdee

Sturdee's Pipistrelle *Pipistrellus sturdeei* **Thomas,** 1915 extinct?

Admiral Sir Frederick Charles Doveton Sturdee R.N., C.M.G., C.V.O., K.C.M.G., G.C.B., 1st Baronet (1859–1925), was Admiral of the Fleet. He was educated at the Royal Naval School, New Cross, from 1870 to 1871, then entered HMS *Britannia* as a naval cadet from 1871 to 1873. He served as a Midshipman in the Channel Squadron and on the East India Station, 1873–1878. He was promoted to Sub-Lieutenant and took various courses in the gunnery school ship HMS *Excellent* from 1878 to 1880. He was promoted to Lieutenant in 1880 and served in HMS *Hecla* on the Mediterranean Station from 1881 to 1882. He was then a student in the torpedo school ship HMS *Vernon*, 1882–1885; served as a torpedo officer in HMS *Bellerophon* on the North American and West Indies Station, 1885–1889; was on the staff of HMS *Vernon*, 1889–1893; and gained extensive experience of torpedo boats. He was promoted to Commander in 1893 and worked as a torpedo specialist in the Naval Ordnance Department of the Admiralty from 1893 to 1897. His first command was HMS *Porpoise* on the Australian Station, 1897–1899, then taking command of the British force in Samoa during hostilities between Germany and the USA. He was promoted to Captain and returned to the Admiralty to serve as Assistant Director of Naval Intelligence from 1900 to 1902. He commanded various ships in home waters, 1902–1905. He was then Chief of Staff to Lord Charles Beresford, Commander in Chief of the Mediterranean Fleet, 1905–1907, and the Channel Fleet, 1907–1908. He commanded the battleship HMS *New Zealand*, 1908–1909, the First Battle Squadron in 1910, and various cruiser squadrons, 1912–1913, and was promoted to Vice Admiral in 1913. He became Chief of the War Staff in 1914 and served as Commander in Chief of the Atlantic and South Pacific Stations from 1914 to 1915, scoring a decisive victory in the Falkland Islands in 1914. He commanded the Fourth Battle Squadron from 1915 until 1918, during which time he was present at the Battle of Jutland, 1916. He was promoted to full Admiral in 1918 and became Commander in Chief at the Nore, serving from 1918 until 1921. He was promoted to Admiral of the Fleet in 1921. He retired to Camberley, in Surrey, and died there on 7 May 1925. The bat is known only from Hahajima Island in the Bonin Islands, Japan. Some authorities think this type locality is in error, and that the species' true origin is unknown. It is sometimes listed as being extinct, as no further records have been documented since the species was described.

Styan

Styan's Squirrel *Callosciurus erythraeus styani* **Thomas,** 1894
Styan's Water Shrew *Chimarrogale styani* **de Winton,** 1899
Styan's Red Panda *Ailurus fulgens styani* Thomas, 1902

Frederick William Styan (1838–1934) spent 27 years in China as a tea trader and collector, corresponding from Kiukiang. He was a Fellow of the Zoological Society and was elected as a member of the British Ornithologists' Union in 1887. He is remembered in the common names of two birds: Styan's Bulbul *Pycnonotus taivanus* and Styan's Grasshopper Warbler *Locustella pleskei*. The squirrel is found in southeast China, and the water shrew in northern Myanmar and China (Sichuan and Shensi provinces). The *styani* race of Red Panda occurs in the Gongshan Mountains of Yunnan and the Hengduan Mountains of Sichuan (China).

Sumichrast

Cacomistle *Bassariscus sumichrasti* **Saussure,** 1860
Sumichrast's Vesper Rat *Nyctomys sumichrasti* Saussure, 1860
Sumichrast's Harvest Mouse *Reithrodontomys sumichrasti* Saussure, 1861

Adrien Jean Louis François de Sumichrast (1828–1882) was a Swiss naturalist who accompanied Saussure on his travels from 1854 to 1856 in the West Indies, the USA, and Mexico. They made considerable collections of specimens, which Saussure took back to Geneva in 1856. Sumichrast stayed on in Mexico for the rest of his life, marrying a local woman. He appears to have variously used different parts of his name, depending upon who he was dealing with. He may have taken Mexican nationality, though sources in Mexico record him as a Swiss naturalist called A. L. François Sumichrast. But he was known as Professor

François Sumichrast to the Smithsonian Institution, which also has him down as being a French naturalist, under which guise he undertook an expedition in Mexico for the Smithsonian. (He applied to Baird at the Smithsonian for a second expedition in 1870, but funding was refused. However, he did undertake several other collecting expeditions under the institution's auspices.) He is also commemorated in the names of many other taxa, such as Sumichrast's Wren *Hylorchilus sumichrasti* and Sumichrast's Garter Snake *Thamnophis sumichrasti*. All three mammals named after him are found from southern Mexico to Panama.

Sundevall

Sundevall's Jird *Meriones crassus* Sundevall, 1842 [Alt. Gentle Jird]
Sundevall's Roundleaf Bat *Hipposideros caffer* Sundevall, 1846

Professor Carl Jakob Sundevall, sometimes Sundewall (1801–1875), was a Swedish zoologist and ornithologist who had a strong interest in spiders as well as birds. He graduated with a doctorate in zoology from Lund University in 1823. He then went on a journey to East Asia before returning to Lund, where he qualified as a physician in 1830. From 1833 to 1871 he was employed by the Swedish Natural History Museum in Stockholm, being Professor and Keeper of the Vertebrates Section from 1839. He wrote *Svenska füglarna* in 1856, and *Tentamen (Methodi naturalis avium disponendarum tentamen)* in 1889. He is also remembered in the name of Sundevall's Garter Snake *Elapsoidea sundevallii*, among other taxa. The jird is found across North Africa and southwest Asia, from Morocco to Iran and Afghanistan. The bat occurs over much of sub-Saharan Africa, the southwest Arabian Peninsula, and in Morocco.

Swainson

Dusky Antechinus *Antechinus swainsonii* **Waterhouse,** 1840

William Swainson (1789–1855) was a naturalist and illustrator. He was born in Liverpool, the son of a collector of customs duty. After elementary education he worked as a junior clerk, and then in the army commissariat in Malta and Sicily. Before going abroad he drew up, at the request of the Liverpool Museum, the *Instructions for Collecting and Preserving Subjects of Natural History*, privately printed in Liverpool in 1808. He served for eight years from 1807 to 1815 with the army commissariat and amassed a large collection of zoological specimens. At the end of the Napoleonic wars he retired on half-pay. In 1816 he left for Brazil and traveled, collecting specimens, through Pernambuco to the Rio São Francisco and then on to Rio de Janeiro. On his return he published a sketch of his journey in the *Edinburgh Philosophical Journal*, in 1819, very briefly describing the voyage without any scientific detail. He then endeavored to sort his zoological specimens. He learned the new technique of lithography and produced *Zoological Illustration* (published in three volumes from 1820 to 1823), the *Naturalists Guide* (1822), and *Exotic Conchology* (1841), among other works. In 1828 he visited museums in Paris under the guidance of Cuvier and St. Hilaire, meeting the other great French naturalists. In the same year he moved to the English countryside and worked as a full-time artist and author. In 1840 he left for New Zealand and became the country's first Attorney General. Unfortunately most of his specimen collection was lost on the voyage. He remained in New Zealand for the rest of his life. Swainson was a Fellow of the Linnean Society and of the Royal Society, as well as of numerous foreign academies. He published many papers, as well as *Birds of Brazil*, which appeared in five parts between 1834 and 1835. He wrote the bird section of Sir John Richardson's *Fauna Boreali-Americana* and contributed to the 11 volumes of Lardner's *Cabinet Encyclopaedia* (1834–1840) and the 3 volumes of Jardine's *Naturalist's Library* (1833–1846). Swainson was primarily interested in birds, and no fewer than 19 are named after him in their common names alone. The antechinus is found in Tasmania and in eastern Australia as far north as southeast Queensland.

Swarth

Santiago Galápagos Mouse *Nesoryzomys swarthi* Orr, 1938

Harry Schelwald Swarth (1878–1935) was an American zoologist. He began collecting birds in 1894. In 1896, only 18 years old, he made a trip to the Huachuca Mountains in Arizona, where he developed such an interest in the area that he took five more trips to different regions of the state and became the leading authority on Arizona ornithology. In 1904 he became Assistant in the Department of Zoology at the Field Museum in Chicago, and in 1908 he became Assistant in Ornithology at the Museum of Vertebrate Zoology at the University of California, Berkeley, becoming Curator two years later. In 1912 he became a member of the California Academy of Sciences and was appointed Curator of the Department of Ornithology and Mammalogy there in 1927. He left in 1913 to become Assistant Secretary of another institution but returned three years later. He wrote *The Avifauna of the Galapagos Islands* (1931) and *A New Bird Family (Geospizidae) from the Galapagos Islands* (1929). In 1932 he led the Templeton Crocker expedition of the California Academy of Sciences to the Galápagos. There is an account of the voyage during which Swarth collected and studied the Galápagos fauna: *Log of the Schooner "Academy" on a Voyage of Scientific Research to the Galapagos Islands 1905–1906*, written by Joseph R. Stevin. The mouse is confined to Santiago Island, Galápagos. It was thought to be extinct until rediscovered in 1997.

Swayne

Swayne's Hartebeest *Alcelaphus buselaphus swaynei* **Sclater,** 1892 [Alt. Korkay]
Swayne's Dik-dik *Madoqua (saltiana) swaynei* **Thomas,** 1894

Colonel Harald George Carlos Swayne (1860–1940) served in Somaliland. He wrote *Seventeen Trips through Somaliland: A Record of Exploration and Big Game Shooting, 1885 to 1893—Being the Narrative of Several Journeys in the Hinterland . . . Notes on the Wild Fauna of the Country,* published in 1895. The hartebeest is found in Ethiopia (and formerly in Somalia, but now extinct there). The dik-dik occurs in the Juba Valley of southern Somalia.

Swinderen

Greater Cane Rat *Thryonomys swinderianus* **Temminck,** 1827 [Alt. Marsh Cane Rat, Cutting Grass]

Professor Dr. Theodorus van Swinderen (1784–1851) was a Dutch naturalist. He was also a Doctor of Linguistics and of Law, and a school inspector for 40 years. He was appointed Professor of Natural History at the University of Groningen in 1814. He founded the Museum of Natural History in Groningen and ensured a flow of specimens for the collection by means of his links with a network of scholars and students in France, Germany, England, the South African Cape, Java, and China as well as in the Netherlands. He sent the type specimen of the cane rat to Temminck. He is also remembered in the scientific name of the Black-collared Lovebird *Agapornis swindernianus.* The rat is found in much of Africa south of the Sahara, except for the southwest corner.

Swinhoe

Swinhoe's Deer *Cervus unicolor swinhoii* **Sclater,** 1862 [Alt. Formosan Sambar; Syn. *Rusa unicolor swinhoii*]
Taiwan Serow *Naemorhedus swinhoei* **Gray,** 1862
Swinhoe's Striped Squirrel *Tamiops swinhoei* **Milne-Edwards,** 1874
Swinhoe's Jird *Meriones crassus swinhoei* **Scully,** 1881

Robert Swinhoe F.R.S. (1836–1877) was born in Calcutta, India, but in 1852 he was sent to London to be educated. In 1854, while at the University of London, he was recruited into the China consular corps by the Foreign Office. Before he left for Hong Kong he deposited a small collection of British birds' nests and eggs with the British Museum. His time in China as a diplomat gave him a terrific opportunity as a naturalist; he explored a vast area that had not been open previously to any other collector. As a result he discovered new species at the rate of about one per month throughout the more than 19 years he was there. He was primarily an ornithologist, and so the majority of his discoveries were birds, but his name is also associated with dozens of Chinese animals including mammals, fish, and insects. He first returned to London in late 1862. He was a prolific writer with, for example, 37 papers in the *Proceedings of the Zoological Society* just between 1861 and 1870. He brought part of his vast collection of specimens to meetings of the Zoological Society, as well as to those of its counterparts in France and Holland. He was somewhat taken aback by having to allow someone else to name the 200-plus new bird species he had discovered. He has at least 13 birds named after him in their common names, from Swinhoe's Bush Warbler *Cettia acanthizoides* to Swinhoe's Snipe *Gallinago megala*. The deer and serow are found on Taiwan. The squirrel is found in southern China (Sichuan, Yunnan, and Tibet) and northern Burma. The jird (a race of Sundevall's Jird) comes from Afghanistan and northwest Pakistan.

Swinny

Swinny's Horseshoe Bat *Rhinolophus swinnyi* Gough, 1908

H. H. Swinny (1876–1958) was a collector in Africa. His main areas of interest were insects and spiders, but we can find little else about him. The bat is found from southern Zaire and Zanzibar south to Cape Province (South Africa).

Swynnerton

Swynnerton's Bush Squirrel *Paraxerus vexillarius* Kershaw, 1923 [Alt. Lushoto Mountain Squirrel]

Charles Francis Massy Swynnerton (1877–1938) was principally an entomologist. He was born in India and worked in Africa, becoming the first game warden in Tanganyika (now Tanzania) between 1919 and 1929. He then spent 10 years between 1929 and 1938 as head of tsetse research in East Africa. He published papers on many aspects of natural history, including "On the Birds of Gazaland, Southern Rhodesia" (1907). He was killed in an air-crash. He is also commemorated in the names of two birds, Swynnerton's Robin *Swynnertonia swynnertoni* and Swynnerton's Thrush *Turdus olivaceus swynnertoni*. The squirrel has a very restricted range in the Usambara Mountains, Tanzania. Some authorities include *Paraxerus byatti* in this species, which then extends the species' distribution into montane forests of eastern Tanzania.

Sykes

Sykes' Monkey *Cercopithecus albogularis* Sykes, 1831

Colonel William Henry Sykes (1790–1872) saw plenty of action in the 20 years after he joined the Bombay army, a part of the armed forces of the Honourable East India Company, at the age of 14. In 1824 he was appointed to a statistician's position, and later his statistical researches involved him in natural history. In 1832 he published *Catalogue of Birds of the Raptorial and Incessorial Orders Observed in the Dukkan*. He retired from the military in 1837 and became a director of the East India Company, Rector of Aberdeen University, and Member of Parliament for Aberdeen. He is remembered in the common names of at least three bird species, all of which he described himself: Sykes' Crested Lark *Galerida deva*, Sykes' Nightjar *Caprimulgus mahrattensis*, and Sykes' Warbler *Hippolais rama*. The typical form of Sykes' Monkey is found on Zanzibar. Other subspecies occur from southern Somalia to South Africa.

Taczanowski

Mountain Paca *Cuniculus taczanowskii* **Stolzmann,** 1865 [Syn. *Agouti taczanowskii*]

Taczanowski's Oldfield Mouse *Thomasomys taczanowskii* **Thomas,** 1882

Dr. Wladyslaw Taczanowski (1819–1890) was Curator for the Zoological Cabinet of the Royal University of Warsaw, later the Branicki Museum, which he transformed from a teaching institution into a scientific center. An outstanding zoologist and ornithologist, he took part in collecting expeditions with Kalinowski in South America in 1884, and with A. S. Waga in North Africa in 1866–1867. He wrote *Ornithologie du Peru,* which was published in French, in three volumes, between 1884 and 1886. This was the first handbook of the birds of a Neotropical country and is still considered an important benchmark for South American ornithology. There are other taxa named after him, including fish such as Taczanowski's Gudgeon *Ladislavia taczanowskii* and birds such as Taczanowski's Tinamou *Nothoprocta taczanowskii*. The paca is found in the Andes from northwest Venezuela through Colombia and Ecuador to Peru. The mouse occurs in northwest Peru.

Taddei

Taddei's Bat *Eptesicus taddeii* Miranda, Bernardi, and Passos, 2006

Dr. Valdir Antonio Taddei (1942–2004) was a Brazilian biologist and zoologist who published widely on bats. He was awarded his doctorate in 1973 by the State University of São Paulo, where he later became Professor of Zoology. He also acted as the Curator of the university museum's bat collection. He died from lung cancer. This bat is found in the Atlantic Forest of southeast Brazil.

Talazac

Talazac's Shrew-Tenrec *Microgale talazaci* **Major,** 1896

The Rev. Père Talazac (dates not found) was a Jesuit priest, a missionary at Tandrakazo, South Betsileo, in Madagascar. The shrew-tenrec is found in northern and eastern Madagascar.

Tancré

Zaisan Mole Vole *Ellobius tancrei* **Blasius,** 1884 [Alt. Eastern Mole-Vole]

Rudolph Tancré (1842–1934) employed a man called Ruckbeil to do his actual collecting in central Asia. He was interested in both entomology and ornithology. His important collection is in the Zoological Museum in Amsterdam, as part of the larger Sillem van Marle collection. He described a few taxa himself, sometimes with, or after, the German naturalist Alexander von Homeyer, such as the butterfly *Limenitis homeyeri*. The vole is found from northeast Turkmenistan, Uzbekistan, and Kazakhstan to northern China and Mongolia.

Tantalus

Tantalus Monkey *Chlorocebus tantalus* **Ogilby,** 1841

Tantalus was King of Sipylos and a son of Zeus. He was invited to eat and drink with the gods on Olympus, and repaid this great honor by stealing nectar and ambrosia to take back to his people. He also sacrificed his son, Pelops, to the gods, boiled his body, and served pieces of it as food to them. The gods were neither deceived nor forgiving; Tantalus was sentenced to eternal punishment in Hades. He was condemned to stand forever in a pool of water beneath a fruit tree. Whenever he tried to pick a fruit, the branch raised itself beyond his grasp, and whenever he bent to drink, the water receded. It is the proverbial punishment of temptation without satisfaction. The monkey was first named from a specimen held at the London Zoo, and the mythological name was presumably applied fancifully. It is found from Ghana east to Sudan and Uganda.

Tarabul

Tarabul's Gerbil *Gerbillus tarabuli* **Thomas,** 1902

Tarabul (or Tarabulus) is the Arabic name of the Libyan capital, Tripoli, so this rodent is not named after a person, and references to "Tarabul's Gerbil" arise from a misunderstanding. It would be better simply to call it the Tarabul Gerbil. First recorded in Libya, the species has also been found in southeast Niger, close to Lake Chad. It may be more widely distributed than is currently realized.

Tate

Sombre Bat *Eptesicus tatei* **Ellerman** and Morrison-Scott, 1951
Tate's Triok *Dactylopsila tatei* Laurie, 1952 [Alt. Fergusson Island Striped Possum]
Tate's Fat-tailed Mouse-Opossum *Thylamys tatei* **Handley,** 1956
Tate's Shrew-Rat *Tateomys rhinogradoides* **Musser,** 1969
Tate's Rice Rat *Hylaeamys tatei* Musser et al., 1998

Dr. George Henry Hamilton Tate (1894–1953) was an American zoologist and ecologist with a particular interest in marine mammals. He worked at the American Museum of Natural History from 1921 onward but was widely traveled, collecting, for example, in Ecuador between 1921 and 1924, Venezuela between 1925 and 1928, and Australia in 1952. He wrote many scientific papers published from the 1920s to the 1950s, and he published several larger works, including *Mammals of Eastern Asia,* published in 1947, and *Mammals of the Pacific World,* written with two co-authors. The bat is known only from the Darjeeling area of northeast India, and its taxonomic status is uncertain. The triok is endemic to Fergusson Island, off southeast Papua New Guinea. The mouse-opossum comes from the central coastal areas of Peru, the shrew-rat from Sulawesi, and the rice rat from the eastern Andean piedmont of Ecuador.

Tattersall

Tattersall's Sifaka *Propithecus tattersalli* Simons, 1988 [Alt. Golden-crowned Sifaka]

Professor Dr. Ian Tattersall (b. 1945) is a British anthropologist, zoologist, archeologist, geologist, and paleontologist. He was born in England but brought up in East Africa. He trained in archeology and anthropology at Cambridge, and in geology and vertebrate paleontology at Yale, where he was awarded his Ph.D. in 1971. He has undertaken fieldwork in Madagascar, Vietnam, Suriname, Yemen, and Mauritius. He is acknowledged as a leader in the analysis of the human fossil record and the study of the ecology and systematics of the lemurs of Madagascar. He is currently Curator in the Department of Anthropology of the American Museum of Natural History, New York City; Adjunct Professor, Department of Anthropology, Columbia University; and Adjunct Professor, Ph.D. Program in Anthropology, CUNY. Among his publications are *The Primates of Madagascar* (1982), *Extinct Humans* (with Jeffrey Schwartz, 2000), *Becoming Human: Evolution and Human Uniqueness* (1998), and *The Last Neanderthal: The Rise, Success and Mysterious Extinction of Our Closest Human Relatives* (1999). He has published many articles in the *Scientific American* and co-edited the *Encyclo-*

pedia of Human Evolution and Prehistory. The sifaka has a very restricted range in northern Madagascar.

Taylor, E. H.

Taylor's Flying Fox *Pteropus pumilus tablasi* Taylor, 1934

Professor Edward Harrison Taylor (1889–1978) was a zoologist whose major interest was herpetology. He went to the Philippines in 1912 as a teacher in a village school, returning briefly to Kansas University to finish his M.A. After going back to Manila he was appointed Chief of Fisheries. From 1923 he was head of the Zoology Department of the University of the Philippines. He returned to Kansas University in 1927 and became a Professor in 1934. His book *Recollections of an Herpetologist* was published in 1975. He described the bat as *Pteropus tablasi,* but it has now been downgraded to a subspecies of the Little Golden-mantled Flying Fox *Pteropus pumilus.* He is remembered in the names of more than 30 reptiles, including Taylor's Spiny Lizard *Sceloporus edwardtaylori* and Edward Taylor's Gecko *Cyrtodactylus edwardtaylori* (the latter named after him in 2005—a measure of his enduring reputation among herpetologists). He is also remembered in the names of other taxa such as a species of Mexican salamander, *Ambystoma taylori.* The flying fox is found on Tablas Island in the Philippines.

Taylor, W.

Northern Pygmy Mouse *Baiomys taylori* **Thomas,** 1887

William Taylor (dates not found) was a collector. We know very little about him, although in naming the mouse after him, Thomas wrote that the Natural History Museum was "indebted [to Taylor] for many rare Rodents." The mouse is found in the eastern half of Texas and southwest Oklahoma (USA), and from southeast Arizona to central Mexico.

Telfair

Lesser Hedgehog-Tenrec *Echinops telfairi* **Martin,** 1838 [Alt. Pygmy Hedgehog-Tenrec]

Dr. Charles Telfair (1778–1833) was an Ulsterman who was a physician, a naval surgeon, a botanist, a sugar planter, and probably also a rum smuggler. He lived on Mauritius but traveled widely in the Indian Ocean and further afield. He was in China in 1826 when he acquired some banana plants and sent them to a friend in England. From him they passed to the Duke of Devonshire, who successfully grew them in the glass houses at Chatsworth. This variety was formally named in 1836 after the Duke of Devonshire's family name, Cavendish. From Chatsworth, horticulturalists spread the Cavendish variety far and wide, always by vegetative propagation of the suckers. Today it is the most popular variety you can buy in British shops and accounts for about 99 percent of consumption. Telfair was able to obtain some bones of an extinct bird, the Rodrigues Solitaire, in 1831 and presented them to the Zoological Society of London. There is a charming garden in Mauritius founded by and now named after him, and to improve its attractiveness, he imported Nile crocodiles from Madagascar. He clearly spent some time in Madagascar, as Bennett in 1832 reported on "Characters of a New Genus of Lemuridae, Presented by Mr. Telfair." A lizard, the Round Island or Telfair's Skink *Leiolopisma telfairii,* and a genus of plants, *Telfairia,* are also named after him. The lesser hedgehog-tenrec comes from southwest Madagascar.

Telford

Telford's Shrew *Crocidura telfordi* **Hutterer,** 1986

Sam Rountree Telford Jr. (b. 1932) is an American zoologist. He graduated from the University of Florida and took his M.A. at the University of Virginia and his Ph.D. at the University of California in 1964, after which he briefly took up a post as Lecturer there. Between 1965 and 1967 he undertook a postdoctoral fellowship at the Medical Sciences Department of the University of Tokyo, then taking up a post as Vertebrate Ecologist at the Gorgas Memorial Laboratory in Panama until 1970. He went on to the Florida State University, where he was an Assistant Professor until 1973 and is still a "Courtesy Curator." From 1973 until 1980 he worked at the World Health Organization in Geneva as a medical zoologist, then at the Danish International Development Agency until 1985. He returned to the University of Florida and undertook various research posts, finally retiring in 1993. His main interests remain herpetology, veterinary medicine, and parasitology. He has published one book in two volumes, *The Ecology of a Symbiotic Community*, and over 200 articles such as study, published in 1986, dealing with saurian malaria in relation to the African flying lizard *Holaspis guentheri* (Lacertidae) from the Uluguru and the Udzungwe Mountains, Tanzania, from whence hails the shrew.

Temchuk

Temchuk's Bolo Mouse *Necromys temchuki* Massoia, 1980 [Alt. Temchuk's Akodont; Syn. *Bolomys temchuki*]

Eduardo Temchuk is an Argentine agronomist who worked at an experimental agricultural research center, INTA (Instituto Nacional de Technología Agropecuaria), in Misiones Province, Argentina. The type specimen of the bolo mouse was captured on the grounds of that research center. Massoia, who described the species, was also associated with INTA. The mouse is confined to northeast Argentina. A study published in 2008 concluded that *Necromys temchuki* was not a valid species but should instead be regarded as a junior synonym of *N. lasiurus*.

Temminck

Temminck's Red Colobus *Piliocolobus badius temminckii* **Kuhl,** 1820

Temminck's Golden Cat *Catopuma temminckii* Vigors and Horsfield, 1827 [Alt. Asian Golden Cat]

Temminck's Ground Pangolin *Manis temminckii* Smuts, 1832 [Alt. Cape Pangolin]

Temminck's Trident Bat *Aselliscus tricuspidatus* Temminck, 1835

Temminck's Tailless Fruit Bat *Megaerops ecaudatus* Temminck, 1837

Temminck's Mole *Mogera wogura* Temminck, 1842 [Alt. Japanese Mole]

Temminck's Flying Squirrel *Petinomys setosus* Temminck, 1844

Temminck's Mouse *Mus musculoides* Temminck, 1853 [Alt. Subsaharan Pygmy Mouse]

Temminck's Giant Forest Squirrel *Epixerus ebii* Temminck, 1853 [Alt. Ebian Squirrel]

Temminck's Spotted Squirrel *Heliosciurus punctatus* Temminck, 1853 [Alt. Small Sun Squirrel]

Temminck's Striped Mouse *Hybomys trivirgatus* Temminck, 1853

Dwarf Dog-faced Bat *Molossops temminckii* **Burmeister,** 1854

Temminck's Flying Fox *Pteropus temmincki* **Peters** 1867

Coenraad Jacob Temminck (1778–1858) was a Dutch zoologist, illustrator, and collector. He became the first Director of the Rijksmuseum van Natuurlijke Historie in Leiden in 1820 and held that post until his death. He was a wealthy man who had a very large collection of specimens, including live birds. His first task as an ornithologist was to catalogue his father's very extensive collection. (His father was Jacob Temminck, for whom Le Vaillant collected specimens.) In 1815 he issued his *Manuel d'ornithologie, ou Tableau systématique des oiseaux qui se trouvent en Europe,* and he wrote *Nouveau recueil de planches coloriées d'oiseaux* in 1820. He is also commemorated in the common names of no fewer than 18 birds, such as Temminck's Stint *Calidris temminckii* and Temminck's Tragopan *Tragopan temminckii.* The colobus is found in Senegal, Gambia, and western Guinea. The golden cat occurs from Nepal and Tibet east to Vietnam and south to Sumatra. The pangolin's distribution covers eastern and southern Africa, except for the far south. The trident bat is found from the Moluccas (Indonesia) east through New Guinea to the Solomon Islands and Vanuatu. The tailless fruit bat inhabits Thailand, peninsular Malaysia, Sumatra, and Borneo. The flying squirrel has a similar distribution. The mole comes from Japan. Temminck's Mouse occurs widely in sub-Saharan Africa. The forest squirrel, spotted squirrel, and striped mouse are all West African endemics. The dog-faced bat lives in tropical South America, as far south as Paraguay and Uruguay. The flying fox comes from the Moluccan islands of Ambon, Buru, and Seram.

Teusz

Teusz's Dolphin *Sousa teuszii* Kükenthal, 1892 [Alt. Atlantic Hump-backed Dolphin]

Edward Teusz (dates not found) was a German naturalist who found the first known specimen of the dolphin, partially eaten by sharks and floating in the sea in the Bay of Warships, the Cameroon Estuary. He donated some of his specimens to the Jena Natural History Museum. The dolphin is found in coastal waters from southern Morocco to Gabon.

Thaeler

Thaeler's Pocket Gopher *Orthogeomys thaeleri* Alberico, 1990

Professor Dr. Charles S. Thaeler Jr. (1932–2004) was a biologist who spent most of his career at the New Mexico State University in Las Cruces. In 1962 he was in California, attached to the Museum of Vertebrate Zoology at Berkeley, where he was awarded his doctorate in 1964. He divided his time between teaching and field research in mammalogy. He was a member of the American Society of Mammalogists. In 1989 he wrote, jointly with E. P. Lessa, "A Reassessment of Morphological Specializations for Digging in Pocket Gophers." He also wrote on spiny rats. The gopher is found only in northwest Colombia.

Thalia

Thalia's Shrew *Crocidura thalia* Dippenaar, 1980

In Greek mythology Thalia was one of the three Graces and was regarded as the goddess of festivity. Another Thalia (or Thaleia) was one of the Muses and was sometimes called the goddess of comedy. The original description of the shrew does not specify which of these Thalias is intended or give any reason why this name was chosen. The shrew is found in the Ethiopian highlands.

Theobald

Theobald's Tomb Bat *Taphozous theobaldi* **Dobson,** 1872

William Theobald (1829–1908) was Deputy Superintendent of the Geological Survey of India. He was also a naturalist with an interest in herpetology, publishing *Catalogue of Reptiles of British Burma* in 1868 and *Descriptive Catalogue of the Reptiles of British India* in 1876. Theobald was interested in malacology as well; with S. Hanley, he produced *Conchologia Indica* (1876). He also commented, with a degree of skepticism, on reports of "wolf-children" (human children supposedly raised by wolves). The tomb bat is found from central India to Vietnam, and also on Java, Borneo, and Sulawesi.

Theresa

Theresa's Short-tailed Opossum *Monodelphis theresa* **Thomas,** 1921

The type specimen of this opossum was collected near Rio de Janeiro at a mountain town called Theresopolis. Combined with the fact that the binomial *theresa* is not in the genitive, which would be *theresae,* we are confident that *theresa* derives from the town and not directly from a person. The species comes from the Atlantic Rainforest of southeast Brazil. Reported occurrence in the Andes of Peru probably refers to a different taxon.

Therese

Therese's Shrew *Crocidura theresae* **Heim de Balsac,** 1968

Madame Marie-Thérèse Guilbot (dates not found) had this shrew named after her by Heim de Balsac as thanks for her services over a long period. We could not find out any more about her. The species is found in West African savanna regions, from Guinea to Ghana.

Thersites

Railer Bat *Mops thersites* **Thomas,** 1903

Thersites is the only common soldier mentioned in the *Iliad* for whom Homer gives any description. He also gives him a speaking part, in which Thersites tries to persuade the Greek army to return home without taking Troy. For this treasonable speech and other episodes, Achilles killed him. Thersites was reputed to be lame in one foot, incredibly ugly, and prone to "railing" against anyone and anything that did not please him. He appears as a character in Aeschylus' play *Seven against Thebes* and is mentioned in Shakespeare's *Troilus and Cressida.* The bat occurs from Sierra Leone east to Uganda.

Thetis

Red-necked Pademelon *Thylogale thetis* **Lesson,** 1828

The pademelon is not named after a person but a ship, which in turn was named after a character from Greek mythology. Thetis was a goddess of the sea and mother of the hero Achilles. The name became attached to this wallaby because the first known specimen was brought to Europe by the frigate *La Thetis* with Lesson (q.v.) on board. So the pademelon is indirectly named after a goddess. It is found in eastern Australia.

Thierry

Thierry's Genet *Genetta thierryi* **Matschie,** 1902 [Alt. Haussa Genet]

Gaston Thierry (1866–1904) was, despite his French names, an Oberleutnant (First Lieutenant) in the Imperial German Army. He was one of those sent to Togoland in 1896 to establish a series of bases, after the French and German governments had reached agreement on the border between German Togoland (now partly in Ghana and partly the state of Togo) and French Dahomey (now Benin), to enforce German control over the country. The government in Berlin expected its officers to collect examples of the local fauna, and Thierry appears as one of those who made such a collection. He left Togo in 1899 and was killed in 1904 in Cameroon (then a German colony) by a poisoned arrow. The genet is found in the savannah zone of West Africa, from Senegal to northern Cameroon.

Thollon

Thollon's Red Colobus *Piliocolobus tholloni* **Milne-Edwards,** 1886

François-Romain Thollon (1855–1896) was a French collector in the Congo (now Zaire) and a member of the de Brazza mission, in Gabon, in 1884, which was noted for its fish collection. He is also celebrated in the scientific names of a wide range of animals, including two fish, *Tilapia tholloni* and *Alestes tholloni;* the elephant tick *Amblyomma tholloni* (which carries the cattle disease heartwater); and a bird, the Congo Moor-Chat *Myrmecocichla tholloni*. The colobus is found in DRC (Zaire), south of the Congo River.

Thomas, M.

Arata and Thomas' Yellow-shouldered Bat *Sturnira aratathomasi* Peterson and Tamsitt, 1968 [Alt. Aratathomas's Yellow-shouldered Bat, Giant Andean Fruit Bat]

The bat is named after Andrew Arata (q.v.) and Dr. Maurice Thomas, who collected the holotype. Dr. Thomas is a Professor of Biology at Palm Beach Atlantic University. He took his master's degree at Cornell University and his doctorate at Tulane. At the end of 2002 he was in Panama to study bat populations as part of a sabbatical from teaching. The bat is known from northwest Venezuela, Colombia, Ecuador, and Peru.

Thomas, M. R. O.

The Neotropical mouse genus *Thomasomys* **Coues,** 1884 [approx. 35 species]

Thomas' Oldfield Mouse *Thomasomys pyrrhonotus* Thomas, 1886

Thomas' Rope Squirrel *Funisciurus anerythrus* Thomas, 1890

Thomas' Langur *Presbytis thomasi* **Collett,** 1893 [Alt. Northern Sumatran Leaf Monkey]

Thomas' Shrew-Tenrec *Microgale thomasi* **Major,** 1896

Giant Atlantic Tree Rat *Phyllomys thomasi* **Ihering,** 1897

Thomas' Rock Rat *Aethomys thomasi* **de Winton,** 1897

Thomas' Small-eared Shrew *Cryptotis thomasi* **Merriam,** 1897

Thomas' Giant Deer Mouse *Megadontomys thomasi* Merriam, 1898

Thomas' Broad-nosed Bat *Platyrrhinus dorsalis* Thomas, 1900

Thomas' Flying Squirrel *Aeromys thomasi* **Hose,** 1900

Thomas' Fruit-eating Bat *Artibeus watsoni* Thomas, 1901 [Alt. Watson's Fruit-eating Bat]

Thomas' Mastiff Bat *Eumops maurus* Thomas, 1901 [Alt. Guianan Bonneted Bat]

Thomas' Pine Vole *Microtus thomasi* Barrett-Hamilton, 1903

Thomas' Yellow Bat *Rhogeessa io* Thomas, 1903

Thomas' Nectar Bat *Lonchophylla thomasi* **J. A. Allen,** 1904

Thomas' Sac-winged Bat *Balantiopteryx io* Thomas, 1904

Thomas' Horseshoe Bat *Rhinolophus thomasi* **K. Andersen,** 1905

Thomas' Rice Rat *Oryzomys dimidiatus* Thomas, 1905

Thomas' Galago *Galago thomasi* Elliot, 1907

Thomas' Moustached Tamarin *Saguinus labiatus thomasi* **Goeldi,** 1907

Thomas' Pygmy Mouse *Mus sorella* Thomas, 1909

Thomas' Spiny Rat *Trinomys iheringi* Thomas, 1911 [Alt. Ihering's Spiny Rat]

Thomas' Melomys *Paramelomys mollis* Thomas, 1913 [Alt. Thomas' Mosaic-tailed Rat]

Thomas' Pipistrelle *Pipistrellus paterculus* Thomas, 1915 [Alt. Mount Popa Pipistrelle]

Thomas' Night Monkey *Aotus miconax* Thomas, 1927 [Alt. Peruvian Night Monkey, Andean Owl Monkey]

Thomas' Pygmy Jerboa *Salpingotus thomasi* **Vinogradov,** 1928

Thomas' Water Mouse *Rheomys thomasi* **Dickey,** 1928

Thomas' Pika *Ochotona thomasi*
Argyropulo, 1948

Michael Rogers Oldfield Thomas (1858–1929) has been called the greatest expert on mammalian taxonomy of his time. He certainly devoted his life to the study of mammals and by the time of his death had described no fewer than 2,900 genera, species, and races of mammals. He spent part of his childhood in South Africa, where his father was Archdeacon of Cape Town, and he developed his love of natural history while collecting insects on Table Mountain. On his return to England he studied at Haileybury College, but in 1876, showing little interest in further education, he secured a clerkship in the Museum Secretary's office at the British Museum. In 1878 he transferred to the Zoological Department, serving a short time as a clerk. Later he was appointed Assistant Curator in Charge of Mammals, in which post he served until his retirement in 1923. He worked with many of the leading zoologists of that period and was a frequent correspondent with many others such as Wilhelm Peters and Angel Cabrera. He wrote *Catalogue of Marsupialia and Monotremata* in 1888. In 1891 Thomas married an heiress to a small fortune, which gave him the finances to hire mammal collectors to provide specimens for the museum. (Some people have posited that he was involved in the "Piltdown Man" affair and may have contributed the lower jaw to the skull hoax.) He committed suicide in 1929 while at the height of his career, a few months after his wife had died, as he could not face life without her. His outstanding services to mammalogy are reflected in the large number of species named after him. The speciose genus *Thomasomys* includes a form that has had Thomas's name attached to it in the vernacular: Thomas' Oldfield Mouse occurs in southern Ecuador and northern Peru. The rope squirrel is found in tropical Africa, from Benin east to Uganda. The langur is restricted to northern Sumatra, and the shrew-tenrec to eastern Madagascar. The tree rat is endemic to Isla de São Sebastião, off the coast of São Paulo, Brazil. The rock rat comes from Angola, and the shrew from Colombia, Ecuador, and northern Peru. The giant deer mouse is found in southwest Mexico. The broad-nosed bat occurs from Panama to Bolivia. The flying squirrel is a Borneo endemic. The fruit-eating bat comes from Central America, and the mastiff bat from eastern Venezuela, Guyana, Suriname, and adjacent parts of northeast Brazil. The pine vole occurs in the southern Balkans (Herzegovina to Greece). The range of the yellow bat extends from Nicaragua to central Brazil. The nectar bat occurs widely in tropical South America. The sac-winged bat is confined to southern Mexico, Belize, and Guatemala. The horseshoe bat is found from Myanmar east to Vietnam. The rice rat is a little-known species from southern Nicaragua. Thomas' Galago is known with certainty only from Uganda and northeast DRC (Zaire); similar forms have been reported as far west as Cameroon, but it is not yet known whether these represent undescribed species or subspecies. The tamarin is found in western Amazonian Brazil. The pygmy mouse occurs in eastern DRC (Zaire), Uganda, Kenya, and northern Tanzania; it has also been recorded in Angola and eastern Cameroon, so the full extent of its range is uncertain. The spiny rat is confined to eastern Brazil. The melomys inhabits the montane forests of New Guinea. The pipistrelle has been recorded in northern India, Myanmar, Yunnan (China), Thailand, and Vietnam. The night monkey has a very restricted distribution in northern Peru. The jerboa is known only from the type specimen, said to come from Afghanistan. The water mouse is found in southern Mexico, Guatemala, and El Salvador. The pika comes from central China (Sichuan, Gansu, and eastern Qinghai).

Thomas, R.

Thomas' Yellow-shouldered Bat *Sturnira thomasi* **de la Torre** and Schwartz, 1966

Professor Dr. Richard Thomas is a zoologist and is Professor at the Department of Biology, University of Puerto Rico. He obtained his doctorate in 1976 from the University of Louisiana. His main interest is herpetology, and a number of reptiles, including Thomas' Blind Snake *Leptotyphlops pyrites* and Thomas' Galliwasp *Celestus marcotus*, are named after him. Thomas collected the type specimen of the bat on Guadeloupe in January 1963. De la Torre and Schwartz name him as one of the persons thanked for "the valuable and interested assistance rendered . . . in the course of these studies on the fauna of the West Indies." The species is found on the islands of Guadeloupe and Montserrat.

Thompson

Thompson's Pygmy Shrew *Sorex thompsoni* **Baird,** 1858 [Alt. Northeastern Pygmy Shrew; Syn. *S. hoyi thompsoni*]

The Rev. Zadock Thompson (1796–1856) was an American naturalist, amateur botanist, and geologist. He wrote an almanac on the weather and geography of Vermont and used the money he raised to pay for an education at the University of Vermont. He graduated in 1823, later being appointed as Professor there and teaching until his death. He published his most important work, *The History of the State of Vermont, Natural, Civil and Statistical,* in 1842. In 1849 some laborers digging near Lake Champlain uncovered the fossilized bones of what turned out to be a whale. Thompson reassembled the skeleton and thought he had discovered a new species, which he named *Delphinus vermontanus.* This was the only species he named in his whole career, but it is now regarded as being the same species as the Beluga Whale *Delphinapterus leucas.* In 1853 he was appointed as State Naturalist and Geologist for Vermont. He was a prolific author on the natural history of the state. Baird mentions his "gazetteer" in correspondence with Audubon. The shrew is found from Nova Scotia south into New England, and along the Appalachians to northern Georgia. It is sometimes treated as a full species but often as a race of *Sorex hoyi.*

Thomson

Thomson's Gazelle *Gazella thomsonii* **Günther,** 1884

Joseph Thomson (1858–1895) has been described as "one of the most colourful and prudent of 19th century African explorers." In 1879, at only 20 years of age, after studying at Edinburgh University, he was placed as second-in-charge on his very first exploration for the Royal Geographic Society. The leader, the 34-year-old Scottish cartographer Keith Johnston, succumbed to dysentery very early in the expedition. Thomson buried him under a tree on which he carved Johnston's initials and the date of his death (28 June 1879), just outside the village of Behobeho in southern Tanzania. He said that because of his inexperience he asked himself, "Should we simply turn back?" and answered, "I feel I must go forward, whatever might be my destiny." He carried on and successfully led the expedition to Lake Nyasa (now Lake Malawi) and Lake Tanganyika. Thereafter, he made many more safe and successful explorations on behalf of the Royal Geographic Society. He explored routes through Kenya and Tanzania from 1878 until 1884, Nigeria in 1885, and Morocco in 1888. He was the first to explore Masai land successfully, simply because he was not confrontational. He said, "In my opinion the travellers' strength would lie more in his manner toward and treatment of the natives than in his guns and revolvers."

Despite his relaxed style of leadership, he always put the safety of the party first, and his motto was "He who goes slowly, goes safely; he who goes safely, goes far." In 1885 he wrote *Through Masai Land: A Journey of Exploration among the Snowclad Volcanic Mountains and Strange Tribes of Eastern Equatorial Africa, Being the Narrative of the Royal Geographical Society's Expedition to Mount Kenya and Lake Victoria Nyanza, 1883–1884.* The diseases he caught in tropical Africa led to his early death from pneumonia. He is buried at Morton Cemetery in Thornhill, Dumfries and Galloway, Scotland. Interestingly for Victorian times, he openly lived with the Scottish poet Anderson and another man, described as a watercolorist, as "intimate friends." He was first to find the eponymous gazelle, which is found in southeast Sudan, Kenya, and northern Tanzania.

Thonglongya

Kitti's Hog-nosed Bat *Craseonycteris thonglongyai* **Hill,** 1974 [Alt. Bumblebee Bat, Khun Kitti Bat]

Kitti Thonglongya first discovered this tiny species of bat in 1973. It occurs in Thailand and Myanmar. See **Kitti** for biographical details.

Thornicroft

Thornicroft's Giraffe *Giraffa camelopardalis thornicrofti* **Lydekker,** 1903

Harry Scott Thornicroft (b. 1868) was District Commissioner in Northern Rhodesia (now Zambia) for 17 years, having been sent there by the Africa Lakes Corporation in 1898. He married a local African woman and had 11 children. He died when a tractor fell on him. In 1902 he sent the skin of a giraffe, which he had shot, to the British Museum to confirm that this was indeed a unique race. This subspecies of giraffe occurs in eastern Zambia.

Thorold

Thorold's Deer *Cervus albirostris* **Przewalski,** 1883 [Alt. White-lipped Deer; Syn. *Przewalskium albirostris*]

Dr. W. G. Thorold (dates not found) was a Surgeon-Captain in the Indian Medical Service of the British army and an explorer, collector, and naturalist. From 1891 to 1892 he accompanied the intelligence officer Major General Sir Hamilton Bower on the latter's trip from Ladakh across the Tibetan Plateau (where the deer is found) to China—the first crossing by Europeans. He collected about 115 species belonging to 69 genera in Tibet in 1891, including two specimens of this deer, although it turned out that Przewalski had discovered it previously, in 1879. Blanford named it *Cervus thoroldi* in 1893. Thorold certainly continued to collect and corresponded with the British Museum (Natural History) in 1894.

Thyone

Northern Little Yellow-eared Bat *Vampyressa thyone* **Thomas,** 1909

Thyone is a character from Greek mythology. She was originally a mortal, a Princess of Thebes called Semele. Zeus seduced her and by him she became the mother of the wine-god Dionysus. It was Dionysus who later fetched Semele up from the underworld to join the gods of Olympus, under the new name of Thyone. She was also regarded as the goddess of the inspired frenzy of the Bacchic orgy, especially the religious ecstasy experienced by female disciples of that god. Thomas was quite fond of using names from mythology when he described mammal species; often there is no obvious reason for his choices. The bat is found from southern Mexico to western Brazil and northern Bolivia.

Tickell

Tickell's Bat *Hesperoptenus tickelli* **Blyth,** 1851 [Alt. Tickell's False Serotine]

Colonel Samuel Richard Tickell (1811–1875) was a British army officer, artist, and ornithologist in India and Burma (now Myanmar). He has been described by Sir Norman Kinnear as "one of the best field naturalists India has known." He made important early contributions to Indian ornithology and mammalogy through field observations and the collection of specimens while he was stationed in several localities in the 1830s and 1840s. He planned to publish a book on the birds and mammals of India but never did. However, his manuscript notes and illustrations are preserved in the library of the Zoological Society of London. These notes contain many references to observations of birds in Bihar, Orissa, Darjeeling, and Tenasserim. He also published studies on the structure and vocabulary of the Ho language. He is commemorated in the common names of at least eight birds, including Tickell's Thrush *Turdus unicolor* and Tickell's Laughingthrush *Garrulax strepitans.* The bat is found in India, Sri Lanka, Nepal, Bhutan, Myanmar, Thailand, Laos, and Cambodia.

Tilda

Tilda's Yellow-shouldered Bat *Sturnira tildae* **de la Torre,** 1959

Tilda Brandt. The original etymology reads, "It is a pleasure to name this species for Tilda Brandt, of Essen, Germany, in appreciation of her valuable help in translating the critical German literature pertaining to the genus *Sturnira.*" The bat was first discovered on Trinidad but has since been found to occur quite widely in tropical South America.

Tim Ealey

Pilbara Ningaui *Ningaui timealeyi* **Archer,** 1975

Dr. Eric Herbert Mitchell "Tim" Ealey (b. 1927) is a retired Australian environmental scientist and ecologist who has researched marine and coastal environments as well as a wide-ranging number of subjects from kangaroo conservation to logging. He was Director of the Monash University Graduate School of Environmental Science and in retirement devotes his time to reviving mangrove colonies. This small carnivorous marsupial is found in northwest Western Australia.

Timmins

Annamite Striped Rabbit *Nesolagus timminsi* Averianov, Abramov, and Tikhinov, 2000

Dr. Robert "Rob" J. Timmins is a British biologist and keen birdwatcher. For some years in the late 1990s he worked at New York's Bronx Zoo on the Wildlife Conservation Society's Lao Program. He spent most of his time helping to establish the conservation priorities for a national park in Laos. He has published many articles, alone and with others, such as "Status and Conservation of the Giant Muntjac *Megamuntiacus vuquangensis,* and Notes on Other Muntjac Species in Laos," published in *Oryx.* Writing in 1997, he noted that "a distinctive forest rabbit found in the area (probably *Nesolagus* sp.) awaits formal description," referring to three rabbits he found dead and sent for description. The rabbit is found in the Annamite Mountains of Laos and Vietnam.

Tippelskirch

Masai Giraffe *Giraffa camelopardalis tippelskirchi* **Matschie,** 1898

Lieutenant General Ernst Ludwig von Tippelskirch (1774–1840) was a German explorer as well as a professional soldier. He fought in one of Marshal Blücher's Prussian Brigades at both Ligny and at Waterloo and was awarded the highest Prussian military decoration, Pour le Mérite, in 1815. From 1827 to 1840 he was Commandant and Military Governor of Berlin. This race of giraffe is found in Kenya and Tanzania.

Titania

Titania's Woolly Bat *Kerivoula titania* Bates et al., 2007

Titania is the Queen of the Fairies in Shakespeare's *A Midsummer Night's Dream.* In traditional folklore the Queen of the Fairies has no name, and Shakespeare took the name from Ovid's *Metamorphoses,* where it is used as a term for the daughters of the Titans. According to the bat's describers, "The name is chosen to reflect the nymph-like nature of this forest bat." The species has been recorded from Myanmar, Thailand, Laos, Cambodia, and Vietnam.

Tokuda

Guam Flying Fox *Pteropus tokudae* **Tate,** 1934 extinct

Tokuda's Mole *Mogera tokudae* Kuroda, 1940

The Ryukyu spiny rat genus *Tokudaia* Kuroda, 1943 [3 species: *muenninki, osimensis,* and *tokunoshimensis*]

Dr. Mitoshi Tokuda (dates not found) wrote a number of scientific papers in the 1930s and 1940s, including "A Revised Monograph of the Japanese and Manchou-Korean Muridae" (1941), and also translated Darwin's *Origin of Species* into Japanese. Furthermore, he published a book in 1944, *Animals of East Asia, Mammals in the Southern Area.* The flying fox was endemic to Guam (Marianas Islands), but no specimens have been found since 1968. The mole is found only on Sado Island, off Honshu, Japan. The *Tokudaia* genus of spiny rats is confined to islands in the Ryukyu chain, south of the main islands of Japan.

Tomes

Tomes' Rice Rat *Nephelomys albigularis* Tomes, 1860

Tomes' Spiny Rat *Proechimys semispinosus* Tomes, 1860

Tomes' Long-eared Bat *Lonchorhina aurita* Tomes, 1863 [Alt. Tomes' Sword-nosed Bat]

Large False Serotine *Hesperoptenus tomesi* **Thomas,** 1905

Robert Fisher Tomes (1823–1904) was a British farmer and a zoologist with a special interest in bats. He wrote *Panama in 1855: An Account of the Panama Rail-road, of the Cities of Panama and Aspinwall, with Sketches of Life and Character on the Isthmus,* and scientific papers such as "Report of a Collection of Mammals Made by Osbert Salven, Esq., at Dueñas, Guatemala" (1861). The rice rat is found from the Panama-Colombia border south to northern Peru. The spiny rat's distribution extends from southern Honduras to Ecuador. The long-eared bat occurs from southern Mexico to northern Bolivia and eastern Brazil. The false serotine is known from the Malay Peninsula and Borneo.

Torre

Torre's Cave Rat *Boromys torrei* **Allen,** 1917 extinct

Professor Carlos de la Torre y la Huerta (1858–1950) of Havana University was regarded as the foremost Cuban naturalist of his generation. He was closely associated with the Smithsonian Institution in Washington, DC, and was a leading figure in the Havana Academy of Medical, Physical, and Natural Sciences. He is also commemorated in the name of a bird genus, *Torreornis,* which includes *T. inexpectata,* the Zapata Sparrow. The cave rat was formerly found on Cuba and the Isle of Pines. No precise date of extinction is known, but some remains of *Boromys* are so fresh that the genus may have lingered into the 19th century.

Torres

Torres' Crimson-nosed Rat *Bibimys torresi* Massoia, 1979

Norberto Torres and his family were honored by Elio Massoia when he named this rodent, as they helped him to obtain the first known specimens. Massoia mentions Adrian, Eduardo, and Marcos Torres by name, but it is not clear what their relationship was to Norberto (were they sons?) or each other (were they brothers?). The family lived close to the Agricultural Experimental Station of the Paraná Delta, which is run by the National Institute of Farming Technology, for which Massoia worked. The rat is found in the region of the Paraná Delta, Argentina.

Townsend, C.

Guadelupe Fur Seal *Arctocephalus townsendi* **Merriam,** 1897

Charles Haskins Townsend (1859–1944) was an American zoologist. From 1883 to 1902 he spent much of his time aboard the *Albatross,* working for the U.S. Fish Commission. During 1883–1884 he explored northern California and, in 1885, the Kobuk River in Alaska. In 1896 he was an expert before the Russo-American fisheries arbitration at The Hague. In 1912–1913 he was President of the American Fisheries Society. He was Director of the New York Aquarium from 1902 until he retired in 1937. He published a number of scientific papers, including "Field Notes on the Mammals, Birds, and Reptiles of Northern California" and "The Distribution of Certain Whales as Shown by Logbook Records of American Whale Ships." He also published a book called *The Public Aquarium,* a sort of how-to book still in print today. He is commemorated in the name of Townsend's Shearwater. This fur seal was hunted to the very edge of extinction by the early 1890s, but numbers have gradually recovered. It breeds on Guadelupe Island, off Baja California, Mexico.

Townsend, J.

Townsend's Big-eared Bat *Corynorhinus townsendii* **Cooper,** 1837
Townsend's Chipmunk *Tamias townsendii* **Bachman,** 1839
Townsend's Ground Squirrel *Spermophilus townsendii* Bachman, 1839
Townsend's Hare *Lepus townsendii* Bachman, 1839 [Alt. White-tailed Jackrabbit, Prairie Hare]
Townsend's (Shrew) Mole *Scapanus townsendii* Bachman, 1839
Townsend's Pocket Gopher *Thomomys townsendii* Bachman, 1839
Townsend's Vole *Microtus townsendii* Bachman, 1839

John Kirk Townsend (1809–1851) was an American naturalist, ornithologist, writer, and collector. He sent many of his specimens to Audubon, who gave him free reign in naming anything he found. He wrote *Narrative of a Journey across the Rocky Mountains,* which tells the story of Wyeth's expedition across the Rockies to the Pacific Ocean between 1834 and 1835. Ironically, Townsend died of arsenic poisoning, arsenic being the "secret" ingredient of the powder he had formulated to use in his taxidermy. He is also commemorated in the names of such birds as Townsend's Warbler *Dendroica townsendi.* The bat is found from southern British Columbia (Canada) through the western USA to southern Mexico. The chipmunk occurs in southwest British Columbia and the northwest USA, both the mole and the vole having similar distributions. The ground squirrel is endemic to southern Washington State (northwest USA). The jackrabbit is found in southern Canada and the western and central USA, as far south as eastern California and Colorado. The gopher is found in the Snake River Valley of Idaho, south and west to southeast Oregon, northeast California, and northern Nevada.

Tranier

Tranier's Gerbil *Taterillus tranieri* Dobigny et al., 2003

Professor Dr. Michel Tranier is a zoologist and the Director of the Natural History Museum in Paris. He has been working for many years on the systematics and cytotaxonomy of rodents. The gerbil is known from western Mali and southeast Mauritania.

Travers

Spade-toothed Beaked Whale *Mesoplodon traversii* **Gray,** 1874 [Alt. Bahamonde's Beaked Whale; Syn. *M. bahamondi*]

Henry H. Travers (1844–1928) was a New Zealand naturalist who collected widely in his native land. He was mostly known as a botanist and made a special study of the Chatham Islands from 1863 onward. In 1872 he wrote a pamphlet with his father on the *Birds of the Chatham Islands*. His father was an explorer and has a number of geographical features, including a mountain, named after him. He is remembered in the name of what was probably the only truly flightless passerine bird, the extinct Travers' Wren *Traversia lyalli*. Travers is also commemorated in the names of several plants, such as*Archeria traversii*, *Hebe traversii*, and *Pimelia traversii*. The Spade-toothed Beaked Whale was overlooked for many decades after its original description, being considered conspecific with *Mesoplodon layardii*. In 1995 a "new" species of beaked whale was described as *Mesoplodon bahamondi*, but this was then shown to be a junior synonym of *M. traversii*—which was "resurrected" as a valid taxon in 2002 after genetic research. Specimens of this whale have been found in New Zealand and on the Juan Fernández Islands, west of Chile.

Trevelyan

Giant Golden-Mole *Chrysospalax trevelyani* **Günther,** 1875

Herbert Trevelyan (dates not found) obtained the type specimen of this golden-mole from a native South African and presented it to the British Museum. The species is found in Eastern Cape Province, South Africa.

Trevor

Trevor's Free-tailed Bat *Mops trevori* **J. A. Allen,** 1917

John Bond Trevor (1878–1956) was a trustee of the American Museum of Natural History and Chairman of the Committee on African Exploration. He was a Harvard-educated lawyer and industrialist descended from one of the signatories of the Declaration of Independence. He served as an officer in military intelligence just after WW1, a role in which he "made his own rules, gave himself his own assignments," according to a colleague at the time. For example, Trevor developed a plan to suppress a mass uprising of Jewish "subversives" in New York City, going so far as to order 6,000 rifles and a machine gun battalion for deployment in Jewish neighborhoods in anticipation of a disturbance that never took place. Keeping secret ties to military intelligence even after his return to civilian life, Trevor became, according to one historian, "one of the most influential unelected individuals affiliated with the U.S. Congress." According to others, he was a racist. Certainly he devised the national-origins quota system for immigration that was enacted in 1924, and he was a prominent Nazi sympathizer in the 1930s. This African bat has been recorded in various locations from Ghana east to Uganda.

Tristram

Tristram's Jird *Meriones tristrami* **Thomas,** 1892

The Rev. Henry Baker Tristram F.R.S. (1822–1906) was canon of Durham Cathedral and a traveler, archeologist, naturalist, and antiquarian. He was so fond of collecting bird specimens by shooting them that he became known as "the Great Gun of Durham." Despite being a churchman, he was an early supporter of Darwin. He wrote a number of accounts of his explorations, including *A Journal of Travels in Palestine and the Great Sahara: Wanderings South of the Atlas Mountains,* published in 1860. His writings provided interesting details on the indigenous peoples and their customs as well as on the natural history of the region, including its birds, reptiles, molluscs, and plants as well as mammals. He originally went to Palestine and North Africa because of ill health. Salvin, who named a species of storm-petrel in his honor, was Tristram's cousin by marriage. Despite an early penchant for collecting with a gun, he went on to be a Vice President of the Royal Society for the Protection of Birds from 1904 until his death. He is commemorated in the names of several birds, including Tristram's Starling *Onychognathus tristramii.* The jird is found from eastern Turkey and Armenia to northwest Iran.

Triton

Greater Long-tailed Hamster *Tscherskia triton* **de Winton,** 1899
Grey-bellied Pygmy Mouse *Mus triton* **Thomas,** 1909

Triton was a sea-god in Greek mythology. Often depicted as a merman, he had a conch-shell "trumpet" that he blew to both calm and stir up the waves. What brought him to mind when two great zoologists were naming small rodents is difficult to conjecture. The hamster is found in northeast China, Korea, and the Ussuri region of the Russian Far East. The mouse occurs in northern and eastern DRC (Zaire), Uganda, Kenya, Tanzania, Malawi, Zambia, and Angola.

Trouessart

Trouessart's Trident Bat *Triaenops furculus* Trouessart, 1906

Édouard Louis Trouessart (1842–1927) was a French naturalist who was associated with both the Museum of Natural History of Angers and the Natural History Museum in Paris. From the 1870s to the 1900s he described a number of species endemic to Madagascar, as evidenced by his paper of 1906, "Description de mammifères nouveaux d'Afrique et de Madagascar." In 1921 he was elected an honorary member of the American Society of Mammalogists. He published much on halacarids (aquatic mites) and other Acarina and is commemorated in some of their scientific names, such as *Scaptognathus trouessarti* and *Copidognathus trouessarti.* The bat is found on Madagascar and Aldabra Island.

Troughton

Troughton's Pouched Bat *Saccolaimus mixtus* Troughton, 1925
Troughton's Tomb Bat *Taphozous troughtoni* **Tate,** 1952
Troughton's Forest Bat *Vespadelus troughtoni* Kitchener, Jones, and Caputi, 1987 [Alt. Eastern Cave Bat]

Ellis Le Geyt Troughton (1893–1974) was an Australian zoologist. He was Curator of Mammals at the Australian Museum, Sydney, from 1921 to 1958. During WW2 he was involved in field investigations into scrub typhus in New Guinea. His best-known work is *Furred Animals of Australia,* published in 1941. He is also commemorated in other taxa such as the scientific name of the Australian Sea-Star *Nepanthia troughtoni.* The pouched bat is found in Papua

New Guinea and in northeast Queensland. The tomb bat is a poorly known species from Queensland. The forest bat comes from eastern Australia.

Trowbridge

Trowbridge's Shrew *Sorex trowbridgii* **Baird,** 1857

Professor William P. Trowbridge (1828–1892) was an American army officer who contributed a significant collection of fishes to the Smithsonian in Washington, DC. Baird, who described the shrew, acknowledged the importance of this collection, which had added about 50 new fish species to the North American fauna. His first appointment in the army was with the U.S. Coast Survey, and his first assignment involved the exploration of the Atlantic Coast, particularly the Appomattox and James rivers in Virginia. In 1853, with the rank of Lieutenant, Trowbridge was sent to the U.S. Pacific coast to install tidal gauges at a number of locations, including Astoria in Oregon, where the type specimen of the shrew was collected. He also conducted astronomical, tidal, meteorological, and magnetic investigations from San Diego to Puget Sound and took part in the Williamson and Abbott expedition of the Pacific Railroad Surveys. During the American Civil War he was in charge of the army engineering unit supplying materials for Unionist fortifications. From 1871 to 1877 he served as Professor of Dynamic Engineering at Yale University, and from 1877 until his death in 1892 as Professor of Engineering at Columbia University. The shrew is found in southwest British Columbia (Canada) and south through the Pacific states of the USA to California.

True

Piñon Deer Mouse *Peromyscus truei* Shufeldt, 1885
True's Shrew-Mole *Urotrichus pilirostris* True, 1886 [Syn. *Dymecodon pilirostris*]
True's Vole *Hyperacrius fertilis* True, 1894 [Alt. Subalpine Kashmir Vole]
Mindanao Gymnure *Podogymnura truei* **Mearns,** 1905
True's Porpoise *Phocoenoides dalli truei* **Andrews,** 1911
True's Beaked Whale *Mesoplodon mirus* True, 1913

Frederick William True (1858–1914) was Head Curator of the Department of Biology, U.S. National Museum. After graduating in 1878 he worked for the U.S. Government Fish Commission for a couple of years before returning to New York University to earn his master's degree. After this he was Librarian at the Smithsonian for two years until 1883, when he took over there as Curator of Mammals until 1909. He was Assistant Secretary of the Smithsonian from 1911 until his death. He was an acknowledged expert on whales and wrote much on them from 1885 until 1913. He also wrote parts of some government reports, such as "The Useful Aquatic Reptiles and Batrachians" (1884). The deer mouse is found in the western USA and Baja California (Mexico). The shrew-mole occurs in Japan, and the vole in the mountains of Kashmir and northern Pakistan. The gymnure (a relative of the hedgehog) is endemic to the island of Mindanao in the Philippines. True's Porpoise was formerly regarded as a distinct species, *Phocoenoides truei*, but is now viewed as either a subspecies or a color-morph of Dall's Porpoise. The *truei* type of Dall's Porpoise is mainly found around Japan. The beaked whale appears to have two distinct populations, one in the North Atlantic and one in the Southern Hemisphere (stranded specimens have been found in South Africa and Australia).

Trumbull

Trumbull's Bonneted Bat *Eumops trumbulli* **Thomas,** 1901

J. Trumbull (dates not found) presented the type specimen of this bat, and all that Thomas tells about him is his surname and initial. It is likely that he was the same J. Trumbull who is recorded as having collected birds in Brazil between 1895 and 1904, and who sent a number of his specimens to the National Museum of Ireland. This bat is found in the Amazonian region of South America.

Tschudi

Tschudi's Pygmy Rice Rat *Oligoryzomys destructor* Tschudi, 1844 [Alt. Destructive Pygmy Rice Rat, Destructive Colilargo]
Tschudi's Yellow-shouldered Bat *Sturnira oporaphilum* Tschudi, 1844
Tschudi's Slender Opossum *Marmosops impavidus* Tschudi, 1845 [Alt. Andean Slender Mouse-Opossum]
Grey Brocket *Mazama tschudii* Wagner, 1855 [Alt. Peruvian Brown Brocket; Syn. *M. gouazoubira tschudii*]
Montane Cavy *Cavia tschudii* Fitzinger, 1857 [Alt. Montane Guinea-Pig]

Baron Dr. Johann Jacob von Tschudi (1818–1889) was a Swiss explorer who traveled in Peru, Brazil, Argentina, and Chile. He was also a physician, a diplomat, a naturalist, a student of South America, a hunter, an anthropologist, a cultural historian, a language researcher, and a statesman. He wrote *Untersuchungen uber die Fauna Peruana Ornithologie,* published in 1844. He is remembered in the scientific names of many different taxa, including the Desert Coral Snake *Micrurus tschudii* and fish such as the Apron Ray *Discopyge tschudii.* The rice rat is found from southern Colombia to western Bolivia and northwest Argentina. The bat has a similar distribution. The opossum occurs from eastern Panama south through western South America to Bolivia. The brocket deer was originally described as a full species but has been demoted to a race of the Brown/Grey Brocket, which is widely distributed in tropical South America. The cavy is found in Peru, northern Chile, Bolivia, and northwest Argentina.

Tsolov

Tsolov's Mouse-like Hamster *Calomyscus tsolovi* Peshev, 1991

Peshev originally described this hamster as a subspecies of *Calomyscus bailwardi,* without giving any information about the "Tsolov" he was naming it after. One possible candidate is Valentin Tsolov, a Bulgarian scientist who worked at the Faculty of Veterinary Medicine at Trakia University and published on veterinary matters. The hamster is known only from southwest Syria.

Tullberg

Tullberg's Soft-furred Mouse *Praomys tullbergi* **Thomas,** 1894

Professor Dr. Tycho Fredrik Hugo Tullberg (1842–1920) was a Swedish zoologist. His great-great grandfather was no less a person than Linnaeus (q.v.). He gained a doctorate in philosophy at Uppsala in 1869 and became a lecturer there in 1871. He was instrumental in the development of modern zoological veterinary medicine in Sweden. Between 1882 and 1907 he was Professor of Zoology at Uppsala, and after 1902 he was the Chairman of the Linnean Society at Hammarby. As far as we can ascertain, he was an academic as opposed to someone active in the field. He published in Swedish, English, and German, the latter being the preferred learned language for Swedes of his day. His works included *Neomeni, a New Genus of Invertebrate Animals* (1875), *Djurriket* (The Animal Kingdom; 1885), and his major work, *Über das System der Nagethiere, eine phylogenetische Studie* (1899). He is also commemorated in the name of a bird, Tullberg's Woodpecker *Campethera tullbergi*. The mouse is found from Gambia east to northern DRC (Zaire) and on the island of Bioko (Equatorial Guinea). It has also been recorded in northwest Angola and northwest Kenya.

Tunney

Pale Field Rat *Rattus tunneyi* **Thomas,** 1904

John Thomas Tunney (1871–1929) was a professional collector. He worked as such for the Western Australian Museum between 1895 and 1906. He led the Tunney expedition of 1903 to the Northern Territory of Australia. Thomas wrote a paper on the expedition's findings in 1904 entitled "On a Collection of Mammals Made by Mr. J. T. Tunney in Arnhem Land, Northern Territory of South Australia." This trip was for Lord Rothschild of the Tring Museum. Tunney is also commemorated in the trinomial of a subspecies of bird, the Yellow Chat *Epthianura crocea tunneyi*. The rat was formerly widespread in Australia, but its range has contracted. It is now found mainly in northern Australia, and in southeast Queensland and northeast New South Wales.

Tweedy

Tweedy's Crab-eating Rat *Ichthyomys tweedii* **Anthony,** 1921 [Alt. Northern/ Tweedie's Crab-eating Rat]

Andrew Mellick Tweedy (dates not found) was the manager of mines belonging to the South American Development Company in Ecuador. Tweedy originally went to Portovelo in 1916 and spent the next 30 years there. In 2004 his grandson, John Tweedy, made a film called *Streams of Gold* about his grandparents and the town that grew around the mines. The original etymology reads, "I take pleasure in naming this fine species in honor of Mr. A. M. Tweedy, the resident manager of the mine at Portovelo, who extended to the Museum's expedition all the assistance it lay in his power to give." The museum referred to is the American Museum of Natural History. The type specimen of the crab-eating rat was taken at Portovelo in western Ecuador. The species is also known from central Panama.

Tyler

Tyler's Mouse-Opossum *Marmosa tyleriana* **Tate,** 1931 [Alt. Tyleria Mouse-Opossum]

Sidney F. Tyler Jr. (d. 1937) was an American historian and photographer. He was wealthy enough to support the American Museum of Natural History's expedition of 1928–1929 to Mount Duida, at the headwaters of the Orinoco River; the expedition is now referred to as the Tyler-Duida expedition. Tate, who described the mouse-opossum, was a member of that expedition. The somewhat unusual form of the scientific name—i.e. *tyleriana* rather than simply *tyleri*, as one might expect for a species named after a person—is explained by the opossum being found in *Tyleria* forest on the plateau of Mount Duida. This plant is named after S. F. Tyler; the opossum is thus named after him "at one remove."

Uchida

Ryukyu Mole *Mogera uchidai* **Abe,** Shiraishi, and Arai, 1991 [Alt. Senkaku Mole; Syn. *Nesoscaptor uchidai*]

Professor Dr. Toru (sometimes Tohru) Uchida (1897–1981) was a Japanese zoologist and a professor at Hokkaido University. He edited several popular zoological encyclopedias and books on Japanese fauna, as well as writing more serious zoological works on systematics and papers on a wide variety of fauna. He is commemorated in the scientific names of a number of different taxa, such as the hydrozoan *Phialidium uchidai* and the fish *Zoarchias uchidai*. The mole is known only from Uotsurijima Island, in the Senkaku Islands, northeast of Taiwan. (These islands are currently controlled by Japan but are also claimed by Taiwan and by the People's Republic of China.)

Underwood

Underwood's Long-tongued Bat *Hylonycteris underwoodi* **Thomas,** 1903
Underwood's Water Mouse *Rheomys underwoodi* Thomas, 1906
Underwood's Pocket Gopher *Orthogeomys underwoodi* **Osgood,** 1931
Underwood's Bonneted Bat *Eumops underwoodi* **Goodwin,** 1940

Cecil F. Underwood (1873–1943) was a British ornithologist and collector. He went to Costa Rica in 1889 and stayed there until his death. He was an all-round naturalist who collected for a number of foreign museums, combining this with his job as a taxidermist at Costa Rica's National Museum. He described many new mammal taxa from Central America, often in association with George Goodwin. The long-tongued bat occurs from west-central Mexico to western Panama. The mouse inhabits central Costa Rica and western Panama. The gopher is a Costa Rican endemic. The bonneted bat is found from southern Arizona through western Mexico and south to Nicaragua.

Urich

Northern Bolo Mouse *Necromys urichi* **J. A. Allen** and **Chapman,** 1897 [formerly Northern Grass Mouse, *Akodon urichi*]
Sucre Spiny Rat *Proechimys urichi* J. A. Allen, 1899

Frederick William Urich (1872–1936) was a Trinidadian naturalist and one of the co-founders of the Trinidad Field Naturalists' Club (now the Trinidad and Tobago Field Naturalists' Club). He is also commemorated in the name of a bird, Urich's Tyrannulet *Phyllomyias urichi*. The grass mouse is found in eastern Colombia, Venezuela, northern Brazil, and Trinidad and Tobago. The spiny rat comes from northern Venezuela.

Uta Hick

Uta Hick's Bearded Saki *Chiropotes utahickae* **Hershkovitz,** 1985 [formerly *C. satanas utahicki*]

Frau Uta Hick(-Ruempler) was a Curator at the Cologne Zoo and the longtime editor of the zoo's journal, *Zeitschrift des Kölner Zoo*. The original citation reads, "Much is owed Miss Hick for her contributions to our knowledge of pitheciines. The largest assemblage of captive living pitheciines ever gathered flourished in the Kölner Zoo under Curator Hick's personal care." The original spelling of the scientific name was *utahicki*, but because the species is named after a woman this has now been corrected to the feminine *utahickae*. This monkey has been elevated to full-species status, having previously been regarded as a race of the Black Bearded Saki *Chiropotes satanas*. It is found in eastern Amazonian Brazil, between the rivers Xingu and Tocantins.

Val

Val's Gundi *Ctenodactylus vali* **Thomas,** 1902 [Alt. Sahara Gundi]

Thomas did not explain his choice of scientific name in this case, which suggests that this rodent is not named after a person. It may be that *vali,* like the word "gundi," is taken from a local name for the species. The gundi is found in Morocco, Algeria, southern Tunisia, and northwest Libya. The type specimen came from "Wadi Bey" in Libya.

Van Beneden

Van Beneden's Colobus *Procolobus verus* Van Beneden, 1838 [Alt. Olive Colobus]

Dr. Pierre-Joseph Van Beneden (1809–1894) was a Belgian physician who was appointed Curator of the Natural History Museum at the University of Louvain in 1831. In 1836 he became Professor of Zoology and Comparative Anatomy in the Catholic University at Louvain, an appointment he held until his death. Early in his career he specialized in invertebrates, particularly marine ones, and in 1843 he established a marine laboratory and aquarium in Ostend, which is believed to have been one of the first examples of its kind. Later he turned his attention to the vertebrates. When the port of Antwerp was being fortified, a number of bones of fossil whales were dug up, and Van Beneden became interested, undertaking a detailed study of the group. On the subject of the Cetacea, living and extinct, he published a number of papers and several large works; the most important was *Ostéographie des cétacés vivants et fossils,* which was jointly written with Gervais and published between 1868 and 1880. By the end of his life he had a Europe-wide reputation and was a foreign member of the Royal Society and also of the Linnean, Geological, and Zoological societies of London. The colobus has a patchy distribution in West Africa, from Sierra Leone to southern Nigeria.

Van Breda

Rough-toothed Dolphin *Steno bredanensis* **Cuvier,** 1828

Jacob Gijsbert Samuel Van Breda (1788–1867) was a Dutch academic. He was Professor of Botany, Zoology, and Comparative Anatomy at the University of Ghent. He had wide interests and was a leading geologist, a cartographer, editor, and illustrator, and the Curator of the Botanical Gardens in Ghent. After Belgium became independent he returned to Holland and became Professor of Zoology at Leiden. He was also Secretary of the Hollandsche Maatschappij der Wetenschappen (the Royal Holland Society of Sciences and Humanities) and Director of Teyler's Museum. In the Natural History Museum in Paris, a skull belonging to a Rough-toothed Dolphin had apparently been erroneously matched with a skin from a Neotropical river dolphin, *Inia geoffrensis.* While visiting the French capital, Van Breda pointed out this error to Cuvier, who later named the species to which the skull belonged after Van Breda. The dolphin is found worldwide in warm-temperate and tropical waters.

Van der Decken

Van der Decken's Sifaka *Propithecus deckenii* **Peters,** 1870

This species of lemur is endemic to western Madagascar. See **Decken** for biographical details.

Van Deusen

Van Deusen's Rat *Stenomys vandeuseni* Taylor and **Calaby,** 1982 [Syn. *Rattus vandeuseni*]

Hobart Merritt Van Deusen (1910–1976) was an American zoologist, author, and mammalogist who worked at the American Museum of Natural History in New York. He was a member of the fourth, sixth, and seventh Archbold expeditions to New Guinea in 1953, 1959, and 1964 respectively. He published widely, alone and with others, including for example with Taylor and Calaby, who described the rat, "A Revision of the Genus *Rattus* (Rodentia, Muridae) in the New Guinean Region." This species is known only from the region of Mount Dayman, Papua New Guinea.

Van Gelder

Van Gelder's Bat *Bauerus dubiaquercus* Van Gelder, 1959

Dr. Richard George Van Gelder (1928–1994) was an American zoologist who for many years served as Curator of Mammalogy at the American Museum of Natural History in New York. In addition to his academic works on mammals, he wrote several books aimed at children, such as *The Professor and the Mysterious Box.* The bat was first described from the Tres Marias Islands off western Mexico but has also been found on the mainland of Central America from southern Mexico and Belize to Costa Rica.

Van Heurn

Lesser Forest Wallaby *Dorcopsulus vanheurni* **Thomas,** 1922

Willem Cornelis Van Heurn (1887–1972) was a Dutch taxonomist and biologist who worked for a period at the National Natural History Museum in Leiden. He came from a wealthy family but chose to work all his life. He went to Suriname in 1911, to Simaloer (off Sumatra) in 1913, and to Dutch New Guinea between 1920 and 1921. He then lived in the Dutch East Indies (mostly Java) for 15 years, where he ran a laboratory for sea research; studied rat control on Java, Timor, and Flores; was a schoolteacher; and served as head of the Botany Department at the Netherlands Indies Medical School in Java. He returned to Holland in 1939. He collected natural history specimens wherever he traveled or settled, which he meticulously prepared and labeled. Most he sent to the Leiden Museum, where Van Heurn himself worked as an Assistant Curator for Fossil Mammals from 1941 to 1945. He was a prolific writer. He published around 100 articles on a wide range of topics, including such gems as "The Safety Instinct in Chickens" (1927), "Cannibalism in Frogs" (1928), "Do Tits Lay Eggs Together as the Result of a Housing Shortage?" (1955), and "Wrinkled Eggs" (1958). It was said of him in a memorial booklet published by the museum, "He made natural history collections wherever he went and gave his attention to almost all animal groups. He was an excellent shot, and a competent preparator; his mammal and bird skins are exemplary." He is also commemorated in the name of a fish, Van Heurn's Rainbowfish *Melanotaenia vanheurni.* The wallaby is native to New Guinea. See also **Heuren.**

Van Roosmalen

Van Roosmalens' Hairy Dwarf Porcupine *Coendou roosmalenorum* **Voss,** 2001

Dr. Marc Van Roosmalen (b. 1947) is a Dutch zoologist, primatologist, and ecologist. When he collected the type specimen of the porcupine he was accompanied by his son, Tomas, so the scientific name *roosmalenorum* (meaning "of the Roosmalens") honors them both. In 1977 he published the two-volume *Surinaams Vruchtenboek.* In 1980 he was awarded his doctorate by the University of Amsterdam. In 1985 he published *Guide to the Fruits of the Guianan Flora,* with illustrations by his wife Betty. He spent years in Suriname, a former Dutch colony, studying spider monkeys. On the strength of his doctoral research into tropical ecology he was appointed by the Brazilian government to a scientific post in Manaus in 1987. He is a leading advocate of an environmental-protection law that enables Brazilian nongovernment organizations to buy rainforest tracts for ecotourism and research. As the discoverer of new species, he has the right to choose their scientific names; he hatched the very interesting plan to auction off this privilege to the highest bidder and use the proceeds to protect the species' native habitat. The porcupine is known from both banks of the Rio Madeira, in Amazonian Brazil. See also **Roosmalen.**

Van Sung

See **Caovansung.**

Van Swinderen

See **Swinderen.**

Vanzolini

Black Squirrel Monkey *Saimiri vanzolinii* **Ayres,** 1985
Vanzolini's Bald-faced Saki *Pithecia irrorata vanzolinii* **Hershkovitz,** 1987

Professor Dr. Paulo Emilio Vanzolini (b. 1924) is a Brazilian zoologist and herpetologist. He was appointed as a biologist to the Zoological Museum of the University of São Paulo in 1946. He studied at Harvard under Alfred Romer from 1948 to 1951 and was awarded his doctorate there. While in the USA he became acquainted at the American Museum of Natural History in New York with the herpetologists Ernest Williams and Charles Bogert, with whom he became lifelong friends. He returned to the museum in São Paulo, as Curator, in 1951. He was Director from 1962 until 1993, when he was compulsorily retired at the age of 70. Nevertheless, at the time of writing he is still working at the museum. In 1993 he published *Elementary Statistical Methods in Zoological Systematics.* He is remembered in the names of a number of reptiles, including Vanzolini's Ground Snake *Liophis vanzolinii* and Vanzolini's Worm Lizard *Amphisbaena vanzolinii.* He is also famous in Brazil as a composer of samba music. The squirrel monkey is known only from a small area of Amazonian Brazil, in a triangle of land between the Japurá and Solimões rivers. The saki monkey also has a very restricted range in Brazil, between the upper Rio Juruá and its south bank tributary, the Rio Tarauacá.

Van Zyl

Van Zyl's Golden-Mole *Cryptochloris zyli* **Shortridge** and Carter, 1938

Gideon van Zyl was a South African landowner whose family farmed on the southern bank of the Olifants River. The family still own a farm and vineyard there, and it is clear that the Christian name has passed down generation to generation, as the present proprietor is also called Gideon. We think it likely that the man after whom the mole was named was Major Gideon Brand van Zyl (1873–1956), who was Governor General of South Africa from 1946 to 1950. The property was known as "Van Zyl's" in the 18th century. R. J. Gordon, in his *Journal of the Third Journey through a Part of Southern Africa, Carried out between 28th Aug. 1778 and 25th Jan. 1779,* mentions the place several times. Immediately to the west of Van Zyl's was a place called Compagnies Drift, where salt could be collected. In 1937 the type specimen of this rare golden-mole was collected at Compagnies Drift, northwest Cape Province, South Africa. The original description contains the following etymology: "I have named this very distinct Golden Mole in honour of Mr. Gideon van Zyl, Compagnies Drift, to whom I am indebted for much assistance during my visit to Lamberts Bay." The golden-mole is known from the type locality and one other location about 150 km (90 miles) further north along the Namaqualand coast.

Vardon

Puku *Kobus vardonii* **Livingstone,** 1857

Major Frank Vardon (dates not found) was an English elephant hunter. He entered the Madras army of the Honourable East India Company as a Cadet in 1831 and was commissioned in 1832 into the 25th Regiment, Madras Native Infantry. He was promoted to Captain in 1845 and in 1846 was in Africa, hunting along the Limpopo River with William Cotton Oswell. He was in England in 1848 but back in Africa in 1850, when he met and became a friend of David Livingstone. In his book *The Zambesi Expedition,* which was published in 1865, Livingstone wrote, "We venture to call the poku after the late Major Vardon, a noble-hearted African traveller." This species of antelope is found in southern DRC (Zaïre), northeast Angola, Zambia, Malawi, and southern Tanzania.

Varian

Giant Sable Antelope *Hippotragus niger variani* **Thomas,** 1916

Henry Francis "Frank" Varian (1876–1960) was an engineer, big-game hunter, and naturalist. He was Chief Engineer on the Benguela Railway in Angola and spent 50 years working on engineering projects in Mozambique, Rhodesia, and other parts of southern Africa. He wrote *Some African Milestones*—which, while about the development of the railways, is full of personal anecdotes and observations as well as zoological notes. He first described the giant sable in 1909, calling it "the finest antelope in Africa, bearing the most magnificent horns of all." This subspecies is now critically endangered and is restricted to Angola between the Cuango and Luando rivers.

Varona

Varona and Garrido's Hutia *Mesocapromys sanfelipensis* Varona and **Garrido,** 1970 [Alt. San Felipe Hutia]

Luis S. Varona-Calvo (d. 1988) was a Cuban zoologist. He published a great number of books and articles over a long period, from *The Mammals of Cuba* in 1923 to "The Distribution of *Crocodylus acutus* in Cuba" in 1985. The hutia was found on the Cayos de San Felipe (small islands off southwest Cuba), but a survey in 2003–2004 failed to find any trace of surviving animals, and the species may well now be extinct.

Veldkamp

Veldkamp's Bat *Nanonycteris veldkampii* **Jentink,** 1888 [Alt. Little Flying Cow, Veldkamp's Dwarf Epauletted Fruit Bat]

Antonie Veldkamp (1853–1927) was a Dutchman who worked in West Africa for H. Muller and Co., a trading company. He went to the Gold Coast (now Ghana) in 1879 and in 1883 was appointed as Dutch Consular Agent at Elmina in the Gold Coast. In 1887 Muller and Co. transferred Veldkamp to Monrovia in Liberia. From 1887 until 1890 he also acted as Dutch Consul in Liberia, and in 1891 he returned to the Netherlands. The bat is found in West Africa from Guinea east to the Central African Republic.

Veloz

Veloz's Hutia *Plagiodontia velozi* **Rimoli,** 1977 extinct

Professor Marcio Veloz Maggiolo (b. 1936) is a prominent figure in the fields of Hispaniolan archeology, literature, and anthropology. He graduated in archeology from the University of Madrid. He was Director in charge of excavations (archeological and anthropological) for the Museum of the Dominican Republic.

He later served as Ambassador to Mexico and to Italy and was the first Director of the Independent University of Santo Domingo, serving from 1964 to 1968. He was a colleague of Renato Rimoli, who described the hutia, at the Museum of Dominican Anthropology between 1973 and 1975. He published novels, essays, and poetry in addition to works on pre-Columbian ceramics, archeology, and the like. *Plagiodontia velozi* may well be only a junior synonym of *P. ipnaeum*, rather than a distinct species. It lived on the island of Hispaniola (Haiti and Dominican Republic) and probably survived until after Europeans reached the New World.

Verhagen

Verhagen's Brush-furred Rat *Lophuromys verhageni* **Verheyen,** Hulselmans, Dierckx, and Verheyen, 2002

Professor Ronald Verhagen is a Belgian biologist and mammalogist who works at the University of Antwerp. He is particularly interested in hedgehogs and water shrews and is an expert on rodent biology. He has published widely on small mammals—for example, in 2003, with Adil Baghli, "The Distribution and Status of the Polecat, *Mustela putorius*, in Luxembourg." He has also published jointly with Verheyen, who was one of the describers of the brush-furred rat. The species is known only from Mount Meru, northern Tanzania.

Verheyen

Verheyen's Multimammate Mouse *Mastomys verheyeni* Robbins and Van der Straeten, 1989

Verheyen's Shrew *Congosorex verheyeni* **Hutterer,** Barriere, and Colyn, 2002 [Alt. Lesser Congo Shrew]

Verheyen's Wood Mouse *Hylomyscus walterverheyeni* Nicolas et al., 2008

Professor Dr. Walter N. Verheyen (1932–2005) was a Belgian zoologist who was a Professor in the Biology Department of the University of Antwerp. He published widely on rodents and other small mammals of Africa—for example, with Leirs and Verhagen, "The Basis of Reproductive Seasonality in *Mastomys* Rats (Rodentia: Muridae) in Tanzania." The mouse occurs only in the region of Lake Chad, in northeast Nigeria and northern Cameroon. However, it is now believed that this "species" is invalid and only a junior synonym of *Mastomys kollmannspergeri*. Nicolas said in his etymology of the wood mouse, "The species is named in honor of our colleague the late Walter Verheyen, who initiated the taxonomic study on the genus *Hylomyscus*, and in recognition of his significant contribution to the systematics and biogeographic research on African small mammals." The shrew and the wood mouse are both found in the western Congo basin, central Africa.

Verhoeven

Verhoeven's Giant Tree Rat *Papagomys theodorverhoeveni* **Musser,** 1981 extinct

Father Dr. Theodor Verhoeven (1907–1990) was a Dutch archeologist who was also a Catholic missionary in Indonesia. He worked in Timor and Java, but he is best known for the 20 years he spent in Flores. He spent some years in the 1960s excavating in a cave where, in more recent times, the small hominid dubbed "hobbit" (*Homo floresiensis*) has been found. Verhoeven found stone tools there and remains of small elephants. After 20 years as a priest on Flores, he left the priesthood, married his secretary, and returned to Europe. The tree rat is known only from subfossil remains on Flores. It is included here because the species might have survived until recent times (or even still be extant, alongside its close relative *Papagomys armandvillei*).

Vernay

Red Climbing Mouse *Vernaya fulva* **G. Allen,** 1927

Vernay's Climbing Mouse *Dendromus vernayi* Hill and Carter, 1937

Vernay's Lion *Panthera leo vernayi* **Roberts,** 1948 [Alt. Kalahari Lion]

Arthur Stannard Vernay (1877–1960) was an English businessman and philanthropist. As a young man he went to New York and set up a business dealing in antiques. As his business flourished, he was able to indulge his other interests: natural history and travel in exotic places. As a Trustee of the American Museum of Natural History, he, along with British Colonel John C. Faunthorpe, mounted six collecting expeditions to Burma (now Myanmar), India, and Thailand between 1922 and 1928. He also financed a British Museum collecting trip to Tunisia in 1925 and another U.S. trip in the same year to Angola. In 1940 he sold his collection of English furniture, porcelains, silver, glassware, and other art objects and retired to the Bahamas. The red climbing mouse was originally described as a species of *Chiropodomys,* but in 1941 it was placed in its own unique genus, *Vernaya.* It is found in northern Myanmar and southern China (Sichuan and Yunnan). To make things confusing, this species is sometimes known as "Vernay's Climbing Mouse," which is the generally accepted common name for *Dendromus vernayi.* The latter is known only from the type locality in east-central Angola. Vernay's Lion, from the Kalahari, is not now generally recognized as a valid subspecies.

Verreaux

Verreaux's Mouse *Myomyscus verreauxii* **A. Smith,** 1834 [Alt. Verreaux's White-footed Rat]

Verreaux's Sifaka *Propithecus verreauxi* **Grandidier,** 1867

Jean Baptiste Edouard Verreaux (1810–1868) and Jules Pierre Verreaux (1807–1873) were French naturalists, collectors, and dealers. They both worked in China and in South Africa's Cape Colony. There was a third brother, Joseph Alexis Verreaux, yet another naturalist, who lived in Cape Town. The Verreaux family traded in Paris from a huge emporium for feathers and stuffed birds, which they called the Maison Verreaux. They were clearly ambitious taxidermists and gained notoriety on account of having once attended the funeral of a tribal chief, whose body they then disinterred, took to Cape Town, and stuffed. The Catalán veterinarian Francisco Darder, then Curator of the zoo in Barcelona, purchased the "specimen" from one of the brothers' sons, Edouard Verreaux, in 1888. This controversial exhibit was on show in Barcelona until the end of the 20th century, when the man's descendants demanded that the body be returned for a decent burial. Jules was employed as an ornithologist and plant collector for the Natural History Museum in Paris, which sent him to Australia in 1842. He returned to France around 1851 with a collection of natural history specimens reported to number 115,000 items. He also assisted Andrew Smith in founding the South African National Museum at Cape Town. He is commemorated in the common names of at least 10 birds, including Verreaux's Eagle Owl *Bubo lacteus* and Verreaux's Turaco *Tauraco macrorhynchus.* The mouse comes from western Cape Province, South Africa. The sifaka is found in southern and southwest Madagascar.

Verschuren

Verschuren's Swamp Rat *Malacomys verschureni* **Verheyen** and Van der Straeten, 1977

Dr. Jacques Verschuren (b. 1926) is a Belgian zoologist and biologist who worked in Zaire for many years and then at the Royal Institute of Natural Sciences of Belgium in Brussels. It was he who first told Dian Fossey about the gorillas on the Virunga volcanoes. In 1960, with François Bourlière as co-author, he wrote *Enquetes sur les grands mammifères du parc national Albert*, and in 2002 he published his autobiography, *Ma vie sauver la nature*. The rat is found in northeast DRC (Zaire).

Vespucci

Vespucci's Rat *Noronhomys vespuccii* Carleton and Olson, 1999 extinct

Amerigo Vespucci (1454–1512) was an Italian navigator after whom the Americas are named. In 1503 he discovered the island of Fernando de Noronha, off the coast of Brazil. In *Lettera di Amerigo Vespucci delle isole nuovamente in quattro suoi viaggi*, he makes reference to "very big rats" living there. Later explorers found no native mammals there, and as Vespucci was one of the first Europeans to reach the island, the rats he saw were unlikely to have been an introduced European species. In 1973, fossils of a large, previously undescribed rodent were found on the island. Carleton and Olson named the species after Vespucci because his *Lettera* was the only document "suggesting the existence of an indigenous rodent on the island." Vespucci's Rat seems to have been the only native land mammal of the archipelago.

Victoria

Savanna Hare *Lepus victoriae* **Thomas,** 1893 [Syn. *L. microtis* **Heuglin,** 1865]
Giant Genet *Genetta victoriae* Thomas, 1901

Queen Victoria (1819–1901) was Queen of the United Kingdom from 1837 and Empress of India from 1876. It seems likely, although Thomas is not forthcoming in his original descriptions, that the two mammals were named after Lake Victoria, which in turn was named after the British Queen by John Hanning Speke. The type specimen of the hare was collected on Speke Gulf, Lake Victoria. The species is, however, widely distributed in African savannah regions, as far south as Natal (South Africa). There is also an isolated population in western Algeria. The "correct" (i.e. oldest valid) scientific name for the Savanna Hare appears to be *Lepus microtis*, but *victoriae* is found widely in references. The first known specimen of the genet was obtained at Entebbe, Uganda, on the shores of Lake Victoria. However, it is not native to that area, and the skin must have been brought from elsewhere. The genet's distribution is centered on eastern DRC (Zaire); it may extend into western Uganda.

Vieira

Vieira's Long-snouted Bat *Xeronycteris vieirai* Gregorin and Ditchfield, 2005
Vieira's Spiny Rat *Echimys vieirai* Iack-Ximenes, de Vivo, and Percequillo, 2005

Dr. Carlos Octaviano da Cunha Vieira (d. 1958) was a mammalogist and Curator in charge of the Mammal Collection at the Zoological Museum, University of São Paulo, Brazil, from the early 1940s until his death. Through his efforts the museum's mammal collection was greatly improved in both size and geographic coverage. Vieira wrote *Ensaio monografico sobre os quirópteros do Brazil* in 1942, which is cited in the etymology for the long-snouted bat as the major reference work for Brazilian bats. The bat is found in northeast Brazil. The spiny rat is known from the south bank of the Amazon in the Brazilian states of Amazonas and Pará.

Vigne

Urial *Ovis vignei* **Blyth,** 1841

Godfrey Thomas Vigne (1801–1863) was a British explorer, traveler, artist, and adventurer who also played the occasional game of first-class cricket for the Marylebone Cricket Club or Hampshire when he happened to be in England. However, much of his time seems to have been spent in the mountainous areas of Asia. He was in Kabul in the 1830s and gave an early account of the game now known as polo. In 1844 he published *Travels in Kashmir, Ladak and Iskardo.* Between 1800 and 1841 the Great Trigonometrical Survey of India took place. Vigne was with this survey around 1835 and left posterity a sketch of Sir George Everest (after whom the mountain is named) and an assistant overseeing some workmen erecting a pole. The Urial, a wild sheep, is found in Uzbekistan, Tadzhikistan, northeast Iran, Afghanistan, Pakistan, and northwest India.

Villa

Villa's Pocket Gopher *Thomomys bottae villai* Baker, 1953
Villa's Grey Shrew *Notiosorex villai* Carraway and Timm, 2000

Dr. Bernardo Villa-Ramirez (1911–2006) obtained a doctorate in biology at the National Autonomous University in Mexico City. He was head of the Department of Zoology at the Institute of Biology, University in Mexico. He was also a leading member of the Association of Mexican Mammalogists, an honorary member of the Society for Marine Mammalogy, and an honorary trustee of the Smithsonian. He published a number of scientific papers including, with Hall, "Systematics of the Smooth-Toothed Pocket Gopher *Thomomys umbrinus* in the Mexican Transvolcanic Belt" (1948). The Bernardo Villa Student Fund was established in 2002 to support participation by students at the North American Symposium on Bat Research. The pocket gopher, one of almost 200 named subspecies of *Thomomys bottae,* is found in the state of Coahuila, northern Mexico. The shrew occurs in the state of Tamaulipas, northeast Mexico.

Vincent

Vincent's Bush Squirrel *Paraxerus vincenti* **Hayman,** 1950

Colonel Jack Vincent (1904–1999) was a zoologist, conservationist, farmer. and soldier. He was born in London, moved to South Africa at the age of 21, and worked on farms in Natal. He then returned to England and worked for the British Museum. In the late 1920s and early 1930s he went on a number of bird-collecting expeditions in eastern and southern Africa, but his monumental work was a study of the birds of northern Portuguese East Africa (now Mozambique), leading to the publication of *The Birds of Northern Portuguese East Africa* in 1934. Several subspecies of birds bear his name. After his last expedition in 1934 he was sent to Zanzibar by Jardine Mathieson Co. to start a clove distillery there. He stayed for three years before being transferred to a sisal plantation in Tanganyika. In 1937 he bought a farm in Natal. He was awarded the M.B.E. for his war service, and in 1949 he became the first Director of the Natal Parks Board, which was instrumental in saving the White Rhinoceros. In the late 1940s and early 1950s he served as editor of *The Ostrich,* the journal of the South African Ornithological Society. He then took a job with the International Council for Bird Preservation and moved to Switzerland, working in international conservation for four years. During this time he was awarded the World Wildlife Fund Gold Medal and the Order of the Golden Ark by

Prince Bernhard of the Netherlands. He rejoined the Parks Board in 1967. He retired to his farm in 1974. In 1990 he self-published his autobiography, *Web of Experience*. The squirrel is now considered to be critically endangered, as it is only known from Mount Namuli, northern Mozambique.

Vinogradov

Vinogradov's Jird *Meriones vinogradovi* **Heptner,** 1931
Vinogradov's Jerboa *Allactaga vinogradovi* Argyropulo, 1941
Wrangel Collared Lemming *Dicrostonyx vinogradovi* **Ognev,** 1948

Professor Boris Stepanovich Vinogradov (1891–1958) was the head of the Department of Terrestrial Vertebrates at the St. Petersburg Institute from 1928. He was also one of the leading specialists in rodent systematics. He supervised investigations into the systematics of carnivores, ungulates, insectivores, lagomorphs, rodents, whales, and seals—all carried out at the laboratory of theriology (mammalogy). The jird is found in eastern Turkey, Armenia, Azerbaijan, Syria, and northern Iran. The jerboa occurs in southern Kazakhstan, Uzbekistan, and Kyrgyzstan. The lemming is endemic to Wrangel Island in the Arctic Ocean, north of the Siberian mainland.

Vinson

Vinson's Slit-faced Bat *Nycteris vinsoni* **Dalquest,** 1965

Jerry Vinson (dates not found) of Wichita Falls, Texas, was a wealthy independent oilman. During WW2 he organized the collecting of scrap metal for the war effort. He invited Dalquest to join him on a hunting trip to Mozambique in 1963, so that Dalquest could collect mammals for the vertebrate collections of Midwestern University. As this was an inaccessible and little-studied area, Dalquest jumped at the chance, and this is why he named the bat— the type of which was collected during the trip—after Vinson. Vinson sponsored a second trip in 1965. A reptile, Vinson's Gecko *Phelsuma vinsoni,* is also named after him. The bat is known from only one locality in southern Mozambique. Its taxonomic status is uncertain. Some regard it as a subspecies of *Nycteris macrotis.*

Virginia

Virginia Dunnart *Sminthopsis virginiae* de Tarragon, 1847 [Alt. Red-cheeked Dunnart]

The type specimen of this small marsupial is lost, and de Tarragon did not specify a type locality. The significance of the scientific name *virginiae* is not known. The species is found in northern Australia, southern New Guinea, and the Aru Islands.

Visagie

Visagie's Golden-Mole *Chrysochloris visagiei* Broom, 1950

I. H. J. Visagie (dates not found) was a landowner in Cape Province, South Africa. The type—and only known—specimen of this golden-mole was captured "by Captain Shortridge at Gouna, with the assistance of Mr. I. H. J. Visagie to whom the estate belongs." The species has not been found again, despite expeditions to the type locality to look for it.

Vives

Fishing Bat *Myotis vivesi* Ménégaux, 1901 [Alt. Fish-eating Myotis; Syn. *Pizonyx vivesi*]

Auguste Ménégaux (1857–1937), the describer of the bat, was a zoologist at the Paris Natural History Museum from 1901 to 1926. He gave no indication as to who the "Vives" honored in the binomial might have been. We do know that Vives was not the name of the person who discovered the species: that was Leon Diguet.

Some authorities think that *vivesi* does not refer to a person but derives from the Latin *vivere* (to be alive). The bat is found on the coasts and offshore islands of northwest Mexico (Baja California and Sonora).

Vleeschouwers

Vleeschouwers' Talapoin *Miopithecus talapoin vleeschouwersi* **Poll,** 1940 [Alt. Zaire Talapoin]

C. Vleeschouwers (fl. 1925–1956) spent at least two decades in the Belgian Congo. He appears to have concentrated on ichthyology and collected a number of different fish species in the early 1940s from the River Luie in the Kwango region, and these fish were also described by Poll, who was working in the Congo in the same period. Vleeschouwers was a member of the colonial administration with responsibilities encompassing local hunting and related matters. In 1953 he commented on the growth in the use by local peoples of automatic firearms, and in 1956 he published, in the Congo's *Agricultural Bulletin,* a paper entitled "Contribution à l'étude cynégétique du District du Kwango" (Contribution to the Study of Hunting in the Kwango District). This small monkey is found in western DRC (Zaire) on the south bank of the lower Congo River. The subspecies is no longer generally recognized as being valid.

Voeltzkow

Pemba Flying Fox *Pteropus voeltzkowi* **Matschie,** 1909

Professor Dr. Alfred Voeltzkow (1860–1947) was a German traveler and zoologist who spent many years in East Africa. Between 1889 and 1895 he traveled in Zanzibar and Pemba. In 1906 he was in the Comoro Islands, and at some stage he visited the island of Aldabra. His East African collection is in the Natural History Museum in Vienna. He is remembered in the scientific names of other taxa, such as the ant *Cataulacus voeltzkowi.* The flying fox is endemic to Pemba Island, Tanzania.

Volnuchin

Volnuchin's Shrew *Sorex volnuchini* **Ognev,** 1922 [Alt. Caucasian Pygmy Shrew]

D. Volnuchin (dates not found) was a Russian collector. He, together with D. Filatov, collected the type specimen of the shrew in the Ukraine. We have been unsuccessful in our attempts to expand his biography. Ognev originally described the shrew as a subspecies of the Eurasian Pygmy Shrew *Sorex minutus,* but it has since been elevated to a full species. It is found in Crimea, the Caucasus, Turkey, and northern Iran.

Vordermann

Vordermann's Flying Squirrel *Petinomys vordermanni* **Jentink,** 1890

Vordermann's Pipistrelle *Hypsugo vordermanni* Jentink, 1890 [formerly *Pipistrellus vordermanni*]

Dr. Adolphe Guillaume Vordermann (1844–1902) was a Dutch physician who was a naval surgeon in the Dutch East Indies in 1866. From 1890 until his death he worked as a government physician of the Dutch health inspectorate in Indonesia, principally as an inspector of prisons. Clearly enlightened, he was concerned about disease among prisoners, and his observations helped link beriberi to vitamin deficiency. Vordermann found that in the prisons using mostly brown rice, less than 1 prisoner in 10,000 had developed beriberi, while in those using mainly white rice the proportion was 1 in 39. He was also a keen naturalist and mounted several brief expeditions, notably to Belitung, an island west of Borneo, in 1891. There is a record of the "List of the Lepidopterous Insects" he collected. Vordermann was the great grandfather of the British television personality Carol Vordermann. He was very interested in bird life and corresponded with others who used his observations. The squirrel is found in southern Myanmar, the Malay Peninsula, and Borneo. The bat is known from Borneo and Belitung (Indonesia).

Vosmaer

Banka Shrew *Crocidura vosmaeri* **Jentink,** 1888

Jan Hendrik Gabriel Vosmaer (1830–1885) was born in Makassar, Celebes (now Sulawesi), Indonesia. He worked as administrator of the tin mines on the island of Banka (Bangka) off eastern Sumatra, and was subsequently Assistant Resident of (the island of) Belitung, which is where he died. In his spare time he collected natural history specimens, most of which he sent to the Leiden Museum in Holland. The shrew is known with certainty only from Banka, but it may also occur on Sumatra.

Voss

Voss' Slender Opossum *Marmosops creightoni* Voss et al., 2004

Dr. Robert S. Voss is an Associate Curator (Mammalogy) at the American Museum of Natural History. To quote the AMNH website, he studies "the systematics and biogeography of neotropical mammals that inhabit moist-forest habitats in Amazonia and the Andes. He is actively involved in long-term revisionary taxonomic research on several clades that have radiated extensively in lowland and montane rainforests, including didelphid marsupials, caviomorph rodents (erethizontids and dasyproctids), and murid rodents." He was awarded his doctorate by the University of Michigan in 1983 and has written numerous scientific papers. He named this species of opossum after Ken Creighton, who had collected, but not identified, the taxon in Bolivia in 1979. It is known from the valley of the Rio Zongo, La Paz Province, Bolivia.

Wagner, E. R. and D. L.

Wagner's Peccary *Catagonus wagneri* Rusconi, 1930 [Alt. Chacoan Peccary, Taagua]

Emilio Roger Wagner (1868–1949) and Duncan Ladislao Wagner (1863–1937) were French archeologists, entomologists, anthropologists, and naturalists. We cannot be sure which of the two brothers the peccary was named after. Their father was a diplomat who served in Brazil, Argentina, and Peru, so when the brothers left Europe for South America they were returning to familiar ground. In 1904 they settled in Santiago del Estero, Argentina, where they became interested in early civilizations and investigated and excavated the pre-Hispanic Chaco-Santiagueña culture. Emilio, after his return from fighting in the French army in WW1, was invited to run a museum, which he did until his death. The museum is today known as the Emilio and Duncan Wagner Museum of Natural and Anthropological Sciences. The peccary was first described from subfossil remains, but to the amazement of zoologists it was discovered as a living animal in 1974 by Dr. Ralph Wetzel. Until then it had been thought extinct since pre-Hispanic times. It is found in the Gran Chaco (dry, thorny scrubland) of Paraguay, southern Bolivia, and northwest Argentina.

Wagner, J. A.

Wagner's Gerbil *Gerbillus dasyurus* J. A. Wagner, 1842 [Syn. *Dipodillus dasyurus*]
Wagner's Marsh Rat *Holochilus sciureus* J. A. Wagner, 1842 [Alt. Amazonian Marsh Rat]
Wagner's Bonneted Bat *Eumops glaucinus* J. A. Wagner, 1843
Wagner's Moustached Bat *Pteronotus personatus* J. A. Wagner, 1843
Wagner's Sac-winged Bat *Cormura brevirostris* J. A. Wagner, 1843 [Alt. Chestnut Sac-winged Bat]

Johann Andreas Wagner (1797–1861) was a German paleontologist, zoologist, and archeologist. He became Professor of Zoology and Assistant Curator at the University of Munich Zoological Museum, where he examined specimens sent from Brazil by such notables as Natterer and Spix. This led to the publication of *Diagnosen neuer Arten brasilischer Säugethiere* in 1842 and *Diagnosen neuer Arten brasilischer Handflugler* in 1843. The gerbil is found in the Sinai Peninsula and in the Arabian Peninsula, Iraq, Syria, Jordan, Israel, and Lebanon. The marsh rat occurs widely in the Amazon and Orinoco river basins; it may be a "composite" of more than one species. The bonneted bat is found from southern Mexico to northern Argentina and on Cuba and Jamaica; populations in southern Florida are now considered to be a distinct species, *Eumops floridanus*. The moustached bat's distribution ranges from Mexico south to Bolivia, and that of the sac-winged bat from Nicaragua to Peru and Amazonian Brazil.

Wahlberg

Wahlberg's Epauletted Bat *Epomophorus wahlbergi* **Sundevall,** 1846

Johan August Wahlberg (1810–1856) was a Swedish naturalist and collector. He studied chemistry at Uppsala in 1829 and worked in a chemist's shop in Stockholm while studying at the Skogsinstitutet (Forestry Institute). He traveled and collected widely in southern Africa between 1838 and 1856, sending thousands of specimens home to Sweden. He returned briefly to Sweden in 1853 but soon decided to go back to Africa—a decision that would cost him his life. While exploring the headwaters of the Limpopo River, he was killed by a wounded elephant. He is also commemorated in the common names of three birds, Wahlberg's Cormorant *Phalacrocorax neglectus,* Wahlberg's Eagle *Aquila wahlbergi,* and Wahlberg's Honeyguide *Prodotiscus regulus.* The bat is found from Cameroon to Sudan and southern Somalia, south to Angola and South Africa.

Walker

Pygmy Nyctophilus *Nyctophilus walkeri* **Thomas,** 1892 [Alt. Pygmy Long-eared Bat]

James John Walker (1851–1939) was a British naval officer, entomologist, and naturalist. He was on board HMS *Penguin* as Chief Engineering Officer during her cruise in the Australian and China seas in the period 1890–1893, during which he collected the type specimen of the bat. He retired from the navy in 1904, went to live in Oxford, and collected insects in his local area. He was President of the Royal Entomological Society of London from 1919 to 1920. The bat is found in northern Australia.

Wallace

Wallace's Striped Dasyure *Myoictis wallacii* **Gray,** 1858
Stripe-faced Fruit Bat *Styloctenium wallacei* Gray, 1866

Alfred Russel Wallace (1823–1913) was an English naturalist, evolutionary scientist, geographer, and anthropologist and one of the giants of Victorian science. He has claims to be regarded as the father of zoogeography. He was also a social critic and theorist, a follower of the utopian socialist Robert Owen. His interest in natural history began while working as an apprentice surveyor, at which time he also attended public lectures. He went to Brazil in 1848, on a self-sustaining natural history collecting expedition. Even on his first expedition he was very interested in how geography limited or facilitated the extension of species' ranges. He not only collected but also mapped, using his surveying skills. His return to England in 1852 was a near disaster; his ship, the brig *Helen,* caught fire and sank with all his specimens, and he was lucky to be rescued by a passing vessel. He spent the next two years writing *Palm Trees of the Amazon and Their Uses* and *A Narrative of Travels on the Amazon and Rio Negro,* and organizing another collecting expedition to the Indonesian archipelago. He managed to get a grant covering his passage to Singapore, in 1862. He spent nearly eight years there, during which he undertook about 70 different expeditions involving a total of around 22,500 km (14,000 miles) of travel. He visited every important island in the archipelago at least once, some many times. He collected a remarkable 125,660 specimens, including more than 1,000 new taxa. He wrote *The Malay Archipelago* in 1869, which is the most celebrated of all writings on Indonesia and ranks as one of the 19th century's best scientific travel books. He also published *Contributions to the History of Natural Selection* in 1870 and *Island Life* in 1880. His essay "On the Law Which Has Regulated the Introduction of New Species," which encapsulated his most profound theories on evolution, was sent to Darwin. He later sent Darwin his essay "On the Tendency of Varieties to Depart Indefinitely from the Original Type," presenting the theory of "survival of the fittest." Darwin and Lyell presented this essay, together with Darwin's own work, to the Linnean Society. Wallace's thinking spurred Darwin to encapsulate these ideas in *The Origin of Species;* the rest is history. Wallace developed the theory of natural selection, based on the differential survival of variable individuals, halfway through his stay in Indonesia. He remained for four more years, during which he continued his systematic exploration and recording of the region's fauna, flora, and people. For the rest of his life he was known as the greatest living authority on the region and its zoogeography. His legacy includes his discovery and description of the faunal discontinuity that now bears his name: Wallace's Line. This natural boundary runs between the islands of Bali and Lombok in the south and Borneo and Sulawesi in the north, and separates the Oriental and Australasian faunal regions. He is commemorated in the common names of at least 12 birds, including Wallace's Standardwing *Semioptera wallacii* (a bird-of-paradise) and Wallace's Fruit Dove *Ptilinopus wallacii*. The dasyure is found in southern New Guinea and the Aru Islands; it was originally described as a species, then demoted to a subspecies of *Myoictis melas*, and has now been re-elevated to a full species. The bat occurs on Sulawesi and the Togian Islands (Indonesia).

Waller

Waller's Gazelle *Litocranius walleri* **Brooke,** 1879 [Alt. Gerenuk]

Brooke described this gazelle on the basis of three skulls in the possession of Gerald Waller, who expressed a wish, in the words of the original etymology, "that the species should be named after his brother, who lost his life in Africa." It is sometimes said that the gazelle was named in honor of the Rev. Horace Waller, who was a missionary in Africa and a friend of Dr. Livingstone, but since he died at his parish in Hampshire in 1896 he cannot be the brother in question. Unfortunately we have not been able to identify the third brother. Gerald Waller was a naturalist who collected in East Africa. He was probably also employed in some capacity by the Imperial British East Africa Company and was involved in negotiations with the Sultan of Zanzibar, who agreed to a concession by which land on the African mainland was ceded to Great Britain. The gazelle is found in eastern Ethiopia, Somalia, Djibouti, Kenya, and northeast Tanzania.

Wallich

Wallich's Deer *Cervus (elaphus) wallichi* **Cuvier,** 1823 [Alt. Shou, Tibetan Red Deer; Syn. *C. affinis*]

Dr. Nathaniel Wallich (1786–1854) was a Danish physician and botanist. In 1806 he graduated from the Royal Academy of Surgeons in Copenhagen, and in the same year he was appointed surgeon to the Danish settlement at Serampore, in Bengal, then known as Frederischnagor, taking up the post in 1807. By the time he arrived the British had annexed the area and he was interned, but he was later paroled into the service of the East India Company. In 1814 he was instrumental in establishing a museum in Calcutta with many of his own botanical specimens. He supervised the gardens belonging to the East India Company in Calcutta, between 1815 and 1846, giving up his museum post around 1819. He prepared a catalogue of more than 20,000 specimens, in addition to publishing *Tentamen flora nepalensis illustratae* in 1824 and *Plantae asiaticae rariories* in 1830. He also collected in Nepal from 1820 to 1822. Hardwicke sent seeds that he had collected to Wallich at the gardens. The scientific name of the Cheer Pheasant *Catreus wallichi* also honors him. The deer, from southern Tibet, has traditionally been regarded as a race of the Red Deer *Cervus elaphus*, but its taxonomy is currently in question; it has been suggested that it should be elevated to a full species, whereas other authorities use the name *C. affinis* for central Asian populations of red deer.

Ward, A.

Ward's Field Mouse *Apodemus wardi* **Wroughton,** 1908 [Syn. *A. pallipes* Barrett-Hamilton, 1900]
Ward's Long-eared Bat *Plecotus wardi* **Thomas,** 1911

Colonel A. E. Ward (dates not found) was an amateur naturalist and member of the Bombay Natural History Society. He was active as a collector in the Ladakh region of northern India, which is where the type specimens of both the above mammals originated. In 1924, in the *Journal of the Bombay Natural History Society,* he published an article, "The Mammals and Birds of Kashmir and the Adjacent Hill Provinces." The taxonomy of the genus *Apodemus* is not fully resolved; it seems that the "correct" (i.e. oldest valid) scientific name for Ward's Field Mouse is *A. pallipes.* It is found in Kyrgyzstan, Tajikistan, Afghanistan, northern Pakistan, northwest India, and Nepal. The bat was long considered to be a subspecies of *Plecotus austriacus* but has been re-elevated to a full species. It is known from the Hindu Kush, Karakoram, and along the southern slopes of the western Himalayas.

Ward, F.

Ward's Red-backed Vole *Eothenomys wardi* **Thomas,** 1912
Ward's Short-tailed Shrew *Blarinella wardi* Thomas, 1915 [Alt. Burmese Short-tailed Shrew]

Captain Francis "Frank" Kingdon-Ward (1885–1958) was an English botanist, collector, and explorer. He wrote *The Land of the Blue Poppy* in 1913 and *Plant Collecting on the Edge of the World* in 1930. He was one of those intrepid late Victorians who went everywhere, enduring the most enormous perils, in his case storms, torrents, and an earthquake measuring over 9.5 on the Richter scale. He traveled extensively in Assam, Burma (now Myanmar), China, and Tibet. He served in the Indian army during WW1 and went on to teach jungle survival techniques to Allied Forces during WW2, after previously avoiding capture by Japanese forces and having made his way alone through the Burmese jungle to India. He was awarded the Founders' Gold Medal of the Royal Geographical Society in 1930. Ward died of a stroke. The vole occurs in the Mekong and Salween valleys of northwest Yunnan (China). The shrew is found in northern Myanmar and Yunnan.

Ward, R.

Greenland Musk-ox *Ovibos moschatus wardi* **Lydekker,** 1900
Ward's Reedbuck *Redunca redunca wardi* **Thomas,** 1900 [Alt. Eastern Bohor Reedbuck, Highland Reedbuck]
Ward's Zebra "*Equus wardi*" Ridgeway, 1910

Rowland Ward (1835–1912) was probably the most famous taxidermist of the Victorian age. He was the son of a taxidermist who had at one time been employed by John James Audubon and ran a shop in Piccadilly in London called The Jungle. He was particularly well known for preserving and mounting big-game trophies. He wrote a famous book, *The Sportsman's Handbook,* which ran to many editions. His sister, Jane, married a man called Tost, by whom she had at least six children; one of them, Ada, married Henry Stewart Rohu and owned a taxidermy and furrier's business in Sydney. The musk-ox is found in Greenland and some of the Arctic islands of Canada. The subspecies has also been introduced into Alaska. The reedbuck comes from Kenya and Tanzania. The story of "Ward's Zebra" is a rather interesting one. A zebra skin was bought from Rowland Ward in the last years of the 19th century by James Cossar Ewart (a zoologist with an interest in hybridizing horses and zebras). This skin was said to have come from Somaliland and appeared distinct from all known forms of zebra. Ewart referred to it informally as Ward's Zebra. Now move on to 1909 and the writings of Professor William Ridgeway in the *Proceedings of the Zoological Society of London*. At first he accepted the validity of the taxon but later learned from Messrs. Ward and Co. that the taxidermy firm did *not* get this unique specimen from Somaliland; it was, in fact, the skin of a specimen from Barnum and Bailey's Menagerie and was most probably a hybrid between a Mountain Zebra and a Chapman's Zebra. With this revelation, "Ward's Zebra" quietly vanished from the zoology textbooks.

Warren

Warren's Spiny Rat *Proechimys warreni* **Thomas,** 1905

S. B. Warren (dates not found) collected the type specimen of the spiny rat in British Guiana (now Guyana). Unfortunately Thomas gives no other information about him, and it has not proved possible to trace him. The rodent is found in Guyana and Suriname. It is perhaps synonymous with *Proechimys guyannensis* (E. Geoffroy, 1803).

Washington

Washington Ground Squirrel *Spermophilus washingtoni* Howell, 1938

This squirrel is named after Washington State, northwest USA, and not directly after President George Washington. It occurs in eastern Washington State and northeast Oregon.

Waterhouse

Waterhouse's Leaf-nosed Bat *Macrotus waterhousii* **Gray,** 1843

George Robert Waterhouse (1810–1888) was a British naturalist and architect whose major interest was entomology. Darwin deputed to him the task of describing the mammals and coleoptera collected on the *Beagle* expedition. He was Curator of the Zoological Society of London's museum from 1836 to 1843. Then from 1851 to 1880 he was at the British Museum, first as Assistant Keeper of the mineralogical and geological branch and later as Keeper of the Department of Geology. The bat is found in Mexico, Belize, and Guatemala, and on various Caribbean islands: Cuba, Jamaica, Hispaniola, the Cayman Islands, and the Bahamas.

Waters

Waters' Gerbil *Gerbillus watersi* **de Winton,** 1901

A. W. Waters (dates not found) worked as a collector for Rothschild. The original etymology reads, "At the request of Mr. Rothschild I name this species in honor of his assistant, Mr. A. W. Waters." The Rothschild referred to was probably Nathaniel Charles Rothschild (1877–1923), whose elder brother was Lord Rothschild of Tring fame. Charles Rothschild was a banker and entomologist who led his own expedition to Sudan in 1901. We surmise that Waters, about whom we could find nothing more recorded, must have been on that expedition. The gerbil is known from Sudan, Djibouti, and Somalia.

Watson

Watson's Climbing Rat *Tylomys watsoni* **Thomas,** 1899
Watson's Fruit-eating Bat *Artibeus watsoni* Thomas, 1901 [Alt. Thomas' Fruit-eating Bat]

H. J. Watson (dates not found) was a collector of zoological specimens in Panama at the end of the 19th century and the beginning of the 20th. Unfortunately we can find nothing more about him. The rat is found in Costa Rica and western Panama. The bat occurs from southern Mexico to Colombia.

Watts

Watts' Pipistrelle *Pipistrellus wattsi* Kitchener, Caputi, and Jones, 1986
Watts' Spiny Rat *Maxomys wattsi* **Musser,** 1991

Dr. Christopher "Chris" Henry Stuart Watts B.Sc. (Hons.), D.Phil. (b. 1939) is an Australian mammalogist who has studied rodents and published on them since the late 1960s, with particular emphasis on the taxonomy and evolution of murids. He was formerly head of the Evolutionary Biology Unit, South Australian Museum, and Chief Scientist at the museum. Although now retired, he is still based at the South Australian Museum in Adelaide as a Honorary Researcher. He is the author of a number of articles on rodents, such as "*Leggadina lakedownensis,* a New Species of Murid Rodent from North Queensland." He appears as co-author of many others, such as, with H. J. Aslin in 1981, "The Rodents of Australia." The bat is found in Papua New Guinea. The spiny rat is known only from the highlands of east-central Sulawesi.

Webb

Webb's Tufted-tailed Rat *Eliurus webbi* **Ellerman,** 1949

Cecil Stanley Webb (b. 1895), known as "Webbie," was a British natural history collector who was Curator-Collector for the Zoological Society of London and responsible for the restocking of both the London and Whipsnade zoos after WW2. He became Curator of Birds and Mammals at the London Zoo and was appointed Superintendent of the Dublin Zoo in 1952 but resigned his position later that year. He wrote a number of articles, particularly in the 1930s, such as "Collecting Waterfowl in Madagascar" (1935), and authored at least one book, *The Odyssey of an Animal Collector* (1954). His collections were sent to the British Museum of Natural History. In 1949 J. R. Ellerman published an article in the *Proceedings of the Zoological Society of London* entitled "Notes on the Rodents from Madagascar in the British Museum, and on a Collection from the Island Obtained by Mr. C. S. Webb." Webb's Madagascar Frog *Mantidactylus webbi* is also named after him. The rat is found in lowland forests of eastern Madagascar.

Weber

Weber's Dwarf Squirrel *Prosciurillus weberi* **Jentink,** 1890
Weber's White-toothed Shrew *Crocidura weberi* Jentink, 1890

Max Wilhelm Carl Weber van Bosse (1852–1937) was a German-Dutch physician and zoologist. He was Director at the Zoological Museum in Amsterdam from 1883, when at the age of 30 he became a naturalized Dutch citizen. He was educated in Germany at Bonn and Berlin. He was in military service in the German army—half the time as a doctor and half as a hussar. In 1881 he made a voyage in a small schooner called *Willem Barents,* appropriately to the Barents Sea. He combined the roles of watch-keeping officer, ship's doctor, and naturalist. His wife, Anna van Bosse, was a skilled botanist, and after their marriage the Webers spent three summers in Norway, where he could dissect whales and she could collect algae—her speciality. They made a number of other voyages: in 1888 to Sumatra, Java, Celebes, and Flores, and in 1894 to South Africa. He was coauthor, with De Beaufort, of the authoritative *The Fishes of the Indo-Australian Archipelago.* He also wrote of "Weber's Line," an important zoogeographical line between Sulawesi and the Moluccas, which is often preferred over Wallace's Line (between Sulawesi and Borneo) as the dividing line between the Oriental and Australasian faunas. He dedicated his great work on fishes to his wife, "who has been always a joyful and helpful traveling-companion to me, in the extreme North, in South Africa, in the Indo-Australian Archipelago and also during the Siboga Expedition." (The "Siboga" expedition was carried out under Weber's personal leadership between 1899 and 1900.) He is also commemorated in the name of a bird, Weber's Lorikeet *Trichoglossus haematodus weberi.* The squirrel is found only in central Sulawesi. The shrew is found in Sumatra, but this "species" is regarded by many mammalogists as a synonym of either *Crocidura beccarii* or *C. malayana.*

Weddell, H. A.

Weddell's Saddle-back Tamarin *Saguinus fuscicollis weddelli* Deville, 1849

Hugh Algernon Weddell (1819–1877) was a botanist and physician, born in England but raised and educated in France. He explored parts of South America from 1843 to 1847 as a member of the Castelnau expedition to Brazil. In particular, he made a study of the cinchona plant, the source of quinine. He wrote *Voyage dans le nord de la Bolivie et dans les parties voisins du Pérou* in 1853 and *Chloris Andina—Essai d'une flore de la région alpine des cordillières de l'Amerique du Sud* in 1855. He was not related to the sealer Captain James Weddell (1787–1834), after whom the Weddell Seal is named (see below). A bird, Weddell's Conure *Aratinga weddellii,* is also named after him. The tamarin occurs in Bolivia, southeast Peru, and adjacent parts of western Brazil south of the Rio Purus.

Weddell, J.

Weddell Seal *Leptonychotes weddellii* **Lesson,** 1826

Captain James Weddell (1787–1834) was a British seaman and explorer, after whom the Weddell Sea is named. He joined the navy in 1796 and by 1815 had risen to the rank of Master and had trained in Antarctic waters. In 1819 he joined the merchant service and persuaded a shipowner to give him command of a brig on a sealing expedition to the newly discovered South Shetland Islands. He returned in 1821 with cargo insufficient for profitability. In 1822 he commanded another ship with a remit to explore further south in search of seals to exploit commercially. The ship had poor provisions but plenty of rum, and this, together with Weddell's leadership skills, enabled it to be successful. Weddell was also skilled at mapping and charting the areas he explored. He had sailed further south than anyone ever had before, and no one managed to repeat his feat for over 90 years. He took home a sketch and a skeleton of the seal in 1824. Weddell continued captaining various merchant vessels but was wrecked in 1829 in the Azores and was saved by lashing himself to a rock. His last voyage, between 1830 and 1832, was to New South Wales and Tasmania. He died in London in poverty. The seal is found around the Antarctic and on such islands as South Georgia, the South Shetlands, and the South Orkney Islands.

Welwitsch

Welwitsch's Bat *Myotis welwitschii* **Gray,** 1866
Kaokoveld Rock Hyrax *Procavia welwitschii* Gray, 1868 [Syn. *P. capensis welwitschii*]

Dr. Friedrich Martin Josef Welwitsch (1806–1872) was an Austrian who worked first as a theater critic in his native country before fleeing to Portugal to escape the consequences of an act of youthful indiscretion. The Portuguese gave him a commission as a plant collector and eventually sent him to Angola to explore and collect. He was in Angola for 12 years and collected over 5,000 specimens, many of which were new to science. He also had the commendable habit of including large amounts of useful information on the labels. He was among the botanists of the day who urged the establishment of the botanical gardens on Madeira. He discovered the plant *Welwitschia mirabilis,* a so-called living fossil, which is named after him. He caused an international quarrel by sending a large proportion of his collection to the Natural History Museum in London instead of to Lisbon. The Portuguese authorities took the view that they had paid him and so the collection should belong to them. In the end the dispute was settled fairly amicably. The collection's duplicate specimens were split off, so both museums got something out of it. Welwitsch himself died in London and is buried in Kensal Green Cemetery, where his tombstone, decorated with an engraving of *Welwitschia,* can be seen to this day. The bat was first found in Angola but has a wide distribution from South Africa to Ethiopia. The hyrax is often treated as a subspecies of *Procavia capensis.* It is found in coastal areas of Angola and northwest Namibia.

Werner

Werner's Guenon *Cercopithecus werneri* **I. Geoffroy,** 1850 [now regarded as a junior synonym of the Green Monkey *Chlorocebus sabaeus* **Linnaeus,** 1766]

Jean-Charles Werner (1798–1856) was a natural history illustrator for the Museum of Natural History in Paris. He produced illustrations for Dumont d'Urville during and after his important scientific expeditions to Australia and the South Pacific. He was also a student of comparative anatomy. Isidore Geoffroy Saint-Hilaire named this species of monkey based on living specimens in the Paris Menagerie, which had an unknown place of origin in Africa. The name Werner's Guenon is hardly ever encountered these days, as it is now known to be merely a junior synonym of the Green Monkey, which is found in West Africa from Senegal to Ghana.

Wetmore

White-collared Fruit Bat *Megaerops wetmorei* **Taylor,** 1934

Frank Alexander Wetmore (1886–1978), always known as Alexander, was an American ornithologist and avian paleontologist who conducted extensive fieldwork in Latin America. His first job was as a bird taxidermist at the Denver Museum of Natural History, Colorado, in 1909. He spent 1911 in Puerto Rico studying birdlife. Later he traveled throughout South America for two years, investigating bird migration between continents, while working for the U.S. Bureau of Biological Survey. In 1925 he was appointed Assistant Secretary of the Smithsonian Institution, the U.S. National Museum, where he worked for 20 years. He was President of the American Ornithologists' Union from 1926 to 1929, and he became the Smithsonian's sixth Secretary, serving from 1945 until 1952. Wetmore made a number of short trips to Haiti, the Dominican Republic, Guatemala, Mexico, Costa Rica, and Colombia. He also conducted a research program in Panama every year from 1946 to 1966, during which he made an exhaustive survey of the birds of the isthmus. He wrote *A Systematic Classification for the Birds of the World* in 1930, which he revised in 1951 and again in 1960. Therein he devised the Wetmore Order, a sequence of bird classification that had widespread acceptance until very recently and is still in use. His other publications included *Birds of Haiti and the Dominican Republic* (1931) and *The Birds of the Republic of Panamá* (1965). Numerous taxa are named in his honor, including the Plain-flanked Rail *Rallus wetmorei* and the lizard *Ameiva wetmorei*. Wetmore also has a glacier and a canopy bridge in the Bayano River rainforest of Panama named after him. He wrote the first descriptions of 189 species and subspecies of living birds, mostly from Central and northern South America. The bat is found on the Malay Peninsula, Sumatra, Borneo, and the island of Mindanao (Philippines).

Wetzel

Wetzel's Climbing Mouse *Rhipidomys wetzeli* **Gardner,** 1989

Dr. Ralph M. Wetzel (fl. 1930–1983) was a Professor of Zoology at the University of Connecticut from 1950 until he retired in 1983; a fund for research into vertebrate zoology has been set up in his memory there. He collected in Paraguay initially, and then in other areas of South America, sending his collections to the University of Connecticut. He published such scientific papers as "Systematics, Distribution, Ecology and Conservation of South American Edentates" (1982). The mouse is found in the highlands of southern Venezuela.

Weyns

Weyns' Duiker *Cephalophus weynsi* **Thomas,** 1901

Lieutenant Colonel Auguste F. G. Weyns (1854–1944) was a Belgian explorer who collected in central Africa from 1888 until 1903. He served as Governor, from 1900 to 1903, of the semiautonomous state of Katanga within the Congo (now Zaire), as "representative" of the Special Committee of Katanga. He is also remembered in the name of a bird, Weyns' Weaver *Ploceus weynsi*. The duiker is found in DRC (Zaire), Uganda, extreme southern Sudan (Imatong Mountains), Rwanda, western Kenya, and possibly also in southwest Ethiopia.

Whitaker

Whitaker's Shrew *Crocidura whitakeri* **de Winton,** 1898

Joseph Isaac Spadafora Whitaker (1850–1936) was a member of a British family that had settled in Palermo, Sicily, in 1806 and had become influential and extremely wealthy as one of the three leading producers of Marsala wine. Joseph was a grandson of the founder of the company and had no interest at all in matters commercial but devoted himself to ornithology, archeology, and botany. He traveled in Tunisia and in 1905 published *Birds of Tunisia*. He bought the island of Mozia, off the Sicilian coast, and excavated there. His villa in Palermo was a center for high society, and its gardens were adorned with exotic plants, including the only example in Europe of *Araucaria rulei* and a huge banyan tree. His collection of bird skins is housed in Edinburgh. The shrew is found in Morocco, Algeria, and Tunisia.

Whitehead

Tufted Pygmy Squirrel *Exilisciurus whiteheadi* **Thomas,** 1887
Whitehead's Spiny Rat *Maxomys whiteheadi* Thomas, 1894
Whitehead's Woolly Bat *Kerivoula whiteheadi* Thomas, 1894
Luzon Striped Rat *Chrotomys whiteheadi* Thomas, 1895
Harpy Fruit Bat *Harpyionycteris whiteheadi* Thomas, 1896

John Whitehead (1860–1899) was a British explorer who collected in Borneo between 1885 and 1888, in the Philippines between 1893 and 1896, and finally on the island of Hainan (China) in 1899. He wrote *Explorations of Mount Kina Balu, North Borneo,* published in 1893, and he may have been the first European to reach the summit of the mountain. He died of fever on Hainan when only 38 years old. He is commemorated in the names of several birds, including Whitehead's Trogon *Harpactes whiteheadi.* The squirrel is confined to the mountains of Borneo. The spiny rat is found on the Malay Peninsula, Sumatra, and Borneo. The woolly bat occurs in the Philippines, Borneo, and the Malay Peninsula. The striped rat is confined to Luzon. The fruit bat is widespread in the Philippines.

Whiteside

Whiteside's Guenon *Cercopithecus ascanius whitesidei* **Thomas,** 1909 [Alt. Yellow-nosed Red-tailed Monkey]

The Rev. H. M. Whiteside (dates not found) was a Baptist missionary in the Lulonga district in the Congo (now Zaire) in 1909. The American ornithologist and naturalist James Paul Chapin, who was in the Congo from 1909 to 1915, records that Whiteside provided him with some ornithological specimens. Chapin also records that Whiteside was in the habit of sending specimens to the British Museum. The monkey is found in DRC (Zaire), south of the Congo River.

Whitley

Bini Free-tailed Bat *Myopterus whitleyi* Scharff, 1900

Dr. J. C. Whitley (dates not found) lived in Benin City, Nigeria. He supplied the type specimen of the bat, which is found from Ghana east to Uganda.

Whyte

Whyte's Hare *Lepus whytei* **Thomas,** 1894 [Alt. Malawi Hare; Syn. *L. microtis whytei*]
Whyte's Mole-Rat *Fukomys whytei* Thomas, 1897 [formerly *Cryptomys hottentotus whytei*]
Whyte's Vervet *Chlorocebus pygerythrus whytei* **Pocock,** 1907

Alexander Whyte F.L.S. (1834–1908) was a government naturalist in Nyasaland (Malawi), where he collected extensively under the patronage of Sir Harry Johnston between 1891 and 1897. Britten wrote *The Plants of Milanji, Nyasa-land, Collected by Mr. Alexander Whyte* (1894), and Thomas wrote "On the Mammals Obtained by Mr. A. Whyte in Nyasaland, and Presented to the British Museum by Sir H. H. Johnston" (*Proceedings of the Zoological Society of London,* 1897). All the mammals named after him come from Malawi. The hare is sometimes listed as a distinct species but is now often considered to be a race of the widespread African Savanna Hare *Lepus microtis.* The mole-rat, conversely, was long regarded as a subspecies of *Cryptomys* (now *Fukomys*) *hottentotus,* but genetic studies have shown it to be a distinct species. The vervet monkey subspecies occurs only on Mount Chiradzulu, Malawi, but this race is probably not distinct from *Chlorocebus pygerythrus rufoviridis.*

Wied

Margay *Leopardus wiedii* Schinz, 1821
Wied's Long-legged Bat *Macrophyllum macrophyllum* Schinz, 1821
Wied's Marmoset *Callithrix kuhlii* Wied-Neuwied, 1826 [Alt. Kuhl's Marmoset, Wied's Black-tufted-ear Marmoset]

Prince Alexander Philipp Maximilian II of Wied-Neuwied (1782–1867) was a German explorer and naturalist. The margay is found from the Texas-Mexico border south to Uruguay and northern Argentina. The bat occurs from southern Mexico to Bolivia and southern Brazil. The marmoset is endemic to Brazil, in southern Bahia State and northwest Minas Gerais. See **Maximilian** for biographical details.

Wilfred

Wilfred's Mouse *Wilfredomys oenax* **Thomas,** 1928 [Alt. Greater Wilfred's Mouse]
Lesser Wilfred's Mouse *Wilfredomys pictipes* **Osgood,** 1933 [Alt. Contreras' Juliomys; Syn. *Juliomys pictipes*]

Wilfred H. Osgood (1875–1947). The Greater Wilfred's Mouse is found in southeast Brazil and Uruguay, the Lesser in extreme northeast Argentina (Misiones Province) and southeast Brazil. The latter species has now been moved into the genus *Juliomys,* leaving *oenax* as the sole member of the genus *Wilfredomys.* See **Osgood** for biographical details.

Wilhelmina

Short-haired Water Rat *Paraleptomys wilhelmina* **Tate** and **Archbold,** 1941 [Alt. Short-haired Hydromyine]
Lesser Antechinus *Antechinus wilhelmina* Tate, 1947 [Syn. *Murexechinus melanurus*]

Both these species are named after Mount Wilhelmina in New Guinea, near to which the type specimens were taken. The mountain was named after Queen Wilhelmina of the Netherlands in the early years of the 20th century, so the mammals are named after her "at one remove." Queen Wilhelmina (1880–1962) was still queen in 1947 and the Netherlands still owned Indonesia, including the western half of New Guinea. She abdicated in 1948 in favor of her daughter, the late Queen Juliana of the Netherlands. The water rat and antechinus are both endemic to central New Guinea. The antechinus is now regarded as being synonymous with *"Antechinus" melanurus*—which has been moved into its own genus, *Murexechinus.*

Williams

Williams' Jerboa *Allactaga williamsi* **Thomas,** 1897

Major W. H. Williams (dates not found) was the British Consul at Van in Kurdistan (Van is in present-day Turkey) in 1896, when he reported on the fighting between Turks and Armenians in the neighborhood. He must have retired or been relieved of his duties quite shortly afterward, as Thomas, in his description of the jerboa, describes Williams as having been "recently H.M. Consul at Van, Kurdistan." The jerboa is found in Turkey.

Williamson

Williamson's Mouse-Deer *Tragulus williamsoni* **Kloss,** 1916

Sir W. J. F. Williamson (1867–1954) was the financial adviser to the government of Siam (now Thailand) from 1904 to 1925. He was coeditor of the *Journal of the Natural History Society of Siam* and published a number of articles in this journal, such as "The Birds of Bangkok." The mouse-deer is known only from the holotype, from northern Thailand.

Wilson, A.

Wilson's Meadow Mouse *Microtus pennsylvanicus pennsylvanicus* **Ord,** 1815 [Alt. Meadow Vole]

Alexander Wilson (1766–1813) was a pioneering American ornithologist, and the first to study American birds in their native habitats. As such he is often called the father of American ornithology. Wilson was born in Paisley, Scotland, where he earned a meager livelihood as an itinerant poet and peddler of muslin. His narrative poem "Watty and Meg" was published anonymously in 1792, attaining great popularity, but was ascribed to the Scottish poet Robert Burns. Subsequently, during a labor dispute in Paisley, Wilson wrote satiric verses lampooning the manufacturers and was imprisoned for libel. Following his release in 1794 he left Scotland for the USA, where he worked as a village schoolmaster in Pennsylvania. Wilson began to collect material for a comprehensive work, illustrated with his own drawings, on the birds of America. From 1808 to 1813 he published seven volumes of his *American Ornithology;* an additional two volumes were edited and published after his death. Ord's original scientific name for this vole was *Mus pennsylvanica.* The vernacular name Wilson's Meadow Mouse is now obsolete and encountered only in historical accounts. The vole has a wide distribution in Alaska and Canada, extending south in the USA to northern New Mexico and northern Georgia. Isolated populations occur in Florida and northern Chihuahua (Mexico).

Wilson, D. E.

Little Blue-eyed Cuscus *Spilocuscus wilsoni* Helgen and **Flannery,** 2004

Dr. Don Ellis Wilson (b. 1944) is a senior scientist and Curator of Mammals at the Smithsonian's National Museum of Natural History. Jointly with D. M. Reeder, he edited *Mammal Species of the World,* which was published by the Smithsonian in 1993 and had reached its third edition by 2005. He is now working on another major project, *Mammals of China.* His other writings include, with K. M. Helgen (who is a joint describer of the cuscus), *The History of the Raccoons of the West Indies* (2002). The cuscus is endemic to the twin islands of Biak and Supiori in Geelvink Bay, northern New Guinea.

Wilson, D. J.

Wilson's Spiny Mouse *Acomys wilsoni* **Thomas,** 1892

D. J. Wilson (dates not found) was an employee of the British East Africa Company, which was a charter company that also acted as the colonial administrator of British possessions in East Africa until 1906, when the Crown assumed direct control. Thomas wrote that Wilson had sent the British Museum various natural history specimens from Mombasa and showed his gratitude by naming the spiny mouse after him. Very little else is known about him, but a D. J. Wilson was recorded as being in Zanzibar around 1894 and 1895, and we believe this to be the same person. The spiny mouse is found from Ethiopia and Somalia south to Tanzania.

Wilson, E. A.

Wilson's Dolphin *Lagenorhynchus cruciger* Quoy and **Gaimard,** 1824 [Alt. Hourglass Dolphin]

Dr. Edward Adrian Wilson (1872–1912) was a naturalist and Antarctic explorer. He was appointed as the Assistant Surgeon and Vertebrate Zoologist to the British National Antarctic Expedition of 1901–1904 aboard *Discovery,* under Commander Robert Falcon Scott. He was also a gifted artist and was probably the last "exploration artist," as after his time photography became the major medium for recording explorations. In 1910 he returned to the Antarctic with Captain Scott aboard *Terra Nova* as Chief of the Scientific Staff. He died with his comrades on the return from the South Pole in 1912. The dolphin occurs in Antarctic and sub-Antarctic waters. Though this species is traditionally placed in the genus *Lagenorynchus*, molecular analysis indicates that it is closer to *Cephalorhyncus*. The dolphin was described in 1824, long before Wilson was born, but in 1915 the name *Lagenorhynchus wilsoni* was coined for a supposed "new" species. This was partly based on a description and drawings of dolphins seen by Wilson in the Antarctic. It was later realized that *wilsoni* was a junior synonym of *cruciger*, but "Wilson's Dolphin" is still occasionally used as a vernacular name.

Wilson, J. L.

Biafran Palm Squirrel *Epixerus wilsoni* Du Chaillu, 1860 [Syn. *E. ebii wilsoni*]

The Rev. John Leighton Wilson D.D. (1809–1896) was an American missionary in West Africa. He was ordained in 1833 and in that year went on a voyage of exploration to West Africa. In early 1834 he returned to the USA, married, then returned to Africa to establish a mission at Cape Palmas, where he and his wife stayed until 1841. During that period they created a written form of the local language, Grebo, and translated two of the Gospels, St. Matthew and St. John, into it. In 1842 Wilson established the headquarters station of the American Board of Foreign (Presbyterian) Missions on the Gaboon River. Again Wilson converted the spoken language of the local Mpongwe people into written form, and parts of the Bible were translated into it. In 1853 his health was failing, so he returned to the USA, where he was appointed as Secretary of the American Presbyterian Board of Foreign Missions. With the outbreak of the American Civil War he resigned his post and returned to his home state, South Carolina, part of the Confederacy. He was appointed to be Secretary of Foreign Missions of the Southern Presbyterian Church and continued in this role until 1885. A pamphlet he wrote in 1852 against the slave trade in West Africa so impressed Lord Palmerston that he was able to use it to silence opposition in Parliament to the presence of the Royal Navy off the West African coast, where it was engaged in trying to eradicate the slave trade, an endeavor in which it was successful in less than five years. Wilson published a number of tracts on his missionary activities in Africa and at least one book, *Western Africa, Its History, Condition, and Prospects* (1857). Of the seaport town of Whydah he says, "There is no place where there is more intense heathenism; and to mention no other feature in their superstitious practices, the worship of snakes at this place fully illustrates this remark. A house in the middle of the town is provided for the exclusive use of these reptiles, and they may be seen here at any time in very great numbers. They are fed, and more care is taken of them than of the human inhabitants of the place. If they are seen straying away they must be brought back; and at the sight of them the people prostrate themselves on the ground, and do them all possible reverence. To kill or injure one of them is to incur the penalty of death. On certain occasions they are taken out by the priests or doctors, and paraded about the streets, the bearers allowing them to coil themselves around their arms, necks, and bodies." The squirrel is found in Cameroon, Equatorial Guinea, and Gabon.

Wimmer

Wimmer's Shrew *Crocidura wimmeri* **Heim de Balsac** and **Aellen,** 1958

Eugène Wimmer (dates not found) was a Swiss civil engineer who went to the Ivory Coast in 1926 to start a construction company that would also provide a place for young Swiss nationals to emigrate to and work in. He was Switzerland's Consul in the Ivory Coast and also an enthusiastic naturalist who was a close friend of two French botanists, George Mangenot, Professor at the Sorbonne, and Claude Favarger, Professor and Director of the Institute of Botany at the University of Neuchâtel. Together the three of them set up a research station in the Ivory Coast, of which the first directors were Ursula Rahm and her husband, Urs. The shrew is known only from southern Ivory Coast.

Winkelmann

Winkelmann's Deer Mouse *Peromyscus winkelmanni* Carleton, 1977 [Alt. Coalcomán Deer Mouse]

Dr. John R. Winkelmann (b. 1931) is Professor of Biology at Gettysburg College, Pennsylvania, where he has taught since 1963. He took his first degree at the University of Illinois and his master's and doctorate at the University of Michigan. Recently he has spent several summers studying bats in Papua New Guinea, Ecuador, and South Africa, and has published numerous papers. Winkelmann collected the type specimen of this mouse in Mexico. Carleton later identified it as being a new species and named it after Winkelmann, a fact Winkelmann discovered only in the 1990s. The species occurs in the states of Michoacan and Guerrero, western Mexico.

Winston

Winston Churchill's Flying Squirrel *Hylopetes winstoni* **Sody,** 1949 [Alt. Sumatran Flying Squirrel]

Sir Winston Leonard Spencer Churchill (1874–1965) was Prime Minister of the UK from 1940 to 1945 and again from 1951 to 1955. It seems unnecessary to give a lengthy précis of the life of so well known a figure. In 2002 the BBC conducted a poll to determine whom the British public regarded as the 100 greatest Britons of all time; Winston Churchill came top of the poll. The flying squirrel is known only from the type locality in northern Sumatra.

Witherby

Steppe Field Mouse *Apodemus witherbyi* **Thomas,** 1902

Henry Forbes (Harry F.) Witherby M.B.E. (1873–1943) was an English ornithologist who was an expert on British and Spanish birds. His varied activities had an enduring influence in most fields of British ornithology. He founded *British Birds* in 1907, a monthly journal that has become an institution. He started the first bird-banding scheme in Britain in 1909. The family firm of H. F. and G. Witherby and Co. began specializing in bird books in the early 20th century. They were merely printers until Harry published his own works. He edited the *Handbook of British Birds,* which was published in five volumes between 1938 and 1941. Witherby was awarded the Godman-Salvin Medal of the British Ornithologists' Union in 1938, a signal honor for distinguished ornithological work. The type specimen of the field mouse was collected by Witherby, and Thomas says of him, "to whom the National Museum owes examples of several interesting Persian mammals." The taxonomy of the genus *Apodemus* is not fully resolved; *witherbyi* may include the taxa known as *hermonensis* and *fulvipectus,* thus giving it a range extending from the Caucasus and Turkey to Pakistan.

Woermann

Woermann's Bat *Megaloglossus woermanni* Pagenstecher, 1885 [Alt. Woermann's Long-tongued Fruit Bat]

Adolph Woermann (1847–1911) entered the family firm of C. Woermann Ltd., which his father had founded in 1837. Its main trade was in spirits and armaments to West Africa, traded for India rubber and palm oil. He became a partner in 1874 and took control of the company in 1880 and, from 1884 to 1890, was a member of the German Parliament and an influential member of the committee dealing with the affairs of Germany's overseas colonies.

In 1885 he founded the Woermann Steamship Line in conjunction with the German East Africa Line (now Deutsches Afrika Linje). He owned properties in a number of West African ports, including in the German colony of Cameroon, where he himself was based for a time. Pagenstecher gives no reason for his choice of *woermanni* as the bat's scientific name, but as Woermann was such an influential figure in Hamburg at the time and Pagenstecher was the Director of the Natural History Museum there, we think he probably assumed it was unnecessary to specify further. The bat is found from Guinea to Uganda and northern Angola and on the island of Bioko.

Wolf

Wolf's Monkey *Cercopithecus wolfi* **Meyer,** 1891

Dr. Ludwig Wolf (1850–1889). In 1883 King Leopold of Belgium authorized what is now know as the Wissmann expedition, to which Wolf was attached as its medical officer, to explore the Congo River system in the Belgian Congo (now Zaire). They sailed from Hamburg at the end of 1883 and explored extensively until 1887, during which period Wolf explored on his own for part of the time. He was one of the first Europeans to reach and enter the Kingdom of the Kuba, a people heavily involved in the slave trade. In 1887 Wolf returned temporarily to Germany, after which he was in West Africa (Togo and Benin), where he died, reportedly following a fall from his horse. Meyer wrote in his description of the monkey, "In the Zoological Garden of Dresden, since the year 1887, there has been a living specimen of a *Cercopithecus,* brought hither from Central West-Africa by Dr. Ludwig Wolf. This specimen so obviously represents an undescribed species of monkey that I need not hesitate any longer in describing it shortly, though this can be done but imperfectly during its life-time." At the end of his description Meyer wrote, "I name this beautiful species *Cercopithecus Wolfi*, in honour of its discoverer, whose early death, which took place in Africa the 26th of June 1889, the scientific world has to deplore. The decease of Dr. Wolf prevents me from ascertaining the exact locality where this specimen was procured." The monkey is found in DRC (Zaire), south of the Congo River.

Wolffsohn

Wolffsohn's Leaf-eared Mouse *Phyllotis wolffsohni* **Thomas,** 1902

Wolffsohn's Viscacha *Lagidium wolffsohni* Thomas, 1907

John A. Wolffsohn (1856–1928) was a British zoologist born in Yorkshire. He worked in Chile from 1891 and collected in some other South American countries. He wrote a number of scientific papers, such as "Contribuciones a la mamalogía Chilena" (1908), "Notas sobre el huemul" (*Revista Chilena de Historia Natural,* 1910), and "Catálogo de cráneos de mamíferos de Chile colectados entre los años 1896 y 1918" (1926). He died in Chile. The mouse inhabits the eastern slopes of the Andes in central Bolivia, while the viscacha is found in southwest Argentina and adjacent parts of Chile.

Wollaston

Wollaston's Roundleaf Bat *Hipposideros wollastoni* **Thomas,** 1913

Mount Everest Pika *Ochotona macrotis wollastoni* Thomas and Hinton, 1922

Dr. Alexander "Sandy" Frederick Richmond Wollaston (1875–1930) was a physician, naturalist, and explorer. He led an expedition to Dutch New Guinea in 1912–1913, having been a participant in an earlier expedition of 1910–1911, which had been deliberately misdirected by the Dutch authorities. On his return he was awarded the Gill Memorial Medal of the Royal Geographical Society. During WW1 he served as a naval surgeon in East Africa, where he met his future brother-in-law, Richard Meinertzhagen (q.v.). He took part in an Everest expedition of 1922 as team doctor and botanist. He wrote a number of papers and longer works, including *Pygmies and Papuans: The Stone Age To-day* (1912) and *An Expedition to Dutch New Guinea* (1914). In 1930 he was teaching in Cambridge when he was shot dead in his rooms by a deranged student, who then shot and killed the policeman who had come to arrest him and finally committed suicide. The bat is found in New Guinea. The pika (a subspecies of the Large-eared Pika) was first found at an altitude of 5,300 m (17,400 feet) on Mount Everest.

Wollebaek

Galapagos Sea-Lion *Zalophus wollebaeki* Sivertsen, 1953

Alf Wollebaek (1879–1960) was a zoologist and Curator at the Oslo Museum who conducted the Norwegian Zoological Expedition to the Galápagos Islands in 1925. He traveled to South America well in advance of the rest of the expedition and rejoined them when their ship, the *Floreana*, arrived. His stay coincided with political unrest, and he spent several weeks of patient waiting at Guayaquil. He relates that during one and a half days of shooting and street fighting he remained in hiding in the hotel. Fortunately for him the revolution ended as quickly as it started. Wollebaek published a number of scientific papers, including "Remarks on Decapod Crustaceans of the North Atlantic and the Norwegian Fjords" (1909). At the time, he was his nation's best-known zoologist and was involved in many bodies such as the Fisheries Board. He also has a creek on one of the Galápagos Islands named after him. This sea-lion, endemic to the Galápagos, was formerly treated as a race of the Californian Sea-Lion but is now often given full-species status.

Wood

Wood's Slit-faced Bat *Nycteris woodi* **K. Andersen,** 1914

Rodney Carrington Wood (1889–1962) traveled widely, particularly in Africa, farmed at times, worked as a game warden in Nyasaland (now Malawi), became a schoolteacher in Natal, and was also employed by the railways in Nyasaland. His real love was natural history, and he was an enthusiastic collector. Although he spent much of his life in Africa, he was in the Seychelles when he died. The bat is recorded from southwest Tanzania, Malawi, Mozambique, Zambia, Zimbabwe, and northeast South Africa.

Woodford

Woodford's Blossom Bat *Melonycteris woodfordi* **Thomas,** 1887 [Alt. Woodford's Fruit Bat]

Dwarf Flying Fox *Pteropus woodfordi* Thomas, 1888

Charles Morris Woodford (1852–1927) was the Resident Commissioner in the Solomon Islands Protectorate between 1896 and 1914. He was an adventurer, a naturalist, and also a philatelist. He established the first postal service in the islands and issued their first stamps, personally franking the envelopes. He wrote *A Naturalist among the Headhunters* (1890), which is referred to in a letter by his friend the novelist Jack London, and *Notes on the Solomon Islands* (1926). He is also commemorated in the name of a bird, Woodford's Rail *Nesoclopeus woodfordi*. Both species of bat are found in the Solomon Islands.

Woodhouse

Woodhouse's Arvicola

Dr. Samuel Washington Woodhouse (1821–1904) was an American surgeon, explorer, and naturalist. He qualified as a physician in 1847 and joined the army. He was a member of the

Creek Indian Boundary Survey under the command of James William Abert in 1849 and 1850. He was doctor and naturalist on the Sitgreaves expedition to the Colorado and Zuni rivers in 1851. This proved to be quite an eventful and hazardous affair, involving fighting hostile Indians, and Woodhouse was wounded by an arrow in his leg. In 1853 and 1854 he was part of an expedition to Nicaragua to investigate the possibility of building a railway line from the Atlantic to the Pacific. When he returned from Central America he went back to the practice of medicine. He was the author of *A Naturalist in Indian Territory: The Journal of S. W. Woodhouse, 1849–50.* In 1856 he resigned his commission in the army, and from then until 1859 he was employed as a surgeon in a military prison. In 1859 he became a ship's doctor on a line trading between Philadelphia and Liverpool, and after the American Civil War he practiced medicine in Philadelphia. Between 1852 and 1853 he published articles in which he described three new species of birds and four new species of mammals. The Woodhouse's Toad *Bufo woodhousii* is also named after him. The name Woodhouse's Arvicola is an obsolete vernacular for a species of North American vole, and we have not been able to find a recognized scientific name for it.

Woods

Woods' Solenodon *Solenodon paradoxus woodi* Ottenwalder, 2001

Professor Dr. Charles Arthur Woods is a biologist who was Curator of Mammalogy at the Florida State Museum, University of Florida. His studied for his bachelor's degree at the University of Denver; his doctorate is from the University of Massachusetts. He worked at the University of Vermont from 1970 to 1979, in which year he took up his position in Florida. In 2000 he returned to the University of Vermont as Adjunct Professor of Biology. He has worked in the Caribbean for many years, in particular in Haiti in connection with the establishment of national parks there. He has published quite extensively on West Indian mammals. The solenodon—a subspecies of the Hispaniolan Solenodon—occurs in parts of southern Haiti and the Dominican Republic. (It is curious that the spelling *woodi* was used in the scientific name, rather than *woodsi.*)

Woodward

Black Wallaroo *Macropus bernardus* **Rothschild,** 1904
Kimberley Rock Rat *Zyzomys woodwardi* **Thomas,** 1909

Bernard Henry Woodward (1846–1912) was born in London but emigrated from there to Western Australia in 1889, where he became a government analyst and Curator of the Geological Museum in Perth. In 1907 he published "National Parks and the Fauna and Flora Reserves in Australia," which appeared in the *West Australian Natural History Society Journal.* He is also remembered in the scientific names of the White-throated Grasswren *Amytornis woodwardi* and the Sandstone Shrike-Thrush *Colluricincla woodwardi.* The wallaroo and rat are found in northern Australia.

Woolley

Woolley's Pseudantechinus *Pseudantechinus woolleyae* Kitchener and Caputi, 1988

Dr. Pat Woolley (b. 1932) is an Australian zoologist. After graduating from the University of Western Australia, in 1955, she worked there as Research Assistant to Professor Harry Waring, investigating marsupial biology. Her lifelong interest in dasyurid marsupials began in 1960 when she moved to Canberra, where she lectured at the Australian National University and completed her Ph.D. in 1966. From 1967 until 2000 she worked at La Trobe University, Melbourne, eventually becoming Reader and Associate Professor. During this time she further researched dasyurid life history and reproduction. In 1991 she started searching for the Julia Creek Dunnart *Sminthopsis douglasi*, which was thought extinct; she caught the first live animals in 1992 and bred them in captivity. During the 1980s and 1990s she made a number of trips to Papua New Guinea searching for and studying dasyurids. She was made an honorary life member of the Australian Mammal Society in 1999 and in 2001 was elected to honorary life membership of the American Society of Mammalogists. In 2000 she received an Outstanding Achievement Award from the Society of Women Geographers and in 2003, the Ellis Troughton Memorial Award and Medal from the Australian Mammal Society. She is currently Emeritus Scholar at La Trobe University. The pseudantechinus is found in Western Australia.

Woosnam

Woosnam's Broad-headed Mouse *Zelotomys woosnami* Schwann, 1906
Woosnam's Brush-furred Rat *Lophuromys woosnami* **Thomas,** 1906

Richard Bowen Woosnam (1880–1915) was a British soldier, traveler, and naturalist. In 1903, after serving in the army during the South African War, he started collecting animals in Africa for the British Museum of Natural History and the Zoological Society of London. In 1905 Woosnam accompanied Colonel A. C. Bailward on a journey from the Persian Gulf to the Black Sea. In 1910 he was appointed Game Warden in the British East African Protectorate but rejoined the army on the outbreak of WW1. He was killed in action at Gallipoli. There is a paper in the *Proceedings of the Zoological Society of London* of 1906 entitled "A List of the Mammals Obtained by Messrs. R. B. Woosnam and R. E. Dent in Bechuanaland." He is also commemorated in the scientific name of a bird, the Trilling Cisticola *Cisticola woosnami*. The mouse is found in Botswana, Namibia, and Northern Cape Province (South Africa). The rat occurs in Rwanda, Burundi, eastern DRC (Zaire), and western Uganda.

Wrangel

Wrangel Collared Lemming *Dicrostonyx vinogradovi* **Ognev,** 1948

Wrangel Island in the Arctic Ocean off Siberia is named after Ferdinand Petrovich von Wrangel (1797–1870), a Russian explorer. Wrangel searched for, but did not find, the island that today bears his name. The lemming is known only from Wrangel Island and is named after the island, not directly after the man.

Wright

Wright's Sportive Lemur *Lepilemur wrighti* Louis et al., 2006

Dr. Patricia C. Wright (b. 1952) is an American primatologist. She is Professor of Anthropology at the State University of New York at Stony Brook, having previously been an Assistant Professor at Duke University. As her dissertation for her doctorate, she wrote what has become a classic on the behavior and ecology of the world's only nocturnal monkeys, the owl monkeys or douroucoulis. She made her first visit to Madagascar in 1986 and was one of the party that discovered the Golden Bamboo Lemur *Hapalemur aureus*. In 1995 she was awarded Madagascar's national Medal of Honor. The sportive lemur was named after her "for her long term dedication and contributions to conservation in Madagascar and tropical environments throughout the world." The binomial of the scientific name should be *wrightae* (feminine), and this lapse in grammar will probably be corrected. The lemur is known from a small area of southern Madagascar.

Wroughton

Wroughton's Free-tailed Bat *Otomops wroughtoni* **Thomas,** 1913 [Alt. Wroughton's Giant Mastiff Bat]

Robert Charles Wroughton (1849–1921) was a member of the Bombay Natural History Society and worked as a forest officer in the Bombay Presidency. In 1911 he organized a survey of mammals making use of the society's membership throughout continental India to provide specimens—perhaps the first collaborative natural history study in the world, resulting in 50,000 specimens (including several new species), 47 publications, and a better understanding of India's biogeography. The results were published in 1919 as "Summary of the Results from the Indian Mammal Survey of the Bombay Natural History Society," with some preliminary results being published in 1917 as "Bombay Natural History Society's Mammal Survey of India, Burma and Ceylon." In 1920 he undertook a survey of the bats of Bangalore and subsequently wrote a paper entitled "Indian Mammal Survey," published in the *Journal of the Bombay Natural History Society,* the journal in which he published the descriptions of most of the mammals he described. This bat was long thought to be endemic to India but has now also been recorded in Cambodia.

Wulsin

Wulsin's Ebony Leaf Monkey *Trachypithecus auratus ebenus* Brandon-Jones, 1995 [Alt. Indochinese Black Langur; Syn. *T. francoisi ebenus*]

Frederick R. Wulsin Jr. (1891–1961) was originally trained as an engineer but regarded himself more as an anthropologist, archeologist, and explorer. Between 1921 and 1925 he and his wife, Janet Elliott Wulsin, traveled (much of the time by bicycle) through Vietnam, Tibet, China, and Inner Mongolia. Between 1931 and 1932 he excavated an archeological site at Turang Tappeh (in modern-day Iran) for the University Museum of the University of Pennsylvania. He collected the type specimen of this monkey in Vietnam in 1924, but it was 70 years before Brandon-Jones described it from its skin, which had been waiting all that time in the American Museum of Natural History. Rather oddly, Brandon-Jones described it as a subspecies of *Trachypithecus auratus* (a species found in Java and Bali). Later it was considered to be a race of François' Langur *T. francoisi.* More recently, DNA work has indicated that *ebenus* may only be a melanistic morph of the Hatinh Leaf Monkey *T. laotum hatinhensis.*

Wynne

Murree Vole *Hyperacrius wynnei* **Blanford,** 1881

Arthur Beevor Wynne (dates not found) was a geologist employed by the Survey of India. He also surveyed in Ireland. In 1864 he described large fossil corals in Sligo as looking "like stumps in a cabbage garden." Blanford, who described the vole, was one of his colleagues. Wynne wrote on geological matters connected with his work—for example, *Memoir on the Geology of Kutch* (1872) and *On the Geology of the Salt Range in the Punjab* (1878). The vole's common name derives from the hill station of Murree in northern Pakistan. The species occurs at high altitudes (1,850–3,050 m; 6,000–10,000 feet) in this region.

Xantippe

Xantippe's Shrew *Crocidura xantippe* **Osgood,** 1910 [Alt. Vermiculate Shrew]

Xantippe was the wife of Socrates (ca. 470–399 B.C.). She was reputed to be a "shrewish" woman, and so her name came to signify a shrew or scolding wife. She was apparently considerably younger than her husband; when Socrates was compelled to commit suicide by drinking hemlock, he was at least 70 years old and she was left with their three sons, one of whom was still an infant. The shrew is found in southeast Kenya and the Usambara Mountains of Tanzania.

Yalden

Yalden's Rat *Desmomys yaldeni* Lavrenchenko, 2003

Dr. Derek William Yalden (b. 1940) is a British zoologist who received his doctorate in 1966 in London. In 2005 he retired from the School of Biological Sciences at the University of Manchester after 40 years' service. In 1968 he was one of the zoologists on the Great Abbai expedition. He has written many scientific papers and longer works, including *History of British Mammals.* In 1975 he co-authored, with P. A. Morris, *The Lives of Bats.* He was the leading author of the "Catalogue of the Mammals of Ethiopia," published in seven parts in *Monitore Zoologico Italiano,* which took from 1976 to 1997 to complete and involved five collecting trips to Ethiopia. The rat is endemic to southwest Ethiopia.

Yamashina

Taiwanese Mole-Shrew *Anourosorex yamashinai* Kuroda, 1935

Marquis Dr. Yoshimaro Yamashina (1900–1989) was the second son of Prince Kikumaro Yamashina and developed a keen interest in birds as a child. He was an ornithology graduate at Tokyo University and was awarded his Ph.D. following research on avian cytology, in affiliation with the University of Hokkaido. In 1932, after army service, he built a museum in his backyard to house his collection of bird specimens and books. Wanting to share it with others, he opened it up to the public, and later it became the Yamashina Institute for Ornithology, moved to larger premises and administered under the auspices of the Department for Education. He was also honored in the trinomial of a bird, *Chalcophaps indica yamashinai,* and received many awards including the Jean Delacour Prize, which some consider the "Nobel Prize in Ornithology." He was co-author of the *Handlist of the Japanese Birds* and author of *Birds in Japan* (1961). In 1981 he described a new species of flightless rail from Okinawa Island. As the mole-shrew's common name implies, the species is from Taiwan where Orii (q.v.) worked as a collector for Marquis Yamashina during 1932 and 1933.

Yepes

Yepes' Long-nosed Armadillo *Dasypus yepesi* Vizcaíno, 1995

Dr. Jose Yepes (1897–1976) was an Argentinean zoologist. He wrote various articles, most prolifically in the 1930s and 1940s. With Angel Cabrera he co-wrote *Mamíferos Sud Americanos* (1940) and *Catálogo de los mamíferos de América del Sur* (1957–1961). The long-nosed armadillo is known from Jujuy and Salta provinces, northwest Argentina.

Yolanda

Santa Fé Tuco-tuco *Ctenomys yolandae* **Contreras** and Berry, 1984

Yolanda Davis is a staff member at the Bernardino Rivadavia Museum in Buenos Aires, Argentina, where she works closely with Professor Julio Contreras, one of those who described the tuco-tuco. This rodent is found in Santa Fé Province, northeast Argentina.

Yonenaga-Yassuda

Yonenaga-Yassuda's Spiny Rat *Trinomys yonenagae* Rocha, 1995 [Alt. Yonenaga's Atlantic Spiny-rat]

Professor Dr. Yatiyo Yonenaga-Yassuda is a Brazilian biologist of Japanese descent. She took three degrees at São Paulo University, culminating in a doctorate in genetic biology in 1973. She became an Assistant Professor in the Biology Department of the Biological Science Institute of the University of São Paulo, Brazil, in 1969. She specializes in vertebrate genetics, in particular those of rodents. She has written numerous articles on "the Brazilian fauna of rodents, marsupials, lizards and amphibians under cytogenetical, molecular and morphological aspects." The rat is found in the state of Bahia, eastern Brazil.

Yoshiyuki

Yoshiyuki's Myotis *Myotis yesoensis* Yoshiyuki, 1984

Dr. Mizuko Yoshiyuki is a Japanese zoologist. For many years she worked at the Department of Zoology, National Science Museum, Tokyo, as a Research Fellow and was Professor at the Tokyo University of Agriculture. She also described the Taiwan Big-eared Bat *Plecotus taivanus* in 1991. She has written a number of papers, such as, with Imaizumi, "Taxonomic Status of the Japanese Otter (Carnivora, Mustelidae), with a Description of a New Species" (1989). The bat is endemic to Hokkaido, Japan.

Young

White-winged Vampire Bat *Diaemus youngi* **Jentink,** 1893

Dr. Charles Grove Young (1849–1934) was in the Medical Service in British Guiana (now Guyana) from 1873 to 1898. He clearly had an overall interest in natural history, as in 1900 he published "Stalk-eyed Crustacea of British Guiana, West Indies, and Bermuda" and, in three parts in *The Ibis* between October 1928 and April 1929, "A Contribution to the Ornithology of the Coastland of British Guiana." He had a relative who was also a keen observer, with whom he co-wrote *Some Birds of Guiana* (1921), illustrated by Charles Grove Young and with the descriptions written by Charles Gore Young. We assume that it was the former after whom the bat was named; the original description says that the type specimen was one of a large collection of bats, preserved in alcohol and "presented to the Leyden Museum by our well known correspondent Dr. C. G. Young from Berbice, New Amsterdam, British Guyana." The bat is found from Mexico to northern Argentina and on Trinidad.

Youngson

Lesser Hairy-footed Dunnart *Sminthopsis youngsoni* **McKenzie** and **Archer,** 1982

W. K. Youngson is an Australian zoologist at the Western Australian Museum. He has written a number of articles since the 1970s, such as, with McKenzie, Burbidge, and Chapman, "The Islands of the North-west Kimberley, Western Australia" (1978). The dunnart is found in desert areas of Australia (Western Australia, Northern Territory, and western Queensland).

Yvonne

Yvonne (Kitchener's) Ningaui *Ningaui yvonneae* Kitchener, Stoddart, and Henry, 1983 [Alt. Southern Ningaui]

Yvonne Carole Kitchener is, we presume, the wife of the senior author of the species description, D. J. Kitchener, though their relationship is not made clear in the original text. This small carnivorous marsupial is found in southern Australia.

Zaitsev

Zaitsev's White-toothed Shew *Crocidura zaitsevi*. **Jenkins,** Abramov, Rozhnov, and Makarova, 2007

Dr. Mikhail V. Zaitsev (1954–2005) worked at the Zoological Institute of the Russian Academy of Sciences, St. Petersburg, and was noted for his studies of the taxonomy of recent and fossil insectivores. With Anna Bannikova, Vladimir Lebedev, and Dmitri Kramerov as coauthors, he wrote "Phylogeny and Systematics of the *Crocidura suaveolens* Species Group: Corroboration and Controversy between Nuclear and Mitochondrial DNA Markers." It was published in 2006, after his death. The shrew is known only from Ngoc Linh Mountain, in Vietnam.

Zakaria

Zakaria's Gerbil *Gerbillus zakariai* Cockrum, Vaughn, and Vaughn, 1976 [Alt. Kerkennah Dipodil; Syn. *Dipodillus zakariai*]

Zakaria Ben Mustapha was President both of the Research Department of the Tunisian Ministry of Agriculture and of TAPNE (Tunisian Association for the Protection of Nature and Environment). The describers named this species in his honor as "only his continued sponsorship, interest and help made our studies of Tunisian mammals possible." The gerbil is found on the Kerkennah Islands off the coast of Tunisia.

Zammarano

Zammarano's White-throated Guenon *Cercopithecus albogularis zammaranoi* **De Beaux,** 1924

Lieutenant Colonel Vittorio Tedesco Zammarano (dates not found) was an Italian traveler and hunter who was involved with the Civic Museum of Milan. He wrote *Azanagò non piange* (Azanagò Does Not Cry) in 1934 and *Auhér mi sogno* (Aughé Dreams about Me) in 1935, both of which were published in Milan. He also wrote books on big-game hunting in Somalia, such as *Hic sunt leones. Un anno di esplorazione e di caccia in Somalia* (Here Be Lions: A Year of Exploration and Hunting in Somalia), published in 1924. In 1924 de Beaux wrote an article entitled "Mammiferi della Somalia Italiana racconta del Maggiore Vittorio Tedesco Zammarano nel Museo Civico de Milano" (Mammals of Italian Somalia Described by Major Vittorio Tedesco Zammarano of the Civic Museum of Milan). He appears to have spent much of the decade 1915–1925 in Africa, and a film he made on location of his time there still exists. Zammarano observed that Ethiopian lions were "more wary and cowardly" than those in other parts of Africa, with the result that hunting in Ethiopia was difficult—hardly surprising, as he had helped to significantly reduce their numbers. The guenon is found in southern Somalia.

Zapadokanad

Zapadokanad's Bear *Ursus arctos pervagor* **Merriam,** 1914

This is a classic example of how a "personalized" vernacular name can be born. Puzzled by mentions on certain websites of "Zapadokanad's Bear," we eventually traced a Czech website using the words "medved západokanadský *Ursus arctos pervagor*"—and all became clear. *Západokanadský* means simply "western Canada." Somebody somewhere in the English-speaking world must have misunderstood this and thought that Zapadokanad was a person, and thus was Zapadokanad's Bear born. The type specimen of this bear was taken at Pemberton Lake in British Columbia. Like many taxa of Brown Bear described by Merriam, the validity of this subspecies is questionable.

Zaphiro

Zaphir's Shrew *Crocidura zaphiri* **Dollman,** 1915

Philip Photious Constantine Zaphiro (1879–1933) was employed by Macmillan as a collector during his Sudan expedition of 1903–1904. He was born in Constantinople (Istanbul) and educated in Cairo. He went to Ethiopia and worked there as a medical dispenser (one reference calls him a "charlatan doctor") before working from 1904 until 1911 for the British East Africa Protectorate on the Kenyan-Ethiopian border. Between 1909 and 1919 he was employed at the British Legation as interpreter for the minister, Wilfred Thesiger, whose son, also Wilfred, was a famous explorer and travel writer. Zaphiro became Vice Consul in 1915, and in 1921 he became Oriental Secretary, working at the British Embassy in Addis Ababa until his death. The shrew is found in southern Ethiopia and Kenya.

Zarudny

Zarudny's Shrew *Crocidura zarudnyi* **Ognev,** 1928
Zarudny's Jird *Meriones zarudnyi* **Heptner,** 1937

Nikolai Alekseyivich Zarudny, sometimes spelled Zarudnyi (1859–1919), was a Russian zoologist, traveler, and ornithologist. From 1879 to 1892 he was a teacher at the Military High School in Orenburg, during which time he undertook five expeditions through the Trans-Caspian region (now Turkmenistan). Then from 1892 until 1906 he was a teacher of natural history at the Pskov Military School and undertook a further four journeys through eastern, central, and western Persia (now Iran), for which he was awarded the Russian Geographical Society's Przhevalski Medal. From 1906 he worked in Tashkent, continuing his exploration of middle Asia. During his expeditions Zarudny collected extensively, and his specimens are now housed by the Zoological Museum of the Russian Academy of Sciences. In 1903 he published *Les reptiles, amphibiens, et poissons de la Perse orientale.* In 1916 he published *Third Excursion over Eastern Persia (Horassan, Seistan and Persian Baluchistan) in 1900–1901.* Zarudny is also commemorated in the scientific name of the Asian Desert Sparrow *Passer zarudnyi* and a subspecies of the Common Pheasant, *Phasianus colchicus zarudnyi.* The shrew comes from southeast Iran, southern Afghanistan, and southwest Pakistan. The jird is found in southern Turkmenistan, northeast Iran, and northern Afghanistan.

Zech

Togo Mole-Rat *Fukomys zechi* **Matschie,** 1900 [Alt. Ghana Mole-Rat; formerly *Cryptomys zechi*]

Count Johann (some say Julius) Nepomuk von Zech auf Neuhofen (1868–1914) was governor of Togo, at that time a German colony, from 1904 until 1910. He had followed a career in the military and was sent to Togo in 1896 as part of the administration there. He rose from District Commissioner to Deputy Governor and finally Governor. In November 1910 he requested retirement from his position on health grounds and returned to Europe. The mole-rat is found in eastern Ghana and western Togo.

Zenker

Pygmy Scaly-tailed Flying Squirrel *Idiurus zenkeri* **Matschie,** 1894
Zenker's Fruit Bat *Scotonycteris zenkeri* Matschie, 1894
Flightless Scaly-tailed Squirrel *Zenkerella insignis* Matschie, 1898 [Alt. Cameroon Scaly-tail]

Georg August Zenker (1855–1922) was a German botanist, ornithologist, and gardener who collected in central Africa from 1895 (or earlier) onward. He had significant land holdings around Bipindi. He devoted much time to collecting plants, termites, fish, and apparently also human bones. He made a particular study of the "pygmies" and other native peoples. He is commemorated in the names of two birds, Zenker's Honeyguide *Melignomon zenkeri* and Zenker's Lovebird *Agapornis swindernianus zenkeri*. The scaly-tailed "squirrels" are more correctly known as anomalures (a rodent family confined to Africa). The pygmy scaly-tail is found from southern Cameroon to Uganda. The flightless scaly-tail occurs in southern Cameroon, Gabon, Equatorial Guinea, and the Central African Republic. Zenker's Fruit Bat is found from Liberia east to the DRC (Zaire) and on the island of Bioko.

Zetta

Red Brocket *Mazama zetta* **Thomas,** 1913 [now *M. americana zetta*]

Thomas sometimes gave mammals a scientific name sounding like a woman's name, but without any explanation for his choice. This is one of those cases. Zetta is a Portuguese name meaning "rose." It is also the last letter of the Latin alphabet. Why the word was applied to a race of deer from northern Colombia was not recorded.

Zhitkov

Greater Fat-tailed Jerboa *Pygeretmus shitkovi* Kuznetsov, 1930

Boris Mikhailovich Zhitkov (1872–1943) was an outstanding Russian zoologist and ornithologist. He was a Professor of Vertebrate Zoology at the Moscow State University, the first scientific "projecter" (manager at the planning stage) of the Astrakhan Nature Reserve, the founder of a scientific school, and author of many publications, including, with S. G. Schtecher, *The Ornithology of the Commander Islands.* He was particularly interested in the biology of the extreme northern regions of Russia and undertook expeditions and research in Novaya Zemla, as well as the White Sea, the River Volga, and central Asia. The All-Russian Scientific Research Institute of Fur Farming and Hunting in Kirov (previously Vyatka) was named after him in 1973. He is known to have been on an expedition to western Siberia with the Kuznetsov brothers. The jerboa is found in the Lake Balkhash region of eastern Kazakhstan.

Ziegler

East Sepik Water Rat *Hydromys ziegleri* Helgen, 2005

Dr. Alan Conrad Ziegler (1930–2003) lived on Hawaii for over 30 years and wrote many books about Hawaiian natural history. He spent 15 years as head of the Vertebrate Zoology Division of the Bishop Museum in Hawaii, followed by another 15 years as an independent zoological consultant, and taught in colleges and at the Zoology Department of the University of Hawaii. In 2002 he published *Hawaiian Natural History, Ecology, and Evolution.* He is also remembered in the scientific name of an extinct bird, the Small Oahu Rail *Porzana ziegleri.* The water rat is found in the East Sepik Province of Papua New Guinea.

Zimmer

Zimmer's Shrew *Crocidura zimmeri* **Osgood,** 1936 [Alt. Upemba Shrew]

Dr. John Todd Zimmer (1889–1957) was an American ornithologist who was Curator of Birds at the Field Museum of Natural History, Chicago, from 1921 to 1930. He wrote *Birds of the Marshall Field Peruvian Expedition* (1930) and *Studies of Peruvian Birds* (1931). He also wrote a catalogue of the Edward E. Ayer Ornithological Library, a part of the Field Museum, in 1926. He was co-author of the *Check-list of Birds of the World,* volume 8, which was published in 1979. He is also commemorated in the names of several birds, including Zimmer's Tapaculo *Scytalopus zimmeri.* The shrew is known only from Upemba National Park, southern DRC (Zaire).

Zimmermann

Zimmermann's Shrew *Crocidura zimmermanni* Wettstein, 1953

Professor Dr. Klaus Zimmermann (d. ca. 1975) worked at the Natural History Museum of the Humboldt Institute, Berlin. In 1966 he was given honorary membership in the American Society of Mammalogists. The shrew is found only in the highlands of Crete.

Zink

Kilimanjaro Mouse-Shrew *Myosorex zinki* **Heim de Balsac** and **Lamotte,** 1956

Dr. Gerhardt Zink (1919–2003) was an ornithologist and zoologist at the State Museum of Natural History, Stuttgart, in the early 1950s. In 1938 he had to enter compulsory military service in the German army, and subsequent events meant he was in the army until 1945. In 1946 he went to university at Tübingen and Munich, where he received his doctorate in natural sciences in 1951. Dr. Zink collected the type specimen of the shrew during the 1951–1952 German East Africa expedition. In 1952 he became a member of the staff of the bird observatory at Radolfzell, where he remained until his retirement in 1984. The shrew was originally described as a subspecies of *Myosorex blarina* but has subsequently been raised to full-species status It is known only from Mount Kilimanjaro.

Zinser

Zinser's Pocket Gopher *Cratogeomys zinseri* **Goldman,** 1939

Juan Zinser (dates not found) was a Mexican civil servant. In 1936, while he was chief of the Wildlife Division of the Mexican Department of Forestry, Game, and Fish, he negotiated a treaty with Major E. A. Goldman of the U.S. Biological Survey for the protection of migratory birds and game mammals. The gopher is endemic to the state of Jalisco, western Mexico.

Zuniga

Zuniga's Dark Rice Rat *Melanomys zunigae* **Sanborn,** 1949

Enrique Zuñiga (dates not found) was a Peruvian zoologist on the staff of the Museum of Natural History in Lima. In 1942 he published *Observaciones ecológicas sobre los mamíferos de Las Lomas.* The rice rat is found in west-central Peru.

Zyl

See **Van Zyl.**

Appendix 1
Vernacular Names

Abbott's Duiker	*Cephalophus spadix*
Abbott's Grey Gibbon	*Hylobates muelleri abbotti*
Abe's Whiskered Bat	*Myotis abei*
Aberdare Shrew	*Surdisorex norae*
Abert's Squirrel	*Sciurus aberti*
Adam's Horseshoe Bat	*Rhinolophus adami*
Aders' Duiker	*Cephalophus adersi*
Admiralty Cuscus	*Spilocuscus kraemeri*
Adolf Friedrichs' Angolan Colobus	*Colobus angolensis ruwenzorii*
Aeecl's Sportive Lemur	*Lepilemur aeeclis*
Aellen's Pipistrelle	*Pipistrellus inexspectatus*
Aellen's Roundleaf Bat	*Hipposideros marisae*
Afghan Flying Squirrel	*Eoglaucomys baberi*
African Groove-toothed Rat	*Mylomys dybowskii*
African Linsang	*Poiana richardsonii*
African Water Rat	*Colomys goslingi*
African Wild Dog	*Lycaon pictus*
Afroalpine Vlei Rat	*Otomys orestes*
Agag Gerbil	*Gerbillus agag*
Agricola's Gracile Opossum	*Cryptonanus agricolai*
Agile Wallaby	*Halmaturus binoe*
Ahmanson's Sportive Lemur	*Lepilemur ahmansoni*
Alaska Marmot	*Marmota broweri*
Alaskan Brown Bear	*Ursus arctos dalli*
Alberico's Broad-nosed Bat	*Platyrrhinus albericoi*
Alcathoe's Myotis	*Myotis alcathoe*
Alcorn's Pocket Gopher	*Pappogeomys alcorni*
Alexander's Bush Squirrel	*Paraxerus alexandri*
Alexander's Cusimanse	*Crossarchus alexandri*
Alfaro's Pygmy Squirrel	*Microsciurus alfari*
Alfaro's Rice Rat	*Oryzomys alfaroi*
Alfaro's Rice Water Rat	*Sigmodontomys alfari*
Allenby's Gerbil	*Gerbillus allenbyi*
Allen's Big-eared Bat	*Idionycteris phyllotis*
Allen's Cotton Rat	*Sigmodon alleni*

Allen's Hutia	*Isolobodon portoricensis*
Allen's Mastiff Bat	*Molossus sinaloae*
Allen's Olingo	*Bassaricyon alleni*
Allen's Round-eared Bat	*Lophostoma carrikeri*
Allen's Squirrel	*Sciurus alleni*
Allen's Squirrel Galago	*Galago alleni*
Allen's Striped Bat	*Glauconycteris alboguttatus*
Allen's Swamp Monkey	*Allenopithecus nigroviridis*
Allen's Wood Mouse	*Hylomyscus alleni*
Allen's Woodrat	*Hodomys alleni*
Allen's Yellow Bat	*Rhogeessa alleni*
Alston's Brown Mouse	*Scotinomys teguina*
Alston's Cotton Rat	*Sigmodon alstoni*
Alston's Woolly Mouse-Opossum	*Micoureus alstoni*
Amami Rabbit	*Pentalagus furnessi*
Amazon River Dolphin	*Inia geoffrensis*
American Pygmy Shrew	*Sorex hoyi*
American Shrew-Mole	*Neurotrichus gibbsii*
Anchieta's Antelope	*Cephalophus (melanorheus) anchietae*
Andean Hairy Armadillo	*Chaetophractus nationi*
Andean Mountain Cat	*Leopardus jacobita*
Andean Squirrel	*Sciurus pucheranii*
Andersen's Bare-backed Fruit Bat	*Dobsonia anderseni*
Andersen's Fruit-eating Bat	*Artibeus anderseni*
Andersen's Horseshoe Bat	*Rhinolophus anderseni*
Anderson's Four-eyed Opossum	*Philander andersoni*
Anderson's Gerbil	*Gerbillus andersoni*
Anderson's Oldfield Mouse	*Thomasomys andersoni*
Anderson's Red-backed Vole	*Myodes andersoni*
Anderson's Rice Rat	*Cerradomys andersoni*
Anderson's Shrew	*Suncus stoliczkanus*
Anderson's Shrew-Mole	*Uropsilus andersoni*
Anderson's Squirrel	*Callosciurus quinquestriatus*
Anderson's White-bellied Rat	*Niviventer andersoni*
Andrews' Beaked Whale	*Mesoplodon bowdoini*
Andrews' Hill Rat	*Bunomys andrewsi*
Andrews' Jerboa	*Stylodipus andrewsi*
Angolan Hairy Bat	*Cistugo seabrai*
Angolan Vlei Rat	*Otomys anchietae*
Anita's Leaf-eared Mouse	*Phyllotis anitae*
Annamite Striped Rabbit	*Nesolagus timminsi*
Annandale's Rat	*Rattus annandalei*
Ansell's Epauletted Fruit Bat	*Epomophorus anselli*
Ansell's Mole-Rat	*Fukomys anselli*
Ansell's Shrew	*Crocidura ansellorum*

Ansell's Wood Mouse	*Hylomyscus anselli*
Ansorge's Cusimanse	*Crossarchus ansorgei*
Ansorge's Free-tailed Bat	*Chaerephon ansorgei*
Ansorge's Grass Rat	*Arvicanthis ansorgei*
Antelope Jackrabbit	*Lepus alleni*
Anthony's Bat	*Sturnira ludovici*
Anthony's Pipistrelle	*Hypsugo anthonyi*
Anthony's Pocket Mouse	*Chaetopidus (fallax) anthonyi*
Anthony's Puma	*Puma concolor anthonyi*
Anthony's Woodrat	*Neotoma anthonyi*
Antillean Ghost-faced Bat	*Mormoops blainvillii*
Antillean Giant Rice Rat	*Megalomys desmarestii*
Aquatic Genet	*Osbornictis piscivora*
Arabian Tahr	*Hemitragus jayakari*
Arata and Thomas' Yellow-shouldered Bat	*Sturnira aratathomasi*
Arends' Golden-Mole	*Carpitalpa arendsi*
Arfak Pygmy Bandicoot	*Microperoryctes aplini*
Arfak Ringtail Possum	*Pseudochirulus schlegeli*
Argali	*Ovis ammon*
Arianus' Rat	*Stenomys omichlodes*
Arnhem Land Rock Rat	*Zyzomys maini*
Arnoux's Beaked Whale	*Berardius arnuxii*
Atherton Antechinus	*Antechinus godmani*
Attenborough's Echidna	*Zaglossus attenboroughi*
Attwater's Pocket Gopher	*Geomys attwateri*
Audubon's Bighorn Sheep (extinct)	*Ovis canadensis auduboni*
Australian False Vampire Bat	*Macroderma gigas*
Australian Snubfin Dolphin	*Orcaella heinsohni*
Ávila Pires' Saddle-back Tamarin	*Saguinus fuscicollis avilapiresi*
Aye-Aye	*Daubentonia madagascariensis*
Ayres' Uacari	*Cacajao ayresi*
Azara's Agouti	*Dasyprocta azarai*
Azara's Dog	*Pseudalopex gymnocercus*
Azara's Grass Mouse	*Akodon azarae*
Azara's Night Monkey	*Aotus azarai*
Azara's Tuco-tuco	*Ctenomys azarai*
Azumi Shrew	*Sorex hosonoi*
Babault's Mouse-Shrew	*Myosorex babaulti*
Bachman's Hare	*Sylvilagus bachmani*
Baer's Wood Mouse	*Hylomyscus baeri*
Bahaman Hutia	*Geocapromys ingrahami*
Bahaman Raccoon	*Procyon maynardii*
Bahamonde's Beaked Whale	*Mesoplodon bahamondi*
Bailey's Bobcat	*Lynx rufus baileyi*

Bailey's Pocket Mouse	*Chaetodipus baileyi*
Bailey's Shrew	*Crocidura baileyi*
Baird's Beaked Whale	*Berardius bairdii*
Baird's Dolphin	*Delphinus bairdi*
Baird's Pocket Gopher	*Geomys breviceps*
Baird's Shrew	*Sorex bairdi*
Baird's Tapir	*Tapirus bairdii*
Baker's Harvest Mouse	*Reithrodontomys bakeri*
Balkan Snow Vole	*Dinaromys bogdanovi*
Baluchistan Pygmy Jerboa	*Salpingotus michaelis*
Banana Bat	*Musonycteris harrisoni*
Banana Climbing Mouse	*Dendromus haymani*
Banded Palm Civet	*Hemigalus derbyanus*
Bangs' Mountain Squirrel	*Syntheosciurus brochus*
Banka Shrew	*Crocidura vosmaeri*
Bannister's Melomys	*Melomys bannisteri*
Baoule's Mouse	*Mus baoulei*
Barbados Raccoon	*Procyon gloveralleni*
Barbara Brown's Brush-tailed Rat	*Isothrix barbarabrownae*
Barbara Brown's Titi	*Callicebus (personatus) barbarabrownae*
Barbe's Leaf Monkey	*Trachypithecus barbei*
Barbour's Rock Mouse	*Petromyscus barbouri*
Barbour's Vlei Rat	*Otomys barbouri*
Barnard's Hairy-nosed Wombat	*Lasiorhinus latifrons barnardi*
Barnes' Mastiff Bat	*Molossus barnesi*
Barnes' Pika	*Ochotona princeps barnesi*
Barren Ground Grizzly Bear	*Ursus arctos richardsoni*
Bartels' Flying Squirrel	*Hylopetes bartelsi*
Bartels' Rat	*Sundamys maxi*
Bartels' Spiny Rat	*Maxomys bartelsii*
Barton's Echidna	*Zaglossus bartoni*
Bastard Big-footed Mouse	*Macrotarsomys bastardi*
Bat genus	*Chaerephon*
Bates' Dwarf Antelope	*Neotragus batesi*
Bates' Shrew	*Crocidura batesi*
Bates' Slit-faced Bat	*Nycteris arge*
Baverstock's Forest Bat	*Vespadelus baverstocki*
Bawean Deer	*Axis kuhlii*
Beach Vole	*Microtus breweri*
Beaked whale genus	*Berardius*
Beatrix Oryx	*Oryx beatrix*
Beatrix's Bat	*Glauconycteris beatrix*
Beaufort's Bare-backed Fruit Bat	*Dobsonia beauforti*
Beccari's Margareta Rat	*Margaretamys beccarii*
Beccari's Mastiff Bat	*Mormopterus beccarii*

Beccari's Sheath-tailed Bat	*Emballonura beccarii*
Beccari's Shrew	*Crocidura beccarii*
Bechstein's Bat	*Myotis bechsteini*
Beddard's Olingo	*Bassaricyon beddardi*
Bedford Takin	*Budorcas taxicolor bedfordi*
Beecroft's Scaly-tailed Squirrel	*Anomalurus beecrofti*
Beecroft's Tree Hyrax	*Dendrohyrax dorsalis*
Behn's Big-eared Bat	*Glyphonycteris behni*
Belding's Ground Squirrel	*Spermophilus beldingi*
Bellier's Striped Grass Mouse	*Lemniscomys bellieri*
Bendire's Shrew	*Sorex bendirii*
Bennett's Chinchilla-Rat	*Abrocoma bennetti*
Bennett's Spear-nosed Bat	*Mimon bennettii*
Bennett's Tree Kangaroo	*Dendrolagus bennettianus*
Bennett's Wallaby	*Macropus rufogriseus fruticus*
Berdmore's Palm Squirrel	*Menetes berdmorei*
Berg's Tuco-tuco	*Ctenomys bergi*
Bergman's Bear	*Ursus arctos piscator*
Bernard's Wolf	*Canis lupus bernardi*
Berthe's Mouse-Lemur	*Microcebus berthae*
Betsileo Sportive Lemur	*Lepilemur betsileo*
Betsileo Woolly Lemur	*Avahi betsileo*
Biafran Palm Squirrel	*Epixerus wilsoni*
Biak Giant Rat	*Uromys boeadii*
Big-eared Opossum	*Didelphis koseritzi*
Bini Free-tailed Bat	*Myopterus whitleyi*
Bishop's Fossorial Spiny Rat	*Clyomys bishopi*
Bishop's Slender Mouse-Opossum	*Marmosops bishopi*
Bismarck Bare-backed Fruit Bat	*Dobsonia praedatrix*
Bismarck Blossom Bat	*Melonycteris melanops*
Bismarck Flying Fox	*Pteropus capistratus*
Bismarck Trumpet-eared Bat	*Kerivoula myrella*
Black Flying Fox	*Pteropus alecto*
Black Snub-nosed Monkey	*Rhinopithecus bieti*
Black Squirrel Monkey	*Saimiri vanzolinii*
Black Wallaroo	*Macropus bernardus*
Black-and-Red Bush Squirrel	*Paraxerus lucifer*
Black-and-Rufous Elephant-Shrew	*Rhynchocyon petersi*
Black-footed Tree Rat	*Mesembriomys gouldii*
Black-lipped Pika	*Ochotona curzoniae*
Blainville's Beaked Whale	*Mesoplodon densirostris*
Blainville's Spotted Dolphin	*Stenella pernettensis*
Blanford's False Serotine	*Hesperoptenus blanfordi*
Blanford's Fox	*Vulpes cana*
Blanford's Fruit Bat	*Sphaerias blanfordi*

Blanford's Jerboa	*Jaculus blanfordi*
Blanford's Rat	*Cremnomys blanfordi*
Blanford's Urial	*Ovis vignei blanfordi*
Blasius' Horseshoe Bat	*Rhinolophus blasii*
Bleyenbergh's Lion	*Panthera leo bleyenberghi*
Blick's Grass Rat	*Arvicanthis blicki*
Blue-eyed Spotted Cuscus	*Spilocuscus wilsoni*
Blyth's Clubfooted Bat	*Tylonycteris pachypus fulvida*
Blyth's Horseshoe Bat	*Rhinolophus lepidus*
Blyth's Pouchbearing Bat	*Saccolaimus saccolaimus crassus*
Blyth's Vole	*Phaiomys leucurus*
Bobrinski's Jerboa	*Allactodipus bobrinskii*
Bobrinski's Serotine	*Eptesicus bobrinskoi*
Bocage's Gerbil	*Gerbilliscus validus*
Bocage's Mole-Rat	*Fukomys bocagei*
Bocage's Rock Rat	*Aethomys bocagei*
Bocage's Tree Squirrel	*Funisciurus bayonii*
Bodenheimer's Pipistrelle	*Hypsugo bodenheimeri*
Boeadi's Roundleaf Bat	*Hipposideros boeadii*
Böhm's Bush Squirrel	*Paraxerus boehmi*
Böhm's Gerbil	*Gerbilliscus boehmi*
Böhm's Zebra	*Equus quagga boehmi*
Bokermann's Nectar Bat	*Lonchophylla bokermanni*
Bolam's Mouse	*Pseudomys bolami*
Bole's Douroucouli	*Aotus bipunctatus*
Bonaparte's Weasel	*Mustela cicognanii*
Bonetto's Tuco-tuco	*Ctenomys bonettoi*
Bonhote's Gerbil	*Gerbillus bonhotei*
Bonhote's Mouse	*Mus famulus*
Bontebok	*Damaliscus dorcas*
Bornean Mountain Ground Squirrel	*Dremomys everetti*
Borneo Black-banded Squirrel	*Callosciurus orestes*
Borneo Roundleaf Bat	*Hipposideros doriae*
Bosman's Potto	*Perodicticus potto*
Botta's Gerbil	*Gerbillus bottai*
Botta's Pocket Gopher	*Thomomys bottae*
Botta's Serotine	*Eptesicus bottae*
Bottego's Shrew	*Crocidura bottegi*
Bougainville's Melomys	*Melomys bougainville*
Bourlon's Genet	*Genetta bourloni*
Bourret's Horseshoe Bat	*Rhinolophus paradoxolophus*
Boutourlini's Blue Monkey	*Cercopithecus mitis boutourlinii*
Bouvier's Red Colobus	*Piliocolobus pennantii bouvieri*
Bowdoin's Beaked Whale	*Mesoplodon bowdoini*
Bowers' White-toothed Rat	*Berylmys bowersi*

Brandt's Bat	*Myotis brandtii*
Brandt's Hamster	*Mesocricetus brandti*
Brandt's Hedgehog	*Hemiechinus hypomelas*
Brandt's Vole	*Lasiopodomys brandtii*
Brandt's Yellow-toothed Cavy	*Galea flavidens*
Brants' Climbing Mouse	*Dendromus mesomelas*
Brants' Whistling Rat	*Parotomys brantsii*
Brazilian Big-eyed Bat	*Chiroderma doriae*
Brelich's Snub-nosed Monkey	*Rhinopithecus brelichi*
Brewer's Shrew Mole	*Parascalops breweri*
Bridges' Degu	*Octodon bridgesi*
Bright's Gazelle	*Gazella granti brighti*
Bristle-faced Freetail Bat	*Mormopterus eleryi*
Broad-striped Dasyure	*Paramurexia rothschildi*
Broad-striped Tube-nosed Bat	*Nyctimene aello*
Brockman's Gerbil	*Gerbillus brockmani*
Brockman's Mouse	*Myomyscus brockmani*
Brock's Yellow-eared Bat	*Vampyressa brocki*
Bronze Quoll	*Dasyurus spartacus*
Brooke's Duiker	*Cephalophus brookei*
Brooke's Squirrel	*Sundasciurus brookei*
Brooks' Large-headed Fruit Bat	*Dyacopterus brooksi*
Broom Hare	*Lepus castroviejoi*
Brosset's Big-eared Bat	*Micronycteris brosseti*
Brown Dorcopsis	*Dorcopsis muelleri*
Brown-bearded Sheath-tailed Bat	*Taphozous achates*
Brown's Hutia	*Geocapromys brownii*
Brown's Pademelon	*Thylogale browni*
Bruijn's Echidna	*Zaglossus bruijni*
Bruijn's Pademelon	*Thylogale brunii*
Brumback's Night Monkey	*Aotus brumbacki*
Brush Deer Mouse	*Peromyscus boylii*
Brush-tailed Mulgara	*Dasycercus blythi*
Bryant's Woodrat	*Neotoma bryanti*
Bryde's Whale	*Balaenoptera brydei*
Budgett's Tantalus Monkey	*Chlorocebus tantalus budgetti*
Budin's Chinchilla-Rat	*Abrocoma budini*
Budin's Grass Mouse	*Akodon budini*
Budin's Tuco-tuco	*Ctenomys budini*
Buettikofer's Epauletted Bat	*Epomops buettikoferi*
Buettikofer's Monkey	*Cercopithecus petaurista buettikoferi*
Buettikofer's Shrew	*Crocidura buettikoferi*
Buffon's Kob	*Kobus kob*
Buffon's Tarsier	*Tarsius syrichta*
Buffy Flower Bat	*Erophylla sezekorni*

Buller's Chipmunk *Tamias bulleri*
Buller's Pocket Gopher *Pappogeomys bulleri*
Bulmer's Fruit Bat *Aproteles bulmerae*
Bumback's Night Monkey *Aotus brumbacki*
Bunker's Woodrat *Neotoma bunkeri*
Bunn's Short-tailed Bandicoot Rat *Nesokia bunni*
Bunting's Thicket Rat *Grammomys buntingi*
Bunyoro Rabbit *Poelagus marjorita*
Burchell's Zebra *Equus quagga burchelli*
Burmeister's Porpoise *Phocoena spinipinnis*
Burmese Red Goral *Naemorhedus baileyi cranbrooki*
Burton's Gerbil *Gerbillus burtoni*
Burton's Melomys *Melomys burtoni*
Burt's Deer Mouse *Peromyscus caniceps*
Bushy-tailed Olingo *Bassaricyon gabbii*
Busuanga Squirrel *Sundasciurus hoogstraali*
Butler's Dunnart *Sminthopsis butleri*
Buxton's Bushbuck *Tragelaphus buxtoni*
Buxton's Jird *Meriones sacramenti*
Byatt's Bush Squirrel *Paraxerus (vexillarius) byatti*
Byrne's Marsupial Mouse *Dasyroides byrnei*

Cabrera's Hutia *Mesocapromys angelcabrerai*
Cabrera's Vole *Microtus cabrerae*
Cacomistle *Bassariscus sumichrasti*
Cadena's Nectar Bat *Lonchophylla cadenai*
Cadena's Tailless Bat *Anoura cadenai*
Cadorna's Pipistrelle *Hypsugo cadornae*
Calaby's Pademelon *Thylogale calabyi*
Calaby's Pebble-mound Mouse *Pseudomys calabyi*
Calamian Tree-Shrew *Tupaia moellendorffi*
California Ground Squirrel *Spermophilus beecheyi*
Callewaert's Mouse *Mus callewaerti*
Campbell's Hamster *Phodopus campbelli*
Campbell's Monkey *Cercopithecus campbelli*
Cansdale's Long-eared Flying Mouse *Idiurus macrotis cansdalei*
Cansdale's Swamp Rat *Malacomys cansdalei*
Cantor's Dusky Leaf Monkey *Trachypithecus obscurus halonifer*
Cantor's Roundleaf Bat *Hipposideros galeritus*
Canut's Horseshoe Bat *Rhinolophus canuti*
Cape Rock Elephant Shrew *Elephantulus edwardii*
Cape York Pipistrelle *Pipistrellus adamsi*
Carmen Mountain Shrew *Sorex milleri*
Carriker's Round-eared Bat *Lophostoma carrikeri*
Carriker's Spiny Rat *Pattonomys carrikeri*

Carruthers' Juniper Vole	*Neodon (juldaschi) carruthersi*
Carruthers' Mountain Squirrel	*Funisciurus carruthersi*
Carter's Myotis	*Myotis carteri*
Caucasian Shrew	*Sorex satunini*
Central African Red Colobus	*Piliocolobus foai*
Central American Red Brocket	*Mazama sartorii*
Central American Squirrel Monkey	*Saimiri oerstedii*
Ceram Bandicoot	*Rhynchomeles prattorum*
Chalchalero Viscacha Rat	*Salinoctomys loschalchalerosorum*
Champion's Tree Mouse	*Pogonomys championi*
Chanler's Mountain Reedbuck	*Redunca fulvorufula chanleri*
Chapin's Free-tailed Bat	*Chaerephon chapini*
Chapman's Prehensile-tailed Hutia	*Mysateles gundlachi*
Chapman's Rice Rat	*Oryzomys chapmani*
Chapman's Zebra	*Equus quagga chapmanni*
Checkered Elephant-Shrew	*Rhynchocyon cirnei*
Cheesman's Gerbil	*Gerbillus cheesmani*
Cheng's Jird	*Meriones chengi*
Cherrie's Pocket Gopher	*Orthogeomys cherriei*
Chestnut Dunnart	*Sminthopsis archeri*
Chestnut Long-tongued Bat	*Lionycteris spurrelli*
Child's Rice Rat	*Nephelomys childi*
Chilean Climbing Mouse	*Irenomys tarsalis*
Chimpanzee genus	*Pan*
Chinese Desert Cat	*Felis bieti*
Chinese Forest Musk Deer	*Moschus berezovskii*
Chinese Noctule	*Nyctalus plancyi*
Chinese Serow	*Naemorhedus (sumatraensis) milneedwardsii*
Chinese Zokor	*Eospalax fontanierii*
Chiriqui Olingo	*Bassaricyon pauli*
Chiru	*Pantholops hodgsonii*
Christie's Long-eared Bat	*Plecotus christiei*
Christy's Dormouse	*Graphiurus christyi*
Chudeau's Spiny Mouse	*Acomys chudeaui*
Cinderella Fat-tailed Opossum	*Thylamys cinderella*
Cinderella's Shrew	*Crocidura cinderella*
Ciscaucasian Hamster	*Mesocricetus raddei*
Claire's Mouse-Lemur	*Microcebus mamiratra*
Clara's Echymipera	*Echymipera clara*
Clarke's Gazelle	*Ammodorcas clarkei*
Clarke's Vole	*Volemys clarkei*
Cleber's Arboreal Rice Rat	*Oecomys cleberi*
Cleese's Woolly Lemur	*Avahi cleesei*
Cloet's Vervet	*Chlorocebus aethiops cloetei*

Clymene Dolphin	*Stenella clymene*
Cockrum's Desert Shrew	*Notiosorex cockrumi*
Coimbra-Filho's Titi	*Callicebus coimbrai*
Coke's Hartebeest	*Alcelaphus buselaphus cokii*
Colburn's Tuco-tuco	*Ctenomys colburni*
Collie's Squirrel	*Sciurus colliaei*
Colombian rice rat genus	*Handleyomys*
Colonel Patterson's Eland	*Taurotragus oryx pattersonianus*
Commerson's Dolphin	*Cephalorhynchus commersonii*
Commerson's Roundleaf Bat	*Hipposideros commersoni*
Commissaris' Long-tongued Bat	*Glossophaga commissarisi*
Common Marmoset	*Callithrix jacchus*
Conover's Tuco-tuco	*Ctenomys conoveri*
Contreras' Juliomys	*Juliomys pictipes*
Cook's Mouse	*Mus cookii*
Cookson's Wildebeest	*Connochaetes taurinus cooksoni*
Cooper's Margay	*Leopardus wiedii cooperi*
Cooper's Melomys	*Melomys cooperae*
Cooper's Mountain Squirrel	*Paraxerus cooperi*
Cooper's Shrew	*Sorex cooperi*
Coppery Brushtail Possum	*Trichosurus johnstonii*
Coquerel's Mouse-Lemur	*Mirza coquereli*
Coquerel's Sifaka	*Propithecus coquereli*
Corbett's Tiger	*Panthera tigris corbetti*
Cordeaux's Dik-dik	*Madoqua saltiana cordeauxi*
Cordier's Angolan Colobus	*Colobus angolensis cordieri*
Cosens' Gerbil	*Gerbillus cosensi*
Cotton's Colobus	*Colobus angolensis cottoni*
Cotton's Oribi	*Ourebia ourebi cottoni*
Cotton's Wide-lipped Rhinoceros	*Ceratotherium simum cottoni*
Cotton-top Tamarin	*Saguinus oedipus*
Coues' Climbing Mouse	*Rhipidomys couesi*
Coues' Deer	*Odocoileus virginianus couesi*
Coues' Rice Rat	*Oryzomys couesi*
Count Branicki's Terrible Mouse	*Dinomys branickii*
Cowan's Shrew-Tenrec	*Microgale cowani*
Coxing's White-bellied Rat	*Niviventer coxingi*
Cox's Roundleaf Bat	*Hipposideros coxi*
Crandall's Saddle-back Tamarin	*Saguinus fuscicollis crandalli*
Crawford-Cabral's Marsh Rat	*Dasymys cabrali*
Crawshay's Hare	*Lepus crawshayi*
Crawshay's Zebra	*Equus quagga crawshaii*
Creagh's Horseshoe Bat	*Rhinolophus creaghi*
Crespo's Pampas Cat	*Oncifelis colocolo crespoi*
Crested Rat	*Lophiomys imhausi*

Crimson-nosed rat genus — *Bibimys*
Cross River Gorilla — *Gorilla gorilla diehli*
Crosse's Shrew — *Crocidura crossei*
Crossley's Dwarf Lemur — *Cheirogaleus crossleyi*
Cross' Guenon — *Cercopithecus preussi*
Crowther's Bear (extinct) — *Ursus crowtheri*
Crump's Mouse — *Diomys crumpi*
Cruz Lima's Saddle-back Tamarin — *Saguinus fuscicollis cruzlimai*
Culion Tree Squirrel — *Sundasciurus moellendorffi*
Curio's Giant Rat (extinct) — *Megaoryzomys curioi*
Curry's Bat — *Glauconycteris curryae*
Curry's Red Rock Rabbit — *Pronolagus rupestris curryi*
Cuvier's Beaked Whale — *Ziphius cavirostris*
Cuvier's Fire-footed Squirrel — *Funisciurus pyrropus*
Cuvier's Gazelle — *Gazella cuvieri*
Cuvier's Hutia — *Plagiodontia aedium*
Cuvier's Spiny Rat — *Proechimys cuvieri*
Cuvier's Spotted Dolphin — *Stenella dubia*
Cuvier's Vervet — *Chlorocebus pygerythrus pygerythrus*
Cyclops Long-beaked Echidna — *Zaglossus attenboroughi*
Cyclops Roundleaf Bat — *Hipposideros cyclops*

Dabbene's Mastiff Bat — *Eumops dabbenei*
Dahl's Jird — *Meriones dahli*
D'Albertis Ringtail Possum — *Pseudochirops albertisii*
Dallon's Gerbil — *Gerbillus dalloni*
Dall's Porpoise — *Phocoenoides dalli*
Dall's Sheep — *Ovis dalli*
Dalquest's Pocket Mouse — *Chaetodipus dalquesti*
Dalton's Mouse — *Praomys daltoni*
D'Anchieta's Fruit Bat — *Plerotes anchietai*
D'Anchieta's Pipistrelle — *Hypsugo anchietai*
Danfoss Mouse-Lemur — *Microcebus danfossi*
Dang's Giant Squirrel (extinct) — *Ratufa indica dealbata*
Daniel's Tufted-tailed Rat — *Eliurus danieli*
Daphne's Oldfield Mouse — *Thomasomys daphne*
Dark Tube-nosed Bat — *Nyctimene celaeno*
Darling's Horseshoe Bat — *Rhinolophus darlingi*
Dartmouth's Vlei Rat — *Otomys dartmouthi*
Darwin's Fox — *Pseudalopex fulvipes*
Darwin's Galápagos Mouse — *Nesoryzomys darwini*
Darwin's Leaf-eared Mouse — *Phyllotis darwini*
Darwin's Sheep — *Ovis ammon darwini*
Daubenton's Bat — *Myotis daubentoni*
Daubenton's Free-tailed Bat — *Myopterus daubentonii*

David's Echymipera	*Echymipera davidi*
Davies' Big-eared Bat	*Glyphonycteris daviesi*
Davis' Long-tongued Bat	*Glossophaga alticola*
Davis' Maroon Langur	*Presbytis rubicunda chrysea*
Davis' Pocket Gopher	*Geomys personatus davisi*
Davis' Round-eared Bat	*Tonatia evotis*
Davy's Naked-backed Bat	*Pteronotus davyi*
Dawson's Caribou (extinct)	*Rangifer tarandus dawsoni*
Day's Grass Mouse	*Akodon dayi*
Day's Shrew	*Suncus dayi*
De Balsac's Mouse	*Heimyscus fumosus*
De Beaux's Grivet	*Chlorocebus aethiops zavattarii*
De Brazza's Monkey	*Cercopithecus neglectus*
De Graaf's Soft-furred Mouse	*Praomys degraafi*
De la Torre's Yellow-shouldered Bat	*Sturnira magna*
De Vis' Bare-backed Fruit Bat	*Dobsonia pannietensis*
De Vis' Woolly Rat	*Mallomys aroaensis*
De Vivo's Disk-winged Bat	*Thyroptera devivoi*
De Vivo's Rice Rat	*Cerradomys vivoi*
De Winton's Golden-Mole	*Cryptochloris wintoni*
De Winton's Long-eared Bat	*Laephotis wintoni*
De Winton's Shrew	*Chodsigoa hypsibius*
De Winton's Tree Squirrel	*Funisciurus substriatus*
Decken's Horseshoe Bat	*Rhinolophus deckenii*
Decken's Sifaka	*Propithecus deckenii*
Dekay's Shrew	*Sorex dekayi*
Dekeyser's Nectar Bat	*Lonchophylla dekeyseri*
Delacour's Leaf Monkey	*Trachypithecus delacouri*
Delacour's Marmoset-Rat	*Hapalomys delacouri*
Delany's Swamp Mouse	*Delanymys brooksi*
Demidoff's Galago	*Galago demidoff*
Dent's Horseshoe Bat	*Rhinolophus denti*
Dent's Monkey	*Cercopithecus denti*
Dent's Shrew	*Crocidura denti*
Dent's Vlei Rat	*Otomys denti*
Deppe's Squirrel	*Sciurus deppei*
Derby's Pale-eared Woolly Opossum	*Caluromys derbianus*
Deroo's Mouse	*Praomys derooi*
Desert Bighorn Sheep	*Ovis canadensis nelsoni*
Desert Cottontail	*Sylvilagus audubonii*
Desert Dormouse	*Selevinia betpakdalaensis*
Desert Musk Shrew	*Crocidura smithii*
Desert Shrew	*Notiosorex crawfordi*
Desmarest's Fig-eating Bat	*Stenoderma rufum*
Desmarest's Hutia	*Capromys pilorides*

Desmarest's Spiny Pocket Mouse	*Heteromys desmarestianus*
Diana Monkey	*Cercopithecus diana*
Dian's Tarsier	*Tarsius dianae*
Diard's Cat	*Neofelis diardi*
Dibatag	*Ammodorcas clarkei*
Dice's Rabbit	*Sylvilagus dicei*
Dickey's Deer Mouse	*Peromyscus dickeyi*
Dieter's Myotis	*Myotis dieteri*
Dieterlen's Brush-furred Rat	*Lophuromys dieterleni*
Dingaan's Yellow Bat	*Scotophilus dinganii*
Dobson's Epauletted Bat	*Epomops dobsoni*
Dobson's Fruit Bat	*Dobsonia chapmani*
Dobson's Horseshoe Bat	*Rhinolophus yunanensis*
Dobson's Large-eared Bat	*Micronycteris brachyotis*
Dobson's Long-tongued Fruit Bat	*Eonycteris spelaea*
Dobson's Painted Bat	*Kerivoula africana*
Dobson's Shrew-Tenrec	*Microgale dobsoni*
Doggett's Blue Monkey	*Cercopithecus doggetti*
Dogramaci's Vole	*Microtus dogramacii*
Dolphin genus	*Steno*
Dollman's Mosaic-tailed Rat	*Melomys dollmani*
Dollman's Rock Rat	*Aethomys chrysophilus dollmani*
Dollman's Spiny Rat	*Maxomys dollmani*
Dollman's Tree Mouse	*Prionomys batesi*
Dollman's Vlei Rat	*Otomys dollmani*
Dolorous Grass Mouse	*Akodon dolores*
Don Felipe's Weasel	*Mustela felipei*
D'Orbigny's Round-eared Bat	*Tonatia silvicola*
D'Orbigny's Tuco-tuco	*Ctenomys dorbignyi*
Dorcas Gazelle	*Gazella dorcas*
Doria's Tree Kangaroo	*Dendrolagus dorianus*
Dormer's Pipistrelle	*Pipistrellus dormeri*
Dorothy's Slender Mouse-Opossum	*Marmosops dorothea*
Doucet's Musk Shrew	*Crocidura douceti*
Douglas' Squirrel	*Tamiasciurus douglasii*
Drouhard's Shrew-Tenrec	*Microgale drouhardi*
Dubost's Bristly Mouse	*Neacomys dubosti*
Dudu's Brush-furred Rat	*Lophuromys dudui*
Duke of Abruzzi's Free-tailed Bat	*Chaerephon aloysiisabaudiae*
Duke of Bedford's Vole	*Proedromys bedfordi*
Dunn's Gerbil	*Gerbillus dunni*
Durga Das' Leaf-nosed Bat	*Hipposideros durgadasi*
Dusky Antechinus	*Antechinus swainsonii*
Dusky Rat	*Rattus colletti*
Duthie's Golden-Mole	*Chlorotalpa duthieae*

Dwarf Bharal	*Pseudois schaeferi*
Dwarf Dog-faced Bat	*Molossops temminckii*
Dwarf Flying Fox	*Pteropus woodfordi*
Dybowski's Sika Deer	*Cervus nippon hortulorum*
Earless Water Rat	*Crossomys moncktoni*
Ear-spot Squirrel	*Callosciurus adamsi*
East Sepik Water Rat	*Hydromys ziegleri*
Eastern Amazon Climbing Mouse	*Rhipidomys emiliae*
Eastern Barred Bandicoot	*Perameles gunnii*
Eastern Gorilla	*Gorilla beringei*
Eastern White-eared Giant Rat	*Hyomys goliath*
Ebian Palm Squirrel	*Epixerus ebii*
Ecuador Fish-eating Rat	*Anotomys leander*
Ecuadorian Cougar	*Puma concolor soderstromii*
Eden's Whale	*Balaenoptera edeni*
Edith's Leaf-eared Mouse	*Graomys edithae*
Edward's Long-clawed Mouse	*Notiomys edwardsii*
Eisentraut's Mouse-Shrew	*Myosorex eisentrauti*
Eisentraut's Pipistrelle	*Hypsugo eisentrauti*
Eisentraut's Shrew	*Crocidura eisentrauti*
Eisentraut's Striped Mouse	*Hybomys eisentrauti*
Eld's Deer	*Cervus eldii*
Ellerman's Tufted-tailed Rat	*Eliurus ellermani*
Elliot's Red Colobus	*Piliocolobus foai ellioti*
Elliot's Short-tailed Shrew	*Blarina hylophaga*
Emilia's Gracile Mouse-Opossum	*Gracilinanus emiliae*
Emilia's Marmoset	*Callithrix emiliae*
Emilia's Short-tailed Opossum	*Monodelphis emiliae*
Emily's Tuco-tuco	*Ctenomys emilianus*
Emin's Gerbil	*Taterillus emini*
Emin's Giant Pouched Rat	*Cricetomys emini*
Emma's Giant Rat	*Uromys emmae*
Emmons' Rice Rat	*Euryoryzomys emmonsae*
Ender's Small-eared Shrew	*Cryptotis endersi*
Endo's Pipistrelle	*Pipistrellus endoi*
Entellus Langur	*Semnopithecus entellus*
Erlanger's Dik-dik	*Madoqua saltiana erlangeri*
Erlanger's Gazelle	*Gazella gazella erlangeri*
Ernst Mayr's Water Rat	*Leptomys ernstmayri*
Erxleben's Guenon	*Cercopithecus pogonias grayi*
Ethiopian Hare	*Lepus fagani*
Ethiopian Thicket Rat	*Grammomys minnae*
Eva's Deer Mouse	*Peromyscus eva*
Eva's Red-backed Vole	*Caryomys eva*

Everett's Ferret-Badger	*Melogale everetti*
Everett's Grizzled Langur	*Presbytis hosei everetti*
Eversmann's Hamster	*Allocricetulus eversmanni*
Falanouc	*Eupleres goudotii*
False Serotine Bat	*Hesperoptenus doriae*
Fardoulis' Blossom Bat	*Melonycteris fardoulisi*
Father Basilio's Striped Mouse	*Hybomys basilii*
Fat-tailed Gerbil	*Pachyuromys duprasi*
Fat-tailed Pseudantechinus	*Pseudantechinus macdonnellensis*
Fea's Muntjac	*Muntiacus feae*
Fea's Tree Rat	*Chiromyscus chiropus*
Felten's Vole	*Microtus felteni*
Ferguson Island Mouse	*Chiruromys forbesi shawmayeri*
Fernandez's Sword-nosed Bat	*Lonchorhina fernandezi*
Fernandina Galápagos Mouse	*Nesoryzomys fernandinae*
Field's Mouse	*Pseudomys fieldi*
Fijian Blossom Bat	*Notopteris macdonaldi*
Fijian Mastiff Bat	*Chaerephon bregullae*
Findley's Myotis	*Myotis findleyi*
Finlayson's Cave Bat	*Vespadelus finlaysoni*
Finlayson's Squirrel	*Callosciurus finlaysonii*
Finsch's Tree Kangaroo	*Dendrolagus inustus finschi*
Fischer's Little Fruit Bat	*Rhinophylla fischerae*
Fischer's Shrew	*Crocidura fischeri*
Fishing Bat	*Myotis vivesi*
Fitzroy's Dolphin	*Lagenorhynchus obscurus*
Flamarion's Tuco-tuco	*Ctenomys flamarioni*
Fleurete's Sportive Lemur	*Lepilemur fleuretae*
Flightless Scaly-tailed Squirrel	*Zenkerella insignis*
Flores Giant Tree Rat	*Papagomys armandvillei*
Flores Shrew	*Suncus mertensi*
Flores Warty Pig	*Sus heureni*
Flower's Gerbil	*Gerbillus floweri*
Flower's Shrew	*Crocidura floweri*
Foa's Red Colobus	*Procolobus foai*
Foch's Tuco-tuco	*Ctenomys fochi*
Fontanier's Cat	*Catopuma temmincki tristis*
Forbes' Tree Mouse	*Chiruromys forbesi*
Forest Giant Squirrel	*Protoxerus stangeri*
Forest-steppe Marmot	*Marmota kastschenkoi*
Fornes' Pygmy Rice Rat	*Oligoryzomys fornesi*
Forrest's Mountain Vole	*Neodon forresti*
Forrest's Mouse	*Leggadina forresti*
Forrest's Pika	*Ochotona forresti*

Forrest's Rock Squirrel	*Sciurotamias forresti*
Forster's Fur Seal	*Arctocephalus forsteri*
Forsyth Major's Sifaka	*Propithecus verreauxi verreauxi*
Foster's Shrew	*Sorex fosteri*
Four-striped Ground Squirrel	*Lariscus hosei*
Fox's Shaggy Rat	*Dasymys foxi*
Fox's Shrew	*Crocidura foxi*
Franciscana	*Pontoporia blainvillei*
François' Leaf Monkey	*Trachypithecus francoisi*
Franklin's Ground Squirrel	*Spermophilus franklinii*
Franquet's Epauletted Bat	*Epomops franqueti*
Fraser's Dolphin	*Lagenodelphis hosei*
Fraser's Musk Shrew	*Crocidura poensis*
Fremont's Squirrel	*Tamiasciurus hudsonicus fremonti*
Frith's Tailless Bat	*Coelops frithi*
Fruit bat genus	*Dobsonia*
Fynbos Golden-Mole	*Amblysomus corriae*
Gabriella's Crested Gibbon	*Nomascus gabriellae*
Gairdner's Shrew-Mouse	*Mus pahari*
Gaisler's Long-eared Bat	*Plecotus teneriffae gaisleri*
Galápagos Sea-Lion	*Zalophus wollebaeki*
Gallagher's Free-tailed Bat	*Chaerephon gallagheri*
Gansu Mole	*Scapanulus oweni*
Gapper's Red-backed Vole	*Myodes gapperi*
Gardner's Climbing Mouse	*Rhipidomys gardneri*
Gardner's Spiny Rat	*Proechimys gardneri*
Garlepp's Mouse	*Galenomys garleppi*
Garnett's Galago	*Otolemur garnettii*
Garrido's Hutia	*Mysateles garridoi*
Gaskell's False Serotine	*Hesperoptenus gaskelli*
Gaumer's Spiny Pocket Mouse	*Heteromys gaumeri*
Gebe Cuscus	*Phalanger alexandrae*
Geelvink Bay Flying Fox	*Pteropus pohlei*
Genoways' Yellow Bat	*Rhogeessa genowaysi*
Geoffroy's Bat	*Myotis emarginatus*
Geoffroy's Cat	*Oncifelis geoffroyi*
Geoffroy's Ground Squirrel	*Xerus erythropus*
Geoffroy's Horseshoe Bat	*Rhinolophus clivosus*
Geoffroy's Marmoset	*Callithrix geoffroyi*
Geoffroy's Monk Saki	*Pithecia monachus monachus*
Geoffroy's Pied Colobus	*Colobus vellerosus*
Geoffroy's Rayed Bat	*Platyrrhinus lineatus*
Geoffroy's Rousette	*Rousettus amplexicaudatus*
Geoffroy's Saddle-back Tamarin	*Saguinus fuscicollis nigrifrons*

Geoffroy's Spider Monkey	*Ateles geoffroyi*
Geoffroy's Tailless Bat	*Anoura geoffroyi*
Geoffroy's Tamarin	*Saguinus geoffroyi*
Geoffroy's Woolly Monkey	*Lagothrix cana cana*
Gerenuk	*Litocranius walleri*
German's Melomys	*Melomys paveli*
German's Shrew-Mouse	*Mayermys germani*
Gervais' Beaked Whale	*Mesoplodon europaeus*
Gervais' Fruit-eating Bat	*Artibeus cinereus*
Gervais' Funnel-eared Bat	*Nyctiellus lepidus*
Gervais' Large-eared Bat	*Micronycteris minuta*
Giant Atlantic Tree Rat	*Phyllomys thomasi*
Giant Bandicoot	*Peroryctes broadbenti*
Giant Forest Hog	*Hylochoerus meinertzhageni*
Giant Genet	*Genetta victoriae*
Giant Golden-Mole	*Chrysospalax trevelyani*
Giant Mole-Shrew	*Anourosorex schmidi*
Giant Naked-tailed Rat	*Uromys anak*
Giant Sable Antelope	*Hippotragus niger variani*
Giant Thicket Rat	*Grammomys gigas*
Giant White-tailed Rat	*Uromys sherrini*
Giffard's Shrew	*Crocidura giffardi*
Gilbert's Dunnart	*Sminthopsis gilberti*
Gilbert's Potoroo	*Potorous gilbertii*
Gilded Tube-nosed Bat	*Murina rozendaali*
Giles' Planigale	*Planigale gilesi*
Gill's Bottle-nosed Dolphin	*Tursiops gilli*
Giovanni's Big-eared Bat	*Micronycteris giovanniae*
Glacier Bear	*Ursus americanus emmonsi*
Glacier Rat	*Stenomys richardsoni*
Gland-tailed Free-tailed Bat	*Chaerephon bemmeleni*
Glass' Shrew	*Crocidura glassi*
Glen's Long-fingered Bat	*Miniopterus gleni*
Glen's Wattled Bat	*Glauconycteris gleni*
Glover's Pika	*Ochotona gloveri*
Gmelin's Shrew	*Crocidura gmelini*
Godman's Long-Tailed Bat	*Choeroniscus godmani*
Godman's Rock Wallaby	*Petrogale godmani*
Goeldi's Marmoset	*Callimico goeldi*
Goeldi's Spiny Rat	*Proechimys goeldii*
Goff's Pocket Gopher (extinct)	*Geomys pinetis goffi*
Golden Atlantic Tree Rat	*Echimys blainvillei*
Golden Leaf Monkey	*Trachypithecus geei*
Golden Monkey	*Cercopithecus kandti*
Golden Mouse	*Ochrotomys nuttalli*

Golden Snub-nosed Monkey	*Rhinopithecus roxellana*
Goldman's Nectar Bat	*Lonchophylla mordax*
Goldman's Pocket Gopher	*Cratogeomys goldmani*
Goldman's Pocket Mouse	*Chaetodipus goldmani*
Goldman's Small-eared Shrew	*Cryptotis goldmani*
Goldman's Spiny Pocket Mouse	*Heteromys goldmani*
Goldman's Water Mouse	*Rheomys raptor*
Goldman's Woodrat	*Neotoma goldmani*
Goliath Shrew	*Crocidura goliath*
Goodfellow's Tree Kangaroo	*Dendrolagus goodfellowi*
Goodfellow's Tuco-tuco	*Ctenomys goodfellowi*
Goodman's Mouse-Lemur	*Microcebus lehilahytsara*
Goodwin's Bat	*Chiroderma trinitatum*
Goodwin's Mouse-like Hamster	*Calomyscus elburzensis*
Goodwin's Small-eared Shrew	*Cryptotis goodwini*
Goodwin's Spiny Pocket Mouse	*Heteromys (desmarestianus) nigricaudatus*
Goodwin's Water Mouse	*Rheomys mexicanus*
Gordon's Red Colobus	*Piliocolobus gordonorum*
Gordon's Wild Cat	*Felis silvestris gordoni*
Gorgas' Rice Rat	*Oryzomys gorgasi*
Gould's Mouse	*Pseudomys gouldii*
Gould's Nyctophilus	*Nyctophilus gouldi*
Gould's Rat-Kangaroo	*Bettongia gouldii*
Gould's Wattled Bat	*Chalinolobus gouldii*
Goya Tuco-tuco	*Ctenomys perrensi*
Graells' Tamarin	*Saguinus graellsi*
Graffman Dolphin	*Stenella graffmani*
Grandidier's Free-tailed Bat	*Chaerephon leucogaster*
Grandidier's Mongoose	*Galidictis grandidieri*
Grandidier's Tufted-tailed Rat	*Eliurus grandidieri*
Grant's Caribou	*Rangifer tarandus granti*
Grant's Gazelle	*Gazella granti*
Grant's Golden-Mole	*Eremitalpa granti*
Grant's Rock Rat	*Aethomys granti*
Grant's Shrew	*Sylvisorex granti*
Grant's Zebra	*Equus quagga boehmi*
Grasse's Shrew	*Crocidura grassei*
Grauer's Gorilla	*Gorilla beringei graueri*
Grauer's Shrew	*Paracrocidura graueri*
Gray's Beaked Whale	*Mesoplodon grayi*
Gray's Crowned Guenon	*Cercopithecus pogonias grayi*
Gray's Dolphin	*Stenella coeruleoalba*
Gray's Four-striped Squirrel	*Funisciurus isabella*
Gray's Long-tongued Bat	*Glossophaga leachii*

Gray's Monk Saki	*Pithecia irrorata irrorata*
Gray's Spinner Dolphin	*Stenella longirostris*
Gray's Spotted Dolphin	*Stenella attentuata*
Great Evening Bat	*Ia io*
Great Kai Island Giant Rat	*Uromys siebersi*
Greater Asiatic Yellow Bat	*Scotophilus heathi*
Greater Cane Rat	*Thryonomys swinderianus*
Greater Dog-like Bat	*Peropteryx kappleri*
Greater Fat-tailed Jerboa	*Pygeretmus shitkovi*
Greater Long-tailed Hamster	*Tscherskia triton*
Greater Monkey-faced Bat	*Pteralopex flanneryi*
Greater Musky Fruit Bat	*Ptenochirus jagori*
Greater Ranee Mouse	*Haeromys margarettae*
Greater Wilfred's Mouse	*Wilfredomys oenax*
Green Acouchi	*Myoprocta pratti*
Green Ringtail Possum	*Pseudochirops archeri*
Greenhall's Dog-faced Bat	*Molossops greenhalli*
Greenland Musk-ox	*Ovibos moschatus wardi*
Green's Puma	*Puma concolor greeni*
Greenwood's Shrew	*Crocidura greenwoodi*
Gregory's Red Wolf	*Canis rufus gregoryi*
Gressitt's Melomys	*Paramelomys gressitti*
Grevy's Zebra	*Equus grevyi*
Grewcock's Sportive Lemur	*Lepilemur grewcocki*
Grey Brocket	*Mazama tschudii*
Grey Duiker	*Sylvicapra grimmia*
Grey Leaf-eared Mouse	*Graomys lockwoodi*
Grey Short-tailed Shrew	*Blarinella griselda*
Grey Slender Loris	*Loris lydekkerianus*
Grey Whale	*Eschrichtius robustus*
Grey-bellied Pygmy Mouse	*Mus triton*
Griselda's Striped Grass Mouse	*Lemniscomys griselda*
Grobben's Gerbil	*Gerbillus grobbeni*
Groove-toothed Forest Mouse	*Leimacomys buettneri*
Groove-toothed Shrew-Mouse	*Microhydromys richardsoni*
Guadelupe Fur Seal	*Arctocephalus townsendi*
Guam Flying Fox (extinct)	*Pteropus tokudae*
Guldenstadt's Shrew	*Crocidura gueldenstaedtii*
Gundlach's Hutia	*Mysateles gundlachi*
Gunning's Golden-Mole	*Neamblysomus gunningi*
Gunnison's Prairie Dog	*Cynomys gunnisoni*
Günther's Dik-dik	*Madoqua guentheri*
Günther's Spiny Rat	*Trinomys dimidiatus*
Günther's Vole	*Microtus guentheri*
Guyanan Red Howler Monkey	*Alouatta (seniculus) macconnelli*

Hagenbeck's Mangabey *Cercocebus galeritus hagenbecki*
Hagenbeck's Rhinoceros *Dicerorhinus sumatrensis*
Hagen's Flying Squirrel *Petinomys hageni*
Haggard's Leaf-eared Mouse *Phyllotis haggardi*
Haggard's Oribi *Ourebia ourebi haggardi*
Hahn's Short-tailed Bat *Carollia subrufa*
Haig's Tuco-tuco *Ctenomys haigi*
Hainald's Rat *Rattus hainaldi*
Hairy-footed Flying Squirrel *Belomys pearsonii*
Hairy-legged Myotis *Myotis keaysi*
Halmahera Blossom Bat *Syconycteris carolinae*
Hamadryas Baboon *Papio hamadryas*
Hamilton's Serval *Leptailurus serval hamiltoni*
Hamilton's Tomb Bat *Taphozous hamiltoni*
Hamlyn's Monkey *Cercopithecus hamlyni*
Hammond's Rice Rat *Mindomys hammondi*
Hanak's Pipistrelle *Pipistrellus hanaki*
Hanang Brush-furred Rat *Lophuromys makundii*
Handley's Nectar Bat *Lonchophylla handleyi*
Handley's Red Bat *Lasiurus atratus*
Handley's Short-tailed Opossum *Monodelphis handleyi*
Handley's Slender Mouse-Opossum *Marmosops handleyi*
Handley's Tailless Bat *Anoura cultrata*
Hanuman Langur *Semnopithecus entellus*
Hardwicke's Woolly Bat *Kerivoula hardwickei*
Harlan's Gibbon *Nomascus concolor concolor*
Harlan's Ground Sloth (extinct) *Glossotherium harlani*
Harlan's Musk-ox (extinct) *Bootherium bombifrons*
Harpy Fruit Bat *Harpyionycteris whiteheadi*
Harrington's Gerbil *Taterillus harringtoni*
Harrington's Rat *Desmomys harringtoni*
Harris' Antelope Squirrel *Ammospermophilus harrisii*
Harris' Olingo *Bassaricyon lasius*
Harris' Rice Water Rat *Sigmodontomys aphrastus*
Harrison's Fruit Bat *Lissonycterisgoliath*
Harrison's Tube-nosed Bat *Murina harrisoni*
Hartmann's Water Mouse *Rheomys raptor hartmanni*
Hartmann's Zebra *Equus zebra hartmannae*
Hart's Fruit-eating Bat *Enchisthenes hartii*
Hartwig's Soft-furred Mouse *Praomys hartwigi*
Harvey's Duiker *Cephalophus harveyi*
Harwood's Gerbil *Gerbillus harwoodi*
Hasselt's Myotis *Myotis hasseltii*
Hatt's Vesper Rat *Otonyctomys hatti*
Hawaiian Monk Seal *Monachus schauinslandi*

Hawk's Sportive Lemur	*Lepilemur tymerlachsoni*
Hayden's Shrew	*Sorex haydeni*
Hayman's Dwarf Epauletted Fruit Bat	*Micropteropus intermedius*
Heath Rat	*Pseudomys shortridgei*
Heaviside's Dolphin	*Cephalorhynchus heavisidii*
Heck's Macaque	*Macaca hecki*
Heck's Wildebeest	*Connochaetes taurinus hecki*
Hector's Beaked Whale	*Mesoplodon hectori*
Hector's Dolphin	*Cephalorhynchus hectori*
Heermann's Kangaroo-Rat	*Dipodomys heermanni*
Heinrich's Hill Rat	*Bunomys heinrichi*
Heller's Broad-nosed Bat	*Platyrrhinus helleri*
Heller's Pipistrelle	*Neoromicia helios*
Heller's Rock Rat	*Aethomys helleri*
Heller's Vervet	*Chlorocebus pygerythrus arenarius*
Hellwald's Spiny Rat	*Maxomys hellwaldii*
Hemprich's Long-eared Bat	*Otonycteris hemprichii*
Hendee's Spiny Rat	*Proechimys hendeei*
Hendee's Woolly Monkey	*Oreonax flavicauda*
Henley's Gerbil	*Gerbillus henleyi*
Heptner's Pygmy Jerboa	*Salpingotus heptneri*
Herbert's Rock Wallaby	*Petrogale herberti*
Herman's Myotis	*Myotis hermani*
Hernández-Camacho's Black Tamarin	*Saguinus nigricollis hernandezi*
Hernández-Camacho's Night Monkey	*Aotus jorgehernandezi*
Hernández-Camacho's Short-tailed Bat	*Carollia monohernandezi*
Hershkovitz's Grass Mouse	*Abrothrix hershkovitzi*
Hershkovitz's Marmoset	*Callithrix intermedia*
Hershkovitz's Night Monkey	*Aotus hershkovitzi*
Heude's Pig	*Sus bucculentus*
Heuglin's Gazelle	*Gazella rufifrons tilonura*
Heuglin's Olive Baboon	*Papio anubis heuglini*
Hewitt's Red Rock Hare	*Pronolagus saundersiae*
Highveld Gerbil	*Gerbilliscus brantsii*
Hildebrandt's Horseshoe Bat	*Rhinolophus hildebrandti*
Hildebrandt's Multimammate Mouse	*Mastomys hildebrandtii*
Hildegarde's Broad-headed Mouse	*Zelotomys hildegardeae*
Hildegarde's Shrew	*Crocidura hildegardeae*
Hildegarde's Tomb Bat	*Taphozous hildegardeae*
Hilgendorf's Tube-nosed Bat	*Murina hilgendorfi*
Hillier's Mulgara	*Dasycercus hillieri*
Hills' Horseshoe Bat	*Rhinolophus hillorum*
Hill's Shrew	*Crocidura hilliana*
Hill's Tomb Bat	*Taphozous hilli*
Himalayan Striped Squirrel	*Tamiops macclellandi*

Hinde's Lesser House Bat	*Scotoecus hindei*
Hinde's Rock Rat	*Aethomys hindei*
Hodgson's Brown-toothed Shrew	*Episoriculus caudatus*
Hodgson's Giant Flying Squirrel	*Petaurista magnificus*
Hodgson's Myotis	*Myotis formosus*
Hodgson's White-bellied Rat	*Niviventer niviventer*
Hodson's Puma	*Puma concolor hudsoni*
Hoffmann's Pika	*Ochotona hoffmanni*
Hoffmann's Rat	*Rattus hoffmanni*
Hoffmanns' Titi	*Callicebus hoffmannsi*
Hoffmann's Two-toed Sloth	*Choloepus hoffmanni*
Hokkaido Flying Squirrel	*Pteromys volans orii*
Hollister's Lion	*Panthera leo hollisteri*
Home's Wombat	*Vombatus ursinus ursinus*
Homez's Big-eared Bat	*Micronycteris homezi*
Hoogerwerf's Rat	*Rattus hoogerwerfi*
Hoogstraal's Gerbil	*Gerbillus hoogstraali*
Hoogstraal's Striped Grass Mouse	*Lemniscomys hoogstraali*
Hooker's Sea Lion	*Phocarctos hookeri*
Hook's Duiker	*Cephalophus (nigrifrons) hooki*
Hoolock's Gibbon	*Hylobates hoolock*
Hooper's Deer Mouse	*Peromyscus hooperi*
Hopkins' Groove-toothed Swamp Rat	*Pelomys hopkinsi*
Horn-skinned Bat	*Eptesicus floweri*
Horsfield's Flying Squirrel	*Iomys horsfieldii*
Horsfield's Myotis	*Myotis horsfieldii*
Horsfield's Short-nosed Fruit Bat	*Cynopterus horsfieldi*
Horsfield's Shrew	*Crocidura horsfieldii*
Horsfield's Tarsier	*Tarsius bancanus*
Hose's Hill Rat	*Bunomys fratrorum*
Hose's Leaf Monkey	*Presbytis hosei*
Hose's Palm Civet	*Diplogale hosei*
Hose's Pygmy Flying Squirrel	*Petaurillus hosei*
Hose's Shrew	*Suncus hosei*
Hosono's Myotis	*Myotis hosonoi*
Hotson's Jerboa	*Allactaga hotsoni*
Hotson's Mouse-like Hamster	*Calomyscus hotsoni*
Howell's Shrew	*Sylvisorex howelli*
How's Melomys	*Melomys howi*
Hubbard's Sportive Lemur	*Lepilemur hubbardi*
Hubbs' Beaked Whale	*Mesoplodon carlhubbsi*
Hubert's Multimammate Mouse	*Mastomys huberti*
Huet's Bush Squirrel	*Paraxerus ochraceus*
Huet's Dormouse	*Graphiurus hueti*
Huey's Kangaroo Rat	*Dipodomys antiquarius*

Hugh's Hedgehog	*Mesechinus hughi*
Humboldt's Big-eared Brown Bat	*Histiotus humboldti*
Humboldt's Hog-nosed Skunk	*Conepatus humboldtii*
Humboldt's Squirrel Monkey	*Saimiri sciureus cassiquiarensis*
Humboldt's Woolly Monkey	*Lagothrix lagothricha*
Hume's Argali	*Ovis ammon humei*
Hummelinck's Vesper Mouse	*Calomys hummelincki*
Hun Shrew	*Crocidura attila*
Hunter's Hartebeest	*Damaliscus hunteri*
Husson's Water Rat	*Hydromys hussoni*
Husson's Yellow Bat	*Rhogeessa hussoni*
Hutterer's Brush-furred Rat	*Lophuromys huttereri*
Hutton's Tube-nosed Bat	*Murina huttoni*
Ihering's Brucie	*Brucepattersonius iheringi*
Ihering's Short-tailed Opossum	*Monodelphis iheringi*
Ihering's Spiny Rat	*Trinomys iheringi*
Ikonnikov's Myotis	*Myotis ikonnikovi*
Illiger's Saddle-back Tamarin	*Saguinus fuscicollis illigeri*
Imaizumi's Horseshoe Bat	*Rhinolophus imaizumii*
Imaizumi's Vole	*Myodes imaizumii*
Indian Bush Rat	*Golunda ellioti*
Indian Gazelle	*Gazella bennetti*
Indian Grey Mongoose	*Herpestes edwardsii*
Indian Hairy-footed Gerbil	*Gerbillus gleadowi*
Indochinese Flying Squirrel	*Hylopetes phayrei*
Indochinese Leaf Monkey	*Trachypithecus germaini*
Indochinese Leopard	*Panthera pardus delacouri*
Inez's Red-backed Vole	*Caryomys inez*
Ingram's Planigale	*Planigale ingrami*
Ingram's Squirrel	*Sciurus (aestuans) ingrami*
Inland Broad-nosed Bat	*Scotorepens balstoni*
Iranian Jerboa	*Allactaga firouzi*
Iranian Mouse-like Hamster	*Calomyscus bailwardi*
Isabella Shrew	*Sylvisorex isabellae*
Isabelle's Ghost Bat	*Diclidurus isabellus*
Issel's Groove-toothed Swamp Rat	*Pelomys isseli*
Jackson's Fat Mouse	*Steatomys jacksoni*
Jackson's Hartebeest	*Alcelaphus buselaphus jacksoni*
Jackson's Mongoose	*Bdeogale jacksoni*
Jackson's Shrew	*Crocidura jacksoni*
Jackson's Soft-furred Mouse	*Praomys jacksoni*
Jamaican Monkey (extinct)	*Xenothrix mcgregori*
James' Gerbil	*Gerbillus jamesi*

James' Sportive Lemur	*Lepilemur jamesi*
Jameson's Red Rock Hare	*Pronolagus randensis*
Japanese Grass Vole	*Microtus montebelli*
Javan Tailless Fruit Bat	*Megaerops kusnotoi*
Jeffery's Tamarin	*Saguinus geoffroyi*
Jelski's Altiplano Mouse	*Chroeomys jelskii*
Jenkins' Shrew	*Crocidura jenkinsi*
Jenkins' Shrew-Tenrec	*Microgale jenkinsae*
Jentink's Dormouse	*Graphiurus crassicaudatus*
Jentink's Duiker	*Cephalophus jentinki*
Jentink's Guenon	*Cercopithecus signatus*
Jerdon's Palm Civet	*Paradoxurus jerdoni*
Joffre's Pipistrelle	*Hypsugo joffrei*
Johan's Spiny Mouse	*Acomys johannis*
John Hill's Roundleaf Bat	*Hipposideros edwardshilli*
John's Langur	*Trachypithecus johnii*
Johnson's Hutia (extinct)	*Plagiodontia ipnaeum*
Johnson's Mouse	*Pseudomys johnsoni*
Johnston's Dormouse	*Graphiurus johnstoni*
Johnston's Genet	*Genetta johnstoni*
Johnston's Grey-cheeked Mangabey	*Lophocebus (albigena) johnstoni*
Johnston's Nyassa Wildebeest	*Connochaetes taurinus johnstoni*
Johnston's Pygmy Shrew	*Sylvisorex johnstoni dieterleni*
Johnston's Shrew	*Sylvisorex johnstoni*
Johnstone's Giant Mastiff Bat	*Otomops johnstonei*
Jolly's Mouse-Lemur	*Microcebus jollyae*
Jones' Roundleaf Bat	*Hipposideros jonesi*
José's Hocicudo	*Oxymycterus josei*
Jouvenet's Shrew	*Crocidura jouvenetae*
Juan Fernandez Fur Seal	*Arctocephalus philippii*
Juldasch's Vole	*Neodon juldaschi*
Julia Creek Dunnart	*Sminthopsis douglasi*
Juliana's Golden-Mole	*Neamblysomus julianae*
Julian's Gerbil	*Gerbillus juliani*
Ka'apori Capuchin	*Cebus (olivaceus) kaapori*
Kai Myotis	*Myotis stalkeri*
Kaiser's Rock Rat	*Aethomys kaiseri*
Kalinowski's Agouti	*Dasyprocta kalinowskii*
Kalinowski's Mastiff Bat	*Mormopterus kalinowskii*
Kalinowski's Mouse-Opossum	*Hyladelphys kalinowskii*
Kalinowski's Oldfield Mouse	*Thomasomys kalinowski*
Kangaroo Island Dunnart	*Sminthopsis aitkeni*
Kano's Mole	*Mogera kanoana*
Kaokoveld Rock Hyrax	*Procavia welwitschii*

Kappler's Armadillo	*Dasypus kappleri*
Karimi's Fat-tailed Mouse-Opossum	*Thylamys karimii*
Kashmir Grey Langur	*Semnopithecus (entellus) ajax*
Kataba Mole-Rat	*Fukomys micklemi*
Keay's Rice Rat	*Nephelomys keaysi*
Keen's Myotis	*Myotis keenii*
Keith's Short-tailed Bat	*Carollia benkeithi*
Kelaart's Leaf-nosed Bat	*Hipposideros lankadiva*
Kelaart's Long-clawed Shrew	*Feroculus feroculus*
Kelaart's Pipistrelle	*Pipistrellus ceylonicus*
Kellen's Dormouse	*Graphiurus kelleni*
Kellogg's Rice Rat	*Euryoryzomys kelloggi*
Kemp's Gerbil	*Gerbilliscus kempi*
Kemp's Grass Mouse	*Deltamys kempi*
Kemp's Spiny Mouse	*Acomys kempi*
Kemp's Thicket Rat	*Thamnomys kempi*
Kenneth's White-toothed rat	*Berylmys mackenziei*
Kermode's Bear	*Ursus americanus kermodei*
Kerr's Tree Rat	*Phyllomys kerri*
Khajuria's Leaf-nosed Bat	*Hipposideros durgadasi*
Kihaule's Mouse-Shrew	*Myosorex kihaulei*
Kilimanjaro Mouse-Shrew	*Myosorex zinki*
Kilonzo's Brush-furred Rat	*Lophuromys kilonzoi*
Kimberley Rock Rat	*Zyzomys woodwardi*
Kirk's Dik-dik	*Madoqua kirkii*
Kirk's Galago	*Otolemur crassicaudatus kirkii*
Kitti's Hog-nosed Bat	*Craseonycteris thonglongyai*
Kloss' Gibbon	*Hylobates klossii*
Kloss' Mole	*Euroscaptor klossi*
Kloss' Squirrel	*Callosciurus albescens*
Knight's Tuco-tuco	*Ctenomys knighti*
Knox Jones' Pocket Gopher	*Geomys knoxjonesi*
Kobayashi's Serotine	*Eptesicus kobayashii*
Kodiak Bear	*Ursus arctos middendorffi*
Koepke's Hairy-nosed Bat	*Mimon koepckeae*
Koford's Grass Mouse	*Akodon kofordi*
Koford's Puna Mouse	*Punomys kofordi*
Kolb's White-collared Monkey	*Cercopithecus albogularis kolbi*
Kollmannsperger's Multimammate Mouse	*Mastomys kollmannspergeri*
Kolombatovic's Big-eared Bat	*Plecotus kolombatovici*
Koopman and Hill's Yellow-shouldered Bat	*Sturnira koopmanhilli*
Koopman's Fruit Bat	*Koopmania concolor*
Koopman's Pencil-tailed Tree Mouse	*Chiropodomys karlkoopmani*
Koopman's Pipistrelle	*Pipistrellus westralis*
Koopman's Porcupine	*Coendou koopmani*

Koopman's Rat	*Rattus koopmani*
Kopsch's Deer	*Cervus nippon kopschi*
Korinch's Rat	*Rattus korinchi*
Kouprey	*Bos sauveli*
Kozlov's Long-eared Bat	*Plecotus kozlovi*
Kozlov's Pika	*Ochotona koslowi*
Kozlov's Pygmy Jerboa	*Salpingotus kozlovi*
Kozlov's Shrew	*Sorex kozlovi*
Kreb's Fat Mouse	*Steatomys krebsii*
Kuhl's Marmoset	*Callithrix kuhlii*
Kuhl's Night Monkey	*Aotus (azarai) infulatus*
Kuhl's Pipistrelle	*Pipistrellus kuhlii*
Kuhl's Tree Squirrel	*Funisciurus congicus*
Kulinas' Spiny Rat	*Proechimys kulinae*
Kuril Harbor Seal	*Phoca vitulina stejnegeri*
Lacépède's Bottle-nosed Dolphin	*Tursiops nesarnack*
Lacépède's Tamarin	*Saguinus midas*
Ladew's Oldfield Mouse	*Thomasomys ladewi*
Lado Giraffe	*Giraffa camelopardalis cottoni*
Lady Burton's Rope Squirrel	*Funisciurus isabella*
Lamotte's Roundleaf Bat	*Hipposideros lamottei*
Lamotte's Shrew	*Crocidura lamottei*
Lander's Horseshoe Bat	*Rhinolophus landeri*
Langguth's Rice Rat	*Cerradomys langguthi*
Langheld's Baboon	*Papio cynocephalus langheldi*
Lar's Gibbon	*Hylobates lar*
Large Asian Roundleaf Bat	*Hipposideros lekaguli*
Large False Serotine	*Hesperoptenus tomesi*
Large Forest Bat	*Vespadelus darlingtoni*
Large Tree Mouse	*Pogonomys loriae*
Large-eared Free-tailed Bat	*Otomops martiensseni*
Large-eared Pied Bat	*Chalinolobus dwyeri*
Large-eared Sheath-tailed Bat	*Emballonura dianae*
Lataste's Gerbil	*Gerbillus latastei*
Lataste's Gundi	*Massoutiera mzabi*
Latona's Shrew	*Crocidura latona*
LaTouche's Free-tailed Bat	*Tadarida latouchei*
LaVal's Disk-winged Bat	*Thyroptera lavali*
Lawrance's Dik-dik	*Madoqua phillipsi lawrancei*
Lawrence Island Shrew	*Sorex jacksoni*
Lawrence's Howler Monkey	*Alouatta pigra*
Laxmann's Shrew	*Sorex caecutiens*
Layard's Beaked Whale	*Mesoplodon layardii*
Layard's Palm Squirrel	*Funambulus layardi*

Leach's Single-leaf Bat	*Monophyllus redmani*
Leadbeater's Possum	*Gymnobelideus leadbeateri*
Leconte's Four-striped Tree Squirrel	*Funisciurus lemniscatus*
Leconte's Pine Mouse	*Microtus pinetorum*
Leib's Myotis	*Myotis leibii*
Leighton's Linsang	*Poiana leightoni*
Leisler's Bat	*Nyctalus leisleri*
Lelwel's Hartebeest	*Alcelaphus buselaphus lelwel*
Lemerle's Hippopotamus (extinct)	*Hippopotamus lemerlei*
Leschenault's Rousette	*Rousettus leschenaulti*
Lesser Antechinus	*Antechinus wilhelmina*
Lesser Asiatic Yellow Bat	*Scotophilus kuhlii*
Lesser Brown Horseshoe Bat	*Rhinolophus stheno*
Lesser Cane Rat	*Thryonomys gregorianus*
Lesser Forest Wallaby	*Dorcopsulus vanheurni*
Lesser Hairy-footed Dunnart	*Sminthopsis youngsoni*
Lesser Hedgehog-Tenrec	*Echinops telfairi*
Lesser Japanese Mole	*Mogera imaizumii*
Lesser Mouse-eared Bat	*Myotis blythii*
Lesser Mouse-tailed Bat	*Rhinopoma hardwickii*
Lesser Nyctophilus	*Nyctophilus geoffroyi*
Lesser Pygmy Flying Squirrel	*Petaurillus emiliae*
Lesser Striped Shrew	*Sorex bedfordiae*
Lesser Wilfred's Mouse	*Wilfredomys pictipes*
Lesser Woolly Horseshoe Bat	*Rhinolophus beddomei*
Lesson's Saddle-back Tamarin	*Saguinus fuscicollis fuscus*
LeSueur's Bettong	*Bettongia lesueur*
LeSueur's Hairy Bat	*Cistugo lesueuri*
Lewis' Marmot	*Arctomys lewisii*
Lewis' Tuco-tuco	*Ctenomys lewisi*
Leyte Shrew-Mouse	*Crunomys rabori*
L'Hoest's Monkey	*Cercopithecus lhoesti*
Liberian Mongoose	*Liberiictis kuhni*
Lichtenstein's Hartebeest	*Sigmoceros lichtensteinii*
Lichtenstein's Jerboa	*Eremodipus lichtensteini*
Lindbergh's Grass Mouse	*Akodon lindberghi*
Linnaeus' False Vampire Bat	*Vampyrum spectrum*
Linnaeus' Mouse-Opossum	*Marmosa murina*
Linné's Two-toed Sloth	*Choloepus didactylus*
Lion-tailed Macaque	*Macaca silenus*
Little Broad-nosed Bat	*Scotorepens Greyii*
Little Red Kaluta	*Dasykaluta rosamondae*
Littledale's Argali	*Ovis ammon littledalei*
Littledale's Whistling Rat	*Parotomys littledalei*
Livingstone's Eland	*Taurotragus oryx livingstonii*

Livingstone's Flying Fox	*Pteropus livingstonii*
Livingstone's Suni	*Neotragus moschatus livingstonianus*
Loder's Gazelle	*Gazella leptoceros loderi*
Long-beaked Common Dolphin	*Delphinus frithii*
Long-clawed Vole	*Prometheomys schaposchnikowi*
Long-fingered Myotis	*Myotis capaccinii*
Long-fingered Triok	*Dactylopsila palpator ernstmayri*
Long-footed Tree Mouse	*Lorentzimys nouhuysi*
Longman's Beaked Whale	*Indopacetus pacificus*
Long-nosed Shrew-Mouse	*Sommeromys macrorhinos*
Long-snouted Phalanger	*Tarsipes spencerae*
Long-tailed Mouse	*Pseudomys higginsi*
Long-tailed Pouched Rat	*Beamys hindei*
Long-tailed rat genus	*Leopoldamys*
Lord Derby's Giant Eland	*Taurotragus derbianus*
Lord Derby's Scaly-tailed Squirrel	*Anomalurus derbyanus*
Lord Howe Long-eared Bat (extinct)	*Nyctophilus howensis*
Lorentz's Melomys	*Paramelomys lorentzii*
Loria's Mastiff Bat	*Mormopterus loriae*
Loring's Rat	*Thallomys loringi*
Louis XV's Rhinoceros	*Rhinoceros unicornis*
Louise's Spiny Mouse	*Acomys louisae*
Lovat's Climbing Mouse	*Dendromus lovati*
Lowe's Gerbil	*Gerbillus lowei*
Lowe's Monkey	*Cercopithecus lowei*
Lowe's Otter-Civet	*Cynogale lowei*
Lowe's Servaline Genet	*Genetta servalina lowei*
Lowe's Shrew	*Chodsigoa parca*
Lowland Brush Mouse	*Pogonomelomys bruijni*
Lowland Red Forest Rat	*Nesomys audeberti*
Low's Squirrel	*Sundasciurus lowii*
Lucas' Short-nosed Fruit Bat	*Penthetor lucasi*
Lucifer Titi	*Callicebus (torquatus) lucifer*
Lucina's Shrew	*Crocidura lucina*
Ludia's Shrew	*Crocidura ludia*
Lugard's Mole-Rat	*Fukomys damarensis lugardi*
Luis Manuel's Tailless Bat	*Anoura luismanueli*
Luis' Yellow-shouldered Bat	*Sturnira luisi*
Lumholtz's Tree Kangaroo	*Dendrolagus lumholtzi*
Lund's Amphibious Rat	*Lundomys molitor*
Lund's Atlantic Tree Rat	*Phyllomys lundi*
Luzon Bushy-tailed Cloud Rat	*Crateromys schadenbergi*
Luzon Hairy-tailed Rat	*Batomys granti*
Luzon Shrew	*Crocidura grayi*
Luzon Striped Rat	*Chrotomys whiteheadi*

Lyle's Flying Fox	*Pteropus lylei*
Lynn's Brown Bat	*Eptesicus lynni*
MacArthur's Mouse-Lemur	*Microcebus macarthurii*
MacArthur's Shrew	*Crocidura macarthuri*
MacConnell's Bat	*Mesophylla macconnelli*
MacConnell's Climbing Mouse	*Rhipidomys macconnelli*
MacConnell's Rice Rat	*Euryoryzomys macconnelli*
MacInnes' Mouse-tailed Bat	*Rhinopoma macinnesi*
Mackenzie's False Pipistrelle	*Pipistrellus mackenziei*
Mackilligin's Gerbil	*Gerbillus mackillingini*
Maclaud's Horseshoe Bat	*Rhinolophus maclaudi*
Maclear's Rat (extinct)	*Rattus macleari*
Macleay's Dorcopsis	*Dorcopsulus macleayi*
Macleay's Marsupial Mouse	*Antechinus stuartii*
MacLeay's Moustached Bat	*Pteronotus macleayii*
Macmillan's Shrew	*Crocidura macmillani*
Macmillan's Thicket Rat	*Grammomys macmillani*
MacNeill's (Red) Deer	*Cervus (canadensis) macneilli*
Macow's Shrew	*Crocidura macowi*
Madagascar Straw-colored Fruit Bat	*Eidolon dupreanum*
Madras Tree-Shrew	*Anathana ellioti*
Magdalena Rat	*Xenomys nelsoni*
Maggie Taylor's Roundleaf Bat	*Hipposideros maggietaylorae*
Mahomet's Mouse	*Mus mahomet*
Major's Long-fingered Bat	*Miniopterus majori*
Major's Pine Vole	*Microtus majori*
Major's Sifaka	*Propithecus verreauxi verreauxi*
Major's Tufted-tailed Rat	*Eliurus majori*
Malagasy Mountain Mouse	*Monticolomys koopmani*
Malagasy Mouse-eared Bat	*Myotis goudoti*
Malayan Tailless Leaf-nosed Bat	*Coelops robinsoni*
Mandelli's Mouse-eared Bat	*Myotis sicarius*
Mandrill	*Mandrillus sphinx*
Manipur Bush Rat	*Hadromys humei*
Manus Melomys	*Melomys matambuai*
Marca's Marmoset	*Callithrix (argentata) marcai*
Marcgraf's Capuchin Monkey	*Cebus flavius*
Marcgrave's Capuchin Monkey	*Cebus flavius*
Marco Polo Sheep	*Ovis ammon polii*
Margay	*Leopardus wiedi*
Marie's Vole	*Volemys musseri*
Marinho's Rice Rat	*Cerradomys marinhus*
Marinkelle's Sword-nosed Bat	*Lonchorhina marinkelleii*
Markham's Grass Mouse	*Akodon markhami*

Markhor	*Capra falconeri*
Marley's Golden-Mole	*Amblysomus marleyi*
Marshall's Horseshoe Bat	*Rhinolophus marshalli*
Martin's False Potto	*Pseudopotto martini*
Martin's Guenon	*Cercopithecus nictitans martini*
Martino's Snow Vole	*Dinaromys bogdanovi*
Martins' Bare-faced Tamarin	*Saguinus martinsi*
Masai Giraffe	*Giraffa camelopardalis tippelskirchi*
Mashona Mole-Rat	*Fukomys darlingi*
Matadi Hyrax	*Heterohyrax chapini*
Matschie's Galago	*Galago matschiei*
Matschie's Guenon	*Chlorocebus aethiops matschiei*
Matschie's Guereza	*Colobus guereza matschiei*
Matschie's Tree Kangaroo	*Dendrolagus matschiei*
Matses' Big-eared Bat	*Micronycteris matses*
Matthey's Mouse	*Mus mattheyi*
Maximowicz's Vole	*Microtus maximowiczii*
Max's Shrew	*Crocidura maxi*
Maxwell's Duiker	*Cephalophus maxwellii*
Maxwell's Otter	*Lutrogale perspicillata maxwelli*
Mayor's Mouse	*Mus mayori*
McIlhenny's Four-eyed Opossum	*Philander mcilhennyi*
Mearns' Flying Fox	*Pteropus (speciosus) mearnsi*
Mearns' Grasshopper Mouse	*Onychomys arenicola*
Mearns' Luzon Rat	*Tryphomys adustus*
Mearns' Pocket Gopher	*Thomomys bottae mearnsi*
Mearns' Pouched Mouse	*Saccostomus mearnsi*
Mearns' Squirrel	*Tamiasciurus mearnsi*
Mechow's Mole-Rat	*Fukomys mechowi*
Medem's Titi	*Callicebus medemi*
Mediterranean Horseshoe Bat	*Rhinolophus euryale*
Mehely's Horseshoe Bat	*Rhinolophus mehelyi*
Melck's House Bat	*Neoromicia melckorum*
Melissa's Yellow-eared Bat	*Vampyressa melissa*
Meller's Mongoose	*Rhynchogale melleri*
Melon-headed Whale	*Peponocephala electra*
Menelik's Bushbuck	*Tragelaphus scriptus meneliki*
Mentawai Langur	*Presbytis potenziani*
Menzbier's Marmot	*Marmota menzbieri*
Menzies' Echymipera	*Echymipera echinista*
Menzies' Mouse	*Pogonomelomys sevia*
Merida Brocket	*Mazama bricenii*
Merriam's Chipmunk	*Tamias merriami*
Merriam's Deer Mouse	*Peromyscus merriami*
Merriam's Desert Shrew	*Megasorex gigas*

Merriam's Ground Squirrel	*Spermophilus canus*
Merriam's Kangaroo-Rat	*Dipodomys merriami*
Merriam's Pocket Gopher	*Cratogeomys merriami*
Merriam's Pocket Mouse	*Perognathus merriami*
Merriam's Shrew	*Sorex merriami*
Merriam's Small-eared Shrew	*Cryptotis merriami*
Merriam's Wapiti (Elk) (extinct)	*Cervus canadensis merriami*
Mexican Grizzly Bear (extinct)	*Ursus arctos nelsoni*
Mexican Volcano Mouse	*Neotomodon alstoni*
Meyen's Dolphin	*Stenella coeruleoalba*
Mhorr's Gazelle	*Gazella dama mhorr*
Michie's Tufted Deer	*Elaphodus cephalophus michianus*
Michoacan Deer Mouse	*Osgoodomys banderanus*
Midas Free-tailed Bat	*Mops midas*
Midas Tamarin	*Saguinus midas*
Middendorff's Vole	*Microtus middendorffii*
Millard's Rat	*Dacnomys millardi*
Miller's Andaman Spiny Shrew	*Crocidura andamanensis*
Miller's Grizzled Langur	*Presbytis hosei canicrus*
Miller's Hutia (extinct)	*Isolobodon levir*
Miller's Long-tongued Bat	*Glossophaga longirostris*
Miller's Mastiff Bat	*Molossus pretiosus*
Miller's Monk Saki	*Pithecia monachus milleri*
Miller's Myotis	*Myotis milleri*
Miller's Nesophontes (extinct)	*Nesophontes hypomicrus*
Miller's Striped Mouse	*Hybomys planifrons*
Miller's Water Shrew	*Neomys anomalus*
Millet's Giant Long-tailed Rat	*Leopoldamys milleti*
Millet's Shrew	*Sorex coronatus*
Milne-Edwards' Long-clawed Mouse	*Notiomys edwardsii*
Milne-Edwards' Long-tailed Giant Rat	*Leopoldamys edwardsi*
Milne-Edwards' Potto	*Perodicticus potto edwardsi*
Milne-Edwards' Sifaka	*Propithecus edwardsi*
Milne-Edwards' Sportive Lemur	*Lepilemur edwardsi*
Milne-Edwards' Swamp Rat	*Malacomys edwardsi*
Mindanao Gymnure	*Podogymnura truei*
Mindanao Hairy-tailed Rat	*Batomys salomonseni*
Mindanao Tree-Shrew	*Urogale everetti*
Mindoro Warty Pig	*Sus (philippensis) oliveri*
Miner's Cat	*Bassariscus astutus*
Misonne's Soft-furred Mouse	*Praomys misonnei*
Miss Ryley's Soft-furred Rat	*Millardia kathleenae*
Miss Waldron's Red Colobus Monkey	*Piliocolobus badius waldroni*
Mitchell's Hopping Mouse	*Notomys mitchelli*
Mittendorf's Striped Grass Mouse	*Lemniscomys mittendorfi*

Mittermeier's Mouse-Lemur	*Microcebus mittermeieri*
Mittermeier's Sportive Lemur	*Lepilemur mittermeieri*
Molina's Grass Mouse	*Akodon molinae*
Molina's Hog-nosed Skunk	*Conepatus chinga*
Moloch Gibbon	*Hylobates moloch*
Moloney's Flat-headed Bat	*Mimetillus moloneyi*
Moloney's Monkey	*Cercopithecus albogularis moloneyi*
Monard's Climbing Mouse	*Dendromus leucostomus*
Monard's Dormouse	*Graphiurus monardi*
Monckton's Melomys	*Paramelomys moncktoni*
Mondolfi's Four-eyed Opossum	*Philander mondolfii*
Monjon	*Petrogale burbidgei*
Montane Cavy	*Cavia tschudii*
Montane Long-nosed Squirrel	*Hyosciurus heinrichi*
Monte Gerbil-Mouse	*Eligmodontia moreni*
Moojen's Atlantic Tree Rat	*Phyllomys kerri*
Moojen's Pygmy Rice Rat	*Oligoryzomys moojeni*
Moojen's Spiny Rat	*Trinomys moojeni*
Morgan's Gerbil-Mouse	*Eligmodontia morgani*
Morris' Bat	*Myotis morrisi*
Morris' Flying Squirrel	*Olisthomys morrisi*
Mortlock Flying Fox	*Pteropus phaeocephalus*
Mo's Spiny Rat	*Maxomys moi*
Moss-forest Blossom Bat	*Syconycteris hobbit*
Moss-forest Rat	*Stenomys niobe*
Mossy Forest Shrew	*Crocidura musseri*
Mouflon	*Ovis gmelini*
Mount Elgon Vlei Rat	*Otomys jacksoni*
Mount Isarog Striped Rat	*Chrotomys gonzalesi*
Mountain Anoa	*Bubalus quarlesi*
Mountain Paca	*Cuniculus taczanowskii*
Mouse genera	*Juliomys, Juscelinomys, Thomasomys, Wilfredomys*
Mouse sp.	*Mus ouwensi*
Mozambique Galago	*Galago granti*
Mrs. Gray's Lechwe	*Kobus megaceros*
Mrs. Millard's Flying Squirrel	*Petaurista sybilla*
Muennink's Spiny Rat	*Tokudaia muenninki*
Müller's Giant Sunda Rat	*Sundamys muelleri*
Müller's Gibbon	*Hylobates muelleri*
Murree Vole	*Hyperacrius wynnei*
Musschenbroek's Spiny Rat	*Maxomys musschenbroeki*
Musser's Bristly Mouse	*Neacomys musseri*
Musser's Shrew-Mouse	*Microhydromys musseri*
Musso's Fish-eating Rat	*Neusticomys mussoi*

Muton's Soft-furred Mouse	*Praomys mutoni*
Myers' Grass Mouse	*Akodon philipmyersi*
Nagtglas' Dormouse	*Graphiurus nagtglasii*
Namdapha Flying Squirrel	*Biswamoyopterus biswasi*
Nancy Ma's Night Monkey	*Aotus nancymaae*
Nasarov's Pine Vole	*Microtus nasarovi*
Nash's Titi Monkey	*Callicebus stephennashi*
Nasolo's Shrew-Tenrec	*Microgale nasoloi*
Nathusius' Pipistrelle	*Pipistrellus nathusii*
Natterer's Bat	*Myotis nattereri*
Natterer's Tuco-tuco	*Ctenomys nattereri*
Neave's Mouse	*Mus neavei*
Nehring's Blind Mole-Rat	*Nannospalax nehringi*
Neill's Long-tailed Giant Rat	*Leopoldamys neilli*
Nelson's and Goldman's Woodrat	*Nelsonia goldmani*
Nelson's Antelope Squirrel	*Ammospermophilus nelsoni*
Nelson's Coati	*Nasua (narica) nelsoni*
Nelson's Collared Lemming	*Dicrostonyx (groenlandicus) nelsoni*
Nelson's Giant Deer Mouse	*Megadontomys nelsoni*
Nelson's Kangaroo-Rat	*Dipodomys nelsoni*
Nelson's Ocelot	*Leopardus pardalis nelsoni*
Nelson's Pocket Mouse	*Chaetodipus nelsoni*
Nelson's Rice Rat (extinct)	*Oryzomys nelsoni*
Nelson's Small-eared Shrew	*Cryptotis nelsoni*
Nelson's Spiny Pocket Mouse	*Heteromys nelsoni*
Nelson's Woodrat	*Neotoma nelsoni*
Neotropical rat genus	*Kunsia*
Nereid Horseshoe Bat	*Rhinolophus nereis*
Neumann's Black-and-White Colobus	*Colobus guereza gallarum*
Neumann's Grass Rat	*Arvicanthis neumanni*
Neumann's Hartebeest	*Alcelaphus buselaphus neumanni*
Neumann's Olive Baboon	*Papio anubis neumanni*
New Britain Flying Fox	*Pteropus gilliardorum*
New Guinea Singing Dog	*Canis hallstromi*
New Zealand Fur Seal	*Arctocephalus forsteri*
New Zealand Sea-Lion	*Phocarctos hookeri*
Niceforo's Big-eared Bat	*Micronycteris nicefori*
Niethammer's Dormouse	*Dryomys niethammeri*
Nigerian Mole-Rat	*Fukomys foxi*
Nikolaus' Mouse	*Megadendromus nikolausi*
Nilgiri Marten	*Martes gwatkinsii*
Nimba Otter-Shrew	*Micropotamogale lamottei*
Niobe Ground Squirrel	*Lariscus niobe*
Niobe's Shrew	*Crocidura niobe*

Noack's Roundleaf Bat	*Hipposideros ruber*
Nolthenius' Long-tailed Climbing Mouse	*Vandeleuria nolthenii*
North African Elephant Shrew	*Elephantulus rozeti*
Northern Bat	*Eptesicus nilssoni*
Northern Bolo Mouse	*Necromys urichi*
Northern Broad-nosed Bat	*Scotorepens sanborni*
Northern Glider	*Petaurus abidi*
Northern Hairy-nosed Wombat	*Lasiorhinus krefftii*
Northern Little Yellow-eared Bat	*Vampyressa thyone*
Northern Palm Squirrel	*Funambulus pennantii*
Northern Pudu	*Pudu mephistophiles*
Northern Pygmy Mouse	*Baiomys taylori*
Northern Tree-Shrew	*Tupaia belangeri*
Northwestern Deer Mouse	*Peromyscus keeni*
Novaes' Bald-headed Uacari	*Cacajao calvus novaesi*
Nuttall's Cottontail	*Sylvilagus nuttallii*
Nyala	*Tragelaphus angasii*
O'Connell's Spiny Rat	*Proechimys oconnelli*
Ogilby's Duiker	*Cephalophus ogilbyi*
Ognev's Dormouse	*Myomimus personatus*
Ognev's Long-eared Bat	*Plecotus ognevi*
Okapi	*Okapia johnstoni*
Oku Rat	*Lamottemys okuensis*
Olalla's Titi	*Callicebus olallae*
Olga's Dormouse	*Graphiurus olga*
Olive Baboon	*Papio anubis*
Olivier's Shrew	*Crocidura olivieri*
Olrog's Chaco Mouse	*Andalgalomys olrogi*
Olrog's Four-eyed Opossum	*Philander olrogi*
Omura's Whale	*Balaenoptera omurai*
One-toothed Shrew-Mouse	*Mayermys ellermani*
Opdenbosch's Mangabey	*Lophocebus opdenboschi*
Orces' Chibchan Water Mouse	*Chibchanomys orcesi*
Orces' Nectar Bat	*Lonchophylla orcesi*
Ord's Kangaroo-Rat	*Dipodomys ordi*
Orii's Shrew	*Crocidura orii*
Orinoco River Dolphin	*Inia geoffrensis humboldtiana*
Orion Broad-nosed Bat	*Scotorepens orion*
Osborn's Caribou	*Rangifer tarandus osborni*
Osborn's Key Mouse (extinct)	*Clidomys osborni*
Osgood's Aztec Mouse	*Peromyscus aztecus evides*
Osgood's Horseshoe Bat	*Rhinolophus osgoodi*
Osgood's Leaf-eared Mouse	*Phyllotis osgoodi*
Osgood's Mouse	*Peromyscus gratus*

Osgood's Rat	*Rattus osgoodi*
Osgood's Short-tailed Opossum	*Monodelphis osgoodi*
Osman Hill's Grey-cheeked Mangabey	*Lophocebus (albigena) osmani*
Osvaldo Reig's Tuco-tuco	*Ctenomys osvaldoreigi*
Otter-Civet	*Cynogale bennetti*
Otto's Sportive Lemur	*Lepilemur otto*
Oustalet's Red Colobus	*Piliocolobus foai oustaleti*
Owen's Marsupial "Lion" (extinct)	*Thylacoleo oweni*
Owen's Pygmy Sperm Whale	*Kogia sima*
Owl's Spiny Rat	*Carterodon sulcidens*
Owston's Palm Civet	*Chrotogale owstoni*
Oyapock's Fish-eating Rat	*Neusticomys oyapocki*
Pacarana	*Dinomys branickii*
Painted Ringtail Possum	*Pseudochirulus forbesi*
Palawan Montane Squirrel	*Sundasciurus rabori*
Palawan Stink Badger	*Mydaus marchei*
Pale Field Rat	*Rattus tunneyi*
Pale-bellied Woolly Mouse-Opossum	*Micoureus constantiae*
Palestine Mole-Rat	*Nannospalax ehrenbergi*
Pallas' Cat	*Felis manul*
Pallas' Long-tongued Bat	*Glossophaga soricina*
Pallas' Mastiff Bat	*Molossus molossus*
Pallas' Pika	*Ochotona pallasi*
Pallas' Squirrel	*Callosciurus erythraeus*
Pallas' Tarsier	*Tarsius spectrum*
Pallas' Tube-nosed Fruit Bat	*Nyctimene cephalotes*
Palmer's Chipmunk	*Tamias palmeri*
Pampas Cat	*Oncifelis colocolo budini*
Panay Bushy-tailed Cloud Rat	*Crateromys heaneyi*
Pardelluch's Lynx	*Lynx pardinus*
Pariente's Fork-marked Lemur	*Phaner (furcifer) parienti*
Parnell's Moustached Bat	*Pteronotus parnellii*
Parry's Marmot Squirrel	*Spermophilus parryii*
Parry's Wallaby	*Macropus parryi*
Pascual's Tuco-tuco	*Ctenomys rosendopascuali*
Patagonian Opossum	*Lestodelphys halli*
Patrizi's Trident Bat	*Aselliscus patrizii*
Patton's Nectar-feeding Bat	*Lonchophylla pattoni*
Patton's Spiny Rat	*Proechimys pattoni*
Paula's Long-nosed Rat	*Paulamys naso*
Paulina's Rock Rat	*Saxatilomys paulinae*
Peale's Dolphin	*Lagenorhynchus australis*
Peale's Free-tailed Bat	*Nyctinomops aurispinosus*
Peale's Meadow Mouse	*Microtus montanus*

Pearson's Chaco Mouse	*Andalgalomys pearsoni*
Pearson's Horseshoe Bat	*Rhinolophus pearsonii*
Pearson's Long-clawed Mouse	*Pearsonomys annectens*
Pearson's Long-clawed Shrew	*Solisorex pearsoni*
Pearson's Puma	*Puma concolor pearsoni*
Pearson's Tuco-tuco	*Ctenomys pearsoni*
Peary Caribou	*Rangifer tarandus pearyi*
Pel's Pouched Bat	*Saccolaimus peli*
Pel's Scaly-tailed Squirrel	*Anomalurus pelii*
Pelzeln's Gazelle	*Gazella (dorcas) pelzelni*
Pemba Flying Fox	*Pteropus voeltzkowi*
Pemberton's Deer Mouse	*Peromyscus pembertoni*
Peninsular Horseshoe Bat	*Rhinolophus robinsoni*
Pennant's Marten	*Martes pennanti*
Pennant's Red Colobus	*Piliocolobus pennantii*
Pen-tailed Tree-Shrew	*Ptilocercus lowii*
Percival's Gerbil	*Gerbillus percivali*
Percival's Spiny Mouse	*Acomys percivali*
Percival's Trident Bat	*Cloeotis percivali*
Père David's Deer	*Elaphurus davidianus*
Père David's Macaque	*Macaca thibetana*
Père David's Mole	*Talpa davidiana*
Père David's Rock Squirrel	*Sciurotamias davidianus*
Père David's Vole	*Euthenomys melanogaster*
Pernetty's Dolphin	*Stenella pernettensis*
Perny's Squirrel	*Dremomys pernyi*
Perrier's Sifaka	*Propithecus perrieri*
Perrin's Beaked Whale	*Mesoplodon perrini*
Persian Field Mouse	*Apodemus arianus*
Persian Mole	*Talpa streeti*
Peters' Bat	*Balantiopteryx plicata*
Peters' Angolan Colobus	*Colobus angolensis palliatus*
Peters' Climbing Rat	*Tylomys nudicaudus*
Peters' Disk-winged Bat	*Thyroptera discifera*
Peters' Duiker	*Cephalophus callipygus*
Peters' Dwarf Epauletted Fruit Bat	*Micropteropus pusillus*
Peters' False Serotine	*Hesperoptenus doriae*
Peters' Flat-headed Bat	*Platymops setiger*
Peters' Gazelle	*Gazella granti petersi*
Peters' Gerbil	*Gerbilliscus leucocaster*
Peters' Ghost-faced Bat	*Mormoops megalophylla*
Peters' Mastiff Bat	*Eumops bonariensis*
Peters' Mouse	*Mus setulosus*
Peters' Musk Shrew	*Crocidura gracilipes*
Peters' Pipistrelle	*Falsistrellus petersi*

Peters' Spiny Pocket Mouse	*Liomys adspersus*
Peters' Squirrel	*Sciurus oculatus*
Peters' Striped Mouse	*Hybomys univittatus*
Peters' Tent-making Bat	*Uroderma bilobatum*
Peters' Trumpet-eared Bat	*Phoniscus jagorii*
Peters' Tube-nosed Bat	*Murina grisea*
Peters' Wrinkle-lipped Bat	*Mormopterus jugularis*
Peterson's Chinchilla-mouse	*Euneomys petersoni*
Peterson's Free-tailed Bat	*Mops petersoni*
Petra Fruit Bat	*Lissonycteris petraea*
Petter's Big-footed Mouse	*Macrotarsomys petteri*
Petter's Gerbil	*Taterillus petteri*
Petter's Soft-furred Mouse	*Praomys petteri*
Petter's Sportive Lemur	*Lepilemur petteri*
Petter's Tufted-tailed Rat	*Eliurus petteri*
Peyrieras' Woolly Lemur	*Avahi peyrierasi*
Pfeiffer's Hairy-tailed Bat	*Lasiurus pfeifferi*
Phayre's Leaf Monkey	*Trachypithecus phayrei*
Phayre's Squirrel	*Callosciurus phayrei*
Philippine Forest Rat	*Rattus everetti*
Philippine Large-headed Fruit Bat	*Dyacopterus rickarti*
Philippine Pouched Bat	*Saccolaimus pluto*
Philippine Pygmy Fruit Bat	*Haplonycteris fischeri*
Phillips' Congo Shrew	*Congosorex phillipsorum*
Phillips' Dik-dik	*Madoqua saltiana phillipsi*
Phillips' Gerbil	*Gerbilliscus phillipsi*
Phillips' Kangaroo-Rat	*Dipodomys phillipsii*
Phillips' Mouse	*Mus phillipsi*
Phillips' Short-eared Shrew	*Cryptotis phillipsii*
Piacentini's Dik-dik	*Madoqua piacentinii*
Pilbara Ningaui	*Ningaui timealeyi*
Pinheiro's Slender Mouse-Opossum	*Marmosops pinheiroi*
Piñon Deer Mouse	*Peromyscus truei*
Pitman's Shrew	*Crocidura pitmani*
Pittier's Crab-eating Rat	*Ichthyomys pittieri*
Plain's Gazelle	*Oryx gazella*
Plain's Zebra	*Equus quagga burchelli*
Plush-coated Ringtail Possum	*Pseudochirops corinnae*
Pluto Monkey	*Cercopithecus mitis mitis*
Pluto Tamarin	*Saguinus mystax pluto*
Pocock's New Guinea Rat	*Rattus (Stenomys) pococki*
Pocock's White-throated Guenon	*Cercopithecus erythrogaster pococki*
Pocock's Zebra	*Equus burchelli pococki*
Poeppig's Woolly Monkey	*Lagothrix poeppigii*
Poey's Pallid Flower Bat	*Phyllonycteris poeyi*

Pohle's Fruit Bat	*Scotonycteris ophiodon*
Polia's Shrew	*Crocidura polia*
Poll's Shrew	*Congosorex polli*
Poncelet's Naked-tailed Rat	*Solomys ponceleti*
Portenko's Shrew	*Sorex portenkoi*
Porteous' Tuco-tuco	*Ctenomys porteousi*
Porter's Rock Rat	*Aconaemys porteri*
Pouched Gerbil	*Desmodilliscus braueri*
Pousargues' Fat Mouse	*Steatomys opimus*
Pousargues' Mongoose	*Dologale dybowskii*
Pousargues' White-collared Monkey	*Cercopithecus albogularis albotorquatus*
Prater's Cat	*Felis chaus prateri*
Pratt's Roundleaf Bat	*Hipposideros pratti*
Pratt's Vole	*Euthenomys chinensis*
Preble's Kangaroo-Rat	*Dipodomys microps preblei*
Preble's Meadow Jumping Mouse	*Zapus hudsonius preblei*
Preble's Shrew	*Sorex preblei*
Preuss' Monkey	*Cercopithecus preussi*
Preuss' Mouse-Shrew	*Myosorex preussi*
Preuss' Red Colobus	*Piliocolobus preussi*
Prevost's Squirrel	*Callosciurus prevostii*
Prigogine's Angolan Colobus	*Colobus angolensis prigoginei*
Prince Alfred's Deer	*Cervus alfredi*
Prince Bernhard's Titi	*Callicebus bernhardi*
Pringle's Gerbil	*Gerbilliscus pringlei*
Przewalski's Hare	*Lepus oiostolus przewalskii*
Przewalski's Gazelle	*Procapra przewalskii*
Przewalski's Gerbil	*Brachiones przewalskii*
Przewalski's Horse	*Equus ferus*
Przewalski's Steppe Lemming	*Eolagurus przewalskii*
Puerto Rican Nesophontes (extinct)	*Nesophontes edithae*
Puku	*Kobus vardonii*
Pundt's Tuco-tuco	*Ctenomys pundti*
Pygmy Fruit Bat	*Aethalops alecto*
Pygmy Nyctophilus	*Nyctophilus walkeri*
Pygmy Ringtail Possum	*Pseudochirulus mayeri*
Pygmy Scaly-tailed Flying Squirrel	*Idiurus zenkeri*
Pygmy Short-tailed Opossum	*Monodelphis kunsi*
Pyrenean Pine Vole	*Microtus gerbei*
Queen Charlotte Caribou (extinct)	*Rangifer tarandus dawsoni*
Queen of Sheba's Gazelle	*Gazella bilkis*
Queen Victoria's Ibex	*Capra pyrenaica victoriae*
Queensland Tube-nosed Fruit Bat	*Nyctimene robinsoni*

Rabor's Tube-nosed Bat	*Nyctimene rabori*
Racey's Pipistrelle	*Pipistrellus raceyi*
Radde's Shrew	*Sorex raddei*
Raffles' Banded Langur	*Presbytis femoralis femoralis*
Raffles' Tarsier	*Tarsius bancanus*
Raffray's Bandicoot	*Peroryctes raffrayana*
Raffray's Sheath-tailed Bat	*Emballonura raffrayana*
Rafinesque's Big-eared Bat	*Corynorhinus rafinesquii*
Rahm's Brush-furred Rat	*Lophuromys rahmi*
Railer Bat	*Mops thersites*
Rainey's Gazelle	*Gazella granti raineyi*
Rainey's Shrew	*Crocidura raineyi*
Ramon's Shrew	*Crocidura ramona*
Randrianasolo's Sportive Lemur	*Lepilemur randrianasoli*
Ranjini's Field Rat	*Rattus ranjiniae*
Ratanaworabhan's Fruit Bat	*Megaerops niphanae*
Rayner's Flying Fox	*Pteropus rayneri*
Red Brocket	*Mazama sheila*
Red Brocket	*Mazama zetta*
Red Climbing Mouse	*Vernaya fulva*
Red Goral	*Naemorhedus baileyi*
Red Rock Hare sp.	*Pronolagus barretti*
Red Viscacha Rat	*Tympanoctomys barrerae*
Red-bellied Marsupial-Shrew	*Phascolosorex doriae*
Red-bellied Melomys	*Protochromys fellowsi*
Red-bellied Titi	*Callicebus moloch*
Red-brown Pipistrelle	*Hypsugo kitcheneri*
Red-handed Howler Monkey	*Alouatta belzebul*
Red-necked Pademelon	*Thylogale thetis*
Red-tailed Guenon	*Cercopithecus ascanius*
Reeves' Muntjac	*Muntiacus reevesi*
Reig's Grass Mouse	*Akodon reigi*
Reig's Montane Mouse	*Aepeomys reigi*
Remy's Shrew	*Suncus remyi*
Rendall's Serotine	*Neoromicia rendalli*
Rennell Flying Fox	*Pteropus rennelli*
Réunion Little Mastiff Bat	*Mormopterus francoismoutoui*
Rhesus Monkey	*Macaca mulatta*
Rhoads' Gerbil	*Gerbillus pulvinatus*
Rhoads' Oldfield Mouse	*Thomasomys rhoadsi*
Richardson's Collared Lemming	*Dicrostonyx richardsoni*
Richardson's Ground Squirrel	*Spermophilus richardsonii*
Richardson's Vole	*Microtus richardsoni*
Richmond's Squirrel	*Sciurus richmondi*
Rickett's Big-footed Bat	*Myotis ricketti*

Ridley's Myotis	*Myotis ridleyi*
Ridley's Roundleaf Bat	*Hipposideros ridleyi*
Riggenbach's Gerbil	*Gerbillus riggenbachi*
Rimoli's Hutia (extinct)	*Hyperplagiodontia stenocoronalis*
Rio de Janeiro Spiny Rat	*Trinomys eliasi*
Risso's Dolphin	*Grampus griseus*
Roach's Mouse-tailed Dormouse	*Myomimus roachi*
Robert's Arboreal Rice Rat	*Oecomys roberti*
Roberts' Flat-headed Bat	*Sauromys petrophilus*
Roberts' Gazelle	*Gazella granti robertsi*
Robert's Hocicudo	*Oxymycterus roberti*
Roberts' Lechwe Antelope (extinct)	*Kobus leche robertsi*
Roberts' Serval	*Leptailurus serval robertsi*
Robert's Snow Vole	*Chionomys roberti*
Robert's Spiny Rat	*Proechimys roberti*
Robinson's Banded Langur	*Presbytis femoralis robinsoni*
Robinson's Mouse-Opossum	*Marmosa robinsoni*
Roborovski's Hamster	*Phodopus roborovskii*
Rock Ringtail Possum	*Petropseudes dahli*
Rodent genus ("brucies")	*Brucepattersonius*
Rodolph's Striped Squirrel	*Tamiops rodolphii*
Rodriguez's Harvest Mouse	*Reithrodontomys rodriguezi*
Rohu's Bat	*Philetor brachypterus*
Rohwer's Shrew	*Sorex rohweri*
Roig's Chaco Mouse	*Graomys roigi*
Roig's Tuco-tuco	*Andalgalomys roigi*
Romanian Hamster	*Mesocricetus newtoni*
Ronald's Short-tailed Opossum	*Monodelphis ronaldi*
Roosevelt's Gazelle	*Gazella granti roosevelti*
Roosevelt's Lion	*Panthera leo roosevelti*
Roosevelt's Muntjac	*Muntiacus rooseveltorum*
Roosevelt's Sable Antelope	*Hippotragus niger roosevelti*
Roosevelt's Shrew	*Crocidura roosevelti*
Roosevelt's Wapiti	*Cervus canadensis roosevelti*
Roosmalen's Dwarf Marmoset	*Callithrix (Callibella) humilis*
Rory's Pseudantechinus	*Pseudantechinus roryi*
Rosalinda Gerbil	*Gerbillus rosalinda*
Rosalinda's Oldfield Mouse	*Thomasomys rosalinda*
Rosenberg's Dwarf Squirrel	*Prosciurillus rosenbergii*
Rosevear's Brush-furred Rat	*Lophuromys roseveari*
Rosevear's Striped Grass Mouse	*Lemniscomys roseveari*
Ross Seal	*Ommatophoca rossii*
Rossetti's Wombat	—
Rothschild's Cuscus	*Phalanger rothschildi*
Rothschild's Giraffe	*Giraffa camelopardalis rothschildi*

Rothschild's Porcupine	*Coendou rothschildi*
Rothschild's Rock Wallaby	*Petrogale rothschildi*
Rothschild's Woolly Rat	*Mallomys rothschildi*
Rothschild's Zokor	*Eospalax rothschildi*
Rough-toothed Dolphin	*Steno bredanensis*
Round-tailed Muskrat	*Neofiber alleni*
Royle's Mountain Vole	*Alticola roylei*
Royle's Pika	*Ochotona roylei*
Rudd's Mole-Rat	*Tachyoryctes ruddi*
Rudd's Mouse	*Uranomys ruddi*
Ruddy Mongoose	*Herpestes smithii*
Rudolph's Whale	*Balaenoptera borealis*
Rufous Horseshoe Bat	*Rhinolophus rouxii*
Rufous Mouse-eared Bat	*Myotis bocagei*
Rümmler's Mouse	*Coccymys ruemmleri*
Rungwe Brush-furred Rat	*Lophuromys machangui*
Rüppell's Broad-nosed Bat	*Scoteanax rueppelli*
Rüppell's Fox	*Vulpes rueppelli*
Rüppell's Guereza	*Colobus guereza guereza*
Rüppell's Horseshoe Bat	*Rhinolophus fumigatus*
Rüppell's Pipistrelle	*Pipistrellus rueppelli*
Rupp's Mouse	*Stenocephalemys ruppi*
Ruschi's Rat	*Abrawayaomys ruschii*
Rusty-sided Atlantic Tree Rat	*Phyllomys pattoni*
Ryukyu Mole	*Mogera uchidai*
Ryukyu Mouse	*Mus caroli*
Ryukyu spiny rat genus	*Tokudaia*
Sabana Hutia	*Capromys (pilorides) gundlachianus*
Sage's Rock Rat	*Aconaemys sagei*
Salenski's Shrew	*Chodsigoa salenskii*
Salim Ali's Fruit Bat	*Latidens salimalii*
Salt's Dik-dik	*Madoqua saltiana*
Salvin's Big-eyed Bat	*Chiroderma salvini*
Salvin's Spiny Pocket Mouse	*Liomys salvini*
Sambirano's Lemur	*Microcebus sambiranensis*
San Esteban Island Deer Mouse	*Peromyscus stephani*
San Nicolas Island Fox	*Urocyon littoralis dickeyi*
Sanborn's Big-eared Bat	*Micronycteris sanborni*
Sanborn's Bonneted Bat	*Eumops hansae*
Sanborn's Flying Fox	*Pteropus mahaganus*
Sanborn's Grass Mouse	*Abrothrix sanborni*
Sanborn's Squirrel	*Sciurus sanborni*
Sand Cat	*Felis margarita*
Sand-colored Soft-furred Rat	*Millardia gleadowi*

Sandstone Pseudantechinus	*Pseudantechinus bilarni*
Sanford's Lemur	*Eulemur (fulvus) sanfordi*
Santa Fé Tuco-tuco	*Ctenomys yolandae*
Santa Margarita Island Kangaroo Rat	*Dipodomys margaritae*
Santiago Galápagos Mouse	*Nesoryzomys swarthi*
Sateré-Maués' Marmoset	*Callithrix saterei*
Saunders' Vlei Rat	*Otomys saundersiae*
Saussure's Shrew	*Sorex saussurei*
Savanna Hare	*Lepus victoriae*
Savile's Bandicoot Rat	*Bandicota savilei*
Savi's Pine Vole	*Microtus savii*
Savi's Pipistrelle	*Hypsugo savii*
Savi's Pygmy Shrew	*Suncus etruscus*
Say's Least Shrew	*Cryptotis parva*
Scaglia's Tuco-tuco	*Ctenomys scagliai*
Schaller's Mouse-Shrew	*Myosorex schalleri*
Schaub's Myotis	*Myotis schaubi*
Scheffel's Sand Cat	*Felis margarita scheffeli*
Schelkovnikov's Water Shrew	*Neomys schelkovnikovi*
Schelkovnikov's Pine Vole	*Microtus schelkovnikovi*
Schlegel's Guenon	*Cercopithecus neglectus*
Schlieffen's Bat	*Nycticeinops schlieffeni*
Schmidly's Deer Mouse	*Peromyscus schmidlyi*
Schmidt's Big-eared Bat	*Micronycteris schmidtorum*
Schmidt's Monkey	*Cercopithecus ascanius schmidti*
Schmitz's Caracal	*Caracal caracal schmitzi*
Schneider's Roundleaf Bat	*Hipposideros speoris*
Schomburgk's Deer (extinct)	*Cervus schomburgki*
Schouteden's Blue Monkey	*Cercopithecus mitis schoutedenii*
Schouteden's Shrew	*Paracrocidura schoutedeni*
Schreber's Yellow Bat	*Scotophilus nigrita*
Schreibers' Bat	*Miniopterus schreibersi*
Schulz's Round-eared Bat	*Lophostoma schulzi*
Schwartz's Myotis	*Myotis martiniquensis*
Schweinfurth's Chimpanzee	*Pan troglodytes schweinfurthii*
Schweitzer's Shrew	*Crocidura schweitzeri*
Schwartz's Fruit-eating Bat	*Artibeus schwartzi*
Sclater's Angolan Colobus	*Colobus angolensis angolensis*
Sclater's Black Lemur	*Eulemur macaco flavifrons*
Sclater's Dog	*Atelocynus microtis*
Sclater's Forest Shrew	*Myosorex sclateri*
Sclater's Golden-Mole	*Chlorotalpa sclateri*
Sclater's Monkey	*Cercopithecus sclateri*
Sclater's Shrew	*Sorex sclateri*
Scott's Rice Rat	*Cerradomys scotti*

Scott's Mouse-eared Bat	*Myotis scotti*
Scott's Tree Kangaroo	*Dendrolagus scottae*
Scully's Tube-nosed Bat	*Murina tubinaris*
Seal's Sportive Lemur	*Lepilemur seali*
Seba's Short-tailed Bat	*Carollia perspicillata*
Selangor Pygmy Flying Squirrel	*Petaurillus kinlochii*
Selborne's Hartebeest	*Alcelaphus buselaphus selbornei*
Selous' Mongoose	*Paracynictis selousi*
Selous' Sitatunga	*Tragelaphus spekii selousi*
Selous' Zebra	*Equus quagga selousi*
Semon's Roundleaf Bat	*Hipposideros semoni*
Seri's Sheath-tailed Bat	*Emballonura serii*
Seri's Tree Kangaroo	*Dendrolagus (dorianus) stellarum*
Setzer's Hairy-footed Gerbil	*Gerbillurus setzeri*
Setzer's Mouse-tailed Dormouse	*Myomimus setzeri*
Setzer's Pygmy Mouse	*Mus setzeri*
Seuanez's Rice Rat	*Hylaeamys seuanezi*
Severtzov's Argali	*Ovis ammon severtzovi*
Severtzov's Birch Mouse	*Sicista severtzovi*
Severtzov's Ibex	*Capra ibex severtzovi*
Severtzov's Jerboa	*Allactaga severtzovi*
Shaggy Bat	*Centronycteris maximiliani*
Shamel's Horseshoe Bat	*Rhinolophus shameli*
Sharman's Rock Wallaby	*Petrogale sharmani*
Sharpe's Colobus	*Colobus angolensis sharpei*
Sharpe's Grysbok	*Raphicerus sharpei*
Shawmayer's Ornate Tree Kangaroo	*Dendrolagus goodfellowi shawmayeri*
Shaw-Mayer's Pogonomelomys	*Pogonomelomys mayeri*
Shaw-Mayer's Shrew Mouse	*Mayermys ellermani*
Shaw-Mayer's Water Rat	*Hydromys shawmayeri*
Shaw's Jird	*Meriones shawi*
Shaw's Mastiff Bat	*Eumops auripendulus*
Shaw's Melomys	*Melomys shawi*
Shepherd's Beaked Whale	*Tasmacetus shepherdi*
Sherman's Fox Squirrel	*Sciurus niger shermani*
Sherman's Pocket Gopher (extinct?)	*Geomys pinetis fontanelus*
Sherman's Short-tailed Shrew	*Blarina shermani*
Shield-faced Roundleaf Bat	*Hipposideros lylei*
Shipton's Mountain Cavy	*Microcavia shiptoni*
Shiras' Moose	*Alces alces shirasi*
Short-eared Bat	*Cyttarops alecto*
Short-haired Hydromyine	*Paraleptomys wilhelmina*
Short-nosed Fruit Bat	*Cynopterus sphinx*
Shortridge's Chacma Baboon	*Papio ursinus ruacana*
Shortridge's Free-tailed Bat	*Chaerephon shortridgei*

Shortridge's Horseshoe Bat	*Rhinolophus shortridgei*
Shortridge's Leaf Monkey	*Trachypithecus shortridgei*
Shortridge's Leopard	*Panthera pardus shortridgei*
Shortridge's Mouse	*Mus shortridgei*
Shortridge's Multimammate Mouse	*Mastomys shortridgei*
Shortridge's Rat	*Thallomys shortridgei*
Shortridge's Rock Mouse	*Petromyscus shortridgei*
Shortridge's Rousette	*Rousettus leschenaulti shortridgei*
Shrew-mouse genus	*Archboldomys*
Sibbald's Whale	*Balaenoptera musculus*
Sibree's Dwarf Lemur	*Cheirogaleus sibreei*
Sierra Madre Shrew-Mouse	*Archboldomys musseri*
Silvery Greater Galago	*Otolemur monteiri*
Simmons' Mouse-Lemur	*Microcebus simmonsi*
Simon's Dipodil	*Gerbillus simoni*
Simons' Spiny Rat	*Proechimys simonsi*
Simpson's Duiker	*Cephalophus monticola simpsoni*
Sir David's Long-beaked Echidna	*Zaglossus attenboroughi*
Sitatunga	*Tragelaphus spekii*
Sladen's Rat	*Rattus tanezumi*
Slender-tailed Giant Squirrel	*Protoxerus aubinnii*
Slevin's Deer Mouse	*Peromyscus slevini*
Sloggett's Vlei Rat	*Otomys sloggetti*
Small Asian Sheath-tailed Bat	*Emballonura alecto*
Small White-toothed Rat	*Berylmys berdmorei*
Small-toothed Fruit Bat	*Neopteryx frosti*
Smith's Bush Squirrel	*Paraxerus cepapi*
Smith's Fruit Bat	*Lissonycteris smithii*
Smith's Red Rock Hare	*Pronolagus rupestris*
Smith's Rock Elephant Shrew	*Elephantulus rupestris*
Smith's Shrew	*Chodsigoa smithii*
Smith's Vole	*Myodes smithii*
Smith's Woolly Bat	*Kerivoula smithii*
Smith's Zokor	*Eospalax smithii*
Smoke-bellied Rat	*Niviventer eha*
Smoky Thumbless Bat	*Amorphochilus schnablii*
Snethlage's Marmoset	*Callithrix emiliae*
Sody's Tree Rat	*Kadarsanomys sodyi*
Soemmerring's Gazelle	*Gazella soemmerringii*
Soft-furred rat genus	*Millardia*
Soft-furred spiny rat genus	*Olallamys*
Sokolov's Dwarf Hamster	*Cricetulus sokolovi*
Sokolov's White-toothed Shrew	*Crocidura sokolovi*
Somali Elephant Shrew	*Elephantulus revoili*
Somali Hedgehog	*Atelerix sclateri*

Somali Pygmy Gerbil	*Microdillus peeli*
Sombre Bat	*Eptesicus tatei*
Someren's Girder-backed Shrew	*Scutisorex somereni*
Sonoran Harvest Mouse	*Reithrodontomys burti*
Sorensen's Leaf-nosed Bat	*Hipposideros sorenseni*
Soriano's Yellow-shouldered Bat	*Sturnira sorianoi*
Southern Bog Lemming	*Synaptomys cooperi*
Southern Giant Pouched Rat	*Cricetomys ansorgei*
Southern Long-finned Pilot Whale	*Globicephala melas edwardii*
Southern Luzon Cloud Rat	*Phloeomys cumingi*
Southern Mountain Brushtail Possum	*Trichosurus cunninghamii*
Southern Myotis	*Myotis aelleni*
Southern Ningaui	*Ningaui yvonneae*
Southern Plains Grey Langur	*Semnopithecus (entellus) dussumieri*
Southern Right-Whale Dolphin	*Lissodelphis peronii*
Southern Sea-Lion	*Otaria byronia*
Sowell's Short-tailed Bat	*Carollia sowelli*
Sowerby's Beaked Whale	*Mesoplodon bidens*
Spade-toothed Beaked Whale	*Mesoplodon traversii*
Speckled Dasyure	*Neophascogale lorentzi*
Spegazzini's Grass Mouse	*Akodon spegazzinii*
Speke's Gazelle	*Gazella spekei*
Speke's Pectinator	*Pectinator spekei*
Spiny rat genus	*Pattonomys*
Spiny Seram Rat	*Rattus feliceus*
Spix's Bearded Saki	*Chiropotes israelita*
Spix's Black-headed Uacari	*Cacajao melanocephalus ouakary*
Spix's Black-mantled Tamarin	*Saguinus nigricollis*
Spix's Disk-winged Bat	*Thyroptera tricolor*
Spix's Owl Monkey	*Aotus vociferans*
Spix's Round-eared Bat	*Tonatia bidens*
Spix's Saddle-back Tamarin	*Saguinus fuscicollis fuscicollis*
Spix's Yellow-toothed Cavy	*Galea spixii*
Spurrell's Free-tailed Bat	*Mops spurrelli*
Spurrell's Woolly Bat	*Kerivoula phalaena*
Squirrel sp.	*Callosciurus crumpi*
Sri Lanka Shrew	*Suncus fellowesgordoni*
Sri Lankan Spiny Mouse	*Mus fernandoni*
St. Aignan's Trumpet-eared Bat	*Kerivoula agnella*
Stampfli's Putty-nosed Monkey	*Cercopithecus nictitans stampflii*
Standing's Hippopotamus	*Hippopotamus amphibius standingi*
Stankovic's Mole	*Talpa stankovici*
Stanley's Brush-furred Rat	*Lophuromys stanleyi*
Starck's Hare	*Lepus starcki*
Starrett's Tailless Bat	*Anoura werckleate*

Steere's Spiny Rat *Proechimys steerei*
Steere's Squirrel *Sundasciurus steerii*
Steinbach's Ocelot *Leopardus pardalis steinbachi*
Steinbach's Tuco-tuco *Ctenomys steinbachi*
Stein's Cuscus *Phalanger vestitus*
Stein's Melomys *Paramelomys steini*
Stein's Rat *Rattus steini*
Stejneger's Beaked Whale *Mesoplodon stejnegeri*
Steller's Sea Cow (extinct) *Hydrodamalis gigas*
Steller's Sea-Lion *Eumetopias jubatus*
Stephen Nash's Titi *Callicebus stephennashi*
Stephens' Kangaroo-Rat *Dipodomys stephensi*
Stephens' Woodrat *Neotoma stephensi*
Steppe Field Mouse *Apodemus witherbyi*
Steppe Polecat *Mustela eversmannii*
Stevenson's Collared Lemming *Dicrostonyx stevensoni*
Stirton's Deer Mouse *Peromyscus stirtoni*
Stoliczka's Mountain Vole *Alticola stoliczkanus*
Stoliczka's Trident Bat *Aselliscus stoliczkanus*
Stolzmann's Crab-eating Rat *Ichthyomys stolzmanni*
Stone's Caribou *Rangifer tarandus stonei*
Stone's Sheep *Ovis dalli stonei*
Storey's Mole-Rat *Tachyoryctes storeyi*
Strachey's Mountain Vole *Alticola stracheyi*
Strand's Birch Mouse *Sicista strandi*
Strecker's Pocket Gopher *Geomys personatus streckeri*
Strelkov's Long-eared Bat *Plecotus strelkovi*
Streltzov's Vole *Alticola strelzowi*
Stripe-faced Fruit Bat *Styloctenium wallacei*
Stuart's Antechinus *Antechinus stuartii*
Stuhlmann's Golden-Mole *Chrysochloris stuhlmanni*
Stuhlmann's Monkey *Cercopithecus mitis stuhlmanni*
Sturdee's Pipistrelle *Pipistrellus sturdeei*
Styan's Red Panda *Ailurus fulgens styani*
Styan's Squirrel *Callosciurus erythraeus styani*
Styan's Water Shrew *Chimarrogale styani*
Sucre Spiny Rat *Proechimys urichi*
Sulawesi Free-tailed Bat *Mopssarasinorum*
Sulawesi Palm Civet *Macrogalidia musschenbroekii*
Sulawesi rat genus *Margaretamys*
Sumatran Flying Squirrel *Hylopetes winstoni*
Sumatran Mastiff Bat *Mormopterus doriae*
Sumatran Orangutan *Pongo abelii*
Sumatran Rabbit *Nesolagus netscheri*
Sumichrast's Harvest Mouse *Reithrodontomys sumichrasti*

Sumichrast's Vesper Rat	*Nyctomys sumichrasti*
Sunda Acerodon	*Acerodon mackloti*
Sundevall's Jird	*Meriones crassus*
Sundevall's Roundleaf Bat	*Hipposideros caffer*
Swamp Deer	*Cervus duvaucelli*
Swayne's Dik-dik	*Madoqua (saltiana) swaynei*
Swayne's Hartebeest	*Alcelaphus buselaphus swaynei*
Swinhoe's Deer	*Cervus unicolor swinhoei*
Swinhoe's Jird	*Meriones crassus*
Swinhoe's Striped Squirrel	*Tamiops swinhoei*
Swinny's Horseshoe Bat	*Rhinolophus swinnyi*
Swynnerton's Bush Squirrel	*Paraxerus vexillarius*
Sykes' Monkey	*Cercopithecus albogularis*
Taczanowski's Oldfield Mouse	*Thomasomys taczanowskii*
Taddei's Bat	*Eptesicus taddeii*
Taiwan Serow	*Naemorhedus swinhoei*
Taiwan Vole	*Volemys kikuchii*
Taiwanese Mole-Shrew	*Anourosorex yamashinai*
Talazac's Shrew-Tenrec	*Microgale talazaci*
Tamar Wallaby	*Macropus eugenii*
Tantalus Monkey	*Chlorocebus tantalus*
Tarabul's Gerbil	*Gerbillus tarabuli*
Tarpan (extinct)	*Equus przewalski gmelini*
Tasmanian Bettong	*Bettongia gaimardi*
Tasmanian Devil	*Sarcophilus harrisii*
Tasmanian Long-eared Bat	*Nyctophilus (timoriensis) sherrini*
Tasmanian Pademelon	*Thylogale billardierii*
Tasmanian Tiger	*Thylacinus harrisii*
Tate's Rice Rat	*Hylaeamys tatei*
Tate's Shrew-Rat	*Tateomys rhinogradoides*
Tate's Triok	*Dactylopsila tatei*
Tattersall's Sifaka	*Propithecus tattersalli*
Taylor's Flying Fox	*Pteropus pumilis tablasi*
Tayra	*Eira barbara*
Telford's Shrew	*Crocidura telfordi*
Temchuk's Bolo Mouse	*Necromys temchuki*
Temminck's Flying Fox	*Pteropus temmincki*
Temminck's Flying Squirrel	*Petinomys setosus*
Temminck's Giant Forest Squirrel	*Epixerus ebii*
Temminck's Golden Cat	*Catopuma temminckii*
Temminck's Ground Pangolin	*Manis temminckii*
Temminck's Mole	*Mogera wogura*
Temminck's Mouse	*Mus musculoides*
Temminck's Red Colobus	*Piliocolobus badius temminckii*

Temminck's Spotted Squirrel	*Heliosciurus punctatus*
Temminck's Striped Mouse	*Hybomys trivirgatus*
Temminck's Tailless Fruit Bat	*Megaerops ecaudatus*
Temminck's Trident Bat	*Aselliscus tricuspidatus*
Terai Grey Langur	*Semnopithecus (entellus) hector*
Teusz's Dolphin	*Sousa teuszii*
Thaeler's Pocket Gopher	*Orthogeomys thaeleri*
Thalia's Shrew	*Crocidura thalia*
Theobald's Tomb Bat	*Taphozous theobaldi*
Theresa's Short-tailed Opossum	*Monodelphis theresa*
Therese's Shrew	*Crocidura theresae*
Thick-thumbed Myotis	*Myotis rosseti*
Thierry's Genet	*Genetta thierryi*
Thollon's Red Colobus	*Piliocolobus thollonii*
Thomas' Broad-nosed Bat	*Platyrrhinus dorsalis*
Thomas' Flying Squirrel	*Aeromys thomasi*
Thomas' Fruit-eating Bat	*Artibeus watsoni*
Thomas' Galago	*Galago thomasi*
Thomas' Giant Deer Mouse	*Megadontomys thomasi*
Thomas' Horseshoe Bat	*Rhinolophus thomasi*
Thomas' Leaf Monkey	*Presbytis thomasi*
Thomas' Mastiff Bat	*Eumops maurus*
Thomas' Melomys	*Paramelomys mollis*
Thomas' Moustached Tamarin	*Saguinus labiatus thomasi*
Thomas' Nectar Bat	*Lonchophylla thomasi*
Thomas' Night Monkey	*Aotus miconax*
Thomas' Oldfield Mouse	*Thomasomys pyrrhonotus*
Thomas' Pika	*Ochotona thomasi*
Thomas' Pine Vole	*Microtus thomasi*
Thomas' Pipistrelle	*Pipistrellus paterculus*
Thomas' Pygmy Jerboa	*Salpingotus thomasi*
Thomas' Pygmy Mouse	*Mus sorella*
Thomas' Rice Rat	*Oryzomys dimidiatus*
Thomas' Rock Rat	*Aethomys thomasi*
Thomas' Rope Squirrel	*Funisciurus anerythrus*
Thomas' Sac-winged Bat	*Balantiopteryx io*
Thomas' Shrew-Tenrec	*Microgale thomasi*
Thomas' Small-eared Shrew	*Cryptotis thomasi*
Thomas' Spiny Rat	*Trinomys iheringi*
Thomas' Water Mouse	*Rheomys thomasi*
Thomas' Yellow Bat	*Rhogeessa io*
Thomas' Yellow-shouldered Bat	*Sturnira thomasi*
Thompson's Pygmy Shrew	*Sorex thompsoni*
Thomson's Gazelle	*Gazella thomsoni*
Thornicroft's Giraffe	*Giraffa camelopardelis thornicrofti*

Thorold's Deer	*Cervus albirostris*
Tibetan Antelope	*Pantholops hodgsonii*
Tickell's False Serotine	*Hesperoptenus tickelli*
Tilda's Yellow-shouldered Bat	*Sturnira tildae*
Titania's Woolly Bat	*Kerivoula titania*
Togo Mole-Rat	*Fukomys zechi*
Tokuda's Mole	*Mogera tokudae*
Tomes' Long-eared Bat	*Lonchorhina aurita*
Tomes' Rice Rat	*Nephelomys albigularis*
Tomes' Spiny Rat	*Proechimys semispinosus*
Tonkin Limestone Rat	*Tonkinomys daovantieni*
Toolache Wallaby (extinct)	*Macropus greyi*
Torre's Cave Rat (extinct)	*Boromys torrei*
Torres' Crimson-nosed Rat	*Bibimys torresi*
Townsend's Big-eared Bat	*Corynorhinus townsendii*
Townsend's Chipmunk	*Tamias townsendii*
Townsend's Ground Squirrel	*Spermophilus townsendii*
Townsend's Hare	*Lepus townsendii*
Townsend's Mole	*Scapanus townsendii*
Townsend's Pocket Gopher	*Thomomys townsendii*
Townsend's Vole	*Microtus townsendii*
Tranier's Gerbil	*Taterillus tranieri*
Tree Bat	*Ardops nichollsi*
Trefoil-toothed Giant Rat	*Lenomys meyeri*
Tres Marias Cottontail	*Sylvilagus graysoni*
Trevor's Free-tailed Bat	*Mops trevori*
Tristram's Jird	*Meriones tristrami*
Trouessart's Trident Bat	*Triaenops furculus*
Troughton's Forest Bat	*Vespadelus troughtoni*
Troughton's Pouched Bat	*Saccolaimus mixtus*
Troughton's Tomb Bat	*Taphozous troughtoni*
Trowbridge's Shrew	*Sorex trowbridgii*
True's Beaked Whale	*Mesoplodon mirus*
True's Porpoise	*Phocoenoides dalli*
True's Shrew-Mole	*Urotrichus pilirostris*
True's Vole	*Hyperacrius fertilis*
Trumbull's Bonneted Bat	*Eumops trumbulli*
Tschudi's Pygmy Rice Rat	*Oligoryzomys destructor*
Tschudi's Slender Opossum	*Marmosops impavidus*
Tschudi's Yellow-shouldered Bat	*Sturnira oporaphilum*
Tsolov's Mouse-like Hamster	*Calomyscus tsolovi*
Tufted Grey Langur	*Semnopithecus (entellus) priam*
Tufted Pygmy Squirrel	*Exilisciurus whiteheadi*
Tuft-tailed Spiny Tree Rat	*Lonchothrix emiliae*
Tullberg's Soft-furred Mouse	*Praomys tullbergi*

Tweedy's Crab-eating Rat	*Ichthyomys tweedii*
Tyler's Mouse-Opossum	*Marmosa tyleriana*
Ufipa Brush-furred Rat	*Lophuromys sabunii*
Underwood's Bonneted Bat	*Eumops underwoodi*
Underwood's Long-tongued Bat	*Hylonycteris underwoodi*
Underwood's Pocket Gopher	*Orthogeomys underwoodi*
Underwood's Water Mouse	*Rheomys underwoodi*
Upemba Lechwe	*Kobus anselli*
Urial	*Ovis vignei*
Uta Hick's Bearded Saki	*Chiropotes utahickae*
Val's Gundi	*Ctenodactylus vali*
Van Beneden's Colobus	*Procolobus verus*
Van der Decken's Sifaka	*Propithecus deckenii*
Van Deusen's Rat	*Stenomys vandeuseni*
Van Gelder's Bat	*Bauerus dubiaquercus*
Van Roosmalens' Hairy Dwarf Porcupine	*Coendou roosmalenorum*
Van Sung's Shrew	*Chodsigoa caovansunga*
Van Zyl's Golden-Mole	*Cryptochloris zyli*
Vanzolini's Bald-faced Saki	*Pithecia irrorata vanzolinii*
Varona and Garrido's Hutia	*Mesocapromys sanfelipensis*
Veldkamp's Bat	*Nanonycteris veldkampi*
Veloz's Hutia (extinct)	*Plagiodontia velozi*
Velvet Climbing Mouse	*Dendroprionomys rousseloti*
Verhagen's Brush-furred Rat	*Lophuromys verhageni*
Verheyen's Shrew	*Congosorex verheyeni*
Verheyen's Multimammate Mouse	*Mastomys verheyeni*
Verhoeven's Giant Tree Rat (extinct)	*Papagomys theodorverhoeveni*
Vernay's Climbing Mouse	*Dendromus vernayi*
Vernay's Lion	*Panthera leo vernayi*
Verreaux's Mouse	*Myomyscus verreauxi*
Verreaux's Sifaka	*Propithecus verreauxi*
Verschuren's Swamp Rat	*Malacomys verschureni*
Vespucci's Rat (extinct)	*Noronhomys vespuccii*
Vieira's Long-snouted Bat	*Xeronycteris vieirai*
Vieira's Spiny Rat	*Echimys vieirai*
Villa's Grey Shrew	*Notiosorex villai*
Villa's Pocket Gopher	*Thomomys bottae villai*
Vincent's Bush Squirrel	*Paraxerus vincenti*
Vinogradov's Jerboa	*Allactaga vinogradovi*
Vinogradov's Jird	*Meriones vinogradovi*
Vinson's Slit-faced Bat	*Nycterisvinsoni*
Virginia Dunnart	*Sminthopsis virginiae*
Visagie's Golden-Mole	*Chrysochloris visagiei*

Visayan Leopard Cat	*Prionailurus bengalensis rabori*
Vleeschouwers' Talapoin	*Miopithecus talapoin vleeschouwersi*
Voalavoanaia	*Gymnuromys roberti*
Volcano Rabbit	*Romerolagus diazi*
Vole genera	*Blanfordimys, Prometheomys*
Volnuchin's Shrew	*Sorex volnuchini*
Von der Decken's Sifaka	*Propithecus deckenii*
Vordermann's Flying Squirrel	*Petinomys vordermanni*
Vordermann's Pipistrelle	*Hypsugo vordermanni*
Voss' Slender Opossum	*Marmosops creightoni*
Wagner's Bonneted Bat	*Eumops glaucinus*
Wagner's Gerbil	*Gerbillus dasyurus*
Wagner's Marsh Rat	*Holochilus sciureus*
Wagner's Moustached Bat	*Pteronotus personatus*
Wagner's Peccary	*Catagonus wagneri*
Wagner's Sac-winged Bat	*Cormora brevirostris*
Wahlberg's Epauletted Bat	*Epomophorus wahlbergi*
Wallace's Striped Dasyure	*Myoictis wallacii*
Waller's Gazelle	*Litocranius walleri*
Wallich's Deer	*Cervus elaphus wallichi*
Ward's Field Mouse	*Apodemus wardi*
Ward's Long-eared Bat	*Plecotus wardi*
Ward's Red-backed Vole	*Eothenomys wardi*
Ward's Reedbuck	*Redunca redunca wardi*
Ward's Short-tailed Shrew	*Blarinella wardi*
Ward's Zebra	*Equus wardi*
Warren's Spiny Rat	*Proechimys warreni*
Washington Ground Squirrel	*Spermophilus washingtoni*
Waterhouse's Leaf-nosed Bat	*Macrotus waterhousii*
Waters' Gerbil	*Gerbillus watersi*
Watson's Climbing Rat	*Tylomys watsoni*
Watts' Pipistrelle	*Pipistrellus wattsi*
Watts' Spiny Rat	*Maxomys wattsi*
Webb's Tufted-tailed Rat	*Eliurus webbi*
Weber's Dwarf Squirrel	*Prosciurillus weberi*
Weddell Seal	*Leptonychotes weddellii*
Weddell's Saddle-back Tamarin	*Saguinus fuscicollis weddelli*
Welwitsch's Bat	*Myotis welwitschii*
Werner's Guenon	*Cercopithecus aethiops werneri*
Western Barred Bandicoot	*Perameles bougainville*
Western Brush Wallaby	*Macropus irma*
Western Long-tongued Bat	*Glossophaga morenoi*
Western Pebble-mound Mouse	*Pseudomys chapmani*
Western Quoll	*Dasyurus geoffroii*

Western Red Bat	*Lasiurus blossevillii*
Western Red Forest Rat	*Nesomys lambertoni*
Western Sucker-footed Bat	*Myzopoda schliemanni*
Western White-bearded Wildebeest	*Connochaetes taurinus mearnsi*
Western White-eared Giant Rat	*Hyomys dammermani*
Wetzel's Climbing Mouse	*Rhipidomys wetzeli*
Weyland Ringtail Possum	*Pseudochirulus caroli*
Weyns' Duiker	*Cephalophus weynsi*
Whitaker's Shrew	*Crocidura whitakeri*
White-collared Fruit Bat	*Megaerops wetmorei*
White-fronted Spider Monkey	*Ateles belzebuth*
Whitehead's Spiny Rat	*Maxomys whiteheadi*
Whitehead's Woolly Bat	*Kerivoula whiteheadi*
Whiteside's Guenon	*Cercopithecus ascanius whitesidei*
White-striped Dorcopsis	*Dorcopsis hageni*
White-winged Vampire Bat	*Diaemus youngi*
Whyte's Hare	*Lepus whytei*
Whyte's Mole-Rat	*Fukomys whytei*
Whyte's Vervet	*Chlorocebus pygerythrus whytei*
Wied's Long-legged Bat	*Macrophyllum macrophyllum*
Wied's Marmoset	*Callithrix kuhlii*
Wilfred's Mouse	*Wilfredomys oenax*
Williams' Jerboa	*Allactaga williamsi*
Williamson's Mouse-Deer	*Tragulus williamsoni*
Wilson's Dolphin	*Lagnenorhynchus cruciger*
Wilson's Meadow Mouse	*Microtus pennsylvanicus pennsylvanicus*
Wilson's Spiny Mouse	*Acomys wilsoni*
Wimmer's Shrew	*Crocidura wimmeri*
Winkelmann's Deer Mouse	*Peromyscus winkelmanni*
Woermann's Bat	*Megaloglossus woermanni*
Wolffsohn's Leaf-eared Mouse	*Phyllotis wolffsohni*
Wolffsohn's Viscacha	*Lagidium wolffsohni*
Wolf's Monkey	*Cercopithecus wolfi*
Wollaston's Roundleaf Bat	*Hipposideros wollastoni*
Wondiwoi Tree Kangaroo	*Dendrolagus dorianus mayri*
Wongai Ningaui	*Ningaui ridei*
Woodford's Blossom Bat	*Melonycteris woodfordi*
Woodhouse's Arvicola	—
Wood's Slit-faced Bat	*Nycteris woodi*
Woods' Solenodon	*Solenodon paradoxus woodi*
Woolley's Pseudantechinus	*Pseudantechinus woolleyae*
Woosnam's Broad-headed Mouse	*Zelotomys woosnami*
Woosnam's Brush-furred Rat	*Lophuromys woosnami*
Wrangel Collared Lemming	*Dicrostonyx vinogradovi*

Wrangel Island Lemming	*Lemmus portenkoi*
Wright's Sportive Lemur	*Lepilemur wrighti*
Wroughton's Free-tailed Bat	*Otomops wroughtoni*
Wulsin's Ebony Leaf Monkey	*Trachypithecus auratus ebenus*
Xantippe's Shrew	*Crocidura xantippe*
Yalden's Rat	*Desmomys yaldeni*
Yellow-lipped Bat	*Vespadelus douglasorum*
Yellow-spotted Hyrax	*Heterohyrax brucei*
Yepes' Long-nosed Armadillo	*Dasypus yepesi*
Yonenaga-Yassuda's Spiny Rat	*Trinomys yonenagae*
Yoshiyuki's Myotis	*Myotis yesoensis*
Yucatan Brown Brocket	*Mazama pandora*
Zaisan Mole Vole	*Ellobius tancrei*
Zaitsev's White-toothed Shrew	*Crocidura zaitsevi*
Zakaria's Gerbil	*Gerbillus zakariai*
Zambesi Sitatunga	*Tragelaphus spekii selousi*
Zammarano's White-throated Guenon	*Cercopithecus albogularis zammaranoi*
Zanzibar Leopard	*Panthera pardus adersi*
Zanzibar Red Colobus	*Piliocolobus kirkii*
Zapadokanad's Bear	*Ursus arctos pervagor*
Zaphir's Shrew	*Crocidura zaphiri*
Zarudny's Jird	*Meriones zarudnyi*
Zarudny's Shrew	*Crocidura zarudnyi*
Zenker's Fruit Bat	*Scotonycteris zenkeri*
Zimmerman's Shrew	*Crocidura zimmermanni*
Zimmer's Shrew	*Crocidura zimmeri*
Zinser's Pocket Gopher	*Cratogeomys zinseri*
Zulu Golden-Mole	*Amblysomus iris*
Zuniga's Dark Rice Rat	*Melanomys zunigae*

Appendix 2
Scientific Names

Abrawayaomys ruschii	Ruschi's Rat
Abrocoma bennetti	Bennett's Chinchilla-Rat
Abrocoma budini	Budin's Chinchilla-Rat
Abrothrix hershkovitzi	Hershkovitz's Grass Mouse
Abrothrix sanborni	Sanborn's Grass Mouse
Acerodon mackloti	Sunda Acerodon
Acomys chudeaui	Chudeau's Spiny Mouse
Acomys johannis	Johan's Spiny Mouse
Acomys kempi	Kemp's Spiny Mouse
Acomys louisae	Louise's Spiny Mouse
Acomys percivali	Percival's Spiny Mouse
Acomys wilsoni	Wilson's Spiny Mouse
Aconaemys porteri	Porter's Rock Rat
Aconaemys sagei	Sage's Rock Rat
Aepeomys reigi	Reig's Montane Mouse
Aeromys thomasi	Thomas' Flying Squirrel
Aethalops alecto	Pygmy Fruit Bat
Aethomys bocagei	Bocage's Rock Rat
Aethomys chrysophilus dollmani	Dollman's Rock Rat
Aethomys granti	Grant's Rock Rat
Aethomys helleri	Heller's Rock Rat
Aethomys hindei	Hinde's Rock Rat
Aethomys kaiseri	Kaiser's Rock Rat
Aethomys thomasi	Thomas' Rock Rat
Ailurus fulgens styani	Styan's Red Panda
Akodon azarae	Azara's Grass Mouse
Akodon budini	Budin's Grass Mouse
Akodon dayi	Day's Grass Mouse
Akodon dolores	Dolorous Grass Mouse
Akodon kofordi	Koford's Grass Mouse
Akodon lindberghi	Lindbergh's Grass Mouse
Akodon markhami	Markham's Grass Mouse
Akodon molinae	Molina's Grass Mouse
Akodon philipmyersi	Myers' Grass Mouse

Akodon reigi	Reig's Grass Mouse
Akodon spegazzinii	Spegazzini's Grass Mouse
Alcelaphus buselaphus cokii	Coke's Hartebeest
Alcelaphus buselaphus jacksoni	Jackson's Hartebeest
Alcelaphus buselaphus lelwel	Lelwel's Hartebeest
Alcelaphus buselaphus neumanni	Neumann's Hartebeest
Alcelaphus buselaphus selbornei	Selborne's Hartebeest
Alcelaphus buselaphus swaynei	Swayne's Hartebeest
Alces alces shirasi	Shiras' Moose
Allactaga firouzi	Iranian Jerboa
Allactaga hotsoni	Hotson's Jerboa
Allactaga severtzovi	Severtzov's Jerboa
Allactaga vinogradovi	Vinogradov's Jerboa
Allactaga williamsi	Williams' Jerboa
Allactodipus bobrinskii	Bobrinski's Jerboa
Allenopithecus nigroviridis	Allen's Swamp Monkey
Allocricetulus eversmanni	Eversmann's Hamster
Alouatta (seniculus) macconnelli	Guyanan Red Howler Monkey
Alouatta belzebul	Red-handed Howler Monkey
Alouatta pigra	Lawrence's Howler Monkey
Alticola roylei	Royle's Mountain Vole
Alticola stoliczkanus	Stoliczka's Mountain Vole
Alticola stracheyi	Strachey's Mountain Vole
Alticola strelzowi	Streltzov's Vole
Amblysomus corriae	Fynbos Golden-Mole
Amblysomus iris	Zulu Golden-Mole
Amblysomus marleyi	Marley's Golden-Mole
Ammodorcas clarkei	Clarke's Gazelle, Dibatag
Ammospermophilus harrisii	Harris' Antelope Squirrel
Ammospermophilus nelsoni	Nelson's Antelope Squirrel
Amorphochilus schnablii	Smoky Thumbless Bat
Anathana ellioti	Madras Tree-Shrew
Andalgalomys olrogi	Olrog's Chaco Mouse
Andalgalomys pearsoni	Pearson's Chaco Mouse
Andalgalomys roigi	Roig's Tuco-tuco
Anomalurus beecrofti	Beecroft's Scaly-tailed Squirrel
Anomalurus derbyanus	Lord Derby's Scaly-tailed Squirrel
Anomalurus pelii	Pel's Scaly-tailed Squirrel
Anotomys leander	Ecuador Fish-eating Rat
Anoura cadenai	Cadena's Tailless Bat
Anoura cultrata	Handley's Tailless Bat
Anoura geoffroyi	Geoffroy's Tailless Bat
Anoura luismanueli	Luis Manuel's Tailless Bat
Anoura werckleate	Starrett's Tailless Bat
Anourosorex schmidi	Giant Mole-Shrew

Anourosorex yamashinai	Taiwanese Mole-Shrew
Antechinus godmani	Atherton Antechinus
Antechinus stuartii	Macleay's Marsupial Mouse
Antechinus stuartii	Stuart's Antechinus
Antechinus swainsonii	Dusky Antechinus
Antechinus wilhelmina	Lesser Antechinus
Aotus (azarai) infulatus	Kuhl's Night Monkey
Aotus azarai	Azara's Night Monkey
Aotus bipunctatus	Bole's Douroucouli
Aotus brumbacki	Brumback's Night Monkey, Bumback's Night Monkey
Aotus hershkovitzi	Hershkovitz's Night Monkey
Aotus jorgehernandezi	Hernández-Camacho's Night Monkey
Aotus miconax	Thomas' Night Monkey
Aotus nancymaae	Nancy Ma's Night Monkey
Aotus vociferans	Spix's Owl Monkey
Apodemus arianus	Persian Field Mouse
Apodemus wardi	Ward's Field Mouse
Apodemus witherbyi	Steppe Field Mouse
Aproteles bulmerae	Bulmer's Fruit Bat
Archboldomys	The shrew-mouse genus
Archboldomys musseri	Sierra Madre Shrew-Mouse
Arctocephalus forsteri	Forster's Fur Seal, New Zealand Fur Seal
Arctocephalus philippii	Juan Fernandez Fur Seal
Arctocephalus townsendi	Guadelupe Fur Seal
Arctomys lewisii	Lewis' Marmot
Ardops nichollsi	Tree Bat
Artibeus anderseni	Andersen's Fruit-eating Bat
Artibeus cinereus	Gervais' Fruit-eating Bat
Artibeus schwartzi	Schwartz's Fruit-eating Bat
Artibeus watsoni	Thomas' Fruit-eating Bat
Arvicanthis ansorgei	Ansorge's Grass Rat
Arvicanthis blicki	Blick's Grass Rat
Arvicanthis neumanni	Neumann's Grass Rat
Asellliscus patrizii	Patrizi's Trident Bat
Asellliscus stoliczkanus	Stoliczka's Trident Bat
Asellliscus tricuspidatus	Temminck's Trident Bat
Atelerix sclateri	Somali Hedgehog
Ateles belzebuth	White-fronted Spider Monkey
Ateles geoffroyi	Geoffroy's Spider Monkey
Atelocynus microtis	Sclater's Dog
Avahi betsileo	Betsileo Woolly Lemur
Avahi cleesei	Cleese's Woolly Lemur

Avahi peyrierasi	Peyrieras' Woolly Lemur
Axis kuhlii	Bawean Deer
Baiomys taylori	Northern Pygmy Mouse
Balaenoptera borealis	Rudolph's Whale
Balaenoptera brydei	Bryde's Whale
Balaenoptera edeni	Eden's Whale
Balaenoptera musculus	Sibbald's Whale
Balaenoptera omurai	Omura's Whale
Balantiopteryx io	Thomas' Sac-winged Bat
Balantiopteryx plicata	Peters' Bat
Bandicota savilei	Savile's Bandicoot Rat
Bassaricyon alleni	Allen's Olingo
Bassaricyon beddardi	Beddard's Olingo
Bassaricyon gabbii	Bushy-tailed Olingo
Bassaricyon lasius	Harris' Olingo
Bassaricyon pauli	Chiriqui Olingo
Bassariscus astutus	Miner's Cat
Bassariscus sumichrasti	Cacomistle
Batomys granti	Luzon Hairy-tailed Ra
Batomys salomonseni	Mindanao Hairy-tailed Rat
Bauerus dubiaquercus	Van Gelder's Bat
Bdeogale jacksoni	Jackson's Mongoose
Beamys hindei	Long-tailed Pouched Rat
Belomys pearsonii	Hairy-footed Flying Squirrel
Berardius	The beaked whale genus
Berardius arnuxii	Arnoux's Beaked Whale
Berardius bairdii	Baird's Beaked Whale
Berylmys berdmorei	Small White-toothed Rat
Berylmys bowersi	Bowers' White-toothed Rat
Berylmys mackenziei	Kenneth's White-toothed rat
Bettongia gaimardi	Tasmanian Bettong
Bettongia gouldii	Gould's Rat-Kangaroo
Bettongia lesueur	LeSueur's Bettong
Bibimys	The crimson-nosed rat genus
Bibimys torresi	Torres' Crimson-nosed Rat
Biswamoyopterus biswasi	Namdapha Flying Squirrel
Blanfordimys	The vole genus
Blarina hylophaga	Elliot's Short-tailed Shrew
Blarina shermani	Sherman's Short-tailed Shrew
Blarinella griselda	Grey Short-tailed Shrew
Blarinella wardi	Ward's Short-tailed Shrew
Bootherium bombifrons	Harlan's Musk-ox (extinct)
Boromys torrei	Torre's Cave Rat (extinct)
Bos sauveli	Kouprey

Brachiones przewalskii	Przewalski's Gerbil
Brucepattersonius	The rodent genus ("brucies")
Brucepattersonius iheringi	Ihering's Brucie
Bubalus quarlesi	Mountain Anoa
Budorcas taxicolor bedfordi	Bedford Takin
Bunomys andrewsi	Andrews' Hill Rat
Bunomys fratrorum	Hose's Hill Rat
Bunomys heinrichi	Heinrich's Hill Rat
Cacajao ayresi	Ayres' Uacari
Cacajao calvus novaesi	Novaes' Bald-headed Uacari
Cacajao melanocephalus ouakary	Spix's Black-headed Uacari
Callicebus (personatus) barbarabrownae	Barbara Brown's Titi
Callicebus (torquatus) lucifer	Lucifer Titi
Callicebus bernhardi	Prince Bernhard's Titi
Callicebus coimbrai	Coimbra-Filho's Titi
Callicebus hoffmannsi	Hoffmanns' Titi
Callicebus medemi	Medem's Titi
Callicebus moloch	Red-bellied Titi
Callicebus olallae	Olalla's Titi
Callicebus stephennashi	Nash's Titi Monkey, Stephen Nash's Titi
Callimico goeldi	Goeldi's Marmoset
Callithrix (argentata) marcai	Marca's Marmoset
Callithrix (Callibella) humilis	Roosmalen's Dwarf Marmoset
Callithrix emiliae	Emilia's Marmoset, Snethlage's Marmoset
Callithrix geoffroyi	Geoffroy's Marmoset
Callithrix intermedia	Hershkovitz's Marmoset
Callithrix jacchus	Common Marmoset
Callithrix kuhlii	Kuhl's Marmoset
Callithrix kuhlii	Wied's Marmoset
Callithrix saterei	Sateré-Maués' Marmoset
Callosciurus adamsi	Ear-spot Squirrel
Callosciurus albescens	Kloss' Squirrel
Callosciurus crumpi	Squirrel sp.
Callosciurus erythraeus	Pallas' Squirrel
Callosciurus erythraeus styani	Styan's Squirrel
Callosciurus finlaysonii	Finlayson's Squirrel
Callosciurus orestes	Borneo Black-banded Squirrel
Callosciurus phayrei	Phayre's Squirrel
Callosciurus prevostii	Prevost's Squirrel
Callosciurus quinquestriatus	Anderson's Squirrel
Calomys hummelincki	Hummelinck's Vesper Mouse
Calomyscus bailwardi	Iranian Mouse-like Hamster

Calomyscus elburzensis	Goodwin's Mouse-like Hamster
Calomyscus hotsoni	Hotson's Mouse-like Hamster
Calomyscus tsolovi	Tsolov's Mouse-like Hamster
Caluromys derbianus	Derby's Pale-eared Woolly Opossum
Canis hallstromi	New Guinea Singing Dog
Canis lupus bernardi	Bernard's Wolf
Canis rufus gregoryi	Gregory's Red Wolf
Capra falconeri	Markhor
Capra ibex severtzovi	Severtzov's Ibex
Capra pyrenaica victoriae	Queen Victoria's Ibex
Capromys (pilorides) gundlachianus	Sabana Hutia
Capromys pilorides	Desmarest's Hutia
Caracal caracal schmitzi	Schmitz's Caracal
Carollia benkeithi	Keith's Short-tailed Bat
Carollia monohernandezi	Hernández-Camacho's Short-tailed Bat
Carollia perspicillata	Seba's Short-tailed Bat
Carollia sowelli	Sowell's Short-tailed Bat
Carollia subrufa	Hahn's Short-tailed Bat
Carpitalpa arendsi	Arends' Golden-Mole
Carterodon sulcidens	Owl's Spiny Rat
Caryomys eva	Eva's Red-backed Vole
Caryomys inez	Inez's Red-backed Vole
Catagonus wagneri	Wagner's Peccary
Catopuma temminckii	Temminck's Golden Cat
Catopuma temminckii tristis	Fontanier's Cat
Cavia tschudii	Montane Cavy
Cebus (olivaceus) kaapori	Ka'apori Capuchin
Cebus flavius	Marcgraf's Capuchin Monkey, Marcgrave's Capuchin Monkey
Centronycteris maximiliani	Shaggy Bat
Cephalophus (melanorheus) anchietae	Anchieta's Antelope
Cephalophus (nigrifrons) hooki	Hook's Duiker
Cephalophus adersi	Aders' Duiker
Cephalophus brookei	Brooke's Duiker
Cephalophus callipygus	Peters' Duiker
Cephalophus harveyi	Harvey's Duiker
Cephalophus jentinki	Jentink's Duiker
Cephalophus maxwellii	Maxwell's Duiker
Cephalophus monticola simpsoni	Simpson's Duiker
Cephalophus ogilbyi	Ogilby's Duiker
Cephalophus spadix	Abbott's Duiker
Cephalophus weynsi	Weyns' Duiker
Cephalorhynchus commersonii	Commerson's Dolphin
Cephalorhynchus heavisidii	Heaviside's Dolphin

Cephalorhynchus hectori	Hector's Dolphin
Ceratotherium simum cottoni	Cotton's Wide-lipped Rhinoceros
Cercocebus galeritus hagenbecki	Hagenbeck's Mangabey
Cercopithecus aethiops werneri	Werner's Guenon
Cercopithecus albogularis	Sykes' Monkey
Cercopithecus albogularis albotorquatus	Pousargues' White-collared Monkey
Cercopithecus albogularis kolbi	Kolb's White-collared Monkey
Cercopithecus albogularis moloneyi	Moloney's Monkey
Cercopithecus albogularis zammaranoi	Zammarano's White-throated Guenon
Cercopithecus ascanius	Red-tailed Guenon
Cercopithecus ascanius schmidti	Schmidt's Monkey
Cercopithecus ascanius whitesidei	Whiteside's Guenon
Cercopithecus campbelli	Campbell's Monkey
Cercopithecus denti	Dent's Monkey
Cercopithecus diana	Diana Monkey
Cercopithecus doggetti	Doggett's Blue Monkey
Cercopithecus erythrogaster pococki	Pocock's White-throated Guenon
Cercopithecus hamlyni	Hamlyn's Monkey
Cercopithecus kandti	Golden Monkey
Cercopithecus lhoesti	L'Hoest's Monkey
Cercopithecus lowei	Lowe's Monkey
Cercopithecus mitis boutourlinii	Boutourlini's Blue Monkey
Cercopithecus mitis mitis	Pluto Monkey
Cercopithecus mitis schoutedenii	Schouteden's Blue Monkey
Cercopithecus mitis stuhlmanni	Stuhlmann's Monkey
Cercopithecus neglectus	De Brazza's Monkey, Schlegel's Guenon
Cercopithecus nictitans martini	Martin's Guenon
Cercopithecus nictitans stampflii	Stampfli's Putty-nosed Monkey
Cercopithecus petaurista buettikoferi	Buettikofer's Monkey
Cercopithecus pogonias grayi	Erxleben's Guenon, Gray's Crowned Guenon
Cercopithecus preussi	Cross' Guenon, Preuss' Monkey
Cercopithecus sclateri	Sclater's Monkey
Cercopithecus signatus	Jentink's Guenon
Cercopithecus wolfi	Wolf's Monkey
Cerradomys andersoni	Anderson's Rice Rat
Cerradomys langguthi	Langguth's Rice Rat
Cerradomys marinhus	Marinho's Rice Rat
Cerradomys scotti	Scott's Rice Rat
Cerradomys vivoi	De Vivo's Rice Rat
Cervus (canadensis) macneilli	MacNeill's (Red) Deer
Cervus albirostris	Thorold's Deer
Cervus alfredi	Prince Alfred's Deer
Cervus canadensis merriami	Merriam's Wapiti (Elk) (extinct)

Cervus canadensis roosevelti	Roosevelt's Wapiti
Cervus duvaucelli	Swamp Deer
Cervus elaphus wallichi	Wallich's Deer
Cervus eldii	Eld's Deer
Cervus nippon hortulorum	Dybowski's Sika Deer
Cervus nippon kopschi	Kopsch's Deer
Cervus schomburgki	Schomburgk's Deer (extinct)
Cervus unicolor swinhoei	Swinhoe's Deer
Chaerephon	The bat genus
Chaerephon aloysiisabaudiae	Duke of Abruzzi's Free-tailed Bat
Chaerephon ansorgei	Ansorge's Free-tailed Bat
Chaerephon bemmeleni	Gland-tailed Free-tailed Bat
Chaerephon bregullae	Fijian Mastiff Bat
Chaerephon chapini	Chapin's Free-tailed Bat
Chaerephon gallagheri	Gallagher's Free-tailed Bat
Chaerephon leucogaster	Grandidier's Free-tailed Bat
Chaerephon shortridgei	Shortridge's Free-tailed Bat
Chaetodipus baileyi	Bailey's Pocket Mouse
Chaetodipus dalquesti	Dalquest's Pocket Mouse
Chaetodipus goldmani	Goldman's Pocket Mouse
Chaetodipus nelsoni	Nelson's Pocket Mouse
Chaetophractus nationi	Andean Hairy Armadillo
Chaetopidus (fallax) anthonyi	Anthony's Pocket Mouse
Chalinolobus dwyeri	Large-eared Pied Bat
Chalinolobus gouldii	Gould's Wattled Bat
Cheirogaleus crossleyi	Crossley's Dwarf Lemur
Cheirogaleus sibreei	Sibree's Dwarf Lemur
Chibchanomys orcesi	Orces' Chibchan Water Mouse
Chimarrogale styani	Styan's Water Shrew
Chionomys roberti	Robert's Snow Vole
Chiroderma doriae	Brazilian Big-eyed Bat
Chiroderma salvini	Salvin's Big-eyed Bat
Chiroderma trinitatum	Goodwin's Bat
Chiromyscus chiropus	Fea's Tree Rat
Chiropodomys karlkoopmani	Koopman's Pencil-tailed Tree Mouse
Chiropotes israelita	Spix's Bearded Saki
Chiropotes utahickae	Uta Hick's Bearded Saki
Chiruromys forbesi	Forbes' Tree Mouse
Chiruromys forbesi shawmayeri	Ferguson Island Mouse
Chlorocebus aethiops cloetei	Cloet's Vervet
Chlorocebus aethiops matschiei	Matschie's Guenon
Chlorocebus aethiops zavattarii	De Beaux's Grivet
Chlorocebus pygerythrus arenarius	Heller's Vervet
Chlorocebus pygerythrus pygerythrus	Cuvier's Vervet
Chlorocebus pygerythrus whytei	Whyte's Vervet

Chlorocebus tantalus	Tantalus Monkey
Chlorocebus tantalus budgetti	Budgett's Tantalus Monkey
Chlorotalpa duthieae	Duthie's Golden-Mole
Chlorotalpa sclateri	Sclater's Golden-Mole
Chodsigoa caovansunga	Van Sung's Shrew
Chodsigoa hypsibius	De Winton's Shrew
Chodsigoa parca	Lowe's Shrew
Chodsigoa salenskii	Salenski's Shrew
Chodsigoa smithii	Smith's Shrew
Choeroniscus godmani	Godman's Long-Tailed Bat
Choloepus didactylus	Linné's Two-toed Sloth
Choloepus hoffmanni	Hoffmann's Two-toed Sloth
Chroeomys jelskii	Jelski's Altiplano Mouse
Chrotogale owstoni	Owston's Palm Civet
Chrotomys gonzalesi	Mount Isarog Striped Rat
Chrotomys whiteheadi	Luzon Striped Rat
Chrysochloris stuhlmanni	Stuhlmann's Golden-Mole
Chrysochloris visagiei	Visagie's Golden-Mole
Chrysospalax trevelyani	Giant Golden-Mole
Cistugo lesueuri	LeSueur's Hairy Bat
Cistugo seabrai	Angolan Hairy Bat
Clidomys osborni	Osborn's Key Mouse (extinct)
Cloeotis percivali	Percival's Trident Bat
Clyomys bishopi	Bishop's Fossorial Spiny Rat
Coccymys ruemmleri	Rümmler's Mouse
Coelops frithi	Frith's Tailless Bat
Coelops robinsoni	Malayan Tailless Leaf-nosed Bat
Coendou koopmani	Koopman's Porcupine
Coendou roosmalenorum	Van Roosmalens' Hairy Dwarf Porcupine
Coendou rothschildi	Rothschild's Porcupine
Colobus angolensis angolensis	Sclater's Angolan Colobus
Colobus angolensis cordieri	Cordier's Angolan Colobus
Colobus angolensis cottoni	Cotton's Colobus
Colobus angolensis palliatus	Peters' Angolan Colobus
Colobus angolensis prigoginei	Prigogine's Angolan Colobus
Colobus angolensis ruwenzorii	Adolf Friedrichs' Angolan Colobus
Colobus angolensis sharpei	Sharpe's Colobus
Colobus guereza gallarum	Neumann's Black-and-White Colobus
Colobus guereza guereza	Rüppell's Guereza
Colobus guereza matschiei	Matschie's Guereza
Colobus vellerosus	Geoffroy's Pied Colobus
Colomys goslingi	African Water Rat
Conepatus chinga	Molina's Hog-nosed Skunk
Conepatus humboldtii	Humboldt's Hog-nosed Skunk

Congosorex phillipsorum	Phillips' Congo Shrew
Congosorex polli	Poll's Shrew
Congosorex verheyeni	Verheyen's Shrew
Connochaetes taurinus cooksoni	Cookson's Wildebeest
Connochaetes taurinus hecki	Heck's Wildebeest
Connochaetes taurinus johnstoni	Johnston's Nyassa Wildebeest
Connochaetes taurinus mearnsi	Western White-bearded Wildebeest
Cormora brevirostris	Wagner's Sac-winged Bat
Corynorhinus rafinesquii	Rafinesque's Big-eared Bat
Corynorhinus townsendii	Townsend's Big-eared Bat
Craseonycteris thonglongyai	Kitti's Hog-nosed Bat
Crateromys heaneyi	Panay Bushy-tailed Cloud Rat
Crateromys schadenbergi	Luzon Bushy-tailed Cloud Rat
Cratogeomys goldmani	Goldman's Pocket Gopher
Cratogeomys merriami	Merriam's Pocket Gopher
Cratogeomys zinseri	Zinser's Pocket Gopher
Cremnomys blanfordi	Blanford's Rat
Cricetomys ansorgei	Southern Giant Pouched Rat
Cricetomys emini	Emin's Giant Pouched Rat
Cricetulus sokolovi	Sokolov's Dwarf Hamster
Crocidura andamanensis	Miller's Andaman Spiny Shrew
Crocidura ansellorum	Ansell's Shrew
Crocidura attila	Hun Shrew
Crocidura baileyi	Bailey's Shrew
Crocidura batesi	Bates' Shrew
Crocidura beccarii	Beccari's Shrew
Crocidura bottegi	Bottego's Shrew
Crocidura buettikoferi	Buettikofer's Shrew
Crocidura cinderella	Cinderella's Shrew
Crocidura crossei	Crosse's Shrew
Crocidura denti	Dent's Shrew
Crocidura douceti	Doucet's Musk Shrew
Crocidura eisentrauti	Eisentraut's Shrew
Crocidura fischeri	Fischer's Shrew
Crocidura floweri	Flower's Shrew
Crocidura foxi	Fox's Shrew
Crocidura giffardi	Giffard's Shrew
Crocidura glassi	Glass' Shrew
Crocidura gmelini	Gmelin's Shrew
Crocidura goliath	Goliath Shrew
Crocidura gracilipes	Peters' Musk Shrew
Crocidura grassei	Grasse's Shrew
Crocidura grayi	Luzon Shrew
Crocidura greenwoodi	Greenwood's Shrew
Crocidura gueldenstaedtii	Guldenstadt's Shrew

Crocidura hildegardeae	Hildegarde's Shrew
Crocidura hilliana	Hill's Shrew
Crocidura horsfieldii	Horsfield's Shrew
Crocidura jacksoni	Jackson's Shrew
Crocidura jenkinsi	Jenkins' Shrew
Crocidura jouvenetae	Jouvenet's Shrew
Crocidura lamottei	Lamotte's Shrew
Crocidura latona	Latona's Shrew
Crocidura lucina	Lucina's Shrew
Crocidura ludia	Ludia's Shrew
Crocidura macarthuri	MacArthur's Shrew
Crocidura macmillani	Macmillan's Shrew
Crocidura macowi	Macow's Shrew
Crocidura maxi	Max's Shrew
Crocidura musseri	Mossy Forest Shrew
Crocidura niobe	Niobe's Shrew
Crocidura olivieri	Olivier's Shrew
Crocidura orii	Orii's Shrew
Crocidura pitmani	Pitman's Shrew
Crocidura poensis	Fraser's Musk Shrew
Crocidura polia	Polia's Shrew
Crocidura raineyi	Rainey's Shrew
Crocidura ramona	Ramon's Shrew
Crocidura roosevelti	Roosevelt's Shrew
Crocidura schweitzeri	Schweitzer's Shrew
Crocidura smithii	Desert Musk Shrew
Crocidura sokolovi	Sokolov's White-toothed Shrew
Crocidura telfordi	Telford's Shrew
Crocidura thalia	Thalia's Shrew
Crocidura theresae	Therese's Shrew
Crocidura vosmaeri	Banka Shrew
Crocidura whitakeri	Whitaker's Shrew
Crocidura wimmeri	Wimmer's Shrew
Crocidura xantippe	Xantippe's Shrew
Crocidura zaitsevi	Zaitsev's White-toothed Shrew
Crocidura zaphiri	Zaphir's Shrew
Crocidura zarudnyi	Zarudny's Shrew
Crocidura zimmeri	Zimmer's Shrew
Crocidura zimmermanni	Zimmerman's Shrew
Crossarchus alexandri	Alexander's Cusimanse
Crossarchus ansorgei	Ansorge's Cusimanse
Crossomys moncktoni	Earless Water Rat
Crunomys rabori	Leyte Shrew-Mouse
Cryptochloris wintoni	De Winton's Golden-Mole
Cryptochloris zyli	Van Zyl's Golden-Mole

Cryptonanus agricolai	Agricola's Gracile Opossum
Cryptotis endersi	Ender's Small-eared Shrew
Cryptotis goldmani	Goldman's Small-eared Shrew
Cryptotis goodwini	Goodwin's Small-eared Shrew
Cryptotis merriami	Merriam's Small-eared Shrew
Cryptotis nelsoni	Nelson's Small-eared Shrew
Cryptotis parva	Say's Least Shrew
Cryptotis phillipsii	Phillips' Short-eared Shrew
Cryptotis thomasi	Thomas' Small-eared Shrew
Ctenodactylus vali	Val's Gundi
Ctenomys azarai	Azara's Tuco-tuco
Ctenomys bergi	Berg's Tuco-tuco
Ctenomys bonettoi	Bonetto's Tuco-tuco
Ctenomys budini	Budin's Tuco-tuco
Ctenomys colburni	Colburn's Tuco-tuco
Ctenomys conoveri	Conover's Tuco-tuco
Ctenomys dorbignyi	D'Orbigny's Tuco-tuco
Ctenomys emilianus	Emily's Tuco-tuco
Ctenomys flamarioni	Flamarion's Tuco-tuco
Ctenomys fochi	Foch's Tuco-tuco
Ctenomys goodfellowi	Goodfellow's Tuco-tuco
Ctenomys haigi	Haig's Tuco-tuco
Ctenomys knighti	Knight's Tuco-tuco
Ctenomys lewisi	Lewis' Tuco-tuco
Ctenomys nattereri	Natterer's Tuco-tuco
Ctenomys osvaldoreigi	Osvaldo Reig's Tuco-tuco
Ctenomys pearsoni	Pearson's Tuco-tuco
Ctenomys perrensi	Goya Tuco-tuco
Ctenomys porteousi	Porteous' Tuco-tuco
Ctenomys pundti	Pundt's Tuco-tuco
Ctenomys rosendopascuali	Pascual's Tuco-tuco
Ctenomys scagliai	Scaglia's Tuco-tuco
Ctenomys steinbachi	Steinbach's Tuco-tuco
Ctenomys yolandae	Santa Fé Tuco-tuco
Cuniculus taczanowskii	Mountain Paca
Cynogale bennetti	Otter-Civet
Cynogale lowei	Lowe's Otter-Civet
Cynomys gunnisoni	Gunnison's Prairie Dog
Cynopterus horsfieldi	Horsfield's Short-nosed Fruit Bat
Cynopterus sphinx	Short-nosed Fruit Bat
Cyttarops alecto	Short-eared Bat
Dacnomys millardi	Millard's Rat
Dactylopsila palpator ernstmayri	Long-fingered Triok
Dactylopsila tatei	Tate's Triok

Damaliscus dorcas	Bontebok
Damaliscus hunteri	Hunter's Hartebeest
Dasycercus blythi	Brush-tailed Mulgara
Dasycercus hillieri	Hillier's Mulgara
Dasykaluta rosamondae	Little Red Kaluta
Dasymys cabrali	Crawford-Cabral's Marsh Rat
Dasymys foxi	Fox's Shaggy Rat
Dasyprocta azarai	Azara's Agouti
Dasyprocta kalinowskii	Kalinowski's Agouti
Dasypus kappleri	Kappler's Armadillo
Dasypus yepesi	Yepes' Long-nosed Armadillo
Dasyroides byrnei	Byrne's Marsupial Mouse
Dasyurus geoffroii	Western Quoll
Dasyurus spartacus	Bronze Quoll
Daubentonia madagascariensis	Aye-Aye
Delanymys brooksi	Delany's Swamp Mouse
Delphinus bairdi	Baird's Dolphin
Delphinus frithii	Long-beaked Common Dolphin
Deltamys kempi	Kemp's Grass Mouse
Dendrohyrax dorsalis	Beecroft's Tree Hyrax
Dendrolagus (dorianus) stellarum	Seri's Tree Kangaroo
Dendrolagus bennettianus	Bennett's Tree Kangaroo
Dendrolagus dorianus	Doria's Tree Kangaroo
Dendrolagus dorianus mayri	Wondiwoi Tree Kangaroo
Dendrolagus goodfellowi	Goodfellow's Tree Kangaroo
Dendrolagus goodfellowi shawmayeri	Shawmayer's Ornate Tree Kangaroo
Dendrolagus inustus finschi	Finsch's Tree Kangaroo
Dendrolagus lumholtzi	Lumholtz's Tree Kangaroo
Dendrolagus matschiei	Matschie's Tree Kangaroo
Dendrolagus scottae	Scott's Tree Kangaroo
Dendromus haymani	Banana Climbing Mouse
Dendromus leucostomus	Monard's Climbing Mouse
Dendromus lovati	Lovat's Climbing Mouse
Dendromus mesomelas	Brants' Climbing Mouse
Dendromus vernayi	Vernay's Climbing Mouse
Dendroprionomys rousseloti	Velvet Climbing Mouse
Desmodilliscus braueri	Pouched Gerbil
Desmomys harringtoni	Harrington's Rat
Desmomys yaldeni	Yalden's Rat
Diaemus youngi	White-winged Vampire Bat
Dicerorhinus sumatrensis	Hagenbeck's Rhinoceros
Diclidurus isabellus	Isabelle's Ghost Bat
Dicrostonyx (groenlandicus) nelsoni	Nelson's Collared Lemming
Dicrostonyx richardsoni	Richardson's Collared Lemming
Dicrostonyx stevensoni	Stevenson's Collared Lemming

Dicrostonyx vinogradovi	Wrangel Collared Lemming
Didelphis koseritzi	Big-eared Opossum
Dinaromys bogdanovi	Balkan Snow Vole, Martino's Snow Vole
Dinomys branickii	Count Branicki's Terrible Mouse, Pacarana
Diomys crumpi	Crump's Mouse
Diplogale hosei	Hose's Palm Civet
Dipodomys antiquarius	Huey's Kangaroo Rat
Dipodomys heermanni	Heermann's Kangaroo-Rat
Dipodomys margaritae	Santa Margarita Island Kangaroo Rat
Dipodomys merriami	Merriam's Kangaroo-Rat
Dipodomys microps preblei	Preble's Kangaroo-Rat
Dipodomys nelsoni	Nelson's Kangaroo-Rat
Dipodomys ordi	Ord's Kangaroo-Rat
Dipodomys phillipsii	Phillips' Kangaroo-Rat
Dipodomys stephensi	Stephens' Kangaroo-Rat
Dobsonia	The fruit bat genus
Dobsonia anderseni	Andersen's Bare-backed Fruit Bat
Dobsonia beauforti	Beaufort's Bare-backed Fruit Bat
Dobsonia chapmani	Dobson's Fruit Bat
Dobsonia pannietensis	De Vis' Bare-backed Fruit Bat
Dobsonia praedatrix	Bismarck Bare-backed Fruit Bat
Dologale dybowskii	Pousargues' Mongoose
Dorcopsis hageni	White-striped Dorcopsis
Dorcopsis muelleri	Brown Dorcopsis
Dorcopsulus macleayi	Macleay's Dorcopsis
Dorcopsulus vanheurni	Lesser Forest Wallaby
Dremomys everetti	Bornean Mountain Ground Squirrel
Dremomys pernyi	Perny's Squirrel
Dryomys niethammeri	Niethammer's Dormouse
Dyacopterus brooksi	Brooks' Large-headed Fruit Bat
Dyacopterus rickarti	Philippine Large-headed Fruit Bat
Echimys blainvillei	Golden Atlantic Tree Rat
Echimys vieirai	Vieira's Spiny Rat
Echinops telfairi	Lesser Hedgehog-Tenrec
Echymipera clara	Clara's Echymipera
Echymipera davidi	David's Echymipera
Echymipera echinista	Menzies' Echymipera
Eidolon dupreanum	Madagascar Straw-colored Fruit Bat
Eira barbara	Tayra
Elaphodus cephalophus michianus	Michie's Tufted Deer
Elaphurus davidianus	Père David's Deer
Elephantulus edwardii	Cape Rock Elephant Shrew

Elephantulus revoili	Somali Elephant Shrew
Elephantulus rozeti	North African Elephant Shrew
Elephantulus rupestris	Smith's Rock Elephant Shrew
Eligmodontia moreni	Monte Gerbil-Mouse
Eligmodontia morgani	Morgan's Gerbil-Mouse
Eliurus danieli	Daniel's Tufted-tailed Rat
Eliurus ellermani	Ellerman's Tufted-tailed Rat
Eliurus grandidieri	Grandidier's Tufted-tailed Rat
Eliurus majori	Major's Tufted-tailed Rat
Eliurus petteri	Petter's Tufted-tailed Rat
Eliurus webbi	Webb's Tufted-tailed Rat
Ellobius tancrei	Zaisan Mole Vole
Emballonura alecto	Small Asian Sheath-tailed Bat
Emballonura beccarii	Beccari's Sheath-tailed Bat
Emballonura dianae	Large-eared Sheath-tailed Bat
Emballonura raffrayana	Raffray's Sheath-tailed Bat
Emballonura serii	Seri's Sheath-tailed Bat
Enchisthenes hartii	Hart's Fruit-eating Bat
Eoglaucomys baberi	Afghan Flying Squirrel
Eolagurus przewalskii	Przewalski's Steppe Lemming
Eonycteris spelaea	Dobson's Long-tongued Fruit Bat
Eospalax fontanierii	Chinese Zokor
Eospalax rothschildi	Rothschild's Zokor
Eospalax smithii	Smith's Zokor
Eothenomys wardi	Ward's Red-backed Vole
Episoriculus caudatus	Hodgson's Brown-toothed Shrew
Epixerus ebii	Ebian Palm Squirrel, Temminck's Giant Forest Squirrel
Epixerus wilsoni	Biafran Palm Squirrel
Epomophorus anselli	Ansell's Epauletted Fruit Bat
Epomophorus wahlbergi	Wahlberg's Epauletted Bat
Epomops buettikoferi	Buettikofer's Epauletted Bat
Epomops dobsoni	Dobson's Epauletted Bat
Epomops franqueti	Franquet's Epauletted Bat
Eptesicus bobrinskoi	Bobrinski's Serotine
Eptesicus bottae	Botta's Serotine
Eptesicus floweri	Horn-skinned Bat
Eptesicus kobayashii	Kobayashi's Serotine
Eptesicus lynni	Lynn's Brown Bat
Eptesicus nilssoni	Northern Bat
Eptesicus taddeii	Taddei's Bat
Eptesicus tatei	Sombre Bat
Equus burchelli pococki	Pocock's Zebra
Equus ferus	Przewalski's Horse
Equus grevyi	Grevy's Zebra

Equus przewalski gmelini	Tarpan (extinct)
Equus quagga boehmi	Böhm's Zebra, Grant's Zebra
Equus quagga burchelli	Burchell's Zebra, Plain's Zebra
Equus quagga chapmanni	Chapman's Zebra
Equus quagga crawshaii	Crawshay's Zebra
Equus quagga selousi	Selous' Zebra
Equus wardi	Ward's Zebra
Equus zebra hartmannae	Hartmann's Zebra
Eremitalpa granti	Grant's Golden-Mole
Eremodipus lichtensteini	Lichtenstein's Jerboa
Erophylla sezekorni	Buffy Flower Bat
Eschrichtius robustus	Grey Whale
Eulemur (fulvus) sanfordi	Sanford's Lemur
Eulemur macaco flavifrons	Sclater's Black Lemur
Eumetopias jubatus	Steller's Sea-Lion
Eumops auripendulus	Shaw's Mastiff Bat
Eumops bonariensis	Peters' Mastiff Bat
Eumops dabbenei	Dabbene's Mastiff Bat
Eumops glaucinus	Wagner's Bonneted Bat
Eumops hansae	Sanborn's Bonneted Bat
Eumops maurus	Thomas' Mastiff Bat
Eumops trumbulli	Trumbull's Bonneted Bat
Eumops underwoodi	Underwood's Bonneted Bat
Euneomys petersoni	Peterson's Chinchilla-mouse
Eupleres goudotii	Falanouc
Euroscaptor klossi	Kloss' Mole
Euryoryzomys emmonsae	Emmons' Rice Rat
Euryoryzomys kelloggi	Kellogg's Rice Rat
Euryoryzomys macconnelli	MacConnell's Rice Rat
Euthenomys chinensis	Pratt's Vole
Euthenomys melanogaster	Père David's Vole
Exilisciurus whiteheadi	Tufted Pygmy Squirrel
Falsistrellus petersi	Peters' Pipistrelle
Felis bieti	Chinese Desert Cat
Felis chaus prateri	Prater's Cat
Felis manul	Pallas' Cat
Felis margarita	Sand Cat
Felis margarita scheffeli	Scheffel's Sand Cat
Felis silvestris gordoni	Gordon's Wild Cat
Feroculus feroculus	Kelaart's Long-clawed Shrew
Fukomys anselli	Ansell's Mole-Rat
Fukomys bocagei	Bocage's Mole-Rat
Fukomys damarensis lugardi	Lugard's Mole-Rat
Fukomys darlingi	Mashona Mole-Rat

Fukomys foxi — Nigerian Mole-Rat
Fukomys mechowi — Mechow's Mole-Rat
Fukomys micklemi — Kataba Mole-Rat
Fukomys whytei — Whyte's Mole-Rat
Fukomys zechi — Togo Mole-Rat
Funambulus layardi — Layard's Palm Squirrel
Funambulus pennantii — Northern Palm Squirrel
Funisciurus anerythrus — Thomas' Rope Squirrel
Funisciurus bayonii — Bocage's Tree Squirrel
Funisciurus carruthersi — Carruthers' Mountain Squirrel
Funisciurus congicus — Kuhl's Tree Squirrel
Funisciurus isabella — Gray's Four-striped Squirrel, Lady Burton's Rope Squirrel
Funisciurus lemniscatus — Leconte's Four-striped Tree Squirrel
Funisciurus pyrropus — Cuvier's Fire-footed Squirrel
Funisciurus substriatus — De Winton's Tree Squirrel

Galago alleni — Allen's Squirrel Galago
Galago demidoff — Demidoff's Galago
Galago granti — Mozambique Galago
Galago matschiei — Matschie's Galago
Galago thomasi — Thomas' Galago
Galea flavidens — Brandt's Yellow-toothed Cavy
Galea spixii — Spix's Yellow-toothed Cavy
Galenomys garleppi — Garlepp's Mouse
Galidictis grandidieri — Grandidier's Mongoose
Gazella (dorcas) pelzelni — Pelzeln's Gazelle
Gazella bennetti — Indian Gazelle
Gazella bilkis — Queen of Sheba's Gazelle
Gazella cuvieri — Cuvier's Gazelle
Gazella dama mhorr — Mhorr's Gazelle
Gazella dorcas — Dorcas Gazelle
Gazella gazella erlangeri — Erlanger's Gazelle
Gazella granti — Grant's Gazelle
Gazella granti brighti — Bright's Gazelle
Gazella granti petersi — Peters' Gazelle
Gazella granti raineyi — Rainey's Gazelle
Gazella granti robertsi — Roberts' Gazelle
Gazella granti roosevelti — Roosevelt's Gazelle
Gazella leptoceros loderi — Loder's Gazelle
Gazella rufifrons tilonura — Heuglin's Gazelle
Gazella soemmerringii — Soemmerring's Gazelle
Gazella spekei — Speke's Gazelle
Gazella thomsoni — Thomson's Gazelle
Genetta bourloni — Bourlon's Genet

Genetta johnstoni	Johnston's Genet
Genetta servalina lowei	Lowe's Servaline Genet
Genetta thierryi	Thierry's Genet
Genetta victoriae	Giant Genet
Geocapromys brownii	Brown's Hutia
Geocapromys ingrahami	Bahaman Hutia
Geomys attwateri	Attwater's Pocket Gopher
Geomys breviceps	Baird's Pocket Gopher
Geomys knoxjonesi	Knox Jones' Pocket Gopher
Geomys personatus davisi	Davis' Pocket Gopher
Geomys personatus streckeri	Strecker's Pocket Gopher
Geomys pinetis fontanelus	Sherman's Pocket Gopher (extinct?)
Geomys pinetis goffi	Goff's Pocket Gopher (extinct)
Gerbilliscus boehmi	Böhm's Gerbil
Gerbilliscus brantsii	Highveld Gerbil
Gerbilliscus kempi	Kemp's Gerbil
Gerbilliscus leucocaster	Peters' Gerbil
Gerbilliscus phillipsi	Phillips' Gerbil
Gerbilliscus pringlei	Pringle's Gerbil
Gerbilliscus validus	Bocage's Gerbil
Gerbillurus setzeri	Setzer's Hairy-footed Gerbil
Gerbillus agag	Agag Gerbil
Gerbillus allenbyi	Allenby's Gerbil
Gerbillus andersoni	Anderson's Gerbil
Gerbillus bonhotei	Bonhote's Gerbil
Gerbillus bottai	Botta's Gerbil
Gerbillus brockmani	Brockman's Gerbil
Gerbillus burtoni	Burton's Gerbil
Gerbillus cheesmani	Cheesman's Gerbil
Gerbillus cosensi	Cosens' Gerbil
Gerbillus dalloni	Dallon's Gerbil
Gerbillus dasyurus	Wagner's Gerbil
Gerbillus dunni	Dunn's Gerbil
Gerbillus floweri	Flower's Gerbil
Gerbillus gleadowi	Indian Hairy-footed Gerbil
Gerbillus grobbeni	Grobben's Gerbil
Gerbillus harwoodi	Harwood's Gerbil
Gerbillus henleyi	Henley's Gerbil
Gerbillus hoogstraali	Hoogstraal's Gerbil
Gerbillus jamesi	James' Gerbil
Gerbillus juliani	Julian's Gerbil
Gerbillus latastei	Lataste's Gerbil
Gerbillus lowei	Lowe's Gerbil
Gerbillus mackillingini	Mackilligin's Gerbil
Gerbillus percivali	Percival's Gerbil

Gerbillus pulvinatus	Rhoads' Gerbil
Gerbillus riggenbachi	Riggenbach's Gerbil
Gerbillus rosalinda	Rosalinda Gerbil
Gerbillus simoni	Simon's Dipodil
Gerbillus tarabuli	Tarabul's Gerbil
Gerbillus watersi	Waters' Gerbil
Gerbillus zakariai	Zakaria's Gerbil
Giraffa camelopardalis cottoni	Lado Giraffe
Giraffa camelopardalis rothschildi	Rothschild's Giraffe
Giraffa camelopardalis tippelskirchi	Masai Giraffe
Giraffa camelopardelis thornicrofti	Thornicroft's Giraffe
Glauconycteris alboguttatus	Allen's Striped Bat
Glauconycteris beatrix	Beatrix's Bat
Glauconycteris curryae	Curry's Bat
Glauconycteris gleni	Glen's Wattled Bat
Globicephala melas edwardii	Southern Long-finned Pilot Whale
Glossophaga alticola	Davis' Long-tongued Bat
Glossophaga commissarisi	Commissaris' Long-tongued Bat
Glossophaga leachii	Gray's Long-tongued Bat
Glossophaga longirostris	Miller's Long-tongued Bat
Glossophaga morenoi	Western Long-tongued Bat
Glossophaga soricina	Pallas' Long tongued Bat
Glossotherium harlani	Harlan's Ground Sloth (extinct)
Glyphonycteris behni	Behn's Big-eared Bat
Glyphonycteris daviesi	Davies' Big-eared Bat
Golunda ellioti	Indian Bush Rat
Gorilla beringei	Eastern Gorilla
Gorilla beringei graueri	Grauer's Gorilla
Gorilla gorilla diehli	Cross River Gorilla
Gracilinanus emiliae	Emilia's Gracile Mouse-Opossum
Grammomys buntingi	Bunting's Thicket Rat
Grammomys gigas	Giant Thicket Rat
Grammomys macmillani	Macmillan's Thicket Rat
Grammomys minnae	Ethiopian Thicket Rat
Grampus griseus	Risso's Dolphin
Graomys edithae	Edith's Leaf-eared Mouse
Graomys lockwoodi	Grey Leaf-eared Mouse
Graomys roigi	Roig's Chaco Mouse
Graphiurus christyi	Christy's Dormouse
Graphiurus crassicaudatus	Jentink's Dormouse
Graphiurus hueti	Huet's Dormouse
Graphiurus johnstoni	Johnston's Dormouse
Graphiurus kelleni	Kellen's Dormouse
Graphiurus monardi	Monard's Dormouse
Graphiurus nagtglasii	Nagtglas' Dormouse

Graphiurus olga	Olga's Dormouse
Gymnobelideus leadbeateri	Leadbeater's Possum
Gymnuromys roberti	Voalavoanaia
Hadromys humei	Manipur Bush Rat
Haeromys margarettae	Greater Ranee Mouse
Halmaturus binoe	Agile Wallaby
Handleyomys	The Colombian rice rat genus
Hapalomys delacouri	Delacour's Marmoset-Rat
Haplonycteris fischeri	Philippine Pygmy Fruit Bat
Harpyionycteris whiteheadi	Harpy Fruit Bat
Heimyscus fumosus	De Balsac's Mouse
Heliosciurus punctatus	Temminck's Spotted Squirrel
Hemiechinus hypomelas	Brandt's Hedgehog
Hemigalus derbyanus	Banded Palm Civet
Hemitragus jayakari	Arabian Tahr
Herpestes edwardsii	Indian Grey Mongoose
Herpestes smithii	Ruddy Mongoose
Hesperoptenus blanfordi	Blanford's False Serotine
Hesperoptenus doriae	False Serotine Bat, Peters' False Serotine
Hesperoptenus gaskelli	Gaskell's False Serotine
Hesperoptenus tickelli	Tickell's False Serotine
Hesperoptenus tomesi	Large False Serotine
Heterohyrax brucei	Yellow-spotted Hyrax
Heterohyrax chapini	Matadi Hyrax
Heteromys (desmarestianus) nigricaudatus	Goodwin's Spiny Pocket Mouse
Heteromys desmarestianus	Desmarest's Spiny Pocket Mouse
Heteromys gaumeri	Gaumer's Spiny Pocket Mouse
Heteromys goldmani	Goldman's Spiny Pocket Mouse
Heteromys nelsoni	Nelson's Spiny Pocket Mouse
Hippopotamus amphibius standingi	Standing's Hippopotamus
Hippopotamus lemerlei	Lemerle's Hippopotamus (extinct)
Hipposideros boeadii	Boeadi's Roundleaf Bat
Hipposideros caffer	Sundevall's Roundleaf Bat
Hipposideros commersoni	Commerson's Roundleaf Bat
Hipposideros coxi	Cox's Roundleaf Bat
Hipposideros cyclops	Cyclops Roundleaf Bat
Hipposideros doriae	Borneo Roundleaf Bat
Hipposideros durgadasi	Durga Das' Leaf-nosed Bat, Khajuria's Leaf- nosed Bat
Hipposideros edwardshilli	John Hill's Roundleaf Bat
Hipposideros galeritus	Cantor's Roundleaf Bat
Hipposideros jonesi	Jones' Roundleaf Bat
Hipposideros lamottei	Lamotte's Roundleaf Bat

Hipposideros lankadiva	Kelaart's Leaf-nosed Bat
Hipposideros lekaguli	Large Asian Roundleaf Bat
Hipposideros lylei	Shield-faced Roundleaf Bat
Hipposideros maggietaylorae	Maggie Taylor's Roundleaf Bat
Hipposideros marisae	Aellen's Roundleaf Bat
Hipposideros pratti	Pratt's Roundleaf Bat
Hipposideros ridleyi	Ridley's Roundleaf Bat
Hipposideros ruber	Noack's Roundleaf Bat
Hipposideros semoni	Semon's Roundleaf Bat
Hipposideros sorenseni	Sorensen's Leaf-nosed Bat
Hipposideros speoris	Schneider's Roundleaf Bat
Hipposideros wollastoni	Wollaston's Roundleaf Bat
Hippotragus niger roosevelti	Roosevelt's Sable Antelope
Hippotragus niger variani	Giant Sable Antelope
Histiotus humboldti	Humboldt's Big-eared Brown Bat
Hodomys alleni	Allen's Woodrat
Holochilus sciureus	Wagner's Marsh Rat
Hybomys basilii	Father Basilio's Striped Mouse
Hybomys eisentrauti	Eisentraut's Striped Mouse
Hybomys planifrons	Miller's Striped Mouse
Hybomys trivirgatus	Temminck's Striped Mouse
Hybomys univittatus	Peters' Striped Mouse
Hydrodamalis gigas	Steller's Sea Cow (extinct)
Hydromys hussoni	Husson's Water Rat
Hydromys shawmayeri	Shaw-Mayer's Water Rat
Hydromys ziegleri	East Sepik Water Rat
Hyladelphys kalinowskii	Kalinowski's Mouse-Opossum
Hylaeamys seuanezi	Seuanez's Rice Rat
Hylaeamys tatei	Tate's Rice Rat
Hylobates hoolock	Hoolock's Gibbon
Hylobates klossii	Kloss' Gibbon
Hylobates lar	Lar's Gibbon
Hylobates moloch	Moloch Gibbon
Hylobates muelleri	Müller's Gibbon
Hylobates muelleri abbotti	Abbott's Grey Gibbon
Hylochoerus meinertzhageni	Giant Forest Hog
Hylomyscus alleni	Allen's Wood Mouse
Hylomyscus anselli	Ansell's Wood Mouse
Hylomyscus baeri	Baer's Wood Mouse
Hylonycteris underwoodi	Underwood's Long-tongued Bat
Hylopetes bartelsi	Bartels' Flying Squirrel
Hylopetes phayrei	Indochinese Flying Squirrel
Hylopetes winstoni	Sumatran Flying Squirrel
Hyomys dammermani	Western White-eared Giant Rat
Hyomys goliath	Eastern White-eared Giant Rat

Hyosciurus heinrichi	Montane Long-nosed Squirrel
Hyperacrius fertilis	True's Vole
Hyperacrius wynnei	Murree Vole
Hyperplagiodontia stenocoronalis	Rimoli's Hutia (extinct)
Hypsugo anchietai	D'Anchieta's Pipistrelle
Hypsugo anthonyi	Anthony's Pipistrelle
Hypsugo bodenheimeri	Bodenheimer's Pipistrelle
Hypsugo cadornae	Cadorna's Pipistrelle
Hypsugo eisentrauti	Eisentraut's Pipistrelle
Hypsugo joffrei	Joffre's Pipistrelle
Hypsugo kitcheneri	Red-brown Pipistrelle
Hypsugo savii	Savi's Pipistrelle
Hypsugo vordermanni	Vordermann's Pipistrelle
Ia io	Great Evening Bat
Ichthyomys pittieri	Pittier's Crab-eating Rat
Ichthyomys stolzmanni	Stolzmann's Crab-eating Rat
Ichthyomys tweedii	Tweedy's Crab-eating Rat
Idionycteris phyllotis	Allen's Big-eared Bat
Idiurus macrotis cansdalei	Cansdale's Long-eared Flying Mouse
Idiurus zenkeri	Pygmy Scaly-tailed Flying Squirrel
Indopacetus pacificus	Longman's Beaked Whale
Inia geoffrensis	Amazon River Dolphin
Inia geoffrensis humboldtiana	Orinoco River Dolphin
Iomys horsfieldii	Horsfield's Flying Squirrel
Irenomys tarsalis	Chilean Climbing Mouse
Isolobodon levir	Miller's Hutia (extinct)
Isolobodon portoricensis	Allen's Hutia
Isothrix barbarabrownae	Barbara Brown's Brush-tailed Rat
Jaculus blanfordi	Blanford's Jerboa
Juliomys	The mouse genus
Juliomys pictipes	Contreras' Juliomys
Juscelinomys	The mouse genus
Kadarsanomys sodyi	Sody's Tree Rat
Kerivoula africana	Dobson's Painted Bat
Kerivoula agnella	St. Aignan's Trumpet-eared Bat
Kerivoula hardwickei	Hardwicke's Woolly Bat
Kerivoula myrella	Bismarck Trumpet-eared Bat
Kerivoula phalaena	Spurrell's Woolly Bat
Kerivoula smithii	Smith's Woolly Bat
Kerivoula titania	Titania's Woolly Bat
Kerivoula whiteheadi	Whitehead's Woolly Bat
Kobus anselli	Upemba Lechwe

Kobus kob	Buffon's Kob
Kobus leche robertsi	Roberts' Lechwe Antelope (extinct)
Kobus megaceros	Mrs. Gray's Lechwe
Kobus vardonii	Puku
Kogia sima	Owen's Pygmy Sperm Whale
Koopmania concolor	Koopman's Fruit Bat
Kunsia	The Neotropical rat genus
Laephotis wintoni	De Winton's Long-eared Bat
Lagenodelphis hosei	Fraser's Dolphin
Lagenorhynchus australis	Peale's Dolphin
Lagenorhynchus obscurus	Fitzroy's Dolphin
Lagidium wolffsohni	Wolffsohn's Viscacha
Lagnenorhynchus cruciger	Wilson's Dolphin
Lagothrix cana cana	Geoffroy's Woolly Monkey
Lagothrix lagothricha	Humboldt's Woolly Monkey
Lagothrix poeppigii	Poeppig's Woolly Monkey
Lamottemys okuensis	Oku Rat
Lariscus hosei	Four-striped Ground Squirrel
Lariscus niobe	Niobe Ground Squirrel
Lasiopodomys brandtii	Brandt's Vole
Lasiorhinus krefftii	Northern Hairy-nosed Wombat
Lasiorhinus latifrons barnardi	Barnard's Hairy-nosed Wombat
Lasiurus atratus	Handley's Red Bat
Lasiurus blossevillii	Western Red Bat
Lasiurus pfeifferi	Pfeiffer's Hairy-tailed Bat
Latidens salimalii	Salim Ali's Fruit Bat
Leggadina forresti	Forrest's Mouse
Leimacomys buettneri	Groove-toothed Forest Mouse
Lemmus portenkoi	Wrangel Island Lemming
Lemniscomys bellieri	Bellier's Striped Grass Mouse
Lemniscomys griselda	Griselda's Striped Grass Mouse
Lemniscomys hoogstraali	Hoogstraal's Striped Grass Mouse
Lemniscomys mittendorfi	Mittendorf's Striped Grass Mouse
Lemniscomys roseveari	Rosevear's Striped Grass Mouse
Lenomys meyeri	Trefoil-toothed Giant Rat
Leopardus jacobita	Andean Mountain Cat
Leopardus pardalis nelsoni	Nelson's Ocelot
Leopardus pardalis steinbachi	Steinbach's Ocelot
Leopardus wiedi	Margay
Leopardus wiedii cooperi	Cooper's Margay
Leopoldamys	The long-tailed rat genus
Leopoldamys edwardsi	Milne-Edwards' Long-tailed Giant Rat
Leopoldamys milleti	Millet's Giant Long-tailed Rat
Leopoldamys neilli	Neill's Long-tailed Giant Rat

Lepilemur aeeclis	Aeecl's Sportive Lemur
Lepilemur ahmansoni	Ahmanson's Sportive Lemur
Lepilemur betsileo	Betsileo Sportive Lemur
Lepilemur edwardsi	Milne-Edwards' Sportive Lemur
Lepilemur fleuretae	Fleurete's Sportive Lemur
Lepilemur grewcocki	Grewcock's Sportive Lemur
Lepilemur hubbardi	Hubbard's Sportive Lemur
Lepilemur jamesi	James' Sportive Lemur
Lepilemur mittermeieri	Mittermeier's Sportive Lemur
Lepilemur otto	Otto's Sportive Lemur
Lepilemur petteri	Petter's Sportive Lemur
Lepilemur randrianasoli	Randrianasolo's Sportive Lemur
Lepilemur seali	Seal's Sportive Lemur
Lepilemur tymerlachsoni	Hawk's Sportive Lemur
Lepilemur wrighti	Wright's Sportive Lemur
Leptailurus serval hamiltoni	Hamilton's Serval
Leptailurus serval robertsi	Roberts' Serval
Leptomys ernstmayri	Ernst Mayr's Water Rat
Leptonychotes weddellii	Weddell Seal
Lepus alleni	Antelope Jackrabbit
Lepus castroviejoi	Broom Hare
Lepus crawshayi	Crawshay's Hare
Lepus fagani	Ethiopian Hare
Lepus oiostolus przewalskii	Przewalski's Hare
Lepus starcki	Starck's Hare
Lepus townsendii	Townsend's Hare
Lepus victoriae	Savanna Hare
Lepus whytei	Whyte's Hare
Lestodelphys halli	Patagonian Opossum
Liberiictis kuhni	Liberian Mongoose
Liomys adspersus	Peters' Spiny Pocket Mouse
Liomys salvini	Salvin's Spiny Pocket Mouse
Lionycteris spurrelli	Chestnut Long-tongued Bat
Lissodelphis peronii	Southern Right-Whale Dolphin
Lissonycterisgoliath	Harrison's Fruit Bat
Lissonycteris petraea	Petra Fruit Bat
Lissonycteris smithii	Smith's Fruit Bat
Litocranius walleri	Gerenuk, Waller's Gazelle
Lonchophylla bokermanni	Bokermann's Nectar Bat
Lonchophylla cadenai	Cadena's Nectar Bat
Lonchophylla dekeyseri	Dekeyser's Nectar Bat
Lonchophylla handleyi	Handley's Nectar Bat
Lonchophylla mordax	Goldman's Nectar Bat
Lonchophylla orcesi	Orces' Nectar Bat
Lonchophylla pattoni	Patton's Nectar-feeding Bat

Lonchophylla thomasi	Thomas' Nectar Bat
Lonchorhina aurita	Tomes' Long-eared Bat
Lonchorhina fernandezi	Fernandez's Sword-nosed Bat
Lonchorhina marinkelleii	Marinkelle's Sword-nosed Bat
Lonchothrix emiliae	Tuft-tailed Spiny Tree Rat
Lophiomys imhausi	Crested Rat
Lophocebus (albigena) johnstoni	Johnston's Grey-cheeked Mangabey
Lophocebus (albigena) osmani	Osman Hill's Grey-cheeked Mangabey
Lophocebus opdenboschi	Opdenbosch's Mangabey
Lophostoma carrikeri	Allen's Round-eared Bat, Carriker's Round- eared Bat
Lophostoma schulzi	Schulz's Round-eared Bat
Lophuromys dieterleni	Dieterlen's Brush-furred Rat
Lophuromys dudui	Dudu's Brush-furred Rat
Lophuromys huttereri	Hutterer's Brush-furred Rat
Lophuromys kilonzoi	Kilonzo's Brush-furred Rat
Lophuromys machangui	Rungwe Brush-furred Rat
Lophuromys makundii	Hanang Brush-furred Rat
Lophuromys rahmi	Rahm's Brush-furred Rat
Lophuromys roseveari	Rosevear's Brush-furred Rat
Lophuromys sabunii	Ufipa Brush-furred Rat
Lophuromys stanleyi	Stanley's Brush-furred Rat
Lophuromys verhageni	Verhagen's Brush-furred Rat
Lophuromys woosnami	Woosnam's Brush-furred Rat
Lorentzimys nouhuysi	Long-footed Tree Mouse
Loris lydekkerianus	Grey Slender Loris
Lundomys molitor	Lund's Amphibious Rat
Lutrogale perspicillata maxwelli	Maxwell's Otter
Lycaon pictus	African Wild Dog
Lynx pardinus	Pardelluch's Lynx
Lynx rufus baileyi	Bailey's Bobcat
Macaca hecki	Heck's Macaque
Macaca mulatta	Rhesus Monkey
Macaca silenus	Lion-tailed Macaque
Macaca thibetana	Père David's Macaque
Macroderma gigas	Australian False Vampire Bat
Macrogalidia musschenbroekii	Sulawesi Palm Civet
Macrophyllum macrophyllum	Wied's Long-legged Bat
Macropus bernardus	Black Wallaroo
Macropus eugenii	Tamar Wallaby
Macropus greyi	Toolache Wallaby (extinct)
Macropus irma	Western Brush Wallaby
Macropus parryi	Parry's Wallaby
Macropus rufogriseus fruticus	Bennett's Wallaby

Macrotarsomys bastardi	Bastard Big-footed Mouse
Macrotarsomys petteri	Petter's Big-footed Mouse
Macrotus waterhousii	Waterhouse's Leaf-nosed Bat
Madoqua (saltiana) swaynei	Swayne's Dik-dik
Madoqua guentheri	Günther's Dik-dik
Madoqua kirkii	Kirk's Dik-dik
Madoqua phillipsi lawrancei	Lawrance's Dik-dik
Madoqua piacentinii	Piacentini's Dik-dik
Madoqua saltiana	Salt's Dik-dik
Madoqua saltiana cordeauxi	Cordeaux's Dik-dik
Madoqua saltiana erlangeri	Erlanger's Dik-dik
Madoqua saltiana phillipsi	Phillips' Dik-dik
Malacomys cansdalei	Cansdale's Swamp Rat
Malacomys edwardsi	Milne-Edwards' Swamp Rat
Malacomys verschureni	Verschuren's Swamp Rat
Mallomys aroaensis	De Vis' Woolly Rat
Mallomys rothschildi	Rothschild's Woolly Rat
Mandrillus sphinx	Mandrill
Manis temminckii	Temminck's Ground Pangolin
Margaretamys	The Sulawesi rat genus
Margaretamys beccarii	Beccari's Margareta Rat
Marmosa murina	Linnaeus' Mouse-Opossum
Marmosa robinsoni	Robinson's Mouse-Opossum
Marmosa tyleriana	Tyler's Mouse-Opossum
Marmosops bishopi	Bishop's Slender Mouse-Opossum
Marmosops creightoni	Voss' Slender Opossum
Marmosops dorothea	Dorothy's Slender Mouse-Opossum
Marmosops handleyi	Handley's Slender Mouse-Opossum
Marmosops impavidus	Tschudi's Slender Opossum
Marmosops pinheiroi	Pinheiro's Slender Mouse-Opossum
Marmota broweri	Alaska Marmot
Marmota kastschenkoi	Forest-steppe Marmot
Marmota menzbieri	Menzbier's Marmot
Martes gwatkinsii	Nilgiri Marten
Martes pennanti	Pennant's Marten
Massoutiera mzabi	Lataste's Gundi
Mastomys hildebrandtii	Hildebrandt's Multimammate Mouse
Mastomys huberti	Hubert's Multimammate Mouse
Mastomys kollmannspergeri	Kollmannsperger's Multimammate Mouse
Mastomys shortridgei	Shortridge's Multimammate Mouse
Mastomys verheyeni	Verheyen's Multimammate Mouse
Maxomys bartelsii	Bartels' Spiny Rat
Maxomys dollmani	Dollman's Spiny Rat

Maxomys hellwaldii	Hellwald's Spiny Rat
Maxomys moi	Mo's Spiny Rat
Maxomys musschenbroeki	Musschenbroek's Spiny Rat
Maxomys wattsi	Watts' Spiny Rat
Maxomys whiteheadi	Whitehead's Spiny Rat
Mayermys ellermani	One-toothed Shrew-Mouse, Shaw-Mayer's Shrew Mouse
Mayermys germani	German's Shrew-Mouse
Mazama bricenii	Merida Brocket
Mazama pandora	Yucatan Brown Brocket
Mazama sartorii	Central American Red Brocket
Mazama sheila	Red Brocket
Mazama tschudii	Grey Brocket
Mazama zetta	Red Brocket
Megadendromus nikolausi	Nikolaus' Mouse
Megadontomys nelsoni	Nelson's Giant Deer Mouse
Megadontomys thomasi	Thomas' Giant Deer Mouse
Megaerops ecaudatus	Temminck's Tailless Fruit Bat
Megaerops kusnotoi	Javan Tailless Fruit Bat
Megaerops niphanae	Ratanaworabhan's Fruit Bat
Megaerops wetmorei	White-collared Fruit Bat
Megaloglossus woermanni	Woermann's Bat
Megalomys desmarestii	Antillean Giant Rice Rat
Megaoryzomys curioi	Curio's Giant Rat (extinct)
Megasorex gigas	Merriam's Desert Shrew
Melanomys zunigae	Zuniga's Dark Rice Rat
Melogale everetti	Everett's Ferret-Badger
Melomys bannisteri	Bannister's Melomys
Melomys bougainville	Bougainville's Melomys
Melomys burtoni	Burton's Melomys
Melomys cooperae	Cooper's Melomys
Melomys dollmani	Dollman's Mosaic-tailed Rat
Melomys howi	How's Melomys
Melomys matambuai	Manus Melomys
Melomys paveli	German's Melomys
Melomys shawi	Shaw's Melomys
Melonycteris fardoulisi	Fardoulis' Blossom Bat
Melonycteris melanops	Bismarck Blossom Bat
Melonycteris woodfordi	Woodford's Blossom Bat
Menetes berdmorei	Berdmore's Palm Squirrel
Meriones chengi	Cheng's Jird
Meriones crassus	Sundevall's Jird, Swinhoe's Jird
Meriones dahli	Dahl's Jird
Meriones sacramenti	Buxton's Jird
Meriones shawi	Shaw's Jird

Meriones tristrami	Tristram's Jird
Meriones vinogradovi	Vinogradov's Jird
Meriones zarudnyi	Zarudny's Jird
Mesechinus hughi	Hugh's Hedgehog
Mesembriomys gouldii	Black-footed Tree Rat
Mesocapromys angelcabrerai	Cabrera's Hutia
Mesocapromys sanfelipensis	Varona and Garrido's Hutia
Mesocricetus brandti	Brandt's Hamster
Mesocricetus newtoni	Romanian Hamster
Mesocricetus raddei	Ciscaucasian Hamster
Mesophylla macconnelli	MacConnell's Bat
Mesoplodon bahamondi	Bahamonde's Beaked Whale
Mesoplodon bidens	Sowerby's Beaked Whale
Mesoplodon bowdoini	Andrews' Beaked Whale, Bowdoin's Beaked Whale
Mesoplodon carlhubbsi	Hubbs' Beaked Whale
Mesoplodon densirostris	Blainville's Beaked Whale
Mesoplodon europaeus	Gervais' Beaked Whale
Mesoplodon grayi	Gray's Beaked Whale
Mesoplodon hectori	Hector's Beaked Whale
Mesoplodon layardii	Layard's Beaked Whale
Mesoplodon mirus	True's Beaked Whale
Mesoplodon perrini	Perrin's Beaked Whale
Mesoplodon stejnegeri	Stejneger's Beaked Whale
Mesoplodon traversii	Spade-toothed Beaked Whale
Micoureus alstoni	Alston's Woolly Mouse-Opossum
Micoureus constantiae	Pale-bellied Woolly Mouse-Opossum
Microcavia shiptoni	Shipton's Mountain Cavy
Microcebus berthae	Berthe's Mouse-Lemur
Microcebus danfossi	Danfoss Mouse-Lemur
Microcebus jollyae	Jolly's Mouse-Lemur
Microcebus lehilahytsara	Goodman's Mouse-Lemur
Microcebus macarthurii	MacArthur's Mouse-Lemur
Microcebus mamiratra	Claire's Mouse-Lemur
Microcebus mittermeieri	Mittermeier's Mouse-Lemur
Microcebus sambiranensis	Sambirano's Lemur
Microcebus simmonsi	Simmons' Mouse-Lemur
Microdillus peeli	Somali Pygmy Gerbil
Microgale cowani	Cowan's Shrew-Tenrec
Microgale dobsoni	Dobson's Shrew-Tenrec
Microgale drouhardi	Drouhard's Shrew-Tenrec
Microgale jenkinsae	Jenkins' Shrew-Tenrec
Microgale nasoloi	Nasolo's Shrew-Tenrec
Microgale talazaci	Talazac's Shrew-Tenrec
Microgale thomasi	Thomas' Shrew-Tenrec

Microhydromys musseri	Musser's Shrew-Mouse
Microhydromys richardsoni	Groove-toothed Shrew-Mouse
Micronycteris brachyotis	Dobson's Large-eared Bat
Micronycteris brosseti	Brosset's Big-eared Bat
Micronycteris giovanniae	Giovanni's Big-eared Bat
Micronycteris homezi	Homez's Big-eared Bat
Micronycteris matses	Matses' Big-eared Bat
Micronycteris minuta	Gervais' Large-eared Bat
Micronycteris nicefori	Niceforo's Big-eared Bat
Micronycteris sanborni	Sanborn's Big-eared Bat
Micronycteris schmidtorum	Schmidt's Big-eared Bat
Microperoryctes aplini	Arfak Pygmy Bandicoot
Micropotamogale lamottei	Nimba Otter-Shrew
Micropteropus intermedius	Hayman's Dwarf Epauletted Fruit Bat
Micropteropus pusillus	Peters' Dwarf Epauletted Fruit Bat
Microsciurus alfari	Alfaro's Pygmy Squirrel
Microtus breweri	Beach Vole
Microtus cabrerae	Cabrera's Vole
Microtus dogramacii	Dogramaci's Vole
Microtus felteni	Felten's Vole
Microtus gerbei	Pyrenean Pine Vole
Microtus guentheri	Günther's Vole
Microtus majori	Major's Pine Vole
Microtus maximowiczii	Maximowicz's Vole
Microtus middendorffii	Middendorff's Vole
Microtus montanus	Peale's Meadow Mouse
Microtus montebelli	Japanese Grass Vole
Microtus nasarovi	Nasarov's Pine Vole
Microtus pennsylvanicus pennsylvanicus	Wilson's Meadow Mouse
Microtus pinetorum	Leconte's Pine Mouse
Microtus richardsoni	Richardson's Vole
Microtus savii	Savi's Pine Vole
Microtus schelkovnikovi	Schelkovnikov's Pine Vole
Microtus thomasi	Thomas' Pine Vole
Microtus townsendii	Townsend's Vole
Millardia	The soft-furred rat genus
Millardia gleadowi	Sand-colored Soft-furred Rat
Millardia kathleenae	Miss Ryley's Soft-furred Rat
Mimetillus moloneyi	Moloney's Flat-headed Bat
Mimon bennettii	Bennett's Spear-nosed Bat
Mimon koepckeae	Koepke's Hairy-nosed Bat
Mindomys hammondi	Hammond's Rice Rat
Miniopterus gleni	Glen's Long-fingered Bat

Miniopterus majori	Major's Long-fingered Bat
Miniopterus schreibersi	Schreibers' Bat
Miopithecus talapoin vleeschouwersi	Vleeschouwers' Talapoin
Mirza coquereli	Coquerel's Mouse-Lemur
Mogera imaizumii	Lesser Japanese Mole
Mogera kanoana	Kano's Mole
Mogera tokudae	Tokuda's Mole
Mogera uchidai	Ryukyu Mole
Mogera wogura	Temminck's Mole
Molossops greenhalli	Greenhall's Dog-faced Bat
Molossops temminckii	Dwarf Dog-faced Bat
Molossus barnesi	Barnes' Mastiff Bat
Molossus molossus	Pallas' Mastiff Bat
Molossus pretiosus	Miller's Mastiff Bat
Molossus sinaloae	Allen's Mastiff Bat
Monachus schauinslandi	Hawaiian Monk Seal
Monodelphis emiliae	Emilia's Short-tailed Opossum
Monodelphis handleyi	Handley's Short-tailed Opossum
Monodelphis iheringi	Ihering's Short-tailed Opossum
Monodelphis kunsi	Pygmy Short-tailed Opossum
Monodelphis osgoodi	Osgood's Short-tailed Opossum
Monodelphis ronaldi	Ronald's Short-tailed Opossum
Monodelphis theresa	Theresa's Short-tailed Opossum
Monophyllus redmani	Leach's Single-leaf Bat
Monticolomys koopmani	Malagasy Mountain Mouse
Mops midas	Midas Free-tailed Bat
Mops petersoni	Peterson's Free-tailed Bat
Mopssarasinorum	Sulawesi Free-tailed Bat
Mops spurrelli	Spurrell's Free-tailed Bat
Mops thersites	Railer Bat
Mops trevori	Trevor's Free-tailed Bat
Mormoops blainvillii	Antillean Ghost-faced Bat
Mormoops megalophylla	Peters' Ghost-faced Bat
Mormopterus beccarii	Beccari's Mastiff Bat
Mormopterus doriae	Sumatran Mastiff Bat
Mormopterus eleryi	Bristle-faced Freetail Bat
Mormopterus francoismoutoui	Réunion Little Mastiff Bat
Mormopterus jugularis	Peters' Wrinkle-lipped Bat
Mormopterus kalinowskii	Kalinowski's Mastiff Bat
Mormopterus loriae	Loria's Mastiff Bat
Moschus berezovskii	Chinese Forest Musk Deer
Muntiacus feae	Fea's Muntjac
Muntiacus reevesi	Reeves' Muntjac
Muntiacus rooseveltorum	Roosevelt's Muntjac
Murina grisea	Peters' Tube-nosed Bat

Murina harrisoni	Harrison's Tube-nosed Bat
Murina hilgendorfi	Hilgendorf'sTube-nosed Bat
Murina huttoni	Hutton's Tube-nosed Bat
Murina rozendaali	Gilded Tube-nosed Bat
Murina tubinaris	Scully's Tube-nosed Bat
Mus baoulei	Baoule's Mouse
Mus callewaerti	Callewaert's Mouse
Mus caroli	Ryukyu Mouse
Mus cookii	Cook's Mouse
Mus famulus	Bonhote's Mouse
Mus fernandoni	Sri Lankan Spiny Mouse
Mus mahomet	Mahomet's Mouse
Mus mattheyi	Matthey's Mouse
Mus mayori	Mayor's Mouse
Mus musculoides	Temminck's Mouse
Mus neavei	Neave's Mouse
Mus ouwensi	Mouse sp.
Mus pahari	Gairdner's Shrew-Mouse
Mus phillipsi	Phillips' Mouse
Mus setulosus	Peters' Mouse
Mus setzeri	Setzer's Pygmy Mouse
Mus shortridgei	Shortridge's Mouse
Mus sorella	Thomas' Pygmy Mouse
Mus triton	Grey-bellied Pygmy Mouse
Musonycteris harrisoni	Banana Bat
Mustela cicognanii	Bonaparte's Weasel
Mustela eversmannii	Steppe Polecat
Mustela felipei	Don Felipe's Weasel
Mydaus marchei	Palawan Stink Badger
Mylomys dybowskii	African Groove-toothed Rat
Myodes andersoni	Anderson's Red-backed Vole
Myodes gapperi	Gapper's Red-backed Vole
Myodes imaizumii	Imaizumi's Vole
Myodes smithii	Smith's Vole
Myoictis wallacii	Wallace's Striped Dasyure
Myomimus personatus	Ognev's Dormouse
Myomimus roachi	Roach's Mouse-tailed Dormouse
Myomimus setzeri	Setzer's Mouse-tailed Dormouse
Myomyscus brockmani	Brockman's Mouse
Myomyscus verreauxi	Verreaux's Mouse
Myoprocta pratti	Green Acouchi
Myopterus daubentonii	Daubenton's Free-tailed Bat
Myopterus whitleyi	Bini Free-tailed Bat
Myosorex babaulti	Babault's Mouse-Shrew
Myosorex eisentrauti	Eisentraut's Mouse-Shrew

Myosorex kihaulei	Kihaule's Mouse-Shrew
Myosorex preussi	Preuss' Mouse-Shrew
Myosorex schalleri	Schaller's Mouse-Shrew
Myosorex sclateri	Sclater's Forest Shrew
Myosorex zinki	Kilimanjaro Mouse-Shrew
Myotis abei	Abe's Whiskered Bat
Myotis aelleni	Southern Myotis
Myotis alcathoe	Alcathoe's Myotis
Myotis bechsteini	Bechstein's Bat
Myotis blythii	Lesser Mouse-eared Bat
Myotis bocagei	Rufous Mouse-eared Bat
Myotis brandtii	Brandt's Bat
Myotis capaccinii	Long-fingered Myotis
Myotis carteri	Carter's Myotis
Myotis daubentoni	Daubenton's Bat
Myotis dieteri	Dieter's Myotis
Myotis emarginatus	Geoffroy's Bat
Myotis findleyi	Findley's Myotis
Myotis formosus	Hodgson's Myotis
Myotis goudoti	Malagasy Mouse-eared Bat
Myotis hasseltii	Hasselt's Myotis
Myotis hermani	Herman's Myotis
Myotis horsfieldii	Horsfield's Myotis
Myotis hosonoi	Hosono's Myotis
Myotis ikonnikovi	Ikonnikov's Myotis
Myotis keaysi	Hairy-legged Myotis
Myotis keenii	Keen's Myotis
Myotis leibii	Leib's Myotis
Myotis martiniquensis	Schwartz's Myotis
Myotis milleri	Miller's Myotis
Myotis morrisi	Morris' Bat
Myotis nattereri	Natterer's Bat
Myotis ricketti	Rickett's Big-footed Bat
Myotis ridleyi	Ridley's Myotis
Myotis rosseti	Thick-thumbed Myotis
Myotis schaubi	Schaub's Myotis
Myotis scotti	Scott's Mouse-eared Bat
Myotis sicarius	Mandelli's Mouse-eared Bat
Myotis stalkeri	Kai Myotis
Myotis vivesi	Fishing Bat
Myotis welwitschii	Welwitsch's Bat
Myotis yesoensis	Yoshiyuki's Myotis
Mysateles garridoi	Garrido's Hutia
Mysateles gundlachi	Chapman's Prehensile-tailed Hutia

Mysateles gundlachi	Gundlach's Hutia
Myzopoda schliemanni	Western Sucker-footed Bat
Naemorhedus (sumatraensis) milneedwardsii	Chinese Serow
Naemorhedus baileyi	Red Goral
Naemorhedus baileyi cranbrooki	Burmese Red Goral
Naemorhedus swinhoei	Taiwan Serow
Nannospalax ehrenbergi	Palestine Mole-Rat
Nannospalax nehringi	Nehring's Blind Mole-Rat
Nanonycteris veldkampi	Veldkamp's Bat
Nasua (narica) nelsoni	Nelson's Coati
Neacomys dubosti	Dubost's Bristly Mouse
Neacomys musseri	Musser's Bristly Mouse
Neamblysomus gunningi	Gunning's Golden-Mole
Neamblysomus julianae	Juliana's Golden-Mole
Necromys temchuki	Temchuk's Bolo Mouse
Necromys urichi	Northern Bolo Mouse
Nelsonia goldmani	Nelson's and Goldman's Woodrat
Neodon (juldaschi) carruthersi	Carruthers' Juniper Vole
Neodon forresti	Forrest's Mountain Vole
Neodon juldaschi	Juldaschi's Vole
Neofelis diardi	Diard's Cat
Neofiber alleni	Round-tailed Muskrat
Neomys anomalus	Miller's Water Shrew
Neomys schelkovnikovi	Schelkovnikov's Water Shrew
Neophascogale lorentzi	Speckled Dasyure
Neopteryx frosti	Small-toothed Fruit Bat
Neoromicia helios	Heller's Pipistrelle
Neoromicia melckorum	Melck's House Bat
Neoromicia rendalli	Rendall's Serotine
Neotoma anthonyi	Anthony's Woodrat
Neotoma bryanti	Bryant's Woodrat
Neotoma bunkeri	Bunker's Woodrat
Neotoma goldmani	Goldman's Woodrat
Neotoma nelsoni	Nelson's Woodrat
Neotoma stephensi	Stephens' Woodrat
Neotomodon alstoni	Mexican Volcano Mouse
Neotragus batesi	Bates' Dwarf Antelope
Neotragus moschatus livingstonianus	Livingstone's Suni
Nephelomys albigularis	Tomes' Rice Rat
Nephelomys childi	Child's Rice Rat
Nephelomys keaysi	Keay's Rice Rat
Nesokia bunni	Bunn's Short-tailed Bandicoot Rat
Nesolagus netscheri	Sumatran Rabbit

Nesolagus timminsi	Annamite Striped Rabbit
Nesomys audeberti	Lowland Red Forest Rat
Nesomys lambertoni	Western Red Forest Rat
Nesophontes edithae	Puerto Rican Nesophontes (extinct)
Nesophontes hypomicrus	Miller's Nesophontes (extinct)
Nesoryzomys darwini	Darwin's Galápagos Mouse
Nesoryzomys fernandinae	Fernandina Galápagos Mouse
Nesoryzomys swarthi	Santiago Galápagos Mouse
Neurotrichus gibbsii	American Shrew-Mole
Neusticomys mussoi	Musso's Fish-eating Rat
Neusticomys oyapocki	Oyapock's Fish-eating Rat
Ningaui ridei	Wongai Ningaui
Ningaui timealeyi	Pilbara Ningaui
Ningaui yvonneae	Southern Ningaui
Niviventer andersoni	Anderson's White-bellied Rat
Niviventer coxingi	Coxing's White-bellied Rat
Niviventer eha	Smoke-bellied Rat
Niviventer niviventer	Hodgson's White-bellied Rat
Nomascus concolor concolor	Harlan's Gibbon
Nomascus gabriellae	Gabriella's Crested Gibbon
Noronhomys vespuccii	Vespucci's Rat (extinct)
Notiomys edwardsii	Edward's Long-clawed Mouse, Milne-Edwards' Long-clawed Mouse
Notiosorex cockrumi	Cockrum's Desert Shrew
Notiosorex crawfordi	Desert Shrew
Notiosorex villai	Villa's Grey Shrew
Notomys mitchelli	Mitchell's Hopping Mouse
Notopteris macdonaldi	Fijian Blossom Bat
Nyctalus leisleri	Leisler's Bat
Nyctalus plancyi	Chinese Noctule
Nycteris arge	Bates' Slit-faced Bat
Nycterisvinsoni	Vinson's Slit-faced Bat
Nycteris woodi	Wood's Slit-faced Bat
Nycticeinops schlieffeni	Schlieffen's Bat
Nyctiellus lepidus	Gervais' Funnel-eared Bat
Nyctimene aello	Broad-striped Tube-nosed Bat
Nyctimene celaeno	Dark Tube-nosed Bat
Nyctimene cephalotes	Pallas' Tube-nosed Fruit Bat
Nyctimene rabori	Rabor's Tube-nosed Bat
Nyctimene robinsoni	Queensland Tube-nosed Fruit Bat
Nyctinomops aurispinosus	Peale's Free-tailed Bat
Nyctomys sumichrasti	Sumichrast's Vesper Rat
Nyctophilus (timoriensis) sherrini	Tasmanian Long-eared Bat
Nyctophilus geoffroyi	Lesser Nyctophilus
Nyctophilus gouldi	Gould's Nyctophilus

Nyctophilus howensis Lord Howe Long-eared Bat (extinct)
Nyctophilus walkeri Pygmy Nyctophilus

Ochotona curzoniae Black-lipped Pika
Ochotona forresti Forrest's Pika
Ochotona gloveri Glover's Pika
Ochotona hoffmanni Hoffmann's Pika
Ochotona koslowi Kozlov's Pika
Ochotona pallasi Pallas' Pika
Ochotona princeps barnesi Barnes' Pika
Ochotona roylei Royle's Pika
Ochotona thomasi Thomas' Pika
Ochrotomys nuttalli Golden Mouse
Octodon bridgesi Bridges' Degu
Odocoileus virginianus couesi Coues' Deer
Oecomys cleberi Cleber's Arboreal Rice Rat
Oecomys roberti Robert's Arboreal Rice Rat
Okapia johnstoni Okapi
Olallamys The soft-furred spiny rat genus
Oligoryzomys destructor Tschudi's Pygmy Rice Rat
Oligoryzomys fornesi Fornes' Pygmy Rice Rat
Oligoryzomys moojeni Moojen's Pygmy Rice Rat
Olisthomys morrisi Morris' Flying Squirrel
Ommatophoca rossii Ross Seal
Oncifelis colocolo budini Pampas Cat
Oncifelis colocolo crespoi Crespo's Pampas Cat
Oncifelis geoffroyi Geoffroy's Cat
Onychomys arenicola Mearns' Grasshopper Mouse
Orcaella heinsohni Australian Snubfin Dolphin
Oreonax flavicauda Hendee's Woolly Monkey
Orthogeomys cherriei Cherrie's Pocket Gopher
Orthogeomys thaeleri Thaeler's Pocket Gopher
Orthogeomys underwoodi Underwood's Pocket Gopher
Oryx beatrix Beatrix Oryx
Oryx gazella Plain's Gazelle
Oryzomys alfaroi Alfaro's Rice Rat
Oryzomys chapmani Chapman's Rice Rat
Oryzomys couesi Coues' Rice Rat
Oryzomys dimidiatus Thomas' Rice Rat
Oryzomys gorgasi Gorgas' Rice Rat
Oryzomys nelsoni Nelson's Rice Rat (extinct)
Osbornictis piscivora Aquatic Genet
Osgoodomys banderanus Michoacan Deer Mouse
Otaria byronia Southern Sea-Lion
Otolemur crassicaudatus kirkii Kirk's Galago

Otolemur garnettii	Garnett's Galago
Otolemur monteiri	Silvery Greater Galago
Otomops johnstonei	Johnstone's Giant Mastiff Bat
Otomops martiensseni	Large-eared Free-tailed Bat
Otomops wroughtoni	Wroughton's Free-tailed Bat
Otomys anchietae	Angolan Vlei Rat
Otomys barbouri	Barbour's Vlei Rat
Otomys dartmouthi	Dartmouth's Vlei Rat
Otomys denti	Dent's Vlei Rat
Otomys dollmani	Dollman's Vlei Rat
Otomys jacksoni	Mount Elgon Vlei Rat
Otomys orestes	Afroalpine Vlei Rat
Otomys saundersiae	Saunders' Vlei Rat
Otomys sloggetti	Sloggett's Vlei Rat
Otonycteris hemprichii	Hemprich's Long-eared Bat
Otonyctomys hatti	Hatt's Vesper Rat
Ourebia ourebi cottoni	Cotton's Oribi
Ourebia ourebi haggardi	Haggard's Oribi
Ovibos moschatus wardi	Greenland Musk-ox
Ovis ammon	Argali
Ovis ammon darwini	Darwin's Sheep
Ovis ammon humei	Hume's Argali
Ovis ammon littledalei	Littledale's Argali
Ovis ammon polii	Marco Polo Sheep
Ovis ammon severtzovi	Severtzov's Argali
Ovis canadensis auduboni	Audubon's Bighorn Sheep (extinct)
Ovis canadensis nelsoni	Desert Bighorn Sheep
Ovis dalli	Dall's Sheep
Ovis dalli stonei	Stone's Sheep
Ovis gmelini	Mouflon
Ovis vignei	Urial
Ovis vignei blanfordi	Blanford's Urial
Oxymycterus josei	José's Hocicudo
Oxymycterus roberti	Robert's Hocicudo
Pachyuromys duprasi	Fat-tailed Gerbil
Pan	The chimpanzee genus
Pan troglodytes schweinfurthii	Schweinfurth's Chimpanzee
Panthera leo bleyenberghi	Bleyenbergh's Lion
Panthera leo hollisteri	Hollister's Lion
Panthera leo roosevelti	Roosevelt's Lion
Panthera leo vernayi	Vernay's Lion
Panthera pardus adersi	Zanzibar Leopard
Panthera pardus delacouri	Indochinese Leopard
Panthera pardus shortridgei	Shortridge's Leopard

Panthera tigris corbetti	Corbett's Tiger
Pantholops hodgsonii	Chiru, Tibetan Antelope
Papagomys armandvillei	Flores Giant Tree Rat
Papagomys theodorverhoeveni	Verhoeven's Giant Tree Rat (extinct)
Papio anubis	Olive Baboon
Papio anubis heuglini	Heuglin's Olive Baboon
Papio anubis neumanni	Neumann's Olive Baboon
Papio cynocephalus langheldi	Langheld's Baboon
Papio hamadryas	Hamadryas Baboon
Papio ursinus ruacana	Shortridge's Chacma Baboon
Pappogeomys alcorni	Alcorn's Pocket Gopher
Pappogeomys bulleri	Buller's Pocket Gopher
Paracrocidura graueri	Grauer's Shrew
Paracrocidura schoutedeni	Schouteden's Shrew
Paracynictis selousi	Selous' Mongoose
Paradoxurus jerdoni	Jerdon's Palm Civet
Paraleptomys wilhelmina	Short-haired Hydromyine
Paramelomys gressitti	Gressitt's Melomys
Paramelomys lorentzii	Lorentz's Melomys
Paramelomys mollis	Thomas' Melomys
Paramelomys moncktoni	Monckton's Melomys
Paramelomys steini	Stein's Melomys
Paramurexia rothschildi	Broad-striped Dasyure
Parascalops breweri	Brewer's Shrew Mole
Paraxerus (vexillarius) byatti	Byatt's Bush Squirrel
Paraxerus alexandri	Alexander's Bush Squirrel
Paraxerus boehmi	Böhm's Bush Squirrel
Paraxerus cepapi	Smith's Bush Squirrel
Paraxerus cooperi	Cooper's Mountain Squirrel
Paraxerus lucifer	Black-and-Red Bush Squirrel
Paraxerus ochraceus	Huet's Bush Squirrel
Paraxerus vexillarius	Swynnerton's Bush Squirrel
Paraxerus vincenti	Vincent's Bush Squirrel
Parotomys brantsii	Brants' Whistling Rat
Parotomys littledalei	Littledale's Whistling Rat
Pattonomys	The spiny rat genus
Pattonomys carrikeri	Carriker's Spiny Rat
Paulamys naso	Paula's Long-nosed Rat
Pearsonomys annectens	Pearson's Long-clawed Mouse
Pectinator spekei	Speke's Pectinator
Pelomys hopkinsi	Hopkins' Groove-toothed Swamp Rat
Pelomys isseli	Issel's Groove-toothed Swamp Rat
Pentalagus furnessi	Amami Rabbit
Penthetor lucasi	Lucas' Short-nosed Fruit Bat
Peponocephala electra	Melon-headed Whale

Perameles bougainville	Western Barred Bandicoot
Perameles gunnii	Eastern Barred Bandicoot
Perodicticus potto	Bosman's Potto
Perodicticus potto edwardsi	Milne-Edwards' Potto
Perognathus merriami	Merriam's Pocket Mouse
Peromyscus aztecus evides	Osgood's Aztec Mouse
Peromyscus boylii	Brush Deer Mouse
Peromyscus caniceps	Burt's Deer Mouse
Peromyscus dickeyi	Dickey's Deer Mouse
Peromyscus eva	Eva's Deer Mouse
Peromyscus gratus	Osgood's Mouse
Peromyscus hooperi	Hooper's Deer Mouse
Peromyscus keeni	Northwestern Deer Mouse
Peromyscus merriami	Merriam's Deer Mouse
Peromyscus pembertoni	Pemberton's Deer Mouse
Peromyscus schmidlyi	Schmidly's Deer Mouse
Peromyscus slevini	Slevin's Deer Mouse
Peromyscus stephani	San Esteban Island Deer Mouse
Peromyscus stirtoni	Stirton's Deer Mouse
Peromyscus truei	Piñon Deer Mouse
Peromyscus winkelmanni	Winkelmann's Deer Mouse
Peropteryx kappleri	Greater Dog-like Bat
Peroryctes broadbenti	Giant Bandicoot
Peroryctes raffrayana	Raffray's Bandicoot
Petaurillus emiliae	Lesser Pygmy Flying Squirrel
Petaurillus hosei	Hose's Pygmy Flying Squirrel
Petaurillus kinlochii	Selangor Pygmy Flying Squirrel
Petaurista magnificus	Hodgson's Giant Flying Squirrel
Petaurista sybilla	Mrs. Millard's Flying Squirrel
Petaurus abidi	Northern Glider
Petinomys hageni	Hagen's Flying Squirrel
Petinomys setosus	Temminck's Flying Squirrel
Petinomys vordermanni	Vordermann's Flying Squirrel
Petrogale burbidgei	Monjon
Petrogale godmani	Godman's Rock Wallaby
Petrogale herberti	Herbert's Rock Wallaby
Petrogale rothschildi	Rothschild's Rock Wallaby
Petrogale sharmani	Sharman's Rock Wallaby
Petromyscus barbouri	Barbour's Rock Mouse
Petromyscus shortridgei	Shortridge's Rock Mouse
Petropseudes dahli	Rock Ringtail Possum
Phaiomys leucurus	Blyth's Vole
Phalanger alexandrae	Gebe Cuscus
Phalanger rothschildi	Rothschild's Cuscus
Phalanger vestitus	Stein's Cuscus

Phaner (furcifer) parienti	Pariente's Fork-marked Lemur
Phascolosorex doriae	Red-bellied Marsupial-Shrew
Philander andersoni	Anderson's Four-eyed Opossum
Philander mcilhennyi	McIlhenny's Four-eyed Opossum
Philander mondolfii	Mondolfi's Four-eyed Opossum
Philander olrogi	Olrog's Four-eyed Opossum
Philetor brachypterus	Rohu's Bat
Phloeomys cumingi	Southern Luzon Cloud Rat
Phoca vitulina stejnegeri	Kuril Harbor Seal
Phocarctos hookeri	Hooker's Sea Lion, New Zealand Sea-Lion
Phocoena spinipinnis	Burmeister's Porpoise
Phocoenoides dalli	Dall's Porpoise, True's Porpoise
Phodopus campbelli	Campbell's Hamster
Phodopus roborovskii	Roborovski's Hamster
Phoniscus jagorii	Peters' Trumpet-eared Bat
Phyllomys kerri	Kerr's Tree Rat, Moojen's Atlantic Tree Rat
Phyllomys lundi	Lund's Atlantic Tree Rat
Phyllomys pattoni	Rusty-sided Atlantic Tree Rat
Phyllomys thomasi	Giant Atlantic Tree Rat
Phyllonycteris poeyi	Poey's Pallid Flower Bat
Phyllotis anitae	Anita's Leaf-eared Mouse
Phyllotis darwini	Darwin's Leaf-eared Mouse
Phyllotis haggardi	Haggard's Leaf-eared Mouse
Phyllotis osgoodi	Osgood's Leaf-eared Mouse
Phyllotis wolffsohni	Wolffsohn's Leaf-eared Mouse
Piliocolobus badius temminckii	Temminck's Red Colobus
Piliocolobus badius waldroni	Miss Waldron's Red Colobus Monkey
Piliocolobus foai	Central African Red Colobus
Piliocolobus foai ellioti	Elliot's Red Colobus
Piliocolobus foai oustaleti	Oustalet's Red Colobus
Piliocolobus gordonorum	Gordon's Red Colobus
Piliocolobus kirkii	Zanzibar Red Colobus
Piliocolobus pennantii	Pennant's Red Colobus
Piliocolobus pennantii bouvieri	Bouvier's Red Colobus
Piliocolobus preussi	Preuss' Red Colobus
Piliocolobus thollonii	Thollon's Red Colobus
Pipistrellus adamsi	Cape York Pipistrelle
Pipistrellus ceylonicus	Kelaart's Pipistrelle
Pipistrellus dormeri	Dormer's Pipistrelle
Pipistrellus endoi	Endo's Pipistrelle
Pipistrellus hanaki	Hanak's Pipistrelle
Pipistrellus inexspectatus	Aellen's Pipistrelle

Pipistrellus kuhlii	Kuhl's Pipistrelle
Pipistrellus mackenziei	Mackenzie's False Pipistrelle
Pipistrellus nathusii	Nathusius' Pipistrelle
Pipistrellus paterculus	Thomas' Pipistrelle
Pipistrellus raceyi	Racey's Pipistrelle
Pipistrellus rueppelli	Rüppell's Pipistrelle
Pipistrellus sturdeei	Sturdee's Pipistrelle
Pipistrellus wattsi	Watts' Pipistrelle
Pipistrellus westralis	Koopman's Pipistrelle
Pithecia irrorata irrorata	Gray's Monk Saki
Pithecia irrorata vanzolinii	Vanzolini's Bald-faced Saki
Pithecia monachus milleri	Miller's Monk Saki
Pithecia monachus monachus	Geoffroy's Monk Saki
Plagiodontia aedium	Cuvier's Hutia
Plagiodontia ipnaeum	Johnson's Hutia (extinct)
Plagiodontia velozi	Veloz's Hutia (extinct)
Planigale gilesi	Giles' Planigale
Planigale ingrami	Ingram's Planigale
Platymops setiger	Peters' Flat-headed Bat
Platyrrhinus albericoi	Alberico's Broad-nosed Bat
Platyrrhinus dorsalis	Thomas' Broad-nosed Bat
Platyrrhinus helleri	Heller's Broad-nosed Bat
Platyrrhinus lineatus	Geoffroy's Rayed Bat
Plecotus christiei	Christie's Long-eared Bat
Plecotus kolombatovici	Kolombatovic's Big-eared Bat
Plecotus kozlovi	Kozlov's Long-eared Bat
Plecotus ognevi	Ognev's Long-eared Bat
Plecotus strelkovi	Strelkov's Long-eared Bat
Plecotus teneriffae gaisleri	Gaisler's Long-eared Bat
Plecotus wardi	Ward's Long-eared Bat
Plerotes anchietai	D'Anchieta's Fruit Bat
Podogymnura truei	Mindanao Gymnure
Poelagus marjorita	Bunyoro Rabbit
Pogonomelomys bruijni	Lowland Brush Mouse
Pogonomelomys mayeri	Shaw-Mayer's Pogonomelomys
Pogonomelomys sevia	Menzies' Mouse
Pogonomys championi	Champion's Tree Mouse
Pogonomys loriae	Large Tree Mouse
Poiana leightoni	Leighton's Linsang
Poiana richardsonii	African Linsang
Pongo abelii	Sumatran Orangutan
Pontoporia blainvillei	Franciscana
Potorous gilbertii	Gilbert's Poteroo
Praomys daltoni	Dalton's Mouse
Praomys degraafi	De Graaf's Soft-furred Mouse

Praomys derooi	Deroo's Mouse
Praomys hartwigi	Hartwig's Soft-furred Mouse
Praomys jacksoni	Jackson's Soft-furred Mouse
Praomys misonnei	Misonne's Soft-furred Mouse
Praomys mutoni	Muton's Soft-furred Mouse
Praomys petteri	Petter's Soft-furred Mouse
Praomys tullbergi	Tullberg's Soft-furred Mouse
Presbytis femoralis femoralis	Raffles' Banded Langur
Presbytis femoralis robinsoni	Robinson's Banded Langur
Presbytis hosei	Hose's Leaf Monkey
Presbytis hosei canicrus	Miller's Grizzled Langur
Presbytis hosei everetti	Everett's Grizzled Langur
Presbytis potenziani	Mentawai Langur
Presbytis rubicunda chrysea	Davis' Maroon Langur
Presbytis thomasi	Thomas' Leaf Monkey
Prionailurus bengalensis rabori	Visayan Leopard Cat
Prionomys batesi	Dollman's Tree Mouse
Procapra przewalskii	Przewalski's Gazelle
Procavia welwitschii	Kaokoveld Rock Hyrax
Procolobus foai	Foa's Red Colobus
Procolobus verus	Van Beneden's Colobus
Procyon gloveralleni	Barbados Raccoon
Procyon maynardii	Bahaman Raccoon
Proechimys cuvieri	Cuvier's Spiny Rat
Proechimys gardneri	Gardner's Spiny Rat
Proechimys goeldii	Goeldi's Spiny Rat
Proechimys hendeei	Hendee's Spiny Rat
Proechimys kulinae	Kulinas' Spiny Rat
Proechimys oconnelli	O'Connell's Spiny Rat
Proechimys pattoni	Patton's Spiny Rat
Proechimys roberti	Robert's Spiny Rat
Proechimys semispinosus	Tomes' Spiny Rat
Proechimys simonsi	Simons' Spiny Rat
Proechimys steerei	Steere's Spiny Rat
Proechimys urichi	Sucre Spiny Rat
Proechimys warreni	Warren's Spiny Rat
Proedromys bedfordi	Duke of Bedford's Vole
Prometheomys	The vole genus
Prometheomys schaposchnikowi	Long-clawed Vole
Pronolagus barretti	Red Rock Hare sp.
Pronolagus randensis	Jameson's Red Rock Hare
Pronolagus rupestris	Smith's Red Rock Hare
Pronolagus rupestris curryi	Curry's Red Rock Rabbit
Pronolagus saundersiae	Hewitt's Red Rock Hare
Propithecus coquereli	Coquerel's Sifaka

Propithecus deckenii	Decken's Sifaka, Van der Decken's Sifaka, Von der Decken's Sifaka
Propithecus edwardsi	Milne-Edwards' Sifaka
Propithecus perrieri	Perrier's Sifaka
Propithecus tattersalli	Tattersall's Sifaka
Propithecus verreauxi	Verreaux's Sifaka
Propithecus verreauxi verreauxi	Forsyth Major's Sifaka, Major's Sifaka
Prosciurillus rosenbergii	Rosenberg's Dwarf Squirrel
Prosciurillus weberi	Weber's Dwarf Squirrel
Protochromys fellowsi	Red-bellied Melomys
Protoxerus aubinnii	Slender-tailed Giant Squirrel
Protoxerus stangeri	Forest Giant Squirrel
Pseudalopex fulvipes	Darwin's Fox
Pseudalopex gymnocercus	Azara's Dog
Pseudantechinus bilarni	Sandstone Pseudantechinus
Pseudantechinus macdonnellensis	Fat-tailed Pseudantechinus
Pseudantechinus roryi	Rory's Pseudantechinus
Pseudantechinus woolleyae	Woolley's Pseudantechinus
Pseudochirops albertisii	D'Albertis Ringtail Possum
Pseudochirops archeri	Green Ringtail Possum
Pseudochirops corinnae	Plush-coated Ringtail Possum
Pseudochirulus caroli	Weyland Ringtail Possum
Pseudochirulus forbesi	Painted Ringtail Possum
Pseudochirulus mayeri	Pygmy Ringtail Possum
Pseudochirulus schlegeli	Arfak Ringtail Possum
Pseudois schaeferi	Dwarf Bharal
Pseudomys bolami	Bolam's Mouse
Pseudomys calabyi	Calaby's Pebble-mound Mouse
Pseudomys chapmani	Western Pebble-mound Mouse
Pseudomys fieldi	Field's Mouse
Pseudomys gouldii	Gould's Mouse
Pseudomys higginsi	Long-tailed Mouse
Pseudomys johnsoni	Johnson's Mouse
Pseudomys shortridgei	Heath Rat
Pseudopotto martini	Martin's False Potto
Ptenochirus jagori	Greater Musky Fruit Bat
Pteralopex flanneryi	Greater Monkey-faced Bat
Pteromys volans orii	Hokkaido Flying Squirrel
Pteronotus davyi	Davy's Naked-backed Bat
Pteronotus macleayii	MacLeay's Moustached Bat
Pteronotus parnellii	Parnell's Moustached Bat
Pteronotus personatus	Wagner's Moustached Bat
Pteropus (speciosus) mearnsi	Mearns' Flying Fox
Pteropus alecto	Black Flying Fox
Pteropus capistratus	Bismarck Flying Fox

Pteropus gilliardorum	New Britain Flying Fox
Pteropus livingstonii	Livingstone's Flying Fox
Pteropus lylei	Lyle's Flying Fox
Pteropus mahaganus	Sanborn's Flying Fox
Pteropus phaeocephalus	Mortlock Flying Fox
Pteropus pohlei	Geelvink Bay Flying Fox
Pteropus pumilis tablasi	Taylor's Flying Fox
Pteropus rayneri	Rayner's Flying Fox
Pteropus rennelli	Rennell Flying Fox
Pteropus temmincki	Temminck's Flying Fox
Pteropus tokudae	Guam Flying Fox (extinct)
Pteropus voeltzkowi	Pemba Flying Fox
Pteropus woodfordi	Dwarf Flying Fox
Ptilocercus lowii	Pen-tailed Tree-Shrew
Pudu mephistophiles	Northern Pudu
Puma concolor anthonyi	Anthony's Puma
Puma concolor greeni	Green's Puma
Puma concolor hudsoni	Hodson's Puma
Puma concolor pearsoni	Pearson's Puma
Puma concolor soderstromii	Ecuadorian Cougar
Punomys kofordi	Koford's Puna Mouse
Pygeretmus shitkovi	Greater Fat-tailed Jerboa
Rangifer tarandus dawsoni	Dawson's Caribou, Queen Charlotte Caribou (extinct)
Rangifer tarandus granti	Grant's Caribou
Rangifer tarandus osborni	Osborn's Caribou
Rangifer tarandus pearyi	Peary Caribou
Rangifer tarandus stonei	Stone's Caribou
Raphicerus sharpei	Sharpe's Grysbok
Rattus annandalei	Annandale's Rat
Rattus colletti	Dusky Rat
Rattus everetti	Philippine Forest Rat
Rattus feliceus	Spiny Seram Rat
Rattus hainaldi	Hainald's Rat
Rattus hoffmanni	Hoffmann's Rat
Rattus hoogerwerfi	Hoogerwerf's Rat
Rattus koopmani	Koopman's Rat
Rattus korinchi	Korinch's Rat
Rattus macleari	Maclear's Rat (extinct)
Rattus osgoodi	Osgood's Rat
Rattus pococki	Pocock's New Guinea Rat
Rattus ranjiniae	Ranjini's Field Rat
Rattus steini	Stein's Rat
Rattus tanezumi	Sladen's Rat

Rattus tunneyi	Pale Field Rat
Ratufa indica dealbata	Dang's Giant Squirrel (extinct)
Redunca fulvorufula chanleri	Chanler's Mountain Reedbuck
Redunca redunca wardi	Ward's Reedbuck
Reithrodontomys bakeri	Baker's Harvest Mouse
Reithrodontomys burti	Sonoran Harvest Mouse
Reithrodontomys rodriguezi	Rodriguez's Harvest Mouse
Reithrodontomys sumichrasti	Sumichrast's Harvest Mouse
Rheomys mexicanus	Goodwin's Water Mouse
Rheomys raptor	Goldman's Water Mouse
Rheomys raptor hartmanni	Hartmann's Water Mouse
Rheomys thomasi	Thomas' Water Mouse
Rheomys underwoodi	Underwood's Water Mouse
Rhinoceros unicornis	Louis XV's Rhinoceros
Rhinolophus adami	Adam's Horseshoe Bat
Rhinolophus anderseni	Andersen's Horseshoe Bat
Rhinolophus beddomei	Lesser Woolly Horseshoe Bat
Rhinolophus blasii	Blasius' Horseshoe Bat
Rhinolophus canuti	Canut's Horseshoe Bat
Rhinolophus clivosus	Geoffroy's Horseshoe Bat
Rhinolophus creaghi	Creagh's Horseshoe Bat
Rhinolophus darlingi	Darling's Horseshoe Bat
Rhinolophus deckenii	Decken's Horseshoe Bat
Rhinolophus denti	Dent's Horseshoe Bat
Rhinolophus euryale	Mediterranean Horseshoe Bat
Rhinolophus fumigatus	Rüppell's Horseshoe Bat
Rhinolophus hildebrandti	Hildebrandt's Horseshoe Bat
Rhinolophus hillorum	Hills' Horseshoe Bat
Rhinolophus imaizumii	Imaizumi's Horseshoe Bat
Rhinolophus landeri	Lander's Horseshoe Bat
Rhinolophus lepidus	Blyth's Horseshoe Bat
Rhinolophus maclaudi	Maclaud's Horseshoe Bat
Rhinolophus marshalli	Marshall's Horseshoe Bat
Rhinolophus mehelyi	Mehely's Horseshoe Bat
Rhinolophus nereis	Nereid Horseshoe Bat
Rhinolophus osgoodi	Osgood's Horseshoe Bat
Rhinolophus paradoxolophus	Bourret's Horseshoe Bat
Rhinolophus pearsonii	Pearson's Horseshoe Bat
Rhinolophus robinsoni	Peninsular Horseshoe Bat
Rhinolophus rouxii	Rufous Horseshoe Bat
Rhinolophus shameli	Shamel's Horseshoe Bat
Rhinolophus shortridgei	Shortridge's Horseshoe Bat
Rhinolophus stheno	Lesser Brown Horseshoe Bat
Rhinolophus swinnyi	Swinny's Horseshoe Bat
Rhinolophus thomasi	Thomas' Horseshoe Bat

Rhinolophus yunanensis	Dobson's Horseshoe Bat
Rhinophylla fischerae	Fischer's Little Fruit Bat
Rhinopithecus bieti	Black Snub-nosed Monkey
Rhinopithecus brelichi	Brelich's Snub-nosed Monkey
Rhinopithecus roxellana	Golden Snub-nosed Monkey
Rhinopoma hardwickii	Lesser Mouse-tailed Bat
Rhinopoma macinnesi	MacInnes' Mouse-tailed Bat
Rhipidomys couesi	Coues' Climbing Mouse
Rhipidomys emiliae	Eastern Amazon Climbing Mouse
Rhipidomys gardneri	Gardner's Climbing Mouse
Rhipidomys macconnelli	MacConnell's Climbing Mouse
Rhipidomys wetzeli	Wetzel's Climbing Mouse
Rhogeessa alleni	Allen's Yellow Bat
Rhogeessa genowaysi	Genoways' Yellow Bat
Rhogeessa hussoni	Husson's Yellow Bat
Rhogeessa io	Thomas' Yellow Bat
Rhynchocyon cirnei	Checkered Elephant-Shrew
Rhynchocyon petersi	Black-and-Rufous Elephant-Shrew
Rhynchogale melleri	Meller's Mongoose
Rhynchomeles prattorum	Ceram Bandicoot
Romerolagus diazi	Volcano Rabbit
Rousettus amplexicaudatus	Geoffroy's Rousette
Rousettus leschenaulti	Leschenault's Rousette
Rousettus leschenaulti shortridgei	Shortridge's Rousette
Saccolaimus mixtus	Troughton's Pouched Bat
Saccolaimus peli	Pel's Pouched Bat
Saccolaimus pluto	Philippine Pouched Bat
Saccolaimus saccolaimus crassus	Blyth's Pouchbearing Bat
Saccostomus mearnsi	Mearns' Pouched Mouse
Saguinus fuscicollis avilapiresi	Ávila Pires' Saddle-back Tamarin
Saguinus fuscicollis crandalli	Crandall's Saddle-back Tamarin
Saguinus fuscicollis cruzlimai	Cruz Lima's Saddle-back Tamarin
Saguinus fuscicollis fuscicollis	Spix's Saddle-back Tamarin
Saguinus fuscicollis fuscus	Lesson's Saddle-back Tamarin
Saguinus fuscicollis illigeri	Illiger's Saddle-back Tamarin
Saguinus fuscicollis nigrifrons	Geoffroy's Saddle-back Tamarin
Saguinus fuscicollis weddelli	Weddell's Saddle-back Tamarin
Saguinus geoffroyi	Geoffroy's Tamarin, Jeffery's Tamarin
Saguinus graellsi	Graells' Tamarin
Saguinus labiatus thomasi	Thomas' Moustached Tamarin
Saguinus martinsi	Martins' Bare-faced Tamarin
Saguinus midas	Lacépède's Tamarin, Midas Tamarin
Saguinus mystax pluto	Pluto Tamarin
Saguinus nigricollis	Spix's Black-mantled Tamarin

Saguinus nigricollis hernandezi	Hernández-Camacho's Black Tamarin
Saguinus oedipus	Cotton-top Tamarin
Saimiri oerstedii	Central American Squirrel Monkey
Saimiri sciureus cassiquiarensis	Humboldt's Squirrel Monkey
Saimiri vanzolinii	Black Squirrel Monkey
Salinoctomys loschalchalerosorum	Chalchalero Viscacha Rat
Salpingotus heptneri	Heptner's Pygmy Jerboa
Salpingotus kozlovi	Kozlov's Pygmy Jerboa
Salpingotus michaelis	Baluchistan Pygmy Jerboa
Salpingotus thomasi	Thomas' Pygmy Jerboa
Sarcophilus harrisii	Tasmanian Devil
Sauromys petrophilus	Roberts' Flat-headed Bat
Saxatilomys paulinae	Paulina's Rock Rat
Scapanulus oweni	Gansu Mole
Scapanus townsendii	Townsend's Mole
Sciurotamias davidianus	Père David's Rock Squirrel
Sciurotamias forresti	Forrest's Rock Squirrel
Sciurus (aestuans) ingrami	Ingram's Squirrel
Sciurus aberti	Abert's Squirrel
Sciurus alleni	Allen's Squirrel
Sciurus colliaei	Collie's Squirrel
Sciurus deppei	Deppe's Squirrel
Sciurus niger shermani	Sherman's Fox Squirrel
Sciurus oculatus	Peters' Squirrel
Sciurus pucheranii	Andean Squirrel
Sciurus richmondi	Richmond's Squirrel
Sciurus sanborni	Sanborn's Squirrel
Scoteanax rueppelli	Rüppell's Broad-nosed Bat
Scotinomys teguina	Alston's Brown Mouse
Scotoecus hindei	Hinde's Lesser House Bat
Scotonycteris ophiodon	Pohle's Fruit Bat
Scotonycteris zenkeri	Zenker's Fruit Bat
Scotophilus dinganii	Dingaan's Yellow Bat
Scotophilus heathi	Greater Asiatic Yellow Bat
Scotophilus kuhlii	Lesser Asiatic Yellow Bat
Scotophilus nigrita	Schreber's Yellow Bat
Scotorepens balstoni	Inland Broad-nosed Bat
Scotorepens greyii	Little Broad-nosed Bat
Scotorepens orion	Orion Broad-nosed Bat
Scotorepens sanborni	Northern Broad-nosed Bat
Scutisorex somereni	Someren's Girder-backed Shrew
Selevinia betpakdalaensis	Desert Dormouse
Semnopithecus (entellus) ajax	Kashmir Grey Langur
Semnopithecus (entellus) dussumieri	Southern Plains Grey Langur
Semnopithecus (entellus) hector	Terai Grey Langur

Semnopithecus (entellus) priam	Tufted Grey Langur
Semnopithecus entellus	Entellus Langur, Hanuman Langur
Sicista severtzovi	Severtzov's Birch Mouse
Sicista strandi	Strand's Birch Mouse
Sigmoceros lichtensteinii	Lichtenstein's Hartebeest
Sigmodon alleni	Allen's Cotton Rat
Sigmodon alstoni	Alston's Cotton Rat
Sigmodontomys alfari	Alfaro's Rice Water Rat
Sigmodontomys aphrastus	Harris' Rice Water Rat
Sminthopsis aitkeni	Kangaroo Island Dunnart
Sminthopsis archeri	Chestnut Dunnart
Sminthopsis butleri	Butler's Dunnart
Sminthopsis douglasi	Julia Creek Dunnart
Sminthopsis gilberti	Gilbert's Dunnart
Sminthopsis virginiae	Virginia Dunnart
Sminthopsis youngsoni	Lesser Hairy-footed Dunnart
Solenodon paradoxus woodi	Woods' Solenodon
Solisorex pearsoni	Pearson's Long-clawed Shrew
Solomys ponceleti	Poncelet's Naked-tailed Rat
Sommeromys macrorhinos	Long-nosed Shrew-Mouse
Sorex bairdi	Baird's Shrew
Sorex bedfordiae	Lesser Striped Shrew
Sorex bendirii	Bendire's Shrew
Sorex caecutiens	Laxmann's Shrew
Sorex cooperi	Cooper's Shrew
Sorex coronatus	Millet's Shrew
Sorex dekayi	Dekay's Shrew
Sorex fosteri	Foster's Shrew
Sorex haydeni	Hayden's Shrew
Sorex hosonoi	Azumi Shrew
Sorex hoyi	American Pygmy Shrew
Sorex jacksoni	Lawrence Island Shrew
Sorex kozlovi	Kozlov's Shrew
Sorex merriami	Merriam's Shrew
Sorex milleri	Carmen Mountain Shrew
Sorex portenkoi	Portenko's Shrew
Sorex preblei	Preble's Shrew
Sorex raddei	Radde's Shrew
Sorex rohweri	Rohwer's Shrew
Sorex satunini	Caucasian Shrew
Sorex saussurei	Saussure's Shrew
Sorex sclateri	Sclater's Shrew
Sorex thompsoni	Thompson's Pygmy Shrew
Sorex trowbridgii	Trowbridge's Shrew
Sorex volnuchini	Volnuchin's Shrew

Sousa teuszii	Teusz's Dolphin
Spermophilus beecheyi	California Ground Squirrel
Spermophilus beldingi	Belding's Ground Squirrel
Spermophilus canus	Merriam's Ground Squirrel
Spermophilus franklinii	Franklin's Ground Squirrel
Spermophilus parryii	Parry's Marmot Squirrel
Spermophilus richardsonii	Richardson's Ground Squirrel
Spermophilus townsendii	Townsend's Ground Squirrel
Spermophilus washingtoni	Washington Ground Squirrel
Sphaerias blanfordi	Blanford's Fruit Bat
Spilocuscus kraemeri	Admiralty Cuscus
Spilocuscus wilsoni	Blue-eyed Spotted Cuscus
Steatomys jacksoni	Jackson's Fat Mouse
Steatomys krebsii	Kreb's Fat Mouse
Steatomys opimus	Pousargues' Fat Mouse
Stenella attentuata	Gray's Spotted Dolphin
Stenella clymene	Clymene Dolphin
Stenella coeruleoalba	Gray's Dolphin, Meyen's Dolphin
Stenella dubia	Cuvier's Spotted Dolphin
Stenella graffmani	Graffman Dolphin
Stenella longirostris	Gray's Spinner Dolphin
Stenella pernettensis	Blainville's Spotted Dolphin, Pernetty's Dolphin
Steno	The dolphin genus
Steno bredanensis	Rough-toothed Dolphin
Stenocephalemys ruppi	Rupp's Mouse
Stenoderma rufum	Desmarest's Fig-eating Bat
Stenomys niobe	Moss-forest Rat
Stenomys omichlodes	Arianus' Rat
Stenomys richardsoni	Glacier Rat
Stenomys vandeuseni	Van Deusen's Rat
Sturnira aratathomasi	Arata and Thomas' Yellow-shouldered Bat
Sturnira koopmanhilli	Koopman and Hill's Yellow-shouldered Bat
Sturnira ludovici	Anthony's Bat
Sturnira luisi	Luis' Yellow-shouldered Bat
Sturnira magna	De la Torre's Yellow-shouldered Bat
Sturnira oporaphilum	Tschudi's Yellow-shouldered Bat
Sturnira sorianoi	Soriano's Yellow-shouldered Bat
Sturnira thomasi	Thomas' Yellow-shouldered Bat
Sturnira tildae	Tilda's Yellow-shouldered Bat
Styloctenium wallacei	Stripe-faced Fruit Bat
Stylodipus andrewsi	Andrews' Jerboa
Suncus dayi	Day's Shrew

Suncus etruscus	Savi's Pygmy Shrew
Suncus fellowesgordoni	Sri Lanka Shrew
Suncus hosei	Hose's Shrew
Suncus mertensi	Flores Shrew
Suncus remyi	Remy's Shrew
Suncus stoliczkanus	Anderson's Shrew
Sundamys maxi	Bartels' Rat
Sundamys muelleri	Müller's Giant Sunda Rat
Sundasciurus brookei	Brooke's Squirrel
Sundasciurus hoogstraali	Busuanga Squirrel
Sundasciurus lowii	Low's Squirrel
Sundasciurus moellendorffi	Culion Tree Squirrel
Sundasciurus rabori	Palawan Montane Squirrel
Sundasciurus steerii	Steere's Squirrel
Surdisorex norae	Aberdare Shrew
Sus (philippensis) oliveri	Mindoro Warty Pig
Sus bucculentus	Heude's Pig
Sus heureni	Flores Warty Pig
Syconycteris carolinae	Halmahera Blossom Bat
Syconycteris hobbit	Moss-forest Blossom Bat
Sylvicapra grimmia	Grey Duiker
Sylvilagus audubonii	Desert Cottontail
Sylvilagus bachmani	Bachman's Hare
Sylvilagus dicei	Dice's Rabbit
Sylvilagus graysoni	Tres Marias Cottontail
Sylvilagus nuttallii	Nuttall's Cottontail
Sylvisorex granti	Grant's Shrew
Sylvisorex howelli	Howell's Shrew
Sylvisorex isabellae	Isabella Shrew
Sylvisorex johnstoni	Johnston's Shrew
Sylvisorex johnstoni dieterleni	Johnston's Pygmy Shrew
Synaptomys cooperi	Southern Bog Lemming
Syntheosciurus brochus	Bangs' Mountain Squirrel
Tachyoryctes ruddi	Rudd's Mole-Rat
Tachyoryctes storeyi	Storey's Mole-Rat
Tadarida latouchei	LaTouche's Free-tailed Bat
Talpa davidiana	Père David's Mole
Talpa stankovici	Stankovic's Mole
Talpa streeti	Persian Mole
Tamias bulleri	Buller's Chipmunk
Tamias merriami	Merriam's Chipmunk
Tamias palmeri	Palmer's Chipmunk
Tamias townsendii	Townsend's Chipmunk
Tamiasciurus douglasii	Douglas' Squirrel

Tamiasciurus hudsonicus fremonti	Fremont's Squirrel
Tamiasciurus mearnsi	Mearns' Squirrel
Tamiops macclellandi	Himalayan Striped Squirrel
Tamiops rodolphii	Rodolph's Striped Squirrel
Tamiops swinhoei	Swinhoe's Striped Squirrel
Taphozous achates	Brown-bearded Sheath-tailed Bat
Taphozous hamiltoni	Hamilton's Tomb Bat
Taphozous hildegardeae	Hildegarde's Tomb Bat
Taphozous hilli	Hill's Tomb Bat
Taphozous theobaldi	Theobald's Tomb Bat
Taphozous troughtoni	Troughton's Tomb Bat
Tapirus bairdii	Baird's Tapir
Tarsipes spencerae	Long-snouted Phalanger
Tarsius bancanus	Horsfield's Tarsier, Raffles' Tarsier
Tarsius dianae	Dian's Tarsier
Tarsius spectrum	Pallas' Tarsier
Tarsius syrichta	Buffon's Tarsier
Tasmacetus shepherdi	Shepherd's Beaked Whale
Tateomys rhinogradoides	Tate's Shrew-Rat
Taterillus emini	Emin's Gerbil
Taterillus harringtoni	Harrington's Gerbil
Taterillus petteri	Petter's Gerbil
Taterillus tranieri	Tranier's Gerbil
Taurotragus derbianus	Lord Derby's Giant Eland
Taurotragus oryx livingstonii	Livingstone's Eland
Taurotragus oryx pattersonianus	Colonel Patterson's Eland
Thallomys loringi	Loring's Rat
Thallomys shortridgei	Shortridge's Rat
Thamnomys kempi	Kemp's Thicket Rat
Thomasomys	The mouse genus
Thomasomys andersoni	Anderson's Oldfield Mouse
Thomasomys daphne	Daphne's Oldfield Mouse
Thomasomys kalinowski	Kalinowski's Oldfield Mouse
Thomasomys ladewi	Ladew's Oldfield Mouse
Thomasomys pyrrhonotus	Thomas' Oldfield Mouse
Thomasomys rhoadsi	Rhoads' Oldfield Mouse
Thomasomys rosalinda	Rosalinda's Oldfield Mouse
Thomasomys taczanowskii	Taczanowski's Oldfield Mouse
Thomomys bottae	Botta's Pocket Gopher
Thomomys bottae mearnsi	Mearns' Pocket Gopher
Thomomys bottae villai	Villa's Pocket Gopher
Thomomys townsendii	Townsend's Pocket Gopher
Thryonomys gregorianus	Lesser Cane Rat
Thryonomys swinderianus	Greater Cane Rat
Thylacinus harrisii	Tasmanian Tiger

Thylacoleo oweni	Owen's Marsupial "Lion" (extinct)
Thylamys cinderella	Cinderella Fat-tailed Opossum
Thylamys karimii	Karimi's Fat-tailed Mouse-Opossum
Thylogale billardierii	Tasmanian Pademelon
Thylogale browni	Brown's Pademelon
Thylogale brunii	Bruijn's Pademelon
Thylogale calabyi	Calaby's Pademelon
Thylogale thetis	Red-necked Pademelon
Thyroptera devivoi	De Vivo's Disk-winged Bat
Thyroptera discifera	Peters' Disk-winged Bat
Thyroptera lavali	LaVal's Disk-winged Bat
Thyroptera tricolor	Spix's Disk-winged Bat
Tokudaia	The Ryukyu spiny rat genus
Tokudaia muenninki	Muennink's Spiny Rat
Tonatia bidens	Spix's Round-eared Bat
Tonatia evotis	Davis' Round-eared Bat
Tonatia silvicola	D'Orbigny's Round-eared Bat
Tonkinomys daovantieni	Tonkin Limestone Rat
Trachypithecus auratus ebenus	Wulsin's Ebony Leaf Monkey
Trachypithecus barbei	Barbe's Leaf Monkey
Trachypithecus delacouri	Delacour's Leaf Monkey
Trachypithecus francoisi	François' Leaf Monkey
Trachypithecus geei	Golden Leaf Monkey
Trachypithecus germaini	Indochinese Leaf Monkey
Trachypithecus johnii	John's Langur
Trachypithecus obscurus halonifer	Cantor's Dusky Leaf Monkey
Trachypithecus phayrei	Phayre's Leaf Monkey
Trachypithecus shortridgei	Shortridge's Leaf Monkey
Tragelaphus angasii	Nyala
Tragelaphus buxtoni	Buxton's Bushbuck
Tragelaphus scriptus meneliki	Menelik's Bushbuck
Tragelaphus spekii	Sitatunga
Tragelaphus spekii selousi	Selous' Sitatunga, Zambesi Sitatunga
Tragulus williamsoni	Williamson's Mouse-Deer
Triaenops furculus	Trouessart's Trident Bat
Trichosurus cunninghamii	Southern Mountain Brushtail Possum
Trichosurus johnstonii	Coppery Brushtail Possum
Trinomys dimidiatus	Günther's Spiny Rat
Trinomys eliasi	Rio de Janeiro Spiny Rat
Trinomys iheringi	Ihering's Spiny Rat, Thomas' Spiny Rat
Trinomys moojeni	Moojen's Spiny Rat
Trinomys yonenagae	Yonenaga-Yassuda's Spiny Rat
Tryphomys adustus	Mearns' Luzon Rat
Tscherskia triton	Greater Long-tailed Hamster

Tupaia belangeri	Northern Tree-Shrew
Tupaia moellendorffi	Calamian Tree-Shrew
Tursiops gilli	Gill's Bottle-nosed Dolphin
Tursiops nesarnack	Lacépède's Bottle-nosed Dolphin
Tylomys nudicaudus	Peters' Climbing Rat
Tylomys watsoni	Watson's Climbing Rat
Tylonycteris pachypus fulvida	Blyth's Clubfooted Bat
Tympanoctomys barrerae	Red Viscacha Rat
Uranomys ruddi	Rudd's Mouse
Urocyon littoralis dickeyi	San Nicolas Island Fox
Uroderma bilobatum	Peters' Tent-making Bat
Urogale everetti	Mindanao Tree-Shrew
Uromys anak	Giant Naked-tailed Rat
Uromys boeadii	Biak Giant Rat
Uromys emmae	Emma's Giant Rat
Uromys sherrini	Giant White-tailed Rat
Uromys siebersi	Great Kai Island Giant Rat
Uropsilus andersoni	Anderson's Shrew-Mole
Urotrichus pilirostris	True's Shrew-Mole
Ursus americanus emmonsi	Glacier Bear
Ursus americanus kermodei	Kermode's Bear
Ursus arctos dalli	Alaskan Brown Bear
Ursus arctos middendorffi	Kodiak Bear
Ursus arctos nelsoni	Mexican Grizzly Bear (extinct)
Ursus arctos pervagor	Zapadokanad's Bear
Ursus arctos piscator	Bergman's Bear
Ursus arctos richardsoni	Barren Ground Grizzly Bear
Ursus crowtheri	Crowther's Bear (extinct)
Vampyressa brocki	Brock's Yellow-eared Bat
Vampyressa melissa	Melissa's Yellow-eared Bat
Vampyressa thyone	Northern Little Yellow-eared Bat
Vampyrum spectrum	Linnaeus' False Vampire Bat
Vandeleuria nolthenii	Nolthenius' Long-tailed Climbing Mouse
Vernaya fulva	Red Climbing Mouse
Vespadelus baverstocki	Baverstock's Forest Bat
Vespadelus darlingtoni	Large Forest Bat
Vespadelus douglasorum	Yellow-lipped Bat
Vespadelus finlaysoni	Finlayson's Cave Bat
Vespadelus troughtoni	Troughton's Forest Bat
Volemys clarkei	Clarke's Vole
Volemys kikuchii	Taiwan Vole
Volemys musseri	Marie's Vole

Vombatus ursinus ursinus	Home's Wombat
Vulpes cana	Blanford's Fox
Vulpes rueppelli	Rüppell's Fox
Wilfredomys	The mouse genus
Wilfredomys oenax	Greater Wilfred's Mouse, Wilfred's Mouse
Wilfredomys pictipes	Lesser Wilfred's Mouse
Xenomys nelsoni	Magdalena Rat
Xenothrix mcgregori	Jamaican Monkey (extinct)
Xeronycteris vieirai	Vieira's Long-snouted Bat
Xerus erythropus	Geoffroy's Ground Squirrel
Zaglossus attenboroughi	Attenborough's Echidna, Cyclops Long-beaked Echidna, Sir David's Long-beaked Echidna
Zaglossus bartoni	Barton's Echidna
Zaglossus bruijni	Bruijn's Echidna
Zalophus wollebaeki	Galápagos Sea-Lion
Zapus hudsonius preblei	Preble's Meadow Jumping Mouse
Zelotomys hildegardeae	Hildegarde's Broad-headed Mouse
Zelotomys woosnami	Woosnam's Broad-headed Mouse
Zenkerella insignis	Flightless Scaly-tailed Squirrel
Ziphius cavirostris	Cuvier's Beaked Whale
Zyzomys maini	Arnhem Land Rock Rat
Zyzomys woodwardi	Kimberley Rock Rat
—	Rossetti's Wombat
—	Woodhouse's Arvicola

Bibliography

Agrawal, V. C., and D. K. Ghosal. 1969. "A new field-rat (Mammalia: Rodentia: Muridae) from Kerala, India." *Proceedings of the Zoological Society of Calcutta* 22.

Allen, G. M. 1939. "A checklist of African mammals." *Bulletin of the Museum of Comparative Zoology at Harvard* 83.

Andriantompohavana, R., et al. 2006. "Mouse lemurs of northeastern Madagascar with a description of a new species at Lokobe Special Reserve." Occasional Papers, Museum of Texas Tech University 259.

Annals and Magazine of Natural History. London, 1841–1966.

Attenborough, D. 2002. *The Life of Mammals.* London: BBC.

Baker, R. J., M. B. O'Neill, and L. R. McAliley. 2003. "A new species of desert shrew, *Notiosorex*, based on nuclear and mitochondrial sequence data." Occasional Papers, Museum of Texas Tech University 222.

Bates, P. J. J., F. H. Ratrimomanarivo, D. L. Harrison, and S. M. Goodman. 2006. "A description of a new species of *Pipistrellus* (Chiroptera: Vespertilionidae) from Madagascar with a review of related Vespertilioninae from the island." *Acta Chiropterologica* 8(2).

Bates, P. J. J., et al. 2007. "A new species of *Kerivoula* (Chiroptera: Vespertilionidae) from Southeast Asia." *Acta Chiropterologica* 9(2).

Beasley, I., K. M. Robertson, and P. Arnold. 2005. "Description of a new dolphin, the Australian snubfin dolphin *Orcaella heinsohni* sp. n. (Cetacea, Delphinidae)." *Marine Mammal Science* 21(3).

Benda, P., P. Hulva, and J. Gaisler. 2004. "Systematic status of African populations of *Pipistrellus pipistrellus* complex (Chiroptera: Vespertilionidae), with a description of a new species from Cyrenaica, Libya." *Acta Chiropterologica* 6(2).

Beolens, B., and M. Watkins. 2003. *Whose Bird?* London: Christopher Helm, A. and C. Black.

Bergmans, W. 1997. "Taxonomy and biogeography of African fruit bats (Mammalia, Megachiroptera). 5. The genera *Lissonycteris* Andersen, 1912, *Myonycteris* Matschie, 1899 and *Megaloglossus* Pagenstecher, 1885; General remarks and conclusions; Annex: Key to all species." *Beaufortia* 47(2).

Bonvicino, C. R. 2003. "A new species of *Oryzomys* (Rodentia, Sigmodontinae) of the subflavus group from the Cerrado of central Brazil." *Mammalian Biology (Zeitschrift für Säugetierkunde)* 68(2).

Boubli, J. P., et al. 2008. "A taxonomic reassessment of *Cacajao melanocephalus* Humboldt (1811), with the description of two new species." *International Journal of Primatology* 29(3).

Bradley, R. D., F. Mendez-Harclerode, M. J. Hamilton, and G. Ceballos. 2004. "A new species of *Reithrodontomys* from Guerrero, Mexico." Occasional Papers, Museum of Texas Tech University 231.

Carleton, M. D., and S. M. Goodman. 2007. "A new species of the *Eliurus majori* complex (Rodentia: Muroidea: Nesomyidae) from south-central Madagascar, with remarks on emergent species groupings in the genus *Eliurus*." *American Museum Novitates* 3547.

Cooper, N. K., K. P. Aplin, and M. Adams. 2000. "A new species of false antechinus (Marsupialia: Dasyuromorphis: Dasyuridae) from the Pilbara region, Western Australia." *Records of the Western Australian Museum* 20.

Corbet, G. B., and J. E. Hill. 1991. *A World List of Mammalian Species*. 3rd edition. London: British Museum (Natural History) Publications.

Cotterill, F. P. D. 2005. "The Upemba lechwe, *Kobus anselli*: An antelope new to science emphasizes the conservation importance of Katanga, Democratic Republic of Congo." *Journal of Zoology (London)* 265.

Csorba, G., and P. J. J. Bates. 2005. "Description of a new species of *Murina* from Cambodia (Chiroptera: Vespertilionidae: Murininae)." *Acta Chiropterologica* 7(1).

Dictionary of National Biography. 1992. Oxford: Oxford University Press.

Dobigny, G., L. Granjon, V. Aniskin, K. Ba, and V. Volobouev. 2003. "A new sibling species of *Taterillus* (Muridae, Gerbillinae) from West Africa." *Mammalian Biology (Zeitschrift für Säugetierkunde)* 68(5).

Duff, A., and A. Lawson. 2004. *Mammals of the World—A Checklist*. London: A. and C. Black.

Ellis, R. 2004. *No Turning Back*. New York: Harper Collins.

Flannery, T. 1995. *Mammals of the South-West Pacific and Moluccan Islands*. Melbourne: Reed Books .

Flannery, T., et al. 1996. *Tree Kangaroos: A Curious Natural History*. Melbourne: Reed Books.

Garbutt, N. 1999. *Mammals of Madagascar*. New Haven: Yale University Press.

Garfield, B. 2007. *The Meinertzhagen Mystery: The Life and Legend of a Colossal Fraud*. Washington, DC: Potomac Books.

Gaubert, P. 2003. "Description of a new species of genet (Carnivora: Viverridae: genus *Genetta*) and taxonomic revision of forest forms related to the Large-spotted Genet complex." *Mammalia* 67(1).

Goodman, S. M., and V. Saorimalala. 2004. "A new species of *Microgale* (Lipotyphla: Tenrecidae: Oryzorictinae) from the Forêt des Mikea of southwestern Madagascar." *Proceedings of the Biological Society of Washington* 117(3).

Gotch, A. F. 1979. *Mammals—Their Latin Names Explained—A Guide to Animal Classification*. Poole, Dorset, England: Blandford Press.

———. 1995. *Latin Names Explained—A Guide to the Classification of Reptiles, Birds and Mammals*. Poole, Dorset, England: Blandford Press.

Groves, C. P. 2001. *Primate Taxonomy.* Washington, DC: Smithsonian Institution Press.

———. 2005. "The genus *Cervus* in eastern Eurasia." *European Journal of Wildlife Research* 52(1).

———. 2007. "The endemic Uganda Mangabey, *Lophocebus ugandae,* and other members of the *albigena*-group *(Lophocebus).*" *Primate Conservation* 22.

Groves, C. P., and H. B. Bell. 2004. "New investigations on the taxonomy of the zebras genus *Equus,* subgenus *Hippotigris.*" *Mammalian Biology (Zeitschrift für Säugetierkunde)* 69.

Happold, M. 2005. "A new species of *Myotis* (Chiroptera: Vespertilionidae) from central Africa." *Acta Chiropterologica* 7(1).

Helgen, K. M. 2003. "A review of the rodent fauna of Seram, Moluccas, with the description of a new subspecies of mosaic-tailed rat, *Melomys rufescens paveli.*" *Journal of Zoology (London)* 261.

———. 2005a. "A new species of murid rodent (genus *Mayermys*) from southeastern New Guinea." *Mammalian Biology (Zeitschrift für Säugetierkunde)* 70(1).

———. 2005b. "Systematics of the Pacific monkey-faced bats (Chiroptera: Pteropodidae), with a new species of *Pteralopex* and a new Fijian genus." *Systematics and Biodiversity* 3.

Helgen, K. M., and T. F. Flannery. 2004. "A new species of bandicoot, *Microperoryctes aplini,* from western New Guinea." *Journal of Zoology* 264.

Iack-Ximenes, G. E., M. de Vivo, and A. R. Percequillo. 2005. "A new species of *Echimys* Cuvier, 1809 (Rodentia, Echimyidae) from Brazil." *Papéis Avulsos de Zoologia* 45(5).

Journal of Mammalogy. American Society of Mammalogists. 1919–current.

Kawada, S., et al. 2007. "Revision of the mole genus *Mogera* (Mammalia: Lipotyphla: Talpidae) from Taiwan." *Systematics and Biodiversity* 5(2).

Kingdon, J. 1997. *The Kingdon Field Guide to African Mammals.* New York: Academic Press.

Kitchener, A. C., M. A. Beaumont, and D. Richardson. 2006. "Geographical variation in the clouded leopard, *Neofelis nebulosa,* reveals two species." *Current Biology* 16.

Lavrenchenko, L. A. 2003. "A contribution to the systematics of *Desmomys* Thomas, 1910 (Rodentia: Muridae) with the description of a new species." *Bonner Zoologische Beiträge* 50(4).

Leite, Y. R. L. 2003. *Evolution and Systematics of the Atlantic Tree Rats, Genus* Phyllomys *(Rodentia, Echimyidae), with Description of Two New Species.* University of California Publications in Zoology 132.

Lekagul, B., and J. A. McNeely. 1977. *Mammals of Thailand.* Bangkok: Association for the Conservation of Wildlife, Sahakarnbhat Co.

Lindenmayer, D. B., J. Dubach, and K. L. Viggers. 2002. "Geographic dimorphism in the mountain brushtail possum (*Trichosurus caninus*): The case for a new species." *Australian Journal of Zoology* 50(4).

Louis, E. E., Jr., et al. 2006a. "Molecular and morphological analyses of the sportive lemurs (family Megaladapidae: genus *Lepilemur*) reveals 11 previously unrecognized species." Texas Tech University Special Publications.

———. 2006b. "Revision of the mouse lemurs *(Microcebus)* of Eastern Madagascar." *International Journal of Primatology* 27(2).

Mantilla-Meluk, H., and R. J. Baker. 2006. "Systematics of small *Anoura* (Chiroptera: Phyllostomidae) from Colombia, with description of a new species." Occasional Papers, Museum of Texas Tech University 261.

McCarthy, T. J., V. L. Albuja, and M. S. Alberico. 2006. "A new species of Chocoan *Sturnira* (Chiroptera: Phyllostomidae: Stenodermatinae) from western Ecuador and Colombia." *Annals of the Carnegie Museum* 75(2).

Meijaard, E., and C. P. Groves. 2004. "A taxonomic revision of the *Tragulus* mouse-deer (Artiodactyla)." *Zoological Journal of the Linnean Society* 140.

Miranda, J. M. D., I. P. Bernardi, and F. C. Passos. 2006. "A new species of *Eptesicus* (Mammalia: Chiroptera: Vespertilionidae) from the Atlantic Forest, Brazil." *Zootaxa* 1383.

Musser, G. G., D. P. Lunde, and T. S. Nguyen. 2006. "Description of a new genus and species of rodent (Murinae, Muridae, Rodentia) from the tower karst region of northeastern Vietnam." *American Museum Novitates* 3517.

Musser, G. G., A. L. Smith, M. F. Robinson, and D. P. Lunde. 2005. "Description of a new genus and species of rodent (Murinae, Muridae, Rodentia) from the Khammouan Limestone National Biodiversity Conservation Area in Lao PDR." *American Museum Novitates* 3497.

Nowak, R. M. 1999. *Walker's Mammals of the World.* 6th ed. Baltimore: Johns Hopkins University Press.

Olivieri, G., et al., 2007. "The ever-increasing diversity in mouse lemurs: Three new species in north and northwestern Madagascar." *Molecular Phylogeny and Evolution* 43.

Patterson, B. D., and P. M. Velazco. 2006. "A distinctive new cloud-forest rodent (Hystricognathi: Echimyidae) from the Manu Biosphere Reserve, Peru." *Mastozoología Neotropical* 13(2).

Proceedings of the Zoological Society of London. London, 1830–1965.

Rasoloarison, R. M., S. M. Goodman, and J. U. Ganzhorn. 2000. "Taxonomic revision of mouse lemurs *(Microcebus)* in the western portions of Madagascar." *International Journal of Primatology* 21(6).

Reeves, R. R., et al. 2002. *Sea Mammals of the World.* London: A. and C. Black.

Seebeck, J., P. R. Brown, R. L. Wallis, and C. Kemper (eds.). 1990. *Bandicoots and Bilbies.* Sydney: Surrey Beatty and Sons.

Solari, S. 2004. "A new species of *Monodelphis* (Didelphimorphia: Didelphidae) from southeastern Peru." *Mammalian Biology (Zeitschrift für Säugetierkunde)* 69(3).

Solari, S., and R. J. Baker. 2006. "Mitochondrial DNA sequence, karyotypic, and morphological variation in the *Carollia castanea* species complex (Chiroptera: Phyllostomidae), with description of a new species." Occasional Papers, Museum of Texas Tech University 254.

Spitzenberger, F., P. P. Strelkov, H. Winkler, and E. Haring. 2006. "A preliminary revision of the genus *Plecotus* (Chiroptera, Vespertilionidae) based on genetic and morphological results." *Zoologica Scripta* 35.

Stanley, W. T., and R. Hutterer. 2000. "A new species of *Myosorex* Gray 1832 (Mammalia: Soricidae) from the Eastern Arc Mountains, Tanzania." *Bonner Zoologische Beiträge* 49(1–4).

Stanley, W. T., M. A. Rogers, and R. Hutterer. 2005. "A new species of *Congosorex* from the Eastern Arc Mountains, Tanzania, with significant biogeographical implications." *Journal of Zoology, London* 265.

Sterling, K. B., R. P. Harmond, G. A. Cevasco, and L. F. Harmond. 1997. *Biographical Dictionary of American and Canadian Naturalists and Environmentalists.* Westport, CT: Greenwood Press.

Straeten, E. van der, E. Lecompte, and C. Denys. 2003. "*Praomys petteri*: Une nouvelle espèce des Muridae africains (Mammalia, Rodentia)." *Bonner Zoologische Beiträge* 50(4).

Thalmann, U., and T. Geissmann. 2005. "New species of woolly lemur *Avahi* (Primates: Lemuriformes) in Bemaraha (Central Western Madagascar)." *American Journal of Primatology* 67.

Thomson, K. S. 1995. *HMS Beagle—The Ship That Changed the Course of History.* New York: W. W. Norton and Co.

Van Roosmalen, M. G. M., T. van Roosmalen, and R. A. Mittermeier. 2002. "A taxonomic review of the titi monkeys, genus *Callicebus* Thomas, 1903, with the description of two new species, *Callicebus bernhardi* and *Callicebus stephennashi*, from Brazilian Amazonia." *Neotropical Primates* 10 (supp.).

Velazco, P. M. 2005. "Morphological phylogeny of the bat genus *Platyrrhinus* Saussure, 1860 (Chiroptera: Phyllostomidae), with the description of four new species." *Fieldiana Zoology* 105.

Verheyen, W. N., J. L. J. Huselmans, T. Dierckx, and E. Verheyen. 2002. "A craniometric and genetic description of two new species from the *Lophuromys flavopunctatus* Thomas 1888 species complex (Rodentia—Muridae—Africa)." *Bulletin de l'Institut Royal des Sciences Naturelles de Belgique (Bulletin van het Koninklijk Belgisch Instituut voor Natuurwetenschappen)* 72.

Verheyen, W. N., et al., 2007. "The characterization of the Kilimanjaro *Lophuromys aquilus* True 1892 population and the description of five new *Lophuromys* species (Rodentia, Muridae)." *Bulletin de l'Institut Royal des Sciences Naturelles de Belgique (Bulletin van het Koninklijk Belgisch Instituut voor Natuurwetenschappen)* 77.

Von Beringe, A. 2002. Unpublished biographical note on his grandfather, Major Robert von Beringe.

Voss, R., T. Tarifa, and E. Yensen. 2004. "An introduction to *Marmosops* (Marsupialia: Didelphidae), with the description of a new species from Bolivia and notes on the taxonomy and distribution of other Bolivian forms." *American Museum Novitates* 3466.

Ward, E. R. 1981. *The Mammals of North America.* New York: John Wiley and Sons.

Weksler, M., A. R. Percequillo, and R. S. Voss. 2006. "Ten new genera of Oryzomyine rodents (Cricetidae: Sigmodontinae)." *American Museum Novitates* 3537.

Wilson, D. E., and D. M. Reeder (eds.). 2005. *Mammal Species of the World: A Taxonomic and Geographic Reference.* 3rd ed. Baltimore: Johns Hopkins University Press.

Woodman, N., and R. M. Timm. 2006. "Characters and phylogenetic relationships of nectar-feeding bats, with descriptions of new *Lonchophylla* from western South America (Mammalia: Chiroptera: Phyllostomidae: Lonchophyllini)." *Proceedings of the Biological Society of Washington* 119(4).

Woolley, P. A. 2005. "Revision of the three-striped dasyures, genus *Myoictis* (Marsupialia: Dasyuridae), of New Guinea, with description of a new species." *Records of the Australian Museum* 57.

Wynne, Col. O. E. 1969. *Biographical Key—Names of Birds of the World—to Authors and Those Commemorated.* Privately published.

Zoologischer Anzeiger. Berlin, Leipzig, 1878–current.